P9-AES-107

Biochemistry of Foods

Biochemistry of Foods

Second Edition

N. A. Michael Eskin

Department of Foods and Nutrition
The University of Manitoba
Winnipeg, Manitoba, Canada

Academic Press
San Diego New York Boston
London Sydney Tokyo Toronto

This book is printed on acid-free paper. ∞

Academic Press
525 B Street, Suite 1900, San Diego, California 92101-4495, USA
http://www.apnet.com

United Kingdom Edition published by
Academic Press Limited
24–28 Oval Road, London NW1 7DX

Library of Congress Cataloging-in-Publication Data

Biochemistry of foods. -- 2nd ed. / edited by Michael Eskin.
p. cm.
Rev. ed. of: Biochemistry of foods / N.A.M. Eskin, H.M. Henderson, R.J. Townsend. 1971.
ISBN 0-12-242351-8 (alk. paper)
1. Food. 2. Biochemistry. I. Eskin, N. A. M. (Neason Akivah Michael) II. Eskin, N. A. M. (Neason Akivah Michael) Biochemistry of foods.
TX531.B56 1990
664--dc20 89-37192
CIP

Printed in the United States of America
98 99 00 01 02 BC 7 6 5 4 3 2

This book is dedicated to my wife,
Nella,
and our four sons,
Katriel, Joshua, Ezra, and Daniel,
and in celebration of the
ninetieth year of my mother,
Ethel Eskin

"How much better it is to get wisdom than gold,
And more desirable to get understanding than silver."
Proverbs.

Contents

Preface

Our understanding of food biochemistry has increased substantially since the publication of the first edition of this book. This has necessitated major revisions of a number of chapters plus reorganization with additional sections incorporated in the text. These changes are reflected by the four major parts in this book. Part I deals with those biochemical changes taking place in raw foods and includes four chapters. Chapter 1 discusses postmortem changes in muscle responsible for the production of edible meat and fish and includes an examination of the role of connective tissue and myofibrillar proteins in this process. Chapter 2 covers the postharvest changes in fruits and vegetables and includes a more extensive treatment of flavor and texture. Chapter 3 examines the biochemistry of cereal development with particular emphasis on wheat, and Chapter 4 reviews the complex biochemical processes involved in milk biosynthesis. Part II focuses on biochemical changes associated with processing with four areas selected. Chapter 5 covers nonenzymatic browning reactions in foods during heating and storage. Chapter 6 includes a detailed discussion of the brewing of beer, and Chapter 7 deals with the biochemistry of baking. The final chapter in this part, Chapter 8, covers the biochemistry of cheese and yoghurt. Part III deals with selected areas in the biochemistry of food spoilage with Chapter 9 on enzymatic browning and Chapter 10 on off-flavors in milk. Part IV on Biotechnology includes a detailed coverage of enzymes in the food industry, including immobilized enzymes, enzyme electrodes, and genetic engineering.

The overall organization of this edition of *Biochemistry of Foods* is far more comprehensive than the previous edition. The chapter on biodeterioration was deleted, due in part to the death of Dr. R. J. Townsend, but also because of the availability of a number of specialized books in this area. This book attempts to bridge the gap between the introductory and highly specialized books dealing with aspects of food biochemistry. It is my hope that this book will serve as a text and reference for undergraduate and graduate students, researchers, and professionals in the fields of food science, horticulture, animal science, dairy science, and cereal chemistry.

Acknowledgments

I am indebted to the following people for their review of one or more of the chapters and for their helpful suggestions during the preparation of this manuscript: Professors C. Biliaderis, H. M. Henderson, and W. Bushuk, Department of Food Science, G. Chauhan, an NSERC/CIDA visiting scientist, University of Pantganar, India, and Dr. R. Przybylski, Department of Foods and Nutrition, University of Manitoba. I am particularly indebted to my wife, Nella, for her patience and understanding during the preparation of this book as well as for editing several of the chapters. I appreciate the support of the editorial staff at Academic Press particularly Suzanne Clancy and her predecessor Valeta Gregg, as well as Kerry Pinchbeck. I acknowledge the Lady Davis Committee for awarding me a fellowship in the Department of Food Engineering and Biotechnology at the Technion, Israel, which enabled me to undertake the initial writing of this text.

Part I

Biochemical Changes in Raw Foods

1

Biochemical Changes in Raw Foods: Meat and Fish

I. Introduction

Meat is defined as the flesh of animals used as food. A more precise definition is provided by the U.S. Food and Drug Administration (Meyer, 1964): meat is that derived from the muscles of animals closely related to man biochemically and therefore of high nutritive value. The more conventional animal species include cattle, pig, sheep, and the avian species chicken and turkey. In fish, however, it is the white muscle which provides the main nutritional source. The per capita consumption of muscle foods in the United States has remained fairly stable as shown in Table 1.1. Beef and pork are clearly the most preferred of the muscle foods, followed by chicken and fish. In the developing continents, Africa, Asia, and Latin America, the consumption of meat and fish is still extremely low or nonexistent, as evident by the increasing incidence of malnutrition. This lack of high-grade proteins and the accompanying deficiency in essential amino acids remains the world's most urgent problem.

This chapter will discuss the dynamic changes involved in the conversion of muscle to meat or edible fish. Following the death of the animal or fish, many chemical, biochemical, and physical changes occur leading to the development of postmortem tenderness. A greater understanding of these changes should

TABLE 1.1

PER CAPITA CONSUMPTION (KG/ANNUM) OF MUSCLE FOODS BY SPECIES (USDA, 1960–76)[a]

Species	1960	1965	1970	1975
Animal source				
Beef/veal	41.5	47.5	52.7	42.0
Pork	29.5	26.6	29.7	23.2
Lamb/mutton	2.2	1.7	1.5	0.8
Avian source				
Chicken	12.6	15.1	19.0	18.3
Turkey	2.8	3.4	3.4	3.9
Aquatic source				
Fish	4.7	5.0	5.1	5.5

[a] From Sink (1979). Copyright © by Institute of Food Technologists.

make an important contribution to the production of high-quality meat or fish products.

II. The Nature of Muscle

While muscles are classified into several types, it is the striated or voluntary muscle which constitutes lean meat. The basic unit of the muscle is the fiber, a multinucleate, cylindrical cell bounded by an outer membrane, the sarcolemma. These fibers associate together into bundles, and are enclosed by a sheath of connective tissue, the perimysium. Fiber bundles are held together by connective tissue and covered by a connective tissue sheath, the epimysium. Connective tissues important to the texture and edibility of the meat and fish include fibrous proteins, collagen, reticulin, and elastin. Fish muscle has much less connective tissue, thus providing less of a problem in tenderization.

A. STRUCTURE

Individual muscle fibers are composed of myofibrils which are 1–2 μm thick and are the basic units of muscular contraction. The skeletal muscle of fish differs from that of mammals in that the fibers arranged between the sheets of connective tissue are much shorter. The connective tissue is present as short transverse sheets (myocommata) which divide the long fish muscles into seg-

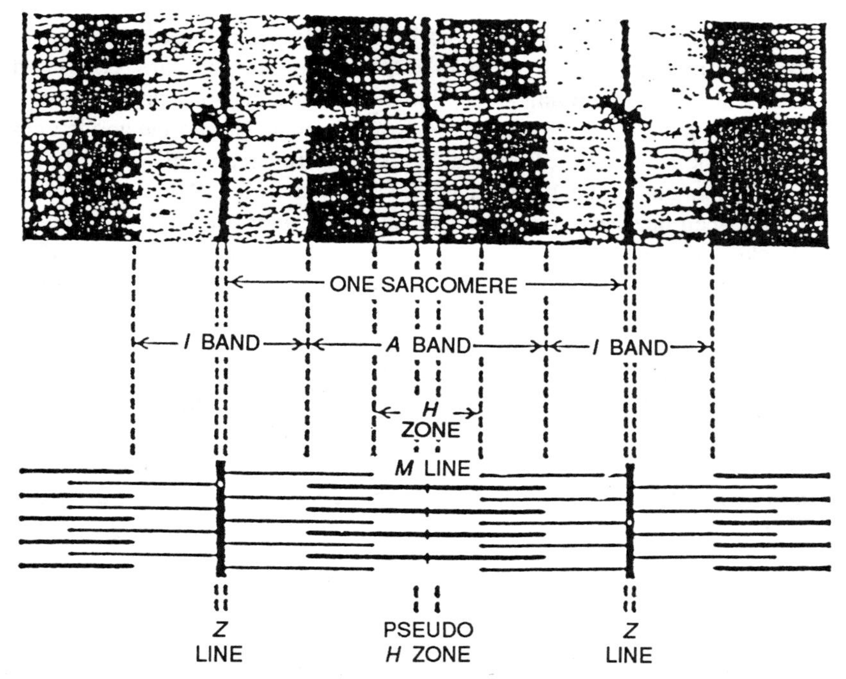

FIG. 1.1. An electron micrograph of a longitudinal section through a frog sartorius muscle is shown at the top of this figure, and a schematic diagram of the longitudinal view of the interdigitating thick and thin filament structure of the myofibril is shown at the bottom (Huxley, 1972a). Reproduced with permission of Academic Press.

ments (myotomes) corresponding in numbers to those of the vertebrae (Dunajski, 1979). The individual myofibrils are separated by a fine network of tubules, the sarcoplasmic reticulum. Within each fiber is a liquid matrix referred to as the sarcoplasm, which contains mitochondria, enzymes, glycogen, adenosine triphosphate, creatine, and myoglobin.

Examination of myofibrils under a phase contrast light microscope shows them to be cross-striated due to the presence of alternating dark or A-bands and light or I-bands. These structures in the myofibrils appear to be very similar in both fish and meat. The A-band is traversed by a lighter band or H-zone, while the I-band has a dark line in the middle known as the Z-line. A further dark line, the M-line, is observed at the center of the H-zone. The basic unit of the myofibril is the sarcomere, defined as the unit between adjacent Z-lines as shown in Figure 1.1. Examination of the sarcomere by electron microscopy reveals two sets of filaments within the fibrils, a thick set consisting mainly of myosin and a thin set containing primarily F-actin.

In addition to the paracrystalline arrangement of the thick and thin set of filaments there appears to be a filamentous "cytoskeletal structure" composed of connectin and desmin (Young *et al.*, 1980–1981). Connectin is now recognized as the major myofibrillar protein in the "gap filaments" in muscle and is present throughout the sarcomere of skeletal muscle (Maruyama *et al.*, 1976a). These gap or G-filaments were reported by Locker and Leet (1976a,b) to span the region between the thick and thin filaments in fibers of overstretched beef muscle. Locker (1984) proposed that each gap filament formed a core to an A-band. Connectin was subsequently characterized as the protein titin, consisting of the three fractions titin-1, -2, and -3, reported by Wang *et al.* (1979) to account for 10–15% of the myofibrillar proteins in chicken breast. Titin-3 is now recognized as a distinct protein and referred to as nebulin (Wang and Williamson, 1980). Desmin, on the other hand, was reported by several researchers to be present in the periphery of each Z-disk in chicken skeletal muscle (Lazarides and Hubbard, 1976; Grainger and Lazarides, 1978). It may have a role in maintaining alignment of adjacent sarcomeres, which unifies the contractile process of the separate myofibrils.

B. Cytoskeleton

Considerable attention has been focused in recent years on the cytoskeleton of muscle. This is composed of two elements, gap filaments and intermediate filaments (Stanley, 1983).

1. *Gap Filaments (G-Filaments)*

Gap filaments were originally identified by Hanson and Huxley (1955) as extremely thin elastic "S-filaments" responsible for keeping the actin filaments together. The model in Figure 1.2, proposed by Hoyle, showed these filaments were located parallel to the fiber axis extending between the Z-disks and referred to them as "gap filaments" (Sjöstrand, 1962). These filaments were found by

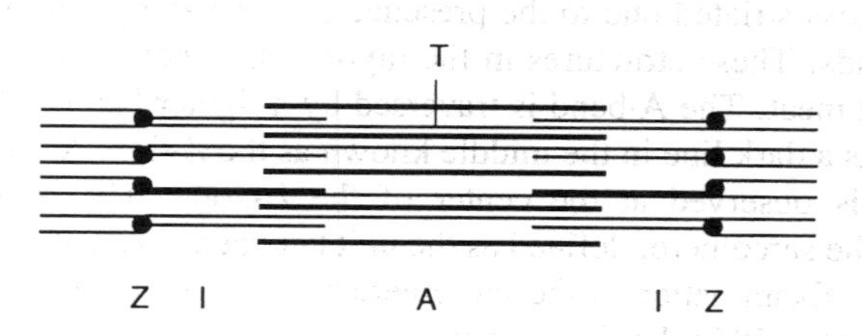

Fig. 1.2. Diagram showing proposed model for muscles including a very thin elastic filament (T) extending between Z-disks (Z) and parallel to the A band (A) and I band (I) (Hoyle, 1967).

Maruyama and co-workers (1976b, 1977) to be composed of a rubbery, insoluble protein called "connectin." Wang and co-workers (1979) identified a high-molecular-weight protein which was referred to as titin. Subsequent research showed the high-molecular-weight components of connectin to be titin (Maruyama *et al.*, 1981). Titin appeared to be the major cytoskeleton protein in the sarcomere responsible for muscle cell integrity (Wang and Ramirez-Mitchell, 1979, 1983a,b). Locker and Leet (1976b) proposed that each gap filament formed a core within an A-band as well as linked the two thick filaments in adjacent sarcomeres through the Z-line. Wang and Ramirez-Mitchell (1984), using four distinct monoclonal antibodies to rabbit titin, showed that titin passed from the M-line through the A-band and into the I-band, thereby discounting the central core model.

Wang and co-workers (1979) also identified a large myofibrillar protein in vertebrate skeletal muscle referred to as nebulin. This protein was later isolated from the myofibrils of rabbit psoas and chicken breast muscles using immunological and electrophoretic techniques. It was found to be distinct from titin (Murayama *et al.*, 1981; Ridpath *et al.*, 1982, 1984; Wang and Williamson, 1980). The location of nebulin in the myofibril was at the N^2 line. Wang and Ramirez-Mitchell (1983b) presented an alternative model for the G-filaments consisting of an elastic filamentous matrix containing both titin and nebulin as additional sarcomere constituents (Locker, 1984).

2. *Intermediate Filaments*

These filaments linking the myofibrils laterally to the sarcolemma are intermediate in size (10 nm in diameter) between the actin (6 nm in diameter) and myosin (14–16 nm in diameter) filaments (Ishikawa *et al.*, 1968). The protein isolated from these filaments, desmin, also referred to as skeletin, is located in the periphery of the Z-disk in the filamentous form (Lazaride and Hubbard, 1976; Richardson *et al.*, 1981). The cytoskeleton role of desmin is to connect Z-lines of adjacent myofibrils (O'Shea *et al.*, 1981; Robson *et al.*, 1984).

C. Connective Tissue

The interstitial space in muscle cells is occupied by three proteins, collagen, reticulin, and elastin, together referred to as connective tissue. The endomysium layer surrounding the muscle fibers is composed of fine reticular and collagenous fibrils, while elastin is sparsely distributed in the muscle with the blood, capillary, and nervous systems (Asghar *et al.*, 1984). Bundles of these muscle fibers are surrounded by a thicker connective tissue, the perimysium. These connective tissues appear to unite at the ends of the muscle in the thick tendon fibers as shown in Figure 1.3 (Etherington and Sims, 1981). During muscle contraction,

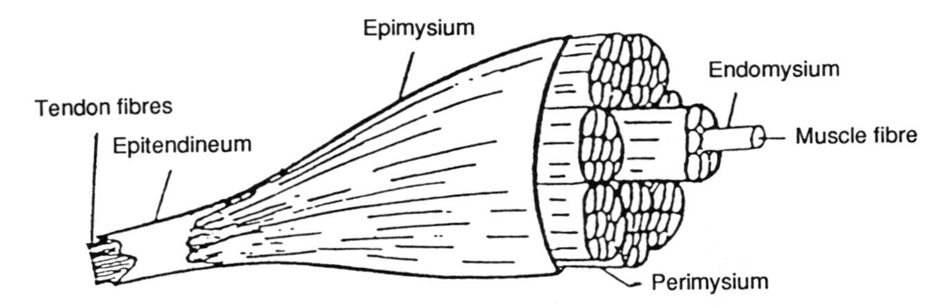

FIG. 1.3. Connective tissues of muscle (Etherington and Sims, 1981).

movement is transmitted via the tendon to the skeleton. The limited elasticity of collagen in the tendon permits the translation of muscle contraction into a high degree of movement.

1. *Collagen*

The major protein of connective tissue is collagen, a glycoprotein. It was originally thought to be composed of two polypeptide chains, α_1- and α_2-chains, which formed a triple helical structure. At least ten different α-chains are now known which appear to be responsible for the different types of collagen so far identified (Types I, II, III, IV, and V). These differ from each other in their primary structure and amino acid composition (Asghar *et al.*, 1984). The subunit of the collagen fiber is the collagen monomer tropocollagen. This is composed of three polypeptide α-chains arranged in a pattern which allows the staggered overlap of one polypeptide chain over the other as shown for the Type I tropocollagen monomer in Figure 1.4 (Asghar and Henrickson, 1982).

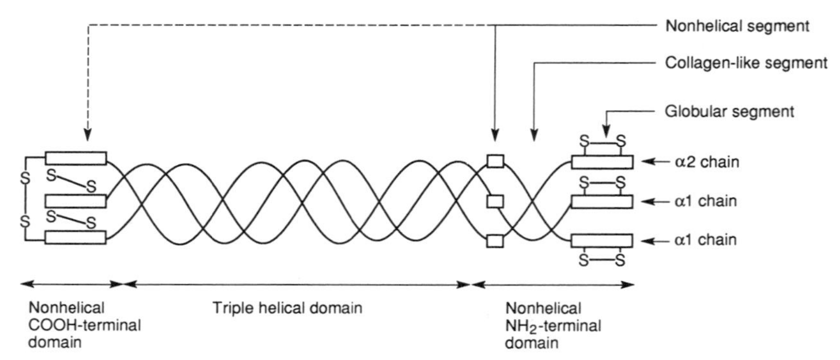

FIG. 1.4. Schematic representation of the Type I tropocollagen monomer, composed of two identical pro-α1(I) chains (solid lines) and one pro-α2(I) chain (dashed line) (Asghar and Henrickson, 1982).

The major collagen component of both the epimysium and perimysium is Type I collagen, while Types III, IV, and V are located in the endomysium (Bailey and Peach, 1968; Bailey and Sims, 1977). Since collagen is the principal component of connective tissue, the texture of meat is greatly influenced by it. Bailey (1972) suggested that an acceptable meat texture requires a certain degree of cross-linkages in collagen. A lack or overabundance of such linkages in collagen produces meat that is either too tender or too tough. The toughness associated with meat from older animals is attributed to the high degree of cross-linkages in the collagen fibers.

a. Collagen and Meat Texture. Attempts to correlate total collagen of muscles with meat texture has resulted in conflicting reports. Dransfield (1977) found a definite relationship between total muscle collagen and toughness. Other studies showed that the qualitative nature of collagen rather than quantity ultimately affected texture (Bailey, 1972; Bailey *et al.*, 1979; Bailey and Sims, 1977; Shinomokai *et al.*, 1972). A recent study by Light and co-workers (1985) examined the role of epimysial, perimysial, and endomysial collagen in the texture of six bovine muscles. These researchers reported a correlation between both collagen fiber diameter and collagen content of perimysial and endomysial connective tissue and meat toughness. A linear plot was obtained when the number of heat-stable cross-links was plotted against the compressive force (kg) (Figure

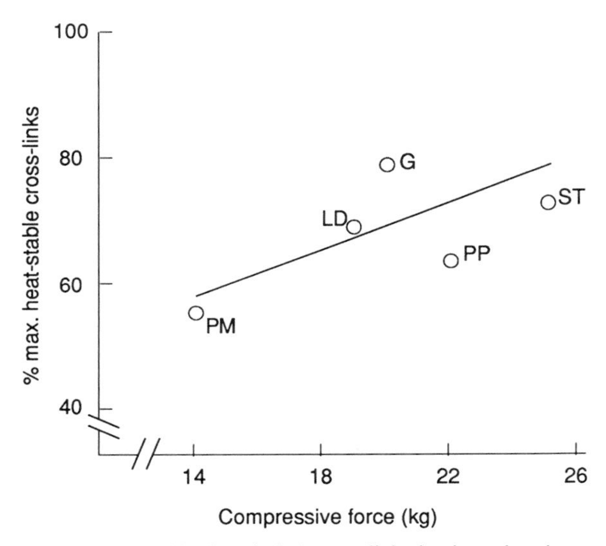

FIG. 1.5. Plot of total heat-stable (keto-imine) cross-links in six perimysia versus compressive force estimated after cooking the muscles for 1 hr at 75°C (from data by Dransfield, 1977). (PM) Psoas major; (LD) longissimus dorsi; (PP) pectoralis profundis; (G) gastrocnemius; (ST) semitendinosus. (Light *et al.*, 1985.)

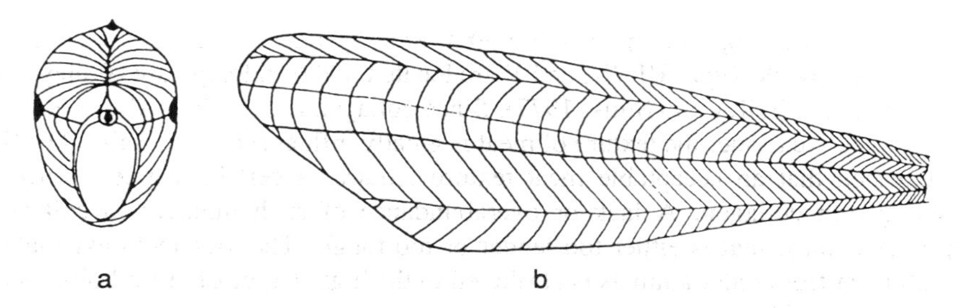

FIG. 1.6. The metameric structure of fish muscles. The patterns of lines on the cross section (a) and longitudinal section (b) represent the arrangement of sheets of connective tissue in the muscles (Dunajski, 1979).

1.5) for six muscle perimysia cooked at 75°C using the data of Dransfield (1977). Although not as clear-cut, similar trends were observed for both epimysial and endomysial muscle samples. Based on these results it was apparent that the cross-links have a crucial role in determining tenderness or toughness of meat. If the primary cause for meat fracture and breakdown, according to Purslow (1985), is *via* the perimysium or at the perimysial–endomysial junctions, the nature of the cross-links between these fibers could be extremely important.

b. Collagen and Fish Texture. Fish muscles generally contain only one-tenth of the collagen found in red meats. They are divided by thin membranes, myocommata, into segments or myotomes as shown in Figure 1.6. The myocommata are composed of connective tissue with each muscle fiber surrounded by a cell wall or basement membrane containing thin collagen fibrils. The integrity of the fish muscles is maintained by the connective tissue of the myocommata and collagen fibers, which together form the endomysial reticulum. If the myotomes are not connected to the myocommata, slits and holes form in the flesh, which is characteristic of gaping. This results in the deterioration of fish quality as the fish fillets fall apart and become quite unacceptable. Love and co-workers (1972) attributed the development of this problem to the rupturing of the endomysial and myocommata connection brought about through rough handling or bending of the stiffened fish. The contribution of connective tissue to the texture of cooked fish remains unclear compared to its role in meat as a consequence of the smaller amount present. For example, the high level of connective tissue in dogfish requires a cooking temperature of 45°C, which is still substantially lower than the cooking temperature of 92°C for 1 hr needed to obtain the same degree of tenderness for beef.

III. Contraction of Muscle

While the majority of studies have been conducted on mammalian muscle, it is apparent that similar changes occur in fish muscle. It is generally accepted that contraction and relaxation of striated muscle occurs by the sliding action of the thick filaments over the thin filaments with the length of the filaments remaining the same (Rowe, 1974). Myosin possesses adenosine triphosphatase (ATPase) activity, which requires the presence of magnesium and calcium ions. It is the regulation of myofibrillar ATPase which determines the contractile response of the muscle. This enzyme catalyzes the hydrolytic cleavage of ATP, thereby providing the most immediate source of energy for muscular contraction:

$$ATP + H_2O \rightarrow ADP + H_3PO_4$$

$$\Delta G_{298} \text{ (Standard free energy change at 25°C)} = -11.6 \text{ kcal/mole}$$

In resting muscle the activity of ATPase is very low, resulting in the slow release of ADP and inorganic phophorous at the active sites of myosin and actin. Once muscle stimulation occurs, the head of myosin, containing the actin-combining and enzymatic sites, interacts with actin with the rapid release of ADP and inorganic phosphate (P_i). The increase in ATP hydrolysis can be several hundred times that observed in the resting state (Perry, 1979). This is accompanied by a conformational change in the myosin head, causing a change in the angle it makes with the actin filament. The overall result is that the actin monomer with the myosin head attached moves forward by approximately 5–10 nm (Huxley, 1969). Once ADP and inorganic phosphate (P_i) are released from the myosin head, the actin monomer detaches itself to permit a fresh molecule of $Mg\text{-}ATP^{2-}$ to be picked up by the enzymatic site on the myosin head and the enzyme–substrate complex is reestablished. Muscular contraction is thus characterized by a rapid conversion of ATP to ADP and inorganic phosphate, and on completion the muscle returns to its resting state. The latter is characterized by the capacity of the substrate–enzyme complex at enzymatic sites in the myosin head to be released once stimulation occurs. The hydrolysis of myosin in the presence of actin has been studied by a number of researchers, although detailed steps remain to be characterized (Chock *et al.*, 1976; Eccleston *et al.*, 1976).

A. Regulation of Muscle Contraction: Troponin and Tropomyosin

The regulation of muscular contraction appears to involve the release of calcium from the vesicles of the sarcoplasmic reticulum, located in the myofibrils. Calcium is released when the stimulus is received at the muscle fiber by

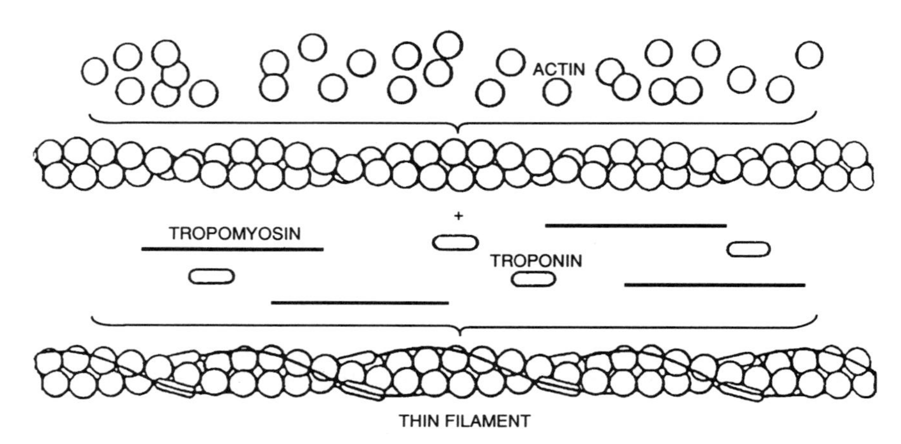

FIG. 1.7. Schematic diagram showing assembly of the thin filament from actin, tropomyosin, and troponin, and the molecular architecture of the assembled thin filament (Murray and Weber, 1974).

way of the central nervous system. It stimulates myosin ATPase, thus releasing the energy necessary for muscle contraction to facilitate the sliding action of actin filaments which form contractile actomyosin (Huxley, 1964). This was confirmed in studies by Goodno *et al.* (1978), who reported up to 100-fold increase in ATPase activity of the myofibril when calcium ion (Ca^{2+}) levels were incrreased from $<10^{-8}$ *M* to $>10^{-5}$ M. Calcium regulates actin–myosin interaction by directly binding to the troponin C component of the I-filament.

Troponin is a myofibrillar protein associated with the thin filaments which appears to control the interaction between actin and myosin. It is an elongated molecule of molecular weight 80,000, which is attached to tropomyosin, another myofibrillar protein. They both provide the regulatory system in muscular contraction (Ebashi, 1974). Tropomyosin, a long, coiled α-helix, is located in each of the two long-pitch helical grooves of the actin filaments (Seymour and O'Brien, 1980). Figure 1.7 shows how troponin and tropomyosin interact with seven actin molecules (Murray and Weber, 1974). In contrast, troponin is found at discrete intervals of 38 nm along the thin actin filaments and is associated with the stripes on the I-band. Troponin is composed of three subunits, troponin C, troponin T, and troponin I. Of these, troponin C binds Ca^{2+} when the muscle is stimulated, which is translated into conformational changes in protein *via* tropomyosin. Skeletal muscles are associated with two types of cells, slow and fast muscle fibers. This is reflected by four calcium binding sites in troponin C of fast muscles compared to three calcium binding sites in slow muscles. In the presence of low calcium levels the formation of cross-links is inhibited by the troponin–tropomyosin complex. An increase in calcium levels following stimulation of the muscle results in the binding of calcium to troponin C and the formation of the actomyosin complex. Accumulation of calcium in the sarcoplasmic reticulum is achieved against a concentration gradient requiring an active transport pumping system involving ATP. This is hydrolyzed by ATPase present in the membranes of the sarcoplasmic reticulum (de Meis and Vianna, 1979).

B. Mechanism of Tropomyosin Action

Extensive studies using electron micrographs of a myosin subfragment (S-1) suggested that tropomyosin regulated muscle contraction by steric blocking and unblocking of the myosin interaction sites in the muscle thin filaments (Haselgrove, 1972; Huxley, 1972b; Parry and Squire, 1973). This theory was questioned by Seymour and O'Brien (1980), who proposed that tropomyosin was located on the opposite side of the thin filament helix axis from the binding sites of myosin S-1 (Moore *et al.*, 1970). Further studies by Taylor and Amos (1981), based on three-dimensional image reconstructions of electron micrographs of thin filaments decorated with myosin S-1, suggested that the location of the binding sites proposed by Moore *et al.* (1970) was incorrect. Taylor and Amos (1981) clearly demonstrated that tropomyosin was located on the same side of the actin helix. Later research by Mendelson (1982) using high-resolution reconstructions of x-ray scattering of myosin S-1 further supported the model for attachment of myosin S-1 to actin presented by Taylor and Amos (1981). Thus regulation of muscle contraction appeared to occur by steric blocking of actin–S-1 myosin interaction. Amos *et al.* (1982) presented structural evidence that the head of myosin S-1 interacted with two sites on F-actin (Figure 1.8). In the active state tropomyosin was thought to occupy a position near the middle of the actin groove, while in the inhibited state it lay on the other side of the groove where it could interface with interaction between S-1 and actin at contact No. 1 or 2 sites.

C. ATP and the Lohmann Reaction

On cessation of the stimulus, calcium ions are removed by a "relaxing factor" with the sarcoplasmic reticulum acting as a calcium pump (Newbold, 1966). While the primary source of ATP in muscle is derived from ADP by aerobic respiration, it can also be resynthesized from ADP and creatine phosphate (CP) by the Lohmann reaction:

$$\text{ADP} + \text{CP} \xrightleftharpoons{\text{ATP:creatine phosphotransferase}} \text{ATP} + \text{creatine}$$

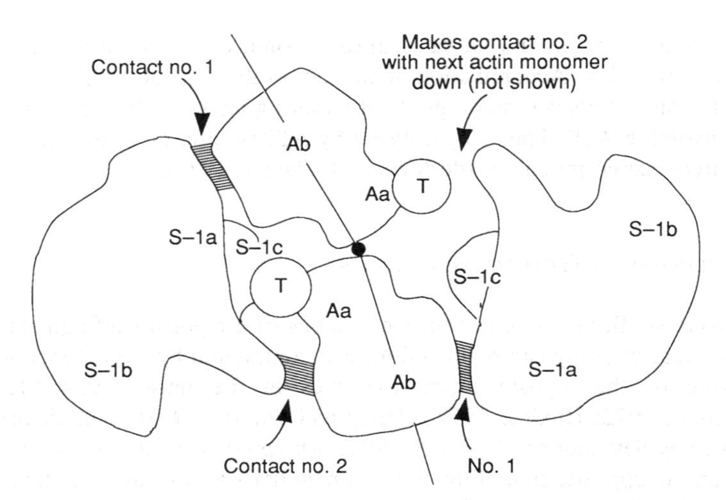

FIG. 1.8. Diagram of the proposed interactions in the rigor state, between myosin heads (S-1), actin monomers (A) in F-actin, and tropomyosin molecules (T), which are thought to lie in the actin helix grooves. Independent sites on S-1 contact two neighboring actin monomers and there is close contact between S-1 and tropomyosin which may account for the stronger binding of S-1 to regulated than to unregulated actin (Amos *et al.*, 1982).

ATP : creatine phosphotransferase and creatine phosphate are both located in the sarcoplasm. This reaction is important in conditions leading to muscle fatigue, representing an immediate pathway for the resynthesis of ATP. Consequently muscular activity can continue until adequate amounts of ATP are generated *via* carbohydrate degradation.

D. FISH MUSCLE CONTRACTION

Fish muscle consists of two types, red and white muscle, in which rigor contractions have been shown to differ. While the proportions of these two muscles vary from one species to the next, red muscle never exceeds 10% of the total muscle for any species, for example, tuna. The dark and white muscle content of 16 species of fish was measured by Obatake and Heya (1985) using a rapid direct gravimetric method on the heated fish. With the exception of saury pike, the dark muscle of all the other species of fishes never exceeded the 10% level. The ratio of dark muscle to whole muscle accounted for over 12% in the so-called red meat species (e.g., sardine, saury pike, frigate herring, and round herring) compared to less than 3% in the white meat species such as yellow sea bream and Silago. The dark or red muscle is characterized by a high myoglobin

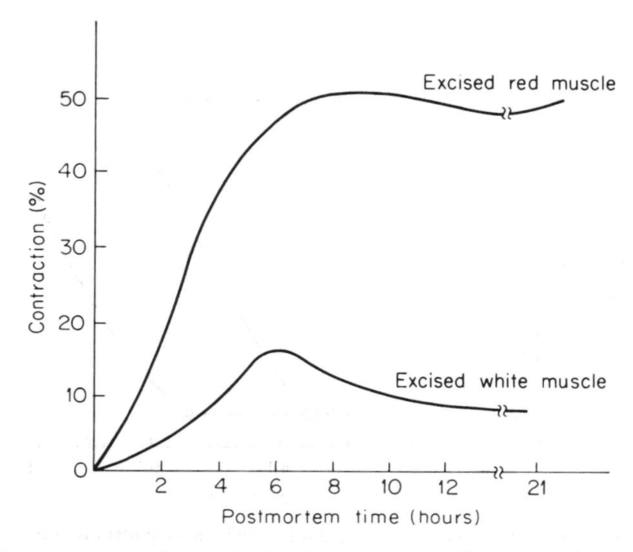

FIG. 1.9. Rigor contraction of excised red and white muscle of lingcod at 20°C (Buttkus, 1963).

content as well as distinct proteins (Hamoir and Konosu, 1965). Obatake *et al.* (1985) found that the dark muscle of fish had higher levels of extractive nitrogen constituents as well as creatine compared to white muscles.

Rigor contraction has been shown to be far greater in red muscle of fish compared to white muscle (Figure 1.9) and corresponded more closely with that of mammalian muscle (Buttkus, 1963). The role of contraction, tension, and elasticity associated with the development of rigor mortis in postmortem fish muscles is still poorly understood. Bate-Smith and Bendall (1956), in studies on rabbit muscle, found considerable shortening during the rigor period which was not concomitant with stiffening and rarely occurred at room temperature. In contrast, the red muscle of lingcod and trout consistently produced a postmortem contraction at 20°C, the rate being indicative of the condition of the fish prior to death (Buttkus, 1963). White muscle is generally regarded as the nutritional flesh of fish, therefore most studies have been confined to this tissue. The importance of red muscle in the postmortem changes of fish cannot be ignored, however, since as yet there is no process capable of separating these two muscles in the fish-processing industry.

Trucco *et al.* (1982) reported that the visual and tactile estimation of rigidity of fish during the prerigor, full rigor, and postrigor stages originally proposed by Cutting (1939) was still the most reproducible method. Their results, shown in Figure 1.10, indicate that for sea bream (*Sparus pagnes*) it took 10 hr for rigor mortis to develop and approximately 30 hr for its resolution compared to 55 hr in

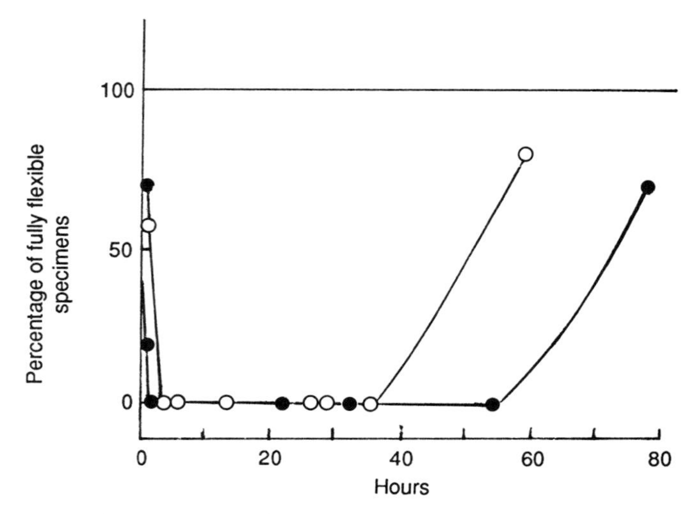

FIG. 1.10. Evolution of rigor mortis in sea bream (○) and anchovy (●) (Trucco *et al.*, 1982).

anchovy. However, resolution of rigor mortis was not apparent until 80 hr. The development of the actomyosin complex during rigor mortis was monitored by the reduced viscosity of the high-strength muscle extract (Crupkin *et al.*, 1979). Thus the course of rigor mortis could be followed by measuring the viscosity of the extract (Figure 1.11).

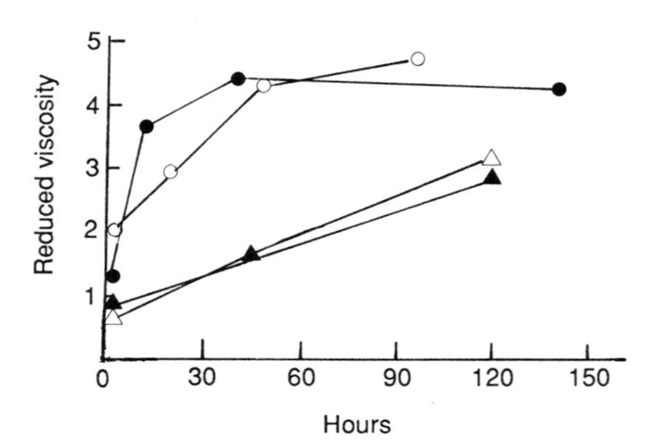

FIG. 1.11. Reduced viscosity of the high ionic strength muscle extract during the development of rigor mortis in hake (○), Patagonian blue whiting (●), tail hake (▲), and grenadier (△) (Trucco *et al.*, 1982).

IV. Conversion of Muscle to Meat and Edible Fish

A vast array of biochemical and physicochemical reactions take place from the time that the animal or fish is killed until it is consumed as meat or edible fish. This period can be divided into three distinct stages:

1. The prerigor state when the muscle tissue is soft and pliable is characterized biochemically by a fall in ATP and creatine phosphate levels as well as by active glycolysis. Postmortem glycolysis results in the conversion of glycogen to lactic acid, causing the pH to fall. The extent of pH change varies from one species to another as well as among different muscles. Nevertheless, in well-fed, rested animals the glycogen reserves are large so that in the postmortem state the meat produced has a lower pH compared to meat produced from animals exhausted at the time of slaughter.
2. The development of the stiff and rigid condition in the muscle known as rigor mortis. This occurs as the pH falls and is associated with formation of actomyosin. The loss of extensibility associated with the formation of actomyosin proceeds slowly at first (*the delay period*) and then extremely rapidly (*fast phase*).The onset of rigor mortis normally occurs at 1–12 hr postmortem and may last for a further 15–20 hr in mammals, depending on a number of factors to be discussed later. Fish generally exhibit a shorter rigor mortis period commencing 1–7 hr after death, with many factors affecting its duration.
3. The postrigor state during which time the meat and fish muscles gradually tenderize, becoming organoleptically acceptable as aging progresses. Mammalian meat usually attains optimum acceptability when stored for 2–3 weeks at 2°C following dissolution of rigor.

The importance of rigor mortis in fish is recognized by the fishing industry, since in addition to retarding microbial spoilage, it affords a stiffness to the fish which is generally recognized by the consumer as a sign of good quality. The rigor period, however, is also a distinct disadvantage with respect to the filleting of fish, as it renders the fish too stiff to process. Thus filleting is carried out following dissolution of rigor, or on trawlers, immediately prior to the development of rigor.

The principal changes following death are summarized in Figure 1.12. Following the death of the animal or fish, circulation of the blood ceases, which results in a complex series of changes in the muscular tissue. Since blood is an ideal medium for spoilage microorganisms, as much as possible is removed from the animal carcass to ensure that the edibility and keeping quality of the meat are maintained. With respect to fish, only some of the larger species are bled; the blood differs from that of warm-blooded animals and coagulates far more rapidly. The advent of the modern fish-processing trawler brought with it a number of

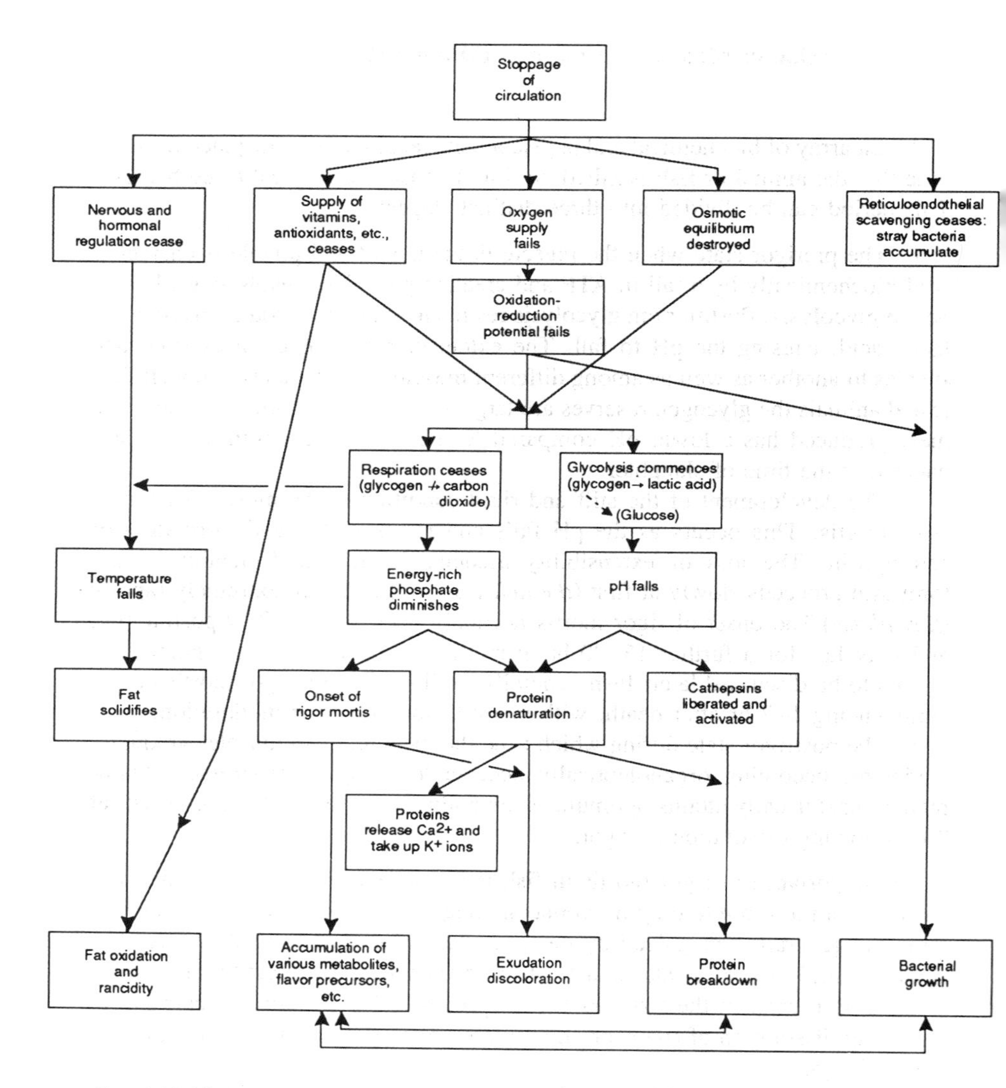

FIG. 1.12. The consequences of stoppage of the circulation in muscular tissue (Lawrie, 1985). Reprinted with permission. Copyright © by Pergamon Press.

problems, particularly surface discoloration in sea-frozen fillets. This discoloration in the prerigor processed cod fillets developed from surface contamination by the fish blood in the form of met-hemoglobin (Kelly and Little, 1966). This phenomenon can be prevented in prerigor fish fillets of ocean perch (*Sebastes marinus*) and cod (*Gadhus morhua*) by correct bleeding methods.

The most immediate effect of the stoppage of blood circulation and the removal of blood from the muscle tissue is depletion of the oxygen supply to the tissue and the subsequent fall in oxidation reduction potential. This results in an inability to resynthesize ATP, as the electron transport chain and oxidative phosphorylation mechanisms are no longer operative.

A. ATP and Postmortem Changes

The major source of ATP supply to the muscle fibers is lost following the death of the animal or fish, since glycogen can no longer be oxidized to carbon dioxide and water. In its place, anaerobic metabolism takes over, resulting in the conversion of glycogen to lactic acid. Under normal aerobic conditions 39 molecules of ATP are produced for each glucosyl unit of glycogen oxidized compared to only 3 molecules of ATP for each hexose unit broken down under anaerobic conditions. The time for the first phase development of rigor mortis is determined by the postmortem level of ATP. The level of ATP is also depleted by the noncontractile ATPase activity of myosin which maintains the temperature and structural integrity of the muscle cell (Bendall, 1973). This results in the production of inorganic phosphate (P_i), which stimulates the degradation of glycogen to lactic acid. Inorganic phosphate is essential for the phosphorolysis of glycogen to glucose l-phosphate by muscle phosphorylase, which is the initial step in the degradation of glycogen. In addition to the ATPase of myosin, the sarcoplasmic reticulum also has ATPase activity.

The level of ATP is maintained in the muscles after death by an active creatine kinase which catalyzes the resynthesis of ATP from ADP and creatine phosphate (CP) (Lawrie, 1966; Newbold, 1966). Thus in the early postmortem or prerigor period, the concentration of ATP remains relatively constant, whereas there is a rapid decline in creatine phosphate levels. In studies with rabbit muscle of well-rested and relaxed animals the creatine phosphate levels were high immediately postmortem but fell rapidly to one-third of the original before any detectable loss in ATP (Bendall, 1951). A rapid fall in creatine phosphate levels was also observed in poultry muscle accompanied by the liberation of free creatine (De Fremery, 1966). The presence of creatine was attributed to the transitory rise in pH in poultry muscle immediately postmortem. Studies by Hamm (1977) on ground beef muscle showed that creatine phosphate was totally degraded within 1–2 hr postmortem.

As discussed earlier, mammalian muscle is capable of maintaining its ATP level for as long as several hours postmortem, compared to fish skeletal muscle, which generally exhibits a rapid decline in ATP levels (Tomlinson and Geiger, 1962). Some species of fish, however, can maintain a constant ATP level but must be in an unexercised state prior to slaughter. The relationship between ATP and creatine phosphate levels in mammalian muscles appears to be similar to that reported in fish skeletal muscle (Partmann, 1965).

The continued activity of the several ATPases in the muscle cell, including in the sarcoplasmic reticulum, mitrochondria, sarcolemma, and the myofibrils, presumably contributes to the depletion of ATP in the muscle. Of these Hamm *et al.* (1973) concluded that it was the myofibrillar ATPase rather than membrane or sarcoplasmic reticulum ATPase which was probably responsible for the degradation of ATP in postmortem skeletal muscle. There is an overall decrease in the ATP level as a consequence of ATPase activity, a decrease in creatine phosphate, and the inability of postmortem glycolysis to synthesize ATP at an effective rate.

The development of rigor mortis in fish is also related to the reduction in ATP. Depletion of creatine phosphate, AMP, and glycogen reserves and the subsequent inability to resynthesize ATP result in the formation of the actomyosin complex. This is accompanied by the muscle becoming tough and inextensible. Unlike land animals, rigor mortis in fish terminates far more rapidly. Jones and Murray (1961) reported that the onset of rigor mortis for cod occurred when ATP dropped to 5% of the original level in the rested fish. This was corroborated for a number of Indian fish by Nazir and Magar (1963), although some species were found to enter rigor at much higher phosphate levels, for example, *Mugul dussumieri* and *Harpodon nehereus*. Jones *et al.* (1965) followed the steady decline in ATP during the postmortem period for cod until the point at which rigor was developed. The ATP levels for unexercised and exercised cod muscle were 2.35 and 0.82 μmole/g, respectively, which showed the effect of exercise in determining the ATP levels at which rigor was established. A reduction in the time required for rigor mortis development is also associated with excessive struggling by fish during capture, which has been correlated with a reduction in creatine phosphate.

B. Postmortem Metabolism of ATP

The development of rigor mortis in animals or fish is a direct response to the decline of ATP. Bendall and Davey (1957) observed that the liberation of ammonia occurred when rabbit voluntary muscle was fatigued or passed into rigor. This was shown to arise from deamination of adenylic acid (AMP) to inosinic acid (IMP). They postulated the direct deamination of adenosine diphosphate in which ammonia was produced in equimolar proportions to the disappearance of adenosine nucleotides, primarily AMP, during the development of rigor. Tsai *et al.* (1972) reported the presence of ATP, ADP, and IMP in prerigor porcine muscle and traces of AMP. The levels of ATP and ADP declined rapidly in the postmortem muscle, while the concentration of IMP, inosine, and hypoxanthine increased markedly.

Fraser *et al.* (1961) reported an increase in ammonia during the resolution of rigor in cod muscle. The postmortem degradation of ATP follows a similar

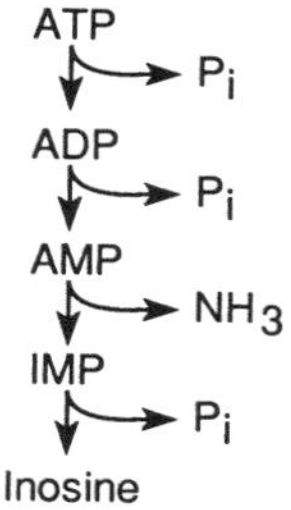

$$\begin{array}{l} ATP \\ \downarrow \rightarrow P_i \\ ADP \\ \downarrow \rightarrow P_i \\ AMP \\ \downarrow \rightarrow NH_3 \\ IMP \\ \downarrow \rightarrow P_i \\ \text{Inosine} \end{array}$$

SCHEME 1.1. Postmortem degradation of ATP to inosine in meat and fish.

pattern to that in mammalian muscles, in which ATP is rapidly degraded to ADP by the sarcoplasmic ATPase, and hydrolyzed by myokinase to AMP. AMP is then converted to IMP by deaminase action (Saito and Arai, 1958). Nucleotides, particularly IMP, are recognized as important contributors to the flavor of good-quality fish (Hashimoto, 1965). Scheme 1.1 summarizes the postmortem degradation of ATP in both fish and meat in which inosinic acid is dephosphorylated to inosine.

Bendall and Davey (1957) demonstrated that these reactions in meat were catalyzed by ATPase, myokinase, and deaminase at 37 and 17°C. In addition to these reactions shown in Scheme 1.1, the presence of ITP and IDP was also reported and attributed to the following reaction:

$$ADP \rightarrow IDP + NH_3$$

$$IDP \rightarrow ITP + IM$$

Small amounts of inosine and hypoxanthine, degradation products of IMP, were also found in postrigor mammalian muscle. The conversion of ATP to IMP occurred by the time the ultimate pH was reached, while the degradation of IMP followed the establishment of the final pH (Lawrie, 1966).

The degradation of ATP in fish muscle also leads to the formation of IMP, which is subsequently hydrolyzed to uric acid (Kassemsarn *et al.*, 1963; Saito *et al.*, 1959; Tarr, 1966) (Scheme 1.2.). The presence of 5′-nucleotidase activity was reported in carp muscle by Tomioka and Endo (1984, 1985).

The level of these nucleotides changes drastically following the death of the fish. Many estimates have been made of fish freshness based on the level of nucleotides (Saito *et al.*, 1959), ammonia (Ota and Nakamura, 1952), amines (Karube *et al.*, 1980), and volatile acids (Suzuki, 1953). Of these the production of nucleotides from ATP appears to be the most reliable indicator in fish. While inosine is comparatively tasteless, its conversion to hypoxanthine gives rise to a bitter substance (Jones, 1965). Conflicting reports refuting this were subsequently presented by a number of Japanese workers, including Hashimoto

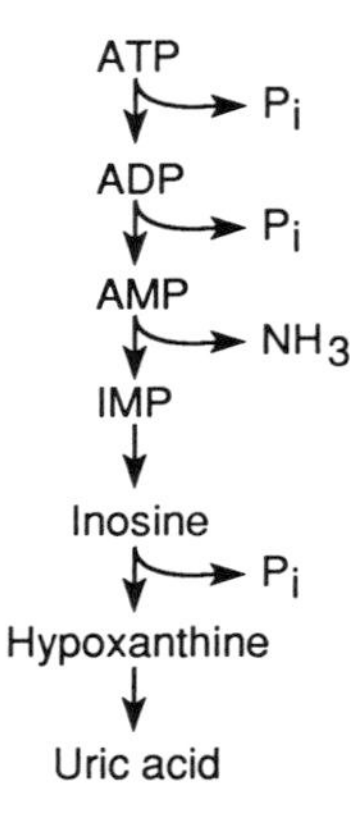

SCHEME 1.2. Degradation of ATP in fish muscle.

(1965), who suggested that hypoxanthine was tasteless. The presence of hypoxanthine in fish muscles was proposed as a chemical index of freshness and quality in fish (Jones *et al.*, 1964; Spinelli *et al.*, 1964). Dugal (1967) suggested that an average rate of hypoxanthine formation could be obtained for a group of fish which would reflect the degree of freshness for a particular species of freshwater fish. Watanabe *et al.* (1984) estimated fish freshness by monitoring the level of IMP by an enzyme sensor. Karube *et al.* (1984) developed a multifunctional enzyme sensor system for assessing fish freshness based on measuring the levels of IMP, inosine, and hypoxanthine. This was based on the changes observed in ATP, ADP, and AMP levels in sea bass, saurel, mackerel, and yellowfish following death. Their results, shown in Figure 1.13, indicate a rapid decrease and reduction in ATP and ADP levels 24 hr after death as well as a drop in AMP to less than 1 μmole/g .

IMP increased sharply during the first 24 hr postmortem and then decreased gradually, accompanied by a rise in inosine and hypoxanthine. The changes in these nucleotides varied with the individual fish species. Since ATP, ADP, and AMP were still present in some of the fish varieties for up to 2 weeks they included these nucleotides with hypoxanthine, inosine, and IMP in the overall equation defining fish freshness.

Saito *et al.* (1959) first proposed the term "*K* value" as an indicator of fresh fish defined as the ratio of inosine plus hypoxanthine to the total amount of ATP-related compounds. The "*K* value" has since been used to express freshness of marine products (Lee *et al.*, 1982; Uchiyama and Kakuda, 1984). Ryder (1985) developed a rapid method for computing "*K* values" based on the quantitative measurement of ATP and its degradation products using HPLC. Surette *et al.* (1988) monitored the postmortem breakdown of ATP-related compounds in Atlantic cod (*Gadus morhua*) and reported that inosine hydrolysis and hypoxanthine formation resulted from both autolytic as well as bacterial enzyme activity.

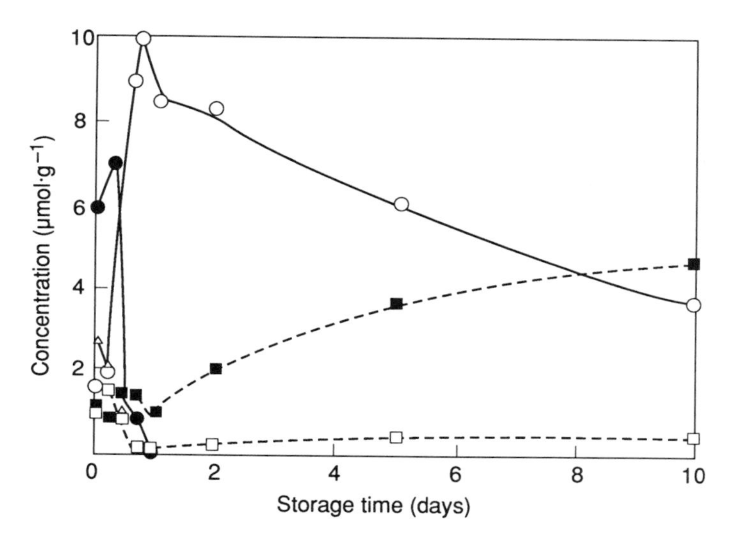

FIG. 1.13. Time course of ATP decomposition and associated reactions in sea bass. (●) ATP; (△) ADP; (□) AMP; (○) IMP; (■) HxR+Hx. Reprinted with permission from Karube *et al.* (1984). Copyright by the American Chemical Society.

These nucleotide catabolites provide a useful index of quality as their presence is affected by spoilage bacteria and mechanical damage during handling. These researchers suggested that a diagnostic kit for measuring catabolites, such as inosine monophosphate (IMP), inosine (HxR), and hypoxanthine (Hx), could provide a useful tool for assessing the fresh quality of cod.

The applicability of the "*K* value" for assessing the freshness of edible meat was reported by Nakatani *et al.* (1986). By monitoring the changes in ATP degradation products in beef and rabbit muscles during cold storage they proposed the following new index K_0:

$$\frac{\text{Inosine} + \text{hypoxanthine} + \text{xanthine} \times 100}{\text{ATP} + \text{ADP} + \text{AMP} + \text{IMP} + \text{adenosine} + \text{inosine} + \text{hypoxanthine} + \text{xanthine}}$$

Recent research by Fujita and co-workers (1988) also found that this index could be used to assess the freshness of both pork and chicken.

C. Adenosine Nucleotides and Protein Denaturation

During frozen storage, deteriorative changes in fish texture have been reported as a consequence of protein denaturation (Andou *et al.*, 1979, 1980; Acton *et al.*, 1983; Dyer, 1951). The possible effect of adenosine nucleotides on protein

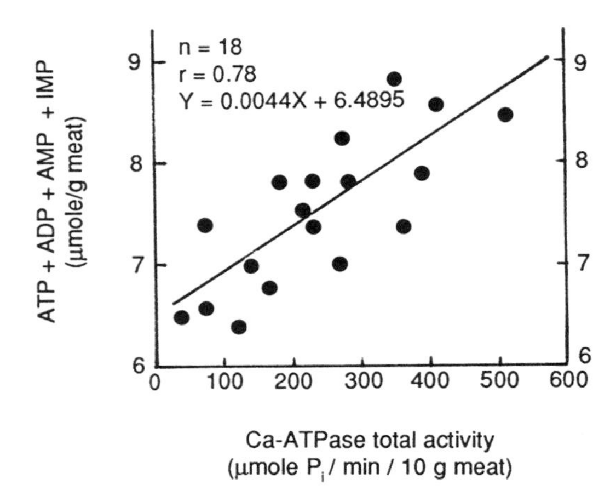

FIG. 1.14. Relationship between the sum of the quantity of adenosine triphosphate (ATP), adenosine diphosphate (ADP), adenosine monophosphate (AMP), and inosine monophosphate (IMP) and Ca-ATPase total activity. Reprinted with permission from Jiang *et al.* (1987). Copyright by the American Chemical Society.

denaturation was recently investigated by Jiang and co-workers (1987). These researchers assessed protein denaturation in fish frozen at −20°C by extractability of actomyosin (AM) and monitored the activities of Ca-ATPase and Mg(EGTA)-ATPase (EGTA=ethylene glycol *bis*(2–aminoethylether)tetraacetic acid) in AM. During frozen storage the molecular weight of the myosin heavy chain and actin decreased. The least stable muscle was associated with the lowest level of ATP, ADP, AMP, and IMP and the highest levels of inosine and hypoxanthine. A correlation of −0.80 was obtained between inosine and hypoxanthine and the Ca-ATPase total activity of AM compared to +0.78 for ATP, ADP, AMP, and IMP (Figs. 1.14 and 1.15). These results point to the possible involvement of adenosine nucleotides with protein denaturation.

D. Postmortem Glycolysis

Once the supply of oxygen to the muscle tissue is depleted, glycogen, the main carbohydrate of animal and fish muscle, undergoes anaerobic glycolysis to lactic acid. Compared to mammalian muscle, the level of glycogen in fish muscle is reported to be much lower. Tomlinson and Geiger (1962), however, found a close similarity between the muscle glycogen levels for many species of both fish and warm-blooded animals. This was attributed to the excessive struggling normally associated with the capture of fish, resulting in depletion of the glycogen level compared to that in the rested fish.

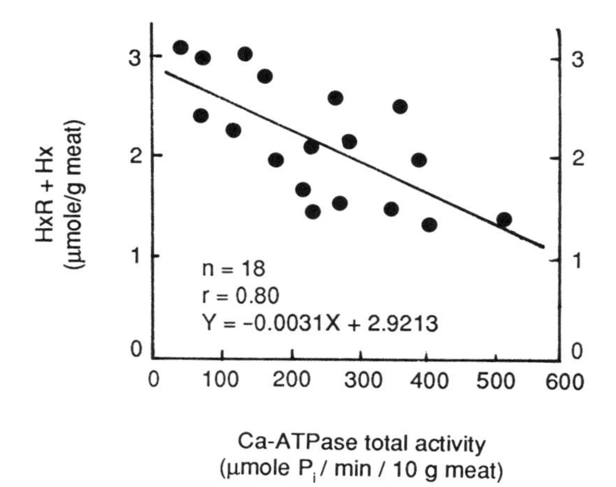

FIG. 1.15. Relationship between the content of inosine (HxR) and hypoxanthine (Hx) and Ca-ATPase total activity. Reprinted with permission from Jiang *et al.* (1987). Copyright by the American Chemical Society.

Postmortem degradation of glycogen in fish muscle suggests that two possible pathways are involved:

1. hydrolytic or amylolytic pathway;
2. phosphorolytic pathway.

These are illustrated in Scheme 1.3.

The postmortem conversion of glucose 6-phosphate to glucose by phosphomonoesterase only occurs to a slight extent in fish muscle. Consequently the hydrolytic pathway appears to be the main one operating in fish. This pathway was first postulated by Ghanekar *et al.* (1956) and since confirmed as the main degradative pathway of glycogen to glucose for most fish (Burt, 1966; Nagayama, 1966; Tarr, 1965). In mammalian muscle, however, it is the phosphoroly-

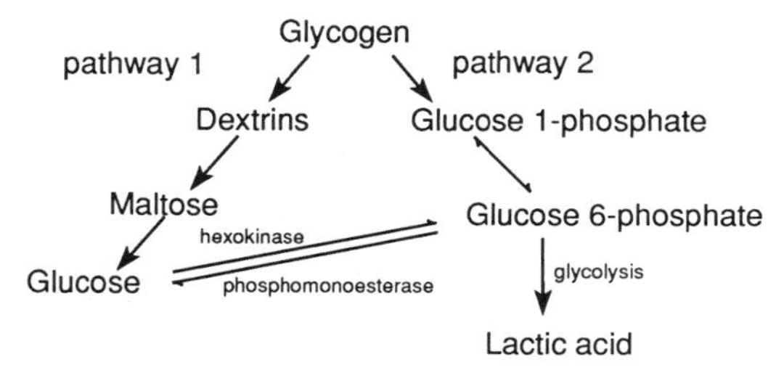

SCHEME 1.3. Postmortem degradation of glycogen.

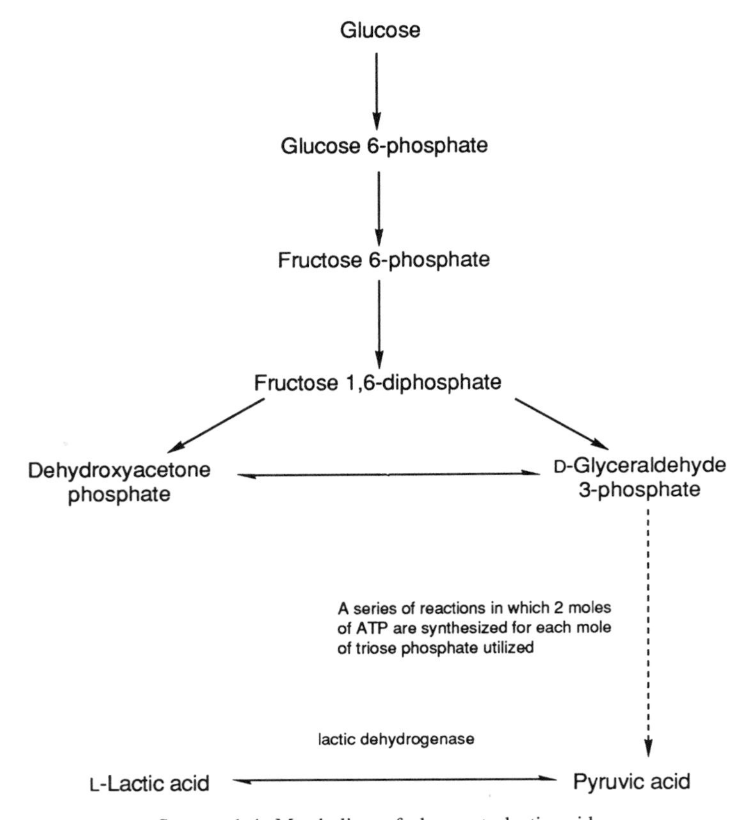

SCHEME 1.4. Metabolism of glucose to lactic acid.

tic pathway that is responsible for glycogen degradation. Irrespective of which pathway is involved in the initial breakdown of glycogen, the final pathway of glycolysis is the same for either animal or fish muscle. The enzymes responsible have, in the main, been characterized and identified in mammalian muscle, with many since reported in many species of fish, including rainbow trout (MacLeod *et al.*, 1963; Tarr, 1968).

The general reactions involved in the glycolytic pathway are outlined in Scheme 1.4.

The rate of postmortem glycolysis in muscles is affected by temperature, muscle fiber type, and hormone secretions, as well as the intensity of the nervous stimuli in the muscle prior to and during slaughter (Bendall, 1973; Beecher *et al.*, 1965; Disney *et al.*, 1967; Tarrant *et al.*, 1972a, b). The effect this has on the pH of the muscle will be discussed in the next section.

E. Postmortem pH

The production of lactic acid causes the pH of the muscle tissue to drop from the physiological pH of 7.2–7.4 in warm-blooded animals to the ultimate postmortem pH of around 5.3–5.5. A direct relationship was demonstrated by Bate-Smith and Bendall (1949) between the rate of fall of pH in postmortem rabbit muscle and lactic acid production. It is particularly important to attain as low a pH as possible in the tissue, since in addition to retarding the growth of spoilage bacteria, it imparts a more desirable color to the meat. In the case of frozen fish, however, a higher pH is more desirable to prevent toughness. The final pH can be attained within the first 24-hr postmortem period, the glycolytic pathway being related to ATP production, the net fall of which is directly responsible for the development of rigor mortis. The interrelationship between the creatine phosphate disappearance, fall in the levels of ATP and pH, and the decrease in extensibility as a measure of rigor mortis is shown in Figure 1.16. ATP is the major source of acid-labile phosphorus, while the fall in pH is a measure of glycolysis.

A postmortem pH of 5.3–5.5 is attained in the muscles of well-rested animals fed just prior to slaughter, when glycogen is at a maximum level. Animals that undergo severe death struggling, however, are fatigued prior to slaughter and characterized by lower glycogen levels as illustrated in Table 1.2. for chicken breast muscle.

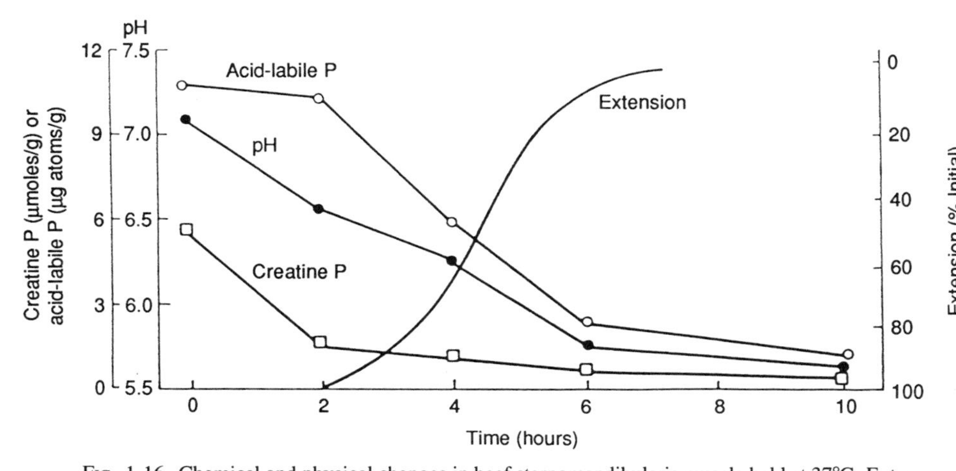

Fig. 1.16. Chemical and physical changes in beef sternomandibularis muscle held at 37°C. Extension changes were recorded on an apparatus similar to that described by Bate-Smith and Bendall (1949) using a load of about 60 g/cm^2 and a loading–unloading cycle of 8 min on and 8 min off. Zero time: 1 hr 45 min postmortem (Newbold, 1966).

TABLE 1.2

THE EFFECT OF SLAUGHTER CONDITIONS ON THE GLYCOGEN CONCENTRATION THREE MINUTES POSTMORTEM IN CHICKEN BREAST MUSCLE[a]

Condition	Initial glycogen level (mg/g)
Anesthetized	8.3
Stunned	6.0
Struggling	3.4

[a] Adapted from De Fremery (1966).

The lower glycogen levels result in a higher final postmortem pH of around 6.0–6.5, producing a dark, dry, and close-textured meat that is much more susceptible to microbial spoilage (Cassens, 1966; Joseph, 1968a,b). This meat, referred to as DFD (dark, firm, and dry), still represents a serious quality problem, particularly in beef (Tarrant, 1981). Improper handling of the cattle prior to slaughter still remains the major cause of physiological stress and exhaustion (Grandin, 1980; Tarrant, 1981). Excessive physical exercise in animals depletes the muscle glycogen although other factors such as fasting, trauma, and psychological stress have also been implicated (Bergstrom *et al.*, 1965; Bergstrom and Hultman, 1966; Conlee *et al.*, 1976; Howard and Lawrie, 1956; Sugden *et al.*, 1976). The incidence of this problem was reported to be 8% in Canada and between 0.3 and 4.8% in the United States (Munns and Burrell, 1966). Tarrant and Sherrington (1980) monitored the final postmortem pH of steer and heifer carcasses at a slaughtering plant in Ireland over a 3-year period. A seasonal effect was noted for the development of DFD with an average incidence of 3.2% in the carcasses examined. Measurement of the ultimate pH of the meat still remained the best method for characterizing this phenomenon. The pH limits reported for the development of DFD ranged from 5.8–5.9 as the lower limit for normal meat to 6.2–6.3 as the upper limit for extreme DFD (Fjelkner-Modig and Ruderus, 1983; Tarrant, 1981). The incidence of DFD in Sweden was examined by Fabiansson and coworkers (1984), who classified beef carcasses with a pH less than 6.2 after 24 hr as DFD. The overall incidence of DFD in electrically stimulated and nonstimulated carcasses was 3.4 and 13.2%, respectively.

The final postmortem pH in meat rarely falls below 5.3, although several exceptions have been reported. For example, in pig longissimus dorsi muscle, pH values ranging from 4.78 to 5.1 have been recorded (Lawrie *et al.*, 1958). Meat with a pH of 5.1–5.5 was found to be in an exudative condition with a whitish color and loose texture, while meat at pH 4.78 had abnormal muscle fibers. This is not unexpected, as the isoelectric point of the major meat proteins is around pH 5.5, which would lead to loss of water-holding capacity.

The lactic acid concentration in fish is similarly dependent on the initial glycogen stores prior to death as well as on the treatment of the fish. Fish muscles have been reported to have lactic acid concentrations ranging from 0.29% in haddock (Ritchie, 1926) to 1.2–1.4% in tuna (Tomlinson and Geiger, 1962) during rigor mortis. Most fish exhibit a higher postmortem pH compared to warm-blooded animals of around 6.2–6.6 at full rigor. An exception is in the case of flatfish, where a final pH of 5.5, similar to that of mammals, has been reported. The struggling of fish during capture substantially depletes the glycogen stores, resulting in a high pH at rigor of around 7.0, giving rise to a condition known as "alkaline rigor." This condition was reported in cod as well as other fish species (Fraser *et al.*, 1961). A recent review by Wells (1987) noted that the capture, transportation, and handling of live fish is accompanied by substantial biochemical and physiological changes. The extent of such changes was dependent on species as well as environmental conditions. Eliminating stress in fish by allowing them to return to a resting condition during captivity prior to death should be the ultimate goal to avoid the abnormal pH changes incurred during struggling.

Low ultimate pH has also been associated with textural problems in fish such as halibut, Alaska pollack, and tuna (Konagaya and Konagaya, 1979; Patashnik and Groninger, 1964; Suzuki, 1981). Love (1975) noted that while low ultimate pH produced a tough texture, high ultimate pH resulted in a "sloppy" soft texture in Atlantic cod (*Gadus morhua*). This condition rendered the fish unfit for filleting and produced a poor frozen product (Love *et al.*, 1982; MacCallum *et al.*, 1967). Postmortem changes in soft-textured cod caught off Newfoundland were examined by Ang and Haard (1985), who reported the lowest ultimate pH in cod which had been feeding heavily on capelin prior to capture. The Atlantic cod do not feed during the months prior to spawning in May and June but feed intensely during the postspawning period. These researchers found that the muscles of cod caught during this intense feeding period were characterized by a persistent low ultimate pH up to 100 hr following rigor. Ang and Haard (1985) suggested that the altered metabolic state as a result of the heavy feeding caused the low and stable ultimate pH in the cod muscle, which was responsible for the soft texture in these fish.

F. Time Course of Postmortem Glycolysis

The ultimate postmortem pH is dependent on the physiological state of the muscle, the type of muscle, as well as the species of animal or fish studied. The different rates of fall of pH with time are shown in Figure 1.17 for three species of animal.

Several postmortem changes observed visually in meat are related to the rate of decline of pH and temperature. For example, a rapid decline in pH in

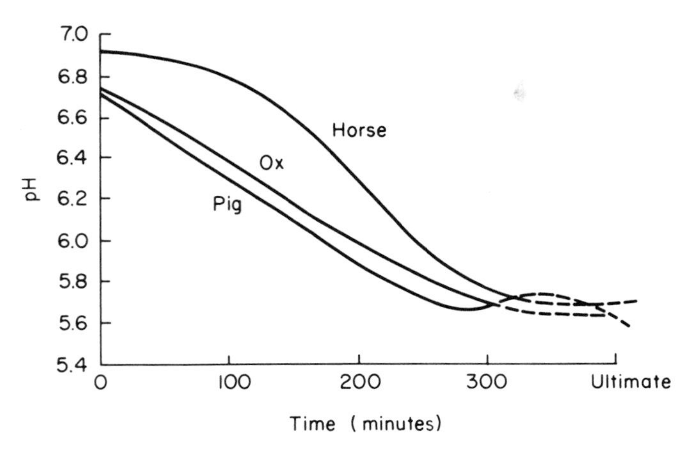

FIG. 1.17. The effect of species in a given muscle (longissimus dorsi) and at a given temperature (37°C) on the progress of glycolysis measured by a fall in pH (Lawrie, 1966). Reprinted with permission. Copyright © by Pergamon Press.

beef held at around body temperature resulted in changes in color, decreased water-holding capacity (WHC), as well as some muscle protein denaturation (Chaudhry *et al.*, 1969; Follett *et al.*, 1974; Lister, 1970; Locker and Daines, 1975; Scopes, 1964). Earlier work by Cassens (1966) and Briskey and coworkers (1966) studied the rate of glycolysis in porcine muscle by monitoring the decline in pH. In addition to the physiological state of the muscle they also observed that certain pigs were predisposed to a rapid postmortem glycolysis. The properties of the meat associated with fall in pH are summarized in Table 1.3.

TABLE 1.3

RELATIONSHIP BETWEEN TYPE OF FALL IN pH AND MEAT PROPERTIES[a]

Final pH	Type of decrease	Properties of meat
6.0–6.5	Slow, gradual	Dark
5.7–6.0	Slow, gradual	Slightly dark
5.3–5.7	Gradual	Normal
5.3–5.6	Rapid	Normal to slightly dark
5.0	Rapid	Dark to pale but exudative
5.1–5.4 then up to 5.3–5.6	Rapid	Pale and exudative

[a] Adapted from Cassens (1966).

Similar changes occur in fish muscle as the pH declines. A low postmortem pH was associated with poor fish texture, low water-holding capacity, and high drip loss (Kelly, 1969). A high water content in cod muscle was identified by Love (1975) in those fish with high postmortem pH, which correlated best with texture of the cooked fish.

G. Effect of Temperature on Postmortem Glycolysis: Cold Shortening

The rate of postmortem glycolysis varies with temperature as evident by differences in the final pH of mammalian muscles (Cassens and Newbold, 1966; Marsh, 1954; Newbold, 1966; Newbold and Scopes, 1967). These researchers all reported the hastening of rigor mortis as temperature is reduced from 5 to 1°C due to increased glycolytic activity and ATP hydrolysis. Stimulation of contractile actomyosin ATPase appeared to be potentiated by the release of Ca^{2+} ions. This phenomenon, known as "cold shortening," results in toughening of cooked meat (Marsh and Leet, 1966). A 30- to 40-fold increase in the level of ionic calcium was reported by Davey and Gilbert (1974) in the myofibril region of beef muscle held at 0°C compared to 15°C. Jeacocke (1977) examined the relationship between temperature and postmortem pH decline in beef sternomandibularis muscle. The results illustrated in Fig. 1.18 indicate a minimum fall in pH over 10–12°C which increased as the temperature dropped to 0°C, characteristic of "cold shortening." This was attributed to an increase in glycolysis due to enhanced ATPase activity in contractile actomyosin. Cornforth and coworkers (1980) subsequently confirmed earlier research by Buege and Marsh (1975) that the mitochondrial content of the muscle was involved in "cold shortening." These researchers also proposed a role for the sarcoplasmic reticulum in the reversibility of this phenomenon. This was attributed to the possible effect of temperature on the membrane of the sarcoplasmic reticulum and the subsequent release of Ca^{2+} ions.

Honikel *et al.* (1983) identified two types of shortening taking place in beef muscle. One which occurred above 20°C was referred to as "rigor shortening" while the other taking place below 15°C was termed "cold shortening." In both cases muscle contraction was explained by the release of Ca^{2+} ions into the myofibrillar space in the presence of of adequate levels of ATP. The uptake of Ca^{2+} ions by the sarcoplasmic reticulum was particularly sensitive to both pH and temperature changes. For example, Honikel (1983) found "rigor shortening" commenced at pH 6.25 in the presence of 2.4 μmole ATP/g muscle. This represented optimum conditions for the uptake of Ca^{2+} ions by the sarcoplasmic reticulum as observed previously by Cornforth and coworkers (1980) and Whiting (1980). The myofibrillar Mg/Ca-ATPase activity was reported by Bendall

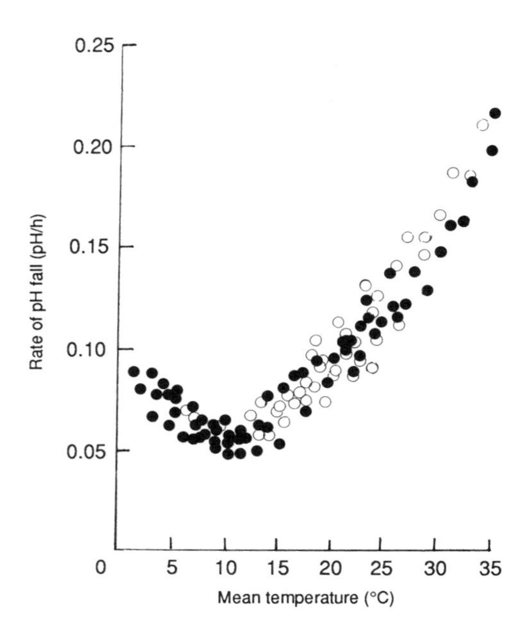

FIG. 1.18. The rate of pH fall in beef sternomandibular muscle as a function of the mean temperature of the adjacent thermocouple junction. The results of 16 different temperature gradients are pooled. (●) Muscles vacuum-packed before insertion into the apparatus; (○) muscles not vacuum-packed (Jeacocke, 1977).

(1969) to be independent of pH between 6 and 7. Both theories explain the development of "rigor shortening" at temperatures of 20°C and higher and at pH below 6.3. In sharp contrast, however, "cold shortening" developed at pH 7 in the presence of full ATP concentration (4 μmole/g) in the muscle. The occurrence of "cold shortening" was attributed by Cornforth *et al.* (1980) to the combined effect of the release of Ca^{2+} ions from the muscle mitochondria and the reduced uptake of Ca^{2+} ions by the sarcoplasmic reticulum.

The development of "cold shortening" is highly undesirable and can be avoided by holding the meat at a minimum temperature of 15°C until the pH drops below 6.0. Lamb carcass, however, should be held for at least 16 hr to ensure that prerigor changes have been completed (McCrea *et al.*, 1971). This represents a delay for the meat-processing industry, which utilizes hot-deboning of beef carcasses which are cut and rapidly refrigerated below 15°C long before the pH falls below 6.0. One technique developed to rapidly reduce the pH of the carcasses to below 6.0 involves the use of electrical stimulation to accelerate postmortem glycolysis.

H. Effect of Electrical Stimulation on Postmortem Glycolysis

Electrical stimulation of muscle has long been known to accelerate postmortem glycolysis and hasten the onset of rigor (De Fremery and Pool, 1960; Forrest and Briskey, 1967; Hallund and Bendall, 1965; Harsham and Detherage, 1951). Carse (1973) prevented "cold shortening" by subjecting freshly slaughtered lamb carcasses to 250-V pulses and attained pH 6.0 within 3 hr compared to 15.4 hr for the unstimulated carcass. This technique facilitated accelerated conditioning of lambs and is commercially used in New Zealand. Similar results were also reported in lamb by Bendall (1976) and Chrystall and Hagyard (1976). The latter researchers monitored the progress of glycolysis in stimulated and unstimulated lamb longissimus dorsi muscles by following the change in pH as shown in Fig. 1.19. Electrical stimulation substantially accelerated postmortem glycolysis, with the final pH 5.5 attained within 8 hr for the stimulated carcass compared to 24 hr for the control. Chrystall and Devine (1978) noted that during electrical stimulation glycolysis was stimulated as much as 150-fold, which resulted in a marked drop in pH. Even following cessation of the stimulus, the rate of glycolysis can increase by as much as threefold. The initial effect was attributed by Newbold and Small (1985) to activation of glycogen phosphorylase, which reached a peak following 30 sec of electrical stimulation. The increased glycolysis and ATP turnover following cessation of electrical stimulation, however, still remains unexplained.

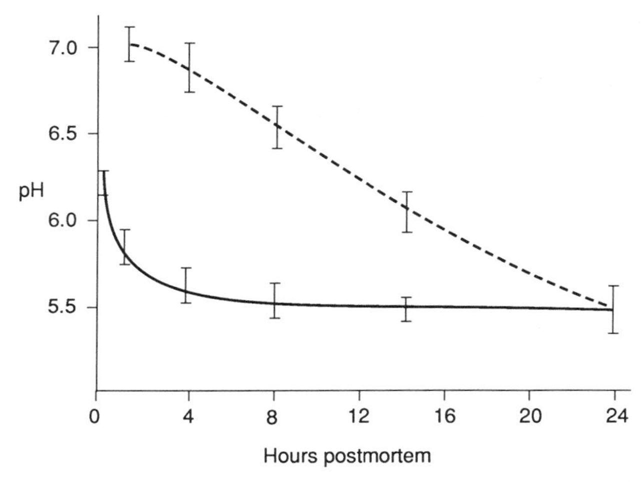

Fig. 1.19. Time course of pH fall in longissimus dorsi muscles of stimulated (——) and unstimulated (-----) animals. Standard deviation is shown by the vertical bar (Chrystall and Hagyard, 1976).

Horgan and Kuypers (1985) examined postmortem glycolysis in rabbit longissimus dorsi muscles following high-voltage and low-voltage stimulation. As expected there was an increase in the fall of pH as well as in, for example, phosphorylase activity following electrical stimulation. This was attributed to an increase in phosphorylase kinase activity and a substantial loss in phosphorylase and phosphatase activities. The yield of sarcoplasmic reticulum was reduced in the electrically stimulated muscles although there was an increase in basal ATPase activity. The different rates of glycolysis, as measured by decline in pH for the low- and high-voltage treatments, were attributed to their effects on myofibrillar ATPase activity.

I. Prerigor Pressurization

Prerigor pressurization is another accelerated procedure for processing meat that has been developed (Elkhalifa *et al.*, 1984a,b; Macfarlane, 1973; Kennick *et al.*, 1980). MacFarlane (1973) observed a rapid fall in pH in pressure-treated meat as well as improved tenderness ratings. This method has some similarities in its effect on meat quality with that described for electrical stimulation. The commercial viability of this procedure remains to be explored.

J. Glycolytic Enzymes

Investigations were conducted to determine the first glycolytic enzyme inhibited as the pH falls, as glycolysis ceases at pH values much higher than 5.3. Newbold and Lee (1965) found that phosphorylase was the limiting enzyme in minced sternomandibular muscle diluted with an equal volume of 0.16 *M* potassium chloride, which was consistent with earlier studies by Briskey and Lawrie (1961). Kastenschmidt and co-workers (1968) examined the metabolism of pig longissimus dorsi muscle and confirmed that phosphorylase was the primary control site in postmortem glycolysis. Phosphorylase is one of the key enzymes in glycolysis present in muscles (Scopes, 1970). In addition to this enzyme, Kastenschmidt and co-workers (1968) also implicated phosphofructokinase and pyruvic kinase in glycolytic control. On the basis of these and related studies, the muscles in pigs were classified as "fast" or "slow" glycolyzing muscles (Briskey *et al.*, 1966; Kastenschmidt *et al.*, 1968).

K. Pale Soft Exudative Condition (PSE)

The pale, soft, and exudative condition associated with porcine muscles, also referred to as PSE, is due to the rapid drop in pH to 5.3–5.8 within 1 hr following death while the muscle temperature is still above 36°C. The combined conditions of high temperature and low pH cause a partial denaturation of muscle proteins (Bendall and Lawrie, 1964; Charpentier, 1969; Goutefongea, 1971). The effect of high temperature and low pH on the properties of phosphorylase was examined by Fischer *et al.* (1979). A combination of high temperature and low pH caused the denaturation of phosphorylase with the resultant loss of activity as well as decreased solubility in PSE muscles. A similar condition was reported in beef muscle, described as pale, watery beef. Fischer and Hamm (1980) studied postmortem changes in fast-glycolyzing muscles of beef. Significant correlations were obtained in which a lower muscle pH was associated with lower water-holding capacity and lower glycogen and higher lactate levels. It was evident that phosphorylase activation occurred in fast-glycolyzing muscles although the overall effect on beef quality was far less severe compared to that in PSE pork.

L. Postmortem Changes in Meat and Fish Proteins

The decline in pH of the muscle to an acidic state, together with the various exothermic reactions, such as glycolysis, has a profound effect on the muscle proteins of both meat and fish. This section will discuss the phenomenon of protein lability within the muscle and its effect on meat and fish quality.

Shortly after death, the body temperature in cattle may rise from 37.6 to 39.5°C (Meyer, 1964). Even during refrigeration meat cools slowly as a result of the various exothermic reactions taking place, such as glycolysis. This phenomenon is known as "animal heat" and was recorded from ancient times. Consequently postmortem changes in muscle proteins are affected very often by a combination of high temperatures and low pH. Such changes include loss of color and decrease in water-holding capacity (Cassens, 1966). In meat it is the sarcoplasmic proteins that are denatured and become firmly attached to the surface of the myofilaments, causing the lightening of meat color (Bendall and Wismer-Pederson, 1962). In fish, sarcoplasmic proteins are generally more stable than the myofibrillar proteins, and are unaffected by dehydration or prolonged cold storage.

1. *Water-Holding Capacity*

The major postmortem change in meat and fish muscle is the loss of water or exudation. In lean meat, which has around 75% water, the majority of the water is somewhat loosely bound by the meat proteins. In the prerigor state, meat has a high water-holding capacity which falls within the first few hours following death to a mimimal level coincident with the establishment of rigor mortis. This minimal level corresponds to the ultimate postmortem pH of 5.3–5.5 which is

TABLE 1.4

DRIP LOSSES IN FOUR MUSCLES IN THE ROUND[a]

Muscle	Depth (cm)	Percentage drip ± SE (n)[b]
M. biceps femoris	1.5	8.1 ± 0.7 (6)
	5	14.9 ± 0.8 (6)
	8	18.2 ± 0.7 (6)
M. semitendinosus	1.5	7.5 ± 1.0 (6)
	8	16.7 ± 1.2 (6)

[a] Adapted from Tanrant and Mothersill (1977).
[b] SE = standard error; n = number of muscles examined.

the isoelectric point of the major muscle proteins. A rapid fall in pH leads to a number of changes including some muscle protein denaturation (Scopes, 1964; Chaudhrey *et al.*, 1969) and loss of water-holding capacity (Disney *et al.*, 1967). Tarrant and Mothersill (1977) determined glycolytic rates at several locations in beef carcasses as they affect the properties of the muscles. The postmortem pH decreased the farther away from the carcass surface, indicating an increased rate of glycolysis which was accompanied by a greater degree of protein denaturation and drip loss. Drip loss was measured by low centrifugation of intact muscle samples from several muscles, as shown in Table 1.4.

The muscles at a depth of 8 cm were paler, softer, and wetter compared to those at 5 or 1.5 cm after 2 days postmortem, which is somewhat characteristic of the PSE (pale, soft, and exudative condition) observed in pork although not as pronounced. This increased drip loss was attributed to a decrease in the water-holding capacity of denatured muscle proteins and sarcolemma disruption. The smallest drip loss at 1.5 cm was attributed to the low temperatures ($\ll$15°C) attained before the pH reached 6.0. This explained why rapid cooling of beef carcasses following slaughter minimized the amount of drip loss.

In the case of fish muscle the ultimate pH tends to be higher than meat, hardly falling below 6.0 even in full rigor. However, considerable water losses were reported from exercised fish muscle similar to that in mammalian skeletal muscle (Partmann, 1965). A rapid rise in expressible fluid was observed in cod stored in ice for a 168-hr period (Banks, 1955). Tomlinson *et al.* (1965) reported that a decrease in pH in halibut caused protein insolubility, which resulted in a pale, soft, exudative condition resembling PSE in pork (Briskey, 1964). This condition, known as chalkiness in halibut, is a problem for the fishing industry in the Pacific Northwest because of rejection by the consumer. To alleviate this condition, fish are allowed to remain alive after capture to remove excess lactic acid, so that a normal postmortem pH is attained following death.

M. Postrigor Tenderness

The most widely used process for tenderizing meat involves postmortem aging of the carcass. An optimum aging period of 8–11 days for choice carcasses from beef was proposed by Smith *et al.* (1978), although longer periods are used by the meat industry. The aging (conditioning or ripening) process is accelerated by raising the temperature. For example, in the Tenderay Process (American Meat Institute Foundation, 1960), beef is held at 15°C for a 3-day period in ultraviolet light to control the surface microbial spoilage. This contrasts with fresh pork, which is not aged because of the rapid onset of fat rancidity even at low temperatures. In addition to temperature, both electrical stimulation and pressurization accelerate postmortem aging as discussed previously (Savell *et al.*, 1981; Koohmaraie *et al.*, 1984).

N. Mechanism of Postrigor Tenderization

Over the past 15 years considerable progress has been made in understanding the mechanism of meat tenderization which is having a great impact on meat quality. For over 50 years increased tenderness was associated with proteolysis of the muscle proteins, which resulted in an increase in insoluble nonprotein nitrogen such as peptides and amino acids. Controversy arose over which of the muscle proteins underwent proteolysis during aging. It is now clear that neither collagen nor the myofibrillar proteins undergo extensive hydrolysis during aging but rather more subtle changes are taking place. These appear to be catalyzed by at least two distinct groups of enzymes, the calcium-activated sarcoplasmic factors and the lysosomal enzymes, the actions of which are summarized in Table 1.5 (Lawrie, 1983).

1. *Calcium-Activated Factor and Myofibril Fragmentation*

A prominent postmortem change observed in the myofibril was the degradation of the Z-disk, which was accelerated by the presence of Ca^{2+} (Busch *et al.*, 1972; Henderson *et al.*, 1970). A calcium-requiring endogenous protease capable of degrading the Z-disk was discovered by Busch *et al.* (1972), which was subsequently referred to as the calcium-activated factor (CAF). This factor was active at neutral pH in the presence of calcium ions (Dayton *et al.*, 1975,

TABLE 1.5

STRUCTURALLY IMPORTANT SITES OF ENZYME ACTION[a] DURING CONDITIONING

Calcium-activated sarcoplasmic factors (CAF)	Lysosomal enzymes (including cathepsins B and D
Troponin T (above pH 6)	Troponin T (below pH 6)
Z-lines (desmin)	Cross-links of nonhelical telopeptide regions of collagen
Connectin ("gap filaments")	Mucopolysaccharides of ground substance
M-line proteins and tropomyosin	Myosin and actin (below pH 5 or above 35°C)

[a] From Lawrie (1983).

1976a,b). It degraded tropomyosin, troponin T, troponin I, filamin, and C-protein with no detectable effect on myosin, actin, α-actinin, or troponin C (Dayton *et al.*, 1975; Dayton and Schollmeyer, 1980). CAF was found by Olson and coworkers (1977) to cause major changes in the myofibrillar proteins. Using a combination of microscopic and SDS–polyacrylamide gel electrophoresis they observed the degradation of the Z-line, fragmentation of the myofibril, and the disappearance of troponin T. This was accompanied by the release of a 30K-dalton myofibril component from aged beef longissimus dorsi muscle. The presence of a 30K-dalton component had been reported earlier during the aging of chicken and beef muscles (Hay *et al.*, 1973; Penny, 1974; Samejima and Wolfe, 1976). Adding calcium ions to minced muscle samples, Cheng and Parrish (1977) increased the rate of troponin T degradation to the 30K-dalton component. These results provided convincing evidence of the importance of CAF in the proteolysis and tenderization of postmortem muscle.

The role of CAF in meat tenderization was shown indirectly by Olson and coworkers (1976), who found postmortem degradation of the Z-disk and myofibril fragmentation in bovine longissimus muscle correlated with tenderness scores as well as Warner-Bratzler shear values (Table 1.6). An objective method was subsequently developed for measuring myofibril fragmentation, referred to as the myofibril fragmentation index (MFI), which correlated with the tenderness scores of beef steaks (Olson and Parrish, 1977). Confirmation of the relationship was provided in later studies on beef loin tenderness and MFI by Culler *et al.* (1978), MacBride and Parrish (1977), and Parrish *et al.* (1979). MacBride and

TABLE 1.6

EFFECT OF POSTMORTEM STORAGE (2°C) ON MYOFIBRIL FRAGMENTATION INDEX (MFI) AND WARNER-BRATZLER (W-B) SHEAR-FORCE OF LONGISSIMUS (L), SEMITENDINOSUS (ST), AND PSOAS MAJOR (PM) MUSCLES[a,b]

	Days of postmortem storage		
	1	3	6
MFI[c]			
L	49.6 ± 1.3	69.8 ± 1.1	76.3 ± 0.9
ST	48.8 ± 0.8	68.2 ± 1.1	77.6 ± 1.0
PM	47.1 ± 0.9	49.3 ± 1.1	54.7 ± 1.0
W-B[d]			
L	2.60 ± 0.20	2.23 ± 0.17	2.13 ± 0.12
ST	3.27 ± 0.11	2.72 ± 0.09	2.64 ± 0.11
PM	2.16 ± 0.12	1.94 ± 0.11	1.86 ± 0.17

[a] From Olson *et al.* (1976). Copyright © by Institute of Food Technologists.

[b] Means ± standard errors of five carcasses. Means not underscored by the same line are significantly different ($P < 0.05$).

[c] Absorbance per 0.5 mg myofibril protein × 200.

[d] Kilograms of shear-force per cm^2.

Parrish (1977) introduced the term "myofibril fragmentation tenderness" to describe tenderness in conventionally aged beef carcasses.

The structural role of troponin T is to bind troponin I and C to tropomyosin and to the thin filaments (Hitchcock *et al.*, 1973; Potter and Gergely, 1975). The dissolution of troponin T was found by Penny and Dransfield (1979) to highly correlate with decreased toughness of the meat during conditioning (Fig. 1.20). This loss reflected the rate and extent of proteolytic activity taking place.

2. *Lysosomal Enzymes*

At one time the lysosomal enzymes, the cathepsins, were considered to be the main agents of postmortem tenderness in meat (Bate-Smith, 1948). However, their precise role in postmortem tenderization still remains inconculsive. Cathepsins exhibit an optimum pH of around 5.5 and are active at 37°C. These enzymes are released from lysosomes under conditions of acid pH at the completion of glycolysis. Schwartz and Bird (1977) showed that rat muscle cathepsins B and D were capable of hydrolyzing myosin. A cathepsin from rabbit muscle was also found to break down myosin, α-actinin, and actin (Okitani *et al.*, 1980). Further

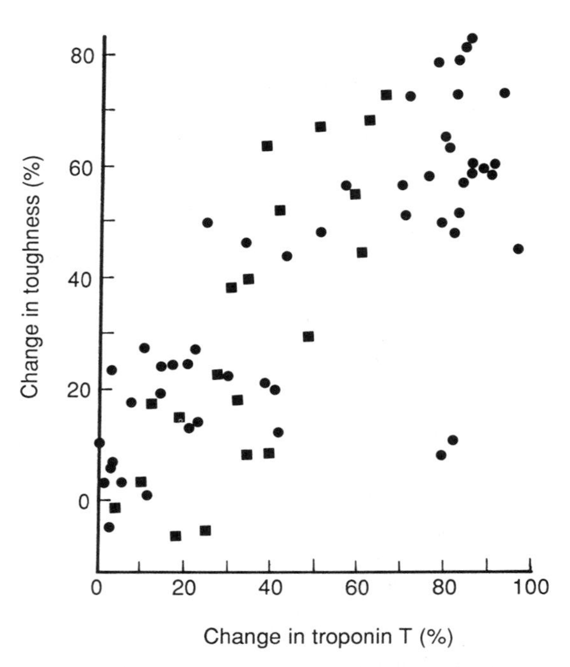

FIG. 1.20. The relationship between the change in toughness and the change in troponin T during conditioning of beef M. semitendinosus (●) and M. longissimus dorsi (■) (Penny and Dransfield, 1979).

research is necessary on the activities of cathepsins B and D which have yet to be demonstrated in postmortem muscle (Parrish and Lusby, 1983). The presence of cathepsin A in fish muscle was first reported by Makinodan and Ikeda (1976). This enzyme was subsequently purified by Toyohara *et al.* (1982) from carp muscle but did not appear to be directly involved in the postmortem proteolysis of the fish muscle.

The combined effects of the calcium-activated factor (CAF) and cathepsin D on the myofibrils of ovine longissimus dorsi muscles were examined by Elgasim and co-workers (1985). In both cases there was loss of the Z-line although in the case of CAF it was totally destroyed. The Z-line proteins, α-actinin, desmin, and actin, were degraded or released by either of the enzymes. CAF degraded desmin and released α-actinin without affecting actin. Cathepsin D primarily affected α-actinin with little or no effect on actin or desmin. Although inconclusive, it appears from this study that these two enzymes might act cooperatively in the degradation of the myofibrillar proteins.

O. Cytoskeleton and Meat Tenderness

1. *G-Filaments*

Davey and Graffhuis (1976) cooked beef neck muscle to maximum stretch and found that the G-filaments remained intact as long as the meat had not been aged. The tenderness of myofibrillar proteins involves weakening of the G-filaments. The rapid degradation of titin, a major protein component of the "G-filaments," was reported by King (1984) to occur at elevated temperatures as well as at 0 and 15°C. Their results differed from those of Locker and Wild (1984), who found that titin was particularly resistant to aging for up to 20 days at 15°C. Nebulin, however, was found to disappear within two days accompanied by an improvement in tenderness. No significant change in titin was reported although a new B-band was detected by polyacrylamide electrophoresis (Locker and Wild, 1986). A later study by Lusby *et al.* (1983) examined the aging of beef longissimus dorsi muscle at three storage temperatures (2, 25, and 37°C) over a 7-day period. Nebulin was also found to be rapidly degraded with the appearance of a new band corresponding to the B-band reported by Locker and Wild (1984). Lusby *et al.* (1983) claimed that titin-1 was converted to titin-2 as the time and temperature increased. However, Locker (1984) pointed out that the photographs of their gels were inconclusive because of lack of resolution of their bands. Paterson and Parrish (1987) noted some minor differences in the titin content between tough and tender muscles after 7 days postmortem. Using a monoclonal antibody method for titin (9 D 10) developed by Wang and Greaser (1985), Ringkob and co-workers (1988) showed that there was an alteration in the shape of titin within 2 days postmortem. This was based on the staining of two bands by the antibody per sarcomere in the 3-hr muscle compared to four bands per sarcomere after 48 hr. This pointed to either proteolysis of titin or another protein to which titin may be attached. The possible relationship between the presence of these bands and tenderness remains to be explored.

It was evident that during the aging of meat the G-filaments are degraded with more subtle changes occurring in titin. This appeared to involve nebulin, which is rapidly degraded in the aging muscle, and based on its role with titin in the model proposed by Wang (1983) this could explain its destabilizing effect on the G-filaments.

2. *Intermediate Filaments*

The degradation of desmin, which plays a role in linking adjacent myofibrils together, may cause physical changes to the postmortem muscle. Robson *et al.* (1984) monitored the changes in three major desmin fractions isolated from bovine skeletal muscle during postmortem aging for up to 7 days at 15°C by electrophoresis on SDS–polyacrylamide gels. Only a slight decrease in the amount of desmin was observed after 1 day although after 7 days storage a substantial loss of intact desmin was evident. This change was paralleled by a decline in troponin T (37K band) which disappeared after 7 days storage and was replaced by a band at around 30K, as observed previously by Olson *et al.* (1977). The degradation of desmin may be carried out by CAF since this has been shown by O'Shea *et al.* (1979) using purified desmin. Thus the degradation of desmin and the connecting intermediate filaments may play an important role in meat tenderness during aging (Yamaguchi *et al.*, 1982, 1983a,b).

P. Effect of Electrical Stimulation on Tenderness

Electrical stimulation has been reported to improve tenderness in meat (Savell *et al.*, 1977; McKeith *et al.*, 1980). Its effect was attributed to several factors, including reduction of cold shortening capacity, myofibril fragmentation, and enhanced activity of acid proteases caused by lowering of the pH (Davey *et al.*, 1976; Chrystall and Hagyard, 1976; Savell *et al.*, 1977, 1978). Sonaiya and co-workers (1982) subjected one side of a cow carcass (semimembranosus, longissimus dorsi, and triceps brachii) to electrical stimulation and monitored the pH, temperature, and myofibril fragmentation index over a period of 1 week following treatment. A significant drop in pH occurred in the treated muscles compared to the corresponding untreated muscles. A higher MFI was observed for the electrically stimulated muscles (Table 1.7), which is normally associated with the degradation of troponin T and the Z-disk by the calcium-activated factor. However, the appearance of the 30,000-D protein, normally associated with CAF action, did not reach a maximum level until 72 hr poststimulation, by which time the MFI had decreased. It was apparent from this study that electrical stimulation did not enhance CAF activity as might have been expected. Since the degradation of troponin T was brought about primarily by the action of cathepsins, the effect of electrical stimulation was attributed to the enhanced activity of these proteolytic enzymes at the low pH.

Q. Effect of Pressurization on Tenderness

1. *Calcium-Activated Factor*

Prerigor pressurization has been reported to increase meat tenderness (Macfarlane, 1973; Elgasim, 1977; Kennick *et al.*, 1980). The effects of pressuriza-

TABLE 1.7

MFI, pH, AND TEMPERATURE OF MUSCLES FROM CONTROL AND ELECTRICALLY STIMULATED COW CARCASSES[e]

	Muscle	Treatment	Time poststimulation (hr)					
			0	3	6	24	72	168
MFI[a,c]	Semimembranosus		59.2 ± 0.8	85.1 ± 3.3	85.7 ± 2.3	113.6 ± 1.6	97.8 ± 2.1	104.2 ± 2.1
	Longissimus dorsi		59.2 ± 1.1	73.3 ± 2.0	92.5 ± 1.1	108.6 ± 22.0	92.9 ± 0.3	87.4 ± 1.3
	Triceps brachii		64.5 ± 1.3	84.0 ± 2.9	93.7 ± 2.9	118.0 ± 2.1	108.6 ± 2.2	102.7 ± 1.0
pH[b,c,d]		Control	6.64 ± 0.03	6.31 ± 0.06	6.09 ± 0.17	5.61 ± 0.04	5.32 ± 0.07	5.18 ± 0.06
		Electrical stimulation	6.35 ± 0.03	5.85 ± 0.03	5.68 ± 0.06	5.55 ± 0.07	5.26 ± 0.05	5.26 ± 0.06
Temperature[b,c,d]		Control	37.6 ± 0.8	30.2 ± 5.7	23.2 ± 7.9	5.9 ± 3.2	2.0 ± 0.1	1.5 ± 0.2
		Electrical stimulation	38.2 ± 1.2	28.4 ± 8.1	22.0 ± 8.3	5.5 ± 3.3	2.0 ± 0.1	1.6 ± 0.2

[a] Means ± standard errors of ten samples.

[b] Means ± standard errors of fifteen samples.

[c] Means not underscored by the same line are significantly different ($P < 0.05$).

[d] Muscle samples were semimembranosus, longissimus dorsi, and triceps brachii.

[e] From Sonaiya *et al.* (1982). Copyright © by Institute of Food Technologists.

tion include the release of calcium ions, weakening of Z-lines, as well as the disappearance of M-lines (Elgasim and Kennick, 1982; Macfarlane and Morton, 1978). This suggests the possible activation of CAF by prerigor pressurization. Koohmaraie and co-workers (1984) examined the effect of prerigor pressurization on CAF activity in bovine muscle. They observed that activation of CAF by this treatment was lower than that in the untreated muscle. The lower CAF activity in the pressurized muscle was attributed to autolysis of this enzyme during the transient rise in Ca^{2+} ions. The results were rather confusing, as on the one hand there was a great similarity between the electrophoretic patterns for myofibrils from pressurized muscles with that for CAF-treated myofibrils, suggesting CAF activation. The absence of the 30K-dalton component in the pressure-treated muscle, characteristic of CAF activity, however, suggested to these researchers that the mechanism of tenderization could be quite different.

2. *Lysosomal Activity*

Elgasim (1977) proposed that the improvements in tenderness induced by prerigor pressurization could be related to lysosomal activity. An examination of bovine longissimus muscle exposed to different pressure levels by Elgasim *et al.* (1983) suggested the early release of lysosomal enzymes when conditions of pH and temperature were still conducive to rapid enzyme activity. This was based on the Z-line degradation observed in the pressurized muscle 24 hr postmortem. The pH of the muscle was less than 6.0 following pressurization, which was shown previously to facilitate the activity of cathepsin D, a lysosomal enzyme (Eino and Stanley, 1973; Robbins and Cohen, 1976). Improvement in tenderness associated with prerigor pressurization may be explained, in part, by the early release of lysosomal enzymes.

R. Lysosomal Enzymes and Collagen Degradation

The possible effects of lysosomal enzymes on the connective tissue of bovine muscles were examined by Wu and co-workers (1981). They showed that high-temperature conditioning (37°C) resulted in the release of lysosomal enzymes and enhancement of collagen solubilization. The increased breakdown of collagen fibers in the presence of lysosomal glycosidases was attributed to the degradation of the proteoglycan components which normally interfere with the degradation of collagen by collagenase (Eyre and Muir, 1974; Osebold and Pedrini, 1976).

IV. Meat Pigments

Consumer acceptance of packaged fresh meat is primarily influenced by its color (Pirko and Ayres, 1957). The bright-red color of fresh meat, caused by the pigment oxymyoglobin, is preferred by most consumers to meat of a darker or browner color. Metmyoglobin, the brown pigment responsible, develops during storage of the meat and is generally recognized by consumers to indicate lack of freshness. Consequently a predominance of this pigment in packaged meat products results in their rejection. This discoloration of packaged meats, referred to by the industry as "loss of bloom," is associated by the consumer with bacterial growth although this may not always be the case. The annual consumption of meat, particularly beef, is well in excess of 100 lb per capita in North America, purchased mainly as raw meat. Any deleterious changes in the nature of these pigments is of great concern to the meat industry as it affects the consumer market.

The major pigment in meat is the purplish-red pigment myoglobin. Hemoglobin, the red blood pigment, was considered at one time to play only a minor role since blood is normally drained from the slaughtered carcass. Solberg (1968) stated, however, that the desirable color of meat is influenced by both myoglobin and hemoglobin. While the amount of hemoglobin reported in meat varies considerably in the literature, the relatively small amount may still be important in terms of color and stability. Myoglobin accounts for 10% of the total iron in an animal prior to slaughter, but in the case of a well-bled carcass it could account for as much as 95% of the total iron (Clydesdale and Francis, 1971). Since the majority of studies carried out on meat color have been concerned primarily with myoglobin, this section will confine its discussion to those changes associated with this pigment.

A. Myoglobin

Myoglobin, the muscle pigment, is composed of a protein moiety, globin, combined with an iron-containing heme group. The latter consists of four pyrrole groups containing a centrally located atom of iron (Fig. 1.21). The primary factors responsible for the color of meat are the valence state of the iron atom and the ligand bond to the free binding site of the heme (Seideman *et al.*, 1984).

1. *Myoglobin Changes in Raw Meat*

Studies on fresh meat discoloration are concerned primarily with the formation of metmyoglobin in postrigor meat. The color of fresh meat is determined by the

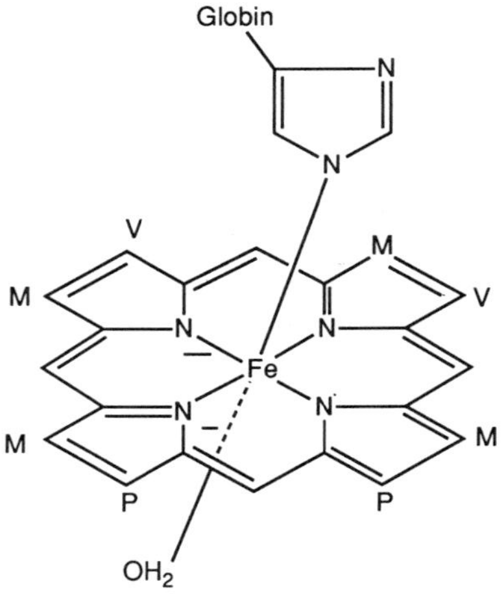

FIG. 1.21. A simplified diagrammatic structure of myoglobin. From "The Science of Meat and Meat Products," edited by J. F. Price and B. S. Schweigert. Copyright © 1960, 1971 by W. H. Freeman and Company.

relative proportions of the three primary forms of myoglobin (Watts *et al.*, 1966). These include oxymyoglobin, metmyoglobin, and reduced myoglobin. When the animal is slaughtered, oxygen is no longer available to the muscle tissue, resulting in the conversion of oxymyoglobin to myoglobin. Other changes during this period include pH, temperature, osmotic pressure, and oxidation–reduction potential (Table 1.8.). Under these conditions the pigment changes occurring in meat are of great concern to the food technologist. Metmyoglobin, the undesirable brown pigment, is in equilibrium with the other pigment forms as shown in Scheme 1.5. Further degradative reactions are mediated by bacterial action,

TABLE 1.8

COMPARISON OF LIVING MUSCLE TISSUE AND MEAT[a]

Condition	Living muscle	Postrigor muscle meat
Myoglobin	Oxymoglobin	Myoglobin
pH	7.35–7.43	5.3–5.5
Temperature (°C)	37.7–39.1	2–5
Osmotic pressure (% NaCl equiv.)	0.936	
Oxidation–reduction potential (mV)	+250	−50

[a] From Solberg (1970).

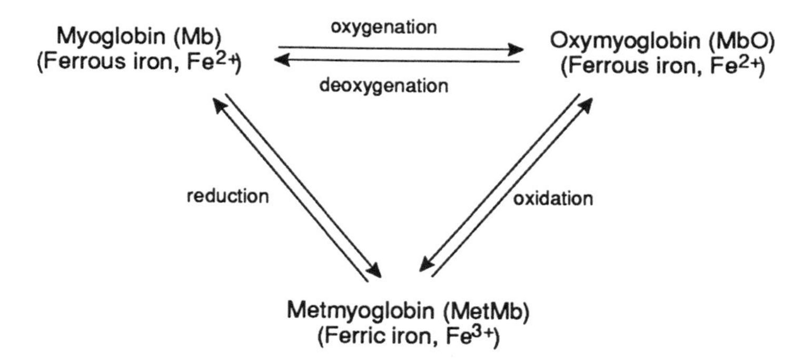

SCHEME 1.5. Myoglobin changes in fresh meat.

causing irreversible damage to the porphyrin ring. This results in the formation of bile pigments, choleglobin, sulfmyoglobin, and oxysulfmyoglobin, as well as other nitrogenous compounds characteristic of meat spoilage.

2. *Myoglobin Changes and Oxygen Tension*

The formation of oxymyoglobin involves the complexing of oxygen with the heme group of myoglobin. This process, referred to as oxygenation, occurs under high oxygen tensions and favors the formation of the desirable red meat pigment. Under low oxygen tensions, however, metmyoglobin is formed (Taylor, 1972). In fresh meat both the reduced and oxygenated forms of myoglobin are present with the predominating form determining the final color of the meat. The formation of oxymyoglobin involves the covalent binding of molecular oxygen to myoglobin (Clydesdale and Francis, 1971). Under conditions of low oxygen tension, oxygen dissociates from the heme to yield myoglobin. The latter, being unstable, is then oxidized to metmyoglobin (Pirko and Ayres, 1957). The formation of the metmyoglobin is accompanied by the loss of an electron in the iron molecule, resulting in a change from the ferrous (Fe^{2t}) to the ferric (Fe^{3t}) state (Giddings, 1977a,b; Giddings and Markarkis, 1973).

3. *Effect of pH, Temperature, and Salt on the Formation of Metmyoglobin*

The formation of metmyoglobin occurs under conditions of high temperature, low pH, ultraviolet light, and in the measure of salt and aerobic bacteria (Seideman *et al.*, 1984). Both high temperatures and low pH cause denaturation of the globin moiety leaving the heme unprotected so that it undergoes rapid oxidation to metmyoglobin (Walters, 1975). The effect of salt is twofold; it lowers the buffering capacity of the meat as well as promotes low oxygen tensions in meat.

Either of these results in oxidation of myoglobin to metmyoglobin (Seideman *et al.*, 1984; Brooks, 1937).

4. Endogenous Meat Enzymes and Metmyoglobin Formation

Freshly stored meat metmyoglobin is formed by two opposing reactions, autoxidation and reduction:

$$\text{Metmyoglobin} \underset{\text{autoxidation}}{\xrightarrow{\text{reduction}}} \text{Myoglobin}$$

These reactions, referred to as "metmyoglobin-reducing activity" (MRA), are responsible for the valence change in myoglobin from ferrous to ferric in the meat tissue. Differences in MRA between muscles contribute to ability of some muscles to retain the bright-red pigment for much longer periods. Stewart *et al.* (1965) separated these two reactions and showed that the reduction of metmyoglobin was carried out by enzymes. They reported considerable variation in the reducing activity (MRA) of their beef samples which increased with a rise in pH and temperature.

Watts and co-workers (1966) concluded that the reduction of metmyoglobin and oxygen in meat was carried out via reduced nicotinamide adenine dinucleotide (NAD^+). Succinic dehydrogenase appeared to be one of the enzymes involved, as addition of succinate to meat increased oxygen utilization. Other intermediates of the citric acid cycle and the amino acid L-glutamate were found to stimulate the reduction of metmyoglobin (Saleh and Watts, 1968). The enzymes involved remained potentially active in the postmortem meat and were capable of resuming activity in the presence of oxygen. This could occur if meat is ground or when cut surfaces are exposed to air provided suitable hydrogen donors are still available (e.g., $NADH^+$). The reduction of metmyoglobin in postmortem meat is due mainly to enzymes in which the mitochondria act as a source of reducing equivalents for the reduction of pyridine nucleotides (MacDougall, 1982). One such enzyme, metmyoglobin reductase, was identified in bovine heart muscle (Hagler *et al.*, 1979). This enzyme essentially reduced metmyoglobin back to myoglobin.

B. Fish Pigments

Oxidation of oxymyoglobin to metmyoglobin also occur in fish. The intensity of the pigment color, however, is considerably less than that in meat. An endogenous enzyme, metmyoglobin reductase, which is capable of reducing metmyoglobin, was reported in blue dolphin by Shimizu and Matsuura (1971). A ferrimyoglobin reductase was later isolated from the skeletal muscle of blue white dolphin by Matsui and co-workers (1975a,b). Purification of metmyoglobin reductase from bluefin tuna was reported by Al-Shaibani (1977).

1. Greening of Tuna

Greening of tuna is a serious problem that sometimes develops during the cooking of tuna. Instead of the normal pink flesh, a greenish-tan color develops throughout the fish flesh. This phenomenon is referred to as greening and develops in a single fish or a whole catch. Since there are no visual means to detect these fish in the raw state, the processor can incur serious financial losses.

The normal pink color of tuna is due to the hemochrome pigments derived from myoglobin, while the green off-color is associated with myoglobin oxidation (Brown and Tappel, 1957; Brown *et al.*, 1958; Naughton *et al.*, 1958). Dollar *et al.* (1961) tried to correlate ferriheme pigment levels (oxidized forms), peroxide values, pH change, total reducing substances, sulfhydryl content, and TBA values with greening but found too much variability in their results.

Considerable research on green tuna by Japanese scientists led to the isolation of a substance from an alcohol extract of tuna which produced greening when cooked with the raw fish (Sasano *et al.*, 1961). This substance was subsequently referred to as "YS" and tentatively identified by Sasano and Tawara (1962) as a peptide. Further studies by Koizumi and Hashimoto (1965a) identified the peptide as trimethylamine oxide. Further verification was provided by the fact that "YS" and trimethylamine oxide were both reduced to trimethylamine when treated with sodium hydrosulfite or titanous chloride. Koizumi and Hashimoto (1965b) boiled trimethylamine oxide (TMAO) or "YS" with normal tuna for 1 hr and reported the development of a tannish-green flesh in both cases. Several degradation products of TMAO were also investigated, including trimethylamine (TMA), dimethylamine (DMA), monomethylamine (MMA), and formaldehyde, but none of these produced a color change, thus demonstrating the importance of the oxidized form of trimethylamine. Earlier reports by Brown *et al.* (1958) noted that sodium hydrosulfite prevented the formation of greening in the affected tuna. They attributed it to the reduction of hemi- to hemochromes, although Koizumi and Hashimoto (1965b) suggested that the effect of sodium hydrosulfite was due to its reduction of TMAO.

The role of TMAO in the greening of tuna makes it a possible predictor of this phenomenon in raw fish. Koizumi and co-workers (1967a) determined the TMAO content of raw fish and found that the level varied at different locations, being highest in the tail area. A closer examination of the tail region showed that a TMAO level less than 7–8 mg% indicated normal color while levels greater than 13 mg% were indicative of greening. Using purified myoglobin, Koizumi *et al.* (1967b) showed that even in the presence of TMAO no greening occurred on heating. This led to the identification of an additional factor, cysteine, which when heated together with TMAO and myoglobin produced a green color. Grosjean *et al.* (1969) demonstrated the importance of denaturing myoglobin in order to expose sulfhydryl groups for the reaction to proceed. They also showed that free cysteine could not be replaced by its oxidative products or other reducing substances but had to be present together with the sulfhydryl groups of the cysteine residues in tuna myoglobin. The formation of greening occurred more readily in the presence of metmyoglobin, so that oxymyoglobin would have to be converted to metmyoglobin. The reaction mechanism responsible for the greening of tuna appeared to involve free cysteine and sulfhydryl groups of heat-denatured tuna myoglobin in the presence of TMAO or air as mild oxidizers. TMAO thus appeared to be involved in the reaction between cysteine and sulfhydryl groups of the protein, forming disulfide bonds in which the ferric ion of metmyoglobin may have a catalytic effect.

$$\text{Native metmyoglobin} \xrightarrow{70^{\circ}\text{C}} \text{Denatured metmyoglobin–SH}$$

$$Fe^{3+} \downarrow \text{RSH}$$

$$\text{Denatured metmyoglobin–S–S–R} + H_2O$$

The level of TMAO in yellow fin tuna (*Thannus albacarres*) was investigated by Yamagata *et al.* (1971). These researchers attempted to confirm earlier findings by Yamagata *et al.* (1969) that the interior portion of the dorsal muscle could be used to assess greening in raw tuna. TMAO and TTMA (total trimethylamine) levels were measured from the epaxial, hypoxial, and tail-end portions. It was possible to predict the greening phenomenon by measuring the TTMA levels at the superficial layer of the central dorsal raw muscle. The accuracy of predicting was reported to be 96% on the lowest level of TMAO-N or TTMA-N for iced and clipper frozen green tuna.

C. Preservation of Meat Pigments

The accelerated conversion of myoglobin to metmyoglobin under conditions of low oxygen partial pressure is extremely important when packaging fresh meat. George and Stratmann (1952) reported that the maximum rate of metmyoglobin formation occurred at around 1–1.4 mm partial pressures of oxygen, decreasing to a constant minimum rate above 30 mm. Consequently the oxygen permeability of the packaging film is of considerable importance when handling

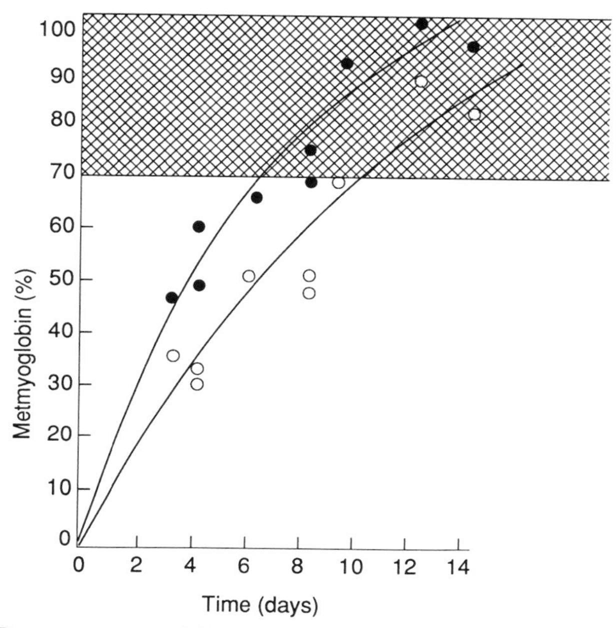

FIG. 1.22. Percentage metmyoglobin at the surface of bovine semimembranosus and adductor muscle slices stored at 4 ± 0.5°C in hermetically sealed semirigid polyvinyl-chloride packages containing an air or an enriched oxygen atmosphere (●, air; ○, oxygen enriched; hatched area denotes unacceptable color) (Daun *et al.*, 1971). Copyright © by Institute of Food Technologists.

fresh meat products. Landrock and Wallace (1955) suggested that packaging should allow an oxygen penetration of 5 liters of O_2/m^2/day/atmosphere to prevent the formation of metmyoglobin and subsequent browning.

Fellers and co-workers (1963), contrary to previous researchers, reported that oxygen levels higher than normally found in the atmosphere preserved oxymyoglobin. This was later confirmed by Bausch (1966). Thus high oxygen partial pressures had considerable potential for enhancing the desirable color of meat. Very little deoxygenation takes place as the excess oxygen recombines with myoglobin to form oxymyoglobin. Daun *et al.* (1971) examined the effect of oxygen-enriched atmospheres on packaged fresh meat using a polyvinyl chloride semirigid tray with a controlled headspace for storing fresh meat. Oxygen flushed into the headspace established an oxygen-enriched atmosphere and the formation of metmyoglobin on the surface of semimembranosus and adductor muscle slices was lower than meat stored under normal air atmosphere. Their results, shown in Figure 1.22, indicated that those samples stored in air were

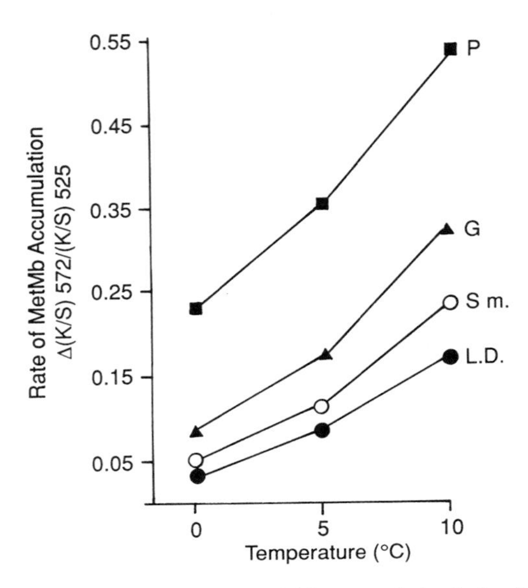

FIG. 1.23. Effect of temperature on metmyoglobin accumulation in four bovine muscles from ten experimental animals. P = *M. psoas major;* G = *M. gluteus medius;* Sm. = *M. semimembranosus;* L. D. = *M. longissimus dorsi* (Hood, 1980).

unacceptable, on the basis of color, after 6 days, while meat stored under oxygen-enriched atmospheres took 10 days to reach the same state. One of the major problems that has limited commercialization of this method was the development of fat rancidity under these conditions.

Of the many factors affecting the discoloration of packaged beef, differences between muscles was shown by Hood (1980) to be the most important single factor. Figure 1.23 shows decreased color stability for all four muscles as the temperature increased from 0 to 10°C. The rate of metmyoglobin accumulation was eight times faster in *M. psoas major* compared to in *M. longissimus dorsi.* Further work by O'Keeffe and Hood (1982) examined the biochemical factors affecting the rate of metmyoglobin discoloration in beef muscles. *M. psoas major* was found to have a higher level of succinic dehydrogenase activity, and, combined with its low myoglobin content, was responsible for a high oxygen consumption rate (OCR), resulting in low oxygen penetration and rapid formation of metmyoglobin compared to in *M. longissimus dorsi.* Based on their research, enzymes responsible for metmyoglobin reduction were the main determinants of metmyoglobin formation in beef muscles, with *M. psoas major* being low in reducing activity and *M. longissimus dorsi* being high.

The interrelationship between lipid and pigment oxidation has been studied since ferric pigments are known to enhance lipid oxidation (Brown *et al.*, 1963;

Younathan and Watts, 1960). Hutchins *et al.* (1967) reported good correlations between lipid oxidation and metmyoglobin formation in raw meat. Greene (1969) attempted to retard rancidity in refrigerated ground raw beef by adding several antioxidants, including propyl gallate (PG) and butylated hydroxyanisole (BHA). The desirable color of the ground beef was extended in the antioxidant-treated samples. Greene *et al.* (1971) treated samples of raw ground beef with a combination of PG or BHA and ascorbic acid (AA). Based on color scores using a trained panel, a consumer panel, and metmyoglobin measurements, the antioxidant-treated meat had a much longer shelf life.

A combination of carbon dioxide and oxygen provides an alternative way of extending the shelf life of meat. While carbon dioxide tended to inhibit spoilage bacteria, oxygen maintained the oxymyoglobin form of the pigment. Taylor and MacDougall (1973) reported the retention of the fresh color in beef for at least 1 week at 1°C in an atmosphere of 60% oxygen and 40% carbon dioxide. Gas packaging of retail cuts of fresh meat in oxygen-free nitrogen or carbon dioxide extended the storage of these products by minimizing oxidative changes and bacterial growth (Huffman, 1974; Partmann and Frank, 1973). O'Keeffe and Hood (1980–1981) examined the effect of anoxic gas packaging of fresh beef in flexible packages or desiccators incorporating a palladium catalyst as an oxygen-scavenging system. No discoloration was observed with meat stored in either carbon dioxide or nitrogen for up to 3 weeks at 0°C. Some differences were observed between the meat samples, which was attributed to animal differences.

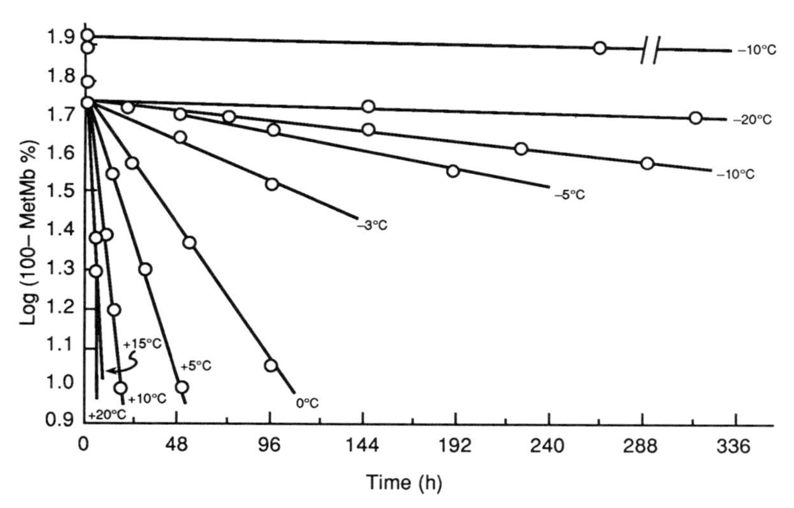

FIG. 1.24. Changes in MetMB percentage of skipjack during storage at different temperatures (Miki and Nishomoto, 1984).

Miki and Nishimoto (1984) examined the relationship between loss of freshness and discoloration in several fish species, including skipjack, mackerel, and sea bream. In red-muscled fish, percentage metmyoglobin has provided a useful index of discoloration. Miki and Nishimoto (1984) plotted the log of (100–metmyoglobin%), which represents the ratio of residual oxymyoglobin to total myoglobin, against storage times at different temperatures. Their results for skipjack in Figure 1.24 clearly indicate a first-order reaction corresponding with changes in freshness. These results were consistent with early work by Matsuura and co-workers (1962), who showed that the rate of autoxidation of isolated myoglobin from fish muscle also followed first-order kinetics.

Bibliography

Acton, J. C., Zieler, G. R., and Burge, D. L., Jr. (1983). Functionality of muscle constituents in the processing of comminuted meat products. *CRC Crit. Rev. Food Sci. Nutr.* **18,** 99.

Al-Shaibani, K. A. (1977). Purification of metmyoglobin from bluefin tuna. *J. Food Sci.* **42,** 1013.

American Meat Institute Foundation (1960). "The Science of Meat and Meat Products." Freeman, San Francisco, California.

Amos, L. A., Huxley, H. E., Holmes, K. C., Goody, R. S., and Taylor, K. A. (1982). Structural evidence that myosin heads may interact with two sites on F-actin. *Nature (London)* **299,** 467.

Andou, S., Takama, K., and Zama, K. (1979). Interaction between lipid and protein during frozen storage. I. Effect of oil dipping on rainbow trout muscle during frozen storage. *Hokkaido Daigaku Kenkyu Iho* **30,** 282.

Andou, S., Takama, K., and Zama, K. (1980). Interaction between lipid and protein during frozen storage. II. Effect off non-polar and polar lipids on rainbow trout myofibrils during frozen storage. *Hokkaido Daigaku Kenkyu Iho* **31,** 201.

Ang, J. F., and Haard, N. F. (1985). Chemical composition and postmortem changes in soft textured muscle from intensely feeding Atlantic cod (*Gadus morhua* L.). *J. Food Biochem.* **9,** 49.

Asghar, A., and Henrickson, R. L. (1982). Chemical, biochemical, functional and nutritional characteristics of collagen in food systems. *Adv. Food Res.* **28,** 231.

Asghar, A., Samejima, K., and Yasui, T. (1984). Functionality of muscle proteins in gelation mechanisms of structured meat products. *CRC Crit. Rev. Food Sci. Nutr.* **22,** 27.

Bailey, A. J. (1972). The basis of meat texture. *J. Sci. Food Agric.* **23,** 995.

Bailey, A. J., and Peach, C. M. (1968). Isolation and structural identification of labile intermolecular crosslink in collagen. *Biochem. Biophys. Res. Commun.* **33,** 812.

Bailey, A. J., and Sims, T. J. (1977). Meat tenderness, distribution of molecular species of collagen in bovine muscle. *J. Sci. Food Agric.* **28,** 565.

Bailey, A. J., Restall, D., Sims, T. J., and Duance, V. C. (1979). Meat tenderness. Immunofluorescent localization of the isomorphic forms of collagen in bovine muscles of varying texture. *J. Sci. Food Agric.* **30,** 203.

Banks, A. (1955). The expressible fluid of fish fillets. IV. The expressible fluid of iced cod. *J. Sci. Food Agric.* **6,** 584.

Bate-Smith, E. C. (1948). The physiology and chemistry of rigor mortis, with special reference to the aging of beef. *Adv. Food Res.* **1,** 1.

Bate-Smith, E. C., and Bendall, J. R. (1949). Factors determining the time course of rigor mortis. *J. Physiol. (London)* **110,** 47.

Bate-Smith, E. C., and Bendall, J. R. (1956). Changes in muscle after death. *Br. Med. Bull.* **12,** 230.

Bausch, E. R. (1966). Color retention of red meat. Canadian Patent 742,165.

Beecher, G. R., Briskey, E. J., and Hockstra, W. G. (1965). Comparison of glycolysis and associated changes in light and dark portions of the porcine semitendinosus. *J. Food Sci.* **30,** 477.

Bendall, J. R. 1951. The shortening of rabbit muscles during rigor mortis: Its relation to the breakdown of adenosine triphosphate and creatine phosphate and to muscular contraction. *J. Physiol. (London)* **114,** 71.

Bendall, J. R. (1969). "Muscles, Molecules and Movement." Heinemanns, London.

Bendall, J. R. (1973). Postmortem changes in muscle. *In* "The Structure and Function of Muscle" (G. H. Bourne, ed.), 2nd ed., vol. 2. Academic Press, New York.

Bendall, J. R. (1976). Electrical stimulation of rabbit and lamb carcasses. *J. Sci. Food Agric.* **27,** 819.

Bendall, J. R., and Davey, C. L. (1957). Ammonia liberation during rigor mortis and its relation to changes in the adenine and inosine nucleotides of rabbit muscle. *Biochim. Biophys. Acta* **26,** 93.

Bendall, J. R., and Lawrie, R. A. (1964). Watery pork. *Anim. Breed. Abstr.* **32,** 1.

Bendall, J. R., and Wismer-Pederson, J. (1962). Some properties of the fibrillar proteins of normal and watery pork muscle. *J. Food Sci.* **27,** 144.

Bergstrom, J., and Hultman, E. (1966). The effect of exercise on muscle glycogen and electrolytes. *Scand. J. Clin. Lab. Invest.* **18,** 16.

Bergstrom, J., Castenors, H., Hultman, E., and Silander, T. (1965). *Acta Chim. Scand.* **130,** 1.

Bodwell, C. E., and McClain, P. E. (1971). Chemistry of Animal Tissues. Proteins. *In* "The Science of Meat and Meat Products" (J. F. Price and B. S. Schweigert, eds.), p. 96. W. H. Freeman, San Francisco.

Briskey, E. J. (1964). Etiological status and associated studies of pale, exudative porcine musculature. *Adv. Food Res.* **13,** 89.

Briskey, E. J., and Lawrie, R. A. (1961). Comparative *in vitro* activities of phosphorylase b and cytochrome oxidase in preparations from two ox muscles. *Nature (London)* **192,** 263.

Briskey, E. J., Kastenschmidt, L. L., Forrest, J. C., Beecher, G. R., Judge, M. D., Cassens, R. G., and Hoekstra, W. G. (1966). Biochemical aspects of post-mortem changes in porcine muscle. *J. Agric. Food Chem.* **14,** 201.

Brooks, J. (1937). Color of meat. *Food Ind.* **9,** 707.

Brown, W. D., and Tappel, A. L. (1957). Identification of the pink pigment of canned tuna. *Food Res.* **22,** 214.

Brown, W. D., Tappel, A. L., and Olcott, H. S. (1958). The pigments of off-colored cooked tuna meat. *Food Res.* **23,** 262.

Brown, W. D., Harris, L. D., and Olcott, H. S. (1963). Catalysis of unsaturated lipid oxidation by iron protoporphyrin derivatives. *Arch. Biochem. Biophys.* **101,** 14.

Buege, D. R., and Marsh, B. B. (1975). Mitochondrial calcium and postmortem muscle shortening. *Biochem. Biophys. Res. Commun.* **65,** 478.

Burt, J. R. (1966). Glycogenolytic enzymes of cod (*Gadus callarias*) muscle. *J. Fish. Res. Board Can.* **23,** 527.

Busch, W. A., Stromer, M. H., Goll, D. E., and Suzuki, A. (1972). Ca^{2+}-specific removal of Z lines from rabbit skeletal muscle. *J. Cell Biol.* **52,** 367.

Buttkus, H. (1963). Red and white muscle of fish in relation to rigor mortis. *J. Fish. Res. Board Can.* **20,** 45.

Carse, W. A. (1973). Meat quality and the acceleration of postmortem glycolysis by electrical stimulation. *J. Food Technol.* **8,** 165.

Cassens, R. G. (1966). General aspects of postmortem changes. *In* "The Physiology and Biochemistry of Muscle as a Food" (E. J. Briskey, R. G. Cassens, and J. C. Trautman, eds.), pp.181–196. Univ. of Wisconsin Press, Madison.

Cassens, R. G., and Newbold, R. P. (1966). Effects of temperature on postmortem metabolism in beef muscle. *J. Sci. Food Agric.* **17,** 254.

Charpentier, J. (1969). Postmortem biochemical characteristics of the sarcoplasmic reticulum of pig muscle. *Ann. Biol. Anim., Biochim., Biophys.* **9,** 101.

Chaudhry, H. M., Parrish F. C., Jr., and Goll, D. E. (1969). Molecular properties of postmortem muscle. VI. *J. Food Sci.* **34,** 183.

Cheng, C-S., and Parrish F. C. Jr. (1977). Effect of Ca^{2+} on changes in myofibrillar proteins of bovine skeletal muscle. *J. Food Sci.* **42,** 1621.

Chock, S. P., Chock, P. B., and Eisenberg, E. (1976). Presteady-state kinetic evidence for a cyclic interaction of myosin subfragment one with actin during the hydrolysis of adenosine-5′-triphosphate. *Biochemistry* **15,** 3244.

Chrystall, B. B., and Devine, C. E. (1978). Electrical stimulation, muscle tension, and glycolysis in bovine sternomandibularis. *Meat Sci.* **2,** 49.

Chrystall, B. B., and Hagyard, C. J. (1976). Electrical stimulation and lamb tenderness. *N. Z. J. Agric. Res.* **19,** 7.

Clydesdale, F. M., and Francis, F. J. (1971). The chemistry of meat color. *Food Prod. Dev.* **5 (3),** 81.

Conlee, R. K., Renna, M. J., and Winder, N. W. (1976). Skeletal muscle glycogen content, diurnal variation and affects of fasting. *Am. J. Physiol.* **231,** 614.

Cornforth, D. P., Pearson, A. M., and Merkel, R. A. (1980). Relationship of mitochondria and sarcoplasmic reticulum to cold shortening. *Meat Sci.* **4,** 103.

Crupkin, M., Barassi, C. A., Martone, C. B., and Trucco, R. E. (1979). Effect of storing hake *(Merluccius merluccius hubbsi)* on ice on the viscosity of the extract of soluble muscle protein. *J. Sci. Food. Agric.* **30,** 911.

Culler, R. D., Parrish, F. C. Jr., Smith, G. C., and Cross, H. R. (1978). MFI relationship of myofibril fragmentation index to palatability and carcass attributes. *J. Food Sci.* **43,** 1177.

Cutting, C. L. (1939). Immediate post mortem changes in trawled fish. *G. B. Dep. Sci. Ind. Res., Rep. Food Invest. Board,* p. 39.

Daun, H., Solberg, M., Franke, W., and Gilbert, S. (1971). Effect of oxygen-enriched atmospheres on storage quality of packaged fresh meat. *J. Food Sci.* **36,** 1011.

Davey, C. L., and Gilbert, K. V. (1974). The mechanism of cold induced shortening in beef muscle. *J. Food Technol.* **9,** 51.

Davey, C. L., and Graafhuis, A. E. (1976). Structural changes in beef muscle during ageing. *J. Sci. Food Agric.* **27,** 301.

Davey, C. L., Gilbert, K. V., and Carse, W. A. (1976). Carcass electrical stimulation to prevent cold shortening toughness in beef. *N. Z. J. Agric. Res.* **19,** 13.

Dayton, W. R., and Schollmeyer, J. R. (1980). Isolation from papaine cardiac muscle of a calcium($^{2+}$)-activated protease that partially degrades myofibrils. *J. Mol. Cell. Cardiol.* **12,** 533.

Dayton, W. R., Goll, D. E., Stromer, M. H., Reville, W. J., Zeece, M. G., and Robson, R. M. (1975). Some properties of a Ca^{2+}-activated protease that may be involved in myofibrillar protein activated protease that may be involved in myofibrillar protein turnover. *In* "Proteases and Biological Control" (E. Reich, D. B. Rifkin, and E. Shaw, eds.), Vol. 2, pp. 551–577. Cold Spring Harbor Lab., Cold Spring Harbor, New York.

Dayton, W. R., Goll, D. E., Zeece, M. G., Robson, R. M., and Reville, W. J. (1976a). A Ca^{2+}-activated protease possibly involved in myofibrillar protein turnover. Purification from porcine muscle. *Biochemistry* **15,** 2150.

Dayton, W. R., Reveille, W. J., Goll, D. E., and Stromer, M. H. (1976b). A Ca^{2+}-activated protease possibly involved in myofibrillar protein turnover. Partial characterization of the purified enzyme. *Biochemistry* **15,** 2159.

De Fremery, D. (1966). Some aspects of post-mortem changes in poultry muscle. *In* "The Physiology and Biochemistry of Muscle as a Food" (E. J. Briskey, R. G. Cassens, and J. C. Trautman, eds.), pp. 205–212. Univ. of Wisconsin Press, Madison.

De Fremery, D., and Pool, M. F. (1960). Biochemistry of chicken muscle as related to rigor mortis and tenderization. *Food Res.* **25,** 73.

de Meis, L., and Vianna, A. L. (1979). Energy interconversions by the Ca dependent ATP'ase of the sarcoplasmic reticulum. *Annu. Rev. Biochem.* **48,** 275.

Disney, J. G., Follett, M. J., and Ratcliff, P. W. (1967). Biochemical changes in beef muscle postmortem. *J. Sci. Food Agric.* **18,** 314.

Dollar, A. M., Goldner, A. M., Brown, W. D., and Olcott, H. S. (1961). Observations on green tuna. *Food Technol.* **15,** 253.

Dransfield, E. (1977). Intramuscular composition and texture of beef muscles. *J. Sci. Food Agric.* **28,** 833.

Dugal, L. C. (1967). Hypoxanthine in iced freshwater fish. *J. Fish. Res. Board Can.* **24,** 2229.

Dunajski, E. (1979). Texture of fish muscle. *J. Texture Stud.* **10,** 301.

Dyer, W. J. (1951). Protein denaturation in frozen and stored fish. *J. Fish. Res. Biol.* **24,** 2229.

Ebashi, S. (1974). Regulatory mechanism of muscle contraction with special reference to the Ca–troponin–tropomyosin system. *Essays Biochem.* **10,** 1.

Eccleston, J. F., Geeves, M. A., Trentham, D. R., Bagshaw, C. R., and Mowa, V. (1976). The binding and cleavage of ATP in the myosin and actomyosin ATPase mechanisms. *Colloq. Ges. Biol. Chem.* **26,** 42.

Eino, M. F., and Stanley, D. W. (1973). Catheptic activity, textural properties and surface ultrastructure of postmortem beef muscle. *J. Food Sci.* **38,** 45.

Elgasim, E. A. (1977). The effect of ultrahydrostatic pressure of prerigor muscle on characteristics of economic importance. M. S. Thesis, Oregon State University, Corvallis.

Elgasim, E. A., and Kennick, W. H. (1982). Effect of high hydrostatic pressure on meat microstructure. *Food Microstruct.* **1,** 75.

Elgasim, E. A., Kennick, W. H., Anglemeir, A. F., Koohmaraie, M., and Elkhalifa, E. A. (1983). Effect of pressurization on bovine lysosomal enzyme activity. *Food Microstruct.* **2,** 91.

Elgasim, E. A., Koohmaraie, M., Anglemeir, A. F., Kennick, W. H., and Elkhalifa, E. A. (1985). The combined effects of the calcium activated factor and cathepsin D on skeletal muscle. *Food Microstruct.* **4,** 55.

Elkhalifa, E. A., Anglemeir, A. F., Kennick, W. H., and Elgasim, E. A. (1984a). Effect of prerigor pressurization on post-mortem bovine muscle lactate dehydrogenase activity and glycogen degradation. *J. Food Sci.* **49,** 593.

Elkhalifa, E. A., Anglemeir, A. F., Kennick, W. H., and Elgasim, E. A. (1984b). Influence of prerigor pressurization on beef muscle creatine phosphokinase activity and degradation of creatine phosphate and adenosine triphosphate. *J. Food Sci.* **49,** 595.

Etherington, D. J., and Sims, T. J. (1981). Detection and estimation of collagen. *J. Sci. Food Agric.* **32,** 539.

Eyre, D. R., and Muir, G. C. (1974). Collagen polymorphism. Two molecular species in pig invertebral disc. *FEBS Lett.* **42,** 192.

Fabiansson, S., Erichsen, I., and Reutersward, M. L. (1984). The incidence of dark cutting beef in Sweden. *Meat Sci.* **10,** 21.

Fellers, D. A., Wahba, I. J., Caldano, J. C., and Ball, C. O. (1963). Factors affecting the color of packaged retail beef cuts. Origin of cuts, package type, and storage conditions. *Food Technol.* **17,** 95.

Fischer, C., and Hamm, R. (1980). Biochemical studies on fast glycolysing bovine muscles. *Meat Sci.* **4,** 41.

Fischer, C., Hamm, R., and Honikel, K. O. (1979). Changes in solubility and enzymic activity of muscle glycogen phosphorylase in PSE-muscles. *Meat Sci.* **3,** 11.

Fjelkner-Modig, S., and Ruderus, H. (1983). The influence of exhaustion and electrical stimulation on the meat quality of young bulls. Part 2. Physical and sensory properties. *Meat Sci.* **8,** 203.

Follett, M. J., Norman, G. A., and Ratcliff, P. W. (1974). The ante-rigor excision and air cooling of beef semimembranosus muscles at temperatures between -5°C and +15°C. *J. Food Technol.* **9,** 509.

Forrest, J. C., and Briskey, E. J. (1967). Response of striated muscle to electrical stimulation. *Meat Sci.* **10,** 35.

Fraser, D. I., Punjamapirom, S., and Dyer, W. J. (1961). Temperature and the biochemical processes occurring during rigor mortis in cod muscle. *J. Fish Res. Board Can.* **18,** 641.

Fujita, T., Hori, Y., Otani, T., Kunita, Y., Sawa, S., Sakai, S., Tanaka, Y., Takagahara, I., and Nakatani, Y. (1988). Applicability of the K_0 value as an index of freshness for porcine and chicken muscles. *Agric. Biol. Chem.* **52,** 107.

George, P., and Stratmann, C. J. (1952). The oxidation of myoglobin to metmyoglobin by oxygen. 2. The relation between the first order rate constant and the partial pressure of oxygen. *Biochem.* J. **51,** 418.

Ghanekar, D. S., Bal, D. V., and Kamala, S. (1956). Enzymes of some elasmobranchs from Bombay. III. Amylases of *Scoliodon sorrakowah* and *Sphyrna blochii. Proc.—Indian Acad. Sci., Sect. B* **43,** 134.

Giddings, G. G. (1977a). The basis of color in muscle foods. *J. Food Sci.* **42,** 288.

Giddings, G. G. (1977b). The basis of color in muscle foods. *CRC Crit. Rev. Food Sci. Nutr.* **9,** 81.

Giddings, G. G., and Markarkis, P. (1973). On the interaction of myoglobin and hemoglobin with molecular oxygen and its lower oxidation states with cytochrome c. *J. Food Sci.* **38,** 705.

Goodno, C. C., Wall, C. M., and Perry, S. V. P. (1978). Kinetics and regulation of the myofibrillar adenosine triphosphatase. *Biochem. J.* **175,** 813.

Goutefongea, R. (1971). Influence du pH et de la température sur le solubilite des protéines musculaires du porc. *Ann. Biol. Anim., Biochim., Biophys.* **11,** 233.

Grainger, B. L., and Lazarides, E. (1978). The existence of an insoluble Z-disc scaffold in chicken skeletal muscle. *Cell. (Cambridge, Mass.)* **15,** 1253.

Grandin, T. (1980). The effect of stress on livestock and meat quality prior to and during slaughter. *Int. J. Stud. Anim. Prob.* **1,** 313.

Greene, B. E. (1969). Lipid oxidation and pigment changes in raw beef. *J. Food Sci.* **34,** 10.

Greene, B. E., Hsin, I., and Zipser, M. W. (1971). Retardation of oxidative color changes in raw ground beef. *J. Food Sci.* **36,** 940.

Grosjean, O., Cob, B. F., III, and Brown, W. D. (1969). Formation of a green pigment from tuna myoglobins. *J. Food Sci.* **34,** 404.

Hagler, L., Coppes, R. I., and Herman, R. H. (1979). Metmyoglobin reductase. Identification and purification of a reduced nicatinamide adenine dinucleotide-dependent enzyme from bovine heart which reduces metmyoglobin. *J. Biol. Chem.* **254,** 6505.

Hallund, O., and Bendall, J. R. (1965). Long-term effect of electrical stimulation on the post-mortem fall of pH in the muscles of Landrace pigs. *J. Food Sci.* **30,** 296.

Hamm, R. (1977). Postmortem breakdown of ATP and glycogen in ground muscle: A review. *Meat Sci.* **1,** 15.

Hamm, R., Dalrymple, R. H., and Honikel, K. O. (1973). *Proc. 19th Meet. Euro. Meat Res. Workers* Vol. 1, p. 73.

Hamoir, G., and Konosu, S. (1965). Carp myogens of white and red muscles. General composition and isolation of low molecular weight components of abnormal amino acid composition. *Biochem. J.* **96,** 85.

Hanson, J., and Huxley, H. E. (1955). The structural basis of contraction in striated muscle. *Symp. Soc. Exp. Biol.* **9,** 228.

Harsham, A., and Detherage, F. E. (1951). Tenderization of meat. U.S. Patent 2,544,681.

Haselgrove, J. C. (1972). X-ray evidence for a conformational change in the actin-containing filaments of vertebrate striated muscle. *Cold Spring Harbor Symp. Quant. Biol.* **37,** 341.

Hashimoto, Y. (1965). Taste-producing substances in marine products. *In* "The Technology of Fish Utilization" (R. Kreuzer, ed.), p. 57. Fishing News (Books), London.

Hay, J. D., Currie, R. W., Wolfe, F. H., and Sanders, E. J. (1973). Effects of post-mortem aging on chicken muscle fibrils. *J. Food Sci.* **38,** 981.

Henderson, D. W., Goll, D. E., and Stromer, M. H. (1970). A comparison of shortening and Z-line degradation in post-mortem bovine, porcine, and rabbit muscle. *Am. J. Anat.* **128,** 117.

Hitchcock, S. E., Huxley, H. E., and Szent-Györgyi, A. G. (1973). Calcium sensitive binding of troponin to actintropomyosin; A two-site model for troponin action. *J. Mol. Biol.* **80,** 825.

Honikel, K. O., Roncales, P., and Hamm, R. (1983). The influence of temperature on shortening and rigor onset in beef muscle. *Meat Sci.* **8,** 221.

Hood, D. E. (1980). Factors affecting the rate of metmyoglobin accumulation in prepackaged beef. *Meat Sci.* **4,** 427.

Horgan, D. J., and Kuypers, R. (1985). Post-mortem glycolysis in rabbit longissimus dorsi muscles following electrical stimulation. *Meat Sci.* **12,** 225.

Howard, A., and Lawrie, R. A. (1956). Beef quality. II. Physiological and biological effects of various preslaughter treatments. *Div. Food Preserv. Tech. Pap. (Aust., C.S.I.R.O.)* **2,** 18.

Hoyle, G. (1967). Diversity of striated muscle. *Am. Zool.* **7,** 435.

Huffman, D. L. (1974). Effect of gas atmospheres on microbial quality of pork. *J. Food Sci.* **39,** 723.

Hutchins, B. K., Liu, T. H. P., and Watts, B. M. (1967). Effect of additives and refrigeration on the reducing activity, metmyoglobin and malonaldehyde of raw ground beef. *J. Food Sci.* **32,** 214.

Huxley, H. E. (1964). Structural arrangements and the contraction mechanism in striated muscle. *Proc. R. Soc. London, Ser. B* **160,** 442.

Huxley, H. E. (1969). The mechanism of muscular contraction. *Science* **164,** 1356.

Huxley, H. E. (1972a). Molecular basis of contraction in cross-striated muscles. *In* "Structure and Function of Muscle," Vol. 1, 2nd ed. (G. H. Bourne, ed.), Academic Press, New York.

Huxley, H. A. (1972b). Structural changes in the actin and myosin-containing filaments [of muscle] during contraction. *Cold Spring Harbor Symp. Quant. Biol.* **37,** 361.

Ishikawa, H., Bischoff, R., and Holtzer, H. (1968). Mitosis and intermediate-sized filaments in developing skeletal muscle. *J. Cell Biol.* **38,** 538.

Jeacocke, R. E. (1977). The temperature dependence of anaerobic glycolysis in beef muscle held in a linear temperature gradient. *J. Sci. Food Agric.* **28,** 551.

Jiang, S.-T., Hwang, B.-S., and Tsao, C.-T. (1987). Protein denaturation and changes in nucleotides of fish muscle during frozen storage. *J. Agric. Food Chem.* **35,** 22.

Jones, N. R. (1965). Interconversions of flavorous catabolites in chilled frozen fish. *Prog Refrig. Sci. Technol., Proc. Int. Congr. Refrig., 11th, 1963, p 917.*

Jones, N. R., and Murray, J. (1961). Nucleotide concentration in codling *(Gadus callarias)* muscle passing through rigor mortis at 0°C. *Z. Vergl. Physiol.* **44,** 174.

Jones, N. R., Murray, J., and Livingstone, E. I. (1964). Rapid estimations of hypoxanthine concentrations as indices of the freshness of chill-stored fish. *J. Sci. Food Agric.* **15,** 763.

Jones, N. R., Burt, J. R., Murray, J., and Stroud, G. D. (1965). Nucleotides and the analytical approach to the rigor mortis problem. *In* "The Technology of Fish Utilization" (R. Kreuzer, ed.), p. 14. Fishing News (Books), London.

Joseph, R. L. (1968a). Biochemistry and quality in beef. Part I. *Process Biochem.* **3 (7),** 20.

Joseph, R. L. (1968b). Biochemistry and quality in beef. Part II. *Process Biochem.* **3 (9),** 32.

Karube, I., Sato, I., Araki, Y., Suzuki, S., and Hideaki, Y. (1980). Monoamine oxidase electrode in freshness testing of meat. *Enzyme Microb. Technol.* **2,** 117.

Karube, I., Maatsuoka, H., Suzuki, S., Watanabe, E., and Toyama, K. (1984). Determination of fish freshness with an enzyme sensor system. *J. Agric. Food Chem.* **32,** 314.

Kassemsarn, B. O., Sang, P., Murray, J., and Jones, N. R. (1963). Nucleotide degradation in the muscle of ice haddock, lemon sole and plaice. *J. Food Sci.* **28,** 28.

Kastenschmidt, L. L., Hoekstra, W. G., and Briskey, E. J. (1968). Glycolytic intermediates and cofactors in "fast" and "slow-glycolyzing" muscles of the pig. *J. Food Sci.* **33,** 151.

Kelly, T. R. (1969). Quality in frozen cod and limiting factors on its shelf life. *J. Food Technol.* **4,** 95.

Kelly, T. R., and Little, W. T. (1966). Brown discolouration in prerigor cut fish fillets. *J. Food Technol.* **1,** 121.

Kennick, W. H., Elgasim, E. A., Holmes, Z. A., and Meyer, P. F. (1980). The effects of ultrahydrostatic pressurization of prerigor muscle on prerigor meat characteristics. *Meat Sci.* **4,** 33.

King, N. L. (1984). Breakdown of connectin during cooking of meat. *Meat Sci.* **11,** 27.

Koizumi, C., and Hashimoto, Y. (1965a). Studies on "green tuna." 1. The significance of trimethylamine oxide. *Bull. Jpn. Soc. Sci. Fish.* **31,** 157.

Koizumi, C., and Hashimoto, Y. (1965b). Studies on "green tuna." 2. Discoloration of cooked tuna meat due to trimethylamine oxide. *Bull. Jpn. Soc. Sci. Fish.* **31,** 439.

Koizumi, C., Kawakani, H., and Howata, J. (1967a). Studies on "green tuna." 3. Relation between greening and trimethylamine oxide concentration in albacore meat. *Bull. Jpn. Soc. Sci. Fish.* **33,** 131.

Koizumi, C., Kawakani, H., and Howata, J. (1967b). Studies on "green tuna." 4. Effect of cysteine on greening of myoglobin in the presence of trimethylamine oxide. *Bull. Jpn. Soc. Sci. Fish.* **33,** 839.

Konagaya, S., and Konagaya, T. (1979). Acid denaturation of myofibrillar protein as the main cause of formation of "yake niku," a spontaneously done meat, in red meat fish. *Nippon Suisan Gakkaishi* **45**(2), 245.

Koohmaraie, M., Kennick, W. H., Elgasim, E. A., and Anglemeir, A. F. (1984). Effect of prerigor pressurization on the activity of calcium-activated factor. *J. Food Sci.* **49,** 680.

Landrock, A. H., and Wallace, G. A. (1955). Discoloration of fresh red meat and its relationship to film oxygen permeability. *Food Technol.* **4,** 194.

Lawrie, R. A. (1966). "Meat Science." Macmillan (Pergamon), New York.

Lawrie, R. A. (1985). "Meat Science," 4th ed. Pergamon, New York.

Lawrie, R. A. (1983). Trends in meat research. *Chem. Ind. (London),* p. 542.

Lawrie, R. A., Gatherum, D. P., and Hale, H. P. (1958). Abnormally low ultimate pH in pig muscle. *Nature (London)* **182,** 807.

Lazarides, E., and Hubbard, B. D .1976. Immunological characterization of the subunit of the 100 Å filaments from muscle cells. *Proc. Natl. Acad. Sci. U. S. A* **73,** 4344.

Lee, E. H., Oshima, T., and Koizumi, C. (1982). High performance liquid chromatographic determination of *K* value as an index of freshness of fish. *Bull. Jpn. Soc. Sci. Fish.* **48,** 255.

Light, N., Champion, A. E., Voyle, C., and Bailey, A. J. (1985). The role of epimysial, perimysial and endomysial collagen in determining texture in six bovine muscles. *Meat Sci.* **13,** 137.

Lister, D. (1970). The physiology of animals and the use of their muscle for food. *In* "The Physiology and Biochemistry of Muscle as a Food" (E. J. Briskey, R. G. Locker, and B. B. Marsh, eds.), p.705. Univ. of Wisconsin Press, Madison.

Locker, R. H. (1984). The role of gap filaments in muscle and meat. *Food Microstruct.* **3,** 17.

Locker, R. H., and Daines, G. J. (1975). Rigor mortis in beef sternomandibularis muscle at 37°C. *J. Sci. Food Agric.* **26,** 1721.

Locker, R. H., and Leet, N. G. (1976a). Histology of highly stretched beef muscle. II. Further evidence on location and nature of gap filaments. *J. Ultrastruct. Res.* **55,** 157.

Locker, R. H., and Leet, N. G. (1976b). Histology of highly stretched beef muscle. IV. Evidence for movement of gap filaments through the Z-line, using the N-line and M-line as markers. *J. Ultrastruct. Res.* **56,** 31.

Locker, R. H., and Wild, D. J. C. (1984). The fate of the large proteins of the myofibril during tenderizing treatments. *Meat Sci.* **11,** 89.

Love, R. M. (1975). Variability in Atlantic cod *(Gadus morhua)* from the northeast Atlantic: A review of seasonal and environmental influences on various attributes of the flesh. *J. Fish. Res. Board Can.* **32,** 2333.

Love, R. M., Roberson, I., Smith, G. L., and Whittle, K. J. (1972). The texture of cod muscle. *J. Texture Stud.* **5,** 201.

Love, R. M., Lavety, J., and Vellas, F. (1982). Unusual properties of the connective tissues of cod *(Gadus murhua L.). In* "Chemistry and Biochemistry of Marine Food Products" (R. E. Martin, G. J. Flick, C. E. Hebard, and D. R. Ward, eds.), pp. 67–73. Avi Publ. Co., Westport, Connecticut.

Lusby, M. L., Ridpath, J. F., Parrish, F. C., Jr., and Robson, R. M. (1983). Effect of post mortem storage on the degradation of the myofibrillar protein titin in bovine longissimus muscle. *J. Food Sci.* **48,** 1787.

MacBride, M. A., and Parrish, F. C., Jr. (1977). The 30,000 dalton component of tender bovine longissimus muscle. *J. Food Sci.* **42,** 1627.

MacCallum, W. A., Jaffray, J. I., Churchill, D. N., Idler, D. R., and Odense, P. H. (1967). Postmortem physicochemical changes in unfrozen Newfoundland trap-caught cod. *J. Fish Res. Board Can.* **24,** 651.

MacDougall, D. B. (1982). Changes in colour and opacity of meat. *Food Chem.* **9,** 75.

Macfarlane, J. J. (1973). Prerigor pressurization of muscle, effect of pH, shear value and taste panel assessment. *J. Food Sci.* **38,** 294.

Macfarlane, J. J., and Morton, D. J. (1978). Effects of pressure treatment on the ultrastructure of striated muscle. *Meat Sci.* **2,** 281.

MacLeod, R. A., Jonas, R. E. E., and Roberts, E. (1963). Glycolytic enzymes in the tissues of a salmonoid fish *(Salmo gairdnerii gaidnerii). Can. J. Biochem. Physiol.* **41,** 1971.

Makinodan, Y., and Ikeda, S. (1976). Studies on fish muscle protease. VI. Separating carp muscle cathepsins A and D and some properties of carp muscle cathepsin A. *Bull. Jpn. Soc. Sci. Fish.* **42,** 239.

Marsh, B. B. (1954). Rigor mortis in beef. *J. Sci. Food Agric.* **5,** 70.

Marsh, B. B., and Leet, N. G. (1966). Studies on meat tenderness. III. The effects of cold-shortening on tenderness. *J. Food Sci.* **31,** 450.

Maruyama, K., Matsubara, S., Natori, R., Nonomura, Y., Kimura, S., Ohashi, K., Murakami, F., Handa, S., and Eguchi, G. (1976a). Connectin, an elastic protein of muscle. Characterization and function. *J. Biochem. (Tokyo)* **82,** 347.

Maruyama, K., Natori, R., and Nonomura, Y. (1976b). New elastic protein from muscle. *Nature (London)* **262,** 58.

Maruyama, K., Matsubara, S., Natori, R., Nonomura, Y., Kimura, S., Ohashi, K., Murakami, F., Harada, S., and Eguchi, G. (1977). Connectin, an elastic protein of muscle. Characterization and function. *J. Biochem. (Tokyo)* **82,** 317.

Maruyama, K., Kimura, M., Kimura, S., Ohashi, K., Suzuki, K., and Katunuma, N. (1981).

Connectin, an elastic protein muscle. Effects of proteolytic enzymes *in situ*. *J. Biochem. (Tokyo)* **89,** 701.

Matsui, T., Shimija, C., and Matsuura, F. (1975a). Studies on metmyoglobin reducing enzyme systems in the muscle of blue white dolphins. Amino acid composition. *Bull. Jpn. Soc. Sci. Fish.* **41,** 877.

Matsui, T., Shimizu, C., and Matsuura, F. (1975b). Studies on metmyoglobin reducing enzyme systems in muscles of blue white dolphins: Purification of some physico-chemical properties of ferrimyoglobin reductase. *Bull. Jpn. Soc. Sci. Fish.* **44,** 761.

Matsuura, F., Hashimoto, K., Kikawada, S., and Yamaguchi, K. (1962). Studies on the autoxidation velocity of fish myoglobin. *Bull. Jpn. Soc. Sci. Fish.* **28,** 210.

McCrea, S. E., Secombe, C. G., Marsh, B. B., and Carse, W. A. (1971). Studies in meat tenderness. 9. The tenderness of various lamb muscles in relation to their skeletal restraint and delay before freezing. *J. Food Sci.* **36,** 566.

McKeith, F. K., Smith, G. C., Dutson, T. R., Savell, J. W., Hostetler, R. L., and Carpenter, Z. L. (1980). Electrical stimulation of intact or split steer and cow carcasses. *J. Food Prot.* **43,** 795.

Mendelson, R. (1982). X-ray scattering by myosin S-1, implications for the steric blocking model of muscle control. *Nature (London)* **298,** 665.

Meyer, L. H. (1964). "Food Chemistry," 3rd ed. Reinhold, New York.

Miki, H., and Nishimoto, J. (1984). Kinetic parameters of freshness-lowering and discoloration based on temperature dependence in fish muscles. *Bull. Jpn. Soc. Sci. Fish.* **50,** 281.

Moore, P. B., Huxley, H. E., and De Roosier, D. J. (1970). Three-dimensional reconstruction of F-actin. Thin filaments and decorated filaments. *J. Mol. Biol.* **50,** 279.

Munns, W. O., and Burrell, D. E. 1966.The incidence of darkcutting beef. *Food Technol.* **20,** 1601.

Murray, J. M., and Weber, A. (1974). The cooperative action of muscle protein. *Sci. Am.* **230** (2), 9.

Nagayama, F. (1966). Mechanisms of breakdown and synthesis of glycogen in tissues of marine animals. *Nippon Suisan Gakkaishi* **32,** 188.

Nakatani, Y., Fujita, T., Sawa, S., Otani, T., Hori, Y., and Takagahara, I. (1986). Changes in ATP-related compounds of beef and rabbit muscles and a new index of freshness of muscle. *Agric. Biol. Chem.* **50,** 1751.

Naughton, J. J., Zeitlin, H., and Frodyma, M. M. (1958). Spectral reflectance studies of the heme pigments in tuna fish flesh. Some characteristics of the pigments and discoloration of tuna meat. *J. Agric. Food Chem.* **6,** 933.

Nazir, D. J., and Magar, N. G. (1963). Biochemical changes in fish muscle during rigor mortis. *J. Food Sci.* **28,** 1.

Newbold, R. P. (1966). Changes associated with rigor mortis. *In* "The Physiology and Biochemistry of Muscle as Food" (E. J. Briskey, R. G. Cassens, and J. C. Trautman, eds.), pp. 213–224. Univ. of Wisconsin Press, Madison.

Newbold, R. P., and Lee, C. A. (1965). Post-mortem glycolysis in skeletal muscle. The extent of glycolysis in diluted preparation of mammalian muscle. *Biochem. J.* **97,** 1.

Newbold, R. P., and Scopes, R. K. (1967). Post-mortem glycolysis in ox skeletal muscle. Effects of temperature on the concentrations of glycolytic intermediates and cofactors. *Biochem. J.* **105,** 127.

Newbold, R. P., and Small, L. M. (1985). Electrical stimulation of post-mortem glycolysis in the semitendinosus muscle of sheep. *Meat Sci.* **12,** 1.

Obatake, A., and Heya, H. (1985). A rapid method to measure dark content in fish. *Bull. Jpn. Soc. Sci. Fish.* **51,** 1001.

Obatake, A., Tsumiyama,, S., and Yamamoto, Y. (1985). Extractive nitrogenous constituents from the dark muscle of fish. *Bull. Jpn. Soc. Sci. Fish.* **5,** 1461.

O'Keeffe, M., and Hood, D. E. (1980–1981). Anoxic storage of fresh beef. 1. Nitrogen and carbon dioxide storage atmospheres. *Meat Sci.* **5,** 27.
O'Keeffe, M., and Hood, D. E. (1982). Biochemical factors influencing metmyoglobin formation on beef from muscles of differing colour stability. *Meat Sci.* **7,** 209.
Okitani, A., Matsukura, U., Kato, H., and Fujimaki, M. (1980). Purification and some properties of a myofobrillar protein-degrading protease, cathepsin rabbit skeletal muscle. *J. Biochem. (Tokyo)* **87,** 1133.
Olson, D. G., and Parrish, F. C., Jr. (1977). Relationship of myofibril fragmentation index to measures of beefsteak tenderness. *J. Food Sci.* **42,** 506.
Olson, D. G., Parrish, F. C., Jr., and Stromer, M. H. (1976). Myofibril fragmentation and shear resistance of three bovine muscles during postmortem storage. *J. Food Sci.* **41,** 1036.
Olson, D. G., Parrish, F. C., Jr., Dayton, W. R., and Goll, D. E. (1977). Effect of postmortem storage and calcium activated factor on the myofibrillar proteins of bovine skeletal muscle. *J. Food Sci.* **42,** 117.
Osebold, W. R., and Pedrini, V. (1976). Pepsin-solubilized collagen of human nucleus pulposus and annuls fibrosus. *Biochim. Biophys. Acta* **434,** 390.
O'Shea, J. M., Robson, R. M., Huiatt, T. W., Hartzer, M. K., and Stromer, M. H. (1979). Purified desmin b from adult mammalian skeletal muscle: A peptide mapping comparison with desmin from adult mammalian and avian smooth muscle. *Biochem. Biophys. Res. Commun.* **89,** 972.
O'Shea, J. M., Robson, R. M., Hartzer, M. K., Huiatt, T. W., Rathbun, W. E., and Stromer, M. H. (1981). Purification of desmin from adult mammalian skeletal muscle. *Biochem. J.* **195,** 345.
Ota, F., and Nakamura, T. (1952). Change of ammonia content in fish meat by heating under pressure. Relation between increase of ammonia and the freshness of fish. *Bull. Jpn. Soc. Sci. Fish.* **18,** 15.
Parrish, F. C., Jr., and Lusby, M. L. (1983). An overview of a symposium on the fundamental properties of muscle proteins important in meat science. *J. Food Biochem.* **7,** 125.
Parrish, F. C., Jr., Vandell, C. J., and Culler, R. D. (1979). Effect of maturity and marbling on the myofibril fragmentation index of bovine longissimus muscle. *J. Food Sci.* **44,** 1668.
Parry, D. A. D., and Squire, J. (1973). Structural role of tropomyosin in muscle regulation, Analysis of the X-ray diffraction patterns from relaxed and contracting muscles. *J. Mol. Biol.* **75,** 33.
Partmann, W. (1965). Changes in proteins, nucleotides and carbohydrates during rigor mortis. *In* "The Technology of Fish Utilization" (R. Kreuzer, ed.), p 4. Fishing News (Books), London.
Partmann, W., and Frank, H. K. (1973). Storage of meat in controlled gaseous atmospheres. *Prog. Refrig. Sci. Technol., Proc. Int. Con. Refrig., 13th, 1971* Vol. **3,** p. 17.
Patashnik, M., and Groninger, H. S. (1964). Observations on the milky condition in some Pacific coast fishes. *J. Fish. Res. Board Can.* **21,** 335.
Paterson, B. C., and Parrish, F. C., Jr. (1987). SDS–PAGE conditions for detection of titin and nebulin in tender and tough bovine muscle. *J. Food Sci.* **52,** 509.
Penny, I. F. (1974). The action of a muscle proteinase on the myofibrillar proteins of bovine muscle. *J. Sci. Food Agric.* **25,** 1273.
Penny, I. F., and Dransfield, E. (1979). Relationship between toughness and troponin T in conditioned beef. *Meat Sci.* **3,** 135.
Perry, S. V. (1979). The regulation of contractile activity in muscle. *Biochem. Soc. Trans.* **7,** 593.
Pirko, P. C., and Ayres, J. C. (1957). Pigment changes in packaged beef during storage. *Food Technol.* **11,** 461.
Potter, J. D., and Gergely, J. (1975). The calcium anmd magnesium binding sites of troponin and their role in the regulation of myofibrillar adenosine triphosphatase. *J. Biol. Chem.* **250,** 4628.
Purslow, P. P. (1985). The physical basis of meat texture: Observations on the fracture behaviour of cooked bovine M. Semitendinosus. *Meat Sci.* **12,** 39.

Richardson, F. L., Stromer, M. H., Huiatt, T. W., and Robson, R. M. (1981). Immunoelectron and fluorescence microscope localisation of desmin in mature avian muscles. *Eur. J. Cell Biol.* **26,** 91.

Ridpath, J. F., Robson, R. M., Huiatt, T. W., Trenkle, A. H., and Lusby, M. L. (1982). Localization and rate of accumulation of nebulin in skeletal and cardiac muscle cell cultures. *J. Cell Biol.* **95,** 361a.

Ringkob, T. P., Marsh, B. B., and Greaser, M. L. (1988). Change in titin position in postmortem bovine muscle. *J. Food Sci.* **53,** 276.

Ritchie, A. D. (1926). Lactic acid and rigor mortis. *J. Physiol. (London)* **6** (1), iv-v.

Robbins, F. M., and Cohen, S. H. (1976). Effects of cathepsin enzymes from spleen on the microstructure of bovine semimembranous muscle. *J. Texture Stud.* **7,** 137.

Robson, R. M., O'Shea, J. M., Hartzer, M. K., Rathbun, W. E., LaSalle, F., Schreiner, P. J., Kasang, L. E., Stromer, M. H., Lusby, M. L., Ridpath, J. F., Pang, Y-Y., Evans, R. R., Zeece, M. G., Parrish, F. C., and Huiatt, T. W. (1984). Role of new cytoskeletal elements in maintenance of muscle integrity. *J. Food Biochem.* **8,** 1.

Rowe, R. W. D. (1974). Collagen fiber arrangement in intramuscular connective tissue. Changes associated with muscle shortening and their possible relevance to raw meat toughness measurements. *J. Food Technol.* **9,** 501.

Ryder, J. M. (1985). Determination of adenosine triphosphate and its breakdown products in fish muscle by high performance liquid chromatography. *J. Agric. Food Chem.* **33,** 678.

Saito, T., and Arai, K. (1958). Further studies of inosinic acid formation in carp muscle. *Nippon Suisan Gakkaishi* **23,** 579.

Saito, T., Arai, A., and Matsuyoshi, M. (1959). A new method for estimating the freshness of fish. *Bull. Jpn. Soc. Sci. Fish.* **24,** 749.

Saleh, B., and Watts, B. M. (1968). Substrates and intermediates in the enzymatic reduction of metmyoglobin in ground beef. *J. Food Sci.* **33,** 353.

Samejima, K., and Wolfe, F. H. (1976). Degradation of myofibrillar protein components during postmortem aging of chicken muscle. *J. Food Sci.* **41,** 250.

Sasano, Y., and Tawara, T. (1962). Studies on the green meat of albacore and yellowfin tuna. 3. Relationship between a substance producing yellow color with ninhydrin and the green meat tuna. *Bull Jpn. Soc. Sci. Fish.* **28,** 722.

Sasano, Y., Tawara, T., and Higashi, K. (1961). Studies on green meat of albacore and yellowfin tuna. *Bull. Jpn. Soc. Sci. Fish.* **27,** 586.

Savell, J. W., Smith, G. C., Dutson, T. R., Carpenter, Z. L., and Suter, D. A. (1977). Effect of electrical stimulation on palatability of beef, lamb and goat meat. *J. Food Sci.* **42,** 702.

Savell, J. W., Dutson, T. R., Smith, G. C., and Carpenter, Z. L. (1978). Structural changes in electrically stimulated beef muscles. *J. Food Sci.* **43,** 1606.

Savell, J. W., McKeith, F. K., and Smith, G. C. (1981). Reducing postmortem aging time of beef with electrical stimulation. *J. Food Sci.* **46,** 1777.

Schwartz, W. N., and Bird, J. N. C. (1977). Degradation of myofibrillar proteins by cathepsins B and D. *Biochem. J.* **167,** 811.

Scopes, R. K. (1964). The influence of post-mortem conditions on the solubilities of muscle proteins. *Biochem. J.* **91,** 201.

Scopes, R. K. (1970). Characterization and study of sarcoplasmic proteins. *In* "The Physiology and Biochemistry of Muscle as a Food" (E. J. Briskey, R. G. Cassens, and B. B. Marsh, eds.), Vol. 2, p. 471. Univ. of Wisconsin Press, Madison.

Seideman, S. C., Cross, H. R., Smith, G. C., and Durland, P. R. (1984). Factors associated with fresh meat colour: A review. *J Food Qual.* **6,** 211.

Seymour, J., and O'Brien, E. J. (1980). The position of tropomyosin in muscle thin filaments. *Nature (London)* **283,** 680.

Shimizu, C., and Matsuura, F. (1971). Occurrence of new enzyme reducing metmyoglobin in dolphin muscle. *Agric. Biol. Chem.* **35,** 468

Shinokomaki, M., Elsden, D. F., and Bailey, A. J. (1972). Meat tenderness: Age related changes in bovine intramuscular collagen. *J. Food Sci.* **37,** 892.

Sink, S. D. (1979). Symposium on meat flavor: Factors influencing the flavor of muscle foods. *J. Food Sci.* **44,** 1.

Sjöstrand, F. S. (1962). The connections between A and I band filaments in striated frog muscle. *J. Ultrastruct. Res.* **7,** 225.

Smith, G. C., Culp, G. R., and Carpenter, Z. L. (1978). Postmortem aging of beef carcasses. *J. Food Sci.* **43,** 823.

Solberg, M. (1968). Factors affecting fresh meat color. *Proc. Meat Ind. Res. Conf.* pp. 32–40.

Solberg, M. (1970). The chemistry of color stability in meat: A review. *Can. Inst. Food Technol. J.* **3,** 55.

Sonaiya, E. B., Stouffer, J. R., and Beerman, D. H. (1982). Electrical stimulation of mature cow carcasses and its effects on tenderness, myofibril protein degradation and fragmentation. *J. Food Sci.* **47,** 889.

Spinelli, J., Eklund, M., and Miyauchi, D. (1964). Measurement of hypoxanthine in fish as a method of assessing freshness. *J. Food Sci.* **29,** 710.

Stanley, D. W. (1983). A review of the muscle cell cytoskeleton and its possible relation to meat texture and sarcolemma emptying. *Food Microstruct.* **2,** 99.

Stewart, M. P., Hutchins, B. K., Zipser, M. W., and Watts, B. M. (1965). Enzymatic reduction of metmyoglobin by ground beef. *J. Food Sci.* **30,** 487.

Sugden, M. C., Sharples, S. C., and Randle, J. (1976). Carcass glycogen as a potential source of glucose during short-term starvation. *Biochem. J.* **160,** 817.

Surette, M. E., Gill, T. A., and LeBlanc, P. J. (1988). Biochemical basis of postmortem nucleotide catabolism in cod *(Gadus morhua)* and its relationship to spoilage. *J. Agric. Food Chem.* **36,** 19.

Suzuki, T. (1953). Determination of volatile acids for judging the freshness of fish. *Bull. Jpn. Soc. Sci. Fish.* **19,** 102.

Suzuki, T. 1981. "Fish and Krill Protein Processing Technology," pp. 31–34. Applied Science Publishers, London.

Tarr, H. L. A. (1965). Pathways of glycogen breakdown. *In* "The Technology of Fish Utilization" (R. Kreuzer, ed.), p. 34. Fishing News (Books), London.

Tarr, H. L. A. (1966). Post-mortem changes in glycogen, nucleotides, sugar phosphates and sugars in fish muscles. A review. *J. Food Sci.* **31,** 846.

Tarr, H. L. A. (1968). Post-mortem degradation of glycogen and starch in fish muscle. *J. Fish. Res. Board Can.* **25,** 1539.

Tarrant, P. J. V. (1981). *In* "The Problem of Dark-Cutting in Beef" (D. E. Hood, and P. J. V. Tarrant, eds.), p. 462. Martinus Nijhoff Publishers, The Hague.

Tarrant, P. V., and Motherstill, C. (1977). Glycolysis and associated changes in beef carcasses. *J. Sci. Food Agric.* **28,** 739.

Tarrant, P. V., and Sherrington, J. (1980). An investigation of ultimate pH in the muscles of commercial beef carcasses. *Meat Sci.* **4,** 287.

Tarrant, P. J. V., McLoughlin, J. V., and Harrington, M. G. (1972a). Anaerobic glycolysis in biopsy and post-mortem porcine longissimus dorsi muscle. *Proc. R. Ir. Acad., Sect. B* **72B,** 55.

Tarrant, P. J. V., Hegarty, P. V. J., and McLoughlin, J. V. (1972b). High-energy phosphates and anaerobic glycolysis in the red and white fibers of porcine semitendinosus muscle. *Proc. R. Ir. Acad., Sect. B* **72B,** 229.

Taylor, A. A. (1972). Gases in fresh meat packaging. *Meat World* **5,** 3.

Taylor, A. A., and Amos, L. A. (1981). A new model for the geometry of the binding of myosin crosslinkages to muscle thin filaments. *J. Mol. Biol.* **147,** 297.

Taylor, A. A., and MacDougall, D. B. (1973). Fresh beef packed in mixtures of oxygen and carbon dioxide. *J. Food Technol.* **8,** 453.

Tomioka, K., and Endo, K. (1984). Purification of 5′-nucleotidase from carp muscle. *Bull. Jpn. Soc. Sci. Fish.* **50,** 1077.

Tomioka, K., and Endo, K. (1985). Zn content and subunit structure of carp muscle 5′-nucleotidase. *Bull. Jpn. Soc. Sci. Fish.* **51,** 857.

Tomlinson, N., and Geiger, S. E. (1962). Glycogen concentration and post-mortem loss of adenosine triphosphate in fish and mammalian muscle: A review. *J. Fish. Res. Board Can.* **19,** 997.

Tomlinson, N., Geiger, S. E., and Dollinger, E. (1965). Chalkiness in halibut in relation to muscle pH and protein denaturation. *J. Fish. Res. Board Can.* **22,** 653.

Toyohara, H., Makinodan, Y., and Ikeda, S. (1982). Purification and properties of carp muscle cathepsin A. *Bull. Jpn. Soc. Sci. Fish.* **48,** 1145.

Trucco, R. E., Lupin, H. M., Giannini, D. H., Crupkin, M., Boeri, R. L., and Barassi, C. A. (1982). Study on the evolution of rigor mortis in batches of fish. *Lebensm.-Wiss. Technol.* **15,** 77.

Tsai, R., Cassens, R. G., Briskey, E. J., and Greaser, M. L. (1972). Studies on nucleotide metabolism in porcine longissimus muscle postmortem. *J. Food Sci.* **37,** 612.

Uchiyama, H., and Kakuda, K. (1984). A simple and rapid method for measuring *K* value, a fish freshness index. *Bull. Jpn. Soc. Sci. Fish.* **50,** 263.

Walters, C. L. (1975). *In* "Meat" (D. J. A. Cole, and R. A. Lawrie, eds.), pp. 385–401.

Wang, K. (1983). Cytoskeletal matrix in striated muscle. The role of titin, nebulin and intermediate filaments. *In* "Cross-Bridge Mechanisms in Muscular and Cellular Control," pp. 439–452. Cold Spring Harbor Lab., Cold Spring Harbor, New York.

Wang, K., and Ramirez-Mitchell, R. (1979). Titin: Possible candidate as putative longtitudinal filaments in striated muscle. *Proc. Natl. Acad. Sci. U.S.A.* **76,** 3698.

Wang, K., and Ramirez-Mitchell, R. (1983a). A network of transverse and longtitudinal intermediate filaments is associated with sarcomeres of adult vertebrate skeletal muscle. *J. Cell Biol.* **83,** 389a.

Wang, K., and Ramirez-Mitchell, R. (1983b). Ultrastructural morphology and epitope distribution of titin, a giant sarcomere-associated cytoskeletal protein. *J. Cell Biol.* **97,** 257a.

Wang, K., and Ramirez-Mitchell, R. (1984). Architecture of titin-containing cytoskeletal matrix in striated muscle. Mapping of distinct epitopes of titin specified by monoclonal antibodies. *Biophys. J.* **45,** 392a.

Wang, K., and Williamson, C. L. (1980). Identification of an N-line protein of striated muscle. *Proc. Natl. Acad. Sci. U.S.A.* **77,** 3254.

Wang, K., McClure, J., and Tu, A. (1979). Titin: Major myofibrillar component of striated muscle. *Proc. Natl. Acad. Sci. U.S.A.* **76,** 3698.

Wang, S. -M., and Greaser, M. L. (1985). Immunocytochemical studies using a monoclonal antibody to bovine cardiac titin on intact and extracted myofibrils. *J. Muscle Res. Cell Motil.* **6,** 293.

Watanabe, E., Toyama, K., Karube, I., Matsuoka, H., and Suzuki, S. (1984). Determination of inosine-5-monophosphate in fish tissue with an enzyme sensor. *J. Food Sci.* **49,** 114

Watts, B. M., Kendrick, J., Zipser, M. W., Hutchins, B., and Saleh, B. (1966). Enzymatic reducing pathways in meat. *J. Food Sci.* **31,** 855.

Wells, R. M. G. (1987). Stress responses imposed by fish capture and handling: A physiological perspective. *Food Technol. Aust.* **39,** 479.

Whiting, R. C. (1980) Calcium uptake by bovine muscle mitochondria and sarcoplasmic reticulum. *J. Food Sci.* **45,** 288.

Wu, J. J., Dutson, T. P., and Carpenter, Z. L. (1981). Effect of postmortem time and temperature on the release of lysosomal enzymes and their possible effect on bovine connective tissue components of muscle. *J. Food Sci.* **46,** 1132.

Yamagata, M., Horimoto, K., and Nagaoka, C. (1969). Assessment of green tuna: Determination of trimethylamine oxide and its derivatives in tuna muscle. *J. Food Sci.* **34,** 156.

Yamagata, M., Horimoto, K., and Nagaoka, C. (1971). Accuracy in predicting occurrence of greening in tuna based on content of trimethylamine oxide. *J. Food Sci.* **36,** 55.

Yamaguchi, M., Robson, R. M., Stromer, M. H., Dahl, D. S., and Oda, T. (1982). Nemaline rod bodies: Structure and composition. *J. Neurol. Sci.* **56,** 35.

Yamaguchi, M., Robson, R. M., and Stromer, M. H. (1983a). Evidence for actin involvement in cardiac Z-line analogs. *J. Cell Biol.* **96,** 435.

Yamaguchi, M., Robson, R. M., Stromer, M. H., Cholvin, N. R., and Izumimoto, M. (1983b). Properties of soleus muscle Z-lines and induced Z-line analogs revealed by dissection with Ca-activated neutral protease. *Anat. Rec.* **206,** 345.

Younathan, M. T., and Watts, B. M. (1960). Oxidation of tissue lipids in cooked pork. *Food Res.* **25,** 538.

Young, O. A., Graafhuis, A. E., and Davey, C. L. 1980–1981. Post-mortem changes in cytoskeletal proteins of muscle. *Meat Sci.* **5,** 41.

2

Biochemical Changes in Raw Foods: Fruits and Vegetables

I. Introduction

Characteristics of fruits and vegetables such as flavor, color, size, shape, and absence of external defects ultimately determine their acceptance by consumers. The development of these characteristics is the result of many chemical and biochemical changes that occur following harvesting and storage. Since harvesting fruits and vegetables at their correct stage of maturity is critical for the development of a highly acceptable product for the fresh market for processing, it is important to more fully understand what changes are taking place. This chapter will highlight those changes occurring within fruits and vegetables during the postharvest period. It is during this period that fruits and vegetables show a gradual reduction in quality concurrent with transpiration and respiration, as well as with other biochemical and physiological changes. Ultimately the plant material deteriorates because of the undesirable enzyme activity and spoilage microorganisms.

The growth and maturation of fruits and vegetables are dependent on photosynthesis and absorption of water and minerals by the parent plant. Once detached, however, they are independent units in which respiratory processes play a major role. This chapter will focus on those changes in postharvest fruits and vegetables which affect quality.

II. Respiration

Respiration is the fundamental process whereby living organisms carry out the exothermic conversion of potential energy into kinetic energy. In higher plants the major storage products are sucrose and starch. These are completely oxidized in the presence of oxygen to carbon dioxide and water, with the production of adenosine triphosphate (ATP):

$$C_6H_{12}O_6 + 6O_2 \rightarrow 6CO_2 + 6H_2O + \text{energy (heat and ATP)}$$

The latter is the form in which energy is stored within the cell. The contribution of proteins and lipids to plant respiration is difficult to assess but can occur *via* the formation of acetyl-CoA. In the absence of oxygen, anaerobic respiration occurs, resulting in only a partial degradation of carbohydrates and a lower ATP production:

The metabolic pathways involved in the respiration of plant tissue result in the conversion of starch or sucrose to glucose-6-P. The latter is then oxidized by glycolysis (Embden–Meyerhoff pathway) or the pentose phosphate pathway to triose phosphate, which enters the tricarboxylic acid cycle by way of pyruvate (Scheme 2.1) (ap Rees, 1977).

The contribution of these two major pathways of carbohydrate oxidation to plant respiration still remains unresolved. Difficulties were encountered with the experimental techniques used in assessing the relative roles of these pathways based on the production of $^{14}CO_2$ or labeled intermediates from labeled hexoses (ap Rees, 1980). Evidence shows that both pathways exist in plant tissues (ap Rees, 1974) and that they change considerably during plant development (ap Rees, 1977). Current evidence supports the glycolytic pathway as the predominant one operating, while the maximum contribution of the pentose phosphate pathway may not exceed 30% of the total (ap Rees, 1980). The relative importance of these pathways probably depends on the particular plant, the organ, and the state of maturity.

A. Fruits

A large number of fruits exhibit a sudden sharp rise in respiratory activity following harvesting, referred to as the climacteric rise in respiration. This phenomenon was first noted by Kidd and West (1922, 1930) as an upsurge in carbon dioxide gas at the end of the maturation phase of apples. Since then there have been numerous reports on this phenomenon in a wide range of fruits. The appropriateness of the term climacteric was questioned by Rhodes (1970), who suggested it should be all-inclusive and describe the "whole of the control phase in the life of fruit triggered by ethylene and the concomitant changes occurring."

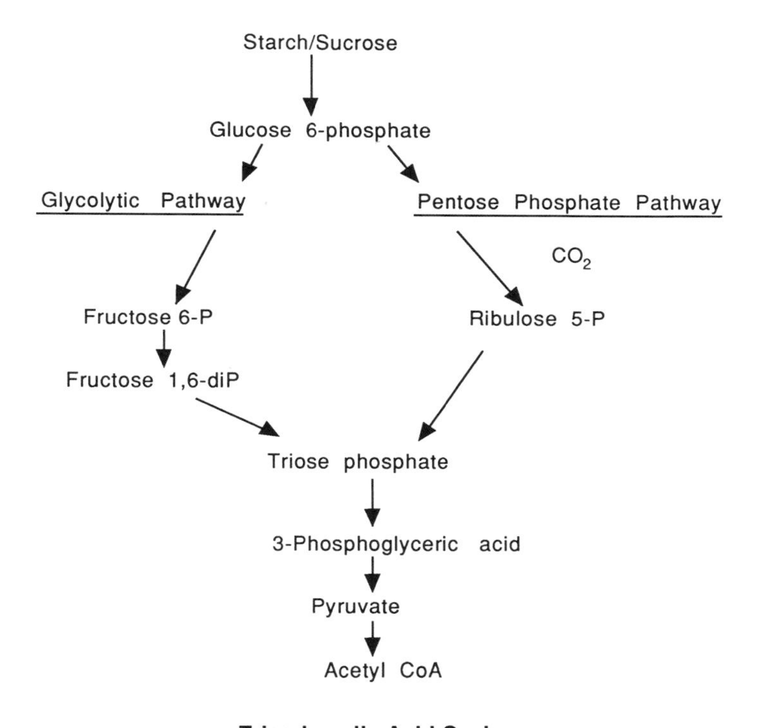

SCHEME 2.1. Glycolytic and pentose phosphate pathways.

McGlasson *et al.* (1978), however, suggested that respiratory climacteric was the more appropriate term to describe this gaseous phenomenon. Biale and Young (1981) nevertheless still preferred the more inclusive description in which climacteric defined those physical, chemical, physiological, and metabolic changes associated with the increased rate of respiration covering the transition phase from growth and maturation to the final stages of senescence. Essentially, climacteric defines the last stages of the fruit at the cellular level which determines the quality of the fruit which is shipped to the consumer.

Biale (1960a,b) tentatively classified fruits as either climacteric or nonclimacteric according to their respiratory rates. A more recent review by Biale and Young (1981), however, suggested a more extensive list of fruits from both groups as shown in Table 2.1. Included are cantaloupe, honeydew melon, and figs, all of which are considered climacteric (Lyons *et al.*, 1962; Pratt and Groeschel, 1968; Marei and Crane, 1971). A number of rare fruits were also included, namely, breadfruit (Biale and Barcus, 1970), guavas, and mammee apples (Akamine and Goo, 1978, 1979a,b).

TABLE 2.1

RESPIRATORY ACTIVITY OF SELECTED FRUITS[a]

Climateric	Nonclimacteric
Apple	Blueberry
Apricot	Grape
Avocado	Grapefruit
Banana	Java plum
Breadfruit	Lemon
Fig	Olive
Guava	Orange
Mammee apple	Pineapple
Muskmelon cantaloupe	Strawberry
	Honeydew

[a] Adapted from Biale and Young (1981).

The period immediately prior to the climacteric rise, when the respiratory level is at a minimum, is known as the preclimacteric. Following the completion of the climacteric rise is the postclimacteric phase, in which a decline in the respiratory rate occurs. Unlike the sudden rise in respiratory activity which characterizes climacteric fruits, nonclimacteric fruits exhibit a steady fall in respiratory activity. This downward trend in the respiratory activity was originally observed for lemons stored at 15°C by Biale and Young (1947) and later in oranges (Biale, 1960a,b). Figure 2.1 illustrates the difference in respiratory activity between climacteric and nonclimacteric fruits, for example, avocadoes and lemons.

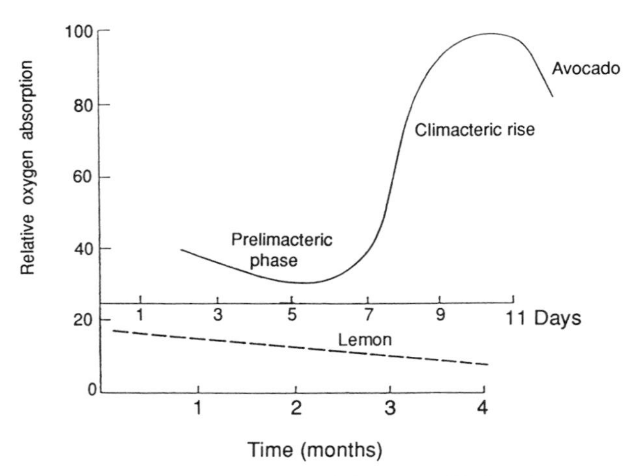

FIG. 2.1. The respiratory trends in climacteric fruits, exemplified by the avocado, compared with nonclimacteric fruit, depicted by the lemon (Biale *et al.*, 1954). Reprinted with permission of copyright owner, American Society of Plant Physiology (ASPP).

In the original classification by Biale (1960a,b), oranges were classified as nonclimacteric fruits since they had a low rate of respiration. It was soon evident that some members of the citrus fruit family had respiratory activities similar to those of climacteric fruits. In spite of the higher respiratory rates, a downward trend was observed for both Valencia and Washington Navel cultivars. This decline in respiration was observed by Bain (1958) in Valencia oranges from fruit set to maturity. Nevertheless, Trout and co-workers (1960) reported a typical respiratory rise in oranges stored at 4.3–10°C. Their results were attributed by Biale and Young (1981) to possible chilling injury at the lower storage temperatures. Only half of those oranges kept at 10°C exhibited a rise in respiration following harvesting, which could be due to the immaturity of some of these fruit. Aharoni (1968) also used the term climacteric to describe the increase in respiratory rate in postharvest, young, unripe Washington Navel, Shamouti, and Valencia oranges as well as in Marsh's seedless grapefruit stored at 16 and 20°C. In contrast, the full-sized and mature fruit did not exhibit any rise in respiratory activity. Eaks (1970) examined the respiratory patterns of several species of citrus fruit throughout ontogeny. The small and immature oranges and grapefruit exhibited a rise in respiration and ethylene activity when stored at 20°C for several days after harvest. As the weight of the fruit increased, characteristic of maturation, the level of CO_2 and ethylene production decreased until full maturity was approached or attained, at which point no change in respiration was noted. Based on this study it was evident that citrus fruits were correctly classified as nonclimacteric. Rhodes (1970) noted that had the climacteric been defined as the period of enhanced metabolic activity during the transition from the growth phase to senescence in fruits, the confusion with citrus fruits could have been avoided. A similar situation was observed in grapes, in which a respiratory rise was reported by Peynaud and Riberau-Gayon (1971) to accompany rapid growth, which was referred to as "rudimentary climacteric." However, this was resolved when the postmaturation change in respiratory activity showed a typical nonclimacteric pattern consistent with earlier work by Geisler and Radler (1963). A similar controversy arose with respect to the pineapple and while Dull *et al.* (1967) found a slight upward respiratory trend it was not typical of climacteric fruit. The identification of several tomato mutants by Herner and Sink (1973) without any climacteric pattern was later reviewed by Tigchelaar and co-workers (1978a,b). These mutants were unable to produce ethylene and had low levels of carotenoids. A particular feature was the extremely low levels of polygalacturonase, which accounted for their prolonged firmness.

B. Vegetables

Once the vegetable becomes detached from the parent plant, metabolism continues to take place, although it is the catabolic reactions that soon become dominant. The climacteric rise in respiration characteristic of certain fruits such as apples and avocadoes is not apparent in vegetables, where there is no clear-cut division between maturation and breakdown.

The intensity and rate of respiration varies with the particular plant, the degree of maturity, and whether the vegetable is actively growing at the time of harvest or functioning as a storage organ. For example, McKenzie (1932) reported higher respiration intensity in freshly harvested immature lettuce *(Lactuca sativa)* during the first 12 hr which then fell to the same level as that of mature lettuce. The deterioration of several vegetables was examined by Platenius (1942), who found that the initial respiratory rate for asparagus *(Asparagus officinalis)* at 24°C was almost 50 times greater compared to potatoes. The respiration rate declined for all vegetables during 60 days storage irrespective of the temperature. The initial rates of respiration appeared to be a useful indicator of the potential storage life of the crop during precooling and early storage. A high respiratory rate, however, was indicative of a short storage life while the reverse was true for crops with a low respiratory rate. This is illustrated for a number of vegetables in Figure 2.2.

Vegetables can be classified as high, low, or intermediate according to the respiration rates. For example, young tissue, such as growing parts of asparagus or developing seeds of green peas, has high respiration rates, while low rates are evident in storage organs such as stems (potatoes), roots (sweet potatoes), and bulbs (onions). Leafy vegetables appear to be intermediate, while some vegetables such as cabbage can be stored at low temperature for considerable periods of time. Other vegetables, including cucumbers *(Cucumis sativa),* are particularly susceptible to chilling injury if stored over a temperature range of of 0–10°C (Eaks and Morris, 1956). Chilling injury is diagnosed by an increase in respiration in which a plateau is reached corresponding to chilling, after which there is a decline in respiration.

1. *Control of the Climacteric Rise*

The dramatic upsurge in respiratory activity associated with the climacteric has been attributed to a number of different factors. One theory proposed was the breakdown in "cell membrane permeability or organization resistance (Solomos and Laties, 1973). While such changes are evident during ripening, the question remains as to whether they are the cause or the consequence of the ripening process (Theologies and Laties, 1978). The second theory was focused on increased protein synthesis as a necessary prerequisite for the ripening process (Brady *et al.*, 1976; Richmond and Biale, 1966) or enhanced ATP turnover with respiratory stimulation (Biale, 1960b). Subsequent research, however, showed there was no difference in the respiratory capacity of mitochondria obtained from preclimacteric and climacteric avocado tissue (Biale, 1969). In addition, the respiratory rate of uncoupled preclimacteric avocado slices was quite sufficient to

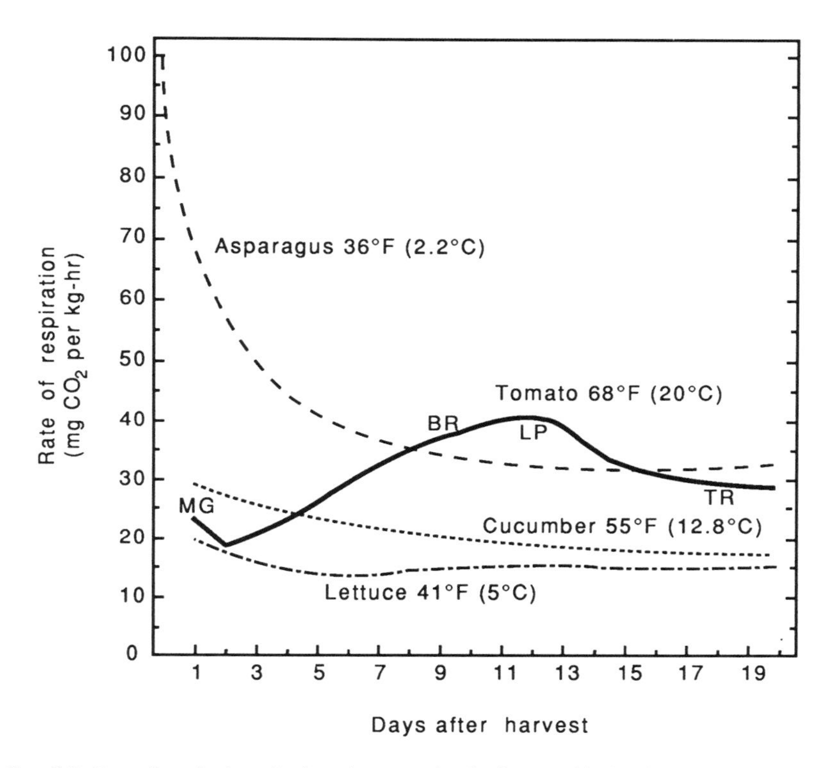

FIG. 2.2. Rate of respiration of a shoot (asparagus), a leafy vegetable (head lettuce), a nonripening fruit (cucumber), and a ripening fruit (tomato) at temperatures commonly encountered during their marketing (for tomato: MG = mature green; BR = breaker; LP = light pink; TR = table ripe) (data taken from Lipton, 1977; Pratt *et al.*, 1954; Workman *et al.*, 1957) (Ryall and Lipton, 1979).

facilitate the climacteric rise (Millerd *et al.*, 1953). This was confirmed in studies using isolated mitochondria from avocado (Biale *et al.*, 1957). Further research by Lance *et al.* (1965) and Hobson *et al.* (1966), using improved techniques, isolated mitochondria from avocado throughout all stages of the climacteric and found that the oxidative and phosphorylating activities remained unchanged provided all cofactors were present. The stimulation of glycolysis in avocado fruit under anaerobic conditions also demonstrated considerable latent glycolytic capacity (Solomos and Laties, 1974). These studies indicated the adequacy of the enzymes in preclimacteric fruits to sustain the respiratory climacteric. This was consistent with studies by Frenkel *et al.* (1968) and McGlasson *et al.* (1971), who found that while inhibiting protein synthesis prevented the ripening of intact peas and banana slices it did not decrease the upsurge in respiration.

The dramatic burst in respiratory activity that accompanies the ripening of climacteric-type fruits appears to be due to some change in the mitochondrial respiratory function *in vivo* (Biale, 1960a,b). A possible explanation for this was attributed to an increase in cyanide-resistant respiration (Solomos and Laties, 1976; Solomos, 1977). Solomos and Laties (1974, 1976) observed that cyanide initiated identical physiological and biochemical changes in avocado and potato tubers. Cyanide is a known inhibitor of cytochrome oxidase, the terminal oxidase in the electron transport system. Thus there is present a cyanide-insensitive pathway which permits the aerobic oxidation of respiratory substrates in the presence of cyanide (Bendall and Bonner, 1971). This cyanide-resistant pathway or alternative pathway was reported to be present in ethylene-responsive fruits (Solomon and Laties, 1974, 1976). Further research by Theologies and Laties (1978) led to an examination of this pathway in the respiration of ripening avocadoes and bananas. These researchers found that the surge in respiration during the climacteric in the intact fruit was cytochrome mediated. The pre-climacteric fruit had the capacity to sustain electron transport through the cytochrome pathway, although it remained unexpressed. During the climacteric rise the alternative pathway appears to remain at a low level of activity and may be involved in the generation of peroxide (Rich *et al.*, 1976). A discussion of the possible regulatory role of cyanide in ethylene biosynthesis can be found in Section IV.

2. *Enzymatic Control*

The possibility of enzymatic activity being the controlling factor in the climacteric has been suggested. Tager and Biale (1957) noted a rise in carboxylase and aldolase activity during the ripening of banana accompanied by a shift from the pentose phosphate to the glycolytic pathway. This may occur during the transition period from the pre- to the postclimacteric phase in fruit ripening.

a. Malic Enzyme. Hulme *et al.* (1963) reported a sharp increase in the activities of malic enzyme and pyruvate carboxylase during the ripening of apples. This explained the slight uptake of oxygen during ripening of apples compared to the marked increase in CO_2 evolution.

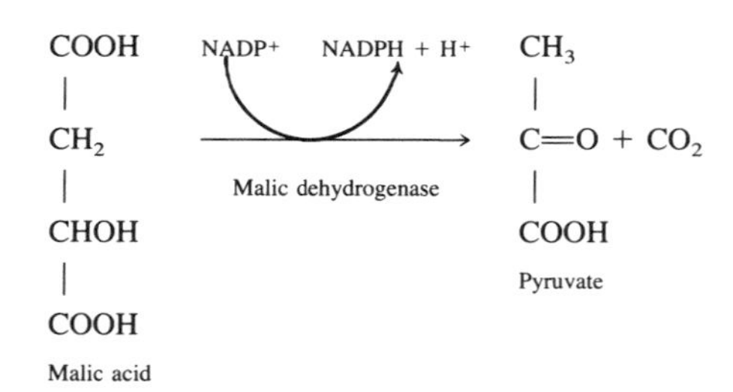

$$\begin{array}{c} CH_3 \\ | \\ C{=}O \\ | \\ COOH \end{array} \xrightarrow[\text{Pyruvate carboxylase}]{\text{TPP } Mg^{2+}} \begin{array}{c} CH_3 \\ | \\ CHO \end{array} + CO_2$$

Several studies showed that malic enzyme exhibited varying degrees of cyanide insensitivity depending on the activity of the enzyme (Coleman and Palmer, 1972; Lance *et al.*, 1967; Macrae, 1971; Neuburger and Douce, 1980). Moreau and Romani (1982) examined the oxidation of malate during the climacteric rise in avocado mitochondria with particular focus on the cyanide-insensitive alternative pathway. The increase in malic enzyme activity paralleled the increase in malate oxidation as ripening advanced through the climacteric. Malate is oxidized by malic dehydrogenase through the cytochrome pathway. It can also be oxidized by the malic enzyme, which involves the alternative pathway via a rotenone-insensitive NADH dehydrogenase located in the inner layer of the mitochondrial membrane (Marx and Brinkmann, 1978; Palmer, 1976; Rustin *et al.*, 1980). These researchers concluded that the malic enzyme and alternate oxidase pathway probably function under conditions of relatively low ATP demands and high energy change characteristics at the later stages of the climacteric. While the regulation of electron transport *via* cytochrome and alternative pathways remains to be clarified in avocado mitochondria, the involvement of the alternative pathway cannot be totally ruled out.

b. Phosphofructokinase and Pyrophosphate: Fructose-6-phosphate Phosphotransferase. Salimen and Young (1975) examined the possibility that the climacteric was regulated by enzyme activation involving phosphofructokinase (PFK) (ATP : D-fructose-6-phosphate-1-phosphotransferase, EC 2. 7. 1. 11). This was based on research by Barker and Solomos (1962), who observed an increase in fructose 1, 6-diphosphate during the ripening of bananas and in tomatoes by Chalmers and Rowan (1971). This increase in fructose 1, 6-diphosphate was attributed to activation of PFK. Salimen and Young (1975) reported that activation of this enzyme accounted for a 20-fold increase of fructose 1, 6-diphosphate during the ripening process. Electrophoretic separation of PFK showed that no new species of the enzyme were produced during the climacteric with the enzyme remaining in the oligomeric form. Rhodes (1971) reported that PFK was present in the oligomeric weight form species in tomato fruits up to the climacteric phase while both the oligomeric and low-molecular-weight species were isolated in the postclimacteric phase. Isaac and Rhodes (1982) later found that PFK existed in the oligomeric form at the "breaker" stage during tomato ripening. Using gel-permeation chromatography Isaac and Rhodes (1987) identified a single peak corresponding to the oligomeric form of PFK at the green and

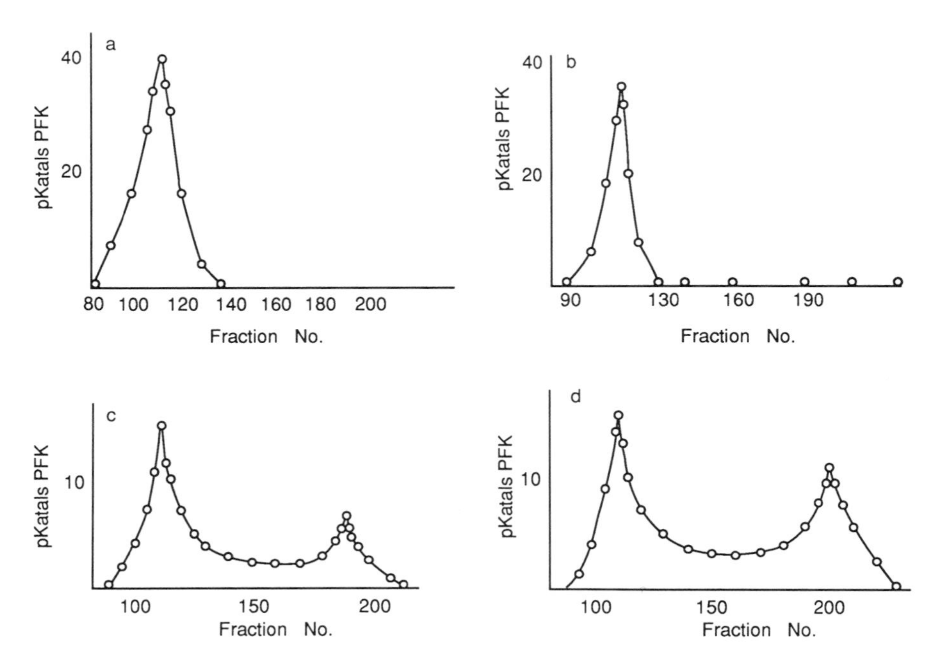

FIG. 2.3. Elution of PFK on Ultrogel AcA 34 for the enzyme preparation from (a) green, (b) breaker, (c) orange, and (d) red tomatoes (Isaac and Rhodes, 1987). Reprinted with permission. Copyright © by Pergamon Press.

breaker stages. Two peaks were separated, however, at the orange and red stages of tomato ripening, which corresponded to oligomeric and monomeric forms of the enzyme (Figure 2.3). To explain the behavior of PFK, these researchers proposed that stimulation of the enzyme occurred because of leakage in P_i from the vacuole as a consequence of permeability changes in the membrane during initiation of the climacteric. The continued leakage of P_i and citrate affected the enzyme at the molecular level by dissociating the oligomeric form of the enzyme into monomeric subunits during the later stages of ripening.

A recent study by Bennett and co-workers (1987) examined the role of glycolytic regulation of the climacteric in avocado fruit. They used *in vivo* ^{31}P nuclear magnetic resonance spectroscopy to monitor the levels of phosphorylated nucleotides. They focused particular attention on pyrophosphate: fructose-6-phosphate phosphotransferase (PFP), an alternative enzyme identified in pineapple by Carnal and Black (1979). This enzyme catalyzes the identical reaction as PFK, utilizing PP_i instead of ATP as phosphate donor, and was activated by fructose 2, 6-biphosphate. An increase in the amount of fructose 2, 6-phosphate concurrently with the rise in respiration suggested to these researchers that PFP may also be involved in the regulation of ripening in avocado fruit.

III. Initiation of Ripening

Ethylene is one of many volatile substances emanating from fruits and vegetables which was subsequently identified by Gane (1934) as the active component for the stimulation of ripening. The application of minute quantities of ethylene, on the order of 1 ppm, stimulates respiratory activity, induces ripening, and hastens the onset of the climacteric. Thus ethylene was soon recognized as a plant hormone that initiates the ripening process as well as regulating many aspects of plant growth, development, and senescence (Abeles, 1973). The response to ethylene in climacteric fruits is only effective if applied during the preclimacteric stage, whereas respiratory activity is stimulated throughout the postharvest period for nonclimacteric fruits (Biale and Young, 1962). In addition to increased respiratory metabolism, ethylene also stimulates its own biosynthesis in ripening climacteric fruits (Burg and Burg, 1965). The application of increased levels of ethylene to climacteric fruits hastens the onset of the climacteric rise accompanied by an increase in oxygen uptake (Figure 2.4). With respect to nonclimacteric fruits, an increase in oxygen absorption accompanied application of ethylene. In the case of climacteric fruits, once ethylene exerts the respiratory rise the process cannot be reversed. This is in sharp contrast to nonclimacteric fruits, in which the respiratory activity returns to the level of the control once ethylene treatment is terminated.

At one time ethylene was considered a by-product rather than a ripening hormone as the amount present during the preclimacteric phase in many fruits was insufficient to stimulate ripening (Biale *et al.*, 1954). This conclusion was based on the amount of ethylene emanating from the fruit rather than the intracellular concentrations and was measured using manometric techniques that were far too insensitive. Subsequent research using gas chromatography provided ample evidence for the presence of ethylene in the intracellular spaces (Burg and Burg, 1965). It is generally accepted that ethylene levels required to stimulate ripening fall within 0.1–1.0 ppm, completely beyond the range of normal manometric techniques. Table 2.2 summarizes the change in internal ethylene levels during the ripening of some climacteric and nonclimacteric fruits.

In the case of avocado, mango, and pears the ethylene level prior to the climacteric rise was lower than the accepted threshold of 0.1 ppm. Biale and Young (1971) noted that the rapid initiation of ripening of avocado required levels of ethylene greater than 1 ppm. These researchers pointed out some ten years later (Biale and Young, 1981) that it was difficult to generalize about the minimum levels of ethylene needed to induce the climacteric rise because of the scant data available. Peacock (1972) suggested that the effectiveness of ethylene was a function of the log of its concentration, length of exposure, and time of application after harvest. As the fruit approached maturity it was evident that there was a decrease in sensitivity to ethylene.

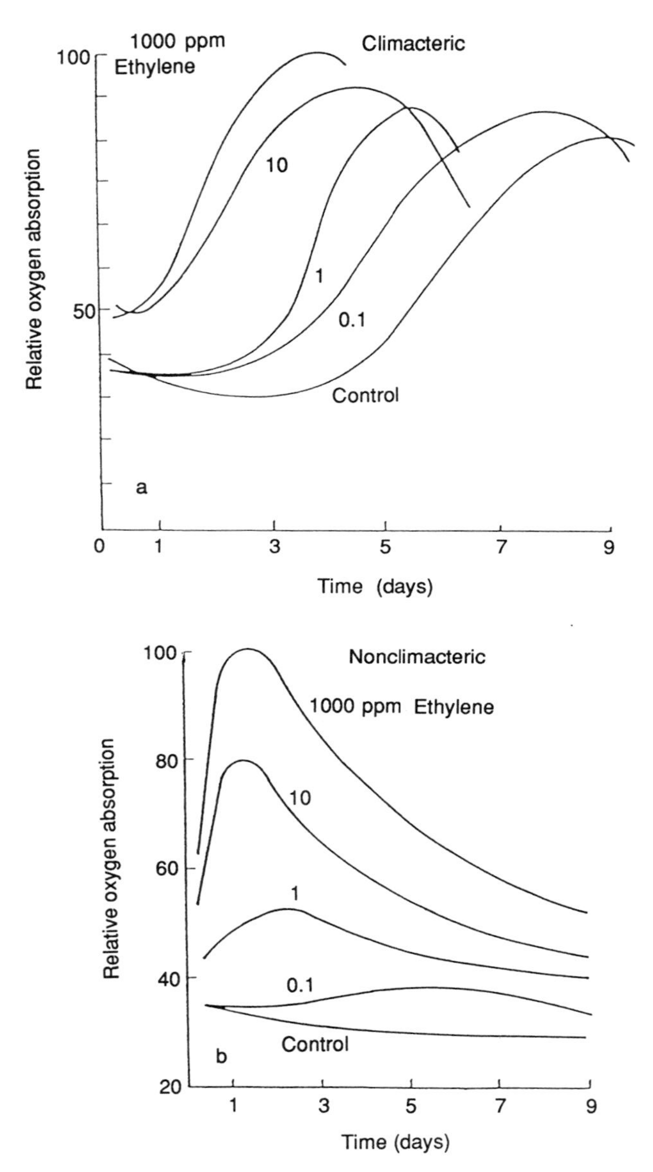

FIG. 2.4. Oxygen uptake by fruits which show the climacteric phenomenon and by fruits which do not , in relation to concentration of external ethylene (Biale, 1964).

TABLE 2.2

INTERNAL ETHYLENE CONTENT (PPM) IN SOME CLIMACTERIC AND NONCLIMACTERIC FRUITS

Fruit	Variety	Preclimacteric	Onset	Climacteric peak
Climacteric				
Avocado[a]	Fuerte	0.03	0.09	25
Banana[b]	Gros-Michel	0.1	1.5	40
Mango[b]	Kent Haden	0.01	0.08	3
Pear[c]	Anjou	0.09	0.4	40
Nonclimacteric			Steady state	
Lemon[a]			0.1–0.2	
Orange[a]			0.1–0.2	
Lime[a]			0.3–2.0	

[a] From Akamine and Goo (1979b).
[b] From Burg and Burg (1962).
[c] From Kosiyachinda and Young (1975).

IV. Biosynthesis of Ethylene

A number of precursors of ethylene have been proposed but of these methionine appears to be the main one in higher plants.

A. METHIONINE AS PRECURSOR OF ETHYLENE

Lieberman and Mapson (1964) initially examined the production of hydrocarbons, including ethane and ethylene, in model systems containing peroxidized linoleic acid, Cu^+, and ascorbic acid. To test whether ethylene production from linoleic acid involved free radicals they added a free radical quencher, methionine. Instead of methionine inhibiting the reaction they found that the production of ethylene was greatly enhanced. Further work showed that ethylene could be produced in the absence of peroxidized lipids as long as methionine–Cu^+–ascorbate was present (Lieberman *et al.*, 1965). It was soon shown that methionine was in fact the biological precursor of ethylene in plants (Yang, 1974; Lieberman *et al.*, 1965). Using ^{14}C-labeled methionine, Lieberman and coworkers (1966) demonstrated its conversion to ethylene in apple fruit tissue. The fact that the C1 from methionine yielded CO_2 and C3 and C4 yielded ethylene in both chemical systems and plant tissue suggested that a common mechanism was involved. The two systems were quite different, however, as methionine was converted via methional with the methyl sulfide group yielding volatile dimethyl sulfide in the model systems. This differed in plant tissue, as methionine is limiting so that the sulfur group is recycled for resynthesis of methionine.

1. *Recycling of Methionine*

The inhibition of ethylene production from methionine in the presence of DNP, an uncoupler of oxidative phosphorylation, suggested the formation of *S*-adenosyl-L-methionine (SAM) as an intermediate in this process (Burg, 1973; Murr and Yang, 1975). Using labeled methionine, Adams and Yang (1977) reported that the CH_3–S group of methionine was released as 5-methylthioadenosine (MTA) during ethylene synthesis in apple slices. The formation of MTA could only be formed as a degradation product if ethylene was synthesized from SAM. In addition to MTA these researchers also detected 5-methylthioribose (MTR), a degradation product of MTA in apple tissue. This suggested that the CH_3–S unit of MTR combined with a four-carbon receptor, such as homoserine, to form methionine, while the ribose group split off. It was subsequently found that the ribose unit of MTA/MTR was directly incorporated into methionine along with the CH_3–S group. Yung and Yang (1980) demonstrated that three MTR molecules were involved in methionine formation with the ribose moiety modified to form the 2, 3-aminobutyrate portion of methionine while the CH_3–S unit remained intact:

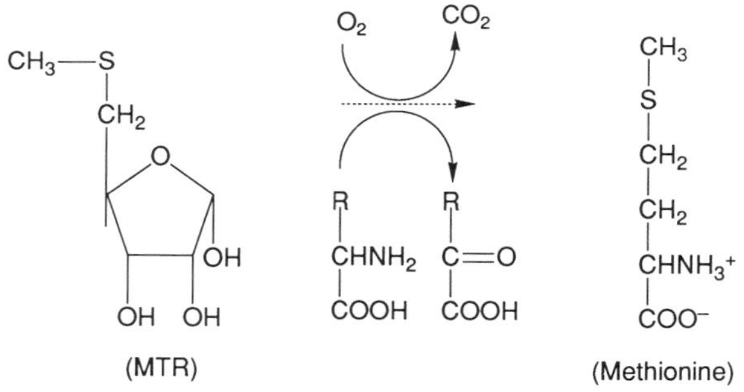

CH_3—S—CH_2 (ribose ring: O, OH, OH, OH)
(MTR)

O_2 → CO_2
R–CHNH$_2$–COOH → R–C=O–COOH

CH_3–S–CH_2–CH_2–CHNH$_3^+$–COO$^-$
(Methionine)

This pathway explains how methionine is recycled and maintained within plants. The overall pathway involved in the resynthesis of methionine from MTA is shown in Scheme 2.2. MTR-1 phosphate is converted to 2-oxo-4-methylthiobutanoic acid from which methionine is re-formed. Miyazaki and Yang (1987) examined the methionine cycle enzymes in a number of fruits and showed that the conversion of MTR to methionine in ripening apples was not a limiting factor in the formation of ethylene.

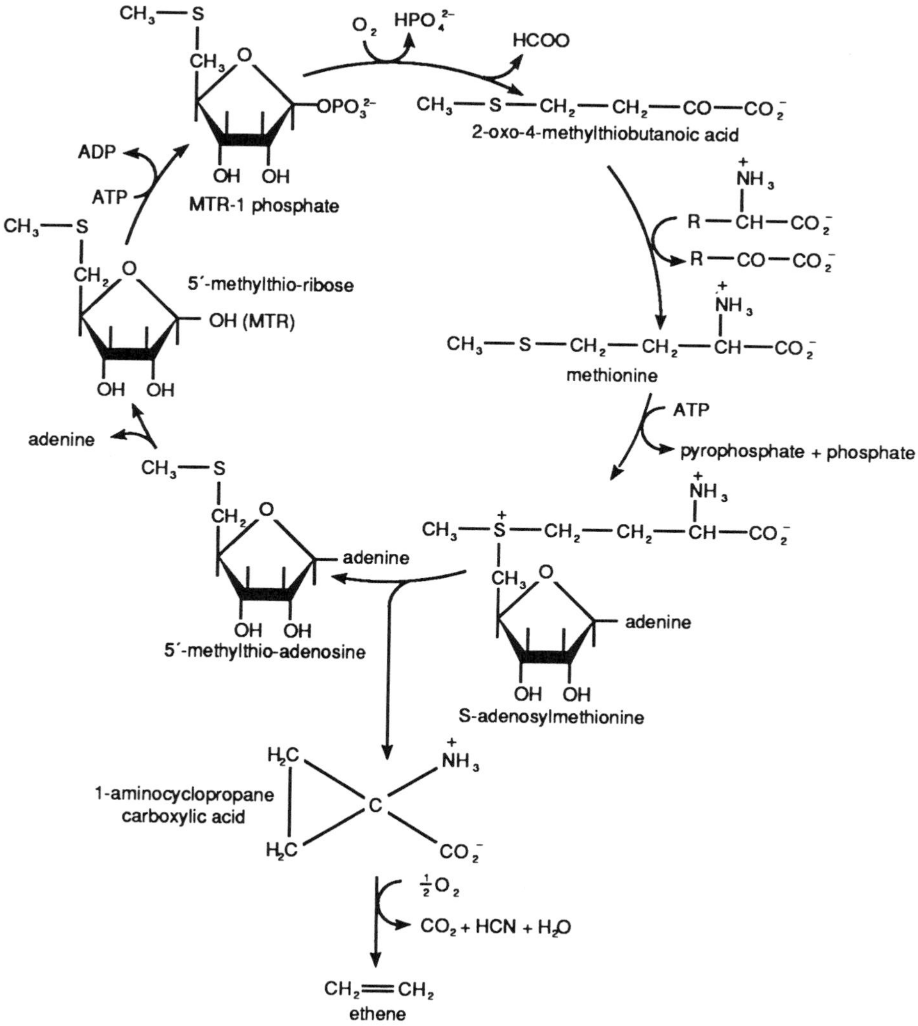

SCHEME 2.2. Ethylene biosynthesis and the methionine cycle (Yang and Hoffman, 1984).

2. *Methionine and Ethylene Biosynthesis*

Early studies by Hansen (1942) and Burg and Thimann (1959) showed that ethylene production ceased when apples and pears were stored in an atmosphere of nitrogen. On reexposure to oxygen, however, the production of ethylene was restored. The rapid production of ethylene suggested the accumulation of an intermediate compound during anaerobic storage. Adams and Yang (1979), using L[U-^{14}C]methionine, identified 1-aminocyclopropane-1-carboxylic acid (ACC) as the intermediate formed in apple fruit stored under nitrogen. It would appear therefore that methionine is first converted to *S*-adenosylmethionine, which then undergoes fragmentation to ACC and MTA. These researchers also found that labeled ACC was converted to ethylene when the apple tissue was incubated in air, which suggested the following sequence:

$$\text{Methionine} \rightarrow \text{SAM} \rightarrow \text{ACC} \rightarrow \text{Ethylene}$$

The conversion of methionine to SAM involves methionine adenosyltransferase (ATP : methionine *S*-adenosyltransferase, EC 2. 5. 1. 6). This enzyme was reported in plant tissues by Konze and Kende (1979) in relation to ethylene production. The addition of aminoethoxylvinylglycine (AVG), an inhibitor of pyridoxal phosphate-mediated enzyme reactions (Rando, 1974), was subsequently shown to inhibit ethylene production from methionine. The part of the reaction sequence affected was SAM to ACC, which involved the participation of pyridoxal phosphate (Adams and Yang, 1979). The enzyme involved, ACC synthase, was identified in tomato preparations and shown to be activated by pyridoxal phosphate (Boller *et al.*, 1979; Yu *et al.*, 1979). ACC synthase has since been identified and studied in apples (Bufler and Bangerth, 1983; Bufler, 1984), tomatoes (Acaster and Kende, 1983), cantaloupe (Hoffman and Yang, 1980), and citrus peel (Riov and Yang, 1982).

The application of ACC to plant organs was shown by Lurssen *et al.* (1979) to enhance ethylene production. These researchers speculated that ACC was derived from methionine via SAM or ACC. The enzyme system involved in the formation of ethylene from ACC appeared to be associated with cellular particles (Imaseki and Watanabe, 1978; Mattoo and Lieberman, 1977). Disruption of the cellular membrane either by treatment with lipophilic compounds or osmotic shock reduced ethylene production in plant tissues (Imaseki and Watanabe, 1978; Odawara *et al.*, 1977). The particular step inhibited was identified as ACC to ethylene (Apelbaum *et al.*, 1981). An enzyme extract capable of converting ACC to ethylene was reported in pea seedlings by Konze and Kende (1979). Similar systems have been reported in a carnation microsomal system (Mayak *et al.*, 1981) and a pea microsomal system (McRae *et al.*, 1982). The enzyme system that is responsible for the conversion of ACC to ethylene remains to be isolated and characterized and is referred to as the ethylene-forming enzyme (EFE). Yang and Hoffman (1984) suggested that ACC might be oxidized by an enzyme, ACC hydroxylase, to *N*-hydroxy-ACC, which is then broken down to ethylene and cyanoformic acid. The latter is extremely labile and spontaneously fragments to carbon dioxide and HCN. Support for this was based on studies by Peiser *et al.* (1983), who reported incorporation of [1-^{14}C] ACC into [4-^{14}C] asparagine in mung bean hypocotyls at levels similar to the production of ethylene. These findings together with work by Miller and Conn (1980), who demonstrated incorporation of Na–CN into asparagine in mung bean, suggested the following pathway:

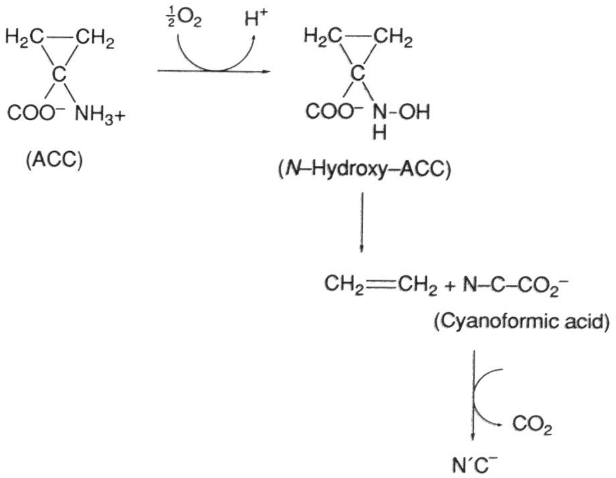

H_2C—CH_2 (C) COO^- NH_3+ (ACC) —($\frac{1}{2}O_2$, H^+)→ H_2C—CH_2 (C) COO^- N-OH H (*N*-Hydroxy–ACC)

↓

CH_2=CH_2 + N–C–CO_2^- (Cyanoformic acid)

↓ → CO_2

$N'C^-$

B. Regulation of Ethylene in Ripening Fruits

1. *ACC Synthase*

The climacteric rise in fruits is associated with enhanced ethylene production at the onset of ripening. The changes in the internal level of ACC were examined by Hoffman and Yang (1980) during the ripening of avocadoes as well as the effect of exogenous ACC on ethylene synthesis in the preclimacteric fruit. Their results in Figure 2.5 show that ACC was present at extremely low levels in the preclimacteric fruit (<0.1 nmole/g) but increased dramatically just prior to the onset of ethylene, then decreasing to 5 nmole/g in the overipe fruit. The low level of ACC in the preclimacteric fruit was attributed to the inability to convert SAM to ACC. Addition of exogenous ACC to the preclimacteric tissue did increase ethylene production but only to a limited degree (Adams and Young, 1977, 1979). Thus the formation of ACC from SAM appeared to be the rate-controlling step in the biosynthesis of ethylene (Yang, 1980). Further confirmation was provided by Liu and co-workers (1985), who studied the effect of ethylene treatment on the production of ethylene in climacteric tomato and can-

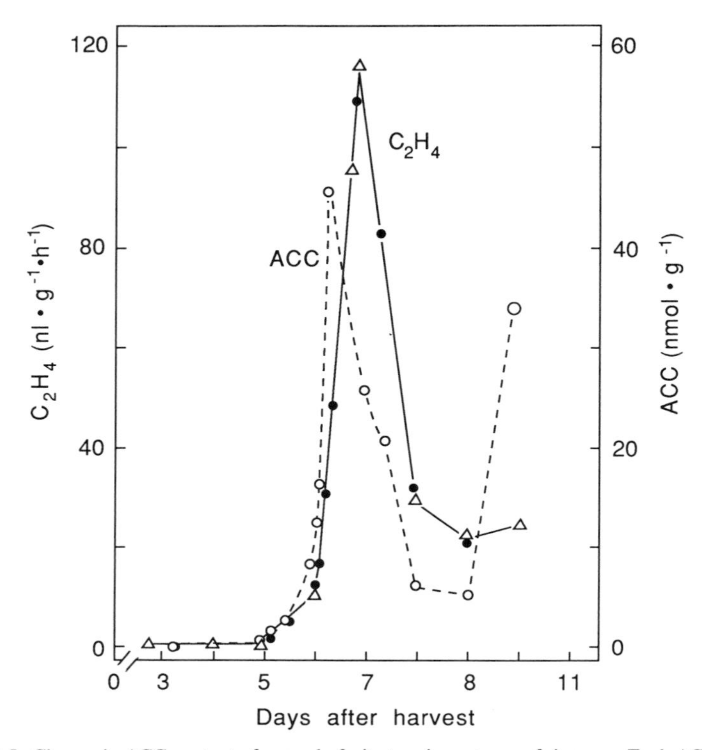

FIG. 2.5. Change in ACC content of avocado fruit at various stages of ripeness. Each ACC value is from a single fruit which had been monitored for ethylene production and assigned an arbitrary stage of ripeness by comparison with the established climacteric patterns of ethylene production (Hoffman and Yang, 1980).

taloupe fruits. They found that when exposed to exogenous ethylene the increased activity of the ethylene-forming enzyme preceded any increase in ACC synthase in these preclimacteric fruit. However, it remains to be established whether this occurs during the normal ripening of fruits. Morin *et al.* (1985) found that cold storage of Passé-Crassane pears was required to initiate ethylene ripening and induce the synthesis of free or conjugated ACC. During cold storage (0°C) both free and conjugated ACC increased together with ribosomes and mRNA. When these pears were then transferred to 15°C, a marked rise in ethylene occurred followed by the climacteric (Hartmann *et al.*, 1987).

A variety of stresses such as wounding or drought can induce ethylene biosynthesis as a result of the increase in ACC synthase activity. A recent study by Bufler (1986) examined the enhancement of ACC synthase in ripening apples and identified two sites in the ethylene pathway which are enhanced by ethylene. These were stimulation of ACC synthase activity and ethylene formation from

ACC. Prior to ripening, ACC synthase is formed which produces sufficient amounts of ACC from which starter ethylene is produced (System 1). The starter ethylene then induces higher activity in the enzyme converting ACC which results in a marked increase in ethylene production (System 2).

2. *Cyanide*

The production of cyanide in the biosynthesis of ethylene from ACC was shown in studies by Peiser et *al.* (1984) and Pirrung (1985). Pirrung and Brauman (1987) suggested that cyanide might regulate ethylene formation during the climacteric period. They proposed that in ethylene biosynthesis the cytochrome and cyanide-resistant respiratory chains were connected *via* cytochrome *c* oxidase. The inhibition of cytochrome *c* oxidase by cyanide during ethylene biosynthesis favored the alternative pathway which, in turn, led to ACC synthesis. Gene expression for ACC synthase in response to the alternative respiratory pathway could explain differences between climacteric and nonclimacteric fruit. The formation of ACC synthase is undoubtedly the prime factor controlling the rate of ethylene biosynthesis (Acaster and Kende, 1983; Yang and Hoffman, 1984).

3. *Organic Acids*

De Pooter *et al.* (1982) observed an increase in CO_2 production and premature ripening when intact Golden Delicious apples were treated with propionic and butyric acids. This change was identical to apples treated with ethylene and suggested a role for these acids in ethylene production. It appeared feasible that ethylene could be produced from these carboxylic acids according to System 1, which then triggered normal ethylene production *via* System 2. Further work by these researchers (De Pooter *et al.*, 1984) confirmed the premature ripening of the intact Golden Delicious apples when treated with acetic or propionic acid vapors. A small part of labeled [2-^{14}C]propionic acid was transformed into [^{14}C]ethylene which then acted as a trigger for ripening (System 2). The degree of fruit maturity was thought to be a major factor, which suggested that in the unripe apples the small amount of ethylene produced was probably derived from simple organic acids. Thus the concentration at which ethylene can trigger ripening depends on the availability of simple organic acids. The ability of CO_2 to delay the onset of ripening in fruits was shown by Bufler (1984) to be due, in part, to inhibition of ACC synthase development (Figure 2.6).

4. *Lipid Peroxidation: Lipoxygenase*

The production of ethylene has been correlated with changes in hydroperoxide levels, peroxidase activity, as well as increase in lipoxygenase activity in ripen-

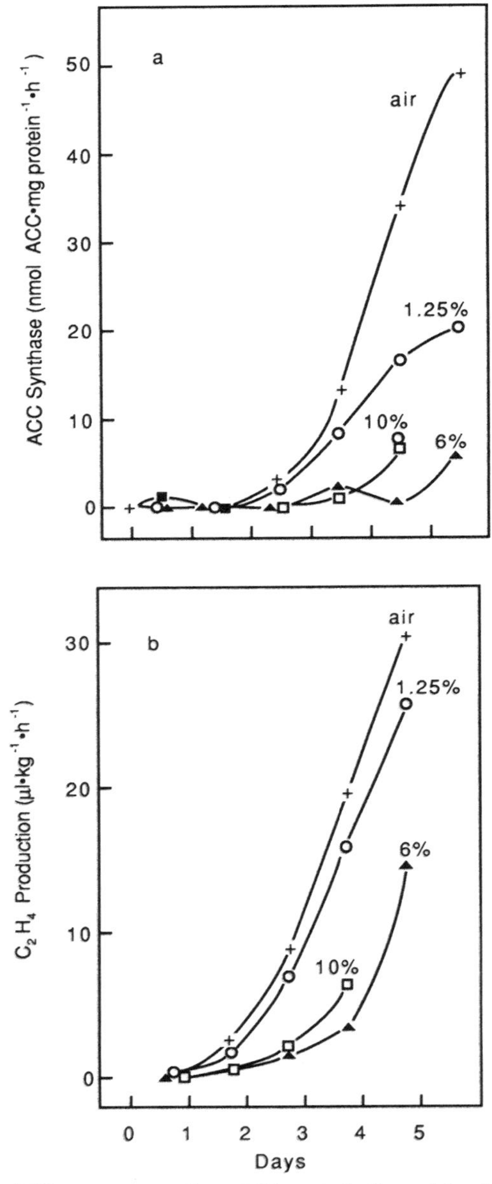

FIG. 2.6. Effect of different concentrations of CO_2 on induction and development of ACC synthase activity (a) and ethylene production (b) in preclimacteric treated apples. Apples were transferred from hypobaric storage to normal pressure and 25°C and immediately treated with air (+), 1.25% (○), 6% (▲), or 10% (□) CO_2 (Bufler, 1984). Reprinted with permission of copyright owner, American Society of Plant Physiology (ASPP).

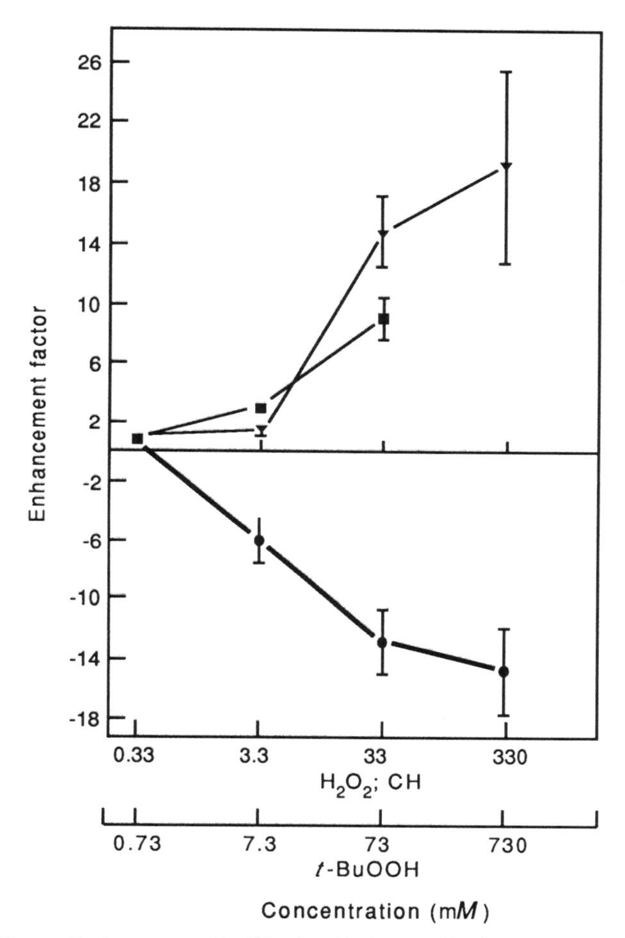

FIG. 2.7. Effects of hydrogen peroxide (●), t-butylhydroperoxide (▲), and cumene hydroperoxide (■) on the conversion of ACC to ethylene by pea microsomal membranes. Final concentrations of hydroperoxide added to the basic reaction mixture are indicated along the abscissa. The enhancement factor is the ratio of ethylene produced in the presence of added hydroperoxide relative to that produced in its absence. Values represent the mean ± SE (n = 3) (Legge and Thompson, 1983). Reprinted with permission. Copyright © by Pergamon Press.

ing fruits (Frenkel and Eskin, 1977; Frenkel, 1979; Meigh *et al.*, 1967). Studies by Adams and Yang (1979) and Konze *et al.* (1980) both suggested peroxidation as one mechanism for the formation of ethylene from ACC, however, the involvement of peroxidase was discounted by later researchers (Machackova and Zmrhal, 1981; Rohwer and Mader, 1981). The enzymatic conversion of ACC to ethylene in a cell-free system was found to be sensitive to catalase and inhibited by hydrogen peroxide (Konze and Kende, 1979; McRae *et al.*, 1982). The effect of hydroperoxides on the enzymatic conversion of ACC to ethylene was investigated by Legge and Thompson (1983) using a model system composed of microsomal membranes from etiolated peas. Addition of hydroperoxides stimulated ethylene production in model systems containing ACC as shown in Figure 2.7. Hydrogen peroxide, a known inhibitor of lipoxygenase, was found to inhibit ethylene formation. Lipoxygenase forms lipid hydroperoxides from linoleic acid (Eskin *et al.*, 1977). A 1.5-fold increase in ethylene production occurred following the addition of linoleic acid to model systems containing this enzyme. There appeared to be an interaction between lipoxygenase activity, a hydroperoxide derivative, and the ethylene-forming enzyme. McRae *et al.* (1982) provided evidence, based on spin-trapping evidence, that oxygen was involved in the formation of ethylene from ACC by pea microsomal membranes. Since hydroperoxides facilitate oxygen activation, the promotion of oxygen via this mechanism could lead to the formation of ethylene. Legge *et al.* (1982) detected free radical formation using a diagnostic spin trap 4-MePyBN which required ACC, oxygen, and hydroperoxides. Their results suggested that free radicals were derived from ACC in the microsomal system producing ethylene. Conversion of ACC to ethylene by pea microsomal membranes is mediated *via* a free radical intermediate requiring hydroperoxides and oxygen. The increase in free radical formation was attributed by Kacperska and Kubacka-Zabalska (1984) to lipoxygenase-mediated oxidation of polyunsaturated fatty acids. This was confirmed with *in vitro* and *in vivo* studies by Kacperska and Kubacka-Zabalska (1985), who found that an increase in ethylene from ACC in winter rape leaf disks resulted from lipoxygenase activity.

5. *Galactose*

A recent study by Jongkee and co-workers (1987) showed that ethylene production was stimulated by galactose during the ripening of tomato fruit. Galactose is a product of reduced cell wall synthesis of galactan and increased activity of cell wall galactosyl residues by β-galactosidase (Lackey *et al.*, 1980; Pressey, 1983). Stimulation of ACC synthase activity by exogenous galactose to the pericarp tissue of green tomatoes suggested a relationship between cell wall turnover and ethylene biosynthesis in ripening tomato fruit (Jongkee *et al.*, 1987).

V. Color Changes

One of the first changes during the ripening of many fruits is the loss of green color. The development of red colors in some fruits and vegetables is due to the

TABLE 2.3

COLOR CHANGES OCCURRING IN SOME FRUITS DURING RIPENING

Fruit	Ripening stage	
	Immature	Ripe
Apple	Green	Yellow/red[a]
Banana	Green	Yellow
Pear	Green	Yellow
Strawbery	Green	Red

[a] Depending on which variety.

formation of anthocyanins. A list of the color changes in some fruits is summarized in Table 2.3. These changes take place immediately following the climacteric rise in respiration and are accompanied by textural changes in the fruit. In the case of leafy vegetables such as cabbage, lettuce, and Brussel sprouts, the loss of chlorophyll is also responsible for the symptom of yellowing during senescence (Lipton, 1987; Lipton and Ryder, 1989).

A. Chlorophyll Changes during Ripening

Ethylene has been reported to promote the degradation of chlorophyll during fruit ripening (Burg and Burg, 1965). A study by Hardy *et al.* (1971) noted the stimulation of chlorophyll biosynthesis in excised cotyledons of cucumber seeds *(Cucumis sativus)* by ethylene and light. Further work by Alscher and Castelfranco (1972) found that stimulation of chlorophyll synthesis occurred only in the dark as exposure to light inhibited chlorophyll synthesis. Little reference can be found on the stimulation of chlorophyll synthesis by ethylene, although the cucumber has provided an excellent system for studying chlorophyll production over the past decade (Chereskin *et al.*, 1982; Hanamoto and Castelfranco, 1983; Pardo *et al.*, 1980; Fuesler *et al.*, 1982). In contrast to stimulation of chlorophyll synthesis by ethylene in the dark, the biosynthesis of anthocyanins in red cabbage was only stimulated by ethylene when exposed to light.

B. Chlorophyll Biosynthesis

Shemin and Russell (1953) demonstrated the role of δ-aminolevulinic acid (ALA) in the biosynthesis of the tetrapyrrole nucleus. The biosynthesis of ALA is poorly understood. Several pathways involving succinyl-CoA and glycine,

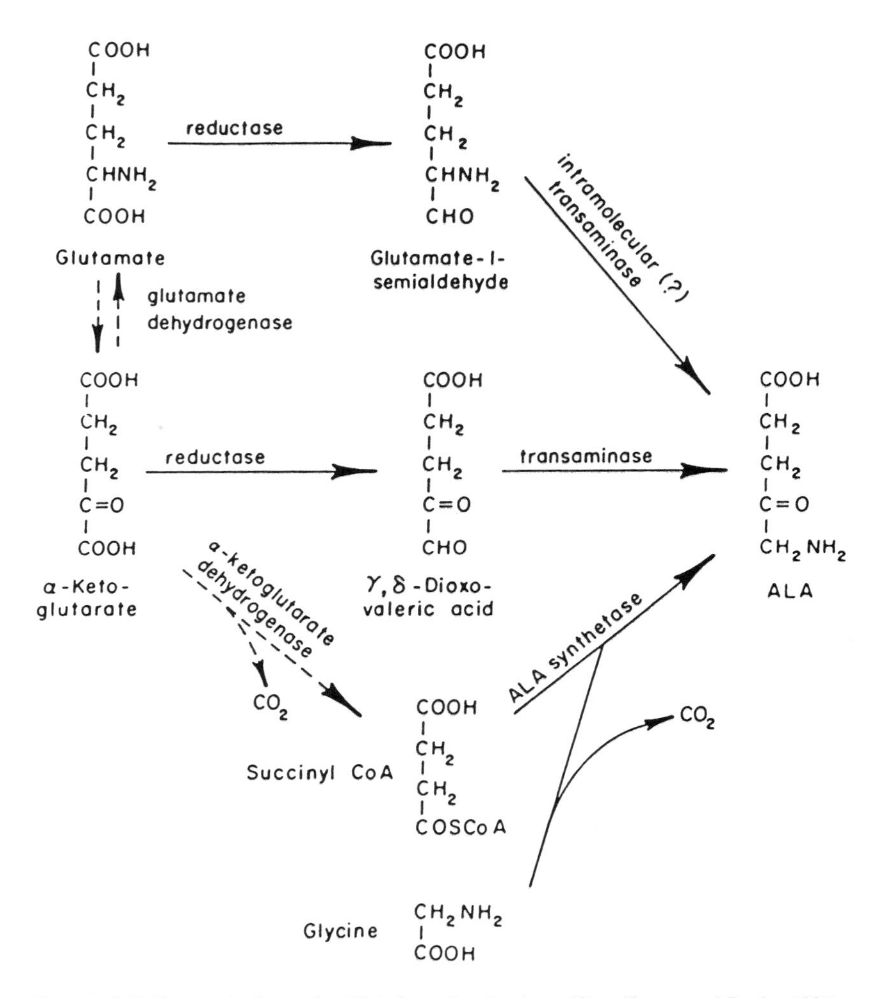

SCHEME 2.3. Proposed scheme for ALA formation in plants (Castelfranco and Beale, 1981).

glutamate, and α-ketoglutarate as starting materials have been proposed. The reactions responsible for the synthesis of ALA are summarized in Scheme 2.3.

The presence of ALA synthase has been reported in green peels of stored potatoes (Ramaswamy and Nair, 1974, 1976). An alternative pathway from glutamate has been found in greening cucumber cotyledons (Weinstein and Castelfranco, 1978), barley (Gough and Kannangara, 1979), wheat (Ford and Friedman, 1979), spinach (Kannangara and Gough, 1979), and maize (Meller *et al.*, 1979). This pathway involves the reduction of glutamate to glutamate-1-semialdehyde followed by removal of the amino group from C2 and the addition or replacement at C1 to yield ALA. The ability to convert GSA or ALA was shown by Kannangara and Gough (1978, 1979) using plastid extracts from greening barley. A third pathway from α-ketoglutarate *via* γ-δ-dioxovaleric acid (DOVA) still remains unconfirmed. Only the transamination of DOVA to ALA has been shown in some higher plants and algae (Foley and Beale, 1982; Gassman *et al.*, 1966; Salvador, 1978).

Once ALA is formed two molecules condense to form porphobilinogen (PBG), catalyzed by ALA dehydrase (5-aminolevulinate hydrolyase EC 4. 2. 1. 24) (Dresel and Falk, 1953; Schmid and Shemin, 1955). It is during this step that an aliphatic compound is converted into an aromatic one.

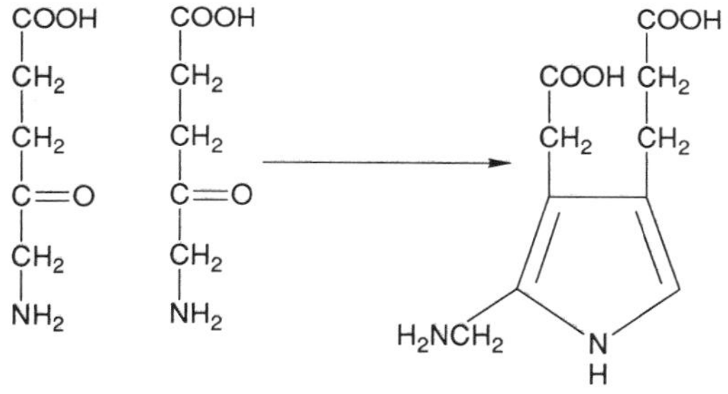

(2) 5–Aminolevulinic acid → Porphobilinogen

ALA dehydrase has been studied extensively in animal tissue and photosynthetic bacteria and in a few plants including wheat (Nandi and Waygood, 1967), soybean tissue culture (Tigier *et al.*, 1968, 1970), and mung bean (Prasad and Prasad, 1987). ALA dehydrase (ALAD) resides in the chloroplasts, where it appears in the soluble form in the plastid stroma or loosely bound to lamella.

The first tetrapyrrole intermediate, a linear hydroxymethylbilane porphyrin precursor, was identified by Battersby *et al.* (1979) and Jordan and Seehra (1979). This results from the head-to-tail condensation of four molecules of PBG catalyzed by PBG deaminase. This linear molecule is enzymatically closed to form the first cyclic tetrapyrrole, uroporphyrinogen III.

The steps leading to the formation of protoporphyrin IX will only be discussed briefly, as a detailed discussion can be found in several excellent reviews (Castelfranco and Beale, 1983; Rebeiz, 1982). Uroporphyrinogen III is converted to co-proporphyrinogen III by decarboxylation of the acetic acid groups on the pyrrole rings A, B, C, and D by a decarboxylase enzyme (Jackson *et al.*, 1976).

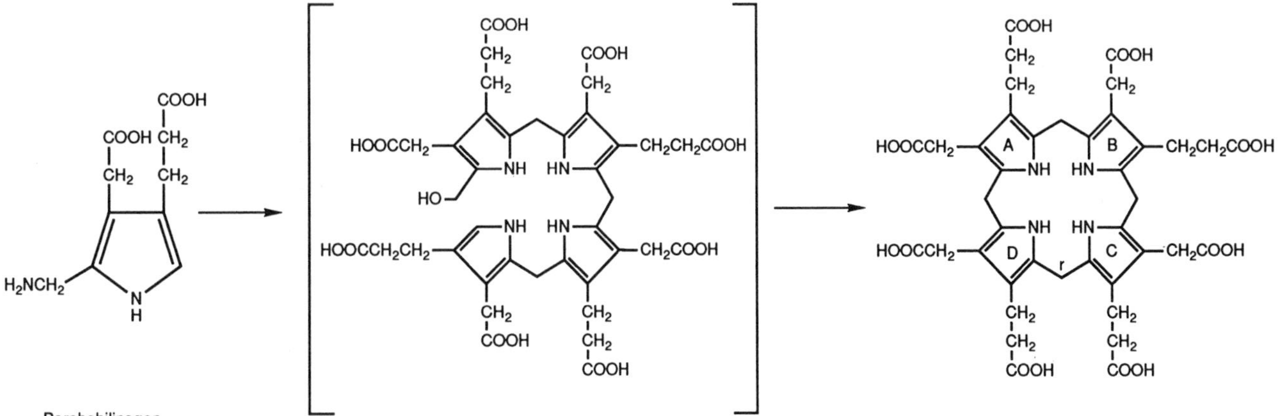

Porphobilinogen

Uroporphyrinogen III

This is followed by oxidative decarboxylation of the propionic acid groups on pyrrole groups A and B by co-proporphyrinogen oxidase to form protoporphyrinogen IX (Games *et al.*, 1976). The final step involves aromatization of protoporphyrin IX in which six electrons are lost (Poulson and Polglase, 1975).

Chelation of protoporphyrin IX is mediated by Mg chelatase and requires a high concentration of ATP (Pardo *et al.*, 1980). This is followed by methylation of one of the propionic acid residues to form Mg-protoporphyrin-*N*-monomethyl ester (Fuesler *et al.*, 1982). The enzyme involved, methyladenosyltransferase (EC 2.5.1.6), requires the presence of *S*-adenosylmethionine (SAM). Fuesler and co-workers (1982) demonstrated the following reaction sequence in which metal chelation preceded methylation, using an HPLC procedure to separate Mg-protoporphyrin and Mg-protoporphyrin-Me ester:

$$\text{Protoporphyrin IX} \xrightarrow{Mg^{2+}\text{-ATP}} \text{Mg-protoporphyrin} \xrightarrow{\text{SAM}} \text{Mg-protoporphyrin-Me}$$

The conversion of Mg-protoporphyrin-Me ester to protochlorophyllide involves the reduction of the vinyl substituent in the side chain of the pyrrole ring B to an ethyl group and oxidation of the methylated propionic acid in ring D to an isocyclic ring. Chereskin *et al.* (1982) proposed the involvement of oxygen in this reaction with formation of β-hydroxypropionate as an intermediate. It differed from the β-oxidation of methylpropionate proposed earlier by Granick (1961).

Protochlorophyllide is converted to chlorophyllide by photoreduction of the D pyrrole ring, the precursor of chlorophyllide *a*. This is the first light-dependent step in which protochlorophyllide is transformed to chlorophyllide by NADPH-protochlorophyllide oxidoreductase (Castelfranco and Beale, 1983; Griffith, 1974). This reaction was recently reported by Dehesh and co-workers (1987) in the green leaves and isolated chloroplasts of barley. The final step in the biosynthesis of chlorophyll is esterification of the proprionate substituent on pyrrole ring D with geranyl geraniol, which is then reduced to phytyl. Chlorophyll *b* is derived from chlorophyllide *b* by oxidation of the methyl groups on ring B to a formyl group. Hanamoto and Castelfranco (1983), however, identified divinyl chlorophyllide as the major intermediate in chlorophyll synthesis rather than the monovinyl derivative. The regeneration of protochlorophyllide was shown by Castelfranco and Beale (1981) to occur during the dark when chlorophyll synthesis ceased. This regeneration of divinyl protochlorophyllide was later reported by Huang and Catelfranco (1986) in isolated developing chloroplasts from greening cucumbers in the presence of glutamate, ATP, reducing power, *S*-adenosyl-L-methionine, and molecular oxygen.

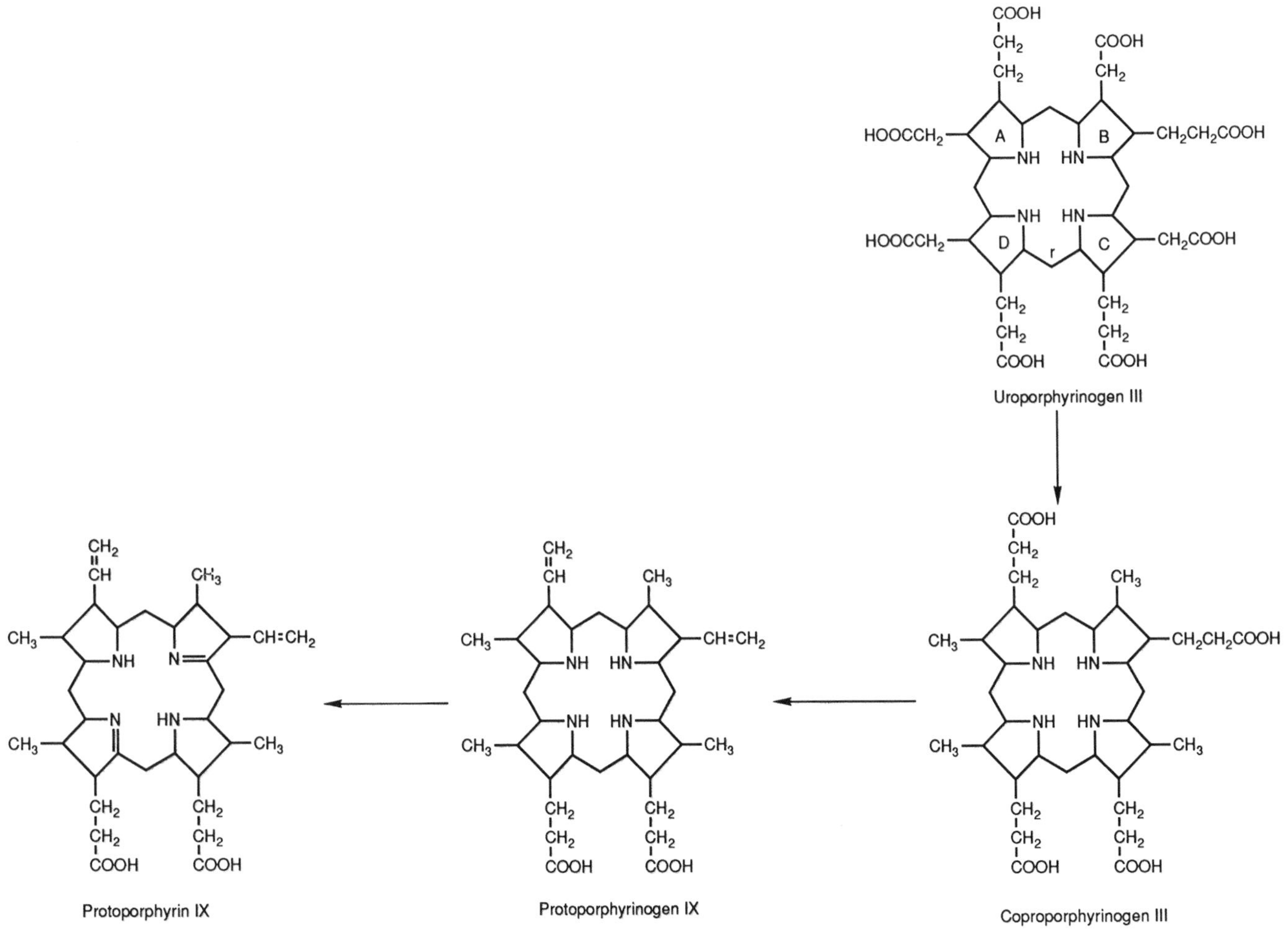

Uroporphyrinogen III

Coproporphyrinogen III

Protoporphyrinogen IX

Protoporphyrin IX

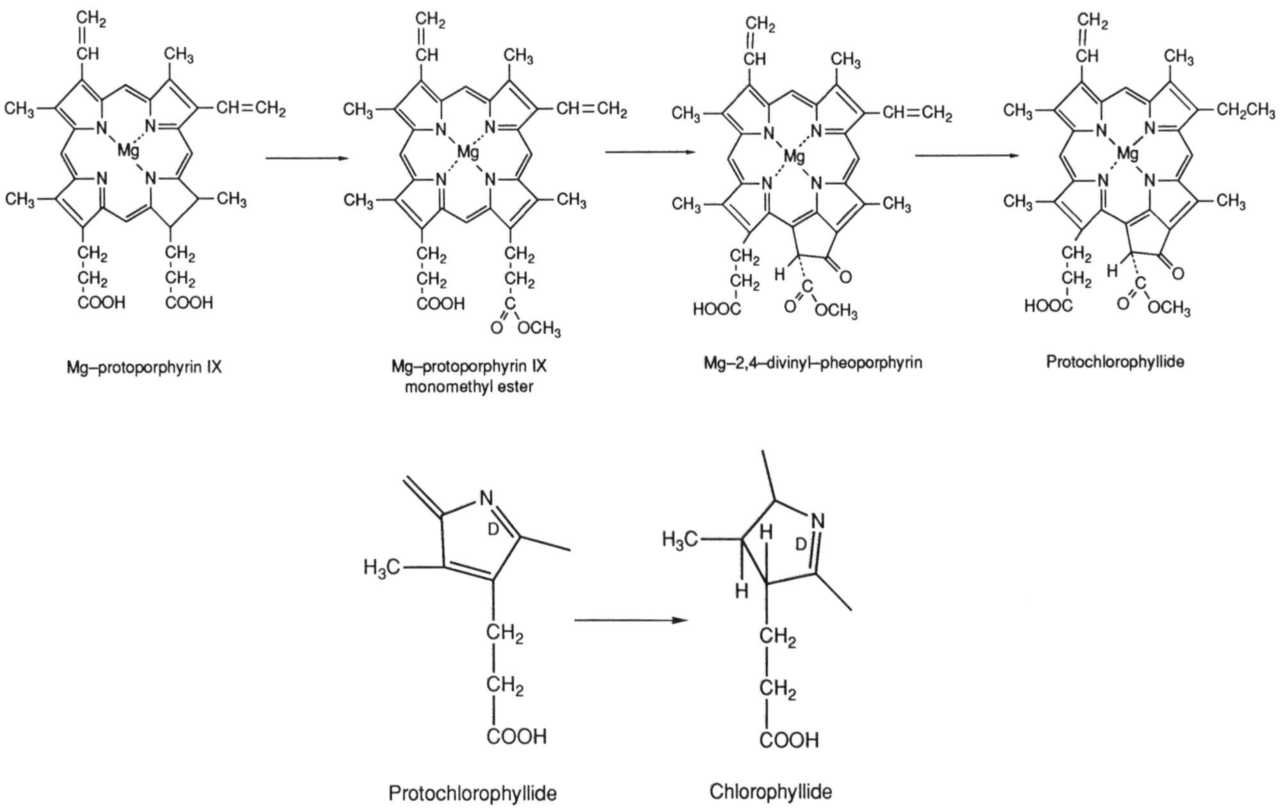

Mg–protoporphyrin IX

Mg–protoporphyrin IX monomethyl ester

Mg–2,4–divinyl–pheoporphyrin

Protochlorophyllide

Protochlorophyllide

Chlorophyllide

Adapted from Castelfranco and Beale (1981).

C. Regulation of Chlorophyll Biosynthesis

The first enzyme involved in tetrapyrrole synthesis, aminolevulinic acid synthetase (ALA), was thought to play a regulatory role in chlorophyll biosynthesis. Subsequent research has implicated δ-aminolevulinic acid dehydrase (ALAD), a metal-sensitive enzyme, as a regulator of chlorophyll synthesis (Hampp *et al.*, 1974; Naito *et al.*, 1980). This was recently confirmed in a study on mung bean seedlings by Prasad and Prasad (1987), who inhibited ALAD in germinating mung beans with lead and mercury, which resulted in a reduction in cholorphyll. ALAD appeared to be located exclusively in the chloroplast, where it regulated the synthesis of chlorophyll.

D. Mechanism of Chlorophyll Degradation

The mechanism of chlorophyll degradation still remains fragmentary. Nevertheless it is generally assumed that chlorophyll is degraded to colorless products, thus exposing the carotenoids. The main steps involved are summarized in Scheme 2.4.

The initial step in chlorophyll breakdown in plant tissues is hydrolysis to chlorophyllide and phytol. This reaction is catalyzed by the enzyme chlorophyllase (chlorophyll chlorophyllidihydrolase, EC 3.1.1.14), an intrinsic membrane glycoprotein located in the lipid envelope of the thylakoid membranes (Bacon and Holden, 1970; Schoch and Vielwerth, 1983). These photosynthetic

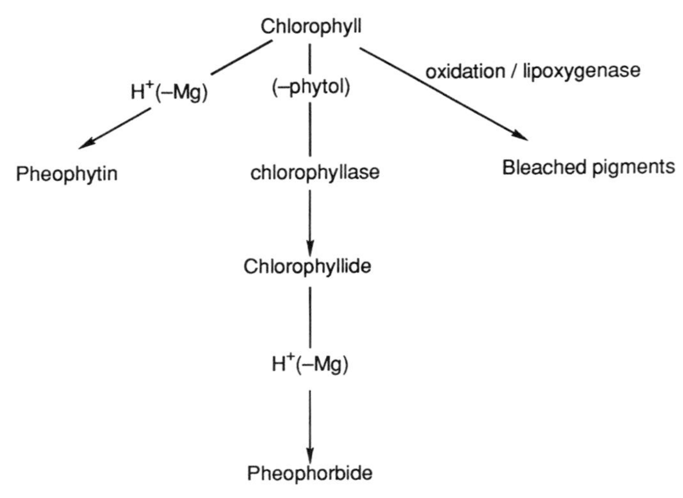

Scheme 2.4. Degradation of chlorophyll.

membranes are composed of lipids and proteins arranged in lipoprotein complexes. Terpstra and Lambers (1983) demonstrated that conversion of chlorophyll *a* by chlorophyllase required the presence of lipids. This was in agreement with earlier work by Terpstra (1981), who noted activation of chlorophyllase by magnesium in the presence of chloroplast lipids. A model for chlorophyllase activity was proposed by Lambers *et al.* (1984a,b) in which chlorophyll is attached to the active site of the enzyme. This site appears to be close to the hydrophobic carbohydrate moiety of the chlorophyllase molecule and its accessibility regulated by Mg^{2+}. Once hydrolyzed, the product chlorophyllide is removed to another site on the enzyme consisting of protein-associated lipid. Thus chlorophyllide can then move from this lipid site to the aqueous medium as the long-chain phytol group which renders the original pigment insoluble has been removed.

The degradation of chlorophyll is completed within a relatively short period of time in senescent plant tissues although the intermediate and final products have not been identified *in vivo*. In fruits and vegetables the loss of chlorophyll is linked to structural changes which release cellular acids and various degradative enzymes. Park and co-workers (1973) reported a 95% loss of chlorophyll during leaf senescence although no colored degradation products were detected as the leaf turned yellow, brown, or red. The rate of decomposition to small products was extremely rapid and no large compounds were detected. Maunders *et al.* (1983) monitored the formation of different chlorophyll derivatives during the senescence of bean (*Phaseolus vulgaris*) and barley (*Hordeum vulgare*) leaves. They identified a number of derivatives, including chlorophyllin, chlorin, and chlorophyll *a*-1 and *b*′. Chlorophyll *b*′ was a C_{10} epimer of chlorophyll *b* while chlorophyll *a*-1 was an oxidized derivative, possibly hydroxychlorophyll. Chlorophyll *a*-1 was not detected in either healthy plant tissue or attached senescent leaves, but was found to increase in excised leaves for up to 10 days, while at the same time chlorophyll *a* decreased. Chlorophyll *b*-1 was not detected. This suggested that chlorophyll *b* was less susceptible to hydroxylation than chlorophyll *a*. The formation of chlorophyll *a*-1 has been demonstrated in excised leaves after boiling or organic solvent diffusion, suggesting that breakdown of proteins or membranes is involved in its formation (Bacon and Holden, 1970; Holden, 1970). The absence of chlorophyll *a*-1 in attached senescent leaves exposed to light was due either to its extremely rapid breakdown or to the fact that this pathway was not involved during natural senescence.

The bleaching of chlorophyll in pea homogenates was observed in the presence of lipoxygenase, peroxidase, and catalase (Wagenknecht and Lee, 1958). Several researchers reported bleaching of chlorophyll by hydrogen peroxide which was catalyzed by peroxidase in the presence of certain phenolics (Aljuburi *et al.*, 1979; Matile, 1980). The peroxidase-catalyzed oxidation of chlorophyll was the subject of a more recent study by Huff (1982). Of a number of phenolic compounds examined, resorcinol and 2, 4-dichlorophenol were found to be the most active in promoting the bleaching of chlorophyll by chlorophyll : H_2O_2 oxidoreductase. The decrease in chlorophyll *a* was far more rapid than that of chlorophyll *b* (Simpson *et al.*, 1976). The magnesium-containing pigments appeared to be destroyed faster than those derivatives without magnesium. Evidence for this mechanism operating during ripening or senescence still remains indirect. Nevertheless peroxide and particulate-bound peroxidase levels were reported to increase during the ripening of pears (Brennan and Frenkel, 1977; Haard, 1973). The decline in catalase activity in senescent tissue accompanying chlorophyll losses suggested that H_2O_2 could penetrate chloroplasts and play a role in chlorophyll degradation (Robinson *et al.*, 1980). The presence of peroxidase and phenols in chloroplasts could facilitate peroxidase-catalyzed oxidation of chlorophylls by H_2O_2 (Henry, 1975; Hurkman and Kennedy, 1977; Kirk and Tilney-Bassett, 1978).

1. *Chlorophyll Degradation: Processing and Storage*

Several pathways have been proposed for loss of chlorophyll during processing and storage of fruits and vegetables. This loss in green color can be undesirable and such changes need to be minimized. One of the main reactions is replacement of the Mg^{2+} atom in chlorophyll by hydrogen under acidic conditions with the formation of pheophytin. The latter pigment is associated with a color change from a bright green to a dull olive green. This reaction was recognized over 50 years ago by Campbell (1937) to cause discoloration in stored frozen peas. The rate of conversion to pheophytin was shown to be first order with respect to the acid concentration (Joslyn and Mackinney, 1938). The formation of pheophytin has been the subject of a large number of studies (Gupte *et al.*, 1964; Hermann, 1970; LaJollo *et al.*, 1971; Roberston and Swinburne, 1981). A linear relationship was reported by Walker (1964) between the appearance and pheophytin formation for frozen beans stored up to 1 year. LaJollo and co-workers (1971) noted that the formation of pheophytin was the predominant reaction at a_W levels greater than 0.32 in freeze-dried, blanched spinach puree stored at 37°C and 55°C under nitrogen and air. Chlorophyll *a* was degraded far more rapidly than chlorophyll *b* by a factor of 2.5–3.0, consistent with earlier reports (Schanderl *et al.*, 1962; Gupte *et al.*, 1964). LaJollo *et al.* (1971) reported a linear relationship between a_W and log time for a 20% loss of chlorophyll (Fig. 2.8).

Besides pheophytinization, chlorophyllase converts chlorophylls to chlorophyllides with the loss of the phytol group. The combined action of chlorophyllase and acid results in the loss of Mg^{2+} and phytol group with the formation of pheophorbides (White *et al.*, 1963). Pheophorbide was also found to be the major degradation product during the brining of cucumbers (Jones *et al.*, 1961, 1963). Several new products were identified by Schwartz and co-workers (1981) in heated spinach puree including pyropheophytins *a* and *b*. These were formed

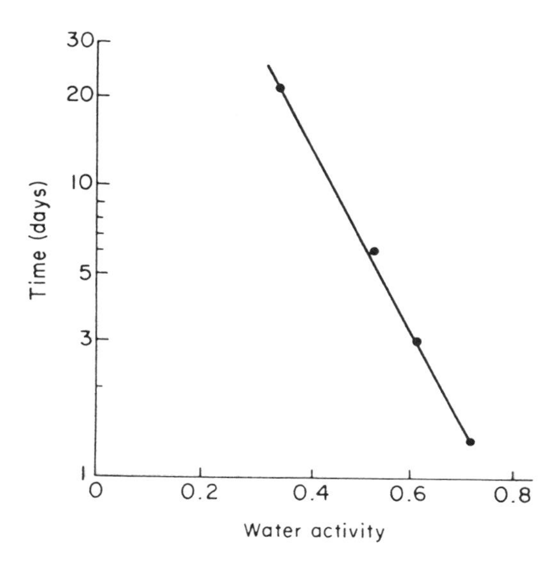

FIG. 2.8. Time required for 20% loss of chlorophyll in spinach at different water activities (37°C in air) (La Jollo *et al.*, 1971). Copyright © by Institute of Food Technologists.

from the corresponding pheophytins as a result of the loss of the carbomethoxy group and separated by HPLC. Schwartz and von Elbe (1983) examined the formation of pyropheophytins in heated spinach. Pheophytins *a* and *b* were both detected after heating at 121°C for up to 15 min, but then a decline in these pigments occurred. Further heating resulted in the formation of pyropheophytin with *a* detected after 4 min while *b* was not observed until after 15 min. A sample of blanched spinach puree heated for almost 2 hr at 126°C produced only pyropheophytins *a* and *b*. No further changes were noted, indicating these were the final products of degradation. The olive-green color associated with canned vegetables was attributed to pyropheophytins. These pigment derivatives probably accounted for the unidentified pigments reported by Buckle and Edwards (1969) and LaJollo *et al.* (1971) to represent 20–30% of the total pigments. Their inability to detect pyropheophytins by these researchers was due to the absence of high-resolution chromatographic techniques required to separate these derivatives. Shwartz and von Elbe (1981) proposed the following kinetic sequence to explain the formation of pyropheophytins:

$$\text{Chlorophyll} \rightarrow \text{Pheophytin} \rightarrow \text{Pyropheophytin}$$

Holden (1965) attributed decoloration of chlorophyll in legume seeds to the oxidation of fatty acids by lipoxygenase. This reaction was coupled with the degradation of fatty acid hydroperoxides and required a thermolabile factor to be present in the crude extract. Subsequent research by Zimmerman and Vick (1970) identified an enzyme, linoleate hydroperoxide isomerase, as the heat-labile factor. The bleaching effect on chlorophyll was attributed to an oxidation–reduction reaction in which the ketohydroxy fatty acid, the isomerized product formed by hydroperoxide isomerase, and a portion of the conjugated double-bond system of chlorophyll were involved. Imamura and Shimizu (1974) disproved the involvement of hydroperoxide isomerase and confirmed the role of lipoxygenase in chlorophyll bleaching. Later researchers confirmed that different lipoxygenase isoenzymes played a role in bleaching chlorophyll and carotenes (Grosch *et al.*, 1976; Ramadoss *et al.*, 1978). The participation of the different lipoxygenase isoenzymes in carotene and chlorophyll bleaching has been the subject of a number of investigations (Cohen *et al.*, 1984; Hilderbrand and Hymovitz, 1982; Reynolds and Klein, 1982). The effect of soybean lipoxygenase-1 on wheat chloroplasts by Kockritz *et al.* (1985) suggested that this enzyme, which selectively attacks free fatty acids, may be involved in senescence and chloroplast breakdown. Lipoxygenase activity has been reported previously in chloroplasts of peas (Borisova and Budnitskaya, 1975; Douillard and Bergeron, 1978).

E. Carotenoids

During the maturation of many fruits there is a change of color from green to orange or red. This is due to the loss in chlorophyll and the unmasking and synthesis of carotenoids (MacKinney, 1961). Such pigmentation changes are accompanied by structural changes in the chloroplasts. The granal–integranal network, in particular, becomes disorganized, resulting in the formation of chromoplasts (Camara and Brangeon, 1981; Spurr and Harris, 1968; Thomson, 1966). The chromoplasts no longer contain chlorophyll or photosynthetic pigments but become the major site for carotenoid biosynthesis (Camara and Brangeon, 1981).

Carotenoids are C_{40} isoprenoid compounds composed of isoprene units joined head to tail to form a system of conjugated double bonds (Eskin, 1979). They are classified into two groups, carotenes and xanthophylls. The former are structurally related to hydrocarbons while the xanthophylls include the corresponding oxidized derivatives (hydroxy, epoxy, and oxy compounds) and are frequently esterified. Examples of carotenes are α- and β-carotenes in carrots and lycopene in tomatoes, while xanthophylls include capsanthin and capsorubin found in red pepper.

Carotenoids are synthesized within the chromoplast from isopentyl pyrophosphate. The latter is synthesized from mevalonic acid (MVA), the basic precursor of these pigments, which itself is formed from acetyl-CoA according to Scheme 2.5 (Britton, 1982).

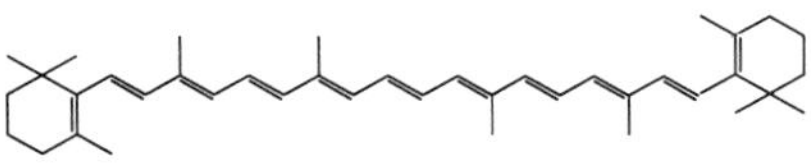

β-Carotene

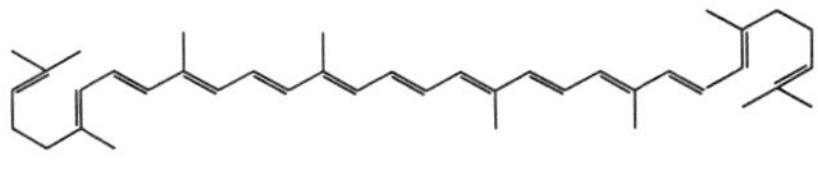

Lycopene

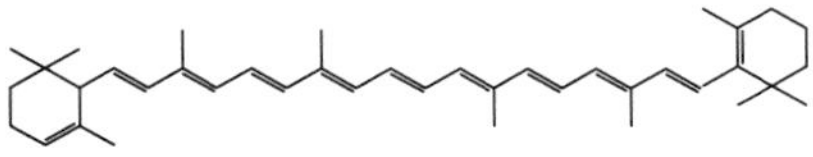

α-Carotene

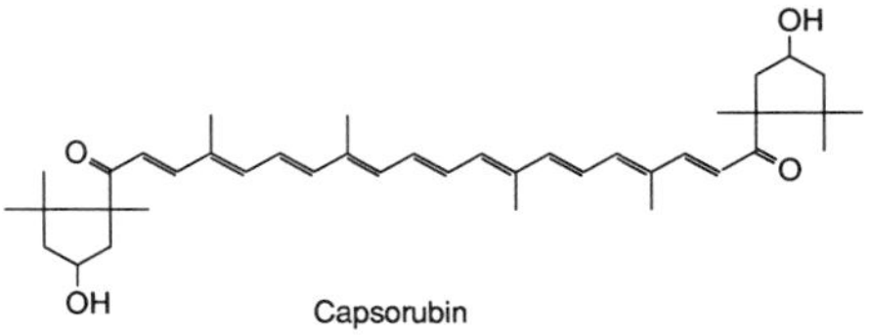

Capsorubin

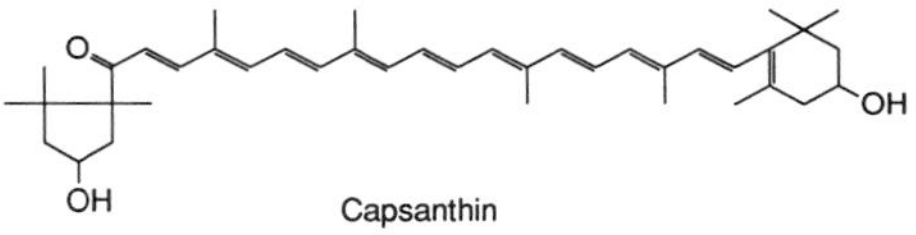

Capsanthin

It remained somewhat unclear whether acetyl-CoA was synthesized within the chloroplast or was of extraplastidic origin. Grumbach and Forn (1980) showed quite clearly that acetyl-CoA was synthesized within the chloroplast and could synthesize carotenoids autonomously. Chromoplasts from both red peppers and red daffodils exhibited ability to synthesize carotenoids from isopentyl phosphates (Beyer *et al.*, 1980; Camara *et al.*, 1982). The presence of enzymes capable of synthesizing acetyl-CoA suggested that a similar autonomy existed in the chromoplasts capable of synthesizing carotenoids.

The formation of geranylgeranyl pyrophosphate from MVA involves phosphorylation by mevalonate kinase (ATP : mevalonate 5-phosphotransferase, EC 27.1.3.6). This enzyme was identified in many plants, including pumkin seedlings (Loomis and Battaille, 1963), green leaves and etiolated cotyledons of French beans (*Phaseolus vulgaris*) (Rogers *et al.*, 1966; Gray and Keckwick, 1969, 1973), orange juice vesicles (Potty and Breumer, 1970), and melon cotyledons (*Cucumis mello*) (Gray and Keckwick, 1972). Phosphorylation of MVA-5P to MVA-5 pyrophosphate (MVA-5 PP) is then catalyzed by 5-phosphomevalonate kinase (ATP : phosphomevalonate phosphotransferase, EC 2.7.4.2). MVA-5 PP is then decarboxylated by pyrophosphomevalonate decarboxylase (ATP : 5-pyrophosphomevalonate carboxylyase (dehydrating), EC 4.1.1.33). This enzyme catalyzes a bimolecular reaction in which ATP and 5-pyrophosphomevalonate are converted to isopentyl pyrophosphate, ADP, phosphate, and CO_2 (Scheme 2.6).

Isomerization of isopentyl pyrophosphate to dimethylallyl pyrophosphate is catalyzed by isopentyl pyrophosphate isomerase (EC 5.3.3.2). The double bond

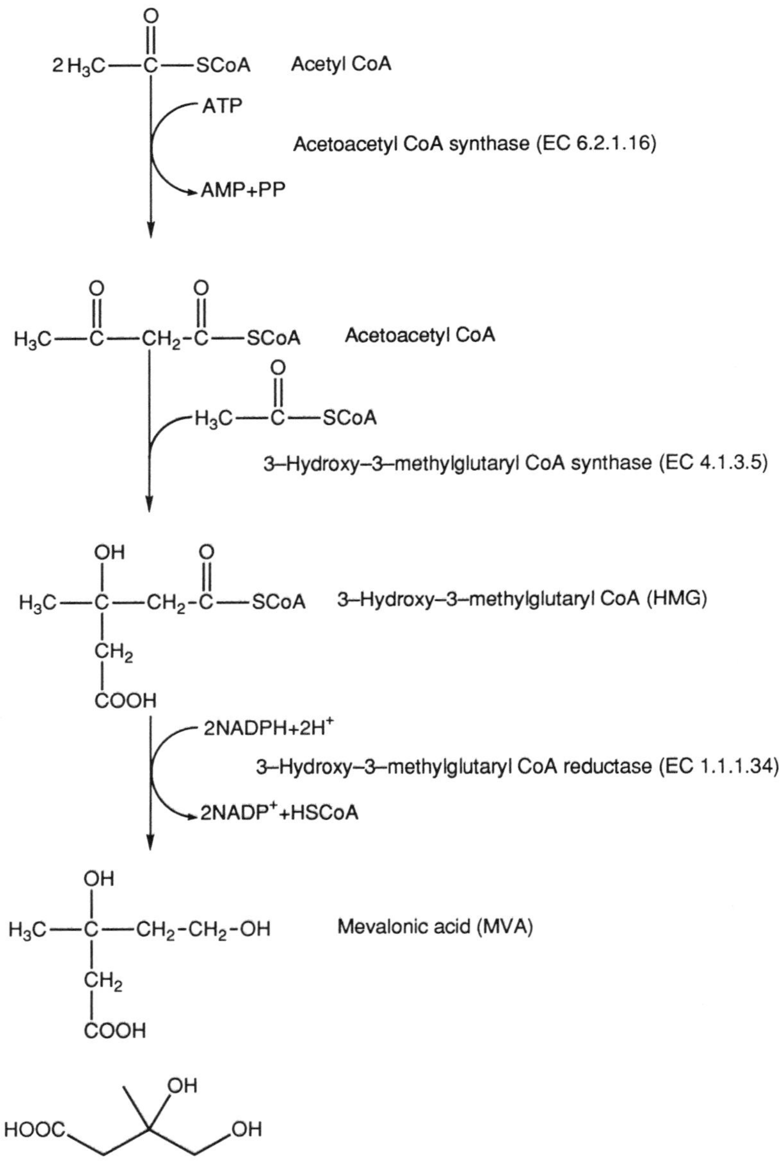

$2\,H_3C{-}C(=O){-}SCoA$ Acetyl CoA

ATP

Acetoacetyl CoA synthase (EC 6.2.1.16)

AMP+PP

$H_3C{-}C(=O){-}CH_2{-}C(=O){-}SCoA$ Acetoacetyl CoA

$H_3C{-}C(=O){-}SCoA$

3–Hydroxy–3–methylglutaryl CoA synthase (EC 4.1.3.5)

$H_3C{-}C(OH)(CH_2COOH){-}CH_2{-}C(=O){-}SCoA$ 3–Hydroxy–3–methylglutaryl CoA (HMG)

$2NADPH+2H^+$

3–Hydroxy–3–methylglutaryl CoA reductase (EC 1.1.1.34)

$2NADP^+ + HSCoA$

$H_3C{-}C(OH)(CH_2COOH){-}CH_2{-}CH_2{-}OH$ Mevalonic acid (MVA)

HOOC, OH, OH

SCHEME 2.5. Mevalonic acid biosynthesis.

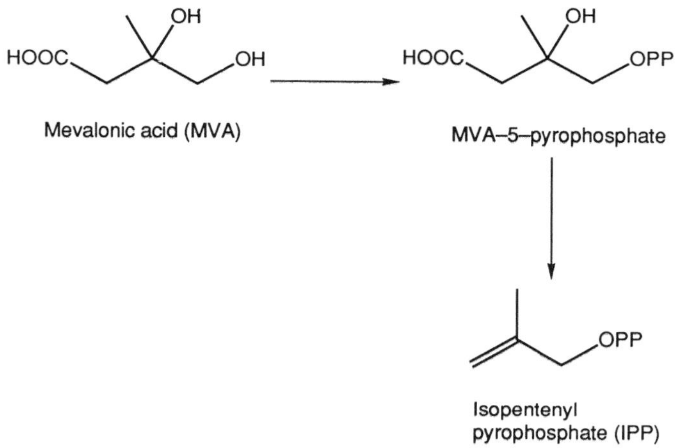

HOOC OH OH

Mevalonic acid (MVA)

HOOC OH OPP

MVA–5–pyrophosphate

OPP

Isopentenyl pyrophosphate (IPP)

SCHEME 2.6. Isopentyl pyrophosphate biosynthesis.

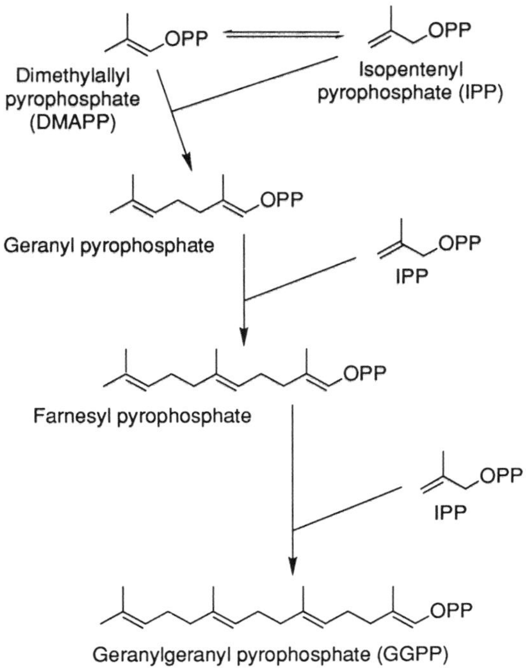

OPP

Dimethylallyl pyrophosphate (DMAPP)

OPP

Isopentenyl pyrophosphate (IPP)

OPP

Geranyl pyrophosphate

OPP

IPP

OPP

Farnesyl pyrophosphate

OPP

IPP

OPP

Geranylgeranyl pyrophosphate (GGPP)

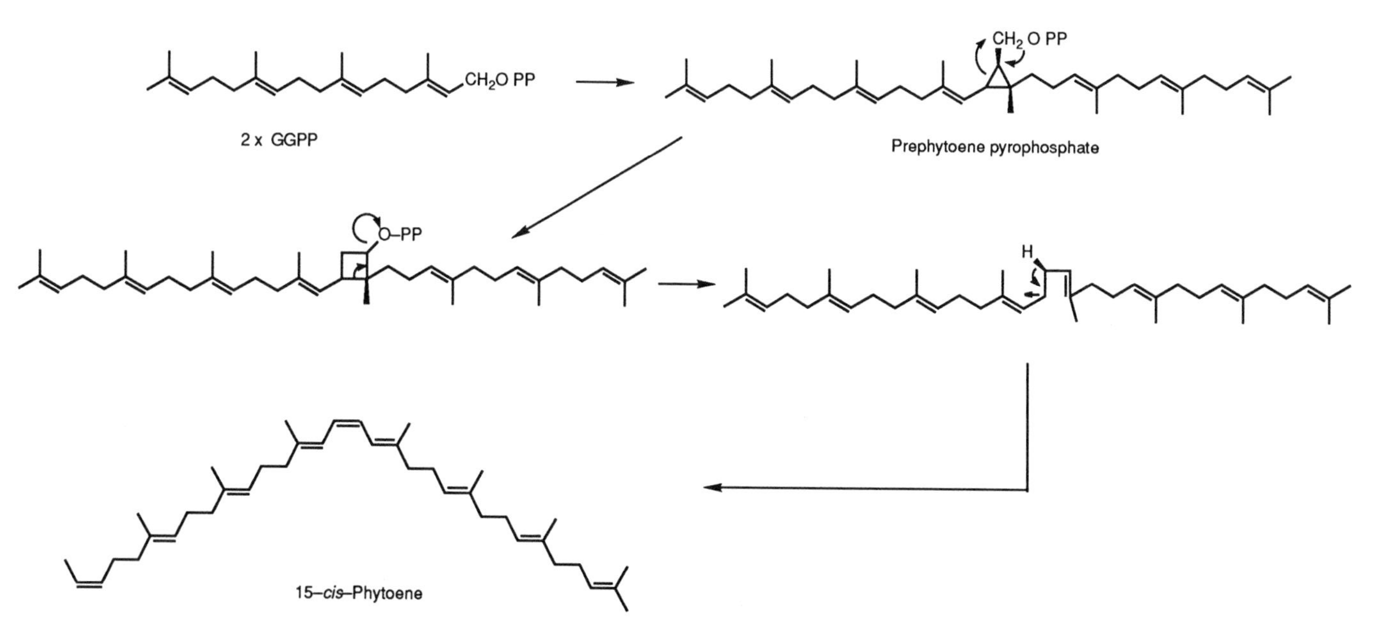

SCHEME 2.7. Mechanism of phytoene biosynthesis (Britton, 1982).

is isomerized from position 3 in isopentyl pyrophosphate to position 2 in dimethylallyl pyrophosphate. This enzyme was isolated from pumpkin fruit by Ogura and co-workers (1968). One molecule of dimethylallyl pyrophosphate then condenses with one, two, or three molecules of isopentyl pyrophosphate, leading to the formation of geranylgeranyl pyrophosphate. The last reactions are catalyzed by a group of enzymes referred to as prenyl transferases. The enzyme responsible for the synthesis of farnesyl pyrophosphate was partially purified from pumpkin seed (Eberhardt and Rilling, 1975).

The formation of the first C_{40} hydrocarbon 15-*cis*-phytoene results from condensation of two molecules of geranylgeranyl pyrophosphate (C_{20}) (Scheme 2.7). The intermediate in this reaction, prephytoene pyrophosphate, loses a proton, which results in a double bond at the C-15 position. Maudinas *et al.* (1975b) identified a soluble enzyme system from tomato fruit plastids which was capable of synthesizing *cis*-phytoene from isopentyl pyrophosphate. Phytoene is quite colorless and is converted to colored carotenoids by a series of desaturation steps which produce a conjugated double-bond system. Unlike phytoene, which is a 15-*cis* isomer, the colored carotenoids are all trans so that isomerization to the trans form must take place during the desaturation process. The mechanism involves loss of hydrogen by trans-elimination and may be mediated by an enzyme complex in the membrane, possibly involving metal ions or cytochromes in a simple electron transfer system (Britton, 1979). The sequential desaturation of phytoene to lycopene was proposed by Porter and Lincoln (1950) and is shown in Scheme 2.8. A similar enzyme system was reported by Qureshi *et al.* (1974) in tangarene mutant tomatoes capable of converting *cis*-β-carotene into all-trans carotene. The conversion of neurosporene into lycopene has been found in fungal systems only (Bramley *et al.*, 1977; Davies, 1973).

The final step in carotenoid biosynthesis is cyclization with the formation of at least one or two cyclic end groups in the carotenoids (Scheme 2.9). The conversion of lycopene to α-, β-, and γ-carotenes was demonstrated in the presence of soluble enzymes from tomato fruit plastids and spinach chloroplasts by Kuwasha *et al.* (1969). Cyclization was shown to be inhibited by nitotine and CPTA [2-(4-chlorophenyltrio)triethyl ammonium chloride] with the accumulation of lycopene in citrus fruits treated with these inhibitors (Britton, 1982). Following cyclization oxygen is then incorporated as a hydroxyl group at C3 or, an epoxide at the 5, 6 position (Britton, 1976; Takeguchi and Yamamoto, 1968). The latter involves a series of reactions referred to as the xanthophyll cycle.

Camara *et al.* (1982) examined the site of carotenoid biosynthesis in chromoplasts of semiripened pepper (*Capsicum annium* L.) fruits. Incubation of [1-^{14}C] isopentyl diphosphate with different chromoplast fractions showed that the membrane was incapable of synthesizing carotenoids while the stroma synthesized the first colorless carotenoid, phytoene (Table 2.4). Thus phytoene synthetase, the enzyme responsible, could be a useful indicator of the chromoplast stroma.

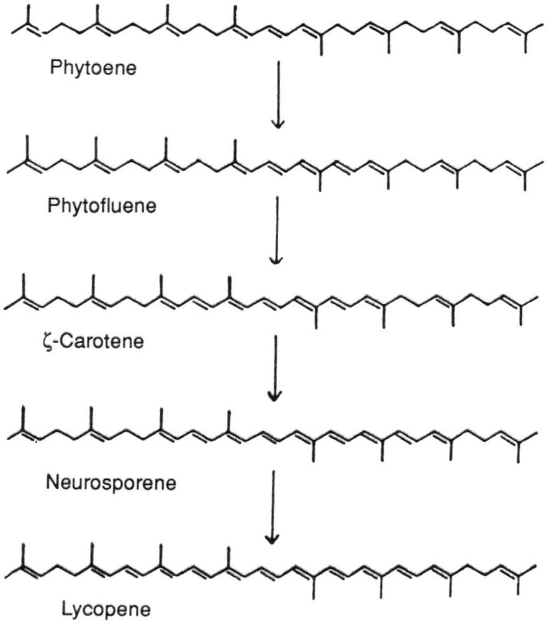

SCHEME 2.8. Sequence of desaturation reactions leading from phytoene to lycopene (Britton, 1982).

Increased incorporation of the labeled substrate in the presence of chromoplast membranes was attributed to their desaturation and cyclization systems, which produced colored carotenoids. This study showed that the enzymes involved in carotenoid biosynthesis were compartmentalized in the chromoplasts. The stroma synthesized phytoene synthesis, which underwent desaturation and cycl-

TABLE 2.4

INCORPORATION OF ^{14}C FROM $[1-^{14}C]$ISOPENTYL DIPHOSPHATE INTO THE UNSAPONIFIABLE BY THE CHROMOPLAST FRACTION[a]

Chromoplast fraction	Isopentyl incorporation (nmole)
Membranes	25
Stroma	450
Membranes + stroma	750

[a] From Camara *et al.* (1982).

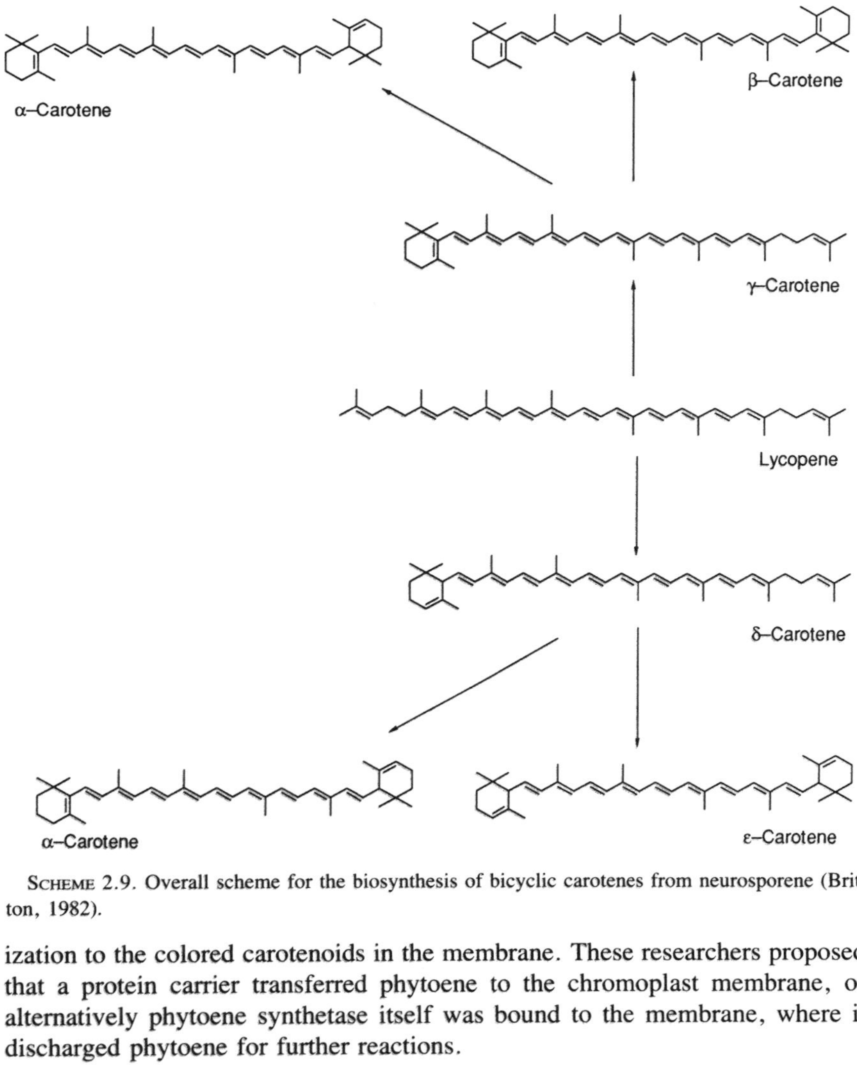

α–Carotene

β–Carotene

γ–Carotene

Lycopene

δ–Carotene

α–Carotene

ε–Carotene

SCHEME 2.9. Overall scheme for the biosynthesis of bicyclic carotenes from neurosporene (Britton, 1982).

ization to the colored carotenoids in the membrane. These researchers proposed that a protein carrier transferred phytoene to the chromoplast membrane, or alternatively phytoene synthetase itself was bound to the membrane, where it discharged phytoene for further reactions.

1. *Carotenoid Changes during Ripening*

Ebert and Gross (1985) examined the carotenoid pigments in the peel of ripening persimmon (*Diospyros kaki* cv. Triumph). A steady decline in chlorophyll (*a* and *b*) was observed during the course of ripening which disappeared in the harvest-ripe fruit (Table 2.5). The chloroplast carotenoids (α- and β-carotenes, lutein, violaxanthin, and neoxanthin) decreased, followed by the

TABLE 2.5

CAROTENOID CHANGES IN THE PEEL OF PERSIMMON (*DIOSPYROS KAKI* CV. TRIUMPH) DURING POSTHARVEST RIPENING[a]

	Ripening stage		
	Harvet-ripe	Intermediate	Fully ripe
Total carotenoids (μg/g fr. wt)	128.0	366.0	491.0
Carotenoid pattern (% of total carotenoids)			
Phytofluene	—	—	0.4
α-Carotene	1.6	1.2	1.0
β-Carotene	9.4	7.6	6.7
Mutatochrome	—	0.7	—
γ-Carotene	—	0.4	—
Lycopene	1.1	0.5	8.2
β-Cryptoxanthin	29.2	50.0	48.2
Cryptoxanthin 5,6-epoxide	0.9	1.2	1.9
Cryptoflavin	0.7	2.1	2.9
Lutein	12.4	5.5	4.1
Zeaxanthin	9.3	9.7	5.9
Mutatoxanthin	0.8	4.7	1.8
Isolutein	0.5	—	0.3
trans-Antheraxanthin	5.4	2.0	4.8
cis-Antheraxanthin	6.2	2.2	2.3
Luteoxanthin	1.7	1.8	1.9
trans-Violaxanthin	6.9	3.7	3.8
cis-Violaxanthin	6.7	1.5	2.0
Neoxanthin	7.2	5.2	3.8

[a] From Ebert and Gross (1985). Reprinted with permission. Copyright © by Pergamon Press.

gradual synthesis of the chromoplast carotenoids (cryptoxanthin, antheraxanthin, and zeaxanthin). Ikemefura and Adamson (1985) monitored changes in the chlorophyll and carotenoid pigments of ripening palm fruit (*Elaeis quineeris* Palmal) and noted a similar degeneration of chloroplasts and formation of chromoplasts. These changes were accompanied by increase in carotenogenesis with the formation of α- and β-carotenes as the major pigments in the ripe fruit.

Farin and co-workers (1983) examined the change in carotenoids during the ripening of the Israeli mandarin hybrid Michal (*Citrus reticulata*). This particular fruit is the most highly colored of the citrus fruits with a bright-reddish color. Total chlorophyll decreased rapidly in the peel and completely disappeared at the ripening stage (Table 2.6). The total carotenoids decreased at the color break stage due to a decline of the chloroplast carotenoids, δ-carotene, lutein, viola-

TABLE 2.6

PIGMENT DISTRIBUTION IN THE FLAVEDO OF A MANDARIN HYBRID (*CITRUS RETICULATA* CV. MICHAL) DURING RIPENING[a]

	Peel		
	Green	Color break	Ripe
Fruit diameter (cm)	4.70	4.85	5.10
Chlorophyll *a* (μg/g fr. wt.)	240.0	52.6	—
Chlorophyll *b* (μg/g fr. wt.)	86.0	15.8	—
Total carotenoids (μg/g fr. wt.)	143.4	51.0	174.1
Carotenoid pattern (% of total carotenoids)			
Phytofluene	—	5.2	3.1
α-Carotene	9.7	2.4	0.2
β-Carotene	6.9	2.5	0.3
ζ-Carotene	—	—	0.4
δ-Carotene	—	—	0.1
Mutatochrome	—	—	0.5
Lycopene	—	—	—
β-Apo-8′-carotenal	0.7	0.7	1.3
α-Cryptoxanthin	1.9	—	—
β-Cryptoxanthin	—	3.1	6.4
Cryptoxanthin 5,6-epoxide	—	—	0.4
Cryptoxanthin 5′,6′-epoxide	—	—	0.3
β-Citraurinene	—	9.5	9.9
β-Citraurin	—	12.3	26.1
Lutein	23.5	12.8	2.6
Zeaxanthin	3.9	1.6	1.0
Mutatoxanthin	1.5	0.2	0.2
trans-Antheraxanthin	3.6	4.3	1.8
cis-Antheraxanthin	—	—	2.5
Luteoxanthin	5.8	5.6	9.1
trans-Violaxanthin	14.0	11.7	9.8
cis-Violaxanthin	11.0	18.2	19.8
trans-Neoxanthin	11.7	6.6	4.2
cis-Neoxanthin	3.6	3.3	—
Neochrome	—	—	—
trans-Trollixanthin	—	—	—
Trollichrome	—	—	—
Unknown	2.2	—	—

[a] Adapted from Farin *et al.* (1983). Reprinted with permission. Copyright © by Pergamon Press.

xanthin, and neoxanthin. A marked rise in carotenogenesis followed with an increase in the chromoplast carotenoids, cryptoxanthin and C_{30} apocarotenoids. The two C_{30} apocarotenoids, β-citraurin and β-citraurinene, accounted for 26.1 and 9.9%, respectively, of the total carotenoids in the ripe fruit. Their formation appeared to require asymmetric degradation of a C_{40} fragment from the side of the C_1 carotenoid, cryptoxanthin.

Gross *et al.* (1983) studied changes in pigments and ultrastructure during the ripening of pummelo (*Citrus grandis* Osbech). Chlorophyll decreased from 90 μg/g at the unripe (green) stage to 11 μg/g in the ripe pale-yellow fruit, totally disappearing by the fully ripe stage. The chloroplast carotenoids, β-carotene, lutein, violaxanthin, and neoxanthin, all decreased during ripening, with the β-carotene totally disappearing. Accumulation of phytofluene was evident at the color break, reaching 67% of the total carotenoids in the ripe fruit. Besides phytofluene, other chromoplast carotenoids were detected, including δ-carotene, neurosporene cryptoxanthin, and cryptoflavin.

These studies showed that compositional changes in carotenoids during ripening reflect the transformation of chloroplasts to chromoplasts. Earlier work by Eilati and co-workers (1975) on ripening Shamouti "orange peel" showed parallel transformations occurred irrespective of whether the fruit was attached or detached from the tree. The transformation of chloroplasts to chromoplasts during degreening of citrus fruits was reported by Huff (1984) to be regulated by the accumulation of sugar in the epicarp. The reverse transformation associated with the regreening of certain citrus species was found to be accompanied by the disappearance of sugars. This regreening phenomenon of *Citrus sinensis* fruit epicarp observed by Thomson and co-workers (1967) was attributed to the reversion of chromoplasts to chloroplasts. Thus, sucrose promoted the formation of chromoplasts while nitrogen stabilized the chloroplasts by retarding the degreening process.

2. *Carotenoid Degradation: Processing and Storage*

Carotenoids impart most of the yellow and orange colors to fruits and vegetables such as pineapples and carrots. The unsaturated nature of carotenoids renders them particularly susceptible to isomerization and oxidation resulting in loss of color which is most pronounced following oxidation. The latter can be brought about by the action of lipoxygenase, which can bleach carotenoids (Eskin *et al.*, 1977). From recent research, it appears that differences in the abilities of lipoxygenases from fruits and vegetables to oxidize carotenoids are due to the isoenzymes present. Lipoxygenase isoenzymes are classified as either type 1 or type 2 depending on their pH optimum and product specificity. Of these isoenzymes, lipoxygenase-2 exhibits a more acidic pH optimum and is involved in cooxidation reactions leading to pigment discoloration (Klein and Grossman, 1985). For example, pea lipoxygenase-2 was shown by Arens *et al.* (1973) to be an effective oxidizer of carotenoids. Chepurenko *et al.* (1978) attributed carotenoid bleaching to the combined effect of the lipoxygenase isoenzymes in peas and not just to lipoxygenase-2. Yoon and Klein (1979) showed definite differences between the rate of carotenoid oxidation for these two pea lipoxygenase isoenzymes. These enzymes were also involved in the biosynthesis of traumatic acid, a wound-healing hormone, as well as the plant growth regulator jasmonic acid (Vick and Zimmerman, 1983; Zimmerman and Coudron, 1979).

Carotenoids are extremely susceptible to nonenzymatic oxidation in dehydrated fruits and vegetables. For example, powdered, dehydrated carrots were reported by MacKinney *et al.* (1958) to lose 21% of their carotenoids when stored in air. Of particular importance to their stability is the amount of moisture present in dehydrated products. The effect of water activity (a_W) on the degradation of β-carotene in model systems was studied by Chou and Breene (1972). These researchers showed that at a_W of 0.44 the oxidative decoloration of β-carotene was reduced compared to the corresponding dry system with or without the presence of the antioxidant butylated hydroxytoluene (BHT) (Fig. 2.9). It is clear from this study that water acts as a barrier to oxygen diffusion.

When carotenoids are heated in the absence of air some of the trans double bonds undergo isomerization to the corresponding cis isomers. Acids also catalyze isomerization from the all-trans form to the corresponding cis isomer.

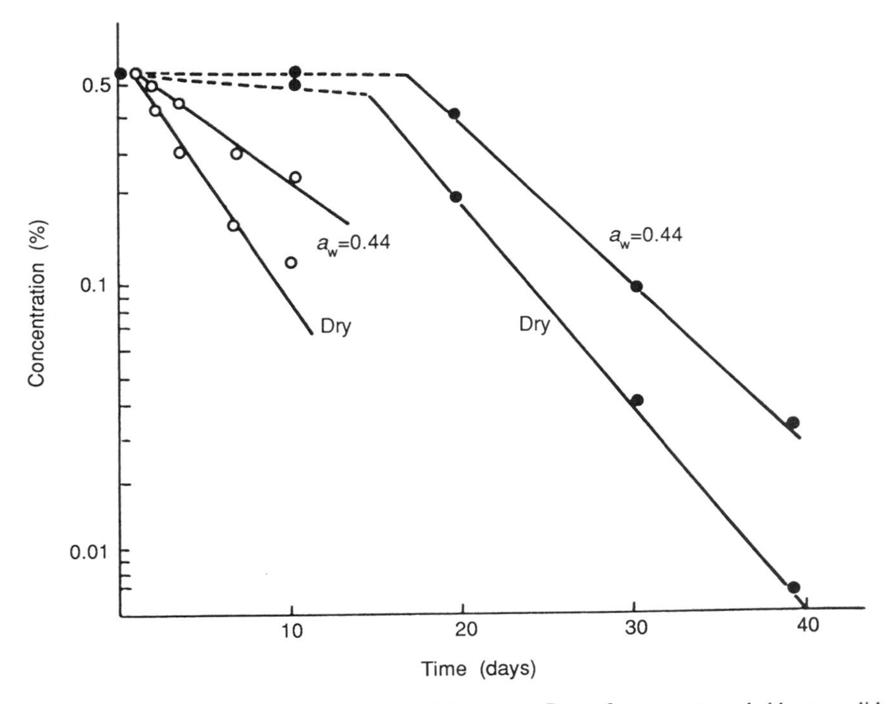

FIG. 2.9. Decoloration of β-carotene in model systems. Dry refers to systems held over solid $CaCl_2$; a_W = 0.44 to systems held over saturated K_2CO_3 (○, control; ●, BHT-butylated hydroxytoluene) (Chou and Breene, 1972). Copyright © by Institute of Food Technologists.

The change in shape associated with the cis isomers reduces the resonance in the molecule as well as the color intensity. These reactions were responsible for the difference in quality between canned and fresh pineapple (Singleton *et al.*, 1961). A change in the spectrum of the extracted carotenoid pigments was reported for canned pineapples, including loss of a peak at 466 nm together with peaks of shorter wavelengths. This spectral shift, although too subtle for the human eye, caused a slight color shift from orange-yellow in the fresh pineapple to a more lemon-yellow in the canned product. This shift was attributed to isomerization of carotenoids with 5, 6-epoxide groups to the corresponding 5, 6-furanoid oxides. In the intact fruit the natural cell vacuole acids have a catalytic effect by coming in contact with the carotenoid-bearing plastids when cell membranes are disrupted during handling or processing. Several studies examined the effect of cooking carrots on the formation of cis isomers. Klaui (1973), however, suggested that the effects of cooking on color and carotene content were insignificant.

F. Anthocyanins

Anthocyanins are responsible for the attractive pink, red, purple, and blue colors of flowers, leaves, fruits, and vegetables (Harborne, 1967). These are water-soluble pigments that accumulate in the epidermal cells of fruits as well as in roots and leaves (Harborne, 1976; Timberlake and Bridle, 1975, 1982). These pigments are formed as the fruit matures and ripens, as is evident for strawberries (Fuleki, 1969), certain varieties of cherries, raspberries, cranberries, and apples (Bishop and Klein, 1975; Cansfield and Francis, 1970; Craker and Wetherbee, 1973; Proctor and Creasy, 1971; Zapsalis and Francis, 1965), black grapes (Liao and Luh, 1970), and blueberries (Suomalainen and Keranen, 1961), as well as in such vegetables as red cabbage, russet potato, radishes, and red onion (Fuleki, 1971; Small and Pecket, 1982). The primary role of anthocyanins is that of an insect or bird attractant for pollination of flowers and for fruit seed dissemination by animals (Brouillard, 1983).

Anthocyanins are flavonoid pigments whose structure is based on the phenylpropanoid nucleus. They occur in nature as glycosides in which the aglycone forms, or anthocyanidins, are substituted flavylium salts. The structural formula of the flavylium cation is as follows:

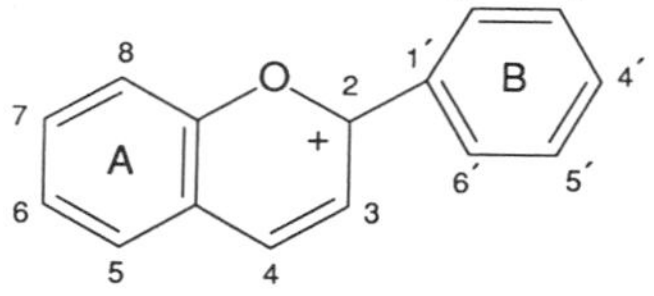

Flavylium cation

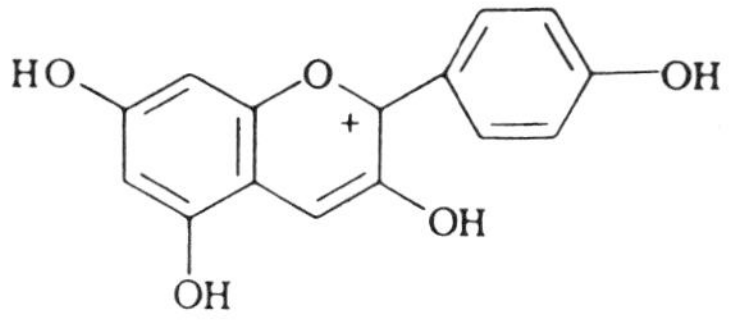

Pelargonidin (3, 5, 7, 4′-tetrahydroxyflavylium cation)

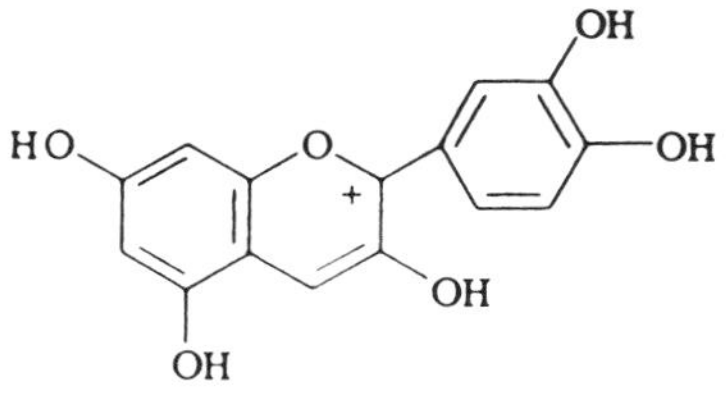

Cyanidin (3, 5, 7, 3′, 4′-pentahydroxyflavylium cation)

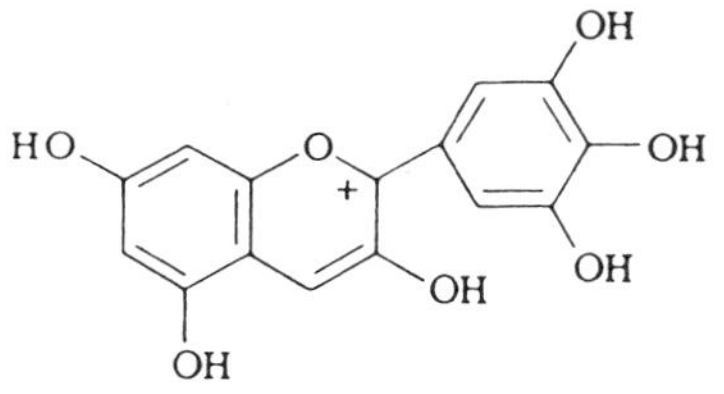

Delphinidin (3, 5, 7, 3′, 4′, 5′-hexahydroxyflavylium cation)

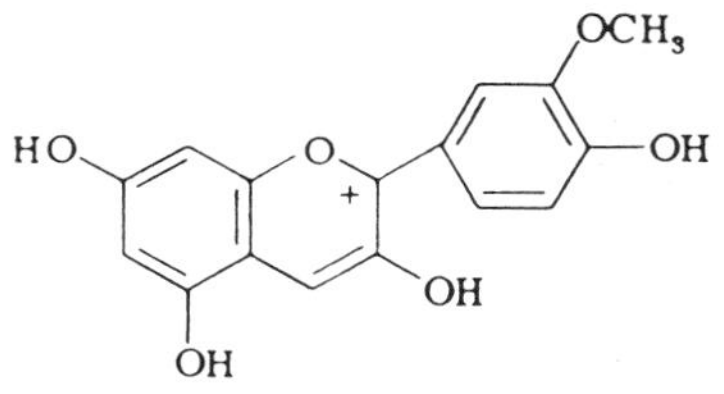

Peonidin (3, 5, 7, 4′-tetrahydroxy-3′-methoxyflavylium cation)

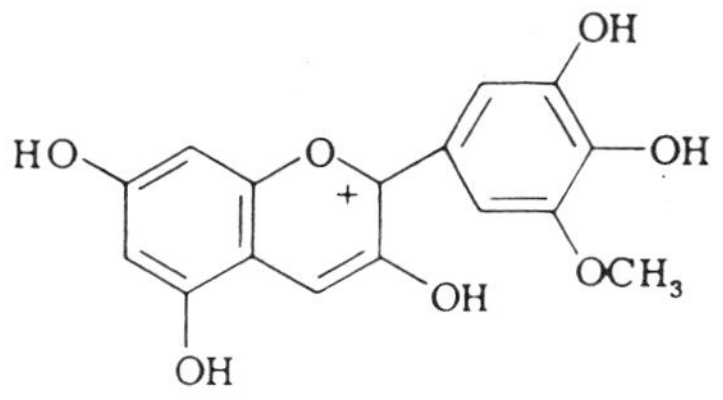

Petunidin (3, 5, 7, 3′, 4′-pentahydroxy-5′-methoxyflavylium cation)

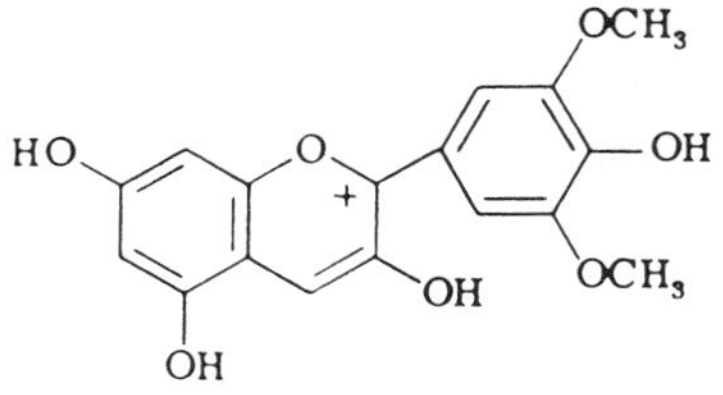

Malvidin (3, 5, 7, 4′-tetrahydroxy-3′, 5′-dimethoxyflavylium cation)

SCHEME 2.10. Major food anthocyanidins.

The anthocyanidins or aglycones are somewhat less stable than the corresponding glycosides. The latter are composed primarily of 3-glycosides and 3,5-glycosides. The major anthocyanidins or aglycones in fruits are listed in Scheme 2.10. D-Glucose, D-lactose, L-rhamnose, and D-xylose are the main sugars linked to the anthocyanidin at the 3-position. The intensity of the color is determined by the nature of the pigment, its concentration, and pH, as well as the presence of pigment mixtures, copigments, and certain metallic ions (Brouillard, 1983). Anthocyanins are dissolved in the aqueous vascular sap in the plant cell, which is slightly acidic. These pigments tend to be more stable under these acidic conditions. The different shades of colors of anthocyanidins reflect the nature of their

TABLE 2.7

DISTRIBUTION OF ANTHOCYANIDINS IN SOME EDIBLE FRUIT

Anthocyanidin	Fruit
Cyanidin	Blackberry, rhubarb
Cyanidin, delphinidin	Blackcurrant
Cyanidin, peonidin	Cherry
Cyanidin, pelargonidin	Strawberry

hydroxylation and methoxylation patterns. An increase in hydroxylation is accompanied by an increase in the blue hue, whereas methoxylation enhances the red color (Braverman, 1963). The distribution of anthocyanidins in some edible fruits is shown in Table 2.7.

1. *Biosynthesis of Anthocyanins*

The biosynthesis of anthocyanins, in common with other phenylpropanoid compounds, involves both acetate and phenylalanine as precursors. The synthesis of aromatic amino acids is achieved in plants via the shikimic acid pathway. The phenylpropane residue, ring B, and carbon atoms 2, 3, and 4 of the heterocyclic ring C are formed from phenylalanine via coumarate, while ring A is produced from three acetate residues condensing head to tail (Hahlbrock and Grisebach, 1975). The first intermediate formed, a chalcone, is common to the biosynthesis of all flavonoids. Unlike the biosynthesis of other flavonoids, however, relatively little is known about the synthesis of anthocyanins beyond the formation of chalcone and cyclization to flavonone (Scheme 2.11).

The enzyme chalcone isomerase, responsible for formation of the six-membered heterocyclic ring C of the flavonone, was first isolated by Moustafa and Wong (1967). It has since been identified in flavonoid-producing plants (Halhlbrock and Grisebach, 1975). Studies on the synthesis of anthocyanins in flowers pointed to the importance of chalcone isomerase and flavonone-3-hydroxylase in the biosynthesis of these pigments. The mechanism whereby dihydroflavonol is converted to anthocyanidins is unknown, although it has been shown to be a precursor of these aglycones.

2. *Phenylalanine Ammonia Lyase*

The activities of phenylalanine ammonia-lyase (PAL), chàlcone synthase, and chalcone isomerase have all been shown to increase with anthocyanin accumulation (Wellman *et al.*, 1976). The key enzyme, however, in the metabolism of phenyl propanoid is phenylalanine ammonia-lyase (PAL). Inhibition of this en-

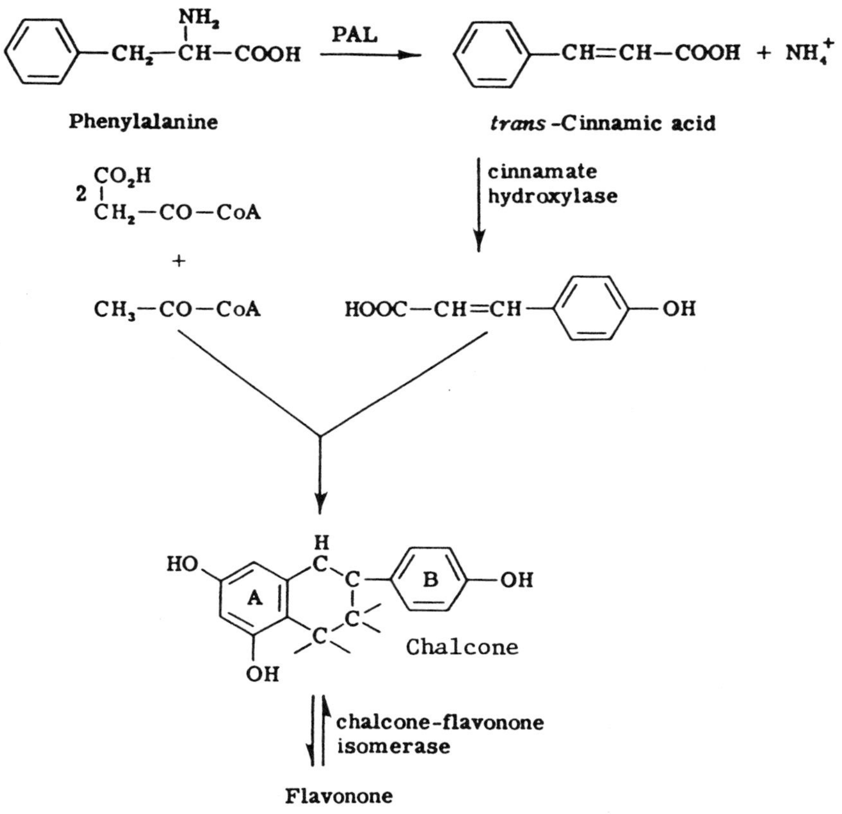

Phenylalanine $\xrightarrow{\text{PAL}}$ *trans*-Cinnamic acid + NH_4^+

trans-Cinnamic acid $\xrightarrow{\text{cinnamate hydroxylase}}$ $HOOC-CH=CH-C_6H_4-OH$

$2\ CO_2H-CH_2-CO-CoA$ + $CH_3-CO-CoA$ + $HOOC-CH=CH-C_6H_4-OH$ → Chalcone

Chalcone $\rightleftharpoons$ Flavonone (chalcone-flavonone isomerase)

SCHEME 2.11. Biosynthesis of flavonoids.

zyme by O-substituted hydroxylamines such as the hydroxylamine analogue of phenylalanine was found to block anythocyanin synthesis in buckwheat as well as other plants (Amrhein, *et al.*, 1978). Stimulation of anthocyanin synthesis in apple skins was attributed to exposure to both low temperature and light. Faragher and Chalmers (1977) reported that PAL activity reached a peak and then declined when Jonathan apples were stimulated with light. Subsequent studies by Faragher (1983) examined the effect of temperature on the regulation of anthocyanins in apple skin. The optimum temperature for accumulation of anthocyanin was 12 and 24°C in unripe and ripe apples, respectively.

3. *Site of Anthocyanin Biosynthesis*

Anthocyanins accumulate in the central vacuole of the plant cell (Saunders and Conn, 1978; Wagner, 1979). Intensely pigmented structures were reported in the vacuoles of anthocyanin-producing cells of over 70 species of plants by Pecket and Small (1980), who referred to them as anthocyanoplasts. On the basis of studies with red cabbage, Small and Pecket (1982) showed that only the later enzymes responsible for anthocyanin synthesis were located in the anthocyanoplast, an intracellular compartment separated from the acidic environment of the vacuolar sap. Hrazdina and co-workers (1980), however, were unable to detect any enzyme activity in the cytoplasmic cell vacuolar fractions, which tended to eliminate the vacuole in the biosynthesis of anthocyanins. Jonsson *et al.* (1983) investigated the presence of methyltransferase, one of the enzymes involved in the latter stages of anthocyanin biosynthesis, in the cell vacuole. They too were unable to detect any biosynthesis of anthocyanins in the cell vacuole and suggested that methyltransferase, like other flavonoid-synthesizing enzymes, were located in the cytosol, where anthocyanin synthesis took place. The presence of anthocyanoplasts in the vacuoles of anthocyanin-containing cells of red cabbage was discounted by Neumann (1984). Using electron microscopy he demonstrated that these were hydrophobic droplets and not organelles and thus discounted the presence of biosynthetically active organelles in the cell vacuole.

4. *Photoregulation of Anthocyanin Biosynthesis*

Anthocyanin biosynthesis is a phytochrome-controlled process. As discussed previously, phytochrome is a photoreceptor in plants which occurs as two photointerconvertible forms; a physiologically inactive red-absorbing (664 nm) form (Pr) and a physiologically active far-red-absorbing (730 nm) form (Pfr):

$$\mathrm{Pfr} \underset{730\ \mathrm{nm}}{\overset{660\ \mathrm{nm}}{\longleftrightarrow}} \mathrm{Pr}$$

The mechanism whereby Pfr triggers biochemical and physiological changes in the plant remains obscure. Its presence in the cell membrane suggests that it could exert a number of effects, including control of active transport, membrane-bound hormones, and membrane-bound proteins. A number of studies have focused on the role of Pf $\rightleftharpoons$ Pfr on anthocyanin biosynthesis in plants (Obrenovic, 1985; Mancinelli, 1984).

5. *Anthocyanins: Effect of Processing*

Anthocyanins are generally unstable during processing with a net loss of color during canning, bottling, and other thermal processing operations. Fruits and vegetables contain many enzymes capable of decolorizing anthocyanins, however, these can be inactivated by blanching. Such enzymes include polyphenol oxidase, anthocyanase, and peroxidase (Grommeck and Markakis, 1964; Peng and Markakis, 1963; Sakamura *et al.*, 1966). In addition to temperature, other factors, including pH, oxygen, and metallic ions, also effect stability of these pigments (Eskin, 1979). Of these, pH is the most important factor affecting the stability of the anthocyanin (Mazza and Brouillard, 1987).

Anthocyanins are stable under acidic conditions although model systems show, in most cases, that they are in a colorless form. The ability to retain their color in plants is attributed to the formation of complexes with other phenolics, nucleic acids, sugars, and amino acids, as well as metallic ions such as calcium, magnesium, and potassium (Brouillard, 1983). Several acylated anthocyanins have been identified in flowers which exhibit remarkable stability in neutral or weakly acidic solutions. One such pigment isolated from the petals of Chinese bellflower (*Platycodon grandiflorium*) is platyconin (Saito *et al.*, 1971). Its structure was subsequently confirmed by ^{1}H-NMR spectroscopy (Gotto *et al.*, 1983). The two acyl groups in this anthocyanin, one located above the pyrilium ring and the other below it, results in its stability in neutral solutions. The presence of two or more acyl residues linked to sugars appear to provide excellent color stability in neutral conditions (Mazza and Brouillard, 1987). The identification of acylated anthocyanins in plant food products has not yet been reported, although their potential as food colorants needs to be established.

The effect of pH on anthocyanins is shown in Scheme 2.12 for pelargonidin chloride (Harper, 1968). Over a pH range of 1–3 the pigment is quite stable, existing as a deep-red oxonium ion (I). As the pH increases toward neutrality a loss of red color occurs because of the formation of the colorless pseudobase (II), which is in equilibrium with its keto form (III). The latter then undergoes ring opening to the α-diketone form (IV), which is present at pH 3–7. Above pH 7, the anhydro base (V) predominates, producing a purple color which rapidly fades below pH 7 owing to the formation of the pseudobase (II) and α-diketone. With an increase in the pH to more alkaline conditions, the purple color changes from mauve to blue because of the formation of the ionized anhydro base (VI). The formation of a brown precipitate is due to the degradation of the ketone, which is responsible for decoloration in anthocyanin-containing fruits during prolonged storage.

VI. Texture

The texture of fruits and vegetables is related to the structure and organization of the plant cell walls and intercellular cementing substances (Eskin, 1979). The architecture of the cell wall has been the subject of numerous studies suggesting that it is composed of cellulose fibrils located in a matrix of pectic substances, hemicellulose, proteins, lignin, low-molecular-weight solutes, and water (Van

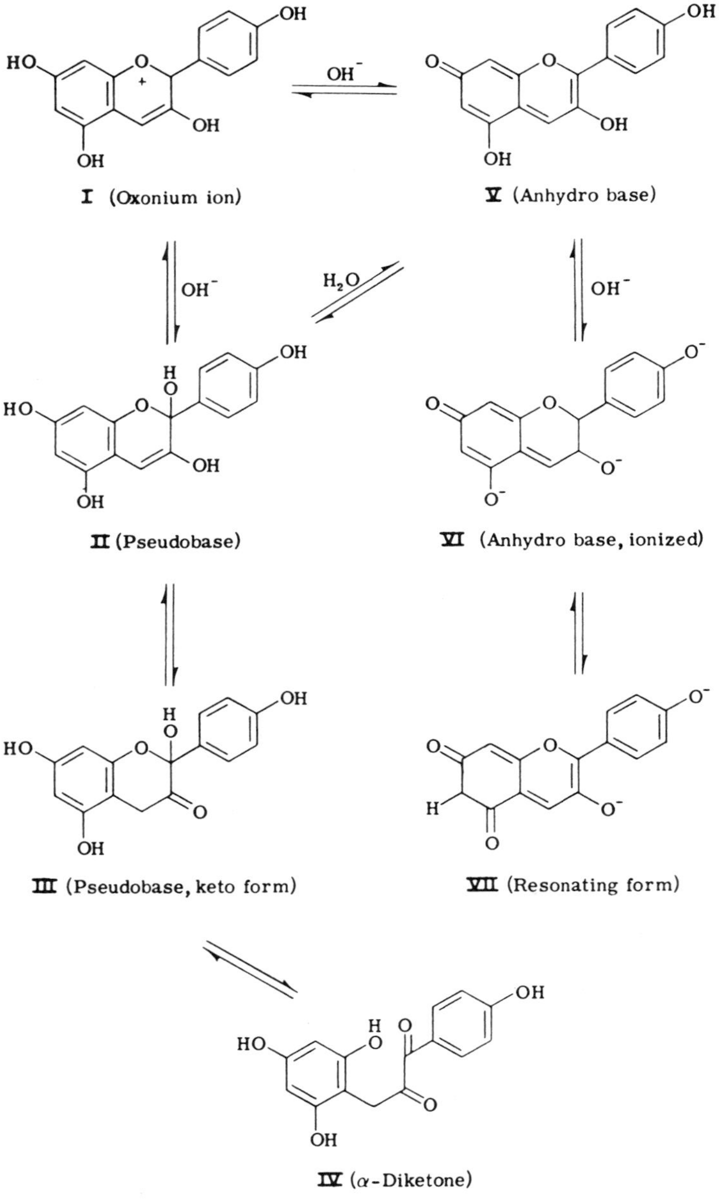

I (Oxonium ion)

OH^-

V (Anhydro base)

OH^-

H_2O

OH^-

II (Pseudobase)

VI (Anhydro base, ionized)

III (Pseudobase, keto form)

VII (Resonating form)

IV (α-Diketone)

SCHEME 2.12. Changes in the molecular structure of pelargonidin chloride with pH (Harper, 1968).

Buren, 1979). In edible plants, the primary cell wall is of central interest as secondary cell walls are virtually absent in mature fruit (Nelmes and Preston, 1968). During the ripening of fruits, a loss of texture results from the deterioration of the primary cell wall constituents. This is in sharp contrast to the maturation of vegetables which is accompanied by a toughening of the texture. The latter is due to the development of the secondary cell walls in vegetables, which contribute to the tough and fibrous texture, as a result of lignin deposition.

A. Cell Wall

The cell walls of plants are composed primarily of complex carbohydrates (Northcote, 1963). The presence of a protein fraction was later established in the primary cell wall of plants (Lambort, 1965). Because of the complexity of the cell wall, numerous models have been proposed including the first complete model by Keegstra and co-workers (1973). These models tended to raise more questions than answers as none described the cell wall structure adequately. The general approach has been to obtain hydrolysed fragments, characterize them, and to reconstruct the individual fragments. This has suggested that the hydroxyproline-rich proteins are covalently linked to the pectic substances through the hemicellulose fraction xyloglucan (Hayashi, 1989). Textural changes occurring during ripening of fruit are due primarily to enzymatic changes on cell wall architecture. Most affected is the middle lamella, an intercellular cement between the primary cell walls of adjacent cells, particularly rich in pectic substances.

1. *Cell Wall Constituents*

a. Polysaccharides. An important constituent of the plant cell wall is cellulose, which is present as linear aggregates or fibrils. They are composed of individual chains held together by hydrogen bonds through the hydroxyl groups at carbon-6 in one cellulose chain with glycosidic oxygens of the adjacent chains (Northcote, 1972). Within the microfibril are highly organized or crystalline regions as well as some amorphous regions. A thorough discussion of cellulose structure can be found in several reviews (Eskin, 1979; Dey and Brinson, 1984).

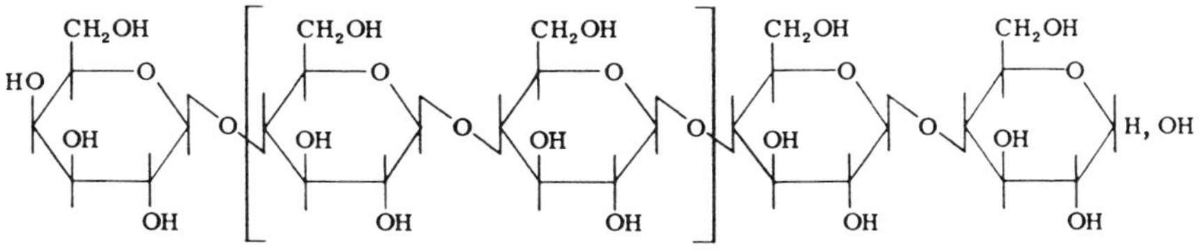

Cellobiose

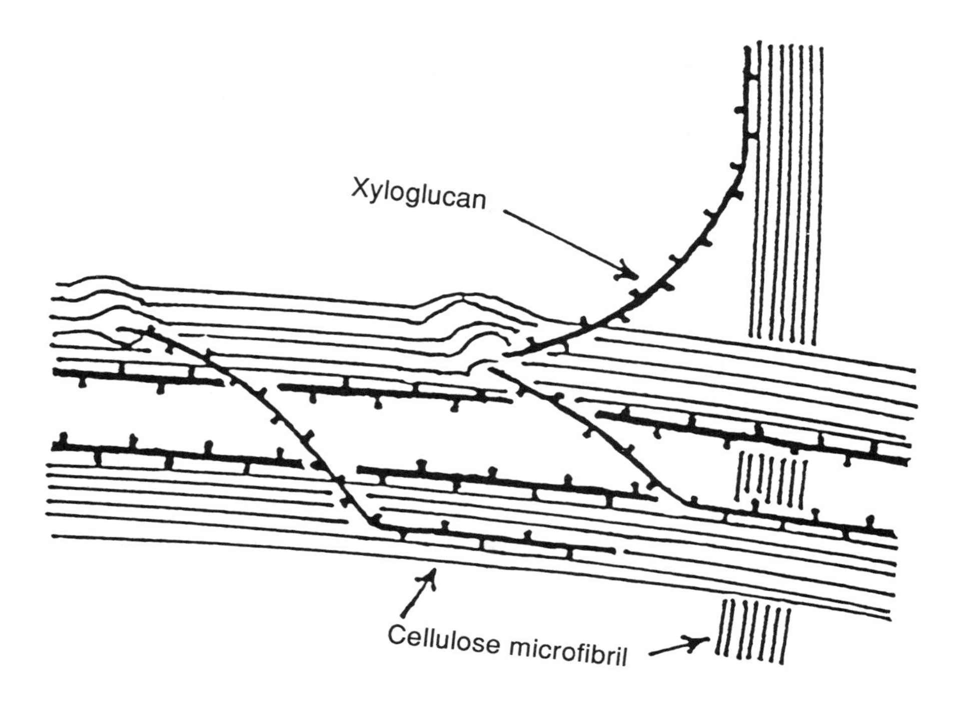

FIG. 2.10. Potential linkages between xyloglucan and cellulose (Hayashi, 1989).

b. Pectic Substances. Pectic substances comprise one-third of the dry substance of the primary cell walls of fruits and vegetables (Van Buren, 1979). The basic structure consists of a backbone of 1,4-linked α-D-galacturonic acid (or the corresponding methyl ester) interspersed with 2-linked L-rhamnosyl residues (Eda and Kato, 1980; McNiel *et al.*, 1980). The predominance of this rhamnogalacturonan was reported in studies on cherry fruit texture by Thibault (1983). A recent study by Lau and co-workers (1985) confirmed the structure of rhamnogalacturonan in the primary cell walls of cultured sycamore cells (*Acer pseudoplatains*). This consisted of a backbone of L-rhamnosyl and D-galacturonic acid with L-arabinosyl and D-galactosyl-rich side chains from O-4 of the branched L-rhamnosyl units (Scheme 2.13).

c. Hemicellulose. The third group of polysaccharides, the hemicelluloses, are a rather poorly defined group of noncellulosic, nonpectic substances extractable by alkaline solution. Three main groups are recognized; xylans, mannans and glucomannans, and galactans and arabogalactan. The hemicellulose xyloglucan

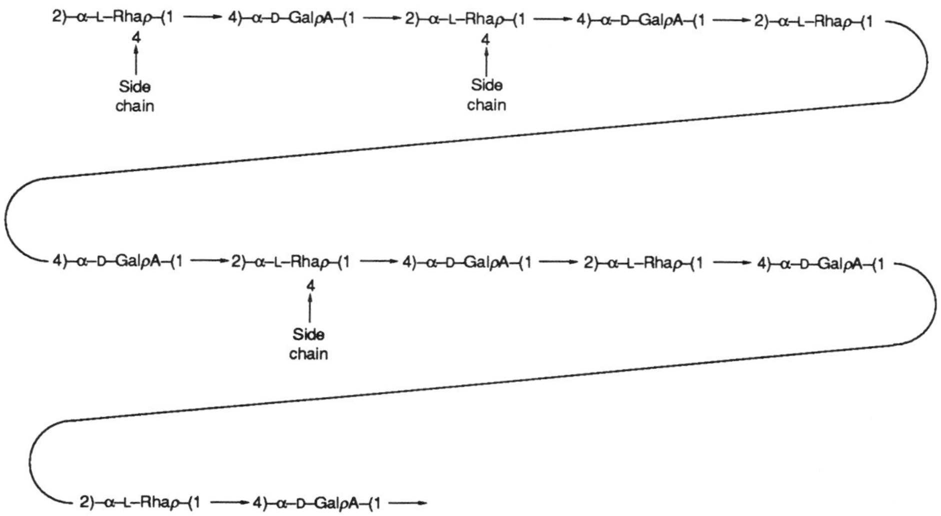

2)-α-L-Rhap-(1 → 4)-α-D-GalpA-(1 → 2)-α-L-Rhap-(1 → 4)-α-D-GalpA-(1 → 2)-α-L-Rhap-(1
4 ↑ Side chain (on first and second Rhap)

4)-α-D-GalpA-(1 → 2)-α-L-Rhap-(1 → 4)-α-D-GalpA-(1 → 2)-α-L-Rhap-(1 → 4)-α-D-GalpA-(1
4 ↑ Side chain (on Rhap)

2)-α-L-Rhap-(1 → 4)-α-D-GalpA-(1 →

SCHEME 2.13. Proposed structure of rhamnogalacturonan (Lau *et al.*, 1985).

appears to be involved in the cross-linking of each cellulose microfibril network through hydrogen bonds as shown in Fig. 2.10.

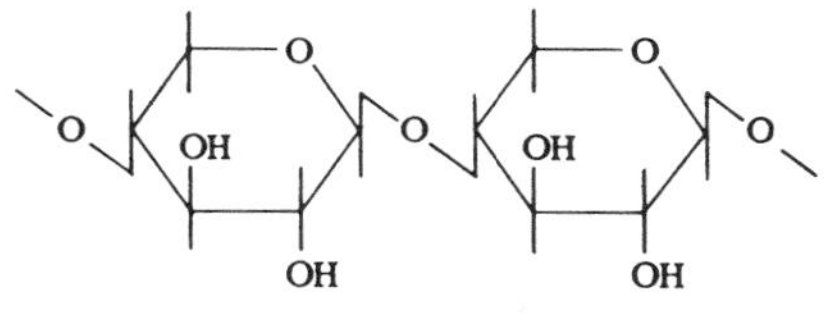

O OH O OH O O OH O OH

Xylan (repeating unit)

d. Lignin. Lignin is synthesized during the formation of the secondary cell wall. It consists of polymers of phenyl propanoid (C_6–C_3) units which provide mechanical strength to the plant cell walls and are responsible for the toughness of such vegetables as celery and asparagus (Herner, 1973; Segerlind and Herner, 1972). A detailed discussion of lignin can be found in several reviews (Eskin, 1979; Grisebach, 1981).

e. Cell Wall Biosynthesis. The biosynthesis of the complex polysaccharides and their deposition in the cell wall is still poorly understood but appears to continue during the senescence period. Evidence for this is based on the incorporation of [^{14}C]methionine into the methyl groups of poly(methyl galacturonate) in ripening apples (Knee, 1978). These results were reinterpreted by Knee and Bartley (1982) in light of research on the methylation of existing pectin. Further studies by Knee (1982) using pear tissue slices showed that incorporation of [$^{14}CH_3$]methionine was associated with new pectin synthesis and not methylation of existing pectin. The new pectin formed was thought to be a direct response to wounding.

The glycosyl donors in the synthesis of cell wall polysaccharides are the sugar nucleotides. These are formed from monosaccharides and ATP in the presence of a nucleoside triphosphate:

$$\text{ATP-glucose} + \text{acceptor} \rightarrow \text{Glycosyl-acceptor} + \text{ADP}$$

These nucleotide esters can be synthesized directly from sucrose and ATP or UTP by the enzyme sucrose synthetase (EC 2.4.1.13). The polymerization of these sugar nucleotides to complex cell wall polysaccharides is poorly understood, although the enzymes involved appear to be membrane bound. Villemez *et al.* (1966) identified a particulate enzyme system from mung bean which could form polygalacturonic acid from UDP-galacturonic acid, which was completely degradable by polygalacturonase. Methylation of the carboxyl groups was demonstrated by Kauss *et al.* (1967, 1969) with *S*-adenosyl-L-methionine as methyl donor. The formation of xylan, galactan, and arabinoxylan from UDP-D-xylose, UDP-D-galactose, and a mixture of UDP-D-xylose and UDP-L-arabinose was observed previously in higher plants (Bailey and Hassid, 1966; McNab *et al.*, 1968; Panyatatos and Villemez, 1973).

The synthesis of a celluloselike (1 → 4) glucan by enzymes is known, although how the cellulose fibril is formed and orients itself remains obscure (Delmer, 1987). Barber *et al.* (1964) identified an enzyme from mung bean that incorporated D-glucose from GDP-D-glucose to a cellulose-like polysaccharide. The formation of the glycoprotein extensin probably occurs via the normal route of protein synthesis in which protein is assembled together on ribosomes (Chrispeels, 1970). The formation of the tetra-L-arabinosyl side chain of extensin involves the sequential transfer of individual arabinose groups to the protein rather than the preformed tetra-arabinosyl unit itself.

2. *Cell Wall Degradation*

The softening of fruits during ripening is attributed to changes in pectin (Kertesz, 1951; Pressey *et al.*, 1971; Hobson, 1967; Tavakoli and Wiley, 1968). This is characterized by a decrease in the level of insoluble pectin (protopectin) with a concomitant increase in soluble pectic substances (Ben-Arie *et al.*, 1979). Protopectin is the generic name ascribed to the water-insoluble parent compound of pectic substances present in unripe fruit. Little is known about this polymer other than that its hydrolysis yields pectin and pectinic acids (Eskin, 1979; Worth,

1967). The latter are polygalacturonic acid polymers varying in the degree of methylation. The pectic material in the middle lamella is quite different from that found in the primary cell walls and consists of pectic substances in the free state or as the corresponding calcium pectate (Eskin, 1979). The degradation of pectin is catalyzed by two groups of enzymes, polygalacturonases (EC 3. 2. 1. 15) and pectin methyl esterases (EC 3. 1. 1. 11) (Eskin, 1979).

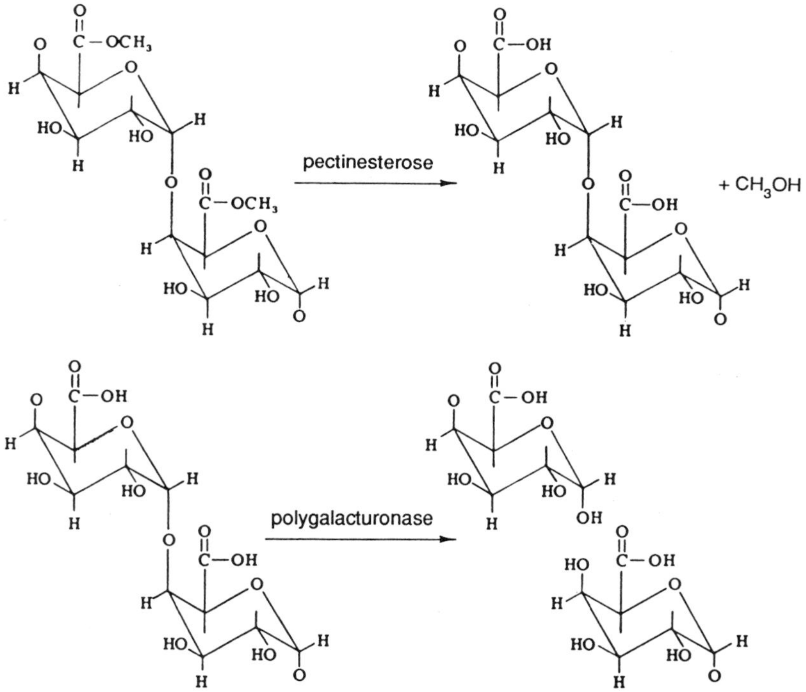

a. Polygalacturonases. Polygalacturonases (PGA) have been implicated in the softening of fruit during ripening. Since the preferred substrate for these enzymes is D-galacturonans, Rexova-Benkova and Markovic (1976) referred to them as D-galacturonases. These enzymes have been reported in many fruits, including peaches (Pressey and Avants, 1973), pears (Bartley *et al.*, 1982; McCready and McComb, 1954; Pressey and Avants, 1976), and tomatoes (Foda, 1957; Hobson, 1964; Patel and Phaff, 1960a,b). The activity of these enzymes increases during the ripening process, when it hydrolyzes pectic material in the middle lamellae and cell walls (Hobson, 1965; Pressey, 1977). The change in polygalacturonase activity during ripening is illustrated in Figure 2. 11 for peaches. Pressey *et al.*

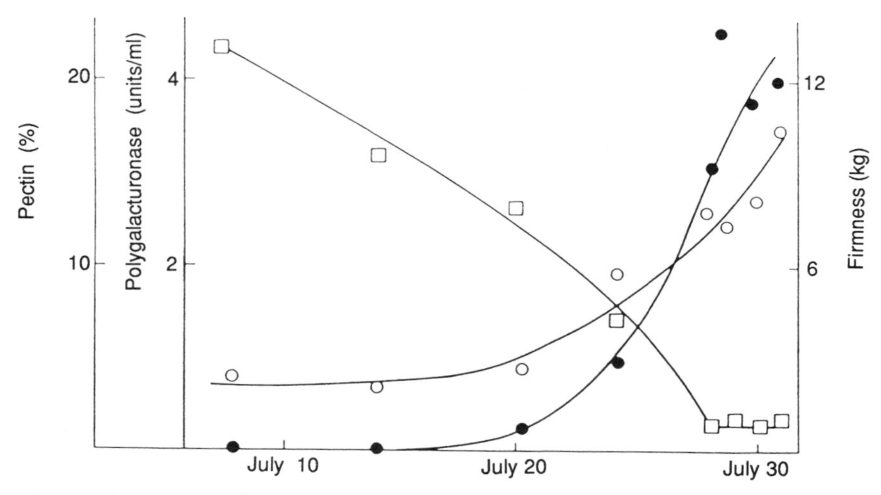

FIG. 2.11. Changes in firmness (□), PGA activity (●), and water-soluble pectin (○) in Elberta peaches during ripening (Pressey *et al.*, 1971). Copyright © by Institute of Food Technologists.

(1971) found the increase in enzyme activity was accompanied by an increase in water-soluble pectin and fruit softening.

Two types of polygalacturonases or D-galacturonases have been identified, endo and exo. The former randomly hydrolyzes the glycosidic bonds in the pectin molecule while the exo-enzyme acts from the terminal end of the pectin molecule (Scheme 2.14). In the presence of endopolygalacturonases the pectin molecules are rapidly degraded into smaller units accompanied by a marked decrease in viscosity. Both forms of the enzyme are found in pears (Bartley and Knee, 1982; Pressey and Avants, 1976) and peaches (Pressey and Avants, 1978). The greater degree of softening observed for Freestone peaches compared to

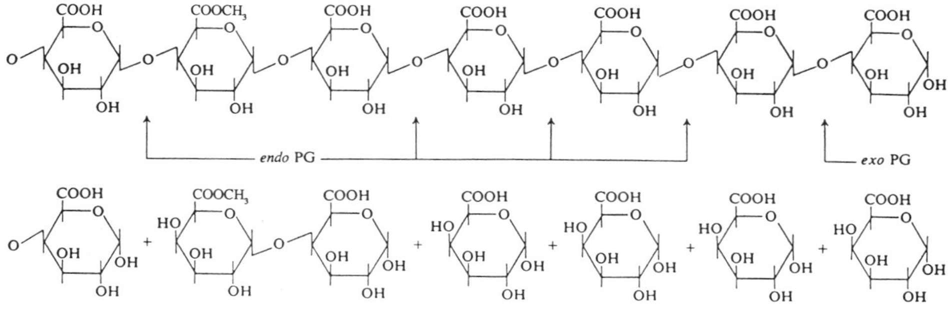

SCHEME 2.14. Action of exo- and endo-polygalacturonases.

Clingstone peaches was attributed to the absence of endopolygalacturonase in the latter fruit. The random degradation of pectin by endopolygalacturonase together with exopolygalacturonase rapidly solubilized the pectin in Freestone peaches. The absence of endopolygalacturonase in Clingstone peaches was evident by the retention of protopectin during ripening.

Exopolygalacturonases (EC 3. 2. 1. 67) have been identified in peaches (Pressey and Avants, 1973), pears (Pressey and Avants, 1976), cucumbers (McFeeters *et al.*, 1980; Pressey and Avants, 1975), and bananas (Markovic *et al.*, 1975). Exopolygalacturonase is the only D-galacturonase found in apples and is responsible for the release of galacturonic acid and polyuronides (Bartley, 1978). Studies using cell wall isolates from tomatoes (Gross and Wallner, 1979; Themmen *et al.*, 1982; Wallner and Bloom, 1977) and pears (Ahmed and Labavitch, 1980) demonstrated that endopolygalacturonases played the major role in pectin degradation during ripening. The release of a water-soluble polymer (WSP) of molecular weight 20, 000 containing galacturonic acid and rhamnose by endopolygalacturonase from the cell walls of red tomatoes was reported by Gross and Wallner (1979) to be identical with the polymer released by the same enzyme from the cell walls of mature green tomatoes. While the activity of polygalacturonases in ripe red tomatoes is predominantly the endo form, it is the exo form which is mainly present in the corresponding green fruit (Pressey and Avants, 1973; Tucker *et al.*, 1980). Although exopolygalacturonase represents only a small fraction of polygalacturonase activity, it was present throughout the ripening of tomatoes at fairly constant levels. Pressey (1987) suggested, therefore, that it was unlikely that exopolygalacturonase played an important part in pectin degradation but might have some role in the growth and development of the tomato fruit.

b. Pectin Methyl Esterase: Pectinesterase. Pectin methyl esterases (PME), or pectinesterases, are widely distributed in many fruits, including bananas (Buescher and Tigchelaar, 1975), peaches (Nagel and Patterson, 1967), and strawberries (Barnes and Patchett, 1976). Considerable confusion surrounds the early studies on changes in PME activity during ripening. For example, Hultin and Levine (1965) noted a rise in PME activity during the ripening of bananas which was not observed by De Swardt and Maxie (1967) when they used polyvinylpyrrolidine (PVP) to remove polyphenols. Brady (1976) also found very little change in PME activity in banana extracts following the addition of 2-mercaptobenzothiazole, a potent inhibitor of banana polyphenol oxidase. PVP was later found by Awad and Young (1980) to suppress PME activity although no inhibitory effects were exerted by the endogenous phenols. PME does not appear to have a major role in fruit softening as it is present at high levels in underdeveloped fruit such as tomatoes and bananas prior to ripening (Barnes and Patchett, 1976; Brady, 1976; Pressey and Avants, 1982; Tucker *et al.*, 1982). In

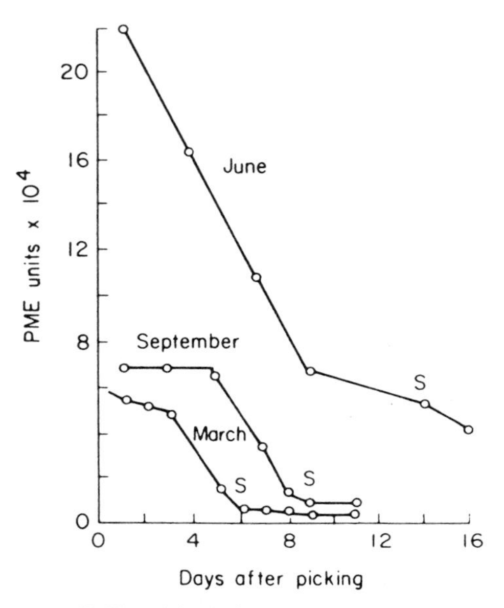

FIG. 2.12. Pectinesterase (PME) activity during storage of avocadoes at various stages of development (S denotes softening of fruits stored at 20°C) (Zaubermann and Schiffman-Nadel, 1972). Reprinted with permission of copyright owner, American Society of Plant Physiology (ASPP).

the case of avocadoes, however, there is a dramatic drop in PME activity just prior to ripening which was reported to be a useful index of fruit maturity (Figure 2.12) (Zauberman and Schiffman-Nadel, 1972). A decrease in PME activity in avocadoes by as much as 50% prior to the climacteric proved to be a reliable indicator of softening time when stored in controlled atmospheres (Barmore and Rouse, 1976). A drop of 80% in PME activity was also reported in avocadoes by Awad and Young (1980).

The traditional view of PME is of de-esterification of the cell wall galacturonans followed by polygalacturonase action. This de-esterification of galacturonans was found by several researchers (Dahodwala *et al.*, 1974; Rexova-Benkova and Markovic, 1976) to enhance the activities of both endo- and exopolygalacturonases. Stimulation of tomato endopolygalacturonases by PME was reported by Pressey and Avants (1980), although these enzymes were capable of hydrolyzing highly esterified substrates at pH 3.5. The presence of a high degree of pectin methylation during the ripening of avocadoes (Dolendo *et al.*, 1966), apples (Knee, 1978), and peaches (Shewfelt *et al.*, 1971) pointed to a rather limited role by PME in fruit softening. Ben-Arie and Sonego (1980) attributed the development of woolly breakdown of peach flesh during cold storage to the inhibition of polygalacturonase activity and enhancement of PME.

This phenomenon was attributed to the inability of peaches to undergo the desirable textural changes associated with normal development. Recent studies by von Mollendorff and De Villiers (1988), however, showed that the primary cause of woolliness in peaches was from the sudden rise in the level of polygalacturonase, while the role of PME was far less clear.

It is evident from these discussions that fruit softening is due to compositional changes in the cell walls of fruit mediated by the combined activity of polygalacturonases and PME. This results in the release of soluble polyuronide with a corresponding decrease in the molecular wight of the polyuronide polymer (Gross and Wallner, 1979; Huber, 1983: Seymour *et al.*, 1987a). The action of polygalacturonase, as discussed previously, may be limited to the demethylated regions of the polygalacturonan, which is brought about by the action of PME. The enhanced production of polyuronides was reported by Pressey and Avants (1982) in isolated cell walls of mature green tomato fruit in the presence of PME. Seymour *et al.* (1987a), using enzyme-inactivated cell wall preparations to eliminate the effect of any endogenous enzymes, found that polyuronide breakdown was much lower for *in vivo* compared to *in vitro* studies. The fact that pectin was not completely de-esterified in spite of high levels of PME suggested that this enzyme may be restricted *in vivo*. Further studies by Seymour *et al.* (1987b) on tomatoes also demonstrated the lower solubilization of polyuronides *in vivo,* which was attributed to the restriction of PME action. The release of two discrete-sized oligomers together with galacturonic acid suggested that the combined pectolytic action was not completely random. These researchers indicated the importance of identifying these oligomers *in vivo* in light of the recent discovery of cell wall elicitors.

c. Cellulase. Cellulose degradation also occurs during the ripening of tomatoes (Babbitt *et al.*, 1973; Pharr and Dickinson, 1973; Sobotka and Stelzig, 1974), strawberries (Barnes and Patchett, 1976), avocadoes (Awad and Young, 1980; Pesis *et al.*, 1978), and Japanese pear fruit (Yamaki and Kakiuchi, 1979). The enzyme involved, cellulase, is composed of several distinct enzymes referred to as the "cellulase complex" (King and Vessal, 1969). These include C_1-cellulase, C_x-cellulase, cellobiase, and exocellulase, which together catalyze the degradation of cellulose as follows:

Insoluble cellulose
$\downarrow$ C_1-cellulase
Soluble cellulose derivatives
$\downarrow$ C_x-cellulase
Cellobiose
$\downarrow$ cellobiase (β-1,4-glucosidase)
Glucose

The degradation of insoluble cellulose to soluble derivatives is poorly understood but appears to involve C_1-cellulase. The breakdown of soluble cellulose is mediated by C_x-cellulase, also referred to by its systematic name 1,4-glucan-4-glucanohydrolase, which randomly cleaves the internal linkages in the cellulose chain. Conflicting reports in the literature suggest that cellulase activity was absent or at very low levels in unripe fruit, whereas others found cellulase activity in immature tomato fruit (Babbitt *et al.*, 1973; Hobson, 1968). Information on the cellulase complex was derived almost exclusively from studies on microbial cellulase. The first study to identify a similar complex in plants was by Sobotka and Stelzig (1974), who partially purified four cellulase fractions from tomato using ammonium sulfate fractionation. These researchers identified C_1-cellulase, C_x-cellulase, β-glucosidase, and cellobiase as the first cellulase complex capable of completely degrading insoluble cellulose in plants. Pharr and Dickinson (1973) were unable to identify an enzyme in tomato fruit that could degrade insoluble cellulose although they did report the presence of C_x-cellulase (EC 3. 2. 1. 4) and cellobiase (EC 3. 2. 1. 21). The presence of a cellulase complex in plants still remains to be established.

The role of cellulase in fruit softening is still somewhat speculative. The only direct evidence with respect to the involvement of cellulase was that reported by Babbitt and co-workers (1973). These researchers investigated the effect of the growth regulators ethephon and gibberellic acid on cellulase and polygalacturonase activities in ripening tomatoes. In the presence of ethephon, cellulase activity increased initially and then declined after 6 days, while polygalacturonase activity increased (Figure 2.13). This contrasted with the almost complete inhibition of polygalacturonase activity by gibberellic acid while cellulase activity continued to increase. The decrease in overall firmness of the tomato fruit in the presence of gibberellic acid pointed to a definite role for cellulase in fruit softening. These researchers proposed that cellulase initiated fruit softening by degrading the cellulose fibrils in the cells walls which permitted pectic enzymes to penetrate the middle lamella. This could explain the observation by Awad and Young (1979), who found that an increase in cellulase activity preceded increases in poylgalacturonase and ethylene production in ripening avocado fruit and subsequent softening (Figure 2.14).

d. β-Galactosidase. The loss of galactose from the cell walls of apples, strawberries, and tomatoes during ripening is the result of β-galactosidase action. The increased activity of this enzyme was correlated with the loss of firmness during the ripening and storage of apples (Bartley, 1974, 1977; Berard *et al.*, 1982; Wallner, 1978). Evidence for this is provided by a decrease in the galactose content of apple cortex cell walls (Knee, 1973), increase in the soluble polyuronide content (Knee, 1975), ability of β-galactosidase to breakdown β-(1 → 4)-linked galactan (Bartley, 1974), and the release of galactose from cell wall preparations (Bartley, 1978). Dick *et al.* (1984) provided preliminary evi-

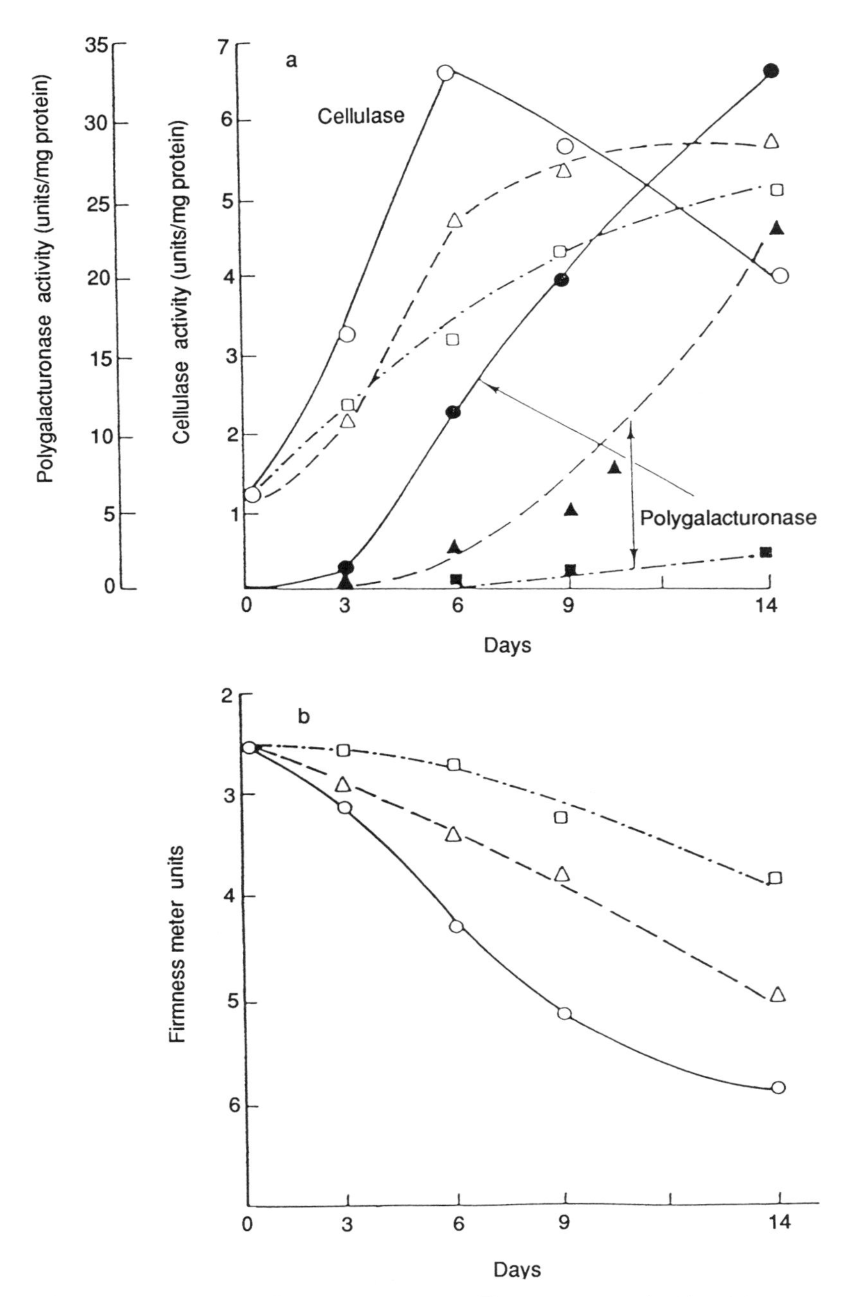

FIG. 2.13. Effect of ethephon (○) and gibberellic acid (□) on enzyme activity (a) and firmness (b) of tomatoes (△, control) (Babbitt *et al.*, 1973).

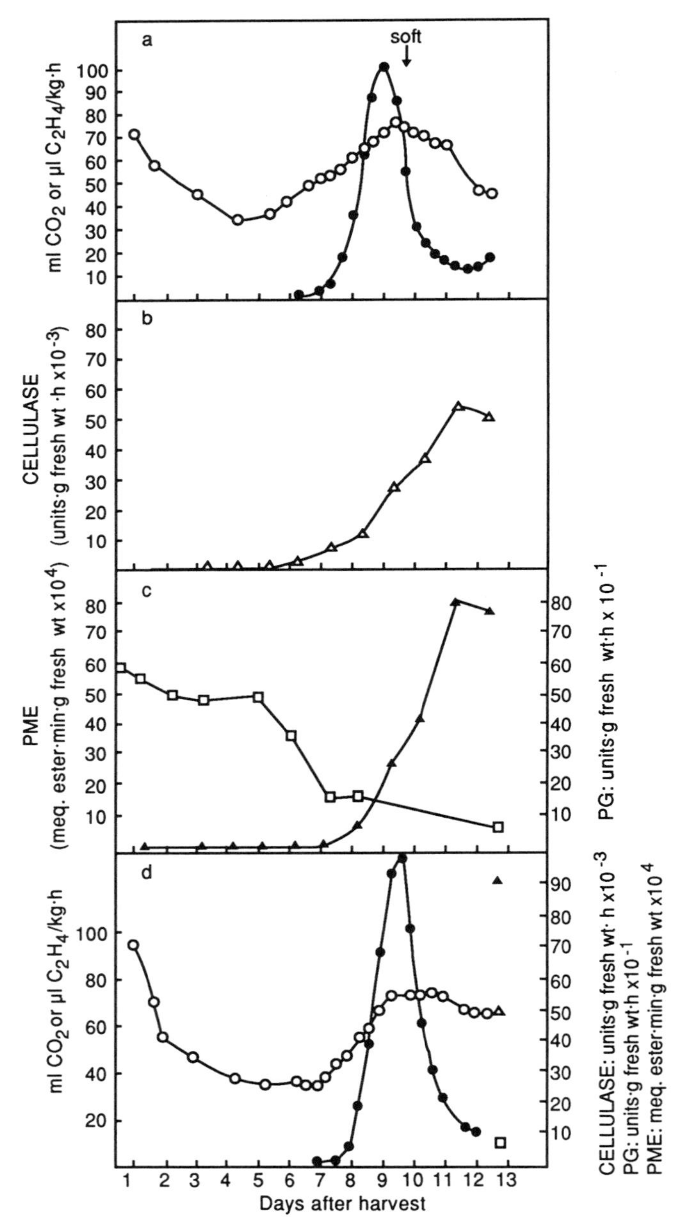

FIG. 2.14. Postharvest trends in cellulase (△), PG (▲), and PME (□) activity and in CO_2 (○) and C_2H_4 (●) production in an individual Fuerte avocado fruit. Fruit in (a) was edibly soft after 9.5 days and in (d) after 10.5 days (Awad and Young, 1979).

dence for the regulation of β-galactosidase activity in McIntosh apples by the presence of an endogenous inhibitor. Unlike β-galactosidase in apples, that found in tomatoes did not appear to be involved in cell wall hydrolysis of galactans (Gross and Wallner, 1979). In fact these researchers were unable to detect any β-galactosidase activity during the ripening of tomatoes. This contrasted with a later study by Pressey (1983), who isolated three enzymes responsible for β-galactosidase activity in tomato. One of these enzymes hydrolyzed tomato galactans and increased in activity during the ripening process. This suggested a possible role for β-galactosidase in fruit softening. Since the galactan polysaccharide in tomato fruit is (1 $\rightarrow$ 4)-linked, the β-galactosidase involved must be β-1, 4-galactosidase. The inability of Gross and Wallner (1979) to detect any galactanase activity was attributed to the preparation of their extracts from frozen fruit. As pointed out by Pressey (1983), the yield of β-galactosidase in frozen fruit is very low compared to that in the corresponding fresh tomato extracts.

VII. Flavor

The flavor of fruits and vegetables is a complex interaction between aroma and taste. Aroma is produced by the volatiles synthesized during fruit ripening and includes aldehydes, alcohols, esters, lactones, terpenes, and sulfur compounds. Taste is provided by many nonvolatile components, including sugars and acids present in the fruit flesh. Vegetables, with a few exceptions, tend to be more bland in flavor.

A. Aroma

Volatiles responsible for aroma originate from proteins, carbohydrates, lipids, and vitamins as shown in Scheme 2.15. The aroma characteristics of individual fruit and vegetable crops develop during ripening and maturation. Many volatiles have been identified in the literature and reviewed by Salunkhe and Do (1976). This section will focus briefly on a few of the biogenic pathways involved in volatile formation.

1. *Aldehydes, Alcohols, and Esters*

Short-chain unsaturated aldehydes and alcohols (C–C) and esters are important contributors to the aroma volatiles of fruits. These are formed during the short ripening period associated with the climacteric rise in respiration (Paillard, 1968; Romani and Ku, 1968; Tressl *et al.*, 1970). Studies on bananas and

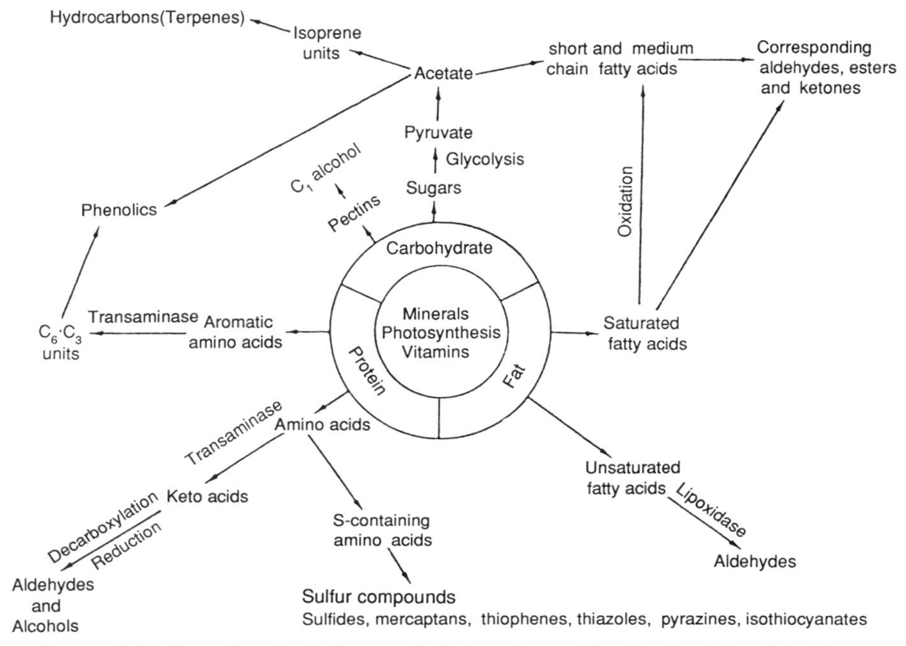

SCHEME 2.15. Formation of volatile aroma in fruits and vegetables. Reprinted from Salunkhe and Do (1976). With permission.

tomatoes have shown them to be synthesized from amino acids or fatty acids (Eskin, 1979; Eskin *et al.*, 1977).

a. Amino Acids. An increase in 3-methyl-1-butanol, isopentyl acetate, isopentyl butyrate, and isovalerate volatiles was reported by Dalal (1965) during tomato ripening. At the same time, 3-methyl-1-butanal increased up to the breaker stage and then decreased. The similarity between the alcohol portion of these esters with the carbon skeleton of 3-methyl-1-butanal suggested to Yu and co-workers (1968a) that they were synthesized from this aldehyde. Since leucine had an identical carbon skeleton with that of 3-methyl-1-butanal, the possible role of this amino acid in the synthesis of this aldehyde was investigated by Yu *et al.*, (1968c). On the basis of their work with L-[^{14}C]leucine, the following pathway was proposed:

$$\underset{\text{L-Leucine}}{CH_3-CH(CH_3)-CH_2-CH(NH_2)-COOH} \longrightarrow \underset{\text{3-Methyl-1-butanal}}{CH_3-CH(CH_3)-CH_2-CHO} \longrightarrow \underset{\text{3-Methyl-1-butanol (isoamyl alcohol)}}{CH_3-CH(CH_3)-CH_2-CH_2OH}$$

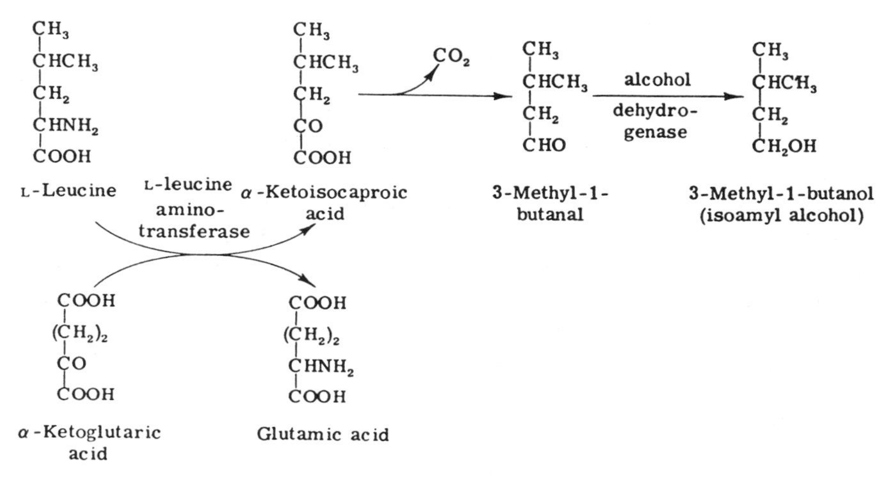

SCHEME 2.16. Biogenesis of isoamyl alcohol from L-leucine in tomato fruit.

In addition to leucine, aspartic acid and alanine were also shown to produce carbonyl compounds when added to tomato extracts (Yu *et al.*, 1968c). The decrease in the levels of these amino acids during ripening pointed to the presence of enzymes capable of utilizing them. The soluble fraction obtained by centrifugation of tomato extracts was particularly active on leucine, while aspartic acid and alanine were metabolized by the mitochondrial fraction. Based on the detection of large amounts of glutamic acid in tomato extracts, Freeman and Woodbridge (1960) and Yu *et al.* (1967) both pointed to the presence of active transaminases. This was confirmed in a subsequent study by Yu *et al.* (1968b), who found a marked production of glutamic acid when extracts of field-grown tomatoes at the green and ripe stages were incubated with these amino acids. Yu and Spencer (1969) incubated L-leucine with fresh tomato extracts and isolated α-keto-isocaproic acid among the products formed. Confirmation with labeled [^{14}C]leucine established the presence of L-leucine : 2 keto-glutarate amino transferase in tomatoes. Scheme 2.16 summarizes the reactions leading to 3-methyl-1-butanol from L-leucine.

b. Fatty Acids. The volatile carbonyls responsible for the aroma of tomatoes and bananas are synthesized from unsaturated fatty acids (Goldstein and Wick, 1969; Jadhav *et al.*, 1972). The major unsaturated fatty acids in the pericarp of tomatoes were shown to be oleic, linoleic, and linolenic acids (Kapp, 1966). As the tomato fruit ripened, Jadhav and co-workers (1972) reported a decrease in the levels of both linoleic and linolenic acids. A marked decrease in linoleic acid was also observed by Goldstein and Wick (1969) in ripe banana pulp, which suggested a possible relationship between this fatty acid and the production of volatiles.

Incorporation of ^{14}C-labeled linoleic and linolenic acids into hexanal using tissue slices or cell-free tomato extracts suggested to Jadhav *et al.* (1972) the involvement of lipoxygenase. This was confirmed by the total inhibition of carbonyls in the presence of hydrogen peroxide, a recognized inhibitor of lipoxygenase. Consequently this enzyme was monitored during tomato ripening and found to increase in activity at the onset of the climacteric. This appeared to explain the increase in volatiles that accompanied maturation of the tomato fruit (Dalal *et al.*, 1968). Kazeniac and Hall (1970) reported the presence of higher levels *cis*-3-hexenal, *trans*-2-hexenal and *n*-hexanol in fully ripened tomato fruit. The formation of *trans*-2-hexenal resulted from the instability of *cis*-3-hexenal to the acidic pulp and juice of the tomato, with isomerization to the *trans* isomer. Stone *et al.* (1975) showed that *cis*-3-hexenal was the major volatile of tomato distillates, while Jadhav and co-workers (1972) found *n*-hexanol to be the major volatile formed. This discrepancy suggested to Stone and co-workers (1975) that *cis*-3-hexenal was a precursor of *n*-hexanol, but this was discounted when only 2% of *cis*-3-[^{14}C]hexenal was incorporated into the alcohol form. Jadhav and co-workers (1972) attributed the presence of hexanol, propanol, 2, 4-decadienal, 2, 6-heptadiene, and *cis*-3-hexenal in tomato volatiles to the formation of 9-, 12-, and 16-hydroperoxides by the action of lipoxygenase on linoleic and linolenic acids. Gaillard and Matthew (1977), however, reported that the major fatty acid hydroperoxides formed from linoleic and linolenic acids were 9- and 13-hydroperoxides in a ratio of 95 : 5. Of these only the 13-hydroperoxide was cleaved to form the nonvolatile compound 12-oxo-dodec-*cis*-9-enoic acid together with hexanal and *cis*-3-hexenal from linoleic and linolenic acids, respectively (Scheme 2.17). Zamora *et al.* (1987) characterized lipoxygenase from tomato fruit and confirmed the 9-hydroperoxide isomer to be the major one formed from linoleic acid. The ratio of 9- to 13-hydroperoxide isomers produced from linoleic acid was found to be 24 : 1, in close agreement with that reported previously by Gaillard and Matthew (1977).

Buttery and co-workers (1987) developed improved trapping methods for the quantitative analysis of the major C_4–C_6 volatiles in tomato fruit. In addition to inactivating tomato enzymes which effected the volatiles during isolation, they also overcame the problem of isomerization of *cis*-3-hexenal to *trans*-2-hexenal reported by Kazeniac and Hall (1970). Using their procedure, which involved Tenax trapping and $CaCl_2$ enzyme deactivation, they identified *cis*-3-hexenal among the major volatiles present. These researchers attributed the lack of flavor in tomatoes purchased in the supermarket to the lower levels of *cis*-3-hexenal present compared to the higher levels present in vine-ripened tomatoes. In addition, they reported that storing tomatoes in the refrigerator caused further loss of flavor in part because of lower levels of *cis*-3-hexenal. This effect of cold storage on tomato flavor was in agreement with previous work by Lammers (1981). In addition to *cis*-3-hexenal other important flavor volatiles were β-ionone, 1-penten-

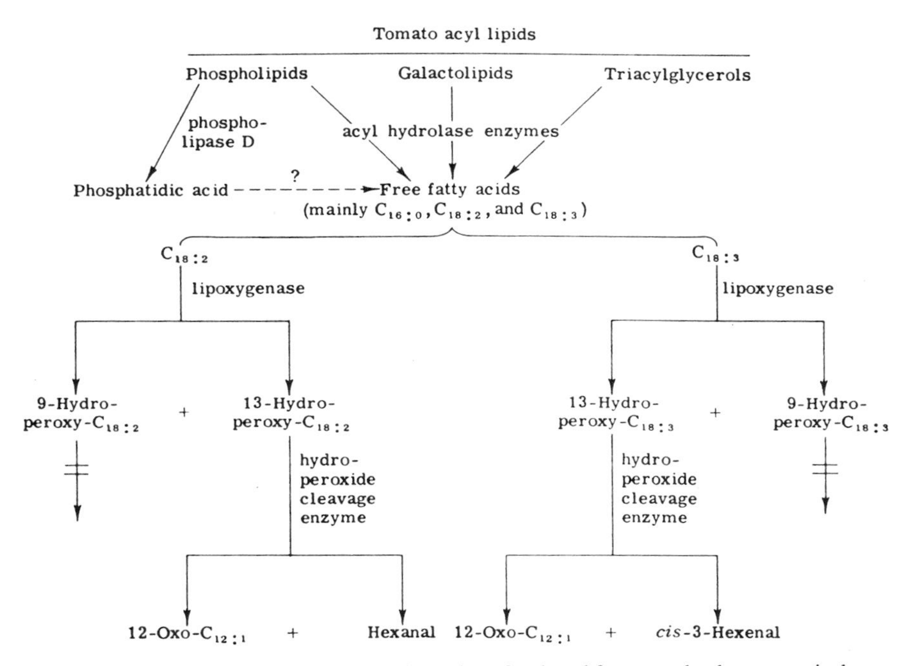

SCHEME 2.17. Proposed pathway for the formation of carbonyl fragments by the enzymatic degradation of acyl lipids in disrupted tomato fruits (Galliard *et al.*, 1977).

3-one, hexanal, *cis*-3-hexanol, *trans*-2-hexanal, 2- and 3-methylbutanol, 2-isobutylthiazole, and 6-methyl-5-hepten-2-one.

trans-2-Hexenal was also identified among the volatiles of *Gros Michel* bananas by Issenberg and Wick (1963). Using a volatile enrichment technique, Tressl and Jennings (1972) confirmed the presence of *trans*-2-hexenal in the headspace of ripening bananas. Separation of volatile fractions from bananas by Palmer (1971) showed the presence of *cis*-3-hexenal, *trans*-2-hexenal, and *n*-hexanal among the aldehydes formed. Tressl and Drawert (1973) found that green banana homogenates produced *trans*-2, *cis*-6-nonadienal, *trans*-2-nonenal, and 9-oxanonanoic acid, similar to that reported in cucumbers (Fleming *et al.*, 1968). Tressl and Drawert (1973) detected hexanal, *trans*-2-hexenal, and 12-oxo-*trans*-10-dodecenoic acid when green bananas were exposed to ethylene and stored for 4 days at 15°C. Incorporation of ^{14}C-labeled linoleic and linolenic acids into these volatiles pointed to the involvement of lipoxygenase. Labeled 13- and 9-hydroperoxyoctadecadienoic acids incubated with crude banana extracts were converted to C_6–C_9 aldehydes as shown in Scheme 2.18. Aldehyde lyase, the enzyme responsible for the breakdown of the hydroperoxy derivatives, was also identified in germinating watermelon seedlings by Vick and Zimmer-

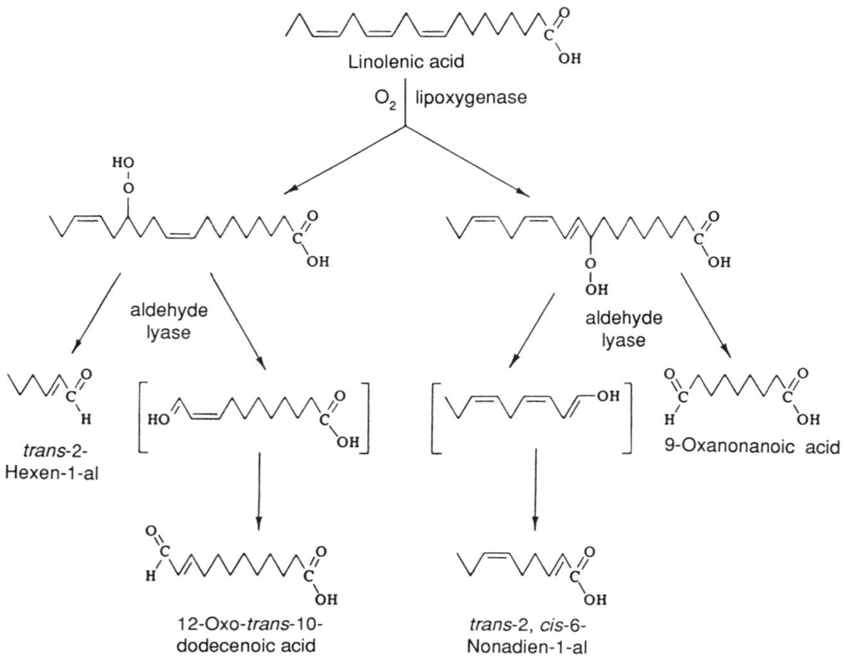

Linolenic acid

O_2 lipoxygenase

aldehyde lyase

aldehyde lyase

trans-2-Hexen-1-al

9-Oxanonanoic acid

12-Oxo-*trans*-10-dodecenoic acid

trans-2, *cis*-6-Nonadien-1-al

SCHEME 2.18. Reaction scheme for enzymatic splitting of linolenic acid into aldehydes and oxo acids. Reprinted with permission from Tressl and Drawert (1973). Copyright by the American Chemical Society.

man (1976). This enzyme catalyzed the formation of 12-oxo-*trans*-10-dodecenoic acid and hexanal from 13-hydroperoxy-*cis*-9, *trans*-11-octadecanoic acid. This enzyme differed from hydroperoxide cleavage enzyme in tomatoes by producing *trans*-2-enals as the primary products.

Studies on apple volatiles by Flath *et al.* (1967) identified 20 compounds from Delicious apple essence, including hexanol and *trans*-2-hexenal. Kim and Grosch (1979) partially purified lipoxygenase from apple homogenates, which produced 13-hydroperoxyoctadeca-9, 11-dienoic acid from linoleic acid. The latter was converted to 2-hexenal and hexanol in a similar manner as described previously.

B. TASTE

The characteristic taste of fruits is determined by the content of sugars and organic acids. The ratio of sugar/acid is particularly useful as an index of ripeness for many fruits. The level of sucrose and L-malic acid together with protein patterns was suggested by Gorin (1973) as parameters for assessing the quality of Golden Delicious apples. Hammett and co-workers (1977) found a high correlation between the ratio of soluble solids to acid content with days from full bloom (DFFB) for Golden Delicious apples. Sugars and acids not only contributed to sweetness and sourness in tomatoes but were responsible for the overall flavor intensity (Jones and Scott, 1983; Kader *et al.*, 1977; Stevens *et al.*, 1979).

In addition to these components, the presence of tannins, phenolic compounds classified into hydrolyzable and nonhydrolyzable, also affects taste. Unlike the hydrolyzable tannins, which yield gallic acid and glucose by enzymatic hydrolysis, the nonhydrolyzable tannins are resistant to enzymatic hydrolysis. The latter appears to be responsible for astringency in many underripe fruits. The loss of astringency in persimmon was attributed by Matsuo and Itoo (1982) to immobilization of tannin with acetaldehyde formed during ripening. No substantial changes in the composition or amount of polyphenols in the fruit of *Rubus* sp. were detected during ripening by several researchers (Haslam *et al.*, 1982; Okuda *et al.*, 1982a,b). On the basis of studies with model systems, Ozawa *et al.* (1987) proposed that the loss of astringency during fruit ripening was due, in part, to the possible interaction between polyphenols and proteins in the fruit.

1. *Starch-Sugar Conversion*

Sugars and transient starch are synthesized in the growing plant by photosynthesis. This is translocated, mainly in the form of sucrose, from the chloroplasts *via* the phloem to the growing cells in the plant, where it is resynthesized into starch. This sucrose–starch conversion appears to involve the sequence of reactions shown in Scheme 2.19.

During the postharvest period, starch is transformed into sucrose, glucose, and fructose. This is affected by the physiological condition of the fruit and vegetables as well as storage temperature and time. Starch hydrolysis is among the more conspicuous changes accompanying ripening of many climacteric fruit. For example, a drop in starch content from 22 to 1% was reported by Palmer (1971) to accompany bananas as they passed from the preclimacteric to the climacteric phase. The presence of phosphorylase and amylase enzymes was observed in the storage tissues of many fruits, although their respective roles in the ripening process remained unclear. Young *et al.* (1974) observed an increase in α-amylase activity during the ripening of bananas but were unable to confirm its involvement in starch hydrolysis. β-Amylase and phosphorylase were also found during ripening but the presence of enzyme inhibitors impeded their assay during the preclimacteric phase. Yang and Ho (1958) suggested that phosphorylase had a role in starch degradation during the climacteric. This was later confirmed in gamma-irradiated Cavendish bananas by Surendranathan and Nair (1973). Three phosphorylases were partially purified from ripe bananas by Singh and Sanwall (1973, 1975, 1976), each exhibiting different biochemical properties.

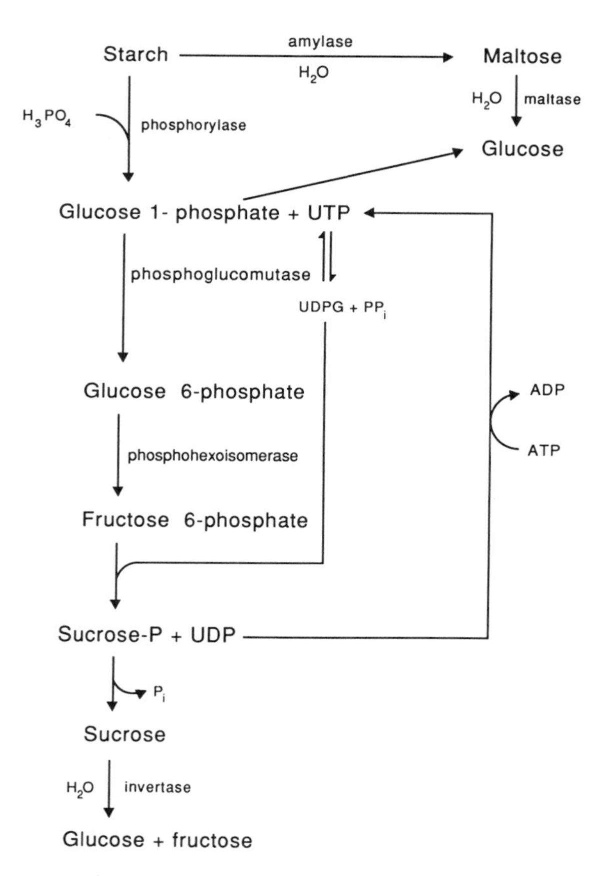

SCHEME 2.19. Starch–sugar conversion.

Starch hydrolysis was examined by Chitarra and LaJollo (1981) during the ripening of hybrid Marmello bananas. This hybrid exhibited abnormal behavior as the peel, color, aroma, and texture did not undergo changes normally associated with ripening. In addition, the starch content dropped to 5% during ripening because of the possible presence of inhibitors. Raising the storage temperature from 20 to 25°C accelerated the climacteric and reduced the time required to stabilize the starch level at 5% from 24.5 to 8 days. The starch level in bananas could be further reduced to 3.3% by storing at 30°C, however, the resulting fruit was overripe and deteriorated rapidly. This contrasted with Dwarf Cavendish bananas, in which complete starch degradation (97.7%) occurred during the climacteric. The lower sugar content in the *Marmello* banana was responsible for the difference in taste compared to the corresponding *Cavendish* variety. Phosphorylase activity in *Marmello* bananas was reported to remain constant during the preclimacteric although it paralleled the change in starch content. This differed from *Cavendish* fruit, in which a 50% increase in phosphorylase activity was reported prior to the climacteric and preceded any starch degradation. These results pointed to the possible involvement of phosphorylase in starch–sucrose transformations as well as sucrose synthetase (Areas and Lajolo, 1981). The exclusive involvement of α-amylase in starch degradation was seriously questioned since it occurred prior to any increase in α-amylase activity. The inhibition of α-amylase synthesis by cycloheximide during banana ripening did not block starch degradation, which supported a role for phosphorylase.

Phosphorylase activity has been associated with the cold storage of potato tubers at 4°C (Hyde and Morrison, 1964). This enzyme is responsible for the initial reaction in the cold-induced sweetening of potatoes (Isherwood, 1976). The overall result is a marked increase in the sugar content of the potato, resulting in the production of chips with unacceptable dark coloration (Talburt and Smith, 1975). To prevent this, potatoes are generally stored at or above 10°C, referred to as "conditioning." The biochemical mechanism involved in these starch/sugar transformations still remains unclear although the effect of storage temperature on a number of enzymes has been reported (Dixon and ap Rees, 1980; Isherwood, 1976; Kennedy and Isherwood, 1975; Pollock and ap Rees, 1975). Increase in sugar phosphates and sucrose was attributed by Pollock and ap Rees (1975) to the cold lability of some of the glycolytic enzymes. One of these, phosphofructokinase, was shown to be a major cause of low-temperature sweetening of potato tubers (Dixon *et al.*, 1981). The cold lability of this enzyme appeared to be due to denaturation of its oligomeric enzyme complex and its dissociation into subunits. This resulted in the inability of the enzyme to oxidize hexose phosphates, with their subsequent accumulation and conversion to sucrose.

2. *Organic Acids*

Fruit ripening is accompanied by changes in organic acids. These reach a maximum during the growth and development of the fruit on the tree, but decrease during storage as well as being highly dependent on temperature. The Krebs cycle is active in the cells of higher plants in generating a variety of organic acids, including citric, malic, and succinic acids. Citric and malic acids are important constituents of most fruits, with oranges, lemons, and strawberries high in citric acid and apples, pears, and plums high in malic acid. During ripening these organic acids decrease, which is accompanied by a decrease in starch content and an increase in sugars responsible for fruit sweetness. Akhavan and Wrolstadt (1980) reported that ripening Bartlett pears attained a maximum sugar content of 13.5% acidity or 6 milliequivalents and flesh firmness of 6 lbs

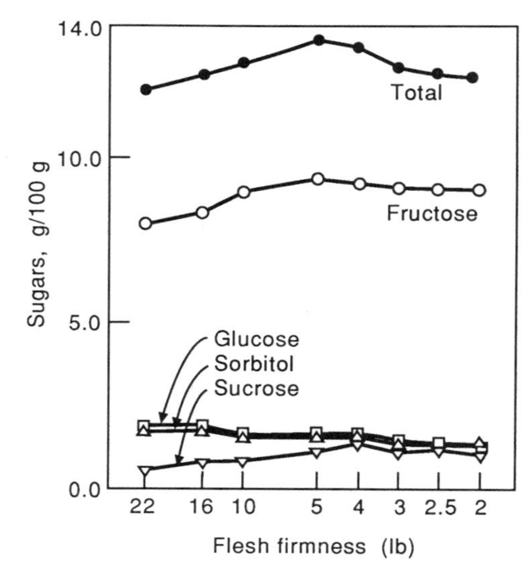

FIG. 2.15. Changes in sugars during the ripening of Bartlett pears (Akhavan and Wrolstad, 1980). Copyright © by Institute of Food Technologists.

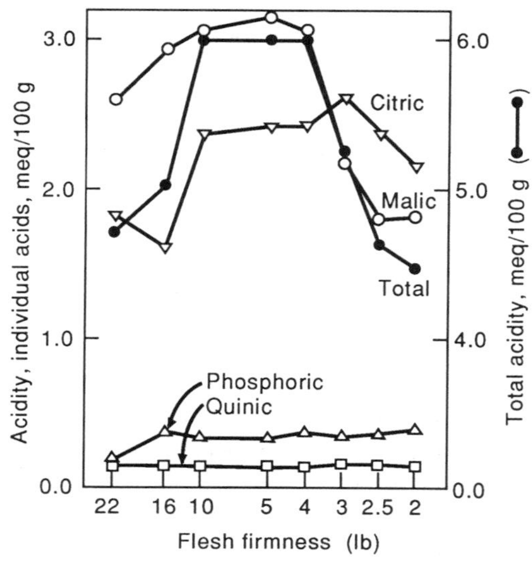

FIG. 2.16. Changes in acids during the ripening of Bartlett pears (Akhavan and Wrolstad, 1980). Copyright © by Institute of Food Technologists.

on the fourth day of ripening (Figure 2.15). Malic and citric acids accounted for the major changes in total acidity both prior to and following ripening (Fig. 2.16). These organic acids contribute to the pH in tomatoes, which is particularly important in processing (Davies and Hobson, 1981). Picha (1987) reported citric and malic acids to be the major organic acids in cherry tomato fruit. During ripening, changes in these organic acids were evident, with citric acid increasing from the immature green to mature green stage while malic acid decreased from the mature green to the table-ripe stage.

In addition to these organic acids, ascorbic acid is also prominent in fruits. It is present in plant tissues, mainly in the reduced form, but can be oxidized to dehydroascorbic acid by the action of ascorbic acid oxidase (see Chapter 10). The presence of L-quinic acid as a minor organic acid has also been reported in a number of fruits including pears (Akhavan and Wrolstad, 1980). Other organic acids reported include oxalic and citramalic acids.

VII. Storage

To ensure that an adequate supply of fruits and vegetables is available for the consumer as well as the food industry, a number of preservation methods have been developed. The oldest of these involves cold storage of the perishable produce, while the other is controlled atmosphere storage. Both these techniques are expensive in terms of storage facilities and maintenance. A new postharvest method for preserving perishable produce was developed recently by Ben-Yehoshua (1985).

A. Cold Storage

The oldest and most popular method for prolonging the shelf life of perishable produce is cold storage. This is a relatively cheap technique which is used for storing potatoes in cooler regions of the world. Cappellinni and co-workers (1984) showed that the recommended storage conditions of 7.2–10°C and 90–95% relative humidity by the U.S. Department of Agriculture (1968) for green bell peppers was still the optimal and cheapest method for maintaining quality. Of particular concern worldwide has been the long-term storage of onions over the winter period. A temperature range between −1 and +3°C has generally been recommended for extending the storage life of onions together with 70–75% relative humidity and good ventilation. Hanaoka and Ito (1957) reported that a high soluble sugar content in onion bulbs was a good indicator of their storage potential. Later researchers also found that the metabolism of carbohydrates in onions was closely linked to their storage performance (Kato, 1966; Toul and Popsilova, 1966). A more detailed study on carbohydrate metabolism in stored onions by Rutherford and Whittle (1982) showed that increased fructose resulted from the hydrolysis of the storage oligosaccharides. A low fructose content in freshly harvested onion bulbs indicated poor storage properties with the reverse true for onion bulbs high in fructose. Measurement of alkaline invertase by Rutherford and Whittle (1984) in onions showed that its activity reflected the level of fructose present. Monitoring fructose content and invertase activity in harvested onions provided a useful indicator for assessing their stability during cold storage.

B. Controlled Atmosphere Storage

Controlled atmosphere storage (CA) is carried out under modified or controlled atmosphere conditions to reduce or retard biochemical processes such as respiration, ripening, and yellowing of fruits and vegetables. The modified environment contains low levels of oxygen and high levels of carbon dioxide, which slow down the catabolic processes and ageing, thereby extending the storage life of fruits and vegetables.

The overall benefit is that produce can retain its freshness and eating quality for much longer periods of time. Not all fruits and vegetables, however, can be stored under these conditions. Considerable research has been conducted on a wide range of produce but apples are still the major fruit handled commercially under controlled atmosphere storage conditions (Blanpied, 1977; Blanpied *et al.*, 1982; Fidler *et al.*, 1977; Padfield, 1975; Porritt, 1977). Other commodities stored or shipped commercially under controlled atmospheres include cabbage, lettuce, avocadoes, asparagus, and bananas (Issenberg *et al.*, 1971; Lipton, 1977). One of the major problems associated with controlled atmosphere storage is that different conditions are required for the storage of the different cultivars. For example, apples have been extensively studied with lists of recommended conditions for specific cultivars available from different regions of the United States, England, and Europe (Blanpied, 1977; Fidler *et al.*, 1977; Porritt, 1977; Stoll, 1970, 1973). This makes controlled atmosphere very expensive as conditions have to be tailor-made for the different cultivars of the same crop as well as for the different fruits and vegetables.

C. New Developments in Postharvest Storage

A new postharvest technique was introduced by Ben-Yehoshua (1978) based on individual packaging of fruit to create a microatmosphere. The particular crop initially studied by Ben-Yohoshua (1969) was citrus fruit, in which a reduction of transpiration was reported which extended its shelf life. This method uses a high-density polyethylene film to seal-package the citrus fruit and so form a barrier resistant to water vapor. This process not only extended the shelf life but reduced shrinkage, weight loss, and incidence of blemishes and represented considerable economic savings compared to refrigerated or controlled atmosphere storage (Ben-Yehoshua, 1985). The applicability of this technique to other fruits and vegetables has considerable potential. The use of plastics to maintain the freshness of fruits was reviewed by Soniso (1986) based on research conducted by Geeson and co-workers at the Institute of Food Research, Norwich, England. These researchers found that packaging fruits in a pack sealed with polyvinylchloride (PVC) or butadiene-styrene copolymer film (20–25 mm thick) resulted in a controlled atmosphere in which oxygen and carbon dioxide accounted for 5%. Under these conditions ripening was retarded, with the result that fruits such as tomatoes and strawberries remained fresh for much longer periods. Smith and co-workers (1987) examined the use of films and coatings to modify the atmospheres in apples for preserving postharvest quality. Coatings included a thin layer of wax or oil on the surface of the fruit as an addition or replacement of the natural waxy coating of the fruit. The films consisted of extruded plastic materials which not only restricted water loss but modified the atmosphere over the fruit. These techniques should help the processor meet the consumer demands for fresh produce.

Bibliography

Abeles, F. B. (1973). "Ethylene in Plant Biology." Academic Press, New York.

Acaster, M. A., and Kende, H. (1983). Properties and partial purification of ACC synthetase. *Plant Physiol.* **72,** 139.

Adams, D. O., and Yang, S. F. (1977). Methionine metabolism in apple tissue: Implication of *S*-adenosylmethionine as an intermediate in the conversion of methionine to ethylene. *Plant Physiol.* **60,** 892.

Adams, D. O., and Yang, S. F. (1979). Ethylene biosynthesis. Identification of 1-aminocyclopropane-1-carboxylic acid as an intermediate in the conversion of methionine to ethylene. *Proc. Natl. Acad. Sci. U.S.A.* **76,** 170.

Aharoni, Y. (1968). Respiration of oranges and grapefruits harvested at different stages of development. *Plant Physiol.* **43,** 99.

Ahmed, A. E., and Labavitch, J. M. (1980). Cell wall metabolism in ripening fruit. II. Changes in carbohydrate-degrading enzymes in ripening "Bartlett" pears. *Plant Physiol.* **65,** 1014.

Akamine, E. K., and Goo, T. (1978). Respiration and ethylene production in Mammee apple (*Mammea americana* L). *J. Am. Soc. Hortic. Sci.* **103,** 308.

Akamine, E. K., and Goo, T. (1979a). Concentrations of carbon dioxide and ethylene in the cavity of attached papaya fruit. *HortScience* **14,** 138.

Akamine, E. K., and Goo, T. (1979b). Respiration and ethylene production in fruits of species and cultivars of *Pridium* and species of *Eugenia*. *J. Am. Soc. Hortic. Sci.* **104,** 632.
Akhavan, I., and Wrolstad, R. E. (1980). Variation of sugars and acids during ripening of pears and in the production and storage of pear concentrate. *J. Food Sci.* **45,** 499.
Aljuburi, H., Huff, A., and Hshieh,M. (1979). Enzymes of chlorophyll catabolism in orange flavedo. *Plant Physiol.* **63,** S-73.
Amrhein, N. (1978). Novel inhibitors of phenylpropanoid metabolism. *Proc. FEBS Meet.* **12.**
Apelbaum, A., Burgoon, A. C., Anderson, J. D., Solomos, T., and Lieberman, M. (1981). Some characteristics of the system converting aminocyclopropane-1-carboxylic acid to ethylene. *Plant Physiol.* **67,** 80.
ap Rees, T. (1974). Pathways of carbohydrate breakdown in higher plants. *MTP Int. Rev. Sci.: Biochem. Ser. One* (D. H. Northcote, ed.), **11,** 51. Butterworth, U.K.
ap Rees, T. (1977). Conservation of carbohydrate by non-photosynthetic cells of higher plants. *Symp. Soc. Exp. Biol.* **31,** 7.
ap Rees, T. (1980). Assessment of the contributions of metabolic pathways to plant respiration. *In* "The Biochemistry of Plants" (D. D. Davies, ed.), Vol. 2, pp. 1–29. Academic Press, New York.
Areas, J. A. C., and LaJollo, F. M. (1981). Determinacao enzimatica especifica de amido, glicose, fructose e sacarose em bananas preclimactericas e climactericas. *An. Farm. Quim. Sao Paulo* **20**(1/2), 307.
Arens, D., Seilmeier, W., Weber, F., Kloos, G., and Grosch, W. (1973). Purification and properties of a carotene cooxidizing lipoxygenase from peas. *Biochim. Biophys. Acta* **327,** 295.
Awad, M., and Young, R. E. (1979). Postharvest variation in cellulase, polygalacturonase, and pectinmethylesterase in avocado (*Persea americana* Mill, cv. Fuerte) fruits in relation to respiration and ethylene production. *Plant Physiol.* **64,** 306.
Awad, M., and Young, R. E. (1980). Avocado pectinmethylesterase activity in relation to temperature, ethylene and ripening. *J. Am. Soc. Hortic. Sci.* **105,** 638.
Babbitt, J. K., Powers, M. J., and Patterson, M. E. (1973). Effects of growth-regulators on cellulase, polygalacturonase, respiration, color and texture of ripening tomatoes. *J. Am. Soc. Hortic. Sci.* **98,** 77.
Bacon, M. F., and Holden, M. (1970). Chlorophyllase of sugarbeet leaves. *Phytochemistry* **9,** 115.
Bailey, B. W., and Hassid, W. Z. (1966). Xylan synthesis from uridine-diphosphate-D-xylose by particulate preparations from immature corncobs. *Proc. Natl. Acad. Sci. U.S.A.* **56,** 1586.
Bain, J. A. (1958). Morphological and physiological changes in the developing fruit of Valencia orange (*Citrus sinesis* L.) Osbeck. *Aust. J. Bot.* **6,** 1.
Barber, G. A., Elbein, A. D., and Hassid, W. Z.1964. The synthesis of cellulose by enzyme systems from higher plants. *J. Biol. Chem.* **239,** 4056.
Barker, J. and Solomos, T. (1962). Mechanisms of carbohydrate breakdown in plants. *Nature (London)* **196,** 189.
Barmore, C. R., and Rouse, A. H. (1976). Pectinesterase activity in controlled atmosphere stored avocadoes. *J. Am. Soc. Hortic. Sci.* **101,** 294.
Barnes, M. F., and Patchett, B. J. (1976). Cell wall degrading enzymes and the softening of senescent strawberry fruit. *J. Food Sci.* **41,** 1392.
Bartley, I. M. (1974). β-Galactosidase activity in ripening apples. *Phytochemistry* **13,** 2107.
Bartley, I. M. (1977). A further study of β-galactosidase activity apples ripening in store. *J. Exp. Bot.* **28,** 943.
Bartley, I. M. (1978). Exo-polygalacturonase of apple. *Phytochemistry* **17,** 213.
Bartley, I. M., and Knee, M. (1982). The chemistry of textural changes in fruit. *Food Chem.* **9,** 47.

Bartley, I. M., Knee, M., and Casimer, M. A. F. (1982). Fruit softening. 1. Changes in cell wall composition and endopolygalacturonase in ripening pears. *J. Exp. Bot.* **33,** 1248.

Battersby, A. R., Fookes, C. J. R., McDonald, E., and Matcham, G. W. J. (1979). Chemical and enzymatic studies on biosynthesis of the natural porphyrin macrocycle, formation and role of unrearranged hydroxymethylbilane and order of assembly of the pyrrole rings. *Bioorg. Chem.* **8,** 451.

Ben-Arie, D. S., and Sonego, L. (1980). Pectolytic enzyme activity involved in woolly breakdown of stored peaches. *Phytochemistry* **19,** 2553.

Ben-Arie, R., Sonego, L., and Frenkel, C. (1979). Changes in pectic substances in ripening pears. *J. Am. Soc. Hortic. Sci.* **104,** 500.

Bendall, D. S., and Bonner, D. W., Jr. (1971). Cyanide-sensitive and cyanide-resistant plant tissues. *Plant Physiol.* **58,** 47.

Bennett, A. B., Smith, G. M., and Nichols, B. G. (1987). Regulation of climacteric respiration in ripening avocado. *Plant Physiol.* **83,** 973.

Ben-Yehoshua, S. (1969). Gas exchange, transpiration and the commercial deterioration in storage of orange fruit. *J. Am. Soc. Hortic. Sci.* **94,** 524.

Ben-Yehoshua, S. (1978). Delayed deterioration of individual citrus fruit by seal-packaging in film of high density polyethylene. 1. General effects. *Proc. Int. Soc. Citric. 1977,* pp. 110–115.

Ben-Yohoshua, S. (1985). Individual seal-packaging of fruit and vegetables in plastic film—A new postharvest technique. *HortScience* **20,** 32.

Berard, L. S., Lougheed, E. C., and Murr, D. P. (1982). β-Galactosidase activity of 'McIntosh' apples in storage. *HortScience* **17,** 660.

Beyer, P., Kreuz, K., and Kleinig, H. (1980). β-Carotene synthesis in isolated chromoplasts from *Narcissus pseudonarcissus*. *Planta* **150,** 435.

Biale, J. B. (1960a). The postharvest biochemistry of tropical and subtropical fruits. *Adv. Food Res.* **10,** 293.

Biale, J. B. (1960b). Respiration of fruits. *In* "Handbuch der Pflanzenphysiologie" (W. Ruhland, ed.), Vol. 12, Part 2, p. 586. Springer-Verlag, Berlin.

Biale, J. B. (1964). Growth, maturation and senescence in fruits. *Science* **146,** 880.

Biale, J. B. (1969). Metabolism at several levels of organization in the fruit of avocado, *Pesea americana*. Mill. *Qual. Plant. Mater. Veg.* **19,** 141.

Biale, J. B., and Barcus, D. E. (1970). Respiration patterns in tropical fruits of the Amazon basin. *Trop. Sci.* **12,** 93.

Biale, J. B., and Young, R. E. (1947). Critical oxygen concentrations for the respiration of lemons. *Am. J. Bot.* **34,** 301.

Biale, J. B., and Young, R. E. (1962). The biochemistry of fruit maturation. *Endeavour* **21,** 164.

Biale, J. B., and Young, R. E. (1971). The avocado pear. *In* "The Biochemistry of Fruits and Their Products" (A. C. Hulme, ed.), Vol. 2, pp. 1–63. Academic Press, London and New York.

Biale, J. B., and Young, R. E. (1981). Respiration and ripening in fruits—Retrospect and prospect. *In* "Recent Advances in the Biochemistry of Fruits and Vegetables" (J. Friend, and M. J. C. Rhodes, eds.), Chapter 1, pp. 1–39. Academic Press, New York.

Biale, J. B., Young, R. E., and Olmstead, A. J. (1954). Fruit respiration and ethylene production. *Plant Physiol.* **29,** 168.

Biale, J. B., Young, R. E., Popper, C. S., and Appleman, W. E. (1957). Metabolic processes in cytoplasmic particles of the avocado fruit. 1. Preparative procedure, cofactor requirements, and oxidative phosphorylation. *Physiol. Plant.* **10,** 48.

Bishop, R. C., and Klein, R. M. (1975). Photo-promotion of anthocyanin synthesis in harvested apples. *HortScience* **10,** 126.

Blanpied, G. D. (1977). Requirements and recommendations for eastern and midwestern apples. *Mich. State Univ. Hortic. Rep.* **28,** 225–230.

Blanpied, G. D., Turk, J. R., and Douglas, J. B. (1982). Low ethylene controlled atmosphere storage for apples. *Symp. Ser.—Oreg. State Univ. Sch. Agric.* **1,** 337.

Boller, T., Herner, R. C., and Kende, H. (1979). An assay for the ethylene precursor 1-aminocyclopropane-1-carboxylic acid and studies on its enzymatic formation. *Planta* **145,** 293.

Borisova, I. G., and Budnitskaya, E. V. (1975). Lipoxygenase of chloroplasts. *Dokl. Akad. Nauk SSSR* **225,** 439; *Chem. Abstr.* **84,** 147888p (1976).

Brady, C. J. (1976). The pectinesterase of the pulp of banana fruit. *Aust. J. Plant Physiol.* **3,** 163.

Brady, C. J., Palmer, J. K., O'Connell, P. B. H., and Smillie, R. M. (1976). An increase in protein synthesising during ripening of the banana fruit. *Phytochemistry* **9,** 1037.

Bramley, P. M., Aung Than, and Davies, B. H. (1977). Alternative pathways of carotene cyclisation in *Phycomyces blakesbeanus. Phytochemistry* **16,** 235.

Braverman, J. B. S. (1963). "Introduction to the Biochemistry of Foods." Elsevier, Amsterdam.

Brennan, T., and Frenkel, C. (1977). Involvement of hydrogen peroxide in the regulation of senescence in pear. *Plant Physiol.* **59,** 411.

Britton, G. (1976). Biosynthesis of carotenoids. *In* "Chemistry and Biochemistry of Plant Pigments" (T. W. Goodwin, ed.), Vol. 1, pp. 262–327. Academic Press, New York.

Britton, G. (1979). Carotenoid biosynthesis—A target for herbicide activity. *Z. Naturforsch., C: Biosci.* **34C,** 979.

Britton, G.1982. Carotenoid biosynthesis in higher plants. *Physiol. Veg.* **20,** 735.

Brouillard, R. (1983). The *in vivo* expression of anthocyanin colour in plants. *Phytochemistry* **22,** 1311.

Buckle, K. A., and Edwards, R. A. (1969). Chlorophyll degradation products from processed pea puree. *Phytochemistry* **8,** 1901.

Buescher, R. W., and Tigchelaar, E. C. (1975). Pectinesterase, polygalacturonase, C_x-cellulase activities and softening of x in the *rin* tomato mutant. *HortScience* **10,** 624.

Bufler, G. (1984). Ethylene-enhanced 1-aminocyclopropane-1-carboxylic acid synthase activity in ripening apples. *Plant Physiol.* **75,** 192.

Bufler, G. (1986). Ethylene-promoted conversion of 1-aminocyclopropane-1-carboxylic acid to ethylene in peel of apple at various stages of fruit development. *Plant Physiol.* **80,** 539.

Bufler, G., and Banzerth, F. (1983). Effects of propylene and oxygen on the ethylene producing system of apples. *Physiol. Plant.* **58,** 486.

Burg, S. P. (1973). Ethylene in plant growth. *Proc. Natl. Acad. Sci. U.S.A.* **70,** 591.

Burg, S. P., and Burg, E. A. (1962). Role of ethylene in fruit ripening. *Plant Physiol.* **37,** 179.

Burg, S. P., and Burg, E. A. (1965). Ethylene action and the ripening of fruits. *Science* **148,** 1190.

Burg, S. P., and Thimann, K. V. (1959). The physiology of ethylene formation in apples. *Proc. Natl. Acad. Sci. U.S.A.* **45,** 335.

Buttery, R. G., Teranishi, R., and Ling, L. C. (1987). Fresh tomato aroma volatiles: A quantitative study. *J. Agric. Food Chem.* **35,** 540.

Camara, B., and Brangeon, J. (1981). Carotenoid metabolism during chloroplast to chromoplast transformation in *Capsicum annuum* fruit. *Planta* **151,** 359.

Camara, B., Bardat, F., and Moneger, R. (1982). Sites of biosynthesis of carotenoids in *Capsicum chromoplasts. Eur. J. Biochem.* **127,** 255.

Campbell, H. (1937). Undesirable color changes in frozen peas stored at insufficiently low temperatures. *Food Res.* **2,** 55.

Cansfield, P. E., and Francis, F. J. (1970). Quantitative methods for anthocyanins. 5. Separation of cranberry phenolics by electrophoresis and chromatography. *J. Food Sci.* **35,** 309.

Cappellini, M. C., LaChance, P. A., and Hudsson, D. E. (1984). Effect of temperature and carbon dioxide atmospheres on the market quality of green bell peppers. *J. Food Qual.* **7,** 17.

Carnal, N. W., and Black, C. C. (1979). Pyrophosphate-dependent 6-phosphofructokinase, a new glycolytic enzyme in pineapple leaves. *Biochem. Biophys. Res. Commun.* **86,** 20.

Castelfranco, P. A., and Beale, S. I. (1981). Chlorophyll biosynthesis. *In* "The Biochemistry of Plants" (M. D. Hatch, and N. K. Boardman, eds.), Vol. 8. Academic Press, New York.

Castelfranco, P. A., and Beale, S. I. (1983). Chlorophyll biosynthesis: Recent advances and areas of interest. *Annu. Rev. Plant Physiol.* **34,** 241.

Chalmers, D. J., and Rowan, K. S. (1971). The climacteric in ripening tomato fruit. *Plant Physiol.* **48,** 235.

Chepurenko, N. V., Borisova, I. G., and Budnitsksaya, E. V. (1978). Isolation and characterization of isoenzymes of lipoxygenase from pea seeds. *Biochemistry (English Transl.)* **43,** 480.

Chereskin, B. M., Wong, Y.-S., and Castelfranco, P. A. (1982). *In vitro* synthesis of the chlorophyll isocyclic ring. Transformation of magnesium–protoporphyrin IX and magnesium–protoporphyrin mono methyl ester into magnesium-2, 4-divinyl pheoporphyrin a. *Plant Physiol.* **70,** 987. 5

Chitarra, A. B., and LaJollo, F. M. (1981). Phosphorylase, phosphatase, α-amylase activity and starch breakdown during ripening of Marmelo banana (*Musa acuminata* (olla) × *musubalbisiani* (olla) ABB group). Whole fruit and thin slices. *J. Am. Soc. Hortic. Sci.* **106,** 579.

Chou, H., and Breene, N. M. (1972). Oxidative discoloration of β-carotene in low moisture foods. *J. Food Sci.* **37,** 66.

Chrispeels, M. J. (1970). Synthesis of hydroxyproline-containing macromolecules in carrots. *Plant Physiol.* **45,** 223.

Cohen, B.-S., Grossman, S., Pinsky, A., and Klein, B. (1984). Chlorophyll inhibition of lipoxygenase in growing plants. *J. Agric. Food Chem.* **32,** 516.

Coleman, J. O. D., and Palmer, J. M. (1972). The oxidation of malate by isolated plant mitochondria. *Eur. J. Biochem.* **26,** 499.

Craker, L. E., and Wetherbee, P. J. (1973). Ethylene, carbon dioxide, and anthocyanin synthesis. *Plant Physiol.* **52,** 177.

Dahodwala, S., Humphrey, A., and Weibel, M. (1974). Pectic enzymes: Individual and concerted kinetic behavior of pectinesterase and pectinase. *J. Food Sci.* **39,** 920.

Dalal, K. B. (1965). Investigation into flavor chemistry with special reference to synthesis of volatiles in developing tomato fruit *(Lycopersicon esculentum)* under field and greenhouse growing conditions. Ph.D Thesis. Utah State University, Logan.

Dalal, K. B., Salunkhe, D. K., Olson, L. E., Do, J. Y., and Yu, M. H. (1968). Volatile components of developing tomato fruit grown under field and greenhouse conditions. *Plant Cell Physiol.* **9,** 389.

Davies, B. H. (1973). Carotene biosynthesis in fungi. *Pure Appl. Chem.* **35,** 1.

Davies, J. N., and Hobson, G. E. (1981). The constituents of tomato fruit—The influence of environment, nutrition, and genotype. *CRC Crit. Rev. Food Sci. Nutr.* **13,** 205.

Dehesh, K., Kreuz, K., and Apel, K. (1987). Chlorophyll synthesis in green leaves and isolated chloroplasts of barley *(Hardeum vulgare)*. *Physiol. Plant.* **69,** 173.

Delmer, D. P. (1987). Cellulose biosynthesis. *Annu. Rev. Plant Physiol.* **38,** 259.

De Pooter, H. L., Montens, J. P., Dirinck, P. J., Willaert, G. A., and Schamp, N. M. (1982). Ripening induced in preclimacteric immature Golden Delicious apples by proprionic and butyric acids. *Phytochemistry* **21,** 1015.

De Pooter, H. L., D'Ydewalle, Y. E., Willaert,G. A., Dirink, P. J., and Schamp, N. M. (1984). Acetic and propionic acids,inducer of ripening in preclimacteric Golden Delicious apples. *Phytochemistry* **23,** 23.

De Swardt, G. H., and Maxie, E. C. (1967). Pectin methylesterase in the ripening banana. S. Africa. *J. Agric. Res.* **10,** 501.

Dey, P. M., and Brinson, K. (1984). Plant cell walls. *Adv. Carbohydr. Chem. Biochem.* **42,** 265.

Dick, A. J., Laskey, G., and Lidster, P. D. (1984). Inhibition of β-galactosidase isolated from 'McIntosh' apples. *HortScience* **19,** 552.

Dixon, W. L., and ap Rees, T. (1980). Identification of the regulatory steps in glycolysis in potatoe tubers. *Phytochemistry* **19,** 1297.

Dixon, W. L., Franks, F., and ap Rees, T. (1981). Cold lability of phosphofructokinase from potatoe tubers. *Phytochemistry* **20,** 969.

Dolendo, A. L., Luh, B. S., and Pratt, H. K. (1966). Relation of pectic and fatty acid changes to respiration rate during ripening of avocado fruits. *J. Food Sci.* **31,** 332.

Douillard, R., and Bergeron, E. (1978). Activite lipoxyginasque de chloroplastes plantules de Ble. *C. R. Hebd. Seances Acad. Sci., Ser. D* **286,** 753.

Dresel, E. J. B., and Falk, J. E. (1953). Conversion of δ-aminolevulinic acid to porphobilinogen in a tissue system. *Nature (London)* **172,** 1185.

Dull, G. G., Young, R. E., and Biale, J. B. (1967). Respiratory patterns in fruit of pineapple, *Ananas comosus*, detached at different stages of development. *Physiol. Plant.* **20,** 1059.

Eaks, I. L. (1970). Respiratory response, ethylene production, and response to ethylene of citrus fruit during ontogeny. *Plant Physiol.* **45,** 334.

Eaks, I. L., and Morris, L. (1956). Respiration of cucumber fruits associated with physiological injury at chilling temperatures. *Proc. Am. Soc. Hortic. Sci.* **78,** 190.

Eberhardt, N. L., and Rilling, H. C. (1975). Phenyltransferase from *Saccharomyces cerevisiae*. Purification to homogeneity and molecular properties. *J. Biol. Chem.* **250,** 863.

Ebert, G., and Gross, J. (1985). Carotenoid changes in the peel of ripening persimmon (*Diospyros Kam*) Triumph. *Phytochemistry* **24,** 29.

Eda, S., and Kato, K. (1980). Pectin isolated from the midrib of leaves of *Nicotiana tabacum*. *Agric. Biol. Chem.* **44,** 2793.

Eilati, S. K., Budowski, P., and Monselise, S. P. (1975). Carotenoid changes in the 'Shamouti' orange peel during chloroplast–chromoplast transformation on and off the tree. *J. Exp. Bot.* **26,** 624.

Eskin, N. A. M. (1979). "Plant Pigments, Flavors and Textures: The Chemistry and Biochemistry of Selected Compounds." Academic Press, New York.

Eskin, N. A. M., Grossman, S., and Pinsky, A. (1977). The biochemistry of lipoxygenase in relation to food quality. *CRC Crit. Rev. Food Sci. Nutr.* **9,** 1.

Faragher, J. D. (1983). Temperature regulation of anthocyanin accumulation in apple skin. *J. Exp. Bot.* **34,** 1291.

Faragher, J.D., and Chalmers, D. J. (1977). Regulation of anthocyanin synthesis in apple skin.III. Involvement of phenylalanine–ammonia lyase. *Aust. J. Plant Physiol.* **4,** 133.

Farin, D., Ikan, R., and Gross, J. (1983). The carotenoid pigments in the juice and flavedo of a mandarin hybrid *(Citrus reticulata)* cv. Michal during ripening. *Phytochemistry* **22,** 403.

Fidler, J. C., Wilkinson, B. B., Edney, K. L., and Sharples, R. O. (1977). The biology of apple and pear storage. *Res. Rev.—Commonw. Agric. Bur. Engl.* **3,** 235.

Flath, R. A., Black, D. R., Guadagni, D. G., McFadden, W. H., and Schultz, T. H. (1967). Identification and organoleptic evaluation of compounds in Delicious apple essence. *J. Agric. Food Chem.* **15,** 29.

Fleming, H. P., Cobb, W. Y., Etchells, J. L., and Bell, T. A. (1968). The formation of carbonyl compounds in cucumbers. *J. Food Sci.* **33,** 572.

Foda, Y. H. (1957). Pectic changes during ripening as related to flesh firmness in tomato. Ph.D. Thesis, University of Illinois, Urbana.

Foley, T., and Beale, S. I. (1982). δ-Aminolevulinic acid formation from γ, δ-dioxovaleric acid in extracts of *Euglena gracilis*. *Plant Physiol.* **70,** 1495.

Ford, S. M., and Friedman, H. C. (1979). Formation of γ-aminolevulinic acid from glutamic acid by a partially purified enzyme system from wheat leaves. *Biochim. Biolphys. Acta* **569,** 153.

Freeman, J. A., and Woodbridge, C. G. (1960). Effect of maturation, ripening and truss position on the free amino acid content in tomato fruits. *Proc. Am. Soc. Hortic. Sci.* **76,** 515.

Frenkel, C. (1979). Role of oxidative metabolism in the onset of senescence in plant storage. *Z. Ernaehrungswiss.* **18,** 209.

Frenkel, C., and Eskin, N. A. M. (1977). Ethylene evolution as related to changes in hydroperoxides in ripening tomato fruit. *HortScience* **12,** 552.

Frenkel, C., Klein, I., and Dilley, D. R. (1968). Protein synthesis in relation to ripening of pome fruits. *Plant Physiol.* **43,** 1146.

Fuesler, T. P., Hanamoto, C. M., and Castelfranco, P. A. (1982). Separation of Mg–protoperphyrin IX and Mg–protoporphytin IX monomethyl ester synthesized *de novo* by developing cucumber etioplasts. *Plant Physiol.* **69,** 421.

Fuleki, T. (1969). The anthocyanins of strawberry, rhubarb, radish and onion. *J. Food Sci.* **34,** 365.

Fuleki, T. (1971). Anthocyanins in red onion, *Allium cepa. J. Food Sci.* **36,** 101.

Gaillard, T., and Matthew, J. A. (1977). Lipoxygenase-mediated cleavage of fatty acids to carbonyl fragments in tomato fruits. *Phytochemistry* **16,** 339.

Galliard, T., Matthew, J. A., Wright, A. J., and Fishweck, M. J. (1977). The enzymic breakdown of lipids to volatile and non-volatile carbonyl fragments in disrupted tomato fruits. *J. Sci. Food Agric.* **28,** 863.

Games, D. E., Jackson, A. H., Jackson, J. R., Belcher, R. V., and Smith, S. G. (1976). Biosynthesis of protoporphyrin-IX from coproporphyrinogen. III. *J. Chem. Soc., Chem. Commun.,* p. 187.

Gane, R. (1934). Production of ethylene by some ripening fruits. *Nature (London)* **134,** 1008.

Gassman, M. L., Pluscec, J., and Bogorad, L. (1966). δ-Aminolevulinic acid transaminase from *Chlorella* and *Phaseolus. Plant Physiol.* **41,** xiv.

Geisler, G., and Radler, F. (1963). Developmental and ripening processes in grapes of *Vitis. Ber Dtsch. Bot. Ges.* **76,** 112.

Goldstein, J. L., and Wick, E. L. (1969). Lipids in ripening banana fruit. *J. Food Sci.* **34,** 482.

Gorin, N. (1973). Several compounds in Golden Delicious apples as possible parameters of acceptability. *J. Agric. Food Chem.* **21,** 671.

Goto, T., Kondo, T., Tamura, H., Kawahori, K., and Hatton, H. (1983). Structure of platyconin, a diacylated anthocyanin isolated from the Chinese bell-flower. *Platycondon grandiflorum. Tetrahedron Lett.* **24,** 2181.

Gough, S. P., and Kannangara, C. G. (1979). Biosynthesis of δ-amino levulinate in greening barley leaves. III. The formation of δ-aminolevulinate in tigrina mutants of barley. *Carlsberg Res. Commun.* **44,** 403.

Granick, S. (1961). Magnesium protoporphyrin monoester and protoporphyrin monomethyl ester in chlorophyll biosynthesis. *J. Biol. Chem.* **236,** 1168.

Gray, J. C., and Keckwick, R. G. O. (1969). Mevalonate kinase from etiolated cotyledons of French beans. *Biochem. J.* **113,** 37.

Gray, J. C., and Keckwick, R. G. O. (1972). The inhibition of plant mevalonate kinase preparations by phenyl pyrophosphates. *Biochim. Biophys. Acta* **279,** 290.

Gray, J. C., and Keckwick, R. G. O. (1973). Mevalonate kinase in green leaves and etiolated cotyledons of the French bean *Phaseolus vulgaris. Biochem. J.* **133,** 335.

Griffith, W. T. (1974). Protochlorophyll and protochlorophyllide as precursors for chlorophyll synthesis *in vitro. FEBS Lett.* **49,** 196.

Grisebach, H. (1981). Lignins. *In* "The Biochemistry of Plants," Vol. 7. Secondary Plant Products. (E. E. Conn, ed.), chap. 15, p. 457. Academic Press, New York.

Grommeck, R., and Markakis, P. (1964). Effect of peroxidase on anthocyanin pigments. *J. Food Sci.* **29,** 53.

Grosch, W., Laskaway, G., and Weber, F. (1976). Formation of volatile carbonyl compounds and cooxidation of β-carotene by lipoxygenase from wheat, potato, flax and beans. *J. Agric. Food Chem.* **24,** 456.

Gross, J., Timberg, R., and Graej, M. (1983). Pigment and ultrastructural changes in the developing *Mammalo Citrus* grandis "Goliath." *Bot. Gaz. (Chicago)* **144,** 401.

Gross, K. C., and Wallner, S. J. (1979). Degradation of cell wall polysaccharides during tomato fruit ripening. *Plant Physiol.* **63,** 117.

Grumbach, K. H., and Forn, B. (1980). Chloroplast autonomy in acetyl coenzyme-A formation and terpenoid synthesis. *Z. Naturforsch., C: Biosci.* **35C,** 645.

Gupte, S. M., El-Bisi, H. M., and Francis, F. J. (1964). Kinetics of thermal degradation of chlorophyll in spinach puree. *J. Food Sci.* **29,** 379.

Haard, N. (1973). Upsurge of particulate peroxidase in ripening banana fruit. *Phytochemistry* **12,** 555.

Hahlbrock, K., and Grisebach, H. (1975). *In* "The Flavonoids" (J. B. Harborne, T. J. Mabry, and H. Mabry, eds.), pp. 866–915. Chapman & Haal, London.

Hammett, L. K., Kirk, H. J., Todd, H. G., and Hale, S. A. (1977). Association between soluble solids/acid content and days from full bloom of "Golden Delicious" apple fruits. *J. Am. Soc. Hortic. Sci.* **102,** 429.

Hampp, R., Kriebitzsch, C., and Zeigler, H. (1974). Effect of lead on enzymes of porphyrine biosynthesis in chloroplasts and erythrocytes. *Naturwissenschaften* **61,** 504.

Hanamoto, M., and Castelfranco, P. A. (1983). Separation of monovinyl and divinyl protochlorophyllides and chlorophyllides from etiolated and phototransformed cucumber cotyledons. *Plant Physiol.* **73,** 79.

Hanaoka, T., and Ito, K. (1957). Studies on the keeping quality of onions. 1. Relation between the characters of bulbs and their sprouting during storage. *J. Hortic. Assoc. Jpn.* **26,** 129.

Hansen, E. (1942). Quantitative study of ethylene production in relation to respiration of pears. *Bot. Gaz. (Chicago)* **103,** 543.

Harborne, J. B. (1967). "Comparative Biochemistry of the Bioflavonoids." Academic Press, New York.

Harborne, J. B. (1976). Functions of flavonoids in plants. *In* "Chemistry and Biochemistry of Plant Pigments" (T. W. Goodwin, ed.), 2nd ed., Vol. 1. p. 525. Academic Press, New York.

Harper, K. A. (1968). Structural changes of flavylium salts. IV. Polarographic and spectrometric examination of perlargonidin chloride. *Aust. J. Chem.* **21,** 221.

Hartmann, C., Drouet, A., and Morin, F. (1987). Ethylene and ripening of apple, pear and cherry fruit. *Plant Physiol.* **25,** 505.

Haslam. E., Gupta, R. K., Al-Shafi, S. M. K., and Layden, K. (1982). *J. Chem. Soc., Perkin Trans. 1,* p. 2525.

Hayashi, T. (1989). Xyloglucans in the primary cell wall. *Ann. Rev. Plant Physiol. Plant Mol. Biol.* **40,** 139.

Henry, E. N. (1975). Peroxidases in tobacco abscission zone tissue. III. Ultrastructural localization in thylakoids and membrane bound bodies of chloroplasts. *J. Ultrastruct. Res.* **52,** 289.

Hermann, J. (1970). Berechnung der chermischen und sensorischen Verandderungen unserer Lebensmittel bei Erhitzungs-und Lageruugsprozessen. *Ernaehrungsforschung* **15,** 279.

Herner, R. C. (1973). Fiber determination. In progress report on asparagus research. *Mich. Agric. Exp. Stn., Rep.* **217,** 11.

Herner, R. C., and Sink, K. C. (1973). Ethylene production and respiratory behaviour of rin tomato mutant. *Plant Physiol.* **52,** 38.

Hilderbrand, D. F., and Hymovitz, T. (1982). Carotene and chlorophyll bleaching by soybeans with and without seed lipoxygenase. 1. *J. Agric. Food Chem.* **30,** 705.

Hobson, G. E. (1964). Polygalacturonase in normal and abnormal tomato fruit. *Biochem. J.* **92,** 324.

Hobson, G. E. (1965). The firmness of tomato fruit in relation to polygalacturonase activity. *J. Hortic. Sci.* **40,** 66.

Hobson, G. E. (1967). The effects of alleles at the "never ripe" locus on the ripening of tomato fruit. *Phytochemistry* **6,** 1337.
Hobson, G. E. (1968). Cellulase activity during the maturation and ripening of tomato fruit. *J. Food Sci.* **33,** 588.
Hobson, G. E., Lance, C., Young, R. E., and Biale, J. B. (1966). Isolation of active subcellular particles from avocado fruit at various stages of ripeness. *Nature (London)* **209,** 1242.
Hoffman, N. E., and Yang, S. F. (1980). Changes of 1-aminocyclopropane-1-carboxylic acid content in ripening fruits in relation to their ethylene production rates. *J. Am. Soc. Hortic. Sci.* **105,** 492.
Holden, M. (1965). Chlorophyll bleaching by legume seeds. *J. Sci. Food Agric.* **16,** 312.
Holden, M. (1970). Chlorophyll bleaching systems in leaves. *Phytochemistry* **9,** 1771.
Hrazdina, G., Alscher-Herman, R., and Kish, V. M. (1980). Subcellular localization of flavonoid synthesising enzymes in *Pisum, Phaseolus, Brassica* and *Spinacia* cultivars. *Phytochemistry* **19,** 1355.
Huang, L., and Castelfranco, P. A. (1986). Regeneration of magnesium-2, 4-divinylpheophorin a (divinyl protochlorophy 5 llide) in isolated chloroplasts. *Plant Physiol.* **82,** 285.
Huber, D. J. (1983). Polyuronide degradation and hemicellulase modifications in ripening tomato fruit. *J. Am. Soc. Hortic. Sci.* **108,** 405.
Huff, A. (1982). Peroxidase-catalyzed oxidation of chlorophyll by hydrogen peroxide. *Phytochemistry* **21,** 261.
Huff, A. (1984). Sugar regulation of plastid interconversion s in epicarp of citrus fruit. *Plant Physiol.* **76,** 307.
Hulme, A. C., Jones, J. D., and Woolworton, L. S. C. (1963). The respiration climacteric in apple fruits. *Proc. R. Soc. London, Ser. B* **158,** 514.
Hultin, H. O., and Levine, A. S. (1965). Pectinmethylesterase in the ripening banana. *J. Food Sci.* **30,** 917.
Hurkman, W. J., and Kennedy, G. S. (1977). Development and cytochemistry of the thylakoidal body in tobacco chloroplasts. *Am. J. Bot.* **64,** 86.
Hyde, R. B., and Morrison, J. W. (1964). The effect of storage temperature on reducing sugars, pH and phosphorylase enzyme activity in potato tubers. *Am. Potato J.* **41,** 163.
Ikemefura, J., and Adamson, J. (1985). Chlorophyll and carotenoid changes in ripening palm fruit, *Elacis guineesis. Phytochemistry* **23,** 1413.
Imamura, M., and Shimizu, S. (1974). Metabolism of chlorophyll in higher plants. IV. Relationship of fatty acid oxidation and chlorophyll bleaching in plant extracts. *Plant Cell Physiol.* **15,** 187.
Imaseki, H., and Watanabe, A. (1978). Inhibition of ethylene inhibition by osmotic shock. Further evidence for control of ethylene production. *Plant Cell Physiol.* **19,** 345.
Isaac, J. E., and Rhodes, M. J. C. (1982). Purification and properties of phosphofructokinase from fruits of *Lycopersicon esculentum. Phytochemistry* **21,** 1553.
Isaac, J. E., and Rhodes, M. J. C. (1987). Phosphofructokinase and ripening in *Lycopersicon esculentum* fruits. *Phytochemistry* **26,** 649.
Isherwood, F. A. (1976). Mechanisms of starch sugar interconversion in *Solanum tuberosum. Phytochemistry* **15,** 33.
Issenberg, F. M. R., Oyer, E. B., and Engst, C. B. (1971). The effect of modified atmospheres plus physiologically active chemicals on cabbage storage life. *Acta Hortic.* **20,** 7.
Issenberg, P., and Wick, E. L. (1963). Volatile components of bananas. *J. Agric. Food Chem.* **11,** 2.
Jackson, A. H., Sancovich, H. A., Ferramola, A. M., Evans, N., and Games, D. E. (1976). Macrocyclic intermediates in the biosynthesis of porphyrins. *Philos. Trans. R. Soc. London, Ser. B* **273,** 191.
Jadhav, S., Singh, B., and Salunkhe, D. K. (1972). Metabolism of unsaturated fatty acids in tomato fruit. Linoleic and linolenic acid as precursors of hexanal. *Plant Cell Physiol.* **13,** 449.

Jones, I. D., White, R. C., and Gibbs, E. (1961). The formation of pheophorbides during brine preservation of cucumber. *Food Technol.* **15,** 172.

Jones, I. D., White, R. C., and Gibbs, E. (1963). Influence of bleaching or brining treatments on the formation of chlorophyllides, pheophytins and pheophorbides in green plant tissue. *J. Food Sci.* **28,** 437.

Jones, R. A., and Scott, S. J. (1983). Genetic potential to improve tomato flavor in commercial F hybrids. *J. Am. Soc. Hortic. Sci.* **109,** 318.1

Jongkee, K., Gross, K. C., and Solomos, T. (1987). Characterization of the stimulation of ethylene production by galactose in tomato (*Lycopersicon esculentum* Mill.) fruit. *Plant Physiol.* **85,** 804.

Jonsson, L. M. V., Donker-Koopman, W. E., Uitslager, P., and Schram, A. W. (1983). Subcellular localization of anthocyanin methyltransferase in flowers of *Petunia hybrida. Plant Physiol.* **72,** 287.

Jordan, P. M., and Seehra, J. S. (1979). The biosynthesis of uroporphyrinogen. III. Order of assembly of the four porphobilinogen molecules in the formation of the tetrapyrrole ring. *FEBS Lett.* **104,** 364.

Joslyn, M. A., and Mackinney, G. (1938). The rate of conversion of chlorophyll to pheophytin. *J. Am. Chem.Soc.* **60,** 1132.

Kacperska, A., and Kubacka-Zabalska, M. (1984). Is lipoxygenase involved in biosynthesis of wound ethylene?. *In* "Structure, Function and Metabolism of Plant Lipids" (P. A. Siegenthaler and W. Eichenberger, eds.), Publ. B. Elsevier, Amsterdam.

Kacperska, A., and Kubacka-Zabalska, M. (1985). Is lipoxygenase involved in the formation of ethylene from ACC? *Physiol. Plant.* **64,** 333.

Kader, A. A., Stevens, M. A., Albright-Holton, M., Morris, C. C., and Algazi, M. (1977). Effect of fruit ripeness when picked on flavor and composition in fresh market tomatoes. *Proc. Am. Soc. Hortic. Sci.* **91,** 486.

Kannangara, C. G., and Gough, S. P. (1978). Biosynthesis of δ-aminolevulinate in greening barley leaves. II: Glutamate 1-semi-aldehyde aminotransferase. *Carlsberg Res. Commun.* **43,** 185.

Kannangara, C. G., and Gough, S. P. (1979). Biosynthesis of δ-aminolevulinate in greening barley leaves. II: Induction of enzyme synthesis by light. *Carlsberg Res.Commun.* **44,** 11.

Kapp, P. P. (1966). Some effects of variety, maturity,and storage on fatty acids in fruit pericarp of *Lycopersicon esculentum* Mill. *Diss. Abstr.* **27,** 77B.

Kato, T. (1966). Physiological studies on the bulbing and dormancy of onion plant. VIII. Relations between dormancy and organic constituents of bulbs. *J. Jpn. Soc. Hortic. Sci.* **35,** 142.

Kauss, H. (1967). Biosynthesis of the glucuronic acid unit of hemicellulose B from UDP-glucuronic acid. *Biochim. Biophys Acta* **148,** 572.

Kauss, H. (1969). Enzymic 4-*O*-methylation of glucuronic acid linked to galactose in hemicellulose polysaccharides from *Phaseolus aureus. Phytochemistry* **8,** 985.

Kazeniac, S. J., and Hall, R. M. (1970). Flavor chemistry of tomato volatiles. *J. Food Sci.* **35,** 519.

Keegstra, K., Talmadge, K. W., Bauer, W. D., and Albershein, P. (1973). The structure of plant cell walls. III. A model of the walls of suspension-cultured cells based on the interconnections of the macromolecular components. *Plant Physiol.* **51,** 188.

Kennedy, M. G. H., and Isherwood, F. A. (1975). Activity of phosphorylase in *Solanum tuberosum* during low temperature storage. *Phytochemistry* **14,** 667.

Kertesz, Z. I. (1951). "The Pectic Substances." Wiley (Interscience), New York.

Kidd, F., and West, C. (1922). "Brown heart," a functional disease of apples and pears. *Food Invest Board Rep.* **1921,** 14.

Kidd, F., and West, C. (1930). Physiology of fruits: Changes in the respiratory activity of apples during senescence at different temperatures. *Proc. R. Soc. London* **106,** 93.

Kim, I., and Grosch, W. (1979). Partial purification of a lipoxygenase from apples. *J. Agric. Food Chem.* **27,** 243.

King, K. J. N., and Vessal, M. J. (1969). Enzymes of the cellulase complex. *Adv. Chem. Ser.* **95,** 7.

Kirk, J. T. D., and Tilney-Bassett, P. A. E. (1978). "The Plastids: Their Chemistry, Structure, Growth and Inheritance." Elsevier/North-Holland Biomedical Press, New York.

Klaui, H. (1973). Carotinoide in Lebensmitteln. *Funct. Prop. Fats Foods, Adv. Study Course, 1971.*

Klein, B. P., and Grossman, S. (1985). Co-oxidation reactions of lipoxygenase in plant systems. *Adv. Free Radical Biol. Med.* **1,** 309.

Knee, M. (1973). Polysaccharide changes in cell walls of ripening apples. *Phytochemistry* **12,** 1543.

Knee, M. (1975). Changes in structural polysaccharides of apples ripening during storage. *Colloq. Int. C. N. R. S.* **238,** 241.

Knee, M. (1978). Metabolism of polymethylgalacturonate in apple cortical tissue during ripening. *Phytochemistry* **17,** 1261.

Knee, M. (1982). Fruit softening. II. Precursor incorporation into pectin by pear tissue slices. *J. Exp. Bot.* **33,** 1256.

Knee, M., and Bartley, I. M. (1982). Composition and metabolism of cell wall polysaccharides in ripening fruits. *In* "Recent Advances in the Biochemistry of Fruits and Vegetables" (J. Friend and M. J. C. Rhodes, eds.), pp. 133–48. Academic Press, London and New York.

Kockritz, A., Schewe, T., Hieke, B., and Hass, N. (1985). The effect of soybean lipoxygenase-1 on chloroplasts from wheat. *Phytochemistry* **24,** 381.

Konze, J. R., and Kende, H. (1979). Ethylene formation from 1-amino-cyclopropane-1-carboxylic acid in homogenates of etiolated pea seedlings. *Planta* **146,** 293.

Konze, J. R., Jones, J. F., Boller, T., and Kende, H. (1980). Effect of 1-aminocyclopropane-1-carboxylic acid on the production of ethylene in senescing flowers of *Ipomoea tricolor* Cav. *Plant Physiol.* **66,** 566.

Kosiyachinda, S., and Young, R. E. (1975). Ethylene production in relation to the initiation of respiratory climacteric in fruit. *Plant Cell Physiol.* **16,** 595.

Kuwasha, S. C., Subbarayan, C., Beeler, D. A., and Porter, J. W. (1969). The conversion of lycopenes-15,15'H to cyclic carotenes by soluble extracts of higher plant plastids. *J. Biol. Chem.* **244,** 3635.

Lackey, G. D., Gross, K. C., and Wallner, S. J. (1980). Loss of tomato cell wall galactan may involve reduced rate of synthesis. *Plant Physiol.* **66,** 532.

LaJollo, F. M., Tannenbaum, S. R., and Labuza, T. P. (1971). Reaction at limited water concentration. 2. Chlorophyll degradation. *J. Food Sci.* **36,** 850.

Lambers, J. J. W., Terpstra, W., and Levine, J. K. (1984a). Studies on the action of the membrane enzyme chlorophyllase. 1. Binding of the fluorescence probe 4,4′-*bis*(1-anilino-8-naphthalene sulfonate)[*Bis*(ans)]. Influence of Mg^{2+} and intrinsic lipids. *Biochim. Biophys. Acta* **789,** 188.

Lambers, J. J. W., Terpstra, W., and Levine, J. K. (1984b). Studies on the action of the membrane chlorophyllase. 2. Interaction of chlorophyllide with chlorophyllase and with chlorophyllase complexed with 4,4′-bis(1-anilino-8-naphthalene sulfonate [Bis(Ans)]. *Biochim. Biophys. Acta* **789,** 197.

Lammers, S. M. (1981). "All About Tomatoes," p. 93. Ortho Books, Chevron Chemical Co., San Francisco, California.

Lamport, D. T. A. (1965). The protein component of primary cell walls. *Adv. Bot. Res.* **2,** 151.

Lance, C., Hobson, G. E., Young, R. E., and Biale, J. B. (1965). Metabolic processes in cytoplasmic particles of the avocado fruit. VII. Oxidative and phosphorylative activities throughout the climacteric cycle. *Plant Physiol.* **40,** 1116.

Lance, C., Hobson, G. E., Young, J. B., and Biale, J. B. (1967). Metabolic processes in cytoplasmic particles in avocado fruit. IX. The oxidation of pyruvate and malate during the climacteric cycle. *Plant Physiol.* **42,** 471.

Lau, J. M., McNiel, M., Darvill, A. G., and Albersheim, P. (1985). Structure of the backbone of rhamnogalacturonan I, a pectic polysaccharide in the primary cell walls of plants. *Carbohydr. Res.* **137,** 111.

Legge, R. L., and Thompson, J. E. (1983). Involvement of hydroperoxides and an ACC-derived free radical in the formation of ethylene. *Phytochemistry* **22,** 2161.

Legge, R. L., Thompson, J. E., and Baker, J. E. (1982). Free radical-mediated formation of ethylene from 1-aminocycloprane-1-carboxylic acid, a spin-trap study. *Plant Cell Physiol.* **23,** 171.

Liao, F. W. H., and Luh, B. S. (1970). Anthocyanin pigments in *Tinto Cao* grapes. *J. Food Sci.* **35,** 41.

Lieberman, M., and Mapson, L. (1964). Genesis and biogenesis of ethylene. *Nature (London)* **204,** 343.

Lieberman, M., Mapson, L. W., Kunishi, A. T., and Wardale, D. A. (1965). Ethylene production from methionine. *Biochem. J.* **97,** 449.

Lieberman, M., Kunishi, A. T., Mapson, L. W., and Wardale, D. A. (1966). Stimulation of ethylene production in apple tissue slices by methionine. *Plant Physiol.* **41,** 376.

Lipton, W. J. (1957). Physiological changes in harvested asparagus (*Asparagus officinalis*) as related to temperature. Ph.D. Thesis. Univ. Calif., Davis.

Lipton, W. J. (1977). Recommendations for CA storage of broccoli, brussel sprouts, cabbage, cauliflower, asparagus and potatoes. *Mich. State Univ. Hortic. Rep.* **28,** 277–280.

Lipton, W. J. (1987). Senescence in leafy vegetables. *HortScience* **22,** 854.

Lipton, W. J., and Ryder, E. J. (1989). Lettuce. *In* "Quality and Preservation of Vegetables" (N. A. M. Eskin, ed.), chap. 4, pp. 212–245. CRC Press.

Liu, Y., Hoffman, N. E., and Yang, S. F. (1985). Promotion by ethylene of the capability to convert 1-aminocyclopropane-1-carboxylic acid to ethylene in preclimacteric tomato and cantaloupe fruits. *Plant Physiol.* **77,** 407.

Loomis, W. D., and Battaile, J. (1963). Biosynthesis of terpenes. III. Mevalonic kinase from higher plants. *Biochim. Biophys. Acta* **67,** 54.

Lurssen, K., Naumann, K., and Schroder, R. (1979). 1-Aminocyclopropane-1-carboxylic acid—An intermediate in the ethylene biosynthesis in higher plants. *Z. Pflanzenphysiol.* **92,** 285.

Lyons, J. M., McGlasson, W. B., and Pratt, H. K. (1962). Ethylene production, respiration, and internal gas concentrations in cantaloupe fruits at various stages of maturity. *Plant Physiol.* **37,** 31.

Machackova, I., and Zmrhal, Z. (1981). Is peroxidase involved in ethylene biosynthesis? *Physiol. Plant.* **53,** 479.

MacKinney, G. (1961). Coloring matters. *In* "The Orange: Its Biochemistry and Physiology" (W. B. Sinclair, ed.), pp. 302–333. University of California Printing Dep., Berkeley.

MacKinney, G., Lukton, A., and Greenbaum, A. (1958). Carotenoid stability in stored dehydrated carrots. *Food Technol.* **12,** 164.

Macrae, A. R. (1971). Malic enzyme activity of plant mitochondria. *Phytochemistry* **10,** 2343.

Mancinelli, A. L. (1984). Red–far red reversibility of anthocyanin synthesis in dark-grown and light-pretreated seedlings. *Plant Physiol.* **76,** 281.

Marei, N., and Crane, J. C. (1971). Growth and respiratory response of fig (*Ficus carica* L. cv. Mission) fruits to ethylene. *Plant Physiol.* **48,** 249.

Markovic, O., Heinrichova, K., and Lenkey, B. (1975). Pectolytic enzymes from banana. *Collect. Czech. Chem. Commun.* **40,** 769.

Marx, R., and Brinkmann, K. (1978). Characteristics of retenone-insensitive oxidation of matrix-NADH broad bean mitochondria. *Planta* **142,** 83.

Matile, P. (1980). Catabolism of chlorophyll, involvement of peroxidase? *Z. Pflanzenphysiol.* **99S,** 475.

Matsuo, T., and Itoo, S. (1982). A model experiment for deastringency of persimmon fruit with high carbon dioxide treatment, *in vitro* gelation of kaki-tannin by reacting with acetaldehyde. *Agric. Biol. Chem.* **46,** 683.

Mattoo, A. K., and Lieberman, M. (1977). Localization of the ethylene-synthesising system in apple tissue. *Plant Physiol.* **60,** 794.

Mattoo, A. K., Baker, J. E., Chalutz, E., and Lieberman, M. (1977). Effect of temperature on the ethylene-synthesising system in apple, tomato and *Penicillium digitatum. Plant Cell Physiol.* **1228,** 715.

Maudinas, B., Bucholtz, M. L., Papastephanou, C., Katigar, S. S., Briedes, A. V., and Porter, J. W. (1975a). Adenosine 5′-triphosphate stimulation of the activity of a partially purified phytoene synthetase complex. *Biochem. Biophys. Res. Commun.* **66,** 430.

Maudinas, B., Bucholz, M. L., and Porter, J. W. (1975b). The partial purification and properties of a phytoene synthetase complex isolated from tomato fruit plastids. *Abstr. Int. Symp. Carotenoids, 4th,* p. 41.

Maunders, M. J., Brown, S. B., and Woodhouse, H. W. (1983). The appearance of chlorophyll derivatives in senescing tissue. *Phytochemistry* **22,** 2443.

Mayak, S., Legge, R. L., and Thompson, J. E. (1981). Ethylene formation from 1-aminocyclopropane-1-carboxylic acid by microsomal membranes from senescing carnation flowers. *Planta* **153,** 49.

Mazza, G., and Brovillard, R. (1987). Recent developments in the stabilization of anthocyanins in food products. *Food Chem.* **25,** 207.

McCready, B. M., and McComb, E. A. (1954). Pectic constituents in ripe and unripe fruit. *Food Res.* **19,** 530.

McFeeters, R. F., Bell, T. A., and Fleming, H. P. (1980). An endopolygalacturonase in cucumber fruit. *J. Food Biochem.* **4,** 1.

McGlasson, W. B., Palmer, J. K., Vendrall, M., and Brady, C. (1971). Metabolic studies with banana fruits. II. Effect of inhibitors on respiration, ethylene production and ripening. *Aust. J. Biol. Sci.* **24,** 103.

McGlasson, W. B., Wade, N. L., and Adato, I. (1978). *In* "Phytohormones and Related Compounds: A Comprehensive Treatise" (D. S. Letham, P. B. Goodwin, and T. J. Higgins, eds.), Vol. 1, pp. 475–519. Academic Press, London and New York.

McKenzie, K. A. (1932). Respiration studies with lettuce. *Proc. Am. Soc. Hortic. Sci.* **28,** 244.

McNab, J. M., Villemez, C. L., and Albersheim, P. (1968). Biosynthesis of galactan by a particulate enzyme preparation from *Phaseolus aureus* seedlings. *Biochem. J.* **106,** 355.

McNiel, M., Darvill, A. G., and Albersheim, P. (1980). Structure of plant cell walls. X. Rhamnogalacturonan 1. A structurally complex pectic polysaccharide in the walls of suspension-cultured sycamore cells. *Plant Physiol.* **66,** 1128.

McRae, D. G., Baker, J. E., and Thompson, J. E. (1982). Evidence for involvement of the superoxide radical in the conversion of 1-aminocyclopropane-1-carboxylic acid to ethylene by pea microsomal membranes. *Plant Cell Physiol.* **23,** 375.

Meigh, D. F., Jones, J. D., and Hulme, A. C. (1967). The respiration climacteric in the apple. Production of ethylene and fatty acids in fruit attached and detached from the tree. *Phytochemistry* **6,** 1507.

Meller, E., Harel, E., and Kannangara, C. G. (1979). Conversion of glutamic-semialdehyde and 4, 5-dioxovaleric acid to 5-aminolevulinic acid by cell-free preparations from greening maize leaves. *Plant Physiol.* **635,** 98.

Miller, C. O., and Conn, E. E. (1980). Metabolism of hydrogen cyanide by higher plants. *Plant Physiol.* **65,** 1199.

Millerd, A., Bonner, J., and Biale, J. B. (1953). The climacteric rise in fruit respiration as controlled by phosphorylative coupling. *Plant Physiol.* **28,** 521.

Miyazaki, J. H., and Yang, S. F. (1987). Inhibition of the methionine cycle enzymes. *Phytochemistry* **26,** 2655.

Moreau, F., and Romani, R. (1982). Malate oxidation and cyanide insensitive respiration in avocado mitochondria during the climacteric cycle. *Plant Physiol.* **70,** 1385.

Morin, F., Rigault, R., and Hartmann, C. (1985). Conséquences d'un séjour au froid sur le métabolism de l'ethylene au cours de la maturation de la poire Passe-Crassane après recolte. *Physiol. Veg.* **23.**

Moustafa, E., and Wong, E. (1967). Purification and properties of chalcone-isomerase from soya bean seed. *Phytochemistry* **6,** 625.

Murr, D. P., and Yang, S. F. (1975). Inhibition of *in vivo* conversion of methionine to ethylene by L-canaline and 2, 4-dinitrophenol. *Plant Physiol.* **55,** 79.

Nagel, C. W., and Patterson, M. E. (1967). Pectic enzymes and development of the pear *(Pyrus communis). J. Food Sci.* **32,** 294.

Naito, K., Ebato, T., Endo, Y., and Shimizu, S. (1980). Effect of benzyladenine on γ-aminolevulinic acid synthetic ability and γ-aminolevulinic acid dehydratase differential response to benzyladenine according to leaf age. *Z. Pflanzenphysiol.* **96,** 95.

Nandi, D. L., and Waygood, G. R. (1967). Biosynthesis of porphyrins in wheat leaves. *Can. J. Biochem.* **45,** 322.

Nelmes, B. J., and Preston, R. D. (1968). Wall development in apple fruit: A study of the life history of a parenchyma. *J. Exp. Bot.* **19,** 496.

Neuburger, M., and Douce, R. (1980). Effect of bicarbonate and oxaloacetate on malate oxidation by spinach leaf mitochondria. *Biochim. Biophys. Acta* **589,** 176.

Neumann, D. (1984). Subcellular localization of anthocyanins in red cabbage seedlings. *Biochem. Physiol. Pflanz.* **178,** 405.

Northcote, D. H. (1963). The biology and chemistry of the cell walls of higher plants, algae and fungi. *Int. Rev. Cytol.* **14,** 223.

Northcote, D. H. (1972). Chemistry of the plant cell wall. *Annu. Rev. Plant Physiol.* **23,** 13.

Obrenovic, S. (1985). Effect of light pretreatments on light induced anthocyanin formation in *Sinapsis alba* L. *Biochem. Physiol. Pflanz.* **180,** 25.

Odawara, S., Watanabe, A., and Imaseki, H. (1977). Involvement of cellular membranes in regulation of ethylene production. *Plant Cell Physiol.* **18,** 5699.

Ogura, K., Nishino, T., and Seto, S. (1968). Purification of prenyltransferase and isopentyl pyrophosphate isomerase of pumpkin fruit and some of their properties. *J. Biochem. (Tokyo)* **64,** 197.

Okuda, T., Hatano, T., and Ogawa, N. (1982a). Rugosin D, E, F and G, dimeric and trimeric hydrolyzable tannins. *Chem. Pharm. Bull.* **30,** 4234.

Okuda, T., Yoshida, T., Kuwaharu, M., Memon, M., and Shingu, T. (1982b). Agrimonin and potentillin, an ellagitannin dimer and monomer having an α-glucose core. *J. Chem. Soc., Chem. Commun.* p. 351.

Ozawa, T., Lilley, T. H., and Haslam, E. (1987). Polyphenol interactions: Astringency and the loss of astringency in ripening fruit. *Phytochemistry* **26,** 2937.

Padfield, C. A. S. (1975). Storage conditions for some of the more recently introduced apple varieties. *Orchadist N. Z.* **48,** 64.

Paillard, N. (1968). Analyse de l'arome de pommes de la variété Calville blanc par chromatographic sur colonne capillaire. *Fruits* **23,** 283.

Palmer, J. K. (1971). The banana. *In* "The Biochemistry of Fruits and Their Products" (A. C. Hulme, ed.), Vol. 2, pp. 65–105. Academic Press, London and New York.

Palmer, J. K. (1973). Separation of components of aroma concentrates on the basis of functional group and aroma quality. *J. Agri. Food Chem.* **21,** 923.

Palmer, J. M. (1976). The organization and regulation of electron transport in plant mitochondria. *Annu. Rev. Plant Physiol.* **27,** 133.

Panyatatos, N., and Villemez, C. L. (1973). The formation of a β-(1–4)-D-galactan chain catalyzed by a *Phaseolus aureus* enzyme. *Biochem. J.* **133,** 263.

Pardo, A. D., Chereskin, B. M., Castelfranco, P. A., Franceschi, V. R., and Wezelman, (1980). ATP requirement for higher plants. *Annu. Rev. Plant Physiol.* **24,** 129.

Park, Y., Morris, M. M., and MacKinney, G. (1973). On chlorophyll breakdown in senescent leaves. *J. Agric. Food Chem.* **21,** 279.

Patel, D. S., and Phaff, H. J. (1960a). Studies on the purification of tomato polygalacturonase. *Food Res.* **25,** 37.

Patel, D. S., and Phaff, H. J. (1960b). Properties of purified tomato polygalacturonase. *Food Res.* **25,** 47.

Peacock, B. C. (1972). Role of ethylene in the initiation of fruit ripening. *Queensl. J. Agric. Sci.* **29,** 137.

Pecket, C., and Small, C. J. (1980). Occurrence, location and development of anthocyanoplasts. *Phyotochemistry* **19,** 2571.

Peiser, G. D., Wang, T.-T., Hoffman, N. E., and Yang, S. F. (1983). Evidence for CN formation from [^{1}C]ACC during *in vivo* conversion of ACC to ethylene. *Plant Physiol. Suppl.* **72,** No. 203 (abstr.).

Peiser, G. D., Wang, T.-T., Hoffman, N. E., Yang, S. F., Liu, H.-W., and Walsh, C. T. (1984). Formation of cyanide from carbon 1 of 1-aminocyclopropane-1-carboxylic acid during its conversion to ethylene. *Proc. Natl. Acad. Sci. U.S.A.* **81,** 3059

Peng, C. Y., and Markakis, P. (1963). Effect of phenolase on anthocyanins. *Nature (London)* **199,** 2571.

Pesis, E., Fuchs, Y., and Zauberman, G. (1978). Cellulase activity in avocado. *Plant Physiol.* **61,** 416.

Peynaud, E., and Riberau-Gayon, P. (1971). The grape. *In* "The Biochemistry of Fruits and Their Products" (A. C. Hulme, ed.), Vol. 2, pp. 171–205. Academic Press, London and New York.

Pharr, D. M., and Dickinson, D. B. (1973). Partial characterization of C cellulase and cellobiase from ripening tomato × fruit. *Plant Physiol.* **51,** 577.

Picha, D. H. (1987). Sugar and organic acid content of cherry tomato fruit at different ripening stages. *HortScience* **22,** 94.

Pirrung, M. C. (1985). Ethylene biosynthesis. 3. Evidence concerning the fate of C1–N1 of ACC. *Bioorg. Chem.* **13,** 219.

Pirrung, M. C., and Brauman, J. I. (1987). Involvement of cyanide in the regulation of ethylene biosynthesis. *Plant Physiol. Biochem. (Paris)* **25,** 55.

Platenius, H. (1942). Effect of temperature on the respiration rate and quotient of some vegetables. *Plant Physiol.* **17,** 179.

Pollock, C. J., and ap Rees, T. (1975). Activities of enzymes of sugar metabolism in cold-stored tubers of *Solanum tuberosum. Phytochemistry* **14,** 613.

Porritt, S. W. (1977). Conditions and practices used in CA of apples in Western US and Canada. *Mich. State Univ. Hortic. Rep.* **28,** 231–232.

Porter, J. W., and Lincoln, P. E. (1950). *Lycopersicon* selection containing a high content of carotenes and colorless polyenes. II. The mechanism of carotene biosynthesis. *Arch. Biochem.* **27,** 390.

Potty, V. H., and Breumer, J. (1970). Formation of isoprenoid pyrophosphates from mevalonate by orange enzymes. *Phytochemistry* **9,** 1229.

Poulson, R., and Polglase, W. J. (1975). The enzymatic conversion of protoporphyrinogen-IX to protoporphyrin IX. Protoporphyrinogen oxidase activity in mitochondrial extracts of *Saccharomyces cerevisiae. J. Biol. Chem.* **250,** 1269.

Prasad, D. D. K., and Prasad, A. R. K. (1987). Effect of lead and mercury on chlorophyll synthesis in mung bean seedlings. *Phytochemistry* **9,** 99.

Pratt, H. K. and Groeschl, J. D. (1968). *In* "Biochemistry and Physiology of Plant Growth Substances" (F. Wrightmann, and G. Setterfield, eds.), pp. 1295–1302. Runge Press, Ottawa, Canada.

Pratt, H. K., Morris, L. L., and Tucker, C. L. (1954). Temperature and lettuce deterioration of broccoli varieties. *Proc. Conf. Transport. Perishables,* Univ. Calif., Davis, pp. 77–83.

Pressey, R. (1977). Enzymes involved in fruit softening. *ACS Symp. Ser.* **47,** 172–191.

Pressey, R. (1983). β-Galactosidases in ripening tomatoes. *Plant Physiol.* **71,** 132.,

Perssey, R. (1987). Exopolygalacturonase in tomato fruit. *Phytochemistry* **26,** 1867.

Pressey, R., and Avants, J. K. (1973). Separation and characterization of endopolygalacturonase and exopolygalacturonase from peaches. *Plant Physiol.* **52,** 252.

Pressey, R., and Avants, J. K. (1975). Cucumber polygalacturonase. *J. Food Sci.* **40,** 937.

Pressey, R., and Avants, J. K. (1976). Pear polygalacturonases. *Phytochemistry* **15,** 1349.

Pressey, R., and Avants, J. K. (1978). Differences in polygalacturonase composition of Clingstone and Freestone peaches. *J. Food Sci.* **43,** 1415.

Pressey, R., and Avants, J. K. (1982a). Pectic enzymes in Long Keeper tomatoes. *HortScience* **17,** 398.

Pressey, R., and Avants, J. K. (1982b). Solubilization of cell walls by tomato polygalacturonases: Effects of pectinesterases. *J. Food Biochem.* **6,** 57.

Pressey, R., Hinton, D. M., and Avants, J. K. (1971). Development of polygalacturonase activity and solubilization of pectin in peaches during ripening. *J. Food Sci.* **36,** 1070.

Proctor, J. T., and Creasy, L. L. (1971). Effects of supplementary light on anthocyanin synthesis in 'McIntosh' apples. *J. Am. Soc. Hortic. Sci.* **96,** 523.

Qureshi, A. A., Manok, K., Qureshi, N., and Porter, J. W. (1974). The enzymatic conversion of *cis*-[C]phyto-14-fluene, *trans*[C] carotene to poly-*cis* acyclic carotenes by a cell-free preparation of tangarene tomato fruit plastids. *Arch. Biochem. Biophys.* **162,** 108.

Ramadoss, C. S., Pistorius, E. K., and Axelrod, B. (1978). Coupled oxidation of carotene by lipoxygenase requires one soybean lipoxygenase isoenzyme. *Arch. Biochem. Biophys.* **190,** 549.

Ramaswamy, N. K., and Nair, P. M. (1974). Temperature and light dependency of chlorophyll synthesis in potatoes. *Plant Sci. Lett.* **2,** 249.

Ramaswamy, N. K., and Nair, P. M. (1976). Pathway for the biosynthesis of delta aminolevulinic acid in greening potatoes. *Indian J. Biochem. Biophys.* **13,** 294.

Rando, R. R. (1974). β γ-Unsaturated amino acids as irreversible inhibitors. *Nature (London)* **250,** 586.

Rebeiz, C. A. (1982). Chlorophyll: Anatomy of a discovery. *Chem-Tech. (Heidelberg)* **15,** 52.

Rexova-Benkova, L., and Markovic, O. (1976). Pectic enzymes. *Adv. Carbohydr. Chem. Biochem.* **33,** 311–367.

Reynolds, P. A., and Klein, B. P. (1982). Purification and characterization of a type-1 lipoxygenase from pea seeds. *J. Agric. Food Chem.* **30,** 1157.

Rhodes, M. J. C. (1970). The climacteric and ripening of fruits. In "The Biochemistry of Fruits and Their Products" (A. C. Hulme, ed.), Vol. 1, p. 524. Academic Press, London and New York.

Rhodes, M. J. C. (1971). Respiration and senescence of plant organs. In "The Biochemistry of Fruits and Their Products" (A. C. Hulme, ed.), Vol. 2. Academic Press, New York.

Rich, P. R., Boveris, A., Bonner, W. D., and Moore, A. L. (1976). Hydrogen peroxide generated by the alternative oxidase of higher plants. *Biochem. Biophys. Res. Commun.* **3,** 695.

Richmond, A., and Biale, J. B. (1966). Protein and nucleic acid metabolism in fruits. Studies of amino acid incorporation during the climacteric rise in respiration of the avocado. *Plant Physiol.* **41,** 1247.

Riov, J., and Yang, S. F. (1982). Effects of exogenous ethylene on ethylene production in citrus leaf tissue. *Plant Physiol.* **70,** 136.

Roberston, G. L., and Swinburne, D. (1981). Changes in chlorophyll to pheophytin. *J. Am. Chem. Soc.* **62,** 755.

Robinson, J. M., Smith, M. G., and Gibbs, M. (1980). Influence of hydrogen peroxide upon carbon dioxide photoassimilation in the spinach chloroplasts. *Plant Physiol.* **65,** 755.

Rogers, L. J., Sha, S. P. J., and Goodwin, T. W. (1966). Mevalonate-kinase isoenzymes in plant cells. *Biochem. J.* **100,** 14c.

Rohwer, F., and Mader, M. (1981). The role of peroxidase in ethylene formation from 1-aminocyclopropane-1-carboxylic acid. *Z. Pflanzenphysiol.* **104,** 363.

Romani, R. J., and Ku, L. (1968). Direct gas chromatographic analysis of volatiles produced by ripening fruit. *J. Food Sci.* **31,** 558.

Rustin, P., Moreu, F., and Lance, C. (1980). Malate oxidation in plant mitochondria via malic enzyme and the cyanide insensitive electron transport pathway. *Plant Physiol.* **66,** 457.

Rutherford, P. P., and Whittle, R. (1982). The carbohydrate composition of onions during long-term cold storage. *J. Hortic. Sci.* **57,** 349.

Rutherford, P. P., and Whittle, R. (1984). Methods of predicting the long-term storage of onions. *J. Hortic. Sci.* **59,** 537.

Ryall, A. L., and Lipton, W. J. (1979). Vegetables as living products—respiration and heat production. *In* "Handling, Transportation, and Storage of Fruits and Vegetables," 2nd ed., Vol. 1, Chapt. 1. Avi Publishing Company, Westport, Connecticut.

Saito, N., Osawa, Y., and Hayashi, K. (1971). Platyconin, a new acylated anthocyanin in Chinese Bell-Flower. *Platycodon grandiflorum. Bot. Mag.* **85,** 105; *Chem. Abstr.* **77,** 149733.

Sakamura, S., Watanabe, S., and Obata, Y. T. (1966). Separation of a polyphenol oxidase for anthocyanin degradation in eggplant. *J. Food Sci.* **31,** 181.

Salimen, S. O., and Young, R. E. (1975). The control properties of phosphofructokinase in relation to the respiratory climacteric in banana fruit. *Plant Physiol.* **55,** 45.

Salunkhe, D. K., and Do, J. Y. (1976). Biogenesis of aroma constituents of fruits and vegetables. *CRC Crit. Rev. Food Sci. Nutr.* **8,** 161.

Salvador, G. F. (1978). δ-Aminolevulinic acid synthesis from γ-, δ-dioxovaleric acid by acellular preparations of *Euglena gracilis. Plant Sci. Lett.* **13,** 351.

Saunders, J. A., and Conn, E. C. (1978). Presence of cyanogenic glucoside dhurrin in isolated vacuoles from sorghum. *Plant Physiol.* **61,** 154.

Schanderl, S. H., Chichester, C. O., and Marsh, B. L. (1962). Degradation of chlorophyll and several derivatives in acid solution. *J. Org. Chem.* **27,** 3865.

Schmid, R., and Shemin, D. (1955). The enzymic formation of porphobilinogen from γ-aminolevulinic acid and its conversion to protoporphyrin. *J. Am. Chem. Soc.* **77,** 506.

Schoch, S., and Vielwerth, F. X. (1983). Chlorophyll degradation in senescent tobacco cell culture (*Nicotiana tabacum* var "Samsun"). *Z. Pflanzenphysiol.* **110,** 317.

Schwartz, S. J., and von Elbe, J. H. (1983). Kinetics of chlorophyll degradation to pyropheophytin in vegetables. *J. Food Sci.* **48,** 1303.

Schwartz, S. J., Wool, S. L., and von Elbe, J. H. (1981). High performance liquid chromatography of chlorophylls and their derivatives in fresh and processed spinach. *J. Agric. Food Chem.* **29,** 533.

Segerlind, L. J., and Herner, R. C. (1972). On the fiber content problem of processed asparagus. *CASAE Publ.* **72,** 882.

Seymour, G. B., Harding, S. E., Taylor, A. J., Hobson, G. E., and Tucker, G. A. (1987a). Polyuronide solubilization during ripening of normal and mutant tomato fruit. *Phytochemistry* **26,** 1871.

Seymour, G. B., Lasslett, Y., and Tucker, G. A. (1987b). Differential effects of pectolytic enzymes on tomato polyuronides *in vivo* and *in vitro*. *Phytochemistry* **26,** 3137.

Shemin, D., and Russell, R. S. (1953). δ-Aminolevulinic acid: Its role in the biosynthesis of porphyrins and purines. *J. Am. Chem. Soc.* **75,** 4873.

Shewfelt, A. L, Panter, V. A., and Jen, J. J. (1971). Textural changes and molecular characteristics of pectin constituents in ripening peaches. *J. Food Sci.* **36,** 573.

Simpson, K. L., Lee, T. C., Rodriguez, D. B., and Chichester, C. O. (1976). Metabolism in senescent and stored tissues. *In* "Chemistry and Biochemistry of Plant Pigments" (T. W. Goodwin, ed.), 2nd ed., Vol. 1, pp. 779–843. Academic Press, New York.

Singh, S., and Sanwall, G. G. (1973). An allosteric α-glucan phosphorylase from banana fruits. *Biochim. Biophys. Acta* **309,** 280.

Singh, S., and Sanwall, G. G. (1975). Characterization of multiple forms of α-glucan phosphorylase from *Musa paradisiaca. Phytochemistry* **14,** 113.

Singh, S., and Sanwall, G. G. (1976). Multiple forms of α-glucan phosphorylase in banana fruits: Properties and kinetics. *Phytochemistry* **15,** 1447.

Singleton, V. L., Gortner, W. A., and Yang, H. Y. (1961). Carotenoid pigments of pineapple fruit. 1. Acid-catalyzed isomerization of the pigments. *J. Food Sci.* **26,** 49.

Small, C. J., and Pecket, R. C. (1982). The ultrastructure of anthocyanoplasts in red-cabbage. *Planta* **154,** 97.

Smith, S., Geeson, J., and Stow, J. (1987). Production of modified (plastic) atmospheres in deciduous fruits by the use of films and coating. *HortScience* **22,** 772.

Sobotka, F. E., and Stelzig, A. A. (1974). An apparent cellulase complex in tomato (*Lycopersicon esculentum* L.) fruit. *Plant Physiol.* **53,** 759.

Solomos, T. (1977). Cyanide-resistant respiration in higher plants. *Plant Physiol.* **28,** 279.

Solomos, T., and Laties, G. G. (1973). Cellular organization and fruit ripening. *Nature (London)* **245,** 390.

Solomos, T., and Laties, G. G. (1974). Similarities between ethylene and cyanide action in triggering the rise in respiration in potato slices. *Plant Physiol.* **54,** 506.

Solomos, T., and Laties, G. G. (1976). Induction by ethylene of cyanide-sensitive and cyanide-resistant plant tissues. *Plant Physiol.* **55,** 47.

Soniso, S. (1986). Plastic keeps ripeness under wraps. *New Sci.* **111**(1521), 35.

Spurr, A. R., and Harris, W. M. (1968). Ultrastructure of chloroplasts and chromoplasts in *Capsicum anuum.* I. Thykaloid membrane changes during fruit ripening. *Am. J. Bot.* **55,** 1210.

Stevens, M. A., Kader, A. A., and Albright, M. (1979). Potential for increasing tomato flavor via increased sugar and acid content. *J. Am. Soc. Hortic. Sci.* **104,** 40.

Stoll, K. (1970). Apfellungerung in kontrollierter atmosphare (CA) unter modifiziernen Konditionen. *Schweiz. Z. Obst- Weinbau* **106,** 334.

Stoll, K. (1973). Tables on the storage of fruits and vegetables in controlled atmospheres. *Eidg. Forschungsanst. Obst. Wein Gartenbau, Wadenswil, Flugschr.*, p. 78.

Stone, E. J., Hall, R. M., and Kazeniac, S. J. (1975). Formation of aldehydes and alcohols in tomato fruit from UC labelled linolenic and linoleic acids. *J. Food Sci.* **40,** 1138.

Suomalainen, H., and Keranen, A. J. A. (1961). The first anthocyanins appearing during the ripening of blueberries. *Nature (London)* **191,** 498.

Surendranathan, K. K., and Nair, P. M. (1973). Alterations in carbohydrate metabolism of γ-irradiated Cavendish banana. *Phytochemistry* **12,** 241.

Tager, J. M., and Biale, J. B. (1957). Carboxylase and aldolase activity in the ripening banana. *Physiol. Plant.* **10,** 79.

Takeguchi, C. A., and Yamamoto, H. (1968). Light-induced oxygen-18 uptake by epoxy xanthophylls in New Zealand spinach leaves *(Tetragonia expansa). Biochim. Biophys. Acta* **150,** 459.

Talburt, W. F., and Smith, O. (1975). "Potato Processing," 3rd ed. Avi Publ. Co., Westport, Connecticut.

Tavakoli, M., and Wiley, R. C. (1968). Relation between trimethylsilyl derivatives of fruit tissue polysaccharides to apple texture. *Proc. Am. Soc. Hortic. Sci.* **92,** 780.

Terpstra, W. (1981). Identification of chlorophyllase as a glycoprotein. *FEBS Lett.* **126,** 231.

Terpstra, W., and Lambers, J. W. J. (1983). Interactions between chlorophyllase, chlorophyll a, plant lipids and Mg. *Biochim. Biophys. Acta* **746,** 23.

Themmen, A. P. N., Tucker, G. A., and Grierson, D. (1982). Degradation of isolated tomato cell walls by purified polygalacturonase *in vitro*. *Plant Physiol.* **69,** 249.

Theologies, A., and Laties, G. G. (1978). Respiratory contribution of the alternative pathway during various stages of ripening in avocado and banana fruit. *Plant Physiol.* **62,** 122.

Thibault, J.-F. (1983). Enzymatic degradation and B-elimination of the pectic substances in cherry fruit. *Phytochemistry* **22,** 1567.

Thomson, W. W. (1966). Ultrastructural development of chromoplasts in Valencia oranges. *Bot. Gaz. (Chicago)* **127,** 133.

Thomson, W. W., Lewis, L. N., and Coggins, C. W. (1967). The reversion of chromoplasts to chloroplasts in Valencia oranges. *Cytologia* **32,** 117.

Tigchelaar, E. C., McGlasson, W. B., and Buescher, R. W. (1978a). Genetic regulation of tomato fruit ripening. *HortScience* **13,** 508.

Tigchelaar, E. C., McGlasson, W. B., and Franklin, M. J. (1978b). Natural and ethephon-stimulated ripening of F hybrids of the ripening inhibitor (*rin*) and non-ripening inhibitor (*nor*) mutants of tomato (*Lycopersicon esculentum* Mill.). *Aust. J. Plant Physiol.* **5,** 449.

Tigier, H. A., Batlle, A. M. del C., and Locascio, G. (1968). Porphyrin biosynthesis in the soybean callus system. II. Improved purification and some properties of δ-aminolevulinic acid dehydratase. *Enzymologia* **38,** 43.

Tigier, H. A., Batlle, A. M. del C., and Locascio, G. (1970). Porphyrin biosynthesis in soybean callus tissue system. Isolation, purification and general properties of γ-aminolevulinate dehydratase. *Biochim. Biophys. Acta* **151,** 300.

Timberlake, C. F., and Bridle, P. (1975). Anthocyanins. *In* "The Flavonoids" (J. B. Harborne, T. J. Mabry, and H. Mabry, eds.), Chapter 5, pp. 214–266. Chapman & Hall, London.

Timberlake, C. F., and Bridle, P. (1982). Distribution of anthocyanins in food plants. *In* "Anthocyanins as Food Colors" (P. Markakis, ed.), p. 125. Academic Press, New York.

Toul, V., and Popsilova, J. (1966). Chemical composition of onion varieties (*Allium cepa* L.) *Bull. Vysk. Zelinarsky Olomouc* **10,** 55.

Tressl, R., and Drawert, F. (1973). Biogenesis of banana volatiles. *J. Agric. Food Chem.* **21,** 560.

Tressl, R., and Jennings, W. G. (1972). Production of volatile compounds in the ripening banana. *J. Agric. Food Chem.* **20,** 190.

Tressl, R., Drawert, F., and Heimann, W. (1970). Uber die Biogenese von Aromastoffen bei Pflanzen und Fruchten. V. Anreicherung, Trennung und Identifizierrung von Banaenaromastoffen. *Z. Lebensm. -Unters. -Forsch.* **142,** 249.

Trout, S. A., Huelin, F. E., and Tindale, G. B. (1960). The respiration of Washingtron navel and Valencia oranges. *Div. Food Preserv. Tech. Pap. (Aust., C.S.I.R.O.)* **14,** 1.

Tucker, G. A., Robertson, N. G., and Grierson, D. (1980). Changes in polygalacturonase isoenzymes during the ripening of normal and mutant tomato fruit. *Eur. J. Biochem.* **112,** 119.

Tucker, G. A., Roberston, N. G., and Grierson, D. (1982). Purification and changes in activities of tomato pectin-esterase isoenzymes. *J. Sci. Food Agric.* **33,** 396.

U.S. Department of Agriculture. (1968). "The Commercial Storage of Fruits, Vegetables, and Florist and Nursery Stocks," Agric. Handb. No. 66. USDA, Washington, D.C.

Van Buren, J. P. (1979). The chemistry of texture in fruits and vegetables. *J. Texture Stud.* **10,** 1.

Vick, B. A., and Zimmerman, D. C. (1976). Lipoxygenase and hydroperoxide lyase in germinating watermelon seedlings. *Plant Physiol.* **57,** 580.

Vick, B. A., and Zimmerman, D. C. (1983). The biosynthesis of jasmonic acid: A physiological role for plant lipoxygenase. *Biochem. Biophys. Res. Commun.* **111,** 470.

Villemez, C. L., Swanson, A. L., and Hassid, W. Z. (1966). Properties of a polygalacturonic acid-synthesizing enzyme system from *Phaseolus aureus* seedlings. *Arch. Biochem. Biophys.* **116,** 446.

von Mollendorff, L. J., and De Villiers, O. T. (1988). Role of pectolytic enzymes in the development of wooliness in peaches. *J. Hortic. Sci.* **63,** 53.

Wagenknecht, A. C., and Lee, F. A. (1958). Enzyme action and off-flavor in frozen peas. *Food Res.* **23,** 25.

Wagner, G. J. (1979). Content and vacuole/extravacuole distribution of neutral sugars, free amino acids, and anthocyanins in protoplasts. *Plant Physiol.* **64,** 88.

Walker, G. C. (1964). Color deterioration in frozen French beans (*Phaseolus vulgaris*). *J. Food Sci.* **29,** 383.

Wallner, S. J. (1978). Apple fruit β-galactosidase and softening in storage. *J. Am. Soc. Hortic. Sci.* **103,** 364.

Wallner, S. J., and Bloom, H. L. (1977). Characteristics of tomato cell wall degradation *in vitro.* Implications for the study of fruit-softening enzymes. *Plant Physiol.* **55,** 94.

Weinstein, J. D., and Castelfranco, P. A. (1978). Mg protoporphyrin-IX and δ-aminolevulinic acid synthesis from glutamate in isolated greening chloroplasts. δ-Aminolevulinic acid synthesis. *Arch. Biochem. Biophys.* **186,** 376.

Wellman, E., Hrazdina, G., and Grisebach, H. (1976). Induction of anthocyanin formation and of enzymes related to its biosynthesis by UV light in cell cultures of *Haplopappus gracilis. Phytochemistry* **15,** 913.

White, R. C., Jones, I. D., and Gibbs, E. (1963). Determination of chlorophylls, chlorophyllides and pheophorbides in plant material. *J. Food Sci.* **28,** 431.

Workman, M., Pratt, H. K., and Morris, L. (1957). Studies on the physiology of tomato fruits. 1. Respiration and ripening behaviour at 20°C as related to date of harvest. *Proc. Am. Soc. Hort. Sci.* **69,** 352.

Worth, H. G. J. (1967). The chemistry and biochemistry of pectic substances. *Chem. Rev.* **67,** 465.

Yamaki, S., and Kakiuchi, N. (1979). Changes in hemi-cellulose degrading enzymes during development and ripening of Japanese pear fruit. *Plant Cell Physiol.* **20,** 311.

Yang, S. F. (1974). Ethylene biosynthesis in fruit tissues. *Colloq. Int. C. N. R. S.* **238.**

Yang, S. F. (1980). Regulation of ethylene biosynthesis. *HortScience* **15,** 238.

Yang, S. F., and Ho, H. K. (1958). Biochemical studies on postripening banana. *J. Chin. Chem. Soc. (Taiwan)* **5,** 71.

Yang, S. F., and Hoffman, N. E. (1984). Ethylene biosynthesis and its regulation in higher plants. *Annu. Rev. Plant Physiol.* **35,** 155.

Yoon, S., and Klein, B. (1979). Some properties of pea lipoxygenase isoenzymes. *J. Agric. Food Chem.* **27,** 955.

Young, R. E., Salimen, S., and Sornrivichai, P. (1974). Enzyme regulation associated with ripening in banana fruit. *Colloq. Int. C. N. R. S.* **238,** 271.

Yu, M. H., and Spencer, M. (1969). Conversion of L-leucine to certain keto acids by a tomato enzyme preparation. *Phytochemistry* **8,** 1173.

Yu, M. H., Olson, D. E., and Salunkhe, D. K. (1967). Precursors of volatile components in tomato fruit. I. Compositional changes during development. *Phytochemistry* **6,** 1457.

Yu, M. H., Salunkhe, D. K., and Olson, L. E. (1968a). Production of 3-methylbutanol from L-leucine by tomato extract. *Plant Cell Physiol.* **9,** 633.

Yu, M. H., Olson, D. E., and Salunkhe, D. K. (1968b). Precursors of volatile components in tomato fruit. III. Enzymatic reaction products. *Phytochemistry* **7,** 555.

Yu, M. H., Olson, L. E., and Salunkhe, D. K. (1968c). Precursors of volatile components in tomato fruit. II. Enzymatic reaction products. *Phytochemistry* **7,** 561.

Yu, U. B., Adams, D. O., and Yang, S. F. (1979). 1-Aminocyclopropane-1-carboxylate synthase, a key enzyme in ethylene biosynthesis. *Arch. Biochem. Biophys.* **198,** 180.

Yung, K. H., and Yang, S. F. (1980). Biosynthesis of wound ethylene. *Plant Physiol.* **66,** 281.

Zamora, R., Olras, J. M., and Mesias, J. L. (1987). Purification and characterization of tomato lipoxygenase. *Phytochemistry* **26,** 345.

Zapsalis, C., and Francis, F. J. (1965). Cranberry anthocyanins. *J. Food Sci.* **30,** 396.

Zaubermann, A., and Schiffman-Nadel, M. (1972). Pectin methylesterase and polygalacturonase in avocado fruit at various stages of development. *Plant Physiol.* **49,** 864.

Zimmerman, D. C., and Coudron, C. A. (1979). Identification of traumatin, a wound hormone, as 12-oxo-trans-10-dodecendic acid. *Plant Physiol.* **63,** 536.

Zimmerman, D. C., and Vick, A. A. (1970). Hydroperoxide isomerase: A new enzyme of lipid metabolism. *Plant Physiol.* 46, 445.

3

Biochemical Changes in Raw Foods: Cereals

I. Introduction

Cereal grains are an important dietary source for the world's population, contributing 70 and 50% of the total calories and protein, respectively. In spite of the decline in consumption of cereals in North America, they still provide 20% of calories, protein, magnesium, and zinc, 30–40% of carbohydrate and iron, 20–30% of riboflavin and niacin, and over 40% of thiamine in the diet (Marston and Welsh, 1980). The five main cereals are wheat, maize, barley, oats and rice, all members of the Gramineae family characterized by starchy seeds (Sylvester-Bradley and Folkes, 1976). The estimated world production of these cereals in 1984–1985 is shown in Table 3.1. This chapter will discuss those biochemical changes taking place during development, germination, and storage of cereal grains with particular attention to wheat.

II. Cereal Grain Structure

The cereal seed is composed of three main tissues; the embryo, the endosperm, and the aleurone layer surrounding the storage endosperm. This is illustrated for wheat in Figure 3.1, in which the endosperm comprises over 80% of

TABLE 3.1

WORLD PRODUCTION OF CEREALS IN 1984–1985[a]

Cereal	3×10^3 metric tons
Wheat	510,029
Maize	490,155
Barley	178,004
Oats	45,562
Rice	465,970

[a] From Food and Agriculture Organization (1986).

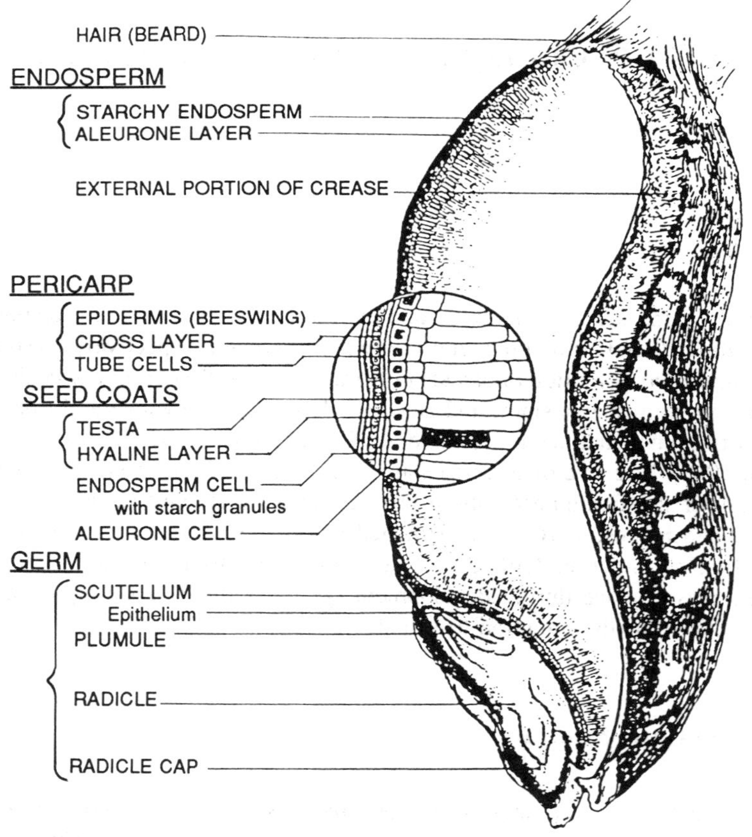

FIG. 3.1. Diagrammatic longitudinal section of wheat grain through crease and germ (based on drawing by E. E. McDermott) (Kent, 1983). Reprinted with permission. Copyright © by Pergamon Press.

the grain weight while the aleurone cells and the germ tissue containing the embryo embedded in the surrounding scutellum account for 15 and 3%, respectively. The anatomical structure of all cereal grains is essentially similar with some minor differences. For instance, wheat and maize are surrounded by a fruit coat or pericarp and seed coat or testa, together referred to as a naked caryopsis. In the case of barley, oats, and rice, however, an additional husk is found surrounding the caryopsis or kernel of the grain. There are several varieties of barley in which the husk is not attached to the caryopsis.

III. Cereal Grain Composition

The largest portion of the seed is the endosperm, which provides the nutrients necessary for embryo development during germination. The nutrients are made available by the release of enzymes from the aleurone layer and embryo which hydrolyze the endosperm reserves. These reserves are contained in discrete storage bodies identified as starch granules and protein bodies.

A. Amyloplasts

Amyloplasts are plastids or organelles in which the starch granules are laid down. The rate of starch synthesis in cereal grains is one of the factors that effects both grain size and yield (Kumar and Singh, 1980). In the mature endosperm of wheat, barley, and rye, starch is found as two distinct fractions based on the size of the granules. The primary or A-type starch granules range in size from 20–45 μm, while the secondary or B-type granules rarely exceed 10 μm in diameter (Evers, 1973). Examination of the particle size distribution in wheat endosperm starch by Evers and Lindley (1977) showed that those starch granules with less than 10 μm diameter accounted for approximately one-third of the total weight of starch. The presence of these two starch granule types in wheat kernels was confirmed in studies by Baruch *et al.* (1979). They found that the size of the starch was affected by seasonal changes in much the same way as grain yield and protein content. The starch granule occupies only a very small part of the total plastid during initial kernel development but accounts for close to 93% at maturity (Briarty *et al.*, 1979). In the mature endosperm, A-type starch granules account for only 3% of the total number of granules although they represent 50–70% of the total weight, due to their larger size (Evers and Lindley, 1977). The smaller B-type granules, however, make up 97% of the total number of starch granules but account for only 25–50% of the overall weight.

Isolated starch granules also contain protein, most of which could be easily removed by washing repeatedly with water. A small part of the protein, however, still remained strongly associated with the granule itself. Lowy *et al.* (1981) found that this protein fraction was readily extracted with salt solution and suggested that it was associated with the starch granule surface. This salt-extractable fraction accounted for 8% of the total protein in the starch granule. The major protein fraction had a molecular weight of around 30,000 and was associated with both A- and B-type starch granules. Based on amino acid analysis this protein was quite different from wheat gluten. An additional protein fraction was extracted from A-type starch granules but only following gelatinization in the presence of sodium dodecyl sulfate. This fraction was quite different and based on electrophoresis, was thought to be part of the internal granule components.

B. The Starch Granule

The shape and size of the starch granule varies with the different cereals (Table 3.2). Starch is composed of amylose and amylopectin with the level of amylose ranging from 20–30% for most cereal starches. In the case of certain varieties of maize, barley, and rice, the starch is composed almost exclusively of amylopectin and these are referred to as "waxy." High-amylose starches are also found, for example, in the case of amylomaize.

C. Biosynthesis of Starch

As described in Chapter 2, starch synthesis is achieved through the action of starch synthase, which can utilize either adenosine diphosphoglucose (ADPG) or

TABLE 3.2

Structure and Amylose Content of Some Whole Granular Cereal Starches[a]

Source	Granule shape	Granule size (nm)	Amylose content (%)
Wheat	Lenticular or round	20–25	22
Maize	Round or polyhedral	15	28
Waxy maize	Round	15(5–15)	1
High-amylose	Round or irregular sausage-shaped	25	52
Barley	Round or elliptical	20–25	22
Rice	Polygonal	3–8	17–19[b] 21–23[c]
Oats	Polyhedral	3–10	23–24

[a] Adapted from Lineback (1984).
[b] Japonica.
[c] Indica.

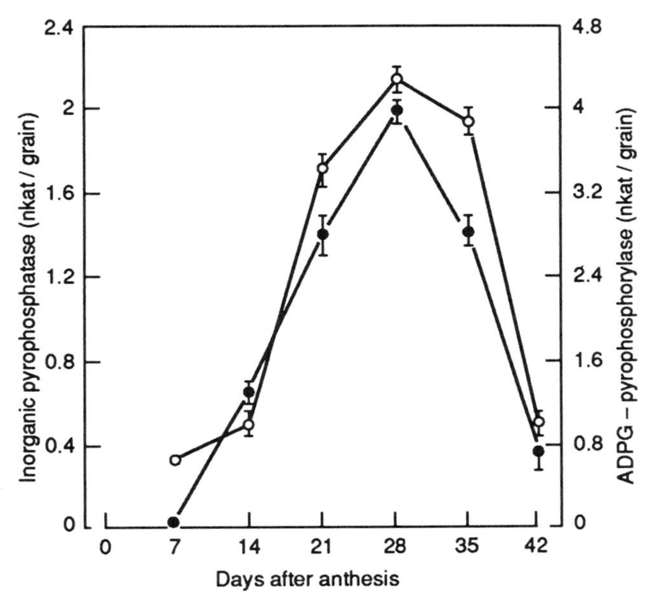

FIG. 3.2. Activity of alkaline inorganic pyrophosphatase (○) and ADPG-pyrophosphorylase (●) during wheat grain development (Kumar and Singh, 1983).

uridine diphosphoglucose (UDPG) as substrates (Recondo and Leloir, 1961). ADPG appeared to be the more active glucosyl donor and was formed by the action of ADPG-pyrophosphorylase (Preiss and Levi, 1979):

$$\text{ATP} + \alpha\text{-glucose-1-P} \rightarrow \text{ADPG} + \text{PP}_i$$

The amount of PP_i in developing grains is controlled by the enzyme alkaline inorganic pyrophosphatase (EC 3. 6. 1. 1). This enzyme limits the accumulation of PP_i and was thought to be the controlling factor in starch synthesis, as PP_i inhibited ADPG-pyrophosphorylase in sweet corn (Amir and Cherry, 1972). The activity of both ADPG-pyrophosphorylase and alkaline pyrophosphatase was monitored by Kumar and Singh (1983) during the development of wheat grain .Their results, shown in Figure 3.2, indicate that both enzymes increased steadily, reaching a maximum 28 days after anthesis, but then declined with maturity. The rapid increase in alkaline pyrophosphatase activity 14 days after anthesis corresponded with the period of rapid starch synthesis. The inability of the intermediate metabolites of sucrose–starch conversion to inhibit the activity of alkaline pyrophosphatase eliminated any possible regulatory role for this enzyme in starch biosynthesis.

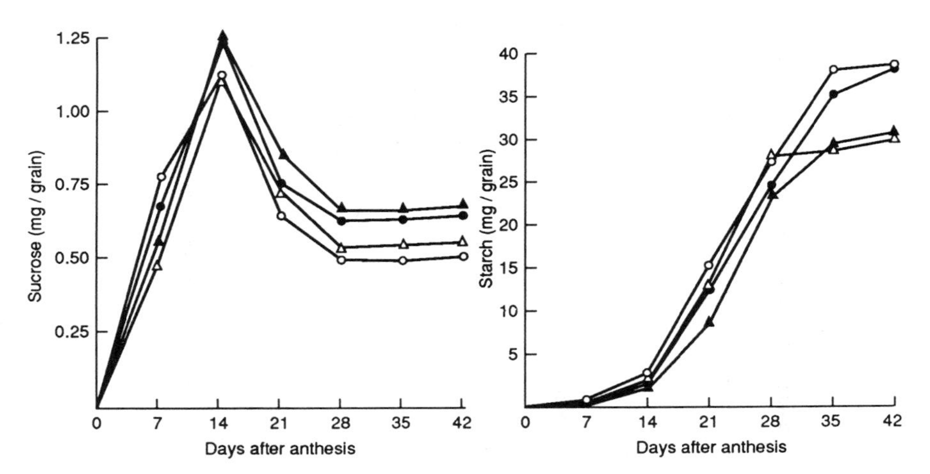

FIG. 3.3. Changes in sucrose and starch (mg per grain) during the development of four wheat grains (Kumar and Singh, 1981). Reprinted with permission. Copyright © by Pergamon Press.

D. SUCROSEESTARCH CONVERSION IN DEVELOPING GRAINS

The amount of free sugars formed during the development of wheat was examined by Kumar and Singh (1981) in relation to grain size and starch content. Their results, summarized in Figure 3.3, indicate that the nonreducing sugar sucrose reached a maximum level 14 days after anthesis, then declined and leveled off after 28 days. Starch synthesis was negligible after 7 days but increased markedly after 14 days, then continued until 35 days after anthesis. The rapid decline in sucrose and reducing sugars once starch synthesis commenced suggested the involvement of hydrolytic enzymes, including invertase. The activity of this enzyme was found by Kumar and Singh (1980) to decrease to negligible levels after 21 days compared to the rather rapid rise in sucrose-UDPG glucosyl tranferase activity (Fig. 3.4). The latter enzyme, also referred to as sucrose synthetase, catalyzes the first step in the formation of starch from sucrose as discussed later in this section. The parallel activities of sucrose-UDPG glucosyl transferase and starch synthesis suggested that this enzyme played a major role in the hydrolysis of sucrose. Kumar and Singh (1984) suggested that the initial role of invertase was to provide substrates for energy liberating respiratory enzymes needed for sustaining active cell division.

Chevalier and Lingle (1983) reported that insoluble invertase was located mainly in the outer pericarp with only slight activity in the endosperm (Figure 3.5). These researchers monitored sucrose synthetase activity which was found

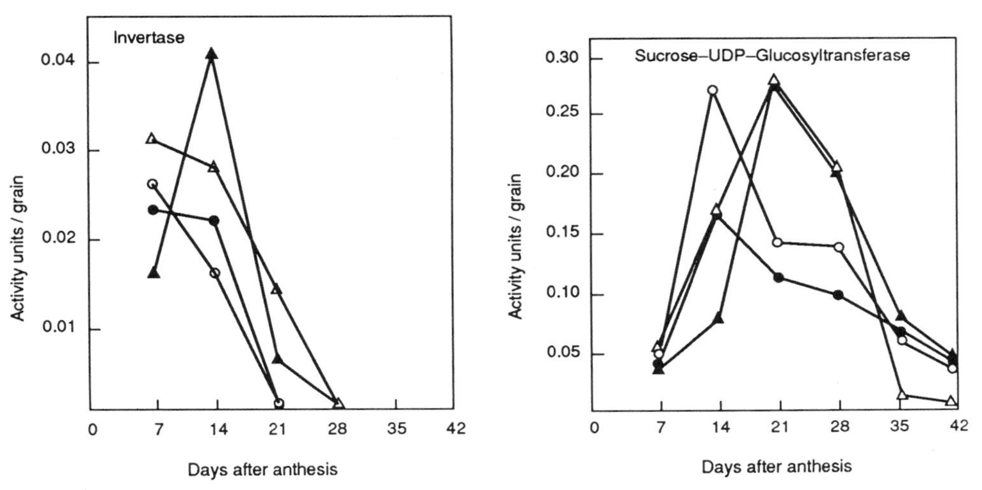

FIG. 3.4. Changes in invertase and sucrose-UDP glucosyl transferase activities during the development of wheat grains (Kumar and Singh, 1980). Reprinted with permission. Copyright © by Pergamon Press.

predominantly in the endosperm. Using whole wheat kernels, Kumar and Singh (1980) monitored invertase activity during the early stages of grain development and that of sucrose synthetase as the grain matured. Chevalier and Lingle (1983) also found an increase in free sucrose in mature wheat and barley kernels, which was consistent with earlier research with wheat (Cerning and Guilbot, 1973), barley (Laberge *et al.*, 1973), and rice (Singh and Juliano, 1977). The marked rise in sucrose observed by Lingle and Chevalier (1980) in the endosperm fraction was accompanied by decreasing sucrose synthetase activity. This decline in sucrose synthetase activity was considered to be an important factor in controlling grain filling. It appeared to be responsible for accumulation of sucrose in extracellular spaces such as the endosperm cavity since the endosperm was now unable to utilize any incoming sucrose. The overall effect was to prevent any more sucrose from entering the kernel.

Kumar and Singh (1984) confirmed the accumulation of sucrose up to 14 days after anthesis, which represented rapid translocation from photosynthetic parts to the wheat endosperm followed by active starch synthesis. Previous work by Chevalier and Lingle (1983) demonstrated the movement of sucrose from the phloem to the endosperm in developing wheat and barley kernels. Using wheat endosperm slices, Rijven and Gifford (1983) also found sucrose to be the preferred substrate for starch synthesis as it was not hydrolyzed prior to its uptake by the endosperm.

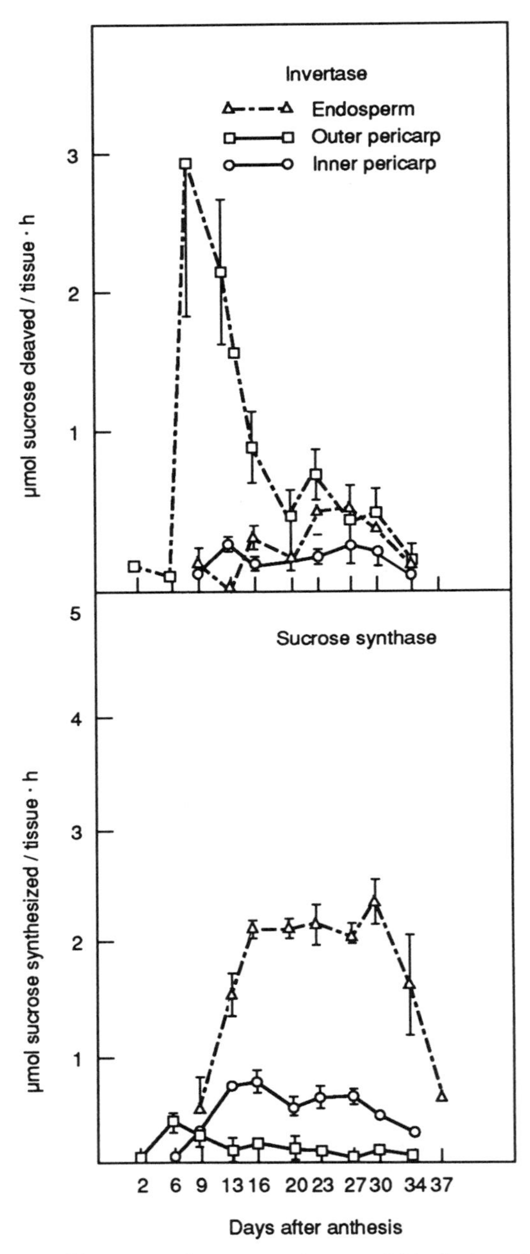

FIG. 3.5. Distribution of invertase and sucrose synthetase activities in the endosperm and pericarp of developing wheat grains. Reproduced from Chevalier and Lingle (1983). With permission of the Crop Science Society of America, Inc.

E. Starch Synthesis

The *in vivo* synthesis of starch involves phosphorylase or synthetase leading to the formation of the linear polymer amylose. Once sucrose enters the endosperm it becomes the starting point for amylose synthesis. The first step involves its conversion to uridine diphosphoglucose by sucrose synthetase (sucrose-UDP glucosyl transferase):

$$\text{Sucrose} + \text{UDP} \xrightarrow{\text{Sucrose—UDPG glucosyl transferase}} \text{UDP-glucose} + \text{fructose-1-P}$$

Following this, fructose-1-P is then converted to glucose-1-P by hexokinase, phosphoglucoisomerase, and phosphoglucomutase. Glucose-1-P is metabolized to ADPG by ADPG-pyrophosphorylase:

$$\text{Glucose-1-P} + \text{ATP} \xrightarrow{\text{ADPG-phosphorylase}} \text{ADPG} + \text{PP}_i$$

The absence of any detectable PP_i suggested that it is rapidly hydrolyzed by pyrophosphatase since, as discussed earlier, it is a potent inhibitor of ADPG-phosphorylase (Amir and Cherry, 1971). Amylose synthesis, as discussed earlier, can be mediated directly by starch synthetase involving UDPG or indirectly by ADPG-starch synthetase to ADPG *via* glucose-1-P:

$$\text{ADPG} + \text{primer (Gn)} \xrightarrow{\text{ADPG-starch synthetase}} \text{Glycosyl primer} \rightarrow (\text{Gn} + 1) + \text{ADP}$$

It appeared, however, that the ADPG reaction was the preferred one for starch biosynthesis in developing wheat grains (Kumar and Singh, 1984). In this reaction glucose is repeatedly transferred from ADPG to a small glucan primer until the elongated starch chain is formed. The extremely small amount of glucose-1-P in the developing grain suggested it was rapidly utilized, pointing to a possible regulatory role for phosphoglucomutase, the enzyme responsible for its formation, in starch biosynthesis. Kumar and Singh (1984) proved conclusively that the termination of starch accumulation in mature wheat grains was due to the loss in synthetic capacity of the endosperm and not due to unavailability of sucrose.

Joshi *et al.* (1980) attempted to explain the regulation of starch biosynthesis in normal and Opaque-2 maize during development of the endosperm. Opaque-2 maize was nutritionally superior although it had decreased grain yield and a lower protein and starch content. These researchers monitored the activities of sucrose-UDPG glucosyl transferase, glucose-6-phosphate ketoisomerase, and soluble and bound ADPG-starch glucosyl transferase in the developing endosperm for 30 days following pollination. Except for sucrose-UDPG glucosyl transferase, all the other enzymes were much lower in Opaque-2 maize, compared to the normal maize during the latter stages of endosperm development. The lower activity of these enzymes was responsible for the reduced amount of starch in Opaque-2 maize, which had 15% less starch content per endosperm. This was accompanied by a decreased protein synthesis in the Opaque-2 endosperm which explained the reduced enzyme synthesis during the later stages of endosperm development.

F. Starch Synthesis: Amylopectin

Biosynthesis of the branched chain amylopectin requires the formation of the amylose *via* phosphorylase or synthase as described in the previous section. The branch points (α-(1, 6)-D-glucosidic linkages) required for amylopectin are introduced by the branching enzyme Q-enzyme (EC 2. 4. 1. 18). Borovsky *et al.* (1979) concluded that the introduction of 1, 6-branch points is a random process in which the Q-enzyme interacts with two 1, 4-glucan chains held together in a possible double helix arrangement.

Amylose and amylopectin are synthesized concurrently in the ratio of 1 : 4 for ordinary starches (Robyt, 1984). A number of hypotheses have been developed to explain the side-by-side occurrence of amylose and amylopectin in the starch granule although our understanding of starch biosynthesis still remains incomplete (Erlander, 1958; Geddes and Greenwood, 1969; Marshall and Whelan, 1970). One such hypothesis suggested that some mechanism was operating which protected the linear polymer from the branching enzyme (Whelan, 1958, 1963). The participation of phospholipids in the regulation of amylopectin was proposed by Vieweg and De Fekete (1976) since they inhibit the action of the branching enzyme. Thus, only amylose without attached phospholipids could theoretically be converted, although this still remains to be verified. Another hypothesis, discussed earlier, is the possible specificity of the branching enzyme for a double helix arrangement involving the shorter amylopectin chains (Borovsky *et al.*, 1979; Robyt, 1984).

G. Protein Bodies

Protein bodies are membrane-bound cellular organelles containing storage proteins located in the starchy endosperm of cereals (Pernollet, 1978, 1982, 1984). They are also found in the aleurone layer, although these differ in composition, structure, and function. While the protein bodies in the endosperm have only a storage function, those in the aleurone layer possess both synthetic and

TABLE 3.3

ULTRASTRUCTURAL DIFFERENCES BETWEEN ALEURONE LAYER AND STARCHY ENDOSPERM PROTEIN BODIES[a]

Species	Aleurone layers		Endosperm	
	Diameter (μm)	Structure	Diameter (μm)	Structure
Wheat	2–3	Two kinds of inclusions	0.1–8	No inclusion. Granular structure
	4–5	One globoid and one crystalloid	1–2	No inclusion Lamellar structure
Barley	2–3	Two kinds of inclusions	2	No inclusion Lamellar structure
	4–5	One globoid and one crystalloid	1–2	No inclusion Lamellar structure
Rice	1.5–4	Globoid	2–5	No inclusion Homogenous
	1–3	Globoid	2–5	No inclusion Homogenous
Maize			1–2	No inclusion Homogenous

[a]Adapted from Pernollet (1978). Reprinted with permission. Copyright © by Pergamon Press.

secretory functions (Simmonds and O'Brien, 1981). Protein bodies in the aleurone layer are 2–4 μm in diameter with globoid and crystalline inclusions, while those in the starchy endosperm have a homogeneous granular structure devoid of inclusions. These differences have been confirmed in wheat, barley, maize, and rice by examination of their ultrastructural differences as indicated in Table 3.3.

In members of the *Triticum* species, these protein bodies vanish as the grain matures, as observed for wheat seeds (Simmonds, 1972; Pernollet and Mossé, 1983; Bechtel *et al.*, 1982a, b; Seckinger and Wolf, 1970), barley seeds (Burgess *et al.*, 1982), and rye seeds (Parker, 1981). This results in the conversion of the spherical protein granules into irregularly shaped protein masses which eventually become the matrix protein which is no longer bound by a membrane between the starch granules.

H. ORIGIN OF PROTEIN BODIES

The origin of protein bodies in the endosperm still remains unclear. Most researchers support their synthesis on the rough endoplasmic reticulum (RER) (Campbell *et al.*, 1981; Miflin *et al.*, 1981; Miflin and Burgess, 1982; Parker and Hawes, 1982), although Bechtel and co-workers (1982a,b) favored secretion of the wheat storage proteins. Irrespective of the mechanism proposed, the initiation and formation of the protein bodies involves the active participation of the Golgi apparatus. Pernollet and Camilleri (1983) examined protein body formation and development in wheat endosperm and found the polypeptides stored in all protein bodies to be similar. Earlier work by Tanaka *et al.* (1980) suggested that only one kind of protein was stored in wheat endosperm. The presence of all cell storage proteins in the protein bodies, however, pointed to a common synthetic pathway operating in wheat seeds. The polypeptides in the protein bodies were similar to those in the endoplasmic reticulum. This suggested the storage proteins were secretory proteins discharged into the endoplasmic reticulum before being translocated to the protein bodies. This model conflicted with the soluble mode of gliadin synthesis proposed by Bechtel *et al.* (1982b), but was in agreement with studies carried out by Greene (1981) and Donovan *et al.* (1982). These researchers reported that messenger RNAs encoding gliadin molecules were translated on polysomes bound to the endoplasmic reticulum.

Three distinct stages were noted by Pernollet and Camilleri (1983) during the development of wheat protein bodies. The initial stage involved protein synthesis of storage proteins and their association as small vesicles into bodies of 5–10 μm in the first month after anthesis. During the next stage, the formation of the small protein bodies slowed down and instead they coalesced into much larger bodies (50–100 μm). The instability of the membrane of these large protein bodies and the mechanical pressure of the developing starch granules resulted in disruption of the membrane with the release of the protein bodies into the matrix protein. This loss of the protein bodies and the matrix protein formation is characteristic of the final stage of development of the mature wheat endosperm. The model proposed by Pernollet and Camilleri (1983) in Fig. 3.6 summarizes the sequence of events leading to the formation of protein bodies in wheat and their eventual disruption.

The protein bodies of barley are similar to those of wheat, but differ quite markedly from those of maize. In maize the membrane is derived from the endoplasmic reticulum, which completely encloses the protein bodies.This differs from wheat and barley, where the endoplasmic reticulum is disrupted by wheat and barley protein body aggregates which are not completely surrounded by this membrane. Oparka and Harris (1982) reported that rice protein bodies were surrounded by a membrane derived from the endoplasmic reticulum.

I. Classification of Plant Proteins

Plant proteins were first classified by Osborne (1895) on the basis of their solubility in different solvents as summarized in the following scheme:

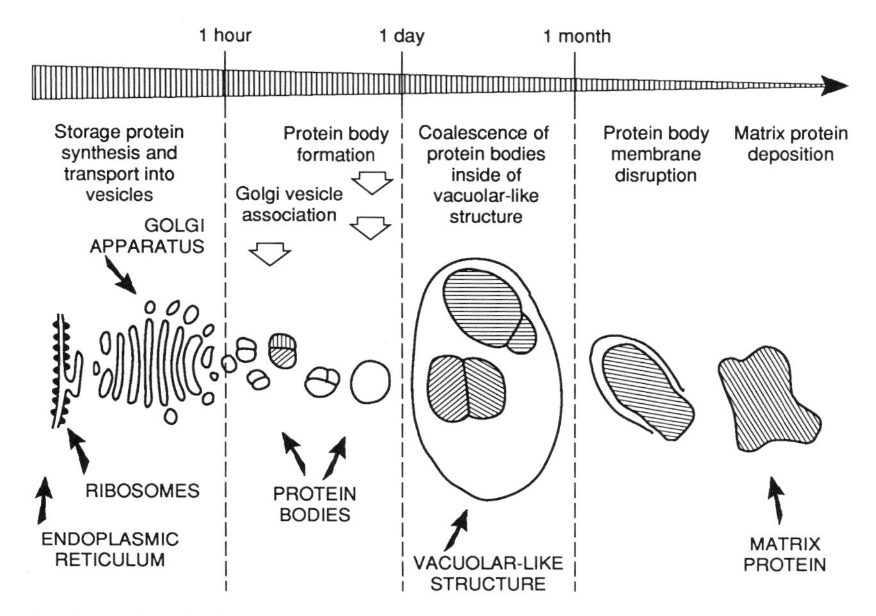

FIG. 3.6. Schematic diagram of wheat endosperm protein body formation and evolution (Pernollet and Camilleri, 1983).

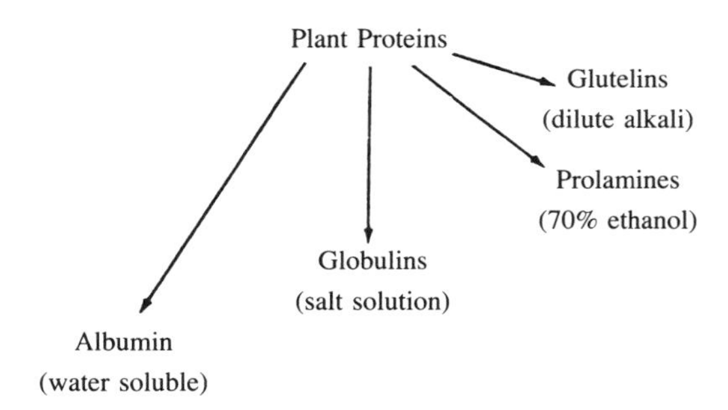

A number of modifications have since been introduced to improve extraction of these fractions. Current practice extracts a combined albumin–globulin fraction as salt-soluble protein while the prolamines are extracted with aqueous propan-1-ol or propan-2-ol plus a reducing agent (Shewry *et al.*, 1980) This method is appropriate for the study of the basic genetic products but quite inappropriate from a technological point of view, as reducing agents result in the reestablishment of new disulfide bonds that change the solubility of the fractions.

TABLE 3.4

RELATIVE PROPORTIONS (%) OF THE OSBORNE PROTEIN FRACTIONS IN CEREAL SEEDS[a]

Cereal	Nonprotein N	Albumins	Globulins	Prolamins	Glutelins	Residues
Barley[b]	11.6		15.6	45.2	18.0	5.0
Wheat[c]		33.1		60.7		6.2
Maize[b]	4.4	0.9	1.5	55.4	22.9	—
Rice[d]		15.7		6.7	61.5	15.4
Oats[e]	11		56	9	23	—

[a] From Bright and Shewry (1983). With permission.
[b] % total seed N.
[c] % recovered seed N.
[d] % total protein.
[e] % recovered protein.

To prevent denaturation of the glutelin fraction by alkali extraction, alternative extractants such as buffers containing the detergent sodium dodecyl sulfate (SDS) at pH 10 are used (Moreaux and Landry, 1968). The relative proportion of the Osborne protein fractions in the seeds of wheat, barley, maize, rye, and oats are summarized in Table 3.4.

J. PROLAMINS

The major storage proteins present in the starchy endosperm of wheat, barley, and maize are the alcohol-soluble proteins, the prolamins. These account for 30–60% of the total grain nitrogen depending on species, nutritional status, and genotype of the plant (Bright and Shewry, 1983; Shewry *et al.*, 1981). The prolamin fractions identified for different cereal species are listed in Table 3.5.

Prolamins derive their name from their unusually high content of proline and

TABLE 3.5

PROLAMIN FRACTIONS OF CEREAL GRAIN

Species	Trivial name
Wheat	Gliadin
Maize	Zein
Barley	Hordein
Oats	Avenin

TABLE 3.6

PROLAMIN FRACTIONS OF WHEAT, BARLEY, AND MAIZE

Wheat	MW	Barley	MW	Maize	MW
α-Gliadin	32,000	B-hordein	35,000–46,000	20 K	20,000–21,000
β-Gliadin	40,000	C-hordein	45,000–72,000	22 K	22,000–23,000
ϖ-Gliadin	40,000–72,000	D-hordein	100,000	9 K	9,000–10,000
HMW subunits	95,000–136,000			14 K	13,000–14,000

amide nitrogen (glutamine). This protein fraction is deficient in the essential amino acid lysine. Oats and rice differ substantially from other cereals in containing very little prolamin (5–10%), with the major storage proteins being globulin and a glutelinlike compound, respectively. Thus these cereals have much more lysine making them nutritionally superior. Electrophoretic separation of the different prolamin fractions on the basis of molecular size is accomplished by polyacrylamide gel electrophoresis in the presence of the detergent sodium dodecyl sulfate (SDS–PAGE). This permitted identification of the different polypeptide patterns in prolamins which vary considerably among different cultivars of the same species. PAGE is a widely used technique for varietal identification of single seeds of wheat and barley. When there are only very minor differences, two-dimensional isoelectric focusing (IEF) and PAGE can be effectively applied. Using these procedures the polypeptides identified for prolamin fractions in wheat, barley, maize, and oats are summarized in Table 3.6.

The wheat gliadins are classified into two groups based on their electrophoretic mobility at low pH. The first group includes the fastest fraction, α-gliadin, followed by β-, γ-, and ω-gliadins, while the second group, with a much higher apparent molecular weight (95,136,000), is referred to as high-molecular-weight units (HMW). All the gliadin fractions are deficient in lysine and threonine. Three groups of hordein protein were separated from barley by SDS–PAGE and are referred to as B, C, and D. They differed from each other in apparent molecular weights and amino acid composition (Miflin and Shewry, 1977). The C fraction had only trace amounts of sulfur amino acids while the D fraction was rich in glycine (13%). Lysine was particularly low in all the hordein protein fractions (<1%) while B and C hordeins were also deficient in threonine.

The zein component of maize protein, although not well defined, was composed of two major and two minor fractions. The two major fractions had apparent molecular weights of 20,000–21,000 and 22,000–23,000, while the minor fractions were 9,000–10,000 and 13,000–14,000, referred to as 20K, 22K, 9K, and 14K zein, respectively. All of these fractions were deficient in lysine. Unlike the other cereal grains, the major storage proteins of oats were 12 S and 7 S globulins as prolamins accounted for less than 15% of the total grain nitrogen (Peterson and Smith, 1976). Burgess and Miflin (1985) showed that 7 S globulin was located mainly in the embryo while 12 S globulin, the larger fraction, was predominant in the endosperm. Based on SDS–PAGE it appeared that the globulin and prolamin fractions were localized in different protein bodies.

K. Protein Synthesis

The development of cereal seed protein is associated with at least three stages. The first stage is characterized by rapid cell division in which protein synthesis remains quite low. When cell division ceases this is followed by an increase in the rough endoplasmic reticulum and accumulation of soluble nucleotides (Briarty *et al.*, 1979; Jenner, 1968). This results in a rapid synthesis of storage proteins which is related to initiation and synthesis of messenger RNA (mRNA) as well as the efficiency of mRNA translation. The accumulation of mRNA in developing wheat seeds was correlated with protein synthesis by Greene (1983). Using labeled [5-^{3}H]uridine and L-[^{3}H]leucine, he studied the synthesis, functioning, and stability of storage protein RNAs. Three developmental stages were apparent:

1. A change from seed protein synthesis of nonstorage to storage proteins.
2. An increase in the rate of accumulation of poly(A) +RNA.
3. An increase in the level of transcription mRNA

A direct relationship between mRNA levels and the rate of protein synthesis is shown in Figure 3.7. Synthesis of the gliadin peptide was predominant from 15 to 25 days following flowering and paralleled the increase in poly(A) + RNA. Thus the storage protein gene expression in wheat endosperm is an mRNA-limiting process based on the amount of storage protein which the mRNA synthesized near the end of endosperm cell division. Okita and Greene (1982) previously identified mRNAs as the major messenger species in Cheyenne seed responsible for gliadin synthesis 20 to 25 days after anthesis. For a more detailed review of cereal proteins the article by Laszity (1984) is recommended.

L. Lipids

Lipids are distributed throughout the cereal grain as part of the intracellular membranes and spherosomes. They are stored as triglyceride-rich droplets in the

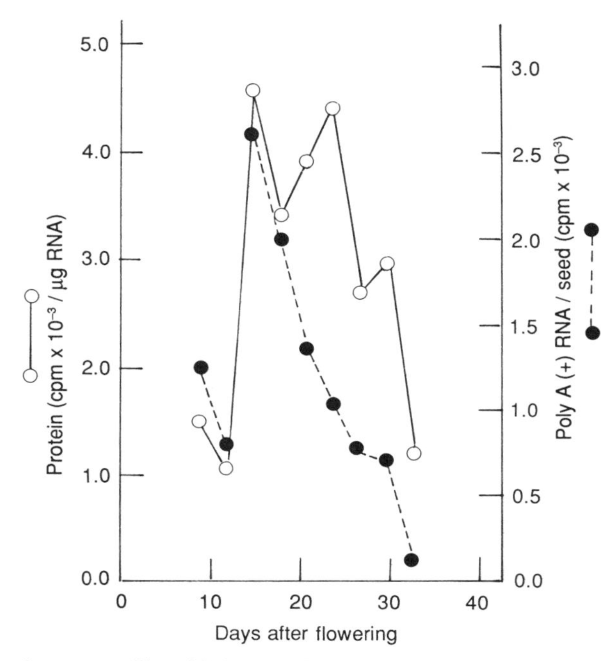

FIG. 3.7. Development profiles of Poly(A) + RNA accumulation and *in vitro* protein synthesis capacity in wheat (Greene, 1983). Reprinted with permission of copyright owner, American Society of Plant Physiology (ASPP).

spherosomes of the aleurone layer which are found clustered around the aleurone grains or with the plasmalemma (Buttrose, 1971; Chamura, 1975; Morrison *et al.*, 1975; Morrison, 1978). Spherosomes are also present in the embryo, scutellum, and coleoptile (Buttrose and Soeffly, 1973; Jelseman *et al.*, 1974; Swift and O'Brien, 1972). Lipids are also found in starch, primarily as monoacyl lipids, lysophosphatidyl ethanolamine, and lysophosphatidylcholine, and as inclusion complexes with amylose inside the starch granule. There appears to be a correlation between amylose and lipid content of cereals, for example, waxy maize has little lipid while high-amylose or amylomaize starch has a higher lipid content than normal maize starch (Acker and Becker, 1971). The distribution of lipids in mature cereal kernels is shown in Table 3.7.

The major fatty acids present in grain lipids are linoleic, oleic, palmitic, and linolenic acids in order of decreasing amounts (Price and Parsons, 1975). The major cereal lipids can be separated into polar and nonpolar lipids by solvent fraction. For example, in the case of a hard red spring wheat, Waldron, the polar and nonpolar lipids accounted for 49.6 and 50.4% of the total lipids, respectively

TABLE 3.7

LIPID CONTENT OF WHOLE GRAIN CEREALS

Cereal	Crude fat (%)
Wheat	1.8
Maize kernel	0.4–1.7
Barley	3.3–4.6
Oats	5.4
Rice	1.9–3.1

(Hargin and Morrison, 1980). The distribution of these lipid fractions within the wheat tissues is shown in Table 3.8.

The germ contains one-third of the total wheat lipids of which 80% are neutral triglycerides. The aleurone lipids account for one-quarter of the total lipids with 80% of these being nonpolar in nature.The endosperm, however, accounts for almost half of the whole kernel lipids. The endosperm starch is associated with 15.6% of the total lipids,of which 96% are phospholipids. The predominant phospholipid in starch endosperm is lysophosphatidyl-choline (Hargin and Morrison, 1980).

The biosynthesis of lipids begins with the formation of fatty acids by a multistep process involving a multienzyme complex, the ACP (acyl protein carrier) fatty acid synthetase. Once formed they are esterified with glycerol to form triglycerides, which serve as an important source of energy during germination of cereals. They are responsible for maintaining the embryo and aleurone layer during the initial stages of germination until sugars are provided from the starchy endosperm.

TABLE 3.8

DISTRIBUTION OF WHEAT LIPIDS WITHIN WHEAT TISSUES[a,b,c]

Total lipids							
Germ (30.4%)		Aleurone layer (24.8%)		Endosperm (44.8%)			
				Nonstarch (29.2%)		Starch (15.6%)	
Nonpolar lipids (24.1%)	Polar lipids (6.3%)	Nonpolar lipids (17.9%)	Polar lipids (6.9%)	Nonpolar lipids (9.7%)	Polar lipids (19.5%)	Nonpolar lipids (0.7%)	Polar lipids (14.9%)

[a] Adapted from Hargin and Morrison (1980).
[b] Calculated and adapted from data by Hargin and Morrison (1980).
[c] Expressed as percentage of total lipids.

IV. Germination of Cereals

Germination of cereals is important in the malting industry, which depends on a certain degree of starch degradation (see Chapter 6). In the production of baked products, however, it is important that most of the starch granules remain intact. Thus germination or sprouting of cereal grains affects the grading of wheat and other cereal grains as a result of the damage it causes. According to the Grain Grading Primer (U.S. Department of Agriculture, 1957), sprouting of wheat is defined as "kernels which have the germ end broken open from germination, and kernels from which sprouts have broken off." This is prevalent during wet weather when the moisture content is increased. The preharvest germination of the wheat reduces grain yield, flour yield, and flour quality. This has an adverse effect on the breadmaking properties of the flour because of the enhanced hydrolysis of the dough starch by α-amylase (Buchanan and Nicholas, 1980). If the activity of α-amylase is excessive it produces a bread product with a wet, sticky crumb.

A. Mobilization of Cereal Starches by α-Amylase

The native starch granule in wheat is attacked by certain α-amylase isoenzymes. Two groups were separated by Sargeant (1979) during germination of wheat, one of which hydrolyzed the starch granules. Halmer (1985) pointed out that since starch hydrolysis is normally carried out with soluble and not native granular starch granules, it is difficult to relate total amylolytic activity, as measured in the laboratory, to the granule-degrading activity of the cereal grain *in vivo*. The hydrolysis of starch by α-amylases is characterized by the endocleavage of amylose and amylopectin (see Chapter 6) (Abbott and Matheson, 1972).

B. Biosynthesis of α-Amylase during Germination

The importance of α-amylase activity in baking and brewing has focused considerable attention on its secretion during germination. A major controversy has centered on whether the site of α-amylase biosynthesis is in the scutellum or aleurone layer (Akazawa and Hara-Nishimura, 1985). In the case of barley grains, α-amylase formation *de novo* has been reported in both the scutellum and aleurone layer (Briggs, 1963, 1964; Chrispeels and Varner, 1967). The biosynthesis and secretion of this enzyme appears to involve the plant hormone gibberellin GA_3. This hormone is produced by the embryo, and triggers the production of α-amylase as well as other hydrolytic enzymes in the aleurone

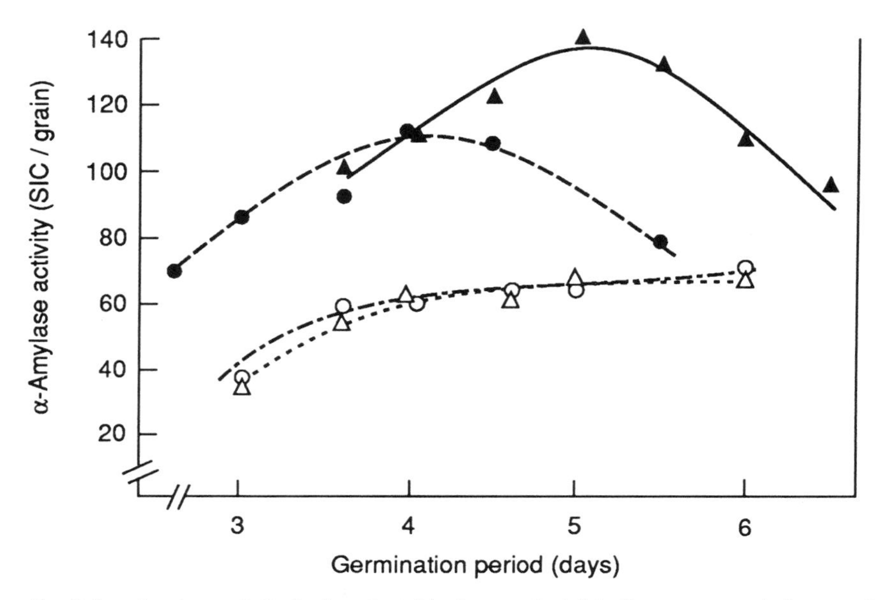

FIG. 3.8. α-Amylase activity in decorticated barley germinated in the presence and absence of K_2SO_4 and GA_3. No additives (○); K_2SO_4 (50 mM) (△); GA_3 (50 μg/ml) (●); GA_3 (50 μg/ml) and K_2SO_4 (▲) (Raynes and Briggs, 1985).

layer (Briggs *et al.*, 1981). The increase in enzyme activity was attributed to an increase in the level of α-amylase mRNA (Bernal-Lugo *et al.*, 1981; Higgins *et al.*, 1976). In the case of wheat, the aleurone layer also becomes the target of hormone-induced enzymes, including increased synthesis of α-amylase by GA (Filmer and Varner, 1967; Melcher and Varner, 1971). Varty and co-workers (1982) found that the plant hormone abscisic acid inhibited both transcription and translation of α-amylase mRNA in isolated wheat aleurone tissue. This explained the ability of abscisic acid to inhibit the induction of α-amylase by GA_3 (Chrispeels and Varner, 1967). Recent studies by Raynes and Briggs (1985) showed increased α-amylase production in decorticated barley grains germinated with or without gibberellic acid. Their results, shown in Figure 3.8, indicate that the onset and amount of enzyme activity were affected by GA_3 and K_2SO_4. The presence of K_2SO_4 appeared to delay the destruction of α-amylase (Briggs, 1968). Based on studies with rice scutellum, calcium also appears to play a role in the biosynthesis and secretion of α-amylase with the possible involvement of calmodulin (Mitsui *et al.*, 1984).

A number of researchers reported that the major isoenzyme form of α-amylase in germinated mature grain or aleurone tissue incubated with GA_3 was α-AMYI (MacGregor, 1983; Marchylo *et al.*, 1981; Sargeant, 1979, 1980). This differed from α-amylase production in pre-mature excised embryo/scutellar tissue in which α-AMY2 was the predominant isoenzyme formed even in the presence of GA_3. While this tissue normally produces little α-amylase activity in pre-mature wheat grain, once removed from the caryopsis it commences synthesizing α-amylase, resulting in the characteristic cytological changes associated with germination. Cornford *et al.* (1987) further examined the production of α-amylase in embryo/scutellar tissue from pre-mature wheat and found that it was influenced by embryo age. While both α-AMY1 and α-AMY2 forms were detected by rocket-line immunoelectrophoresis in the presence of GA_3, it was the production of α-AMY2 that was stimulated by the addition of this growth substance. Abscisic acid inhibited the production of α-AMY1 and several α-AMY2 bands, although four active α-AMY2 bands were still detected. This switching from the developmental to germinative mode by the excised embryos, in terms of α-amylase production, may be due to the loss of abscisic acid from the embryo (Triplett and Quatrano, 1982).

MacGregor and Matsuo (1982) conducted a detailed study on initial starch degradation during germination in endosperms of barley and wheat kernels. The kernels examined were all carefully split longtitudinally through the crease without distorting any of the structural features (Figure 3.9). Using scanning electron microscopy, similar physical changes were evident in both barley and wheat kernels during the initial stages. Starch degradation appeared to commence at the endosperm–embryo junction, then moved along the junction to the dorsal edge of the kernel. This effect was only observed once extensive degradation of the cell wall material and protein matrix in the endosperm had occurred. These results were consistent with earlier work showing that α-amylase synthesis during germination commenced in the embryo (Gibbons, 1979, 1980; Okamoto *et*

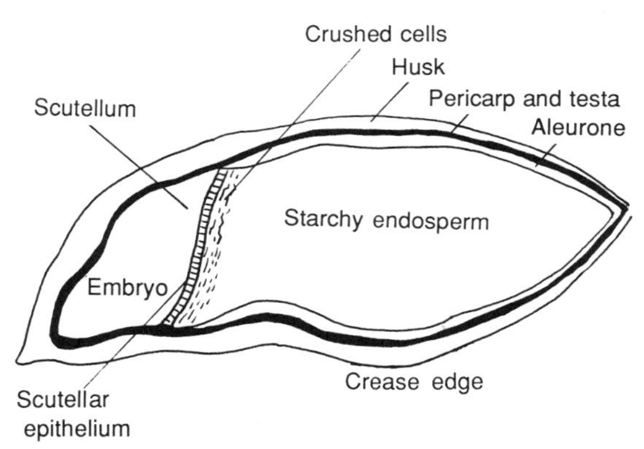

FIG. 3.9. Longitudinal section of barley kernel cracked open through the crease edge (MacGregor and Matsuo, 1982).

al., 1980). Irrespective of where α-amylase is synthesized, it is ultimately discharged into the endosperm, where starch hydrolysis takes place.

C. α-Amylase Activity in Germinated Cereals

During the course of germination starch is degraded by α-amylases and simple sugars are released (Kruger, 1972a,b). Lineback and Ponpipom (1977) monitored the degradation of starch during the germination of cereals, including wheat and oats. They found that an increase in α-amylase activity was accompanied by a rise in free sugars in all cereals examined. The amount produced reflected the degree of damaged starch in the flour milled from the germinated seed. Although the highest α-amylase activity was associated with germinated wheat, the degradation of starch was less compared to that in the other cereals. Starch degradation in wheat was evident by erosion of the granule surface and the equatorial groove. The starch granule from oats was much more resistant to enzyme attack, however, with little damaged starch in the milled flour.

Germination studies conducted on five wheat cultivars by Reddy and coworkers (1984) showed α-amylase development to be temperature dependent. Wheat kernels germinated in growth chambers at 15.5°C developed the highest enzyme activity compared to 20°C for the field-grown kernels. The enzyme activity did not increase significantly until the third day of germination and then rose markedly after 6 days.

D. Effect of Germination on Flour Quality

Lukow and Bushuk (1984) examined the effect of germination on wheat flour quality. Using flours from two cultivars of Canadian hard red spring wheat, they found that α-amylase activity was quite low but increased 1,600- and 3,000-fold during germination. This marked increase in enzyme activity was accompanied by a rise in reducing sugars, which explained the inferior baking characteristics of the germinated flours. The major effect of α-amylase activity was to reduce the water-binding properties of the flour by degradation of the gelatinized starch. The overall result was the production of bread in which the crumbs were damp and sticky (Jongh, 1967; Thomas and Lukow, 1969).

Kruger and Matsuo (1982) studied the effect of preharvest sprouting on the pasta-making quality of durum wheats. α-Amylase activity increased 155- and 320-fold when germinated for 72 and 120 hr, respectively. While cooking during semolina and spaghetti production decreased α-amylase activity it did not destroy the enzyme immediately. These researchers noted that α-amylase was still active during the first 6 min of cooking spaghetti and accounted for the production of reducing sugars, the substantial loss of solids, and the detrimental effect on spaghetti quality.

E. Treatment of Sprouted Grain: Reduction of α-Amylase

Germination of wheat grains commences at harvest time with an adverse effect on grain quality (Meredith and Pomeranz, 1985). The major culprit is α-amylase activity, which increases during germination while β-amylase activity remains unchanged. A variety of different methods have been examined to improve the baking properties of the sprouted grain. Since the starch fraction of sprouted wheat was of good quality, efforts were focused on inhibiting α-amylase activity using heat or chemical agents (Bean *et al.*, 1974; Cawley and Mitchel, 1968; McDermott and Elton, 1971; Westermarck-Rosendahl *et al.*, 1979). An early report by Schultz and Stephan (1960), for example, reported improvement in crumb structure when wheat was treated with acids. Fuller and co-workers (1970) used hydrochloric acid followed by neutralization with ammonia to reduce α-amylase activity, but their method proved impractical. A number of α-amylase inhibitors were examined by Westermarck-Rosendahl *et al.* (1979) to improve the baking qualities of sprouted wheat. The most promising agents were trisodium phosphate, disodium phosphate, sodium polyphosphate, sodium dodecyl sulfate, calcium steoryl lactylate, and citric acid. Evaluations were based on the falling number test values for grain samples in which the optimum value for baking flour was around 200 sec (Greenaway, 1969). These α-amylase inhibitors caused an increase in falling number values well above 200 sec, as shown in Table 3.9 for sodium polyphosphate.

The falling number test measures the time a plunger falls freely in a suspension of the flour in water and the effect of starch amylolytic degradation on the

TABLE 3.9

The Effect of Sodium Polyphosphate on the Falling Number of Sprout-Damaged Wheat[a]

Chemical agent	Concentration[b] (%)	Falling number
Sodium polyphosphate	0.1	147[c]
	0.5	175[d]
	1.0	250[d]

[a] Adapted from Westermarck-Rosendahl *et al.* (1979).
[b] Based on meal weight (moisture content 15.0%).
[c] Difference significant at 5%.
[d] Difference significant at 1%.

viscosity of the flour/water paste. The faster the decrease in viscosity of the flour paste, the lower is the falling number value. Further research by Westermarck-Rosendahl *et al.* (1980) showed that the most promising of the 23 enzyme inhibitors examined were trisodium phosphate and disodium hydrogen phosphate. These were particularly effective in reducing the stickiness problem associated with flours from sprouted wheat as well as improving crumb characteristics. A number of alternative solutions were discussed recently by Meredith and Pomeranz (1985), including the elimination of sprout-susceptible lines through breeding and selection programs.

F. Mobilization of Proteins during Germination

Essential amino acids increase during germination or sprouting of cereal grains (Dalby and Tsai, 1976; Tsai *et al.*, 1975). For example, both lysine and tryptophan increased during the germination of wheat, barley, oats, and rice. The extent of the increase was directly related to the decrease in prolamin content of the grain. A substantial increase in lysine of 50% was noted for wheat, compared to only a slight increase in oats (Figure 3.10). The level of prolamin in oats, however, was much lower than that in wheat. Jones and Tsai (1977) reported an increase in the lysine and tryptophan content of the embryo of normal maize and a corresponding decrease in the endosperm. This indicated that higher levels of lysine were required for embryo growth and development as observed previously by Singh and Axtell (1973) in studies on barley embryo and endosperm proteins. The precursors for lysine biosynthesis in maize may be provided by mobilization of zein reserves in the endosperm.

The release of amino acids during germination of wheat was investigated by Tkachuk (1979). After 122 hr of germination at 16.5°C the proline and glutamine content increased 100- and 80-fold, respectively, while lysine increased only 12-fold (Table 3.10). These results represented the changes taking place in the whole wheat kernels and may not reflect changes occurring in the embryo or aleurone layer. Nevertheless, they do illustrate that considerable proteolysis occurs during germination which could be a method for assessing the extent of germination.

Kruger (1984), using high-performance liquid chromatography in the gel permeation mode, monitored the molecular weight profiles of buffer-soluble (0.5 *M* sodium phosphate buffer, pH 7.0 containing 0.5 *M* sodium chloride) proteins in both sound and germinated wheat kernels. Of the molecular weight protein groups examined, the low-molecular-weight peptides and amino acids exhibited the largest changes. This was further evidence for the increased production of solubilized amino nitrogen, particularly amino acids, during germination. Little change occurred during the first 2 days of germination compared to after 6 days. Further studies by Kruger and Marchylo (1985) examined mobilization of pro-

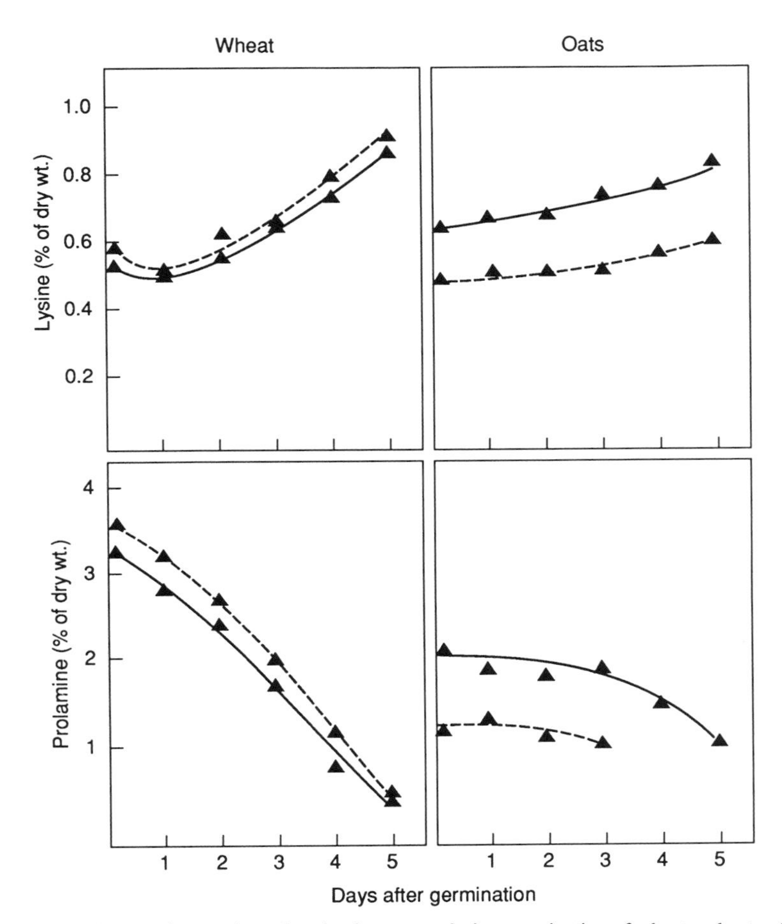

FIG. 3.10. Changes in protein and prolamin content during germination of wheat and oats. Adapted from Dalby and Tsai (1976).

teins during the germination of five wheat cultivars. Six major protein components were eluted, of which only the low-molecular-weight species underwent major changes during germination. These results confirmed earlier work by Kruger (1984) and Lukow and Bushuk (1984) which showed that a very rapid hydrolysis of wheat endosperm proteins occurs following limited endopeptidase activity during the initial period of germination.

Increased release of free amino acids during germination suggests extensive mobilization of the storage proteins during this period. The mechanism controlling this process still remains poorly understood. A number of proteases have

TABLE 3.10

THE EFFECT OF GERMINATION AT 16.5°C ON THE PRODUCTION OF SELECTED FREE AMINO ACIDS IN WHEAT CV. NEEPAWA[a]

	Germination period (hr)	
Amino acid[b]	0	122
Tryptophan	47	50
Lysine	5.7	63
Histidine	2.2	72
Glutamic acid	64	95
Methionine	2.4	27
Isoleucine	5.1	140
Leucine	6.0	170
Tyrosine	4.5	72
Phenylalanine	4.2	150
Proline	7.8	790
Glutamine	12	920

[a] Adapted from Tkachuk (1979).
[b] μmole/g N.

been found in wheat grain, including endopeptidases, carboxypeptidases, and aminopeptidases (Grant and Wang, 1972; Kruger, 1973; Preston and Kruger, 1976a,b, 1977; Kruger and Preston, 1978). Of these, carboxypeptidase is prominent in the endosperm, where it represents one-quarter of the total endopeptidase activity (Preston and Kruger, 1976a). These enzymes have a negligible effect on the endosperm reserves during the first 2 days of germination, possibly because of their compartmentalization, the presence of protease inhibitors, or insolubilization of the substrate. During the course of germination there is limited endopeptidase activity resulting in the formation of intermediate products which are then degraded by carboxypeptidase to amino acids (Kruger and Marchylo, 1985). Only a fraction of the storage proteins are affected at any time, which explains the similarity in protein patterns for sprouted and mature seeds.

G. LIPID MOBILIZATION DURING GERMINATION

Germination and sprouting of cereal grains is accompanied by an increase in total lipid content (Lorenz, 1980; Rahnotra *et al.*, 1977). The presence of lipase in ungerminated wheat and barley seeds is extremely low but develops as soon as germination commences (Huang and Moreau, 1978; Taverner and Laidman,

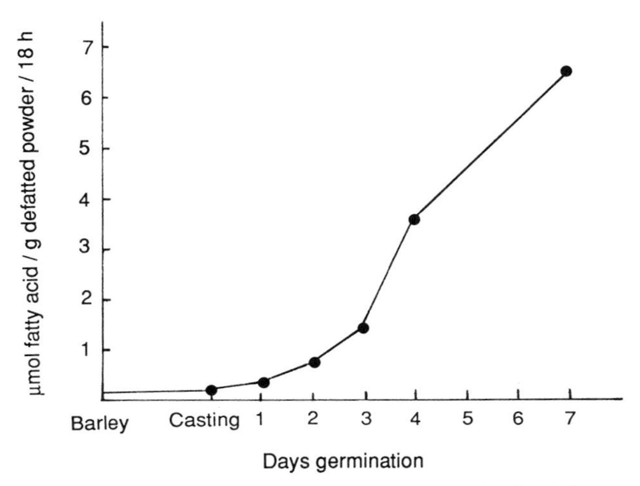

FIG. 3.11. Lipase activity in aqueous extracts of barley (variety Sonja) during germination (Baxter, 1984).

1972). In sharp contrast to these cereals, oats are rich in lipase activity (Matlashewski *et al.*, 1982). Lipase (triacylglycerol lipase, EC 3. 1. 1. 30) hydrolyzes triacylglycerols, diacylglycerols, and possibly monoacylglycerols producing fatty acids. The major difficulty involved in measuring lipase activity is due to the insolubility of the substrate in aqueous solution. This difficulty has been partially overcome using water-soluble substrates such as *p*-nitrophenyl (pNP) acetate or butyrate or by forming a stable emulsion with olive oil. A specific method for assaying lipase was developed by Matlashewski *et al.* (1982) using radioactive triacylglycerols in which the fatty acid moiety was labeled. Using this method, Baxter (1984) examined lipase activity in both germinated and ungerminated barley. The results obtained in Figure 3.11 show that lipase activity increased slowly during the initial 2 days of germination but then rose sharply after 3 days. Two distinct lipase fractions were separated with similar molecular weights (400,000 range) but different ionic properties. The major lipase fraction (I) was associated with the embryo, while the smaller lipase fraction (II) was located in the endosperm. Taverner and Laidman (1972) identified lipases in wheat embryo and endosperm each induced by different factors. Urquardt and co-workers (1984) separated oat embryos from the rest of the kernel and monitored changes in lipase activity during germination. The initial increase in lipase activity appeared to be primarily in the bran layer with little or no activity in the endosperm (Urquardt *et al.*, 1983). While the primary role of lipase is hydrolysis of the storage triacylglycerols, its physiological role remains obscure.

V. Storage of Grains

Following harvest, cereal grains such as wheat are stored in either sacks or bulk silos. These grains are traditionally recognized for their keeping quality, which is affected by moisture, temperature, and invasion by rodents, insects, bacteria, and fungi. Postharvest grain losses worldwide appear to be around 3–10%, sometimes up to 15%, depending on local conditions and resources (Harris, 1984). This section will focus on the effects of moisture and temperature on the quality of grains.

A. Respiration

When cereal grains are dry very little respiration occurs. If the moisture content of the seeds rises above 14%, respiration increases until a critical moisture level is attained. At this point respiration accelerates rapidly with the subsequent heating of the grain. This marked rise in respiration is attributed, in part, to germination and growth of molds such as *Aspergillus* and *Penicillium*. Grain respiration is affected by moisture, temperature, and oxygen tension, although the moisture content is of paramount importance in the commercial storage of cereal grains.

1. *Effect of Moisture Content*

Exposure of grain will result in the uptake of moisture until an equilibrium is reached with the water vapor in the atmosphere. Thus the moisture content of the grain is controlled by the relative humidity in the environment, which in terms of grain storage is the nature of the interstitial atmosphere. When exposed to an atmosphere of uniform relative humidity at a constant temperature, the relative humidity of the stored grain reaches an equilibrium referred to as the equilibrium relative humidity (E. R. H.). The relationship between relative humidity and moisture content is defined by the sorption isotherm, the shape of which is sigmoid. This is due to the larger equilibrium moisture content during desorption as compared to adsorption at a given equilibrium relative humidity. Figure 3.12 shows the moisture isotherm obtained at 30°C for maize with the characteristic sigmoid curve resulting from the greater water content of the desorption isotherm (Denloye and Ade-John, 1985).

The equilibrium moisture content is quite low in grains and only after the isotherm reaches 80% relative humidity does the moisture content rise exponentially with relative humidity (Oxley, 1948). The moisture content that is regarded as safe for grain is that in equilibrium with 70% relative humidity (Pixton and Warburton, 1971). Microbial growth will occur above 75% relative humidity, resulting in extensive deterioration of the grain.

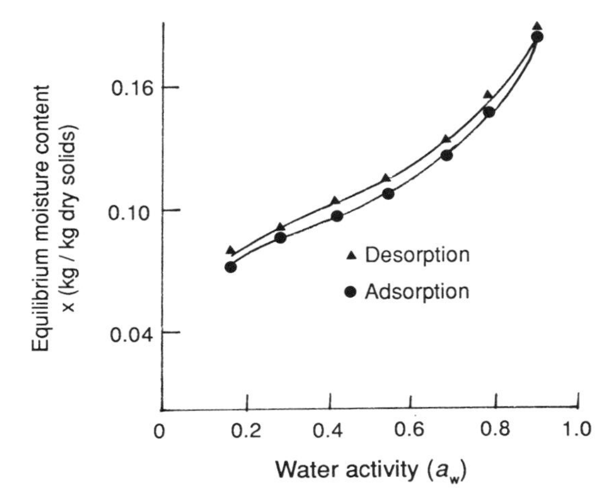

FIG. 3.12. Moisture sorption isotherms for maize (30°C). Reprinted with permission from Denloye and Ade-John (1985). Copyright by Pergamon Press.

Under extreme wet conditions the grain may be harvested at a moisture content which is too high for safe storage. This necessitates the use of drying to reduce the moisture content of the grain which can then be stored with minimal loss in seed viability, nutritive value, and bread-making properties (Bushuk, 1978). Spillane and Pelhate (1982) attempted to bypass the drying step by storing barley harvested with a high moisture content (>30%) under ventilated conditions. Unless the rise in grain temperature, due to respiration, can be controlled, an explosive growth of yeasts and bacteria would take place. This was prevented by continuously ventilating the silo for a month which removed a good portion of the heat generated by respiration and so reduced the final grain temperature to below the critical point of 16°C. The moisture content of the grain was reduced to 16% and the relative humidity of the grain environment to around 80% at the end of the storage period. Under these conditions the growth of yeasts and bacteria was suppressed while quality factors remained intact.

2. *Effect of Temperature*

The equilibrium relative humidity is affected only slightly by changes in temperature. Ayerst (1965) reported that a rise or fall of 10°C resulted in a 3% change in E. R. H. over a relative humidity range of 40– 90%. At higher relative humidities the change never exceeded 1% (Pixton and Warburton, 1975). Using Manitoba wheat, Pixton (1968) showed that at 10% moisture content the E. R. H. increased by 6% when heated at 70°C compared to only 2% when the moisture content was 14%. Prolonging the heating for more than an hour produced little further change. Denloye and Ade-John (1985) noted a decrease in the

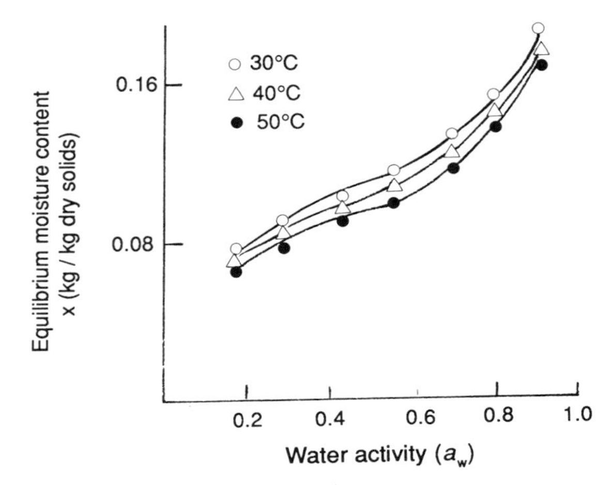

FIG. 3.13. Desorption moisture isotherms for maize at different temperatures. Reprinted with permission from Denloye and Ade-John (1985). Copyright by Pergamon Press.

equilibrium moisture content for maize kept at a constant relative humidity as the temperature was changed from 30 to 50°C (Figure 3.13).

Since grain is stored in bulk, the movement of heat and moisture in the stored grain is extremely important. Anderson *et al.* (1943) first showed that movement of moisture occurred over a temperature gradient from high to low. This process was extremely slow and involved diffusion with some convection currents. The main effect of heating appears to be related to the translocation of moisture brought about by the temperature gradients in the grain.

B. PROLONGED STORAGE OF GRAINS AND FLOUR

Pixton and co-workers (1975) monitored the changes in quality of wheat stored for 16 years under conditions of low temperature (4.5–0.5°C) and low oxygen concentrations (<2% by volume). Two different pest-free dry wheats, Manitoba and Cappelle, with respective moisture contents of 11.9 and 12.6% were placed in bins in 1-ton lots. The moisture content did not change significantly over this period. The crude protein and salt-soluble protein content remained unchanged for both wheats irrespective of the storage conditions. A slight increase in the total fat of 0.5% was observed for both wheat varieties, which was attributed to carbohydrate metabolism during the long storage period. This was based on the slight reduction in total sugars observed in these wheat samples by Pixton and Hill (1967) after 8 years, although maltose and sucrose changed very little during the subsequent storage period. These researchers also monitored vitamin B_1, which remained unchanged throughout the storage period.

A high viability was observed for wheats stored at 4.5°C (96%) compared to only one-third viability when held at ambient temperature for the same period. As long as the wheat was protected from atmospheric moisture, rapid temperature changes, and insects, the baking quality remained intact, although some supplementation with fungal α-amylase was required.

Bibliography

Abbott, I. R., and Matheson, N. K. (1972). Starch depletion in germinating wheat, wrinkled seed peas and senescing tobacco leaves. *Phytochemistry* **11,** 1261.

Acker, L., and Becker, G. (1971). Recent studies on the lipids of cereal starches. II. Lipids of various types of starch and their binding to amylose. *Staerke* **23,** 419.

Aizono, Y., Funatsu, M., Fujiki, Y., and Watanabe, M. (1976b). Biochemical studies of rice bran lipase. Part IV. Purification and characterization of rice bran lipase. II. *Agric. Biol. Chem.* **40,** 317.

Akazawa, T., and Hara-Nishimura, I. (1985). Topographic aspects of biosynthesis, extracellular secretion, and intracellular storage of proteins in plant cells. *Annu. Rev. Physiol.* **36,** 441.

Amir, J., and Cherry, J. H. (1971). Chemical control of sucrose conversion to polysaccharides in sweet corn after harvest. *J. Agric. Food Chem.* **19,** 954.

Amir, J., and Cherry, J. H. (1972). Purification and properties of adenosine diphosphoglucose pyrophosphorylase from sweet corn. *Plant Physiol.* **49,** 893.

Anderson, J. A., Babbitt, J. D., and Meredith, W. O. S. (1943). The effect of temperature differential on the moisture content of wheat. *Can. J. Res., Sect. C* **21,** 297.

Ayerst, G. (1965). Determination of water activity of some hygroscopic food materials by a dew-point method. *J. Sci. Food Agric.* **16,** 17.

Baruch, D. W., Meredith, P., Jenkins, L. D., and Simmonds, L. D. (1979). Starch granules of developing wheat kernels. *Cereal Chem.* **56,** 554.

Baxter, D. E. (1984). Recognition of two lipases from barley and green malt. *J. Inst. Brew.* **90,** 277.

Bean, M. M., Nimmo, C. C., Fullington, J. G., Keagy, P. M., and Mecham, D. K. (1974). Dried Japanese noodles. II. Effect of amylase, protease, salts, and pH on noodle doughs. *Cereal Chem.* **51,** 427.

Bechtel, D. B., Gaines, R. L., and Pomeranz, Y. (1982a). Protein secretion in wheat endosperm. Formation of the protein matrix. *Cereal Chem.* **59,** 336.

Bechtel, D. B., Gaines, R. L., and Pomeranz, Y. (1982b). Early stages in wheat endosperm formation and protein body initiation. *Ann. Bot. (London)* [N.S.] **50,** 507.

Bernal-Lugo, L., Beachy, R. N., and Varner, J. E. (1981). The response of barley aleurone layers to giberellic acid includes the transcription of new sequences. *Biochim. Biophys. Acta* **102,** 617.

Borovsky, D., Smith, E. E., Whelan, W. J., French, D., and Kikumoto, S. (1979). The mechanism of Q-enzyme action and its influence on the structure of amylopectin. *Arch. Biochem. Biophys.* **198,** 627.

Briarty, L. G., Hughes, C. E., and Evers, A. D. (1979). The developing endosperm of wheat. A stereological analysis. *Ann. Bot. (London)* [N.S.] **44,** 641.

Briggs, D. E. (1963). Biochemistry of barley germination: Action of gibberellic acid on barley endosperm. *J. Inst. Brew.* **69,** 13.

Briggs, D. E. (1964). Origin and distribution of α-amylase in malt. *J. Inst. Brew.* **70,** 514.

Briggs, D. E. (1968). α-Amylase in germinating, decorticated barley. III. *Phytochemistry* **7,** 538.

Briggs, D. E., Hough, J. S., Stevens, R., and Young, T. W. (1981). "Malting and Brewing Science 1. Malt and Sweet Wort," 2nd ed. Chapman & Hall, London.

Bright, S. W. J., and Shewry, P. R. (1983). Improvement of protein quality in cereals. *CRC Crit. Rev. Plant Sci.* **1,** 49.

Buchanan, A. M., and Nicholas, E. M. (1980). Sprouting, alpha-amylase, and breadmaking quality. *Cereal Res. Commun.* **8,** 23.

Burgess, S. R., and Miflin, B. J. (1985). The localization of oat (*Avena sativa* L.) seed globulins in protein bodies. *J. Exp. Bot.* **36,** 945.

Burgess, S. R., Turner, R. H., Shewry, P. R., and Miflin, B. J. (1982). The structure of normal and high-lysine barley grains. *J. Exp. Bot.* **33,** 1.

Bushuk, W. (1978). Biochemical changes in edible plant tissue during maturation and storage. *In* "Postharvest Biology and Biotechnology" (H. O. Hultin and M. Miller, eds.), Chapter 1. Foods and Nutrition Press, Westport, Connecticut.

Buttrose, M. S. (1971). Ultrastructure of barley aleurone cells as shown by freeze etching. *Planta* **96,** 13.

Buttrose, M. S., and Soeffly, A. (1973). Ultrastructure of lipid deposits and other contents in freeze-etched coleoptile cells of ungerminated rice grains. *Aust. J. Biol. Sci.* **26,** 357.

Campbell, W. P., Lee, J. W., O'Brien, T. P., and Smart, M. G. (1981). Endosperm morphology and protein body formation in developing wheat grain. *Aust. J. Plant Physiol.* **8,** 5.

Cawley, J. E., and Mitchell, T. A. (1968). Inhibition of wheat α-amylase by bran phytic acid. *J. Sci. Food Agric.* **19,** 106.

Cerning, J., and Guilbot, A. (1973). Changes in the carbohydrate composition during development and maturation of the wheat and barley kernel. *Cereal Chem.* **50,** 220.

Chamura, S. (1975). Histochemical investigation of the accumulation of phosphorised lipid in aleurone cells of rice kernels. *Nippon Sakumotsu Gakkai Kiji* **44,** 243.

Chevalier, P., and Lingle, S. E. (1983). Sugar metabolism in developing kernels of wheat and barley. *Crop Sci.* **23,** 272.

Chrispeels, M. J., and Varner, J. E. (1967). Hormonal control of enzyme synthesis and mode of action of gibberellic acid and abscisin in aleurone layer of barely. *Plant Physiol.* **42,** 1008.

Chung, O. K., and Pomeranz, T. (1981). Recent research on wheat lipids. *Baker's Dig.* **55**(5), 38.

Cornford, C. A., Black, M., Daussant, J., and Murdoch, K. M. (1987). α-Amylase production by pre-mature wheat (*Triticum aestivum* L.) embryos. *J. Exp. Bot.* **38,** 277.

Dalby, A., and Tsai, C. Y. (1976). Lysine and tryptophan increases during germination of cereal grains. *Cereal Chem.* **53,** 222.

Denloye, A. O., and Ade-John, A. O. (1985). Moisture sorption isotherms of some Nigerian food grains. *J. Stored Prod. Res.* **21,** 53.

Donovan, G. R., Lee, J. W., and Longhurst, T. J. (1982). Cellfree synthesis of wheat prolamins. *Aust. J. Plant Physiol.* **9,** 59.

Erlander, S. P. (1958). Proposed mechanism for the synthesis of starch from glycogen. *Enzymologia* **19,** 273.

Evers, A. D. (1973). The size distribution among starch granules in wheat endosperm. *Staerke* **25,** 303.

Evers, A. D., and Lindley, J. (1977). The particle size distribution in wheat endosperm starch. *J. Sci. Food Agric.* **28,** 98.

Filmer, P., and Varner, J. E. (1967). A test for *de novo* 18 synthesis of enzymes, density labelling with H_2O^{18} of barley α-amylase induced by gibberellic acid. *Proc. Natl. Acad. Sci. U.S.A.* **58,** 1520.

Food and Agriculture Organization (1986). "FAO Production Yearbook," FAO Stat. Ser. No. 70, Vol. 39. FAO, Rome.

Fuller, P., Hutchinson, J. B., McDermott, E. E., and Stewart, B. A. (1970). Inactivation of α-amylase in wheat and flour with acid. *J. Sci. Food Agric.* **21,** 27.

Geddes, R., and Greenwood, C. T. (1969). Biosynthesis of starch granules. IV. Observations on the biosynthesis of the starch granule. *Staerke* **21,** 148.

Gibbons, G. C. (1979). On the localisation and transport of α-amylase during germination and early seedling growth of *Hordeum vulgare. Carlsberg Res.Commun.* **44,** 353.

Gibbons, G. C. (1980). On the sequential determination of α-amylase transport and cell wall breakdown in germinating seeds of *Hordeum vulgare. Carlsberg Res. Commun.* **45,** 177.

Grant, D. R., and Wang, C. C. (1972). Dialyzable components resulting in proteolytic activity in extracts of wheat flour. *Cereal Chem.* **49,** 201.

Greenaway, W. T. (1969). The sprouted wheat problem: The search for a solution. *Cereal Sci. Today* **14,** 390.

Greene, F. C. (1981). *In vitro* synthesis of wheat (*Triticum aestivum* L.) storage proteins. *Plant Physiol.* **68,** 778.

Greene, F. C. (1983). Expression of storage protein genes in developing wheat (*Triticum aestivum* L.) seeds. Correlation of RNA accumulation and protein synthesis. *Plant Physiol.* **71,** 40.

Hargin, K. D., and Morrison, W. R. (1980). The distribution of acyl lipids in the germ, aleurone, starch and non-starch endosperm of four wheat varieties. *J. Sci. Food Agric.* **31,** 877.

Harris, L. (1984). Postharvest grain losses in the developing world. *Cereal Foods World* **29,** 456.

Higgins, T. J. V., Zwar, J. A., and Jacobsen, J. V. (1976). Gibberellic acid enhances the level of translatable mRNA for α-amylase in barley aleurone layers. *Nature (London)* **260,** 166.

Huang, A. A. C., and Moreau, R. A. (1978). Lipases in the storage tissue of peanut and other oilseeds during germination. *Planta* **141,** 111.

Jelseman, C. L., Morré, D. J., and Ruddat, M. (1974). Isolation and characterization of spherosomes from aleurone layers of wheat. *Proc. Indiana Acad. Sci.* **84,** 166.

Jenner, C. F. (1968). The composition of soluble nucleotides in the developing wheat grain. *Plant Physiol.* **43,** 41.

Jones, R. A., and Tsai, C. Y. (1977). Changes in lysine and tryptophan content during germination of normal and mutant maize seed. *Cereal Chem.* **54,** 565.

Jongh, G. (1967). Amylase determination. *Getreide Mehl* **17,** 1.

Joshi, S., Lodha, M. L., and Mehta, S. L. (1980). Regulation of starch biosynthesis in normal and opaque-2 maize during endosperm development. *Phytochemistry* **19,** 2305.

Kent, N. L. (1983). "Technology of Cereals," 3rd ed. Pergamon, Oxford.

Kruger, J. E. (1972a). Changes in the amylases of hard red spring wheat during growth and maturation. *Cereal Chem.* **49,** 379.

Kruger, J. E. (1972b). Changes in the amylase of hard spring wheat during germination. *Cereal Chem.* **49,** 391.

Kruger, J. E. (1973). Changes in the levels of proteolytic enzymes from red spring wheat during growth and maturation. *Cereal Chem.* **50,** 122.

Kruger, J. E. (1984). Rapid analysis of changes in the molecular weight distribution of buffer-soluble proteins during germination of wheat. *Cereal Chem.* **61,** 205.

Kruger, J. E., and Marchylo, B. A. (1985). Examination of the mobilization of storage proteins of wheat kernels during germination by high-performance reversed-phase and gel permeation chromatography. *Cereal Chem.* **62,** 1.

Kruger, J. E., and Matsuo, R. R. (1982). Comparison of alpha-amylase and simple sugar levels in sound and germinated durum wheat during pasta processing and spaghetti cooking. *Cereal Chem.* **59,** 26.

Kruger, J. E., and Preston, K. R. (1978). Changes in aminopeptidases of wheat kernels during growth and maturation. *Cereal Chem.* **55,** 360.

Kumar, R., and Singh, R. (1980). The relationship of starch metabolism to grain size in wheat. *Phytochemistry* **19,** 2299.
Kumar, R., and Singh, R.(1981). Free sugars and their relationship with grain size and starch content in developing wheat grains. *J. Sci. Food Agric.* **32,** 229.
Kumar, R., and Singh, R. (1983). Alkaline inorganic pyophosphatase from immature wheat grains. *Phytochemistry* **21,** 2405.
Kumar, R., and Singh, R. (1984). Levels of free sugars, intermediate metabolites, and enzymes of sucrose–starch conversion in developing wheat grains. *J. Agric. Food Chem.* **32,** 806.
Laberge, D. E., MacGregor, A. W., and Meredith, W. O. S. (1973). Changes in the free sugar content of barley kernels during maturation. *J. Inst. Brew.* **79,** 471.
Lasztity, R. (1984). "The Chemistry of Cereal Proteins." CRC Press, Boca Raton, Florida.
Lineback, D. R. (1984). The starch granule; Organization and properties. *Baker's Dig.* **58**(3), 16.
Lineback, D. R., and Ponpipom, S. (1977). Effects of germination of wheat, oats, and pearl millet on alpha-amylase activity and starch degradation. *Staerke* **29,** 52.
Lingle, S. E., and Chevalier, P. (1980). Vascularization of developing barley kernels. *Plant Physiol.* **65,** Suppl., 105.
Lorenz, K. (1980). Cereal sprouts: Composition, nutritive value, food applications. *CRC Crit. Rev. Food Sci. Nutr.* **13,** 353.
Lowy, G. D. A., Sargeant, J. G., and Schofield, J. D. (1981). Wheat starch granule protein: The isolation and characterization of a salt-extractable protein from starch granules. *J. Sci. Food Agric.* **32,** 371.
Lukow, O. M., and Bushuk, W. (1984). Influence of germination on wheat quality. II. Modification of endosperm protein. *Cereal Chem.* **61,** 340.
Lukow, O. M., Bekes, F., and Bushuk, W. (1985). Influence of germination on wheat quality. III. Modification of flour lipid. *Cereal Chem.* **62,** 417.
MacGregor, A. W. (1983). Cereal α-amylases: synthesis and action pattern. *In* "Seed Proteins" (J. Daussant, J. Mossé, and J. Vaughan, eds.), pp. 1–34. Academic Press, New York.
MacGregor, A. W., and Matsuo, R. R. (1982). Starch degradation in endosperms of barley and wheat kernels during initial stages of germination. *Cereal Chem.* **59,** 210.
MacRitchie, F. (1985). Studies on the methodology for fractionation and reconstitution of wheat flours. *J. Cereal Sci.* **3,** 221.
Marchylo, B. A., Lacroix, L. J., and Kruger, J. E. (1981). α-Amylase synthesis in wheat kernels as influenced by seed coat. *Plant Physiol.* **67,** 89.
Marshall, J. J., and Whelan, J. J. (1970). Incomplete conversion of glycogen and starch by crystalline amyloglucosidase and its importance in the determination of amylaceous polymers. *FEBS Lett.* **9,** 85.
Marston, R. M., and Welsh, S. O. (1980). Nutrient content of the national food supply. *Nat. Food Rev., Winter ed.*
Matlashewski, G. J., Urquhart, A. A., Sahasrabudhe, M. R., and Altosaar, I. (1982). Lipase activity in oat flour suspensions and soluble extracts. *Cereal Chem.* **59,** 418.
McDermott, E. E., and Elton, G. A. (1971). Effect of surfactants on the α-amylase activity of wheat flour. *J. Sci. Food Agric.* **22,** 131.
Melcher, V., and Varner, J. E. (1971). Protein release of barley aleurone layers. *J. Inst. Brew.* **77,** 456.
Meredith, P., and Pomeranz, Y. (1985). Sprouted grain. *Adv. Cereal Sci. Technol.* **7,** 239–320.
Miflin, B. J., and Burgess, S. R. (1982). Protein bodies from developing wheat and peas, the effects of protease treatment. *J. Exp. Bot.* **33,** 251.
Miflin, B. J., and Shewry, P. R. (1977). An introduction to the extraction and characterization of barley and maize prolamins. *In* "Techniques for the Separation of Barley and Maize Seed Proteins" (B. J. Miflin and P. R. Shewry, eds.), p. 13. EEC, Luxembourg.

Miflin, B. J., Burgess, S. R., and Shewry, P. R. (1981). The development of protein bodies in the storage tissues of seeds, subcellular separations of homogenates of barley, maize and wheat endosperms of pea cotyledons. *J. Exp. Bot.* **32,** 199.

Mitsui, T., Christeller, J. T., Hara-Nishimura, I., and Akazawa, T. (1984). Possible roles of Ca^{2+} and calmoldulin in the biosynthesis and secretion of α-amylase in rice seed scutellar epithelium. *Plant Physiol.* **75,** 21.

Moreaux, T., and Landry, J. (1968). Extractin selective des protéines du grain de mais et en particulier de la fraction "glutelines." *C. R. Hebd. Seances Acad. Sci.* **266,** 2302.

Morrison, I. N. (1978). Cereal lipids. Chapter 4. *Adv. Cereal Sci. Technol.* **2**, 221–348.

Morrison, I. N., Kuo, J., and O'Brien, T. P. (1975). Histochemistry and fine structure of developing wheat aleurone cells. *Planta* **123,** 105.

Okamoto, K., Kitano, H., and Akazawa, T. (1980). Biosynthesis and excretion of hydrolases in germinating cereal seeds. *Plant Cell Physiol.* **21,** 201.

Okita, T. W., and Greene, F. C. (1982). The wheat storage proteins: Isolation and characterization of gliadin messenger RNAs. *Plant Physiol.* **69,** 834.

Oparka, K. J., and Harris, N. (1982). Rice protein-body formation, all types are initiated by dilation of the endoplasmic reticulum. *Planta* **154,** 184.

Osborne, T. B. (1895).The proteins of barley. *J. Am. Chem. Soc.* **17,** 539.

Oxley, T. A. (1948). "The Scientific Principles of Grain Storage." Northern Publishing, Liverpool.

Parker, M. L. (1981). The structure of mature rye endosperm. *Ann. Bot. (London)* [N.S.] **47,** 181.

Parker, M. L., and Hawes, C. R. (1982). The Golgi apparatus in developing endosperm of wheat (*Triticum aestivum* L.). *Planta* **154,** 277.

Pernollet, J.-C. (1978). Protein bodies of seeds, ultrastructure, biochemistry and degradation. *Phytochemistry* **17,** 1473.

Pernollet, J.-C. (1982). Les corpuscles protéiques des graines, stade transitoire de vacuoles spécialisées. *Physiol. Veg.* **20,** 259.

Pernollet, J.-C. (1984). Protein bodies of seeds, ultrastructure, biochemistry, biosynthesis and degradation. *Phytochemistry* **17,** 1473.

Pernollet, J.-C., and Camilleri, C. (1983). Formation and development of protein bodies in the wheat endosperm. *Physiol.Veg.* **21,** 1093.

Pernollet, J.-C. and Mossé, J. (1983). Structure and location of legume or cereal seed storage proteins. *In* "Seed Proteins" (J. Daussant, J. Mossé, and J. Vaughan, eds.), pp. 155–191. Academic Press, New York.

Peterson, D. M., and Smith, D. (1976). Changes in nitrogen and carbohydrate fractions in developing oat groats. *Crop Sci.* **16,** 67.

Pixton, S. W. (1968). The effect of heat treatment on the moisture content/relative humidity equilibrium relationship of Manitoba wheat. *J. Stored Prod. Res.***4**, 267.

Pixton, S. W., and Hill, S. T. (1967). Long-term storage of wheat. II. *J. Sci. Food Agric.* **18,** 94.

Pixton, S. W., and Warburton, S. (1971). Moisture content relative humidity equilibrium of some cereal grains at different temperatures. *J. Stored Prod. Res.* **6,** 283.

Pixton, S. W., and Warburton, S. (1975). The moisture content equilibrium relative humidity relationship of rice bran at different temperatures. *J. Stored Prod. Res.* **11,** 1.

Pixton, S. W., Warburton, S., and Hill, S. T. (1975). Longterm storage of wheat. III. Some changes in the quality of wheat observed during 16 years of storage. *J. Stored Prod. Res.* **11,** 177.

Preiss, J., and Levi, C. (1979). Metabolism of starch in leaves. *Encycl. Plant Physiol. New Ser.* **6,** 282–312.

Preston, K. R., and Kruger, J. E. (1976a). Location and activity of proteolytic enzymes in developing wheat kernels. *Can. J. Plant Sci.* **56,** 217.

Preston, K. R., and Kruger, J. E. (1976b). Purification and properties of two proteolytic enzymes with carboxypeptidase activity in germinated wheat. *Plant Physiol.* **58,** 516.

Preston, K. R., and Kruger, J. E. (1977). Specificity of two isolated wheat carboxypeptidases. *Phytochemistry* **16,** 525.

Price, P. B., and Parsons, J. G. (1975). Lipids of seven cereal grains. *J. Am. Oil Chem. Soc.* **52,** 490.

Rahnotra, G. S., Loewe, R. J., and Lehmann, T. A. (1977). Breadmaking quality and nutritive value of sprouted wheat. *J. Food Sci.* **42,** 1373.

Raynes, J. G., and Briggs, D. E. (1985). Genotype and the production of α-amylase in barley grains germinated in the presence and absence of gibberellic acid. *J. Cereal Sci.* **3,** 55.

Recondo, E., and Leloir, L. F. (1961). Adenosine diphosphate glucose and starch synthesis. *Biochem. Biophys. Res. Commun.* **6,** 85.

Reddy, L. V., Ching, T. M., and Metzer, R. J. (1984). Alpha-amylase activity in wheat kernels matured and germinated under different temperature conditions. *Cereal Chem.* **61,** 228.

Rijven, A. H. G. C., and Gifford, R. M. (1983). Accumulation and conversion of sugars by developing wheat grains. 3. Non-diffusional uptake of sucrose, the substrate preferred by endosperm slices. *Plant Cell Environ.* **6,** 417.

Robyt, J. F. (1984). Enzymes in the hydrolysis and synthesis of starch. *In* "Starch: Chemistry and Technology" (R. L. Whistler, J. N. BeMiller, and E. F. Paschall, eds.), 2nd ed. Academic Press, New York.

Sargeant, J. G. (1979). The α-amylase isoenzymes of developing and germinating wheat grain. *In* "Recent Advances in the Biochemistry of Cereals" (D. L. Laidman and R. G. Wyn Jones, eds.), pp. 339–343. Academic Press, New York.

Sargeant, J. G. (1980). α-Amylase isoenzymes and starch degradation. *Cereal Res. Commun.* **8,** 77.

Schulz, A., and Stephan, H. (1960). Untersuchungen unber eine zweckmassige Verarbeitung aus wuchsgeschadigter Roggenmehle. *Brot Gebaeck* **14,** 240.

Seckinger, H. L., and Wolf, M. (1970). Electron microscopy of endosperm protein from hard and soft wheat. *Cereal Chem.* **47,** 236.

Shewry, P. R., Field, J. M., Kirkman, M. A., Faulks, A. J., and Miflin, B. J. (1980). The extraction, solubility and characterization of two groups of barley storage polypeptides. *J. Exp. Bot.* **31,** 393.

Shewry, P. R., Miflin, B. J., Forde, B. G., and Bright, S. W. J. (1981). Conventional and novel approaches to the improvement of the nutritional quality of cereal and legume seeds. *Sci. Prog., (Oxford)* **67,** 575.

Simmonds, D. H. (1972). The ultrastructure of the mature wheat endosperm. *Cereal Chem.* **49,** 212.

Simmonds, D.H., and O'Brien, T. P. (1981). Morphological and biochemical development of the wheat endosperm. *Adv. Cereal Sci. Technol.* **4,** 5.

Singh, R., and Axtell, J. D. (1973). High lysine mutant gene (hl) that improves protein quality and biological value of grain sorghum. *Crop Sci.* **13,** 535.

Singh, R., and Juliano, B. O. (1977). Free sugars in relation to starch accumulation in developing rice grain. *Plant Physiol.* **59,** 417.

Spillane, P. A., and Pelhate, J. (1982). Changes in quality and in microbiological activity during extended storage of high-moisture grain. *Cereal Foods World* **27,** 108.

Swift, J. G., and O'Brien, T. P. (1972). The fine structure of wheat scutellum before germination. *Aust. J. Biol. Sci.* **25,** 9.

Sylvester-Bradley, R., and Folkes, B. F. (1976). Cereal grains: Their protein components and nutritional quality. *Sci. Prog. (Oxford)* **63,** 241.

Tanaka, K., Sugimoto, T., Ogawa, M., and Kasai, Z. (1980). Isolation and characterization of two types of protein bodies in rice endosperm. *Agric. Biol. Chem.* **44,** 1633.

Taverner, R. J. A., and Laidman, D. L. (1972). The induction of lipase activity in the germinating wheat grain. *Phytochemistry* **11,** 989.

Thomas, B., and Lukow, G. (1969). Starch degradation during dough raising with regard to α-amylase. *Brot Gebaeck* **23,** 24.

Tkachuk, R. (1979). Free amino acids in germinated wheat. *J. Sci. Food Agric.* **30,** 53.
Triplett, B. A., and Quatrano, R. S. (1982). Timing localization and control of wheat germ agglutin synthesis in developing wheat embryos. *Dev. Biol.* **91,** 491.
Tsai, C. Y., Dalby, A., and Jones, R. A. (1975). Lysine and tryptophan increases during germination of maize seed. *Cereal Chem.* **52,** 356.
Urquardt, A. A., Altosaar, I., and Matlashewski, G. J. (1983). Localization and lipase activity in oat grains and milled oat fractions. *Cereal Chem.* **60,** 181.
Urquardt, A. A., Brumell, C. A., Altosaar, I., Matlashewski, G. J., and Sahasrabudhe, M. R. (1984). Lipase activity in oats during grain maturation and germination. *Cereal Chem.* **61,** 105.
U.S. Department of Agriculture (1957). "Grain Grading Primer," Misc. Publ. No. 740, p. 32. USDA, Washington, D.C.
Varty, K., Arreguin, B., Miguel Gomez, T., Pablo Jaime Lopez, T., and Miguel Angel Gomez, L. (1982). Effects of abscisic acid and ethylene on the gibberellic acid-induced synthesis of α-amylase by isolated wheat aleurone layers. *Plant Physiol.* **73,** 692.
Vieweg, G. H., and De Fekete, M. A. R. (1976). The effect of phospholipids on starch metabolism. *Planta* **129,** 155.
Westermarck-Rosendahl, C., Junnila, L., and Koivistionen, P. (1979). Efforts to improve the baking properties of sprout-damaged wheat by reagents reducing α-amylase activity. I. Screening tests by the falling number method. *Lebensm.-Wiss. -Technol.* **12,** 321.
Westermarck-Rosendahl, C., Junnila, L., and Koivistoinen, P. (1980). Efforts to improve the baking properties of sprout-damaged wheat by reagents reducing α-amylase activity. III. Effects on technological properties of flour. *Lebensm.-Wiss. -Technol.* **13,** 193.
Whelan, W. J. (1958). *In* "Handbuch der Pflanzenphysiologie" (W. Ruhland, ed.), Vol. 6, p. 154. Springer-Verlag, Berlin.
Whelan, W. J. (1963). Recent advances in starch metabolism. *Staerke* **15,** 247.

4

Biochemical Changes in Raw Foods: Milk

I. Introduction

Milk and dairy products are important sources of animal protein, vitamins, and essential fatty acids for infants and young adults (Jensen and Nielsen, 1982). The major source of milk is obtained from Western breeds of dairy cattle (*Bos tauras*), although milk is available from other species of mammals in some countries. According to the Pasteurized Milk Ordinance and Code recommended by the United States Public Health Service (1985), milk is defined as the lacteal secretion, practically free from colostrum, obtained by the complete milking of one or more healthy cows, containing not less than 8.25% milk solids not-fat and not less than 3.25% of milk fat. The exclusion of colostrum, a fluid secreted immediately following parturition, is primarily for esthetic reasons.

This chapter will discuss the dynamic biochemical systems involved in the biosynthesis of milk components.

II. Composition of Milk

The major constituents of milk are water (86–88%), milk fat (3–6%), protein (3–4%), lactose (5%), and phosphorus and minerals (ash) (0.7%), with total

TABLE 4.1

REPRESENTATIVE GROSS COMPOSITION (%) OF MILK FROM COWS OF DIFFERENT BREEDS[a]

Breed	Water	Fat	Protein	Lactose	Ash	Total solids
Guernsey	85.35	5.05	3.90	4.96	0.74	14.65
Jersey	85.47	5.05	3.78	5.00	0.70	14.53
Ayrshire	86.97	4.03	3.51	4.81	0.68	13.03
Brown Swiss	86.97	3.85	3.48	5.08	0.72	13.13
Holstein[b]	87.72	3.41	3.32	4.87	0.68	12.28

[a] Compiled from data from Corbin and Whittier (1965).
[b] Represents predominant dairy breed in the United States.

solids of 11–14%. The composition of milk is affected by a variety of factors, and of particular importance are breed and seasonal changes. Table 4.1 illustrates the influence of six main breeds in the United States, while seasonal changes are shown for fat, protein, lactose, ash, and total solids in Table 4.2. In recent years there has been considerable interest in the percentage of solids-not-fat (SNF) or protein in determining milk prices. This measurement has been facilitated through the development of rapid, automated methods based on infrared analysis. Wilcox *et al.* (1971) completed a large study on milk composition in five breeds and reported that the effect of age on fat, SNF, protein, and total solids

TABLE 4.2

INFLUENCE OF SEASON ON GROSS COMPOSITION (%) OF MILK[a]

Month	Fat	Protein	Lactose	Ash	Total solids
January	4.31	3.67	4.87	0.72	13.57
February	3.22	3.62	4.89	0.72	13.45
March	4.16	3.56	4.98	0.71	13.37
April	4.10	3.54	5.01	0.71	13.37
May	4.10	3.53	5.04	0.71	13.37
June	3.96	3.45	5.02	0.70	13.13
July	3.95	3.46	5.02	0.70	13.12
August	3.95	3.54	5.00	0.69	13.18
September	4.10	3.62	4.96	0.70	13.38
October	4.24	3.66	4.92	0.71	13.53
November	4.27	3.69	4.88	0.72	13.55
December	4.30	3.65	4.92	0.72	13.59

[a] Compiled from Corbin and Whittier (1965) based on 2426 samples from 1482 individual breeds.

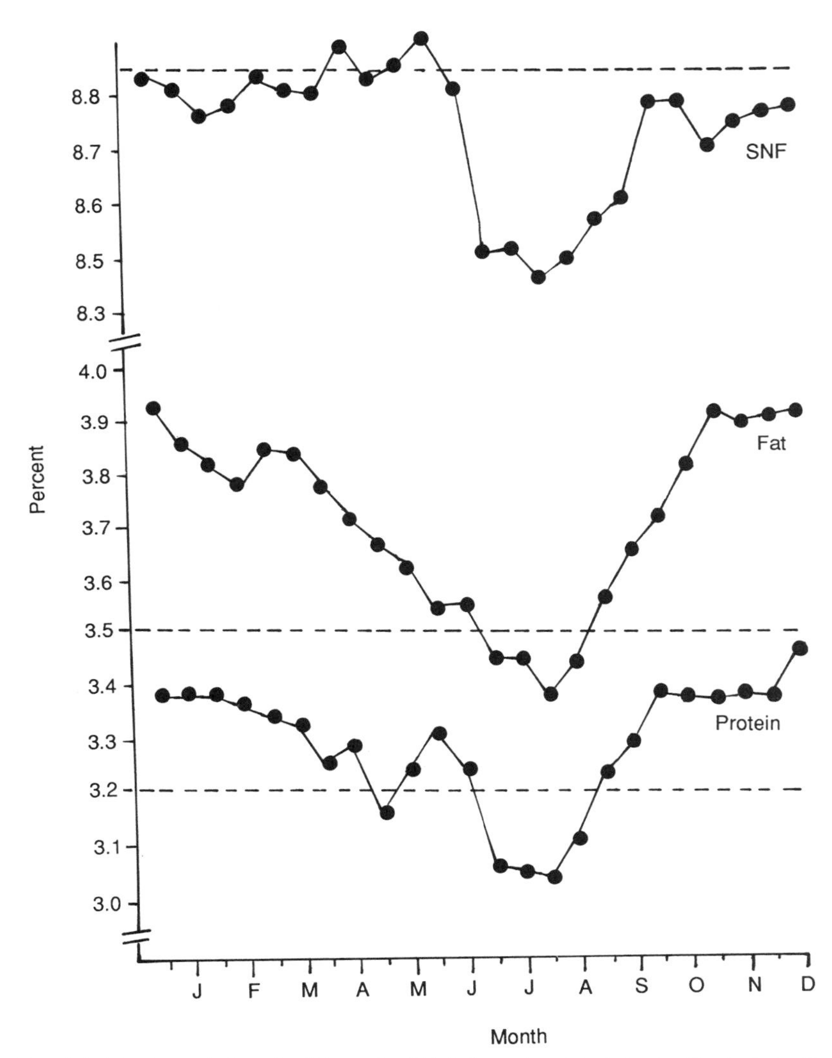

FIG. 4.1. Monthly variation of solids-not-fat (SNF), fat, and protein in herd milks (includes Grade A and manufacturing-grade herds). Dashed lines indicate 8.75% SNF base, 3.5% fat base, and 3.2% protein base (Sommerfeldt and Baer, 1986).

(TS) was as important as its effect on milk yield. A later study by Norman *et al.* (1978) examined the influence of age and month of calving on milk, fat, SNF, and protein yields over a period of nine years. This study was the largest ever conducted and involved 106,411 cows from 2215 herds in 41 states. These researchers found that the levels of milk components were lower than those reported by Wilcox *et al.* (1971) but reflected more closely the average U.S. yields for milk and fat. The average percentages for fat and SNF reported by Norman *et al.* (1978) were 7.8 and 8.6%, respectively. The effect of breed on milk composition was consistent with previous studies. A recent study by Sommerfeldt and Baer (1986) monitored the variation in milk components from 1705 herds from eastern South Dakota, Minnesota, and northern Iowa. Milk was collected biweekly over a 1-year period to evaluate the advantages and disadvantages of milk payment based on SNF. The greatest variability was observed for fat (8.4% coefficient of variation), followed by protein, TS, and SNF with coefficients of variation of 6.3, 4.1, and 3.4%, respectively. The monthly variations for protein, fat, and SNF are shown in Figure 4.1. Sommerfeld and Baer (1986) did not find any consistent relationship between SNF and protein or fat in the milk examined and recommended that testing for SNF should also include the fat content and producer grade. The introduction of a milk component pricing plan was considered to be beneficial to both the producer and processor. For example, a milk high in protein would be particularly attractive to the cheese manufacturer as this will lead to higher yields.

III. Milk Constituents

The overall composition of cow's milk provides useful information to the farmer and processor in determining quality and market value. This section will discuss the chemistry and biochemistry of major components in milk.

A. Lipids

The lipid component of cow's milk is present as small fat globules ranging in size from 0.1 to 20 μm in diameter. Each fat globule is surrounded by an interfacial layer or milk fat globule membrane (MFGM), which separates it from the aqueous milk serum. This layer is composed mainly of triacylglycerols (95%) with small amounts of diglyceride, free fatty acids, monoglycerides, phospholipids, and traces of cholesterol esters. In addition to these components the MFGM layer also contains trace elements, enzymes, protein, and glycoprotein. McPherson and Kitchen (1983) proposed the model shown in Figure 4.2 for the bovine milk fat globule membrane consisting of a layer of high-melting triacylglycerols surrounding the core fat. The main body was composed of phospholipids, cholesterol, and cholesterol esters in which are embedded various

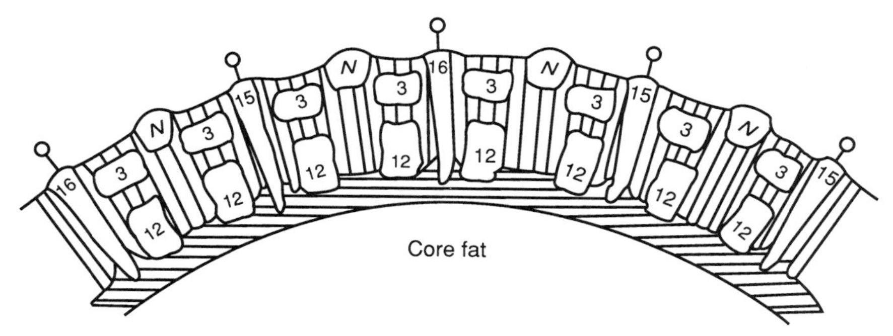

FIG. 4.2. Proposed model for the bovine milk fat globule membrane. Polypeptides are numbered using the nomenclature of Mather and Keenan (1975), carbohydrate side chains are shown as small circles attached to extensions from polypeptides 15 and 16, the polypeptide labeled N is 5′-nucleotidase, horizontal shading indicates the high-melting triglyceride regions, and vertical shading indicates the main body of the membrane, comprising phospholipids, cholesterol, cholesterol esters, and other triglycerides (McPherson and Kitchen, 1983).

polypeptides. Xanthine oxidase appeared to be located just below the outer surface, while the enzyme 5′-nucleotidase was on the outer surface of the membrane. The MFGM layer has an important influence on the stability of the fat phase in milk and the changes that occur during the processing of milk and cream. McPherson and Kitchen (1983) summarized those factors affecting the composition of the MFGM layer as shown in Scheme 4.1. The effect of some of these compositional changes on the properties of the MFGM layer, however, still remains unclear. The outer surfaces of the MFGM are quite labile and can be removed by simple washing procedures and by temperature manipulation. Such losses of the outer surface have an impact on the processing and storage of milk.

The composition of milk fat shown in Table 4.3 indicates that triacylgycerols account for 97–98% of the total. The fatty acid composition of the triacylglycerols is influenced by breed and management practices. Milk fats are not markedly affected by changes in dietary lipids since some lipids are metabolized in the cow's rumen (Garton, 1964). Nevertheless, modification of milk fat can be achieved by feeding the cow triacylglycerol emulsion protected by a protein membrane cross-linked with formaldehyde, to protect the fat from microbial metabolism in the rumen. This technique allowed the fatty acid composition of the milk fat to be manipulated and the degree of unsaturation of fats modified. The fatty acid composition of bovine milk lipids is more varied than any other natural product, with 437 components listed by Patton and Jensen (1975). Of these, the long-chain fatty acids (C_{14}–C_{23}) and short-chain fatty acids (C_4–C_6) account for 75 and 15% of the total fatty acids in the triacylglycerols of milk fat, respectively (Table 4.3).

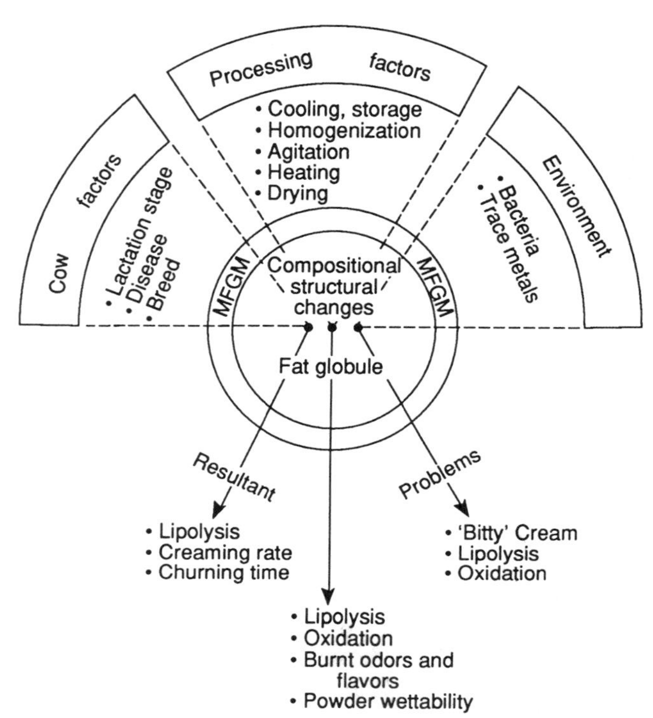

SCHEME 4.1. Factors affecting the milk fat globule membrane (McPherson and Kitchen, 1983).

1. *Biosynthesis of Milk Fat*

The majority of C_{16}–C_{18} fatty acids in bovine milk triglycerides are derived from dietary fat taken up from the blood, while the C_4–C_{14} fatty acids appear to be synthesized *de novo* in the mammary gland. The dietary fatty acids are transported from the intestine to the mammary glands and other tissues by chylomicrons (Zinder *et al.*, 1974). These chylomicrons are formed in the intestinal epithelial cells from lipids absorbed in the intestine and then transported by the lymph in the intestinal lymphatic vessels to the thoracic duct *via* the bloodstream to be taken up by various tissues of the body (Kennedy, 1957; Robinson, 1970). Chylomicrons are spherical particles consisting of a triacylglycerol core with traces of cholesteryl esters. They are enclosed by a monolayer of surface film 25–30 Å in width composed primarily of phospholipids and diglycerides with small amounts of cholesterol, protein, monoglyceride, and fatty acids (Scow *et al.*, 1976). Current evidence supports the hypothesis that triacylglycerols are taken up by the mammary gland from the chylomicrons, when present, and as low-density liprotein ($d < 1.05$) (Moore and Christie, 1979). Glascock *et al.*, (1966) reported that

TABLE 4.3

FATTY ACID COMPOSITION OF HOLSTEIN HERD MILK FAT DURING SUMMER AND WINTER[a]

Fatty acid	Summer	Winter
4:0	3.6	3.5
6:0	1.3	1.4
8:0	0.9	1.1
10:0	2.4	2.7
12:0	2.7	2.9
13:0	0.1	0.1
14:0	9.8	12.7
14:1	—	—
15:0	1.1	1.0
16:0	25.4	34.4
16:1	0.9	1.3
17:0	0.7	0.7
18:0	15.8	11.6
18:1	—	—
18:1 cis	24.3	19.9
18:1 trans	6.4	2.5
18:2 + 18:3	1.9	1.4

[a] Adapted from Patton and Jensen (1975). Figures are wt% by gas chromatography of methyl esters.

only a small amount of total circulating triacylglycerols associated with the serum lipid fraction as lipoprotein ($d > 1.05$) precipitable with dextran sulfate is used for milk fat. This conflicted with earlier work by Glascock and Wright (1962), which excluded the total pool of blood triacylglycerols as sources for milk fat synthesis. In other words, the triacylglycerols circulating in the blood as lipoprotein ($d > 1.05$) do not donate their fatty acids to the mammary gland for milk fat synthesis. Data presented by Brumby and Welch (1970) and Stead and Welch (1975) demonstrated quite clearly that serum lipoproteins from lactating cows contained negligible amounts of triacylglycerols. In spite of all these reports, Bickerstaffe (1971) still claimed that triacylglycerols were present in all lipoprotein fractions in the blood of lactating cows and were taken up by the mammary gland.

Under normal conditions when a standard dietary regimen is fed to the cow, the main fatty acids in the blood triacylglycerols are C16:0, C18:0, and C18:1, which are taken up by the mammary gland. Modification of the diet by feeding sunflower oil particles coated with formaldehyde-treated casein protected the fat from metabolism in the rumen, and resulted in an increase in C18:2 content in the plasma triacylglycerols of lactating cows (Gooden and Lascelles, 1973). This increase was reflected by a corresponding increase in the assimilation of C18 : 2 by the mammary gland. It is clear, however, that the fatty acid composition in the plasma is not the sole determinant of fatty acids in the plasma triacylglycerols. Annison *et al.* (1967) and West *et al.* (1972) both observed that the total triacylglycerols in the arterial plasma was much higher in C18 : 0 and lower in C18 : 1. This selective assimilation was confirmed by Moore *et al.* (1969), who concluded that the mammary glands selectively assimilated a plasma triacylglycerol with low levels of C18 : 2 in the presence of high C18 : 2 plasma triacylglycerols for the synthesis of milk fat.

The distribution of triacylglycerol fatty acids (see Chapter 10, Section I,A,1, for nomenclature) among the different lipoprotein fractions was examined because the triacylglycerol in the chylomicrons and VLDL (very-low-density lipoprotein) fraction was thought to be hydrolyzed during uptake into tissues by lipoprotein lipase. This enzyme was investigated by Askew *et al.* (1970) using homogenates of bovine mammary tissue. Lipoprotein lipase purified from cow's milk was shown by a number of researchers (Morley and Kuksis, 1972; Morley *et al.* 1972, 1975; Paltauf *et al.*, 1974; Paltauf and Wagner, 1976) to have a high specificity for the acyl ester bond in position 1 of the triacylglycerol, hydrolyzing it to 2-monoacyl glycerol according to the following sequence:

$$\begin{array}{l} CH_2{-}O{-}\overset{O}{\overset{\|}{C}}{-}R \\ | \\ CH{-}O{-}\overset{O}{\overset{\|}{C}}{-}R' \\ | \\ CH_2{-}O{-}\overset{O}{\overset{\|}{C}}{-}R'' \end{array} \longrightarrow \begin{array}{l} CH_2{-}OH \\ | \\ CH{-}O{-}\overset{O}{\overset{\|}{C}}{-}R' \\ | \\ CH_2{-}O{-}\overset{O}{\overset{\|}{C}}{-}R'' \end{array} \longrightarrow \begin{array}{l} CH_2{-}OH \\ | \\ CH{-}O{-}\overset{O}{\overset{\|}{C}}{-}R' \\ | \\ CH_2{-}OH \end{array}$$

Triglyceride (TG) 2,3–Diacyl glycerol 2–Monoacyl glycerol

Studies on the uptake of blood triacylglycerol by the mammary gland and the role of lipoprotein lipase have been conducted primarily with nonruminant animals. Nevertheless, the overall results indicate that chylomicrons and VLDL triacylglycerols of mammary tissue are the substrates and apoprotein and phospholipid cofactors for optimum lipoprotein lipase activity. The enzymatic hydrolysis of triacylglycerol (TG) only occurs in the chylomicron attached to the endothelium, possibly by hydrolysis of the acyl ester bond at the 1 position. Further hydrolysis of some of the remaining diglycerides at the lumenal surface may occur, and they are then taken up into the microvesicles and transported across the endothelial cells. The second acyl ester bond of the diglyceride is hydrolyzed as the microvesicle crosses the endothelium where some monoglyceride (MG) is also hydrolyzed to free glycerol. Once the diglycerides enter the alveolar cells of the mammary gland they are resynthesized into milk TG. Under normal dietary conditions, a fraction of plasma unesterified fatty acids equal to that released into the bloodstream by hydrolysis of chylomicron TG is taken up for the synthesis of milk fat (Moore and Christie, 1979).

2. *Fatty Acid Synthesis in the Mammary Gland*

The synthesis of fatty acids in the mammary gland involves the stepwise condensation of C_2 units by reversal of β-oxidation (Hele, 1954):

$$(1)\ CH_3{-}\overset{O}{\overset{\|}{C}}{-}SCoA + HCO_3^- + ATP \longrightarrow HOOC{-}CH_2{-}\overset{O}{\overset{\|}{C}}{-}SCoA + ADP + P_i$$

$$(11)\ CH_3{-}\overset{O}{\overset{\|}{C}}{-}SCoA + (7)\ HOOC{-}CH_2{-}CH_2{-}SCoA + 14NADPH$$

$$\downarrow$$

$$CH_3{-}CH_2\,(CH_2{-}CH_2)_6CH_2COOH + 7CO_2 + 14NADP + 8CoASH + 6H_2O$$

Reaction (1) is catalyzed by acetyl-CoA carboxylase forming malonyl-CoA, while the second reaction (11) is mediated by a group of enzymes known collectively as the fatty acid synthetase reaction sequence. Current evidence indicates that acetate and β-hydroxybutyrate are the two main sources for the *de novo* synthesis of fatty acids in lactating ruminants. The fatty acid synthetase sequence consists of seven enzymes integrated into a multienzyme complex which copurifies as a single protein (Smith, 1976). In animals, fatty acid synthetase requires NADPH and produces C16 : 0 as the main free fatty acid. This is illustrated in Scheme 4.2, in which each cycle is initiated by the transfer of a saturated acyl group from 4–phosphopantatheine thiol to the cysteine site (B_2) with the simultaneous transfer of a malonyl group from the loading site (B_1) to the 4-phosphopantatheine thiol. Subsequent reactions include condensation, ketoreduction, dehydration, and enoyl reduction, with the final end product being a saturated acyl moiety. To synthesize palmityl this cycle has to be completed seven times. The termination of the acyl chain occurs through the action of a deacylase or thioesterase enzyme which hydrolyzes the acyl-4-phosphopantatheine thioester bond.

Ruminant milk fat contains substantial amounts of short- and medium-chain

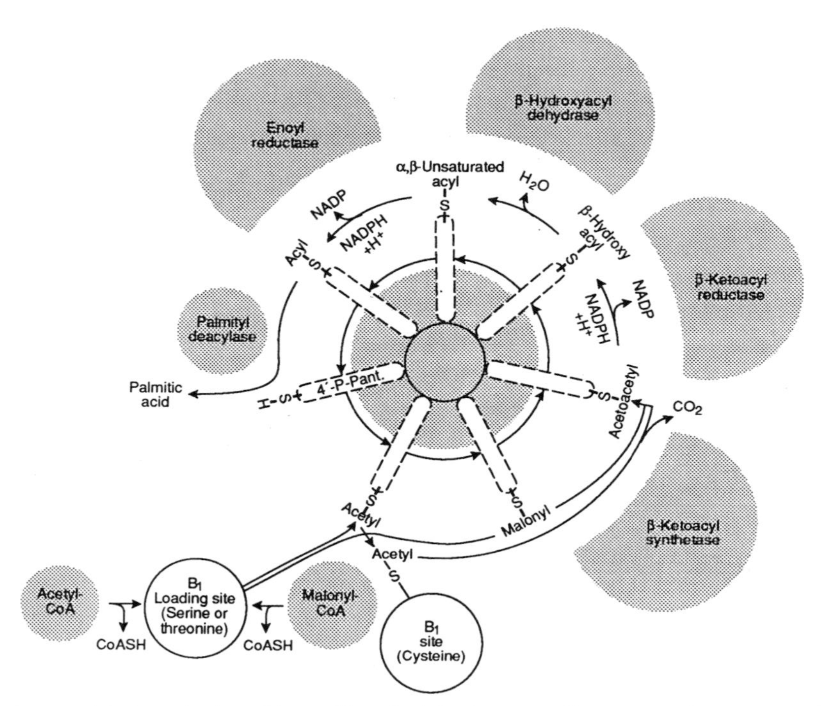

SCHEME 4.2. A mechanism of fatty acid synthesis by the mammalian prototype 1 system according to J. Porter. Note the flow of acetyl-CoA and malonyl-CoA to the loading site and the subsequent movement of the C_2 and C_3 units attached to the central ACP-like protein which serves as the substrate component for the peripherally oriented enzyme systems (Phillips *et al.* 1970).

fatty acids which are synthesized in the mammary gland. The synthesis of fatty acids from [^{14}C]acetate by mammary tissue obtained by biopsy from cows prior to and after parturition was examined with tissue slices (Mellenberger *et al.*, 1973) and homogenates (Kinsella, 1975). These researchers noted that the synthesis of short- and medium-chain fatty acids varied with the physiological state of the mammary glands. Only trace amounts of C4 : 0 to C10 : 0 fatty acids were synthesized in mammary tissue 18 days prior to parturition, with C16 : 0, C14 : 0, C18 : 0, and C12 : 0 accounting for 60, 30, 5, and 4% of the total fatty acids, respectively. This was in sharp contrast to 7 days prior to parturition when C4 : 0–C10 : 0 fatty acids accounted for 40% of the total fatty acids while C16 : 0 represented only 30% of the total. A 30-fold increase in the rate of fatty acid synthesis was observed in both mammary slices and homogenates which commenced 18 days prior to parturition and continued until 20 days post parturition. Based on these studies it was clear that some hormonal control was involved in the synthesis of specific patterns of fatty acid by the mammary glands.

The mechanism involved in chain length termination, particularly in the synthesis of short- and medium-chain fatty acids, remained unclear. Smith (1980) examined the mechanism of chain length termination in milk fat synthesis by selectively cleaving the thioesterase component from the fatty acid synthetase complex. This was based on previous studies which showed that limited proteolysis of the fatty acid synthetase complex from rat liver and mammary gland with chymotrypsin and trypsin produced defective chain termination (Agradi *et al.*, 1976). Trypsin was later shown to selectively cleave the thioesterase component of the fatty acid synthetase complex (Dileepan *et al.*, 1978; Smith *et al.*, 1976). The fatty acid synthetase complex appeared to be composed of two identical subunits each susceptible to tryptic attack at three sites. Thioesterase I occupied a terminal locus at one end of the two polypeptide chains. It had a molecular weight of 25, 000 with a trypsin-susceptible site near the center. The core of the fatty acid synthetase complex remaining after the addition of trypsin retained the 4′-phosphopantatheine moiety as well as the enzyme activities associated with it, with the exception of thioesterase. The complex was still capable of synthesizing long-chain fatty acids but now lacked the ability to terminate this process. Related studies using an inhibitor of thioesterase I activity also showed the ability of the modified multienzyme complex to synthesize a single acyl moiety. Unlike the dimer, fatty acid synthetase monomer catalyzed all of the reaction described previously with the exception of condensation, which is only carried out by the former (Kumar *et al.*, 1970). The separation of the components of fatty acid synthetase involved in chain growth and chain termination provided a unique opportunity to pinpoint chain length regulation. Since fatty acid synthetase from animal tissues preferentially synthesizes C16 : 0, a common mechanism must be responsible for the termination of the acyl chain. Thus removal or inhibition of thioesterase I resulted in the modified multienzyme complex elongating the fatty acid chain to C_{22}, however, the rate of elongation of C16 : 0 to C18 : 0 was much slower than the rate of formation of C16 : 0, with further chain elongation being even slower. If thioesterase I was involved in chain length determination there would be little activity toward short- and medium-chain thioesters with a sharp distinction between C_{14} and C_{16} thioesters. The enzymes from liver and mammary multienzymes were identical, with a preference for long-chain C_{16} and C_{18} acyl moieties.

Studies on the specificity of the condensing enzyme, the chain-elongating enzyme, and the chain-terminating enzyme (thioesterase I) pinpointed where chain lengthening was controlled. The inability of the condensing enzyme to elongate rapidly beyond C_{16} and of thioesterase I to hydrolyze acyl thioesters shorter than C16 : 0 ensured that C16 : 0 would be the main product synthesized. Examination of the great diversity in the fatty acid distribution between species shows that rat, rabbit, rodent, and mice all have large amounts of medium-chain fatty acids while guinea pigs have only long-chain fatty acids in milk fat. Smith

(1980) demonstrated the presence of a tissue-specific thioesterase II in the mammary glands of rabbits, rats, and mice which was responsible for modification of the product specificity of fatty acid synthetase. This enzyme was isolated and purified from rabbit and rat mammary glands and when added to a purified fatty acid synthetase it switched production from C_{16} to medium-chain fatty acids. It was intimately involved with the fatty acid synthetase complex in ruminant mammary glands.

2. *Synthesis of Unsaturated Fatty Acids*

Kinsella (1970) showed that freshly dispersed lactating bovine mammary cells actively desaturated C18 : 0 to C18 : 1. Subcellular fractionation of lactating and nonlactating cow mammary tissues indicated that the desaturase activity occurred exclusively in the microsomes.

B. Milk Proteins

The proteins of milk are a heterogeneous mixture and include two main groups, caseins and whey or serum proteins. These are composed of six major proteins: α_{s1}-casein, α_{s2}-casein, β-casein, κ-casein, β-lactoglobulin, and α-lactalbumin. Other protein fractions are also present but at very low levels, including blood serum albumin, immune globulins, lactoferritin, proteose peptone 3, and ceruloplasmin (Fox and Mulvihill, 1982). Caseins represent the major protein group and were defined by Jennes *et al.* (1956) as phosphoproteins precipitated from raw milk by acidification to pH 4.6 at 20°C. The residual proteins in the serum or whey after the removal of caseins were referred to as whey proteins. The nomenclature of proteins in cow's milk has undergone a number of revisions in response to the improved separation of the proteins by gel electrophoresis. The latest revision of protein nomenclature of cow's milk by Eigel *et al.* (1984) recommended that electrophoresis no longer be used as the basis for classification. Instead they suggested caseins be identified according to the homology of their primary amino acid sequences into four genetic families: α_{s1}-, α_{s2}-, β-, and κ-caseins. Gel electrophoresis was still useful, however, for identifying the different members of these families.

1. *Caseins*

Caseins account for approximately 80% of the total protein in cow's milk. They are present as macromolecular aggregates or "casein micelles" ranging in size from 30 to 300 nm. These micelles scatter light and are responsible for the whitish opaque nature of skim milk. The major casein fractions, α_{s1}-, α_{s2}-, κ-, β-, and γ-caseins, account for 38, 10, 36, 13, and 3% of whole casein, respectively (Davies and Law, 1980). The individual casein fractions differ from each other in their behavior toward calcium ions. Waugh and von Hippel (1956) separated micellar casein into a calcium-sensitive and calcium-insensitive fraction. This was based on the differential solubilities in the presence of specified amounts of calcium ions, resulting in the identification of two fractions referred to as α_s-casein and κ-casein. Aschaffenburg (1961), using alkaline urea paper electrophoresis, demonstrated genetic variants in cow's milk caseins, thus confirming genetic polymorphism. This was consistent with earlier studies by Aschaffenburg and Drewry (1955), who reported the presence of genetic variants in whey proteins. The three genetic variants for α_{s1}-casein were designated A, B, and C based on their decreasing mobility during electrophoresis in urea-containing starch gels by Thompson *et al.* (1962). A fourth genetic variant, α_{s1}-D, was identified by Grosclaude and co-workers (1966). Genetic polymorphism occurs as a result of substitution of one or two amino acids in the same protein, although in the case of α_{s1}-D, eight amino acid residues are deleted. This phenomenon is directly related to breed and genus (Aschaffenburg, 1968; Bell *et al.*, 1981; Swaisgood, 1982). With the exception of α_{s1}-D and κ-caseins, differences between genetic variants, however, do not have any technological importance. Variants of β-casein were reported earlier by Aschaffenburg (1961) and designated A, B, and C. A fourth genetic variant, β-casein D, was identified in *Bos indicus* by Aschaffenburg *et al.* (1968).

The κ-casein fraction originally identified by von Hippel and Waugh (1955) and Waugh and von Hippel (1956) was considered to be identical to the protective colloid Z-casein described by Lindstrom-Lang and Kodoma (1925). It was later found that κ-casein was associated with α_s-casein, stabilizing it from being precipitated by calcium ions. This fraction appeared to stabilize the casein micelles as well as limit their size. κ-Casein was also composed of several genetic variants, including A and B (Neeling, 1964; Schmidt, 1964; Woychik, 1964).

2. *Molecular and Structural Characteristics of Caseins*

The complete amino acid sequence has been determined for the primary structure of all the major bovine caseins (Grosclaude *et al.*, 1973; Jollès *et al.*, 1972; Mercier *et al.*, 1971; Ribadeau-Dumas *et al.*, 1972), which permits their average hydrophobicity to be calculated.

a. α-Casein: α_{s1}-. α-Casein is the largest fraction and includes those phosphoproteins precipitated at low calcium concentrations. Of these, α_{s1}-casein (α_{s1}-CN) contains 199 amino acid residues of which 8.4% are prolyl residues evenly distributed throughout the polypeptide chain. A highly

10 14 20
H.Arg - Pro - Lys - His - Pro - Ile - Lys - His - Gln - Gly - Leu - Pro - Gln - [Glu - Val - Leu - Asn - Glu - Asn - Leu -
Absent in Variant A

26 30 40
Leu - Arg - Phe - Phe - Val - Ala] - Pro - Phe - Pro - Gln - Val - Phe - Gly - Lys - Glu - Lys - Val - Asn - Glu - Leu -

50 53 59 60
Ser(P) - Lys - Asp - Ile - Gly - Ser(P) - Glu - Ser(P) - Thr - Glu - Asp - Gln - [Ala] - Met - Glu - Asp - Ile - Lys - [Gln] - Met -
[P] (α_{s1}-CN B-9P) ThrP (Variant D) Lys (Variant E)

70 80
Glu - Ala - Glu - Ser(P) - Ile - Ser(P) - Ser(P) - Ser(P) - Glu - Glu - Ile - Val - Pro - Asn - Ser(P) - Val - Glu - Gln - Lys - His -

90 100
Ile - Gln - Lys - Glu - Asp - Val - Pro - Ser - Glu - Arg - Tyr - Leu - Gly - Tyr - Leu - Glu - Gln - Leu - Leu - Arg -

110 120
Leu - Lys - Lys - Tyr - Lys - Val - Pro - Gln - Leu - Glu - Ile - Val - Pro - Asn - Ser(P) - Ala - Glu - Glu - Arg - Leu -

130 140
His - Ser - Met - Lys - Glu - Gly - Ile - His - Ala - Gln - Gln - Lys - Glu - Pro - Met - Ile - Gly - Val - Asn - Gln -

150 160
Glu - Leu - Ala - Tyr - Phe - Tyr - Pro - Glu - Leu - Phe - Arg - Gln - Phe - Tyr - Gln - Leu - Asp - Ala - Tyr - Pro -

170 180
Ser - Gly - Ala - Trp - Tyr - Tyr - Val - Pro - Leu - Gly - Thr - Gln - Tyr - Thr - Asp - Ala - Pro - Ser - Phe - Ser -

190 192 199
Asp - Ile - Pro - Asn - Pro - Ile - Gly - Ser - Glu - Asn - Ser - [Glu] - Lys - Thr - Thr - Met - Pro - Leu - Trp.OH
Gly (Variants C & E)

FIG. 4.3. Primary structure of *Bos* α_{s1}-CN B-8P. The amino acid residues enclosed in brackets are situated corresponding to mutational differences in the genetic variants A, C, D, and E. These residues represent the site of additional phosphorylation in α_{s1}-CN B-9P (Eigel *et al.*, 1984).

c. β-Caseins. β-Caseins (β-CN) are also phosphoproteins but differ from α-caseins by their strong temperature-dependent association as well as the temperature dependency of their solubility in the presence of calcium ions. β-Casein has 209 amino acid residues of which 16.7% are proline evenly distributed along the polypeptide, which limits the formation of an α-helix. Seven genetic variants are recognized, which separate differently by electrophoresis depending on whether acidic or alkaline conditions are used. While A can be differentiated from B, C, and D by electrophoresis under alkaline conditions, the A variants can only be separated under acidic conditions. For further details on the nature and nomenclature of these variants the review article by Eigel *et al.* (1984) is recommended. The primary sequence for the β-CN A-5P variant is shown in Figure 4.5. β-Caseins are the most hydrophobic of the casein fractions because of the number of hydrophobic residues present, although it has a strongly charged N-terminal region. Techniques including circular dichroism and spherical rotary dispersion excluded the presence of secondary and tertiary structures, although Andrews *et al.* (1979) calculated the presence of 10% α-helix, 13% sheets, and 77% unordered structure in β-casein.

d. γ-Casein. Until recently γ-casein was considered to be a distinct fraction accounting for 3% of whole casein. It has been shown by electrophoresis to be identical to the C-terminal portion of β-casein (Gordon *et al.* 1972; Groves *et al.*, 1973). Trieu-Cuot and Gripon (1981), using electrofocusing and two-dimensional electrophoresis of bovine caseins, obtained two-dimensional patterns by enzymatic hydrolysis of β-casein with bovine plasma similar to γ-casein. Studies conducted by a number of researchers suggested that β-casein was hydrolyzed by milk proteinase (plasmin) at three sites adjacent to lysyl residues 28, 104, and 106, producing six polypeptides including γ_1, γ_2, and γ_3-caseins as well as protease peptones (heat-stable, acid-soluble phosphoproteins) found in milk (Andrews, 1979; Eigel, 1977; Groves *et al.*, 1973). Thus γ-casein could arise by trypsinlike proteolysis of β-casein prior to or following milking. The latest fifth revision of the nomenclature of the protein of cow's milk no longer categorizes γ-caseins but considers them degradation products of β-casein (Eigel *et al.*, 1984).

The open structure and hydrophobicity of caseins render them extremely susceptible to proteolysis with a high propensity to formation of bitter peptides (Guigoz and Solms, 1976). Caseins are accessible to attack by the indigenous milk or psychotrophic proteinases which do not affect the whey proteins (Fox, 1981).

e. κ-Casein. κ-Casein (κ-CN) contains 169 amino acid residues of which 11.8% are proline (Figure 4.6). The major κ-CN component is carbohydrate-free while the minor κ-CN, a glycoprotein, is thought to be glycosylated forms of the

10 18 20
H.Arg - Glu - Leu - Glu - Glu - Leu - Asn - Val - Pro - Gly - Glu - Ile - Val - Glu - Ser - Leu - Ser - [Ser] - Ser - Glu -
P P [P] P
Lys (Variant D)

28 ↓ 29 30 35 36 37 40
Glu - Ser - Ile - Thr - Agr - Ile - Asn - Lys - Lys - Ile - Glu - Lys - Phe - Gln - Ser - [Glu] - [Glu] - Gln - Gln - Gln -
(Absent in Variant C) [P] Lys Lys (Variant C)
(Variant E)

50 60
Thr - Glu - Asp - Glu - Leu - Gln - Asp - Lys - Ile - His - Pro - Phe - Ala - Gln - Thr - Gln - Ser - Leu - Val - Tyr -

67 70 80
Pro - Phe - Pro - Gly - Pro - Ile - [Pro] - Asn - Ser - Leu - Pro - Gln - Asn - Ile - Pro - Pro - Leu - Thr - Gln - Thr -
(Variants C, A^1, and B) His

90 100
Pro - Val - Val - Val - Pro - Pro - Phe - Leu - Gln - Pro - Glu - Val - Met - Gly - Val - Ser - Lys - Val - Lys - Glu -

105 ↓ 106 107 ↓ 108 110 120
Ala - Met - Ala - Pro - Lys - [His] - Lys - Glu - Met - Pro - Phe - Pro - Lys - Tyr - Pro - Val - Gln - Pro - Phe - Thr -
Gln (Variant A^3)

122 130 140
Glu - [Ser] - Gln - Ser - Leu - Thr - Leu - Thr - Asp - Val - Glu - Asn - Leu - His - Leu - Pro - Pro - Leu - Leu - Leu -
Arg (Variant B)

150 160
Gln - Ser - Trp - Met - His - Gln - Pro - His - Gln - Pro - Leu - Pro - Pro - Thr - Val - Met - Phe - Pro - Pro - Gln -

170 180
Ser - Val - Leu - Ser - Leu - Ser - Gln - Ser - Lys - Val - Leu - Pro - Val - Pro - Glu - Lys - Ala - Val - Pro - Tyr -

190 200
Pro - Gln - Arg - Asp - Met - Pro - Ile - Gln - Ala - Phe - Leu - Leu - Tyr - Gln - Gln - Pro - Val - Leu - Gly - Pro -

209
Val - Arg - Gly - Pro - Phe - Pro - Ile - Ile - Val.OH

FIG. 4.5. Primary structure of *Bos* β-CN A^2-5P. The amino acid residues enclosed in brackets are the sites corresponding to the mutational differences in the genetic variants A^1, A^3, B, C, A, and E. The arrows indicate the points of attack by plasmin, which is responsible for the β-CN fragments present in milk (Eigel *et al.*, 1984).

```
                                                     10                                                        20
PyroGlu - Glu - Gln - Asn - Gln - Glu - Gln - Pro - Ile - Arg - Cys - Glu - Lys - Asp - Glu - Arg - Phe - Phe - Ser - Asp -

                                                     30                                                        40
    Lys - Ile - Ala - Lys - Tyr - Ile - Pro - Ile - Gln - Tyr - Val - Leu - Ser - Arg - Tyr - Pro - Ser - Tyr - Gly - Leu -

                                                     50                                                        60
    Asn - Tyr - Tyr - Gln - Gln - Lys - Pro - Val - Ala - Leu - Ile - Asn - Asn - Gln - Phe - Leu - Pro - Tyr - Pro - Tyr -

                                                     70                                                        80
    Tyr - Ala - Lys - Pro - Ala - Ala - Val - Arg - Ser - Pro - Ala - Gln - Ile - Leu - Gln - Trp - Gln - Val - Leu - Ser -

                                                     90                                                        100
    Asp - Thr - Val - Pro - Ala - Lys - Ser - Cys - Gln - Ala - Gln - Pro - Thr - Thr - Met - Ala - Arg - His - Pro - His -

                           105 ↓ 106                 110                                                       120
    Pro - His - Leu - Ser - Phe - Met - Ala - Ile - Pro - Pro - Lys - Lys - Asn - Gln - Asp - Lys - Thr - Glu - Ile - Pro -

                                                     130                                   136                 140
    Thr - Ile - Asn - Thr - Ile - Ala - Ser - Gly - Glu - Pro - Thr - Ser - Thr - Pro - Thr - [Ile] - Glu - Ala - Val - Glu -
                                                                                           Thr (Variant A)
                                          148        150                                                       160
    Ser - Thr - Val - Ala - Thr - Leu - Glu - [Ala] - Ser - Pro - Glu - Val - Ile - Glu - Ser - Pro - Pro - Glu - Ile - Asn -
                            (Variant A)     Asp      P
                                                     169
    Thr - Val - Gln - Val - Thr - Ser - Thr - Ala - Val.OH
```

FIG. 4.6. Primary structure of *Bos* κ-CN B-1P. The amino acid residues enclosed in brackets are the sites corresponding to the mutational differences in the A variant. The arrow indicates the point of attack by chymosin (rennin) (Eigel *et al.*, 1984).

major κ-CN. The carbohydrate portion contains *N*-neuraminic acid (NANA), galactose (Gal), and *N*-acetylgalactosamine (NeuNAC) and is present as either a trisaccharide or tetrasaccharide (Jollès and Fiat, 1979). The lack of information regarding the structure of the minor κ-CN components makes nomenclature of these casein fractions inconclusive at present (Eigel *et al.*, 1984).

$$\text{NeuNAC} \xrightarrow{\alpha 2,3} \text{Gal} \xrightarrow{\beta 1,3} \text{GalNAC} \xrightarrow{\beta 1} \text{Threonine}$$

$$\downarrow \text{NeuNAC } \alpha 2,6$$

$$\text{NeuNAC} \longrightarrow \text{Gal} \longrightarrow \text{GalNAC} \xrightarrow{\beta 1} \text{Threonine}$$

κ-Casein, an amphiphilic molecule with only one phosphoseryl residue, has charged tri- or tetrasaccharide moieties located in the C-terminal segment. The rest of the molecule, however, is highly hydrophobic in character. This protein is unique in that it is soluble in calcium solutions which would normally precipitate the other casein fractions. κ-Casein exerts a stabilizing effect on the casein fractions by forming colloidal micelles. It is this protein fraction that is specifically hydrolyzed by "rennin," which releases a macropeptide from the C-terminal region containing the carbohydrates. The specific bond hydrolyzed in κ-casein is 105–106, the Phe–Met linkage. The remaining product with N-terminus and two-thirds of the original peptide chain is referred to as para-κ-casein. This reaction, discussed in Chapter 6, destabilizes the casein micelle, causing formation of the curd.

C. Casein Micelle

The dispersed phase of milk consists of coarse colloidal particles, casein micelles, which have an approximate molecular weight of 10,000 and a mean diameter of 100 nm. In addition to protein (94%) the micelles also contain small ions (6.0%) such as calcium, phosphate, magnesium, and citrate, referred to as colloidal calcium phosphate (CCP). These micelles are composed of spherical submicelles (10–15 nm) which give it a porous structure. The stability of the micelles is particularly important as they exert a great influence on the processing properties of the milk. This has resulted in a number of studies on the nature of protein–protein and protein–ion interactions within the micelle structure. A number of models were proposed, including the "core coat," "chain polymer," and subunit of submicelle models (Garnier and Ribadeau-Dumas, 1970; Morr, 1967; Payens, 1966; Rose, 1969; Schmidt, 1980; Schmidt and Payens, 1976; Slattery, 1976; Slattery and Evard, 1973; Talbot and Waugh, 1970; Waugh *et al.*, 1970; Waugh and Noble, 1965). Of these models the submicelle model is now considered to be the most appropriate one.

FIG. 4.7. Artist's conception of a casein micelle containing about 40 subunits. The lighter portions of the subunits represent the α_{s1}- and/or β-caseins and are therefore relatively hydrophobic. The darker portions cover about 20% of the total surface area of the subunits and represent the hydrophilic, carbohydrate-containing parts of associated κ-casein molecules. Interior subunits have the same structure so that hydrophobic association gives rise to pores or channels in the micelle, one of which is shown in the center of the drawing. Further growth is prevented by the fact that such a large percentage of the available surface is hydrophilic. Drawing by Mrs. L. C. Innes (Slattery and Evard, 1973).

1. *Submicelle Model*

The essential feature of this model is that the casein micelles consist of spherical particles of subunits or submicelles 10–15 nm in diameter (Payens, 1979). Since the micelle is composed of three major components, α-casein, β-casein, and κ-casein, it is their distribution within the micelle that is of particular concern. Slattery (1978) and Slattery and Evard (1973) proposed a model (Fig. 4.7) in which the submicelle is amphipathic with κ-casein together with its carbohydrate component, thus forming a polar patch (represented by a dark zone) and hydrophobic interactions between the nonpolar groups of α-casein and β-casein in the submicelle after binding with calcium (represented by the light portion). Figure 4.7 illustrates aggregates of casein micelles of up to 40 units. This model incorporates several features of earlier models in which spherical submicellar particles act as building blocks. Morr's model (Morr, 1967) considered these submicellar spheres to be composed of calcium-α-casein and β-caseinates surrounded by a κ-rich layer. These spheres were believed to be held together by calcium bridges or calcium or phosphate ions and it was thought that κ-casein was evenly distributed on the surface layers of all spheres. In Waugh's model, the micelles were seen as a core of spherical polymers composed of α- and β-casein surrounded by a stabilizing coat of κ-casein which limited the growth of the micelles. Experimental evidence could not support either of these models since at least 30% of κ-casein was in the interior of the micelle and the exterior contained both α-casein and β-casein.

A model proposed by Schmidt (1980) emphasized colloidal phosphate as the cement holding the submicelles together but still incorporated some aspects of the Morr model (Morr, 1967) as well as extending the earlier model of Schmidt and Payens (1976) with certain features of the Slattery model (Slattery, 1976). For instance, the polar portions of the Schmidt model are assumed to be on the surface of the micelle with κ-casein localized in a particular area and the submicelles held together by interaction between the polar portions of α- and β-caseins and colloidal calcium phosphate.

2. *Casein Micelle Structure*

Heth and Swaisgood (1982) examined the casein micelle structure using the technique of reversible covalent immobilization. This involved isolating the native micelles from raw milk by fractionation on Biol columns and immobilizing them covalently by exposure to thioester-derivatized glass beads at room temperature. Once attached, the immobilized micelles were then dissociated with urea to detach the noncovalently bound micelles in order to leave behind those submicelles covalently attached. The covalently bound micelles were released by exposure to hydroxylamine and identified and quantitated by a micro-analytical ion exchange on DEAE-23 cellulose. Approximately 30–40% of the covalently bound monomers were released, which was similar to the percentage released in control experiments using purified α_s-casein and β-casein, thus indicating no preferential release of proteins. Using these analytical methods, profiles were obtained which showed that the monomers released from the immobilized procedure had a ratio of 1 : 1 : 3 for α : β : κ for both whole casein and the casein submicelles. These results suggested that κ-casein was concentrated on the micelle surface and α- and β-caseins were accessible with lysyl residues on the surface of the micelle. This was consistent with either of the two submicelle models proposed and allowed for incomplete exclusion of the polar portions of α- and β-caseins from the micelle surface while not completely excluding κ-casein from the interior of the micelle. The cold lability of the large portions of micellar β-casein and some κ-casein pointed to the importance of hydrophobic interactions in the micellular structure in addition to electrostatic interactions in which colloidal calcium phosphate was involved.

An alternative approach was presented by Chaplin and Green (1982) who examined the location of the casein fractions within the casein micelle after release with enzymes and enzyme–dextran conjugates. The enzymes used were pepsin and carboxypeptidase and their corresponding soluble size fractionated conjugates with dextran. The hydrolysis of isolated κ-casein fractions was independent of the size of the pepsin–dextran conjugates although this was not the

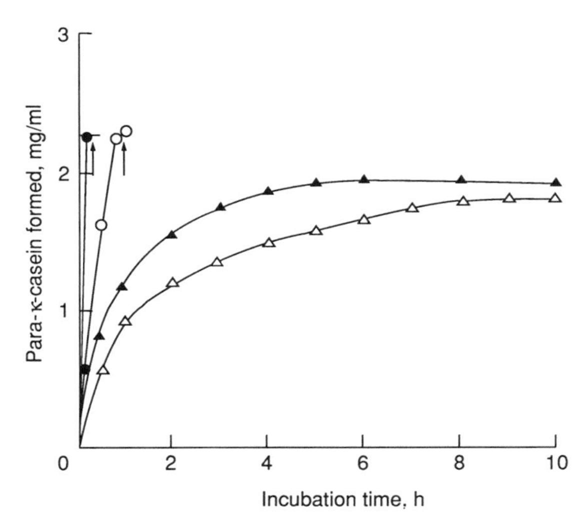

FIG. 4.8. Time course of the action of pepsin (●) and dextran–pepsin conjugates on skim milk, pH 6.6 at 30°C. The enzyme solutions were adjusted to the same activity in hydrolyzing hemoglobin (Chaplin and Green, 1982). Stokes radii of dextran–pepsin conjugates: ○, 5.0 nm; ▲, 8.5nm; △, 11.7 nm; ↑, clotting time.

case for κ-casein in milk. A decrease in the rate of hydrolysis was found evident for acidified skim milks (Fig. 4.8.) as the size of the dextran–pepsin conjugate increased (Stokes radii 5–11.7 nm). Consequently the increased rate of hydrolysis of κ-casein in micelles points to a greater accessibility of this fraction on the surface of the micelle. The β-casein fraction also appeared to be slightly more available for hydrolysis by carboxypeptidase in the native micelle as compared to the disrupted micelle. This suggested that the C-terminal bond of β-casein was on the surface of the micellar units in a hydrolyzable conformation. Based on the Schmidt (1980) model, α- and β-casein were located in the interior of the micelle while κ-casein fraction was on the exterior, thus exposing the proteinase-sensitive bond of κ-casein (phe 105–met 106) and the C-terminus of β-casein. Thomas (1973) reported the proteinase bond was extremely labile and according to work by Walstra (1979) and Walstra *et al.* (1981), the macropeptide portion of κ-casein projected out of the micelle surface. The micelle subunits were thought to consist of a casein component arranged in an orderly manner in which the more hydrophilic portions and C-termini were on the surface with colloidal phosphate-citrate linked primarily to α_{s1}–casein, the latter having a higher net charge than β- or κ-casein at pH 6.8 (Grosclaude *et al.*, 1973; Mercier *et al.*, 1973).

Mehaia and Cheryan (1983) examined the distribution of glyco-κ-casein in bovine casein micelles. By using soluble and immobilized proteases they were able to show that κ-casein was present primarily on the micelle surface. These results supported the Schmidt (1980) model. Pepper and Ferrel (1982) suggested that the casein submicelles could be formed *in vivo* by interaction between sulfhydryl groups on κ-casein monomers with those of α_s- and β-caseins. Neither α_s- nor β-caseins contain free sulfhydryl groups. After milking, oxidation on exposure to air or by the action of sulfhydryl oxidases could result in the random formation of S–S–κ-casein polymers. These researchers supported work by Slattery (1978), who suggested that κ-, α_{s1}-, and β-casein monomers coalesce to submicelles of uniform size and variable composition as the initial step in the aggregation of casein to colloidal casein particles present in the milk. Oxidation of these caseins by sulfhydryl oxidase appears to effect only the κ-casein monomer. A detailed discussion on the integrity of casein micelles can be found in an excellent review by McMahon and Brown (1984). κ-Casein is currently considered to be a "surface" protein which stabilizes the micelle with the "hairy" micelle involving "entropic" stabilization of micellar particles.

D. Biosynthesis of Milk Proteins

The major portion of milk proteins is synthesized by highly specialized mammary secretory cells under genetic control. These produce highly specific proteins which are unique to lactation. The starting materials are free amino acids absorbed from the bloodstream *via* the basal membrane by a process which involves active transport (Christensen, 1975). The possible role of the γ-glutamyl peptidase cycle in mammary amino uptake was suggested by Baumrucker and Pocius (1978). The enzyme involved, glutamyl transpeptidase (EC 2.3.2.2), catalyzes the transfer of the γ-glutamyl residue from glutathione and/or other γ-glutamyl components to amino acids or peptides (Meister *et al.*, 1976). This enzyme is thought to regulate cellular glutathione and amino acid transport via the γ-glutamyl cycle.

$$\begin{array}{ccc} \text{glutathione} & + & \text{amino acid (AA)} \\ & \downarrow & \\ \gamma\text{-glutamyl-AA} & + & \text{cysteinyl glycine} \end{array}$$

The enzyme γ-glutamyl transpeptidase is secreted into the milk where it associates with the milk membranes, including the milk fat globule membrane (MFGM) or another membrane obtained from skim milk. The latter is derived

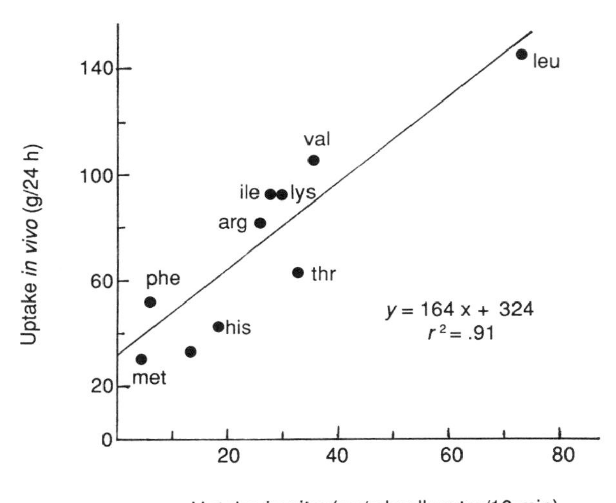

FIG. 4.9. Relationship between amino acid uptake as measured *in vitro* (abscissa) and *in vivo* (ordinate) (Pocius and Baumrucker, 1980). Calculated from data by Clark *et al.* (1975).

from plasma membranes, Golgi apparatus, endoplasmic reticulum, and secretory vessels (Kitchen, 1974). Pocius *et al.* (1981) noted that the level of glutathione was extremely low in the plasma of lactating Holstein cows as compared to in the blood, where it was 200-fold higher. From *in vitro* studies these researchers found that when arteriovenous differences for free amino acids in plasma were quantitated, there was apparent shortage of cysteine for milk protein synthesis. The uptake of glutathione by the mammary gland, however, was more than sufficient to account for any cysteine secreted in milk. Pocius and Baumrucker (1980) studied the *in vitro* uptake pattern of nine essential amino acids by mammary slices compared to the known *in vivo* uptake pattern of the same amino acids in the cow's udder. A significant linear correlation ($r = 0.91$) was evident between these patterns of uptake in spite of the many assumptions made (Figure 4.9).

The biosynthesis of milk proteins is similar to other systems in which the genetic message is transmitted from DNA to messenger RNA and then translated at the ribosomal level into the amino acid sequence of polypeptides. Following synthesis, the export milk protein leaves the ribosomes on the outer surface of the rough endoplasmic reticulum and is transported to the Golgi apparatus, where alteration of some of the export protein as well as synthesis of other major constituents of milk occurs.

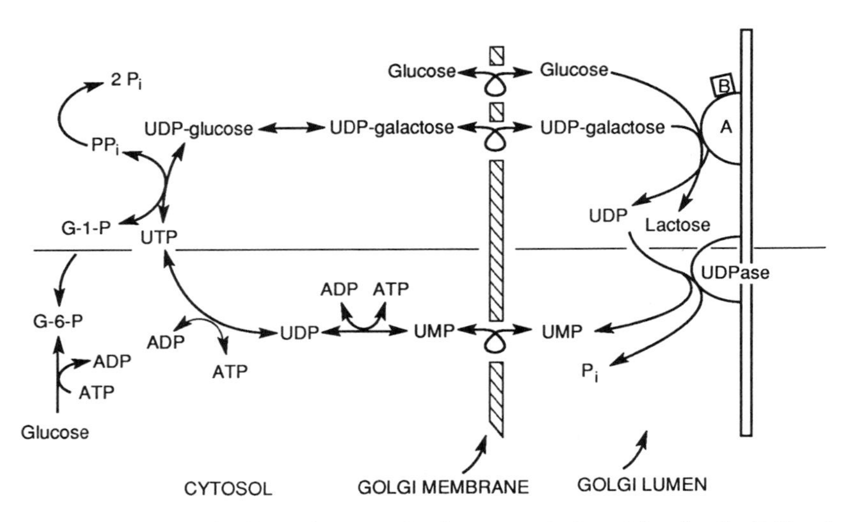

FIG. 4.10. Uridine nucleotide cycle, supporting lactose synthesis and functionally linking the cytosol and Golgi lumen compartments of the mammary secretory cell. A = Galactosyl transferase; B = α-lactalbumin (Kuhn *et al.*, 1980).

E. LACTOSE

The major carbohydrate of milk is lactose, a dissacharide of galactose and glucose linked by an α-(1 → 4) glycosidic bond. It is sometimes referred to as the milk sugar and accounts for 2% of normal cow's milk. Its complete name is lactose α-(1 → 4) galactosylglucopyranose.

The biosynthesis of lactose is catalyzed by the enzyme lactase synthetase (EC 2.4.1.77) in which glucose acts as the galactosyl acceptor (Watkins and Hassid, 1962). This enzyme is located in the luminal face of the Golgi dicytosome membrane, where it receives both glucose and UDP-galactose from the cytosol. The reactions involved are shown in Figure 4.10, in which the uridine nucleotide cycle appears to functionally link the two regions (Kuhn *et al.*, 1980). The transfer of glucose, UDP-galactose, and UMP through the Golgi membrane is probably facilitated by a specific carrier in the membrane. The formation of UMP by the enzyme nucleotide diphosphatase permits the removal of UDP released in the lactose synthetase reaction. This is important as UDP competitively inhibits the lactose synthetase enzyme to form UDP galactose (Kuhn and White, 1975, 1976, 1977). The major steps involved in the biosynthesis of lactose are as follows:

$$\text{(1) UTP + glucose-1-P} \xrightarrow[\text{(EC 2. 7. 7. 9)}]{\text{UDPG-pyrophosphorylase}} \text{UDPG-glucose + PP}_i$$

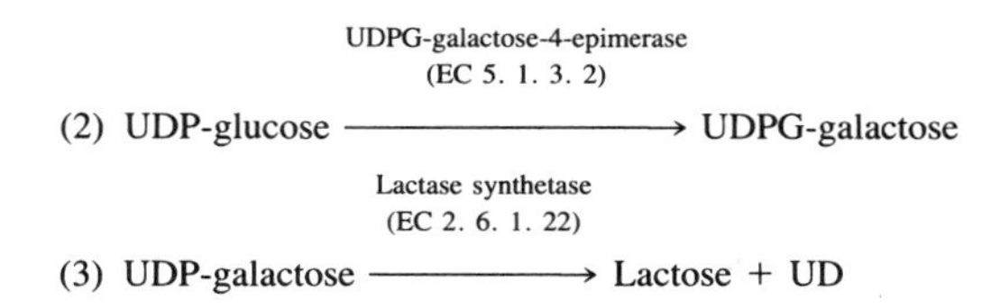

Kuhn *et al.* (1980), from their studies on rat mammary glands, summarized the benefits accrued by the compartmentalization of the lactose synthetase system. The free energy involved could be used to synthesize lactose without creating osmotic problems for the rest of the cells. Second, the concentration of UDP-glucose in the cytosol must not exceed that of UDP-galactose by a factor of three. UDP-glucose must itself be prevented from entering the Golgi lumen as it would inhibit lactose synthetase. The separation of nucleoside diphosphatase prevents hydrolysis of diphosphate in the cytosol and the subsequent depletion of phosphate energy. The role of α-lactalbumin in the biosynthesis of lactose was discovered by Brodbeck and Ebner (1966), who found that lactose synthetase was composed of two components, an "A" protein and "R" protein. The "A" protein was identified by Brew and co-workers (1968) as galactosyltransferase, while the "R" protein was shown by Brodbeck *et al.* (1967) to be α-lactalbumin. While α-lactalbumin has no catalytic function, it acts as a specific carrier protein in facilitating the action of galactosyl transferase. The latter enzyme, in the the absence of α-lactalbumin, could transfer galactose from UDP-galactose to *N*-acetylglucosamine. During the synthesis of lactose, manganese glucose, UDP-galactose, and galactosyl transferase combine with α-lactalbumin to form a dimer which accepts millimolar concentrations of glucose, forming lactose under physiological conditions. α-Lactalbumin appears to be the major regulator during lactogenesis. Other possible regulators identified by these researchers were D-glucose, UDP-galactose, calcium ions, and protein generation within the Golgi lumen, as well as the rate-limiting properties of the Golgi membrane.

Bibliography

Agradi, E., Libertini, L., and Smith, S. (1976). Specific modification of fatty acid synthetase from lactating rat mammary gland by chymotrypsin and trypsin. *Biochem. Biophys. Res. Commun.* **68,** 894.

Andrews, A. L. (1979). The formation and structure of some proteose–peptone components. *J. Dairy Res.* **46,** 215.

Andrews, A. L., Atkinson, D., Evans, M. T. A., Finer, E. G., Green, J. P., Phillips, M. C. and Roberston, R. N. (1979). The conformation and aggregation of bovine β-casein A. I. Molecular aspects of thermal aggregation. *Biopolymers* **18,** 1105.

Annan, W. D., and Manson, W. (1969). A fractionation of the α-casein complex of bovine milk. *J. Dairy Res.* **36,** 259.

Annison, E. F., Linzell, J. L., Fazakerley, S., and Nichols, B. W. (1967). The oxidation and utilization of palmitate, stearate, oleate and acetate by the mammary gland of the fed goat in relation to their overall metabolism, and the role of plasma phospholipids and neutral lipids in milk-fat synthesis. *Biochem. J.* **102,** 637.

Aschaffenburg, R. (1961). Inherited casein variants in cow's milk. *Nature (London).* **176,** 218.

Aschaffenburg, R. (1968). Reviews of the progress of dairy science. Section G. Genetics. Genetic variants of milk proteins, Their breed distribution. *J. Dairy Res.* **35,** 447.

Aschaffenburg, R., and Drewry, J. (1955). Occurrence of different beta-lactoglobulins in cow's milk. *Nature (London)* **176,** 218.

Aschaffenburg, R., Sen, A., and Thompson, M. P. (1968). Genetic variants of casein in Indian and African zebu cattle. *Comp. Biochem. Physiol.* **25,** 177.

Askew, E. W., Emery, R. S., and Thomas, J. W. (1970). Lipoprotein lipase of the bovine mammary gland. *J. Dairy Sci.* **53,** 1415.

Baumrucker, C. R., and Pocius, P. A. (1978). γ-Glutamyl transpeptidase in lactating mammary secretory tissue of cow and rat. *J. Dairy Sci.* **61,** 309.

Bell, K., Hopper, K. E., and McKenzie, H. A. (1981). Bovine α-lactalbumin C and α-, β- and κ-caseins of Bali *(Banteng)* cattle, *Bos (Bibos) javanicus. Aust. J. Biol. Sci.* **34,** 149.

Bickerstaffe, R. (1971). Uptake and metabolism of fat. *In* "Lactation" (I. R. Falconer, ed.), pp. 317–332. Butterworth, London.

Brew, K., Vanaman, T. C., and Hill, R. L. (1968). The role of α-lactalbumin and the A protein in lactose synthetase: A unique mechanism for the control of biological reaction. *Proc. Natl. Acad. Sci. U.S.A* **59,** 491.

Brodbeck, U., and Ebner, K. E. (1966). Resolution of a soluble lactose synthetase into two protein components and solubilization of microsomal lactose synthetase. *J. Biol. Chem.* **241,** 762.

Brodbeck, U., Denton, W. L., Tanahashi, N., and Ebner, K. E. (1967). The isolation and identification of lactose synthetase as α-lactalalbumin. *J. Biol. Chem.* **242,** 1391.

Brumby, P. E., and Welch, V. A. (1970). Fractionation of bovine serum lipoproteins and their characterization by gradient gel electrophoresis. *J. Dairy Res.* **37,** 121.

Chaplin, B., and Green, M. L. (1982). Probing the location of casein fractions in the casein micelle using enzymes and enzyme–dextran conjugates. *J. Dairy Res.* **49,** 631.

Christensen, H. N. (1975). "Biological Transport." Benjamin, London.

Clark, J. H., Derrig, R. G., Davis, C. L., and Spires, H. R. (1975). Metabolism of arginine and ornithine in the cow and rabbit mammary tissue. *J. Dairy Sci.* **58,** 1808.

Corbin, E. A., and Whittier, E. O. (1965). The composition of milk. *In* "Fundamentals of Dairy Chemistry" (B. H. Webb, and A. H. Johnson, eds.), chap. 1, pp. 1–36. Avi Publ. Co., Westport, Connecticut.

Davies, D. T. and Law, A. J. R. (1980). The content and composition of protein in creamery milks in south-west Scotland. *J. Dairy Res.* **47,** 83.

Dileepan, K. N., Lin, C. Y., and Smith, S. (1978). Release of two thioesterase domains from fatty acid synthetase by limited digestion with trypsin. *Biochem. J.* **173,** 11.

Eigel, W. N. (1977). Formation of γ_1-A^2, γ_2-A^2 and γ_3-A caseins by *in vitro* proteolysis of β-casein A^2 with bovine plasmin. *Int. J. Biochem.* **8,** 187.

Eigel, W. N., Butler, J. E., Ernstrom, C. A., Farrell, H. M., Jr., Halwalkar, V. R., Jennes, R., and Whitney, R. McL. (1984). Nomenclature of proteins of cow's milk. Fifth revision. *J. Dairy Sci.* **67,** 1599

Fox, P. F. (1981). Proteinases in dairy technology. *Neth. Milk Dairy J.* **35,** 233.

Fox, P. F., and Mulvihill, D. M. (1982). Milk proteins, molecular, colloidal and functional properties. *J. Dairy Res.* **49,** 679.

Garnier, J., and Ribadeau-Dumas, B. (1970). Structure of the casein micelle. A proposed model. *J. Dairy Res.* **39,** 55.

Garton, G. A. (1964). *In* "Metabolism and Physiological Significance of Lipids" (R. M. C. Dawson and D. N. Rhodes, eds.), p. 335. Wiley, New York.

Glascock, R. F., and Wright, E. W. (1962). *In* "Use of Radioisotopes in Animal Biology and the Medical Sciences" (M. Fried, ed.), Vol. 2. pp. 185–191. Academic Press, London and New York.

Glascock, R. F., Welch, V. A., Bishop, C., Davies, T., Wright, E. W., and Noble, R. C. (1966). An investigation of serum lipoproteins and their contribution to milk fat in the dairy cow. *Biochem. J.* **98,** 149.

Gooden, J. M. and Lascelles, A. K. (1973). Effect of feeding protected lipid on the uptake of precursors of milk fat by the mammary gland. *Aust. J. Biol. Sci.* **26,** 1201.

Gordon, W. G, Groves, M. L., Greenberg, R., Jones, S. B., Kalan, E. B., Peterson, R. F., and Townsend, R. E. (1972). Probable identification of γ,TS-, R- and S-caseins as fragments of β-casein. *J. Dairy Sci.* **55,** 261.

Grosclaude, F., Joudrier, P., and Mahé, M.-F. (1966). Polmorphisime de la caséine α-bovine; étroite liason du locus α-Cn avec les loci délétion dans le variant α-CnD. *Ann. Genet. Sel. Anim.* **10,** 313.

Grosclaude, F., Mahé, M.-F., and Ribadeau-Dumas, B. (1973). Structure primaire de la caséine et de la caséine β-bovine correctif. *Eur. J. Biochem.* **40,** 323.

Groves, M. L., Gordon, W. G., Kalan, E. B., and Jones, S. B. (1973). TS-A^2, TS-B, R- and S-caseins: Their isolation, composition and relationship to the β- and α-casein polymorphs A^2 and B. *J. Dairy Sci.* **56,** 558.

Guigoz, Y., and Solms, J. (1976). Bitter peptides, occurrence and structure. *Chem. Senses Flavour* **2,** 71.

Hele, P. (1954). The acetate activating enzyme of beef heart. *J. Biol. Chem.* **206,** 671.

Heth, A. A., and Swaisgood, H. E. (1982). Examination of casein micelle structure by a method of reversible covalent immobilization. *J. Dairy Sci.* **65,** 2047.

Hoagland, P. D., Thompson, M. P. and Kalan, E. B. (1971). Amino acid composition of α_{s3}-, α_{s4}-, and α_{sr}-caseins. *J. Dairy Sci.* **54,** 1103.

Jennes, R., Larson, B. L., McMeekin, T. L., Swanson, C. H., Whitnah, C. H., and Whitney, R. (1956). Nomenclature of the proteins of bovine milk. *J. Dairy Sci.* **39,** 536.

Jensen, G.-K., and Nielsen, P. (1982). Reviews in the progress of dairy science: Milk powder and recombination of milk and milk products. *J. Dairy Res.* **49,** 515.

Jollès, J., and Fiat, A-M. (1979). The carbohydrate portions of milk glycoproteins. *J. Dairy Res.* **46,** 187.

Jollès, J., Schoentgen, F., Alais, C., and Jollès, P. (1972). Studies on the primary structure of cow κ-casein: The primary structure of cow para-κ-casein. *Chimia* **20,** 148.

Kennedy, E. P. (1957). Metabolism of lipids. *Annu. Rev. Biochem.* **26,** 119.

Kinsella, J. E. (1970). Biosynthesis of lipids from [2-^{14}C]acetate and D(−)-β-hydroxy-(1–3^{14}C)butyrate by mammary cells from bovine and rat. *Biochim. Biophys. Acta* **210,** 28.

Kinsella, J. E. (1975). Coincident synthesis of fatty acids and secretory triglycerides in bovine mammary tissue. *Int. J. Biochem.* **6,** 65.

Kitchen, B. J. (1974). A comparison of the properties of membranes isolated from bovine skim and cream. *Biochim. Biophys. Acta.* **356,** 257.

Kuhn, N. J., and White, A. (1975). Milk glucose as an index of the intracellular glucose concentration of rat mammary gland. *Biochem. J.* **152,** 153.

Kuhn, N. J., and White, A. (1976). Evidence for specific transport of uridine diphosphate galactose across the Golgi membrane of rat mammary gland. *Biochem. J.* **154,** 243.

Kuhn, N. J., and White, A. (1977). The role of nucleoside diphosphatase galactose across the Golgi membrane of rat mammary gland. *Biochem. J.* **168,** 423.

Kuhn, N. J., Carrick, D. T., and Wilde, C. J. (1980). Lactose synthesis: The possibilities of regulation. *J. Dairy Sci.* **63,** 328.

Kumar, S., Dorsey, J. A., Muesing, R. A., and Porter, J. W. (1970). Comparative studies of the pigeon liver fatty acid synthetase complex and its subunits. *J. Biol. Chem.* **245,** 4732.

Lindstrom-Lang, K., and Kodoma, S. (1925). Studies over kasein. *C. R. Trav. Lab. Carlsberg* **16,** 1.

Mather, I. H., and Keenan, T. W. (1975). Studies on the structure of the milk fat globule membrane. *J. Membr. Biol.* **21,** 65.

McMahon, D. J., and Brown, R. J. (1984). Composition, structure and integrity of casein micelles: A review. *J. Dairy Sci.* **67,** 499.

McPherson, A. V., and Kitchen, B. J. (1983). Reviews of the progress of dairy science: The bovine milk fat globule membrane—Its formation, composition, structure and behaviour in milk and dairy products. *J. Dairy Res.* **50,** 107.

Mehaia, M. A., and Cheryan, M. (1983). Distribution of glyco κ-casein micelles. A study using soluble and immobilized proteases. *J. Dairy Sci.* **66,** 2474.

Meister, A., Tate, S. S., and Rose, L. L. (1976). Membrane bound γ-glutamyl transpeptidase. *In* "The Enzymes of Biological Membranes." (A. Martonsi, ed.). Vol, III, p. 315. Plenum Press, New York.

Mellenberger, R. W., Bauman, D. E., and Nelson, D. R. (1973). Metabolic adaptations during lactogenesis; Fatty acid and lactose synthesis in cow mammary tissue. *Biochem. J.* **136,** 741.

Mercier, J.-C., Grosclaude, F., and Ribadeau-Dumas, B. (1971). Structure primaire de la caséine κ β bovine. *Eur. J. Biochem.* **23,** 41.

Mercier, J.-C., Brignon, G., and Ribadeau-Dumas, B. (1973). Structur primaire de la caséine κ β bovine, séquence complete. *Eur. J. Biochem.* **23,** 41.

Moore, J. H., and Christie, W. W. (1979). Lipid metabolism in the mammary gland of ruminant animals. *Prog. Lipid Res.* **17,** 347.

Moore, J. H., Steele, W., and Noble, R. C. (1969). The relationship between dietary fatty acids, plasma lipid composition and milk fat secretion in the cow. *J. Dairy Res.* **36,** 383.

Morley, N. H., and Kuksis, A. (1972). Positional specificity of lipoprotein lipase. *J. Biol. Chem.* **247,** 6389.

Morley, N. H., Kuksis, A., Buchna, D., and Myher, J. (1975). Hydrolysis of diacylglycerols by lipoprotein lipase. *J. Biol. Chem.* **250,** 3414.

Morr, C. V. (1967). Effect of oxalate and urea upon ultracentrifugation properties of raw and heated skim-milk casein micelles. *J. Dairy Sci.* **50,** 1744.

Neeling, J. M. (1964). Variants of κ-casein revealed by improved starch gel electrophoresis. *J. Dairy Sci.* **47,** 506.

Norman, H. D., Kuck, A. L., Cassell, B. G., and Dickinson, F. N. (1978). Effect of age and month-of-calving on solids-not-fat and protein yield for five dairy breeds. *J. Dairy Sci.* **61,** 239.

Paltauf, F., and Wagner, E. (1976). Stereospecificity of lipases. Enzymatic hydrolysis of enantiomeric alkyldiacyl and dialkylacylglycerols by lipoprotein lipase. *Biochim. Biophys. Acta* **431,** 359.

Paltauf, F., Esfandi, F., and Holasek, A. (1974). Stereospecificity of lipases. Enzymic hydrolysis of enantiometric alkyl diacylglycerols by lipoprotein lipase, lingual lipase and pancreatic lipase. *FEBS Lett.* **40,** 119.

Patton, S., and Jensen, R. G. (1975). Lipid metabolism and membrane functions of the mammary gland. *Prog. Chem. Fats Other Lipids* **14,** 167.

Payens, T. A. J. (1966). Association of caseins and their possible relation to the structure of the casein micelle. *J. Dairy Sci.* **49,** 1317.

Payens, T. A. J. (1979). Casein micelles, the colloid–chemical approach. *J. Dairy Res.* **46,** 291.

Pepper, L., and Ferrel, H. M., Jr. (1982). Interactions leading to formation of casein submicelles. *J. Dairy Sci.* **65,** 2259.

Phillips, G. T., Nixon, J. E., Dorsey, J. A., Butterworth, P. H. W., Chesteron, C. J., and Porter, J. W. (1970). The mechanism of sythesis of fatty acids by the pigeon liver enzyme system. *Arch. Biochem. Biophys.* **138,** 380.

Pocius, P. A., and Baumrucker, C. R. (1980). Amino acid uptake by bovine mammary slices. *J. Dairy Sci.* **63,** 746.

Pocius, P. A., Clark, J. H., and Baumrucker, C. R. (1981). Glutathione in bovine blood: Possible source of amino acids for milk protein synthesis. *J. Dairy Sci.* **64,** 1551.

Ribadeau-Dumas, B., Brignon, G., Grosclaude, F., and Mercier, J.-C. (1972). Structure primaire de la caseine β bovine. *Eur. J. Biochem.* **25,** 505.

Robinson, D. S. (1970). The function of plasma triglyceride in fatty acid transport. *Compr. Biochem.* **18,** 51–116.

Rose, D. (1969). A proposed model of micelle structure in bovine milk. *Dairy Sci. Abstr.* **31,** 171.

Rose, D. J., Brunner, J. R., Kalan, E. B., Larson, B. L., Melchnychyn, P., Swaisgood, H. E., and Waugh, D. F. (1970). Nomenclature of the proteins of cow's milk, Third revision. *J. Dairy Sci.* **53,** 1.

Schmidt, D. G. (1964). Starch gel electrophoresis of κ casein. *Biochim. Biophys. Acta* **90,** 411.

Schmidt, D. G. (1980). Colloidal aspects of casein. *Neth. Milk Dairy J.* **34,** 42.

Schmidt, D. G., and Payens, T. A. J. (1976). Micellar aspects of casein. *Surf. Colloid Sci.* **9,** 165.

Scow, R. O., Blanchette-Mackie, E. J., and Smith, L. C. (1976). Role of capillary endothelium in the clearance of chylomicrons: A model for lipid transport from blood by lateral diffusion in cell membranes. *Circ. Res.* **39,** 149.

Slattery, C. W. (1976). Review: Casein micelle structure: An examination of models. *J. Dairy Sci.* **59,** 1547.

Slattery, C. W. (1978). Variation in the glucosylation pattern of bovine κ casein with micelle size and its relationship to a micelle model. *Biochemistry* **17,** 1100.

Slattery, C. W., and Evard, R. (1973). A model for the formation and structure of casein micelles from subunits of variable components. *Biochim. Biophys. Acta* **317,** 529.

Smith, S. (1976). Structural and functional relationships of fatty acid synthetases from various tissues and species. *In* "Immunochemistry of Enzymes and Their Antibodies" (M. R. J. Salton, ed.), Chapter 5. Wiley, New York.

Smith, S. (1980). Mechanism of chain length determination in biosynthesis of milk fatty acids. *J. Dairy Sci.* **63,** 337.

Smith, S., Agradi, E., Libertini, L., and Dileepan, K. N. (1976). Specific release of the thioesterase component of the fatty acid synthetase complex by limited trypsinization. *Proc. Nat. Acad. Sci. U.S.A.* **73,** 1184.

Sommerfeldt, J. L., and Baer, R. J. (1986). Variability of milk components in 1705 herds. *J. Food Prot.* **49,** 729.

Stead, D., and Welch, V. A. (1975). Lipid composition of bovine serum lipoproteins. *J. Dairy Sci.* **58,** 122.

Swaisgood, H. E. (1982). Chemistry of milk protein. *In* "Developments in Dairy Chemistry. 1. Proteins" (P. F. Fox, ed.). Chapt. 1. p. 1. Applied Science Publishers, London.

Talbot, B., and Waugh, D. F. (1970). Micelle-forming characteristics of monomeric and covalent polymeric κ-caseins. *Biochemistry* **9,** 2807.

Thomas, E. L. (1973). The accessibility and lability of a rennin-sensitive bond of bovine κ-casein. *Neth. Milk Dairy J.* **27,** 273.

Thompson, M. P., Kiddy, C. A., Pepper, L., and Zittle, Cs. A. (1962). Casein variants in the milk from individual cows. *J. Dairy Sci.* **45,** 650.

Torneur, C. (1974). The proteolytic activity of lactobacilli. *Int. Dairy Congr., 19th* Vol. 1E, p. 366.

Trieu-Cuot, P., and Gripon, J.-C. (1981). Electrofocusing and two-dimensional electrophoresis of bovine caseins. *J. Dairy Res.* **48,** 303.

von Hippel, P. H., and Waugh, D. F. (1955). Casein, monomers and polymers. *J. Amer. Chem. Soc.* **77,** 4311.

Walstra, P. (1979). The voluminosity of bovine casein micelles and some of its implications. *J. Dairy Res.* **46,** 317.

Walstra, P., Bloomfield, V. A., Wei, G. J., and Jennes, R. (1981). Effect of chymosin action on the hydronamic character of casein micelles. *Biochim. Biophys. Acta* **669,** 258.

Watkins, W. M., and Hassid, W. Z. (1962). The synthesis of lactose by particulate enzyme preparation from guinea pig and bovine mammary glands. *J. Biol. Chem.* **237,** 1432.

Waugh, D. F., and Noble, R. W., Jr. (1965). Casein micelles. Formation and structure. II. *J. Am. Chem. Soc.* **87,** 2246.

Waugh, D. F., and von Hippel, P. H. (1956). κ-Casein and the stabilization of casein micelles. *J. Am. Chem. Soc.* **78,** 4576.

Waugh, D. F., Creamer, L. K., Slattery, C. W., and Dresdner, G. W. (1970). Core polymers of casein micelles. *Biochemistry* **9,** 786.

West, C. E., Bickerstaffe, R., Annison, E. F., and Linzell, J. L. (1972). Studies on the mode of uptake of blood triglycerides by the mammary gland of the lactating goat. *Biochem. J.* **126,** 477.

Whitney, R. McL., Brunner, J. R., Ebner, K. E., Farrell, M. Jr., Josephson, R. V., Morr, C. V., and Swaisgood, H. (1976). Nomenclature of the proteins of cow's milk, Fourth revision. *J. Dairy Sci.* **59,** 795.

Wilcox, C. J., Gaunt, S. N., and Farthing, B. R. (1971). Genetic interrelationship of milk composition and yield. *South. Coop. Ser. Bull.* **155.** Univ. Florida, Gainesville.

Woychik, J. H. (1964). Polymorphism in κ-casein of cow's milk. *Biochem. Biophys. Res. Commun.* **16,** 267.

Zinder, O., Hamosh, M., Clary Fleck, T. R., and Scow, R. O. (1974). Effect of prolactin on lipoprotein lipase in mammary gland and adipose tissue of rats. *Am. J. Physiol.* **226,** 744.

Part II

Biochemistry of Food Processing

5

Biochemistry of Food Processing: Browning Reactions in Foods

I. Introduction

Browning reactions in food are widespread phenomena which take place during processing and storage. These reactions occur during the manufacture of meat, fish, fruit, and vegetable products, as well as when fresh fruits and vegetables are subjected to mechanical injury. Browning affects the flavor, appearance, and nutritive value of the food products involved. However, for certain foods browning is an important part of the preparation process. For example, in the manufacture of coffee, tea, beer, and maple syrup and in the toasting of bread it enhances the appearance and flavor of these products. Browning, to a limited degree, is considered desirable in apple juice, potato chips, and French fries. To control or inhibit these reactions it is important to understand the mechanisms involved.

Three browning mechanisms appear to be involved in foods as shown in Table 5.1. In the case of ascorbic acid browning it can proceed either by the enzyme ascorbic acid oxidase or by direct atmospheric oxygen and oxidation of ascorbic acid.

TABLE 5.1

MECHANISMS OF BROWNING REACTIONS

Mechanism	Requires oxygen	Requires amino group in initial reaction	pH optimum
Maillard	−	+	Alkaline
Caramelization	−	−	Alkaline, acid
Ascorbic acid oxidation	+	−	Slightly acid

II. Nonenzymatic Browning

During the preparation and processing of foods, one soon becomes acquainted with the phenomenon of browning associated with heated and stored products. This phenomenon, referred to as nonenzymatic browning, distinguishes it from the enzyme-catalyzed reactions described in Chapter 9. The importance of this reaction in the production of foods is amply illustrated by its contribution to the flavor, color, and aroma of coffee, caramel, bread, and breakfast cereals. Careful control must be exercised to minimize excessive browning which could lead to unpleasant changes in the food product. In recent years there has been considerable focus on the deleterious effects of nonenzymatic browning reactions in food (Eriksson, 1981). Of particular concern is the toxicity and potential mutagenicity of some of the intermediates formed (Aeschbacher *et al.*, 1981; Gazzani *et al.*, 1987; Grivas *et al.*, 1985; T.-C. Lee *et al.*, 1982; Shinohana *et al.*, 1980; Spingarn and Garvie, 1979; Spingarn *et al.*, 1983). Not all the intermediates formed are deleterious, however, and some appear to exert considerable antioxidant activity (Baltes *et al.*, 1973; Kawashima *et al.*, 1977; Kirigaya *et al.*, 1968, 1969, 1971; Lingnert and Eriksson, 1980; Lingnert and Hall, 1986; Yamaguchi and Fujimaki, 1974).

Since the first edition of this book an enormous number of papers have been published on nonenzymatic browning systems. Nevertheless, our knowledge of this area still remains fragmentary. Current evidence still supports the existence of three major pathways: Maillard reaction, caramelization, and ascorbic acid oxidation.

A. MAILLARD REACTION

The formation of brown pigments and melanoidins was first observed by the French chemist Louis Maillard (1912) following the heating of a solution of glucose and lysine. This reaction was subsequently referred to as the Maillard

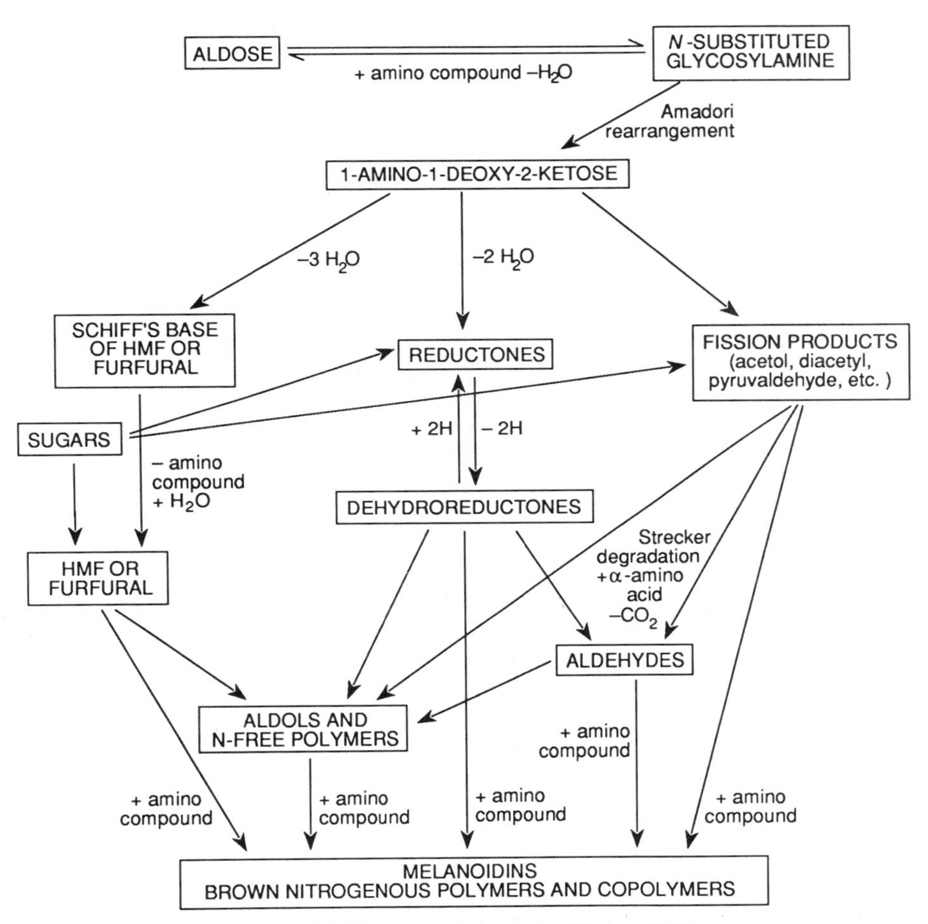

SCHEME 5.1. Nonenzymatic browning (Hodge, 1953).

reaction and essentially covers all those reactions involving compounds with amino groups and carbonyl groups present in foods. These include amines, amino acids, and proteins interacting with sugars, aldehydes, and ketones, as well as with products of lipid oxidation (Feeney *et al.*, 1975; Kwon *et al.*, 1965; Montgomery and Day, 1965). The general mechanism of browning was first proposed by Hodge (1953) and subsequently reviewed by Ellis (1959), Heyns and Paulsen (1960), Reynolds (1963, 1965, 1969), and Baltes (1973). In spite of the volumes of research on this reaction, the original reaction sequence (Scheme 5.1) proposed by Hodge (1953) still remains valid.

1. *Carbonylamino Reaction*

The first step in the Maillard reaction involves condensation between the α-amino groups of amino acids or proteins and the carbonyl groups of reducing sugars: this defines the "carbonylamino" reaction. The initial product is an addition compound which rapidly loses water to form a Schiff base followed by cyclization to the corresponding N-substituted glycosylamine:

D-Glucose $\xrightarrow{+H_2NR}$ Addition compound $\xrightarrow{-H_2O}$ Schiff's base $\xrightarrow{\text{cyclization}}$ *N*-Substituted glycosylamine

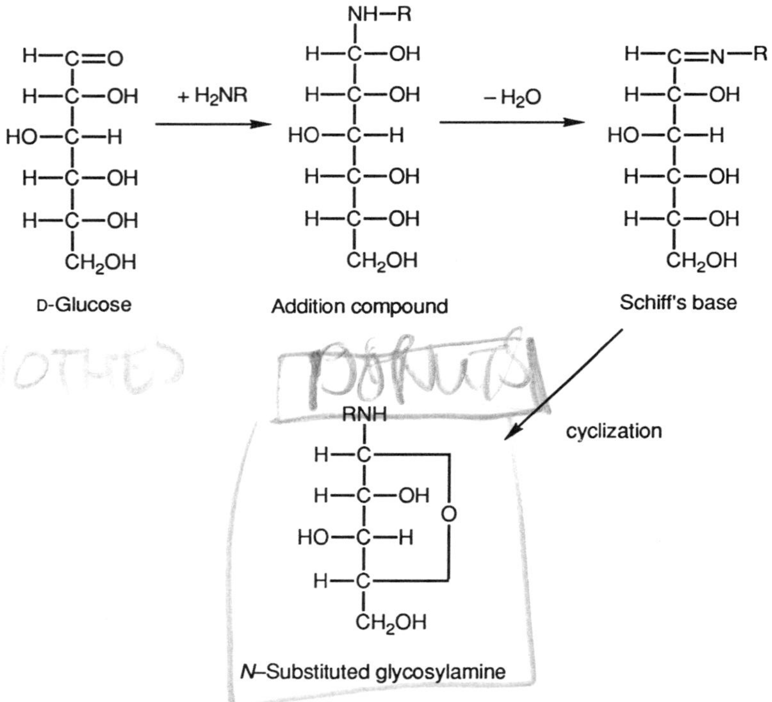

These reactions are all reversible as an equilibrium exists for these compounds in aqueous solution.

2. *Mechanism of the Carbonylamino Reaction*

The formation of the N-substituted glycosylamine involves condensation of the amine group of the amino acid with a carbonyl group of a reducing sugar. This reaction is not necessarily restricted to α-amino acids and can involve the participation of other amino groups found in peptides and proteins. This is facilitated when the pH of the medium is above the isoelectric point of the amino group, thus producing basic amino groups.

The protein molecules are composed of many amino acids joined covalently by peptide bonds, in which the amino acids are presumably unavailable for interaction. Harris and Mattil (1940) observed that lysine provided the majority of free amino groups in proteins, in the form of ϵ-amino groups, which were the main participant in this reaction. While this is true, other amino acids with additional amino groups could also participate, for example, arginine, tryptophan, and histidine. As the temperature is increased many more amino acids are rendered unavailable, which cannot be explained by the cleavage of the peptide bonds, a process which appears to be slight even at fairly high temperatures. Horn *et al.* (1968) found it difficult to explain the rapid and extensive destruction of amino acids in proteins in the presence of sugars simply on the basis of the free amino groups present. A common group such as the imide group of the peptide bond was suggested to be involved, in which the hydrogen of this group was replaced by a carbohydrate moiety. The resulting complex was thought to render the amino acid involved unavailable or prevent enzymatic hydrolysis of the peptide bond itself.

Dworschak and Orsi (1981) examined the Maillard reaction between methionine, lysine, and tryptophan with glucose in the dry state by increasing the temperature up to 700°C in a derivatograph furnace. The derivatograms obtained in the initial stages, corresponding to the condensation step, showed a molar ratio of sugar : amino acid of 1 : 2 for both lysine : glucose and tryptophan : glucose. This illustrated the importance of the bifunctional amine groups of these amino acids in the condensation reaction.

In addition to the mechanisms discussed here, the possibility of anhydride linkages between carboxylic and amino groups was suggested earlier by Harris and Mattil (1940). It was later proposed that these linkages might arise from the interaction between the ϵ-amino groups of lysine and free dicarboxylic acids in the protein chain. Patton *et al.* (1954) proposed that such linkages could involve aspartic and glutamic acids in the protein.

3. *Amadori Rearrangement*

The final condensation product in the carbonylamino reaction is the N-substituted glycosylamine. It was soon evident, however, that this compound was extremely unstable and underwent a series of rearrangements, which explained why the reducing power for a casein–glucose system was of the same order as the original glucose (Lea and Hannan, 1950). These changes involved isomerization of the N-substituted glycosylamine to the corresponding fructose–amino acid. The transition from an aldose to a ketose sugar derivative (Figure 5.1) is referred to as the Amadori rearrangement (Weygand, 1940) and involves protonation of nitrogen at carbon-l. In the case of ketones and amines, ketosylamines are formed which then undergo the "Heyns" rearrangement to form

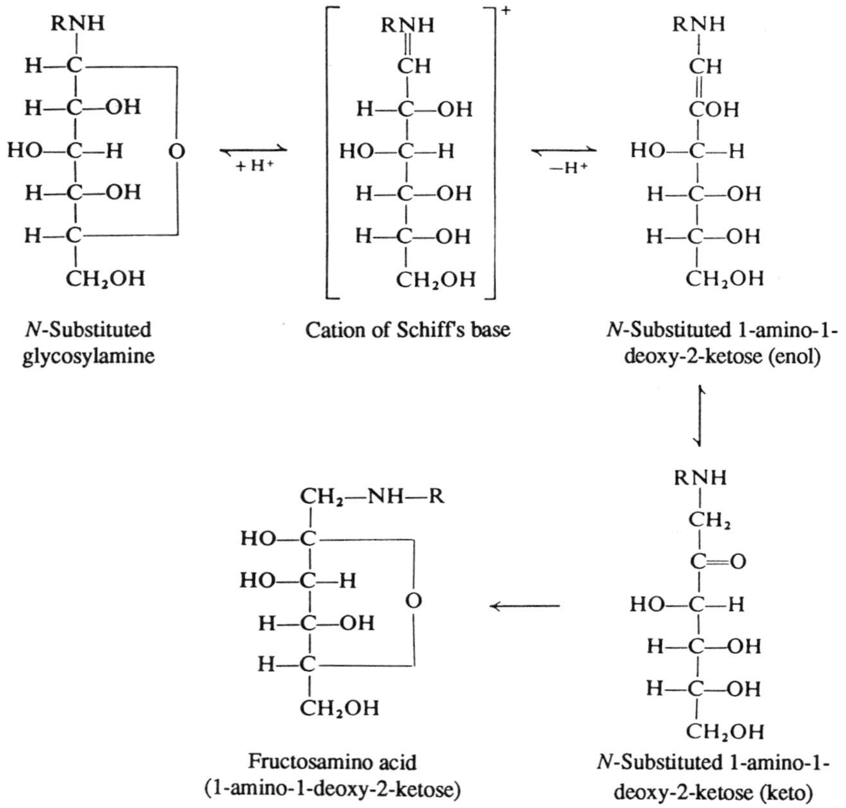

N-Substituted glycosylamine $\xrightleftharpoons{+H^+}$ Cation of Schiff's base $\xrightleftharpoons{-H^+}$ *N*-Substituted 1-amino-1-deoxy-2-ketose (enol)

$\updownarrow$

N-Substituted 1-amino-1-deoxy-2-ketose (keto) $\rightarrow$ Fructosamino acid (1-amino-1-deoxy-2-ketose)

FIG. 5.1. Amadori rearrangement.

2-amino-2-deoxy aldoses (Reynolds, 1965) by protonation of the oxygen at carbon-6 (Kort, 1970).

The Amadori rearrangement has been demonstrated for a series of glucose–amino acid complexes synthesized by Abrams *et al.* (1955). The reactions leading up to the formation of 1-amino-1-deoxy-2 ketone are all reversible. In fact these products are quite stable and have been identified in freeze-dried peaches and apricots (Anet and Reynolds, 1957), soy sauce (Hashiba, 1978), and milk (Finot *et al.*, 1968). Moll *et al.* (1982) isolated and purified a number of Amadori compounds from crude extracts of Maillard reaction systems using HPLC. These included Amadori compounds of alanine–fructose, leucine–fructose, hydroxyproline–fructose and tryptophan–fructose, the structures of which are shown in Scheme 5.2. Lee *et al.* (1979) followed the development of Maillard reaction products during the processing of apricots and found that the level of Amadori

Alanine-fructose

Leucine-fructose

Hydroxyproline-fructose

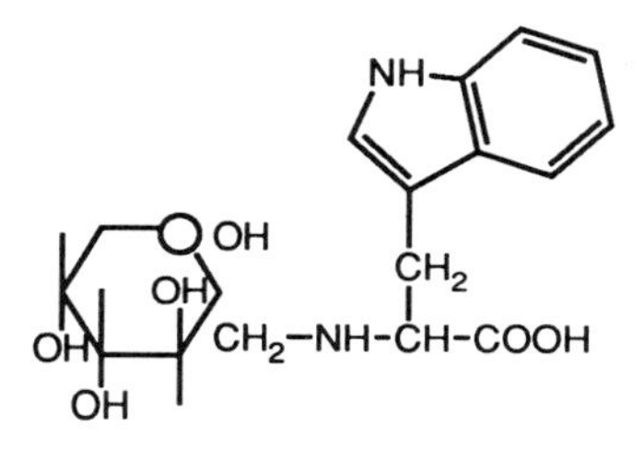

Tryptophan-fructose

SCHEME 5.2. Amadori compounds. Reprinted with permission from Moll *et al.* (1982). Copyright by the American Chemical Society.

compounds reached a maximum prior to the development of any brown color. While these intermediates do not contribute directly to browning or flavor, they result in a loss of nutritional value due in large part to the unavailability of the ϵ-amino group of lysine (Dworschak, 1980; Finot and Mauron, 1972; Friedman, 1982; Hurrell and Carpenter, 1974, 1981; T. C. Lee *et al.*, 1982; Mauron, 1970; Plakas *et al.*, 1988).

4. *Conditions for the Maillard Reaction*

a. pH and Buffers. The carbonylamino reaction can develop in acidic or alkaline media, although it is favored under alkaline conditions, where the amine groups of the amino acids, peptides, and proteins are in the basic form. Increasing the pH also ensures that more of the hexoses are in the open chain or reducing form (Burton and McWeeney, 1963). Several studies have reported an increase in reaction rate as the pH increases (Lea and Hannan, 1949; Underwood *et al.*, 1959). Thus foods of high acidity are less susceptible to these reactions, for example, pickles. An exception to this, however, is the involvement of sucrose in the Maillard reaction. Sucrose, as a nonreducing sugar, will only participate when the glycosidic bond is hydrolyzed and the reducing monosaccharide constituents released. Hydrolysis of the glycosidic bond in sucrose is facilitated by a low pH and high moisture levels, resulting in an increase in the Maillard reaction rate in protein–sucrose systems (Hurrell and Carpenter, 1974).

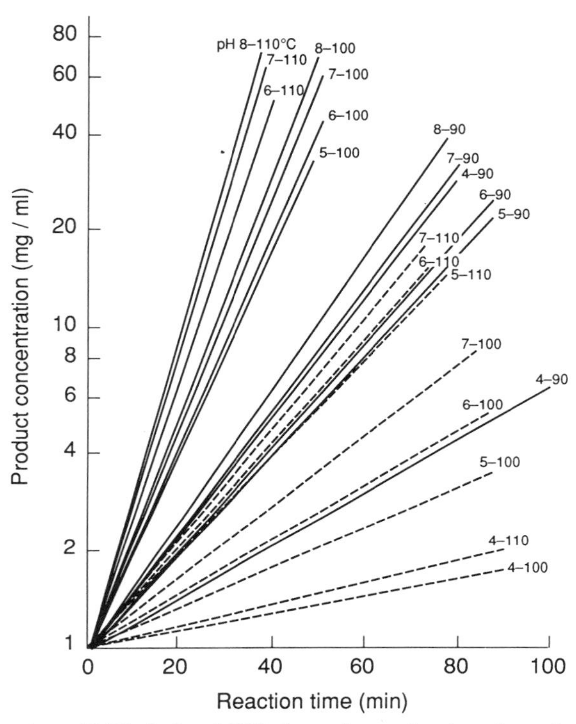

FIG. 5.2. Formation of MFL (—) and DFL (- - - -) as a function of reaction time at various temperatures and pH. All lines were drawn for best fit of the data points. Reprinted with permission from C. M. Lee *et al.* (1984). Copyright by the American Chemical Society.

The role of buffers in nonenzymatic reactions has been shown to increase the rate of browning for sugar–amino acid systems as a result of their influence on the ionic environment in which the reaction takes place. For example, C. M. Lee and co-workers (1984) monitored the formation of the Amadori compounds monofructosyllysine (MFL) and difructosyllysine (DFL) in glucose–lysine mixtures at different temperatures and pH. Their results, shown in Figure 5.2, indicate pseudo-first-order plots for MFL and DFL formation, which increased from pH 4 to 8. This pattern was similar to that observed for pigment formation. A plot of pigment formation as a function of pH, however, showed a parabolic curve with break points at pH 6 and 5 for systems heated at 100 and 110°C, respectively (Figure 5.3).

b. Temperature. The temperature dependence of this reaction has been demonstrated in a number of quantitative studies, where increased rates were reported with rise in temperature. Lea and Hannan (1949) found that the decrease

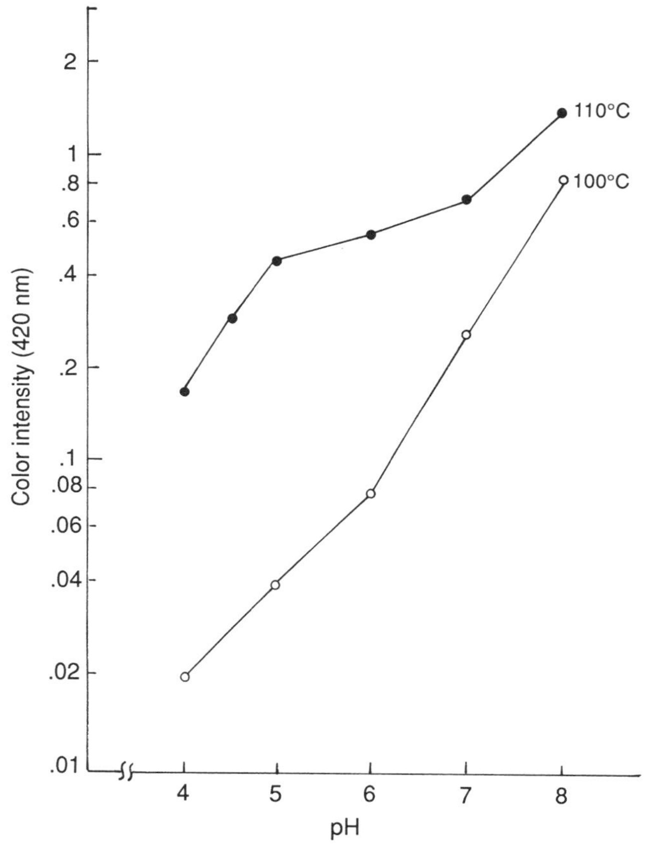

FIG. 5.3. Changes in pigment formation as a function of pH. Reprinted with permission from C. M. Lee *et al.* (1984). Copyright by the American Chemical Society.

in free amino nitrogen for a casein–glucose system conformed to the Arrhenius equation over a temperature range of 0–80°C, where a linear relationship existed between the rate of reaction over this range. The α-amino nitrogen loss possessed 29 cal/mole activation energy in the casein–glucose system while an increase in activation energy from 26 to 36 cal/mole was noted by Hendel *et al.* (1955a) during the browning of dehydrated potato products with increased humidity. Using the formation of hydroxymethylfurfural (HMF) as a measure of progress of the Maillard reaction, Dworschak and Hegedüs (1974) noted an increase in the activation energy for lysine from 29.4 to 34.8 cal/mole when milk powder was heated while increasing the humidity from 2.45 to 5.7%. The amount of HMF produced, however, decreased from 43.6 to 34.6%. Nevertheless, it was appar-

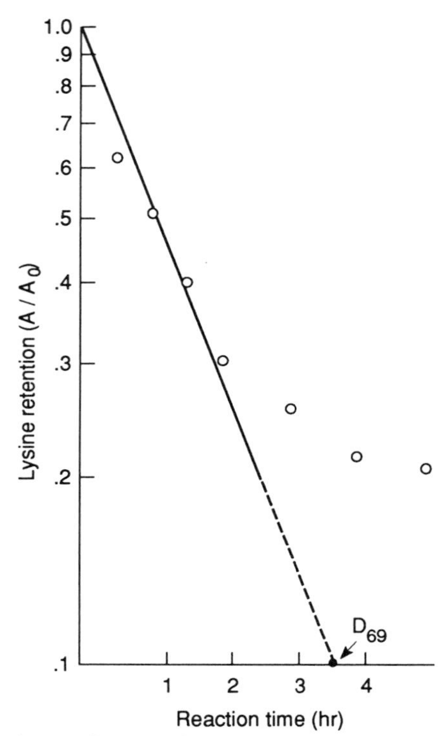

FIG. 5.4. Lysine retention as a function of reaction time. D_{69} = time required for a 90% reduction in the concentration of lysine at 69°C. Reprinted with permission from C. M. Lee *et al.* (1984). Copyright by the American Chemical Society.

ent that the Maillard reaction rate increased fourfold for each 10°C rise in temperature (Gornhardt, 1955).

Hurrell and Carpenter (1974) noted that the loss in ε-amino lysine groups in an albumin–glucose system at 37°C over 30 days was almost equivalent to that in the same system heated at 121°C for 15 min. In both cases the loss in ε-amino groups was 80%, thus emphasizing the importance of duration of storage as well as temperature. A semilog plot of lysine retention during heating of the glucose–lysine model systems at 69°C was shown by C. M. Lee and co-workers (1984) to be linear during the first 2 hr, with an extrapolated decimal reduction time *D* (time required for a 90% reduction of lysine at 69°C) of 3.5 hr (Figure 5.4).

c. Moisture Content. The Maillard reaction proceeds rapidly in solution, although complete dehydration or excessive moisture levels inhibit this process (Wolfrom and Rooney, 1953). Lea and Hannan (1949, 1950) recorded the op-

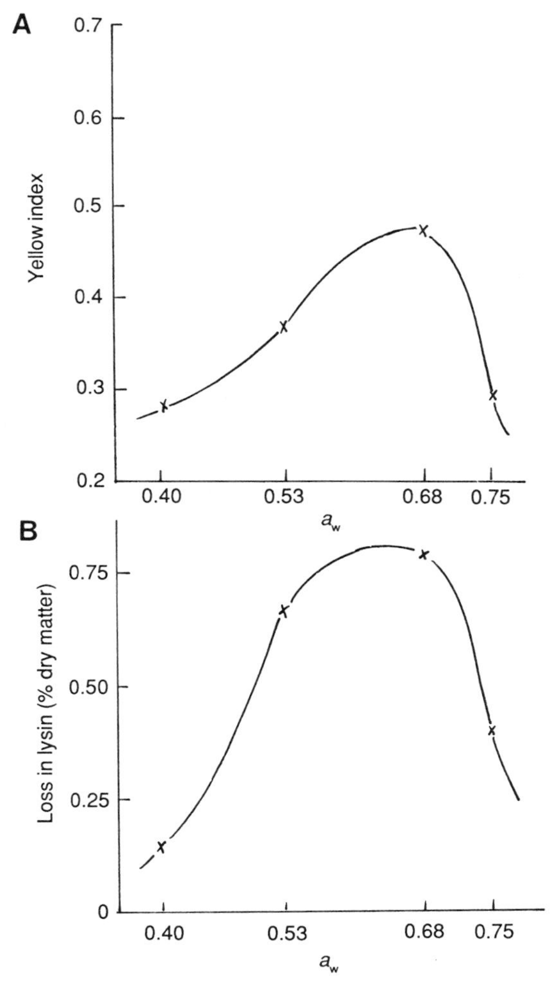

FIG. 5.5. The color change (A) and loss in free lysine (B) of milk powder kept at 40°C for 10 days as a function of a_W (Loncin *et al.*, 1968).

timum moisture level for a casein–glucose system and found that the maximum loss of free amino groups occurred between 65 and 70% relative humidity, which corresponded to a level at which the reactants were still in a comparatively dry state. Loncin *et al.* (1968) monitored browning in milk powder at 40°C as a function of water activity (a_w) and lysine over a period of a day. Their results, shown in Fig. 5.5, show that the loss of lysine paralleled the extent of browning with a maximum between a_w of 0.6 and 0.7. An increase in humidity was also shown by Dworschak and Hegedüs (1974) to cause an increase in the loss of lysine and tryptophan with the concomitant increase in HMF formation and browning. In general, it appears that this reaction is favored at an optimum moisture content corresponding to fairly low moisture levels (Danehy, 1986).

d. Sugars. Reducing sugars are essential ingredients in these reactions, providing the carbonyl groups for interaction with the free amino groups of amino acids, peptides, and proteins. The initial rate of this reaction is dependent on the rate at which the sugar ring opens to the oxo or reducible form. Burton and McWeeny (1963) monitored the concentration of the oxo form of the sugar using polarography and found that it increased with increase in pH. The amount of the oxo form was much higher for pentoses than hexoses, thus explaining the greater reactivity of pentoses in browning systems. This was confirmed by Spark (1969), who found that the order of reactivity was greater for aldopentoses than aldohexoses while reducing disaccharides were considerably less reactive. Of the hexoses examined, reactivity decreased in the order of D-galactose>D-mannose>D-glucose, corresponding to the decreasing rate of ring opening. Tu and Eskin (1973) noted that reducing sugars exerted an inhibitory effect on the hydrolysis of casein by trypsin because of the unavailability of certain essential amino acids resulting from nonenzymatic browning reactions. They found that xylose exerted the greatest inhibitory effect followed by fructose and glucose. Rao and Rao (1972) autoclaved casein with several sugars and found that the reaction was fastest in the presence of arabinose, followed by glucose and lactose. This was accompanied by a significant reduction in lysine availability.

Katchalsky (1941) reported that fructose did not condense with amino acids in dilute solution although scientists have since confirmed that a definite interaction does take place (Heyns and Breuer, 1958; Heyns and Noack, 1962, 1964). D-Fructose has also been reported by Shallenberger and Birch (1975) and Bobbio *et al.* (1973) to brown at a much faster rate than glucose during the initial stages of the browning reaction, but it then falls behind. This was confirmed by Reyes *et al.* (1982) using model systems containing glucose–glycine and fructose–glycine (1 : 1 molar ratio) at 60°C, pH 3.5, and held for 280 hr. The fructose system browned at a faster rate during the first 80-hr period but was subsequently taken over by the glucose system. The consumption pattern of glucose and fructose paralleled the rate of browning (Figure 5.6). The difference in losses was attributed to the greater polymerization of the glucose-derived melanoidins as measured by the formation of a haze after 240 hr of storage compared to the fructose system, which remained essentially clear. A similar haze formation was noted for a sucrose–glycine system (1 : 1 molar ratio) resulting from the hydrolysis of sucrose and the release of glucose.

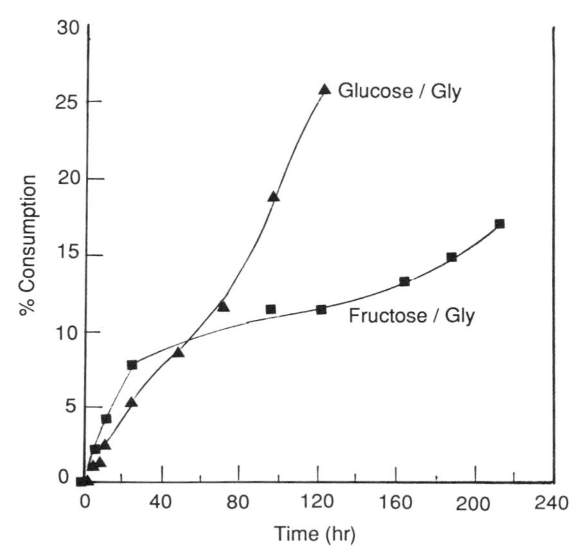

FIG. 5.6. Consumption of glucose and fructose in the glucose and fructose–glycine systems during storage at 60°C and pH 3.5 (% consumption represents the % individual sugar lost) (Reyes *et al.*, 1982). Copyright © by Institute of Food Technologists.

e. Metals. The formation of metal complexes with amino acids can influence the Maillard reaction. This reaction was catalyzed by copper and iron, while manganese and tin inhibited this reaction (Ellis, 1959; Markuze, 1963). Inhibition of browning in glucose–glycine model systems by trace metals was reported by Bohart and Carson (1955). Using an ovalbumin–glucose mixture Kato *et al.* (1981) examined the effect of Na^{+}, Cu^{2+}, Fe^{2+} and Fe^{3+} on the rate of browning at 50°C and 65% relative humidity. Figure 5.7 shows that an acceleration of browning occurred in the presence of Cu^{2+} and Fe^{3+}, while Na^{+} had no effect. Fe^{3+} was more effective than Fe^{2+} in accelerating the browning reaction, which suggested that the first step was an "oxidation activation" resulting in a reduction of the metal. The more rapid browning of a dried egg white–solid glucose system was attributed by both Kato *et al.* (1978) and Watanabe *et al.* (1980) to the presence of trace metals in the egg white. In addition to the catalytic effect of iron on the browning reaction (Hashiba, 1979) it was shown that iron also participated as a chromophore of the pigment (Hashiba, 1986). The possible interaction of iron with hydroxypyridone and hydroxypyranone, both capable of chelating iron, in the melanoidin polymer may be responsible for color formation. The presence of these heterocyclic compounds was reported previously by Tsuchuda and coworkers (1976) after the pyrolysis of nondialyzable melanoidins.

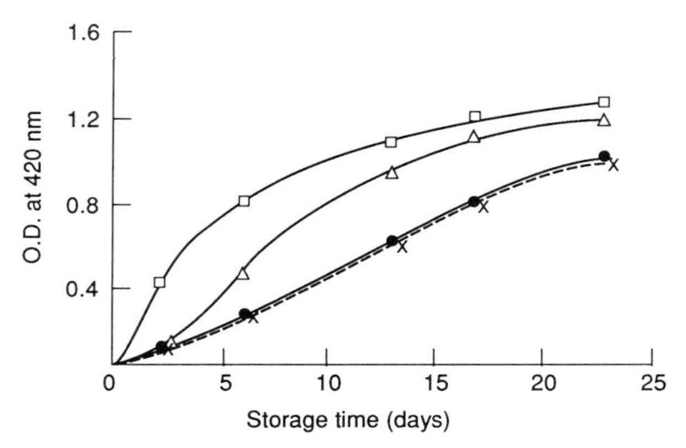

FIG. 5.7. Effect of Na^{+}, Cu^{2+}, and Fe^{3+} additions on the browning color development in ovalbumin–glucose mixtures. (●) OVG, (X) OVG–Na; (□) OVG–Cu; (△) OVG–Fe. Reprinted with permission of Kato *et al.* (1981). Copyright by the American Chemical Society.

III. Pigment Formation

A. Via Amadori Compounds

The reactions involved in the conversion of 1–amino-1-deoxy-1-ketose derivatives to brown pigments or melanoidins are extremely complex and incompletely understood. Nevertheless three distinct pathways have been proposed, two of which are directly involved in pigment formation (Scheme 5.3). They involve different labile intermediates which are the enol forms of the Amadori compounds. In one pathway, enolization of 1–amino-1-deoxy-2-ketose occurs at 2 and 3 positions to irreversibly produce 2,3-enediol. This then undergoes a series of changes including the loss of the amine from C-1 to form a methyl dicarbonyl intermediate (Hodge, 1953; Hodge *et al.*, 1963; Simon and Heubach, 1965). The second pathway involves formation of 1, 2-eneaminol from the Amadori product in which a hydroxyl group is lost at C-3 followed by deamination at C-1 and addition of water to form 3-deoxyhexosulose (Anet, 1960, 1964; Kato, 1962, 1963). The subsequent reactions are complex and little understood but involve a series of aldol condensation and polymerization reactions. The final products are nitrogenous compounds which give rise to the dark-brown pigmentation. A low pH favors the 1, 2-eneaminol pathway while a high pH favors the pathway involving the conversion of 2, 3-enediol to reductones and the subsequent fragmentation to furaneol and pyrones.

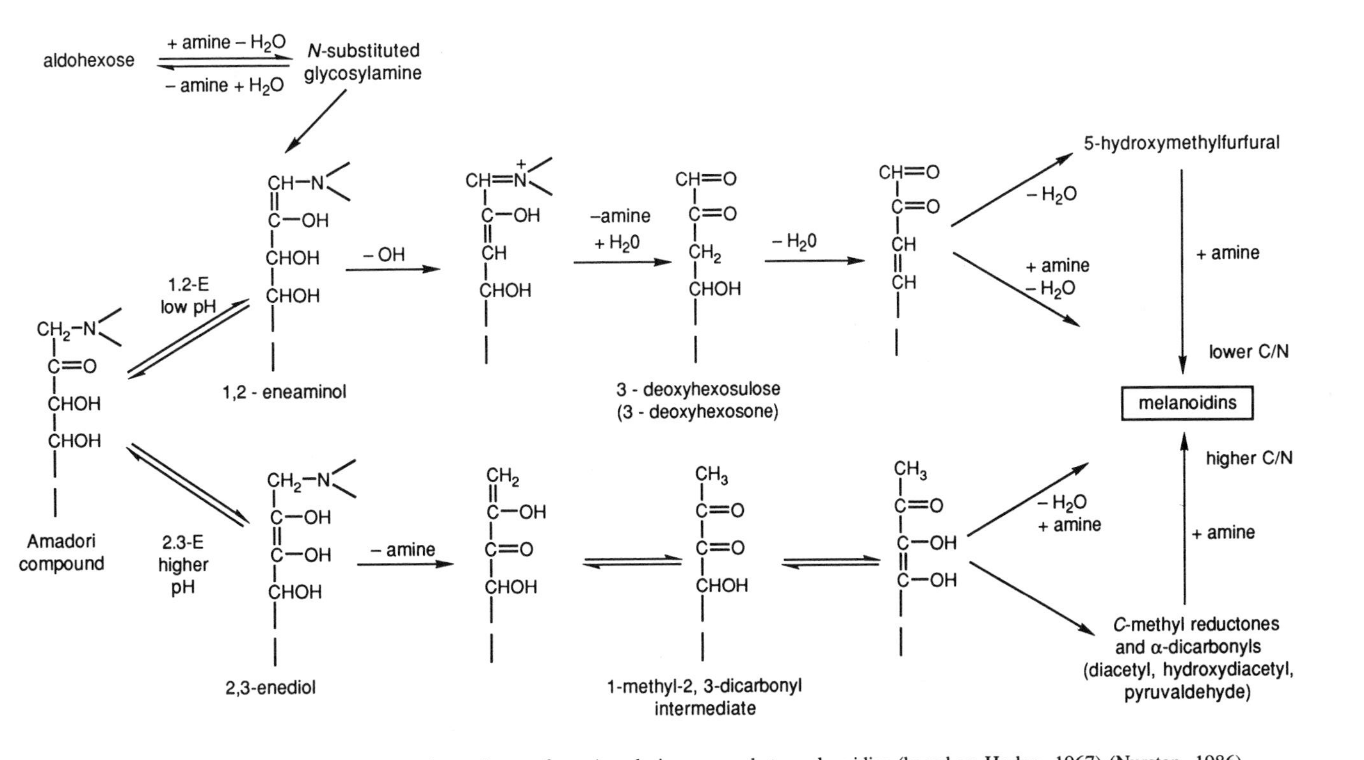

SCHEME 5.3. Maillard reaction: two major pathways from Amadori compounds to melanoidins (based on Hodge, 1967) (Nursten, 1986).

B. Alternative Pathways

The formation of free radicals in browning mixtures of carbonyl compounds and amines or amino acids was first reported 10–15 years ago by Namiki *et al.* (1973), Namiki and Hayashi (1975), and Hayashi *et al.* (1977). The generation of free radicals during the initial stages of the carbonylamino acid reaction for D-glucose–aminobutyric acid isomers was established by Milic and co-workers (1978, 1979, 1980). Namiki and Hayashi (1981) reported that model systems with alanine and arabinose gave rise to ESR (electron spin resonance) spectra with 17 and 23 lines. These simple signals were attributed to the presence of *N*, *N*-dialkyl pyrazine cation radicals, which were detected prior to the formation of Amadori compounds. These researchers proposed the formation of a C_2 sugar fragment as the precursor of this radical, which was confirmed by isolation and identification of glyoxal dialkylimine. This pointed to an alternative pathway for browning in which the sugar moiety of the Schiff base was cleaved prior to the Amadori rearrangement, leading to the formation of glycolaldehyde alkylimine or its corresponding eneaminol (Scheme 5.4) (Namiki and Hayashi, 1983). Further research by Hayashi and Namiki (1986) confirmed the formation of methylglyoxal dialkylimine, a C_2 compound, during the initial stages of the Maillard reaction. The formation of this C_2 compound was thought to arise directly from the Amadori rearrangement. Glycoaldehyde and methylglyoxal, which represented the C_2 and C_3 sugar fragments, exhibited much higher browning rates, corresponding to 2000 and 650 times faster than those of glucose, fructose, or xylose when heated with β-alanine (Table 5.2). Another C_3 compound, glyceraldehyde, also showed close to a 2000-fold increase in the rate of browning compared to the corresponding sugars. Hayashi and Nimiki (1986) presented a summary of the early stages of browning in which they concluded that under acidic conditions the traditionally accepted pathway involved osone formation *via* the Amadori rearrangement. Under alkaline conditions, however, they largely attributed the increase in browning to sugar fragmentation to C_2 and C_3 fragments (Scheme 5.5). Danehy (1986), however, suggested that this pathway be considered as a concomitant one occurring along with the established Maillard reaction scheme.

Several other pathways that also bypass the formation of Amadori compounds were also suggested. Holtermand (1966) proposed migration of a C═N double bond in the Schiff base, which when hydrolyzed, released an oxo acid and a nonreducing sugar. The oxo acid could react with an amino acid and liberate an aldehyde by the Strecker degradation reaction. Another pathway proposed earlier by Burton and McWeeney (1964) suggested a second substitution of the Amadori compound to form a diketone amino compound in which the amino acid was regenerated and the sugar converted to 5-HMF (hydroxymethylfurfural) by dehydration.

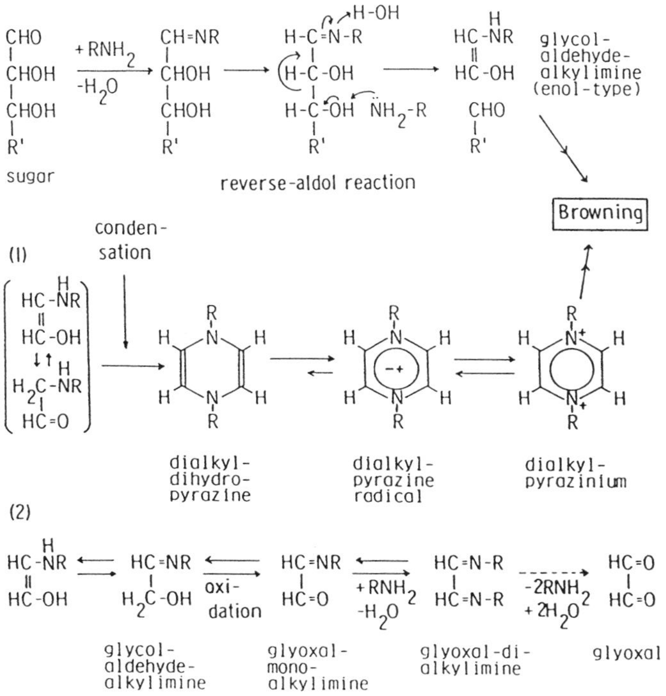

CHO–CHOH–CHOH–R' (sugar) $\xrightarrow[-H_2O]{+RNH_2}$ CH=NR–CHOH–CHOH–R' ⟶ H-C=N-R / H-C-OH / H-C-OH (H-OH, NH_2-R) ⟶ HC-NR (H) ‖ HC-OH, glycol-aldehyde-alkylimine (enol-type) + CHO–R'

reverse-aldol reaction

glycol-aldehyde-alkylimine (enol-type) ⟶ Browning

(1) condensation

[HC-NR (H) ‖ HC-OH ⇅ H_2C-NR (H) | HC=O] ⟶ dialkyl-dihydro-pyrazine ⇌ dialkyl-pyrazine radical ⇌ dialkyl-pyrazinium ⟶ Browning

(2)

HC-NR (H) ‖ HC-OH ⇌ HC=NR | H_2C-OH (glycol-aldehyde-alkylimine) ⇌ oxidation ⇌ HC=NR | HC=O (glyoxal-mono-alkylimine) $\underset{-H_2O}{\overset{+RNH_2}{\rightleftharpoons}}$ HC=N-R | HC=N-R (glyoxal-di-alkylimine) $\xrightarrow[+2H_2O]{-2RNH_2}$ HC=O | HC=O (glyoxal)

SCHEME 5.4. Alternative pathway for browning (Namiki and Hayashi, 1983).

TABLE 5.2

BROWNING RATES OF β-ALANINE WITH SUGAR OR CARBONYL SYSTEMS[a]

Sugar or carbonyl compound	Reaction rate (°C)	Browning activity[b] (liters/min)	Relative value
Glucose	95	0.019	1
Fructose	95	0.014	0.74
Xylose	95	0.166	8.74
Xylose	80	0.037	
Methylglyoxal	80	2.77	654.3
Glyceraldehyde	80	8.33	1967
Glyoxal	80	0.515	121.6
Glycolaldehyde	80	8.93	2109

[a] From Hayashi and Namiki (1986).

[b] Browning rate measured as change in absorbance at 420 nm.

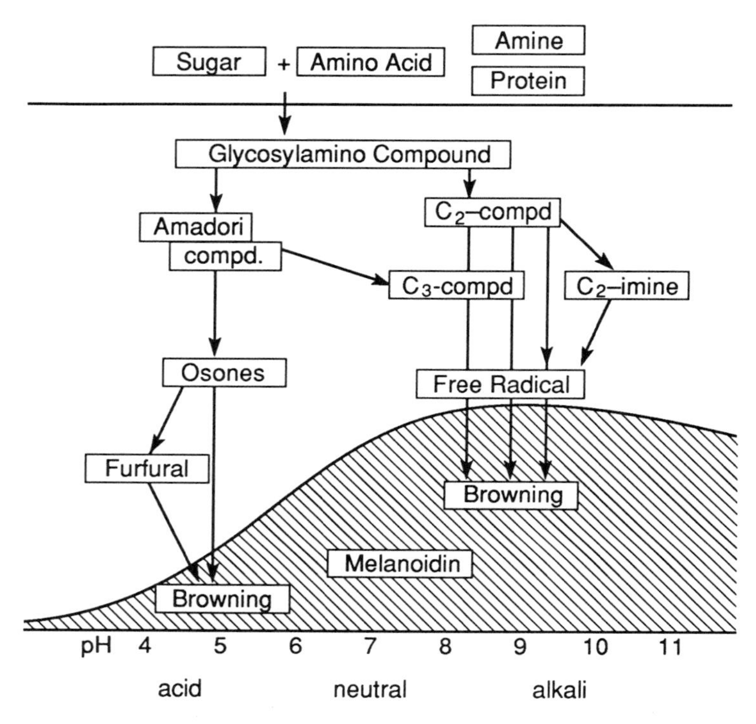

Scheme 5.5. Different pathways for melanoidin formation depending on reaction pH (Hayashi and Namiki, 1986).

C. Strecker Degradation

The third pathway in the Maillard reaction is concerned with the oxidative degradation of amino acids in the presence of α-dicarbonyls or other conjugated dicarbonyl compounds formed from Amadori compounds. This reaction is based on the work originally carried out by Strecker over a century ago in which he observed the oxidation of alanine by alloxan, a weak oxidizing agent. The reaction, now referred to as the Strecker degradation, is not directly concerned with pigment formation but provides reducing compounds essential for its formation. The initial reaction involves the formation of a Schiff base with the amino acid. The tuatomeric end-form then decarboxylates to produce the eneaminol, which then undergoes hydrolysis to the corresponding aldehyde with one carbon less together with a 1–amino-2-keto compound (Scheme 5.6). Schonberg *et al.* (1948) reported that the amino group had to be in the alpha position for this

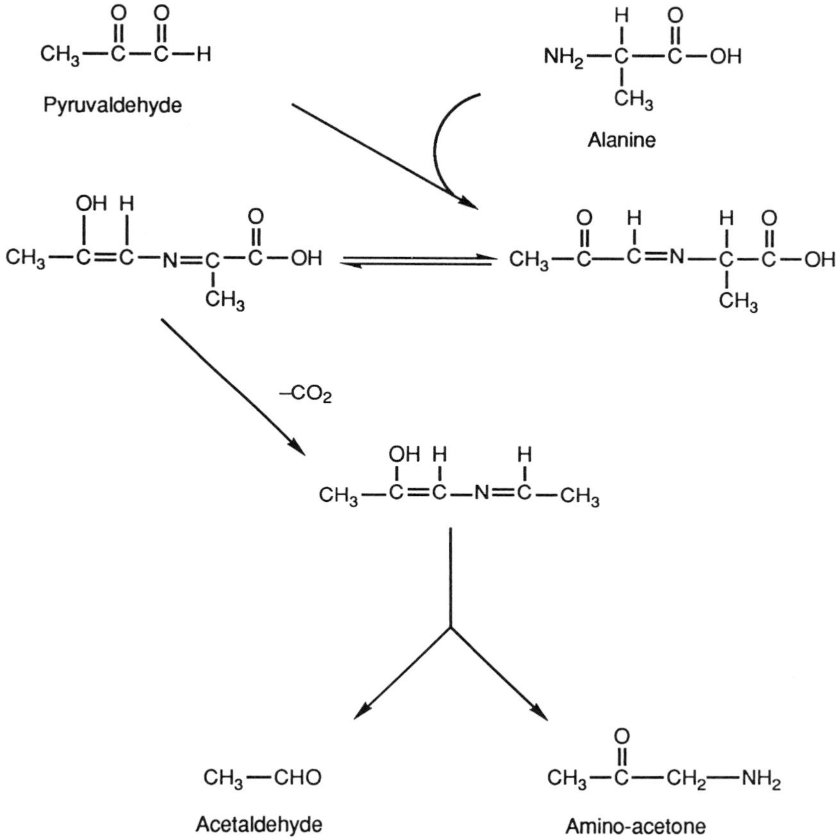

$CH_3-CO-CHO$

Pyruvaldehyde

$NH_2-CH(CH_3)-COOH$

Alanine

$CH_3-C(OH)=CH-N=C(CH_3)-COOH \rightleftharpoons CH_3-CO-CH=N-CH(CH_3)-COOH$

$-CO_2$

$CH_3-C(OH)=CH-N=CH-CH_3$

CH_3-CHO

Acetaldehyde

$CH_3-CO-CH_2-NH_2$

Amino-acetone

SCHEME 5.6. Strecker degradation reaction (Schornberg and Moubacher, 1952).

reaction to proceed in the presence of α-dicarbonyl compounds. The aldehydes formed during the Strecker degradation reaction contribute to flavor. A number of these aldehydes are listed in Table 5.3 together with their flavor characteristics. At one time aldehydes were considered to be directly responsible for the flavor of roasted foods, although they are now recognized to have a contributory role as auxiliary flavor compounds (Hodge *et al.*, 1972; Reynolds, 1970). Van Praag *et al.* (1968) showed it was the secondary products of Strecker aldehydes that were responsible for the strong aroma of cocoa. Hodge *et al.* (1972) similarly noted that the contribution of the Strecker aldehydes isobutyric, isovaleric, and methional to roasted food aroma was only auxiliary. Condensation of the

TABLE 5.3

AROMAS AND VOLATILE COMPOUNDS PRODUCED FROM L-AMINO ACIDS IN MAILLARD REACTION SYSTEMS

Amino acid	Volatile compound	Aroma
Alanine	Acetaldehyde	Roasted barley
Cysteine	Thiol, H_2S	Meaty
Valine	2-Methylpropanal	
Leucine	3-Methylbutanal	Cheesy
Lysine		Breadlike
Methionine	Methional	

intermediates formed by Strecker degradation produced many heterocyclic compounds, pyrazines, pyrrolines, oxazoles, oxazolines, and thiazole derivatives responsible for the flavor of heated foods (Hodge *et al.*, 1972; Maga and Sizer, 1973, Maga, 1982).

IV. Heterocyclic Compounds

A. PYRAZINES

Among the heterocyclic compounds formed from the Strecker degradation products are the pyrazines. These are very potent flavor compounds which have been identified in almost all processed foods, including beef products, soy products, processed cheese, coffee, potatoes, tea, and roasted pecans (Maga, 1981, 1982; Maga and Sizer, 1973). Dawes and Edwards (1960) identified a number of substituted pyrazines in sugar–amino acid model systems including 2, 5-dimethylpyrazine and trimethylpyrazine.

Koehler *et al.* (1969) showed that the C-ring in the substituted pyrazines was derived from the fragmentation of sugars. Koehler and Odell (1970) monitored the formation of methyl- and dimethylpyrazines from sugar–asparagine systems.

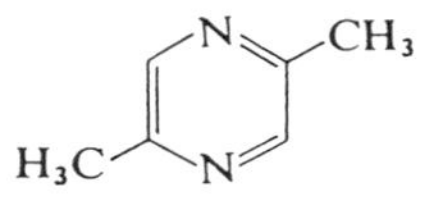

N, CH_3, H_3C, N

2,5-Dimethylpyrazine

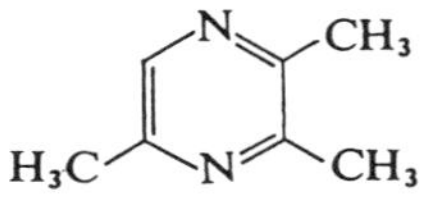

N, CH_3, H_3C, N, CH_3

Trimethylpyrazine

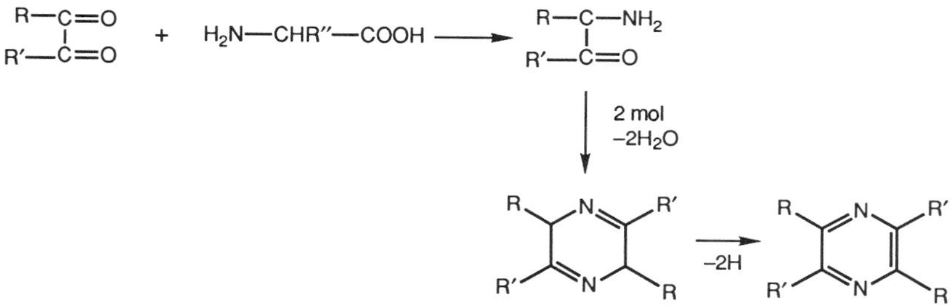

SCHEME 5.7. Formation of pyrazines (Shibamato and Bernard, 1977).

They noted that fructose gave the highest yields while arabinose gave the smallest yields of these compounds. This suggested that the yields and distribution patterns of pyrazine rings were determined by the nature of the sugar. Shibamato and Bernard (1977), using sugar–ammonia model systems, found similar distribution patterns of pyrazines for both pentose and aldose sugars examined. Only in the case of the aldose sugars was the level of unsubstituted pyrazines higher, although higher yields were obtained in the presence of the pentoses. One of the major pathways leading to the formation of pyrazines is illustrated in Scheme 5.7 involving condensation of amino-ketones.

Koehler *et al.* (1969) proposed an alternative pathway involving Strecker degradation in which the bound amino acid nitrogen was the main contributor to nitrogen in the pyrazine rings. Condensation of two 2-carbon sugar fragments with nitrogen produced pyrazine, while a similar reaction involving condensation of a 2-carbon fragment with a 3-carbon sugar fragment produced methylpyrazine. The formation of dimethylpyrazine was attributed to the condensation of two 3-carbon sugar fragments and nitrogen. These researchers showed that the nitrogen was derived from the amino acid and not from ammonia as originally proposed by Newell *et al.* (1967) and Van Praag *et al.* (1968). Milic and Piletic (1984) studied D-glucose and aminobutyric acid model systems using electron spin resonance spectra. D-glucose and 4-aminobutyric acid appeared to give rise to 1, 4-dialkypyrazine radicals which formed pyrazine. The latter was the major constituent produced during the course of nonenzymatic browning in which 27 pyrazine derivatives were identified. These were responsible for the earthy, nutty, baked, cinnamonlike, and caramel-like odors present. Wong and Bernhard (1988) examined five different nitrogen sources (ammonium hydroxide, ammonium acetate, ammonium formate, glycine, and monosodium glutamate) for the formation of pyrazines. They concluded that the nitrogen source had a marked effect on both the amount and types of pyrazines formed during nonenzymatic browning as suggested earlier by Koehler and Odell (1970) and Koehler and co-workers (1969).

B. Pyrroles

An important group of heterocyclic compounds formed during the browning of foods are pyrroles (Hodge, 1953). One of the pathways leading to their formation involves cyclization of methyldicarbonyls to 2, 4-dideoxypentulose-3-ene, which cyclizes to furfural or reacts with an amino acid at C-2 to form a Schiff base which then cyclizes to N-substituted pyrrole-2-aldehyde (Kato and Fujimaki, 1968). The formation of pyrrole derivatives has since been identified in a number of sugar–amino acid systems (Ferretti and Flanagan, 1971, 1973; Rizzi, 1974; Shigematsu *et al.*, 1972). Shaw and Berry (1977) reported the formation of 2-acetylpyrrole and 5-methylpyrrole-2-carboxyaldehyde in fructose–alanine model systems. This pathway is shown in Scheme 5.8, in which the 3-deoxyhexulose derivative underwent Strecker degradation with the amino acid leading to the formation of 1-amino-3-deoxy-2-ketose. Further changes included enolization and dehydration, resulting in the formation of 2-acetylpyrrole.

→ → CHO–C=O–CH=CH–CH_2OH (3,4-Dideoxypentosulos-3-ene) —+ R,NH, $-H_2O$→ CHO–C(=N R)–CH=CH–CH_2OH → → Melanoidins (main product)

3,4-Dideoxypentosulos-3-ene ↓ $-H_2O$ → Furfural (CHO)

CHO–C(=N R)–CH=CH–CH_2OH ↓ $-H_2O$ → *N*-Substituted 2-pyrrole aldehyde (N–R, CHO)

isolated, e.g., R =
—CH_2COOH ex., glycine
—$CHCOOH$ ($CH_2CH(CH_3)_2$) ex., leucine

Furfural *N*-Substituted 2-pyrrole aldehyde

Milic and Piletic (1984) identified seven pyrrole derivatives during the browning of D-glucose–aminobutyric acid systems. Under alkaline conditions the OH ions were thought to cause proton rearrangement in the sugar molecules, resulting in 3-deoxyhexulose and hexosulose-3-ene. These compounds react with amino acids, as described previously, to produce the different pyrrole derivatives. Many pyrroles have extremely powerful flavors which might enhance or exert a detrimental effect on food flavors.

C. Oxazoles and Oxazolines

Oxazoles and oxazolines have been identified among the flavor volatiles of coffee (Stoeffelsman and Pypker, 1968), baked potato (Coleman *et al.*, 1981),

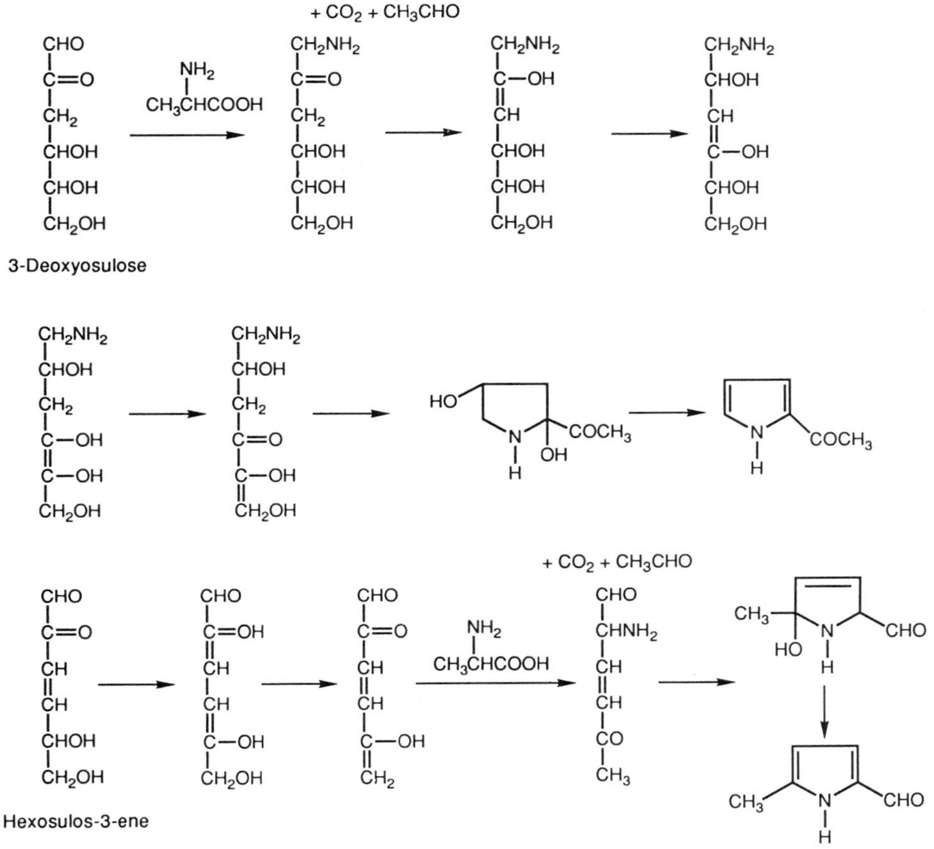

3-Deoxyosulose

Hexosulos-3-ene

SCHEME 5.8. Synthesis of two pyrroles by Strecker degradation. Reprinted with permission from Shaw and Berry (1977). Copyright by the American Chemical Society.

and roasted peanuts (Lee *et al.*, 1981). The role of these compounds in the flavor of foods has been reviewed by Maga (1978, 1981). One such compound, 2, 3, 5-trimethyl-2-oxazole, was identified in the volatiles of boiled beef (Chang *et al.*, 1968) and canned beef stew (Peterson *et al.*, 1975). The latter researchers also reported the presence of the corresponding oxazoline (2, 4, 5-trimethyl-3-oxazoline). The role of the Strecker degradation reaction in the formation of these compounds was first suggested by Rizzi (1969), in which 2-isopropyl-4, 5-dimethyl-3-oxazoline was formed from D-histidine and 2, 3-butadione (diacyl). The formation of oxazoles and oxazolines by Strecker degradation was confirmed by Ho and Hartman (1982) to explain the presence of these com-

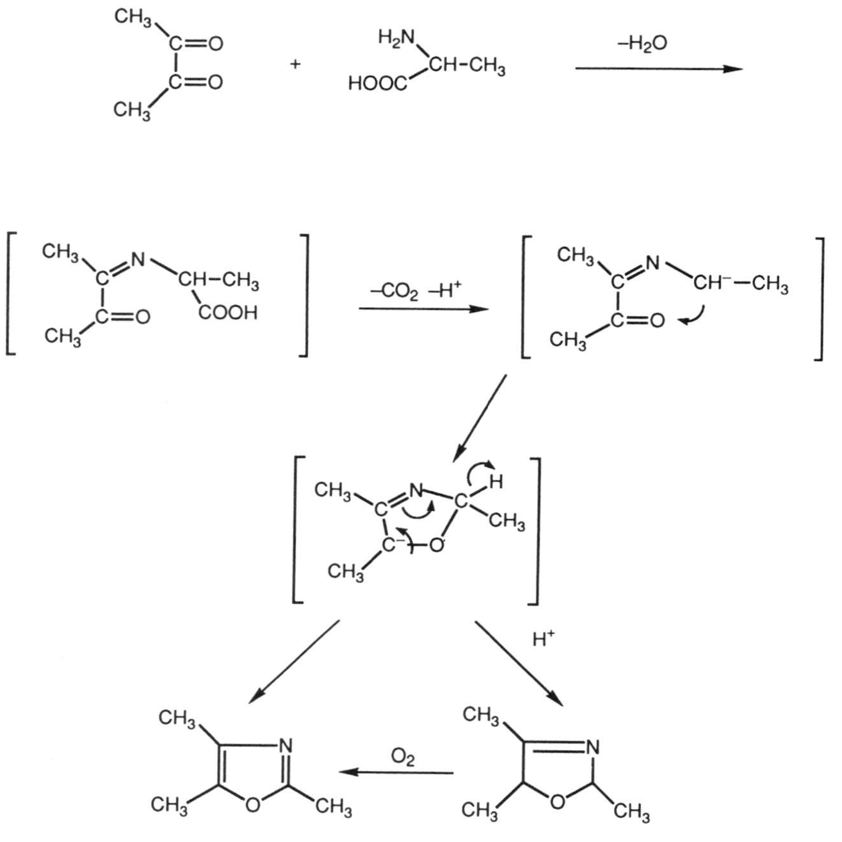

$CH_3-C(=O)-C(=O)-CH_3$ + $H_2N-CH(CH_3)-COOH$ $\xrightarrow{-H_2O}$

$[CH_3-C(=N-CH(CH_3)-COOH)-C(=O)-CH_3]$ $\xrightarrow{-CO_2\ -H^+}$ $[CH_3-C(=N-CH^{-}-CH_3)-C(=O)-CH_3]$

$\rightarrow$ $[$3-oxazolinide ion: CH_3, $C{=}N$, $C(H)(CH_3)$, $C^{-}-O$, $CH_3]$

$\xrightarrow{H^+}$ 2,4,5-trimethyl-3-oxazoline (CH_3, N, O, CH_3, CH_3) $\xrightarrow{O_2}$ 2,4,5-trimethyloxazole (CH_3, N, O, CH_3, CH_3)

SCHEME 5.9. Formation of 2,4,5-trimethyloxazole and 2,4,5-trimethyl-3-oxazoline from the reaction of DL-alanine and 2,3-butanedione. Reprinted with permission from Ho and Hartman (1982). Copyright by the American Chemical Society.

pounds in the flavor volatiles of meat and roasted peanuts. These researchers proposed the pathway in Scheme 5.9 to explain the mechanism for the formation of 2, 4, 5-trimethyloxazole and 2,4,5-trimethyloxazoline from DL-alanine and butanedione. Elimination of water resulted in an unstable Schiff base which then underwent decarboxylation to the corresponding anion followed by cyclization to the 3-oxazolinide ion. Protonation or loss of hydride ion was thought to produce the 2, 4, 5-trimethyloxazoline. The corresponding oxazole was attributed to the oxidation of the oxazoline or loss of hydride.

D. Thiazoles

Thiazoles are formed from sulfur amino acids. These compounds have been identified in coffee, roasted peanuts, cooked beef, and potato chips (Buttery and Ling, 1974; Buttery *et al.*, 1983; Stoll *et al.*, 1967a,b; Walradt *et al.*, 1971). The presence of 2-acetyl-2-thiazoline in beef broth was attributed by Tonsbeek *et al.* (1971) to the Strecker degradation reaction between cysteine and methylglyoxal followed by cyclization.

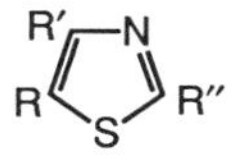

R′ N R R″ S

2-Acetyl-2-thiazoline

The Strecker degradation reaction plays a key role in the production of important flavor compounds by condensation and cyclization of the different aldehydes formed. In addition to the heterocyclic compounds discussed there are many others, including pyrrilodines and pyridines, as well as *o*-heterocyclics such as maltol and isomaltol, whose origins remain unclear. The isolation and chemical synthesis of 2-acetyl-1-pyroline, a key compound responsible for the characteristic smell of cooked rice, provides another example of the role of Strecker degradation in flavor genesis (Buttery *et al.*, 1982, 1983).

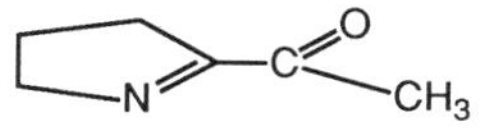

N C O CH_3

2-Acetyl-1-pyroline

In addition to these compounds, other heterocyclic and carbocyclic compounds identified from heated sugar–amine systems, include furanones, pyrrolinones, and cyclopentenones (Ledl and Fritsch, 1984).

V. Protein–Lipid Interactions

The oxidative degradation of polyunsaturated fatty acids produces many compounds capable of interacting with amino groups of amino acids or proteins, including aldehydes and ketohydroxy and epoxy compounds (Lea, 1958). Among the major products of autoxidized polyunsaturated fatty acids is malonaldehyde (Kwon *et al.*, 1965). Malonaldehyde was shown by Buttkus (1967) to interact with myosin, a structural protein, by monitoring the decrease in free ϵ-amino groups. The results obtained are summarized in Figure 5.8, in which a 60% loss in ϵ-amino groups occurred after 4 days at room temperature with 40% interacted after only 8 hr. Considerable reduction in reaction rate was observed at 0°C, whereas that observed at −20°C was of the same order as noted for +20°C.

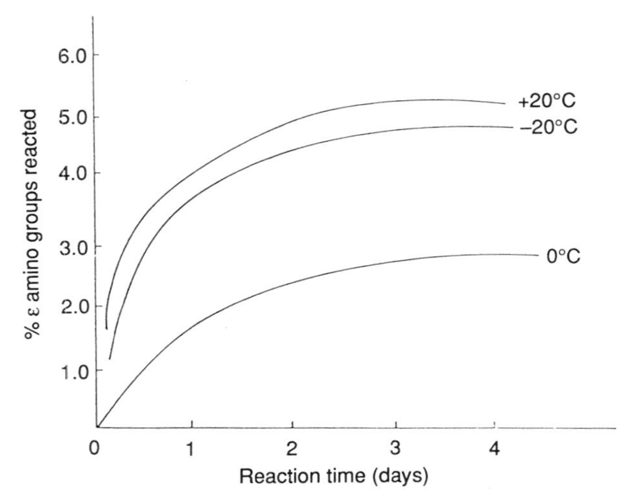

FIG. 5.8. Rate of ϵ-amino groups reacted with malonaldehyde in frozen solution at −20, 0, and +20°C (Buttkus, 1967). Copyright © by Institute of Food Technologists.

This was attributed to a concentration effect resulting from the closer association of the molecules in the reaction mixture due to freezing as well as the result of a catalytic effect in which the ice crystals were thought to participate (Grant *et al.*, 1966).

In addition to the ϵ-amino group of lysine, other amino groups also participated during incubation of myosin and malonaldehyde at -20°C. The order of reactivity was methionine > lysine > tyrosine > arginine. Tannenbaum *et al.* (1969), in studies on the browning of methyl linoleate and casein at 37°C, reported that the intensity of browning was proportional to the loss of methionine. The involvement of histidine with autoxidized lipids was established by Yu and Karel (1978) in which hexanal, a lipid off-flavor component of foods, reacted with histidine, forming a Schiff base. Pokorny *et al.* (1977) reported a linear correlation between the amount of aldehyde that reacted with protein and the degree of browning. Svadlenka and co-workers (1975) examined the binding sites involved in the interaction between malonaldehyde and collagen. They observed a significant interaction between malonaldehyde and lysine and tyrosine residues. The cross-linkages formed between malonaldehyde and collagen altered the structural properties of the protein as well as the ability of pronase to split the protein. Jirousova and Davidek (1975) demonstrated the formation of a Schiff's base when *n*-hexanal interacted with glycine. This reaction occurred best at pH 8.5–9.0, where only nonprotonized amino acids react. The Schiff base was extremely unstable and decomposed rapidly. Salter and co-workers (1988) reported an increase in the total volatiles produced in model systems of glycine and ribose to which phospholipids were added. The major volatiles were lipid degradation products including aliphatic aldehydes, alcohols, and ketones.

VI. Melanoidin–Maillard Polymers

The final products formed in the Maillard reaction are polymers or melanoidins. Unlike the flavor and aroma compounds discussed earlier, the origin and nature of these polymers are poorly understood. A number of studies have attempted to examine melanoidins in model systems, including Barbetti and Chiappini (1976a,b), Ledl (1982 a,b), Ledl and Severin (1982), Velisek and Davidek (1976a,b), Imasato *et al.* (1981), and Bobbio *et al.* (1981). A study by Feather and Nelson (1984) attempted to isolate the Maillard polymers produced in model systems composed of D-glucose/D-fructose/5-(hydroxymethyl)-2-furaldehyde and glycine and D-glucose/D-fructose with methionine. Increasing amounts of water-soluble, nondialyzable polymers with molecular weights greater than 16,000 were obtained for both glycine and methionine systems as a function of time (Figs. 5.9 and 5.10). Elemental analyses (carbon, hydrogen, and nitrogen) of polymers prepared from glycine model systems were similar, which suggested that the amino acid was incorporated into the polymer. The polymer isolated from D-glucose/D-fructose and glycine was composed of sugar and

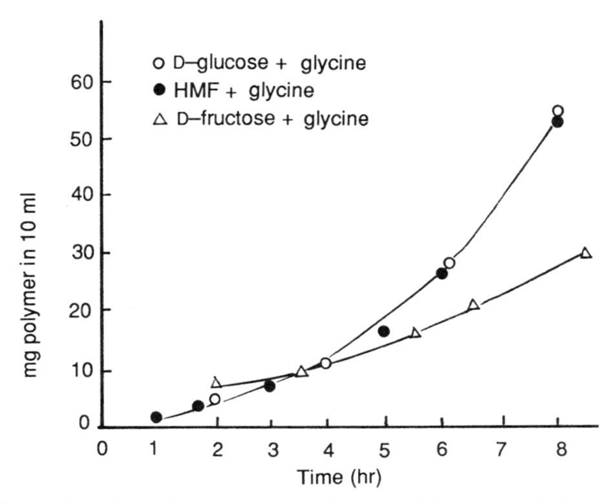

FIG. 5.9. Yields of nondialyzable polymers as a function of time using glycine as the amino acid. Reprinted with permission from Feather and Nelson (1984). Copyright by the American Chemical Society.

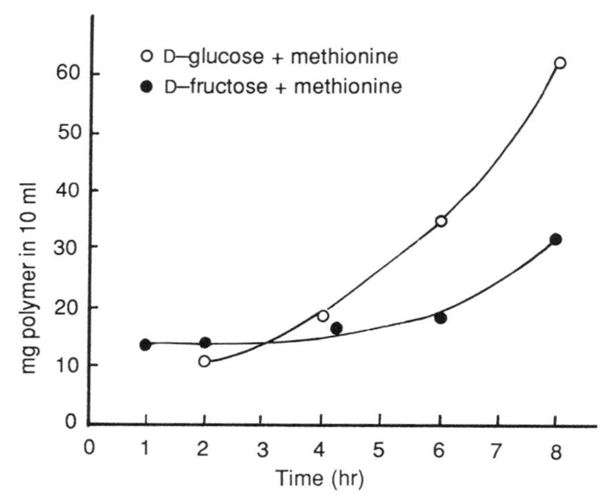

FIG. 5.10. Yields of nondialyzable polymers as a function of time using methionine as the amino acid (Feather and Harris, 1984).

amino acid minus three molecules of water. Detection of sulfur and nitrogen in polymers obtained from D-glucose and methionine also pointed to incorporation of the amino acid intact. The binding of metal ions to these melanoidins was evident by their greater solubility in tap water compared to in distilled water.

The nuclear magnetic resonance (NMR) spectra of these polymers suggested that some aromaticity was present. Feather and Huang (1986) examined the ^{13}C-NMR spectra of water-soluble polymers produced from labeled D-[1-^{13}C] glucose, L-[1-^{13}C]alanine, and L-[2-^{13}C]alanine (90 atom%). Polymers prepared with the C-1 labeled carbon atom in L-alanine had lower activity compared to the C-2 labeled amino acid. This could be due to its degradation to volatile aldehydes, suggesting a more direct role for the Strecker degradation in the Maillard reaction as proposed by Holterman (1966) (Section III,B). Earlier work by Olsson *et al.* (1982) found the aldehyde NMR spectrum was similar to that of an analogous Amadori compound. Based on this study and other NMR spectra it appeared that the nondialyzable polymer was formed by dehydration and polymerization of an Amadori compound as suggested previously by Olsson *et al.* (1982).

Benzing-Purdie and co-workers (1985) examined the effect of temperature on the structure of the melanoidins formed in model systems composed of D-xylose and glycine. In the presence of equimolar amounts of the reactants, an increase in temperature (22, 68, and 100°C) was accompanied by an increase in the aromatic nature of both the low- and high-molecular-weight melanoidin products. These researchers also noted considerable differences in the nature of the melanoidins produced at 22°C compared to those formed at the higher temperatures, with different types of aliphatic carbons and fewer unsaturated carbons.

Ingles and Gallimore (1985) isolated melanoidins from a Maillard reaction by adsorption on a strong anion-exchange resin. The rapid displacement of these melanoidins from the column by acid provided a convenient preparative tool for preparing similar polymers.

The nondialyzable melanoidins produced from glucose–glycine systems heated at 95°C at pH 6.8 were examined by Kato *et al.* (1986). These melanoidins were composed of saturated, aliphatic carbons together with smaller amounts of aromatic carbons. Milic (1987) examined the kinetics of melanoidin formation between D-glucose and 2-, 3-, and 4-aminobutanoic acid isomers using cross-polarization–magic angle spinning ^{13}C nuclear magnetic resonance (CP-MAS ^{13}C-NMR) spectroscopy. The systems were heated in sealed quartz test tubes at 313, 343, and 371 K for 1.0×10^5 to 3.60×10^5 sec under alkaline conditions (pH 9.0) and the brown melanoidins were eluted on an ion-exchange Permutit ES resin with 5% NaCl. The purified melanoidins were then concentrated, dialyzed, and dried under vacuum prior to analysis by CP-MAS ^{13}C-NMR. Based on the CP-MAS spectra, an increase in unsaturation and/or aromaticity was observed at 100 and/or 180 ppm with increasing temperature and time, which leveled off when both glucose and aminobutanoic acid were depleted. Milic (1987) calculated the order of reaction of melanoidin formation, assuming that aromaticity, measured by CP-MAS ^{13}C-NMR spectroscopy, corresponded to the rate of browning temperature (K). The amount of amino acid (A_0) (peak area at 180 ppm) was plotted against the reaction temperature at a constant time of 2.34×10 sec and the straight line obtained showed that melanoidin formation followed first-order kinetics (Figure 5. 11).

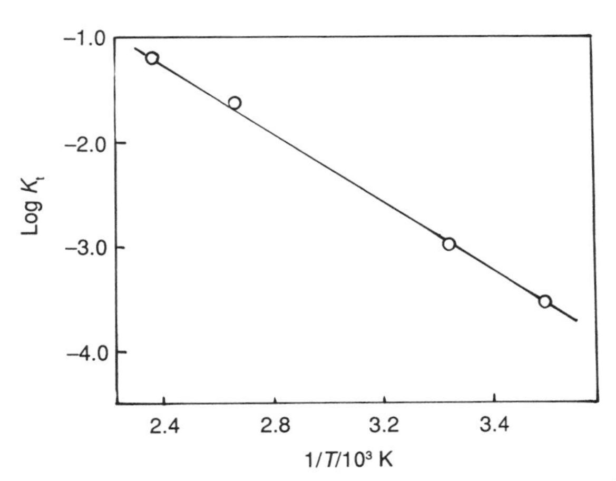

FIG. 5.11. Temperature dependence of approximate rate constant K_t for D(+)-glucose–2-aminobutanoic acid model system (Milic, 1987).

TABLE 5.4

CLASSIFICATION OF CARAMEL COLORS FOR FOOD USE[a]

Classification	Food use
I. Plain (alcohol) caramel	Spirits
II. Caustic sulfite caramel	Spirits
III. Ammonia caramel	Beer
IV. Sulfite ammonia caramel	Soft drinks

[a] From Smolnik (1987).

VII. Caramelization

Caramelization is another example of nonenzymatic browning involving the degradation of sugars. When sugars are heated above their melting points they darken to a brown coloration under alkaline or acidic conditions. If this reaction is not carefully controlled it could lead to the production of unpleasant, burned and bitter products. Consequently it is important to control this reaction during food processing while still retaining the pleasant qualities of caramel. Caramel colors used for coloring foods vary in color from very dark brown to black, syruplike liquids or powders. Smolnik (1987) classified caramel colors into four distinct groups based on differences in functional properties as shown in Table 5.4.

The chemical composition of caramel is extremely complex and still poorly understood, although caramels produced from different sugars all showed similarity in composition. Bryce and Greenwood (1963), using chromatographic techniques, found that pyrolysis of sucrose, glucose, and starch all produced caramels of similar composition. Heyns and Klier (1968), in a series of studies on a whole group of different mono-, di-, and polysaccharides, also found that the volatile products formed at high temperatures were almost identical. Studies by these and other researchers, as reviewed by Feather and Harris (1973), clearly indicated a common pathway for both the acidic and alkaline degradation of sugars. Research conducted since the publication of the first edition of this volume has led to the identification of new groups of compounds formed during this process (Popoff and Theander, 1976; Theander, 1981).

A. ACIDIC DEGRADATION

The first step involves the stepwise conversion of D-glucose to D-fructose and D-mannose, referred to as the Lobry de Bruyn–Alberda van Eckenstein transfor-

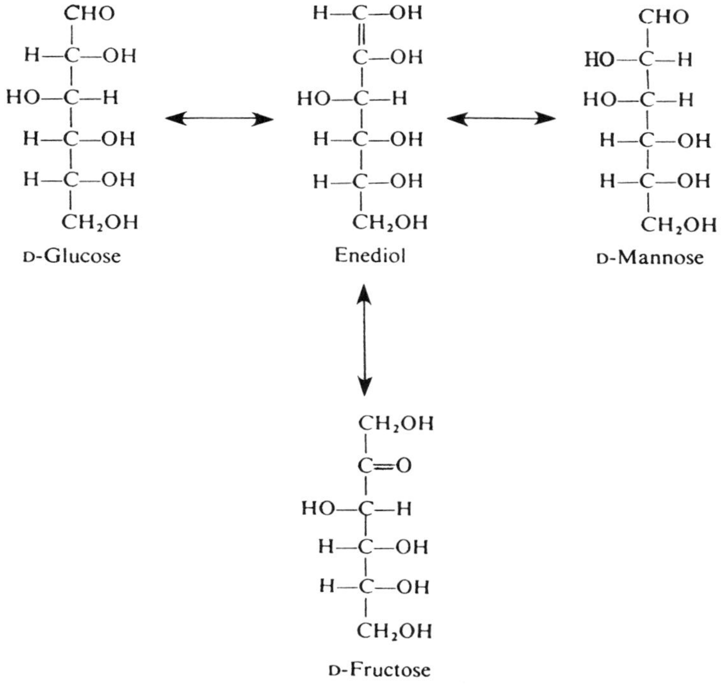

CHO
H—C—OH
HO—C—H
H—C—OH
H—C—OH
CH_2OH
D-Glucose

⟷

H—C—OH
‖
C—OH
HO—C—H
H—C—OH
H—C—OH
CH_2OH
Enediol

⟷

CHO
HO—C—H
HO—C—H
H—C—OH
H—C—OH
CH_2OH
D-Mannose

↕

CH_2OH
C=O
HO—C—H
H—C—OH
H—C—OH
CH_2OH
D-Fructose

SCHEME 5.10. The Lobry de Bruyn–Alberda van Eckenstein transformation (Eskin *et al.*, 1971).

mation (Scheme 5.10). These transformations can be mediated by organic acid catalysts over a pH range of 2.2–2.9 (Hodge and Osman, 1976). The interconversion of these sugars occurs primarily through the 1, 2-enolic form and depends on the ease with which the ring opens. Since D-glucose is the most conformational stable form in both acid or alkaline medium there is much less of the carbonyl (open-chain) form present in solution. This explains the presence of relatively high levels of glucose when fructose is heated at high temperatures over a pH range of 3.0–6.9, whereas only trace amounts of fructose are found when D-glucose is heated under identical conditions. Enolization takes place very slowly under acidic conditions, whereas the hydroxyl group next to the carbonyl group is rapidly removed.

The process of enolization via 1, 2-enediol of sugars under acidic conditions was questioned when Ohno and Ward (1961) reported the presence of only small amounts of fructose when D-glucose was treated with 2.5% sulfuric acid with no mention of mannose. Mawhinney *et al.* (1980) detected the presence of both fructose and mannose when D-glucose was isomerized in acidic solution. Table 5.5, taken from their data, shows that fructose levels off at around 0.8 μg, while

TABLE 5.5

YIELD OF SUGAR OBTAINED FROM 50 MG OF D-GLUCOSE AFTER TREATMENT WITH 2.5% SULFURIC ACID AT 120°C[a]

Reaction time (hr)	Glucose (μg)	Mannose (μg)	Fructose (μg)
0	50.0	0	0
1.0	46.2	4.7	0.6
2.5	43.6	10.8	0.8
5.0	39.7	19.1	0.8
7.5	37.1	27.7	0.9
10.0	35.9	43.8	0.8

[a] From Mawhinney *et al.* (1980).

mannose increased with reaction time. These data show the levels of sugars generated but give no information on their degradation by dehydration, which probably occurred at different reaction rates. The mechanism of this reaction resembled the corresponding isomerase enzyme reaction involving a C-1 → C-2 intramolecular hydrogen transfer in which D-glucose-2-H is converted to D-fructose-1-H (Harris and Feather, 1973, 1975):

H(OH)C=C(OH)–R → CH_2OH–C(=O)–R

Continued heating results in the dehydration of sugars, leading to the formation of hydroxymethylfurfural, levulinic acid, and humin. This process is initiated by the removal of a hydroxyl group from the 1, 2-enediol form located in the α position to the carbonyl group. The initial product, a dicarbonyl, undergoes further degradation. The postulated intermediates in this reaction were thought to be 3-deoxyaldos-2-ene, 3-deoxyosulose, and osulos-3-ene (Isbell, 1944; Wolfrom *et al.*, 1948). These compounds were isolated by Anet (1962) during the acidic degradation of fructose. If the initial sugar was a pentose, the final product was 2-furaldehyde. For example, D-xylose yielded approximately 93% 2-furaldehyde, although the yields from other pentoses were much lower. The dehydration rate of D-glucose was reported to be approximately one-fortieth of that observed for D-fructose with considerably lower product yields (Kuster and Van der Bean, 1977). The mechanism of sugar dehydration from 1, 2-enol to 5-(hydroxymethyl-2-furaldehyde) originally described by Anet (1964) has since been modified. This resulted from work by Feather *et al.* (1972) using isotope exchange in which D-xylose in tritiated water was found to be converted to 2-furaldehyde. By monitoring the proportion and amount of isotope converted

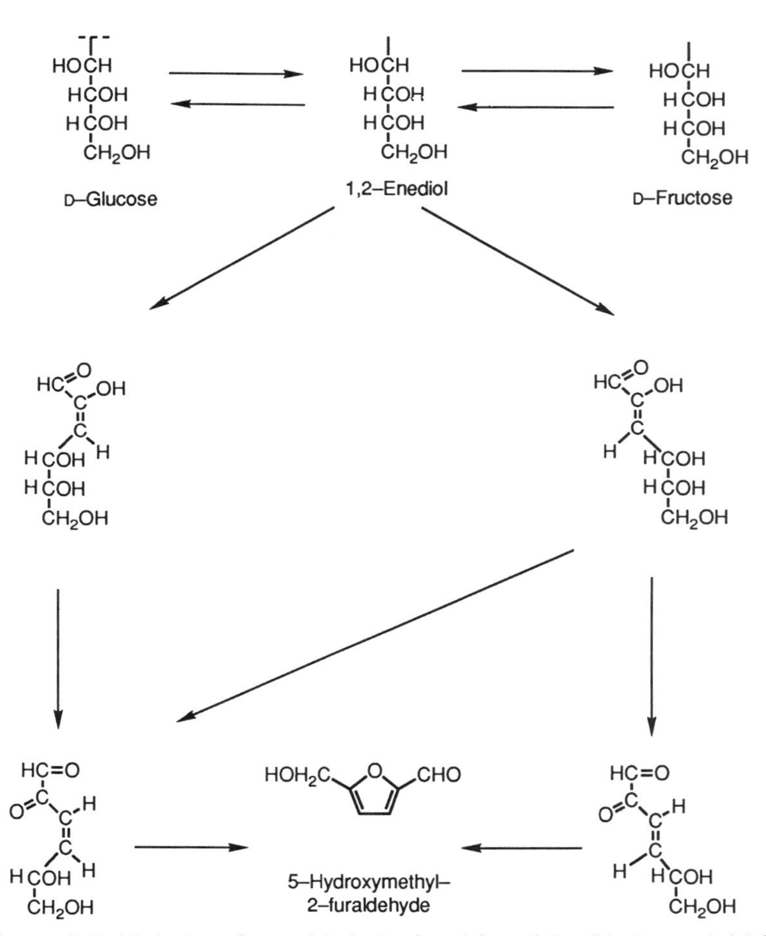

SCHEME 5.11. Mechanism of sugar dehydration from 1,2-enediol to 5-hydroxymethyl-2-furaldehyde. Adapted from Feather and Harris (1973).

into 2-furaldehyde they established the existence of an aldose–ketose 1, 2-enediol equilibrium as incorporation at C-1 of the sugar corresponded to the α-carbon of the 2-furaldehyde. Formation of 3-deoxyglyculose as an intermediate, however, should result in isotope incorporation at the C-3 position in the furan ring. The absence of any isotope exchange eliminated any 1, 2-enediol equilibrium during the reaction, thus making 1, 2-enediol the rate-limiting step. In addition it also eliminated 3-deoxyglyculose as an intermediate and supported the reaction sequence shown in Scheme 5.11.

Isomaltol and 2-(hydroxyacetyl)furan are formed during the acid treatment of D-fructose, suggesting that 2, 3-enediol is the precursor:

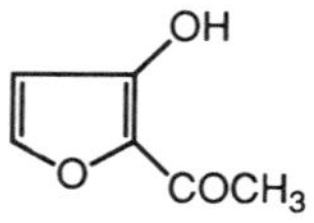

Isomaltol 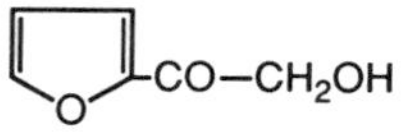 2-Hydroxyacetyl furan

Their formation involved removal of a hydroxyl at C-4 and C-1 followed by dehydration of the furanone ring. Using tritium-labeled D-glucose, D-fructose, and D-mannose, Harris and Feather (1975) found that D-fructose underwent acid-catalyzed dehydration and degradation at a much faster rate than the aldose sugars. Among the major products detected were 5-(hydroxymethyl)-2 furaldehyde, 2-(hydroxyacetyl)furan, and levulinic acid. The difference in rates of degradation explained, in part, why the levels of fructose observed by Mawhinney *et al.* (1980) remained low and constant during the isomerization of acidified D-glucose. Kuster and Temmink (1977) investigated the influence of pH and weak acid anions on the dehydration of D-fructose but were unable to detect 5-hydroxymethyl-2-furaldehyde (HMF) formation from D-fructose at pH > 3.9, while at pH > 2.7 no levulinic acid was formed. Isomerization of D-fructose to D-glucose was observed at pH above 4.5. It was apparent that the formation of HMF by dehydration of D-fructose and rehydration of HMF to levulinic and formic acids were catalyzed by acids, the latter requiring greater acidity. The formation of HMF, one of the major caramelization products produced over a pH range of 6.0–6.7, is presumably a precursor of the pigment. In addition to HMF, a minor product, 2-(2-hydroxyacetyl)furan, also appeared to be formed by 2, 3-enolization instead of 1, 2-enolization of D-fructose.

The pyrolysis of sucrose was reported by Johnson *et al.* (1969) to produce maltol. The formation of maltol and isomaltol, together with ethyl lactate, furfural, 3-hydroxypropionic acid, 5-hydroxymethylfurfural, levulinic acid, and 2-furoic acid, were later detected by Ito (1977) when an aqueous solution of sucrose at pH 2.3 was heated to 120°C. The initial step in the acid-catalyzed browning of sucrose is hydrolytic cleavage with release of the constituent monosaccharides. Several studies have shown that this can occur in freeze-dried sucrose systems at 37°C, where the monolayer of absorbed water is involved in the hydrolysis (Karel and Labuza, 1968; Schoebel *et al.*, 1969). Flink (1983) monitored the development of nonenzymatic browning in sucrose-bound systems during freeze-drying and storage at room temperature. Hydrolysis of sucrose to glucose and fructose occurred following the primary sublimation stage of drying. The increased production of HMF, measured by monitoring absorbance at 280 nm, was evident at the end of the primary freeze-drying. Storage of the freeze-dried samples over an a_W range of 0 to 0.40 showed a reduction in absorbance in the presence of increased levels of water due to the slowing down of the reaction.

The reduction of weight at a_W of 0 was a clear indication that water can be produced in the browning reaction. Following production of HMF, a brown color developed which was monitored at 400 nm. The browning reaction was attributed to the increase in hydrogen ion concentration taking place with passage of ice interface during freeze-drying. Increase in temperature was accompanied by a rapid hydrolysis of sucrose to glucose and fructose in which fructose rapidly underwent dehydration. This study explained the stability of products during freeze-drying and the storage needs of high-acid foods containing sugar, such as fruit juices.

B. Alkaline Degradation

The initial reaction in the degradation of sugars under alkaline conditions follows the Lobry de Bruyn–Alberda van Eckenstein transformation via the 1, 2- and 2, 3-enediol. As discussed previously, enolization is a general reaction for carbonyl compounds with an α-hydrogen atom. Alkalies are much more effective catalysts for the enolization of sugars compared to acids (Pigman and Anet, 1972). Under mild alkaline conditions the series of reactions shown in Scheme 5.12 takes place. Under strong alkaline conditions, continuous enolization progresses along the carbon chain, resulting in a complex mixture of cleavage products including saccharinic acids. The formation of metasaccharinic acid is detailed in Scheme 5.12.

A particular feature of the alkaline degradation of hexoses is the extensive fragmentation which occurs, resulting in the production of 2- and 4-carbon fragments, including saccharinic acids, lactic acid, and 2, 4-dihydroxybutyric acid (Feather and Harris, 1973; Harris, 1972). The recombination of some of these fragments accounts for the formation of a variety of compounds, including 2, 4-dihydroxybutyric acid (Harris, 1972). A detailed discussion of fragmentation and recombination reactions of sugars under alkaline conditions is covered by Feather and Harris (1973).

C. Aromatic Compounds

A number of cyclic compounds have been isolated among the products formed when aqueous solutions of D-glucose and D-fructose were heated to 160°C at pH 4.5. While these compounds were similar for both hexoses, the yields were much lower in the case of D-glucose. The major compound formed was hydroxymethylfurfural, although a large number of phenolic compounds were also produced.

The predominant phenol detected during the acid degradation of hexoses was

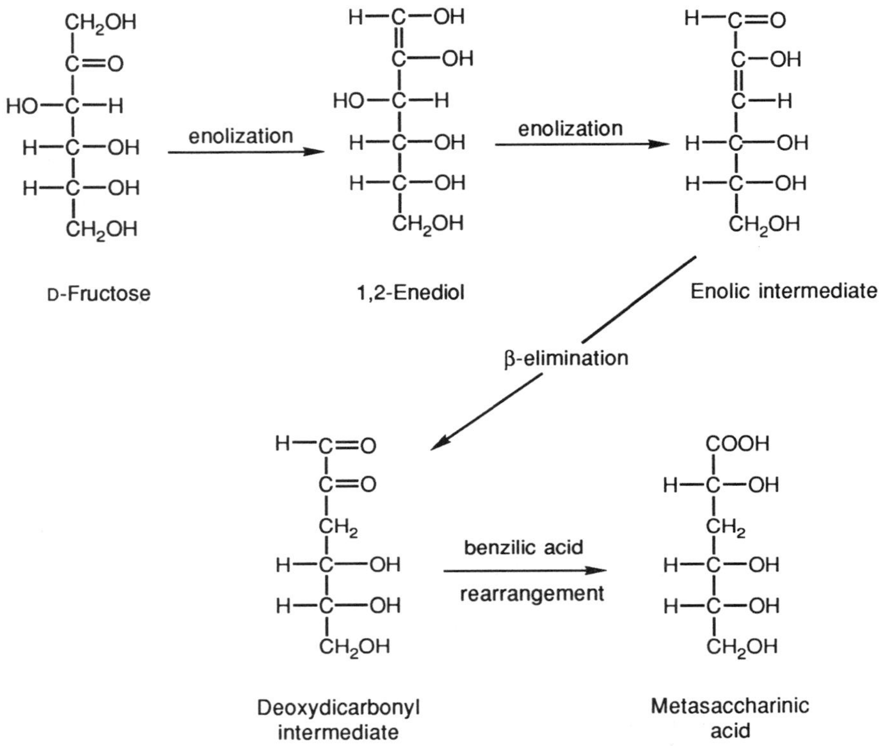

D-Fructose → (enolization) → 1,2-Enediol → (enolization) → Enolic intermediate → (β-elimination) → Deoxydicarbonyl intermediate → (benzilic acid rearrangement) → Metasaccharinic acid

SCHEME 5.12. Alkaline degradation. Adapted from Feather and Harris (1970).

isobenzene furanone, whereas chromone alginetin was the major product formed from pentoses and hexuronic acids (Theander, 1981):

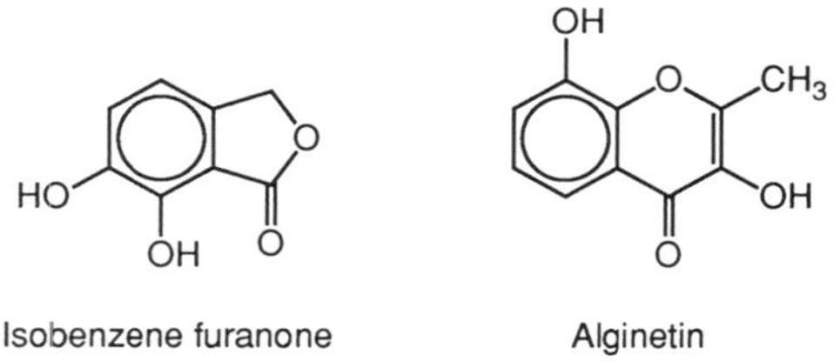

Isobenzene furanone   Alginetin

This difference in specificity was not evident, however, when sugars were degraded under alkaline conditions. Forsskahl *et al.* (1976) noted a similarity in the pattern of phenolics formed under alkaline or neutral conditions. The compounds identified included a number of cyclic enols and phenols. The low yields obtained for these compounds reflected their instability under alkaline conditions. The formation of cyclopentones was reported earlier by Shaw and coworkers (1968) from the alkaline treatment of D-fructose. These compounds were isolated among the aroma components of roasted coffee by Gianturco *et al.* (1963) and had a strong caramel-like odor. The only common phenolic compounds identified from either the acid or alkaline treatments of glucose were catechol, 4-methyl-1, 2-benzene diol, and 3, 4-dihydroxy benzaldedehyde (Popoff and Theander, 1976).

The development of color is extremely complex and involves a series of polymerization reactions. Theander (1981) reported that reductic acid and catechols were much more active color producers than furfurals.

VIII. Ascorbic Acid Oxidation

The browning of citrus juices and concentrates also involves Maillard-type reactions between amino acids and sugars present in citrus products. This was confirmed by Clegg (1969), who demonstrated improved color stability of lemon juice following the removal of amino nitrogen by cation-exchange resins. A patent was subsequently registered by Huffman in 1974 based on improvement of flavor stability when orange concentrate was treated with cation-exchange resins. The acceleration of browning by the addition of amino acids to model systems containing citrus confirmed their role in browning (Curl, 1949; Clegg, 1964; Joslyn, 1957). A recent review of citrus browning by Handwerk and Coleman (1988) suggested that the Maillard reaction was initiated in citrus juice by the formation of hexosamines from amino acids and sugars. The involvement of ascorbic and dehydroascorbic acids occurred at a later stage in this process *via* the formation of α-dicarbonyls, similar to that formed during the degradation of sugars. This section will focus on the degradation of ascorbic acid in citrus products in addition to their role, together with amino compounds in the browning of dehydrated cabbage.

Ascorbic acid plays a central role in the browning of citrus juices and concentrates, for example, lemon and grapefruit. The reaction of ascorbic acid in fruit juices and concentrates is very much dependent on pH, as the browning process is inversely proportional to pH over a range of 2.0–3.5 (Braverman, 1963). Juices with a higher pH are much less susceptible to browning, for example, orange juice at a pH of 3.4. Below pH 4.0, browning is due primarily to decomposition of ascorbic acid to furfural (Huelin, 1953; Huelin *et al.*, 1971).

The degradation of ascorbic acid was investigated by Herrmann and Andrae (1963), who identified 17 decomposition products, including dehydroascorbic acid and 2,3-diketogulonic and oxalic acids:

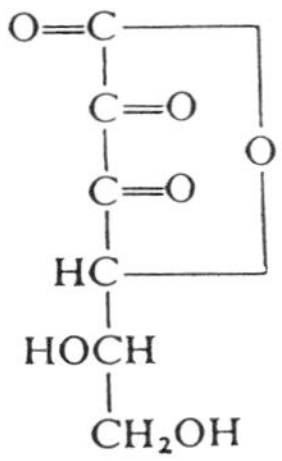

O=C—
C=O
O
C=O
HC—
HOCH
CH_2OH

Dehydroascorbic acid

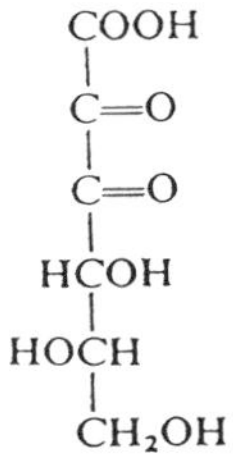

COOH
C=O
C=O
HCOH
HOCH
CH_2OH

2,3-Diketogulonic acid

COOH
COOH

Oxalic acid

Otsuka and co-workers (1986) identified a degradation product of 2, 3-diketogulonic acid by preparative high-performance liquid chromatography. The structure of this compound appeared to be the 3, 4-enediol form of 2, 3-diketogulono-δ-lactone. It was extremely unstable and developed intense brown coloration under mild temperature conditions. These researchers considered 3, 4-enediol to be important in the browning of ascorbic acid:

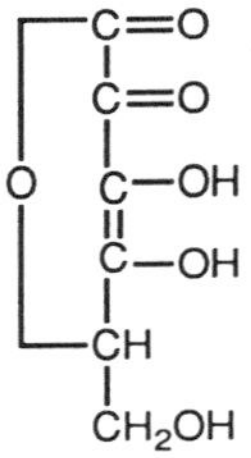

C=O
C=O
O C—OH
C—OH
CH
CH_2OH

3, 4 Enediol

Ascorbic acid degradation can occur under both aerobic and anaerobic conditions. While the level of air in juice is kept as low as possible by the use of vacuum deaeration and live steam injection, there is still some dissolved oxygen in the juice (0.05%) (Nagy, 1980). Only after the oxygen has been used does anaerobic degradation of vitamin C occur, but at a much slower rate. Tatum *et al.* (1967) reported degradation products of ascorbic acid, half of which were identical to the nonenzymatic browning products found in dehydrated orange and grapefruit powders, that is, instant juices. The aerobic and anaerobic degradation of ascorbic acid is outlined in Scheme 5.13 (Bauernfriend and Pinkert, 1970). The dependency of vitamin C degradation on headspace oxygen was recognized over 30 years ago by Bauernfriend (1953). Kefford (1959) reported that the oxidative degradation of ascorbic acid in canned, pasteurized juice occurred during the first few days until the free oxygen was utilized (Nagy and Smoot, 1977). Following this, anaerobic breakdown of ascorbic acid proceeds, but at a tenth of the rate. Improved stability of vitamin C was found in juice sold in tin

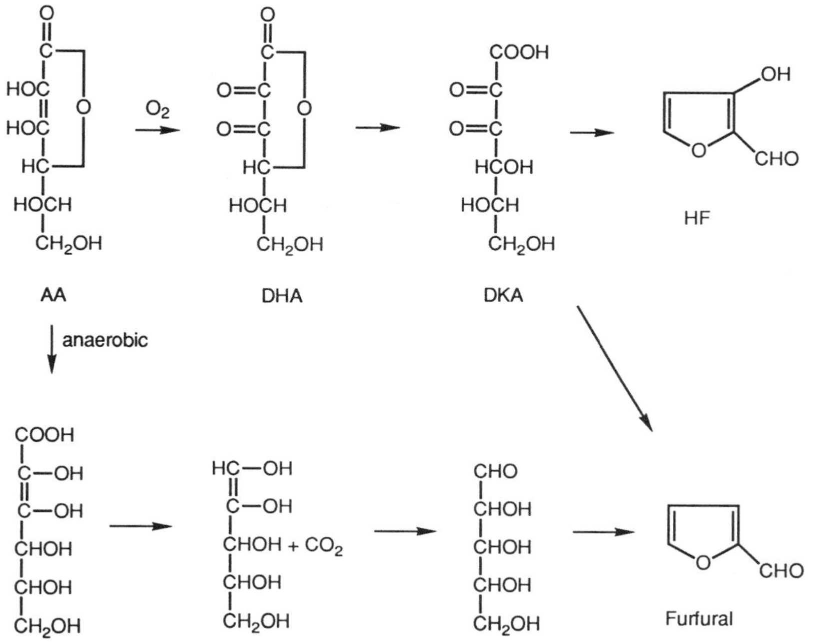

SCHEME 5.13. Possible vitamin C (ascorbic acid) degradation pathways: AA, ascorbic acid; DHA, dehydroascorbic acid; DKA, diketogulonic acid; HF, hydroxyfurfural (Nagy, 1980).

cans compared to enamel-lined cans as a result of oxygen reacting with the tin and competing with ascorbic acid (Riester *et al.*, 1945).

The browning of citrus juices, as discussed earlier, is not due solely to ascorbic acid. Amino acids are also involved *via* the Maillard reaction depending on the pH of the juice and basicity of the amine. This is illustrated by the fact that the main degradation product of juices with pH below 4.0 is furfural (Huelin, 1953; Huelin *et al.*, 1971). At pH above 4.0 this pathway is inoperative and explains the discoloration of dehydrated vegetables, which also involves ascorbic acid. Ranganna and Setty (1968) found that the discolorization of dehydrated cabbage was due to Strecker degradation between ascorbic acid and amino acid. This was facilitated by interactions between the oxidated products of ascorbic acid, dehydroascorbic or 2,3-diketogulonic acids, and amino acids during the final stages of the drying process. The final result was the development of red-to-brown discoloration. Ranganna and Setty (1974) attempted to isolate the condensation product formed in a model system composed of dehydroascorbic acid and glycine ethyl ester. Although unsuccessful they nevertheless demonstrated the presence of a chromogen with color characteristics similar to that formed during the nonenzymatic discoloration of cabbage. Subsequent studies examined the generation of free radicals resulting from the interaction of dehydroascorbic acid with amino acids. Namiki and Hayashi (1974) and Yano and co-workers (1974) observed the formation of stable free radicals from the interaction between dehydroascorbic acid and amino acids or amines. Yano *et al.* (1976) reacted dehydroascorbic acid with a number of amino acids including α-alanine and separated two radical products by thin-layer chromatography. The free radical products were found to be quite stable and could be formed readily in foods. The implications of these reaction products as antioxidants are discussed in the next section.

IX. Antioxidant Activity of Nonenzymatic Browning Products

The ability of Maillard reaction products (MRP) to retard the development of rancidity has been reported by a number of researchers (Anderson *et al.*, 1963; Griffith and Johnson, 1957; Kato, 1973; Lingnert, 1980; Yamaguchi and Fujimaki, 1974). Most of these studies have focused on the effect of these Maillard reaction products in model systems. The antioxidant activity was generated by the heat processing of sugar or sugar and amino acid added to the unprocessed foods. Lingnert and Waller (1983) examined the antioxidant activity of products generated from a histidine–glucose system. Their results, shown in Figure 5.12, indicate a dramatic loss of antioxidant activity was evident in the presence of air compared to storage at 25°C under nitrogen. The loss of antioxidant activity was

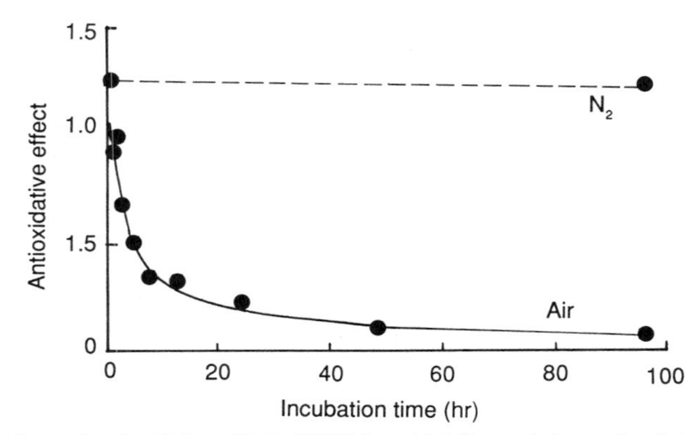

Fig. 5.12. Loss of antioxidative effect of MRP from histidine and glucose incubated at 25°C in an atmosphere of air or nitrogen (Lingnert and Waller, 1983).

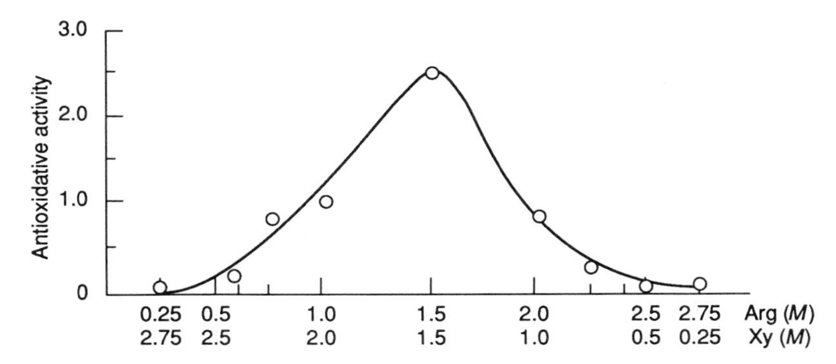

FIG. 5.13. Antioxidative activity versus molar ratio of arginine to xylose (100 μl of 1 : 100 dilution of crude) (Beckel and Waller, 1983). Copyright © by Institute of Food Technologists.

less at low pH (2.0) compared to high pH (8–10). The absence of antioxidant activity Lingnert (1980) when MRP from histidine and glucose were added to dough compared to the addition of histidine and glucose to the dough itself was attributed by these researchers to its instability. The instability of MRP antioxidant activity was also observed by Lingnert and Waller (1983) for arginine–xylose systems. These researchers (Lingnert and Eriksson, 1980) found that neutral or slightly basic conditions favored the formation of antioxidant products from histidine and glucose particularly in the presence of high concentrations of histidine. Beckel and Waller (1983) examined the effect of time, initial pH, and molar ratio of arginine to xylose on antioxidant activity. A pH of 5.0 appeared to be optimal for antioxidant activity with a molar ratio of 1 : 1 producing the maximum effect (Figure 5.13). Repeated dialysis resulted in loss of activity suggesting the involvement of low-molecular-weight antioxidants. This was consistent with studies by Kawashima and co-workers (1977), who examined the antioxidant activity of browning products prepared from low-molecular-weight carbonyl compounds and amino acids. These were produced by reacting methylglyoxal, glyoxal, glyoxylic acid, and dihydroxyacetone, with amino acids. The most potent antioxidants were formed when methylglyoxal and dihydroxyacetone were reacted with the branched-chain amino acids leucine and valine.

The formation of antioxidants during food processing was first reported by Griffith and Johnson (1957). These researchers observed improved stability of cookies to rancidity when 2.5% of the total sucrose was replaced by glucose in cookie dough. The addition of amino acids and protein hydrolysates to cookie dough was also reported by Yamaguchi *et al.* (1980) to improve the storage stability of the cookies. Lingnert (1980) noted that the addition of amino acids and sugars to cookie dough was far more effective than adding the preformed MRP, thereby implicating the role of baking in antioxidant formation. A recent

TABLE 5.6

Antioxidative Effect and Color of Water Extracts from Cookies Baked to a Similar Degree of Darkness[a]

Amino–sugar added	Antioxidative effect[b]	Color[c]
Control	0.1	0.22
His + glucose	0.3	0.33
His + xylose	0.4	0.28
Arg + glucose	0.4	0.27
Arg + xylose	1.2	0.37

[a] From Lingnert and Hall (1986).

[b] Antioxidative effect of cookie extracts measured by polarography.

[c] Color of cookie extracts measured spectrophotometrically at 450 nm.

study by Lingnert and Hall (1986) examined the antioxidative effect of cookies made from doughs containing arginine–glucose/xylose and histidine–glucose/xylose (2 : 1 on a molar basis) or their preformed MRP. Equivalent antioxidative effects were observed irrespective of the addition of the amino acid–sugar or corresponding MRP, although the cookies were considerably darker in the latter case. This experiment was repeated, but in this case the cookies were all baked to the same degree of darkness. Their results in Table 5.6 show that addition of arginine–xylose produced the strongest antioxidative effect.

Farag *et al.* (1982) evaluated the role of Amadori compounds, formed from aldopentoses, aldohexoses, ketohexoses and amino acids, with respect to their antioxidant activity on the oxidation of linoleic acid. Amadori compounds were produced by reacting five different reducing sugars with 16 amino acids. Pentoses produced browning much faster than aldohexoses, which was consistent with studies by Yuichiro (1972) who noted more efficient production of Amadori compounds from pentoses compared to hexoses. Farag *et al.* (1982) observed a higher browning intensity with the straight-chain amino acid lysine compared to the branch-chain amino acids leucine and valine. The rate of reaction reached a maximum at moderate pH with Amadori compounds formed from D-glucose or D-xylose and amino acids exhibiting the highest lipid oxidation retarding activity. The relationship between antioxidant activity and color intensity during the Maillard reaction is confusing. For example, Kirigaya *et al.* (1968) reported that antioxidant activity increased in proportion to color intensity. This contrasted with studies by Hwang and Kim (1973) and Lee *et al.*(1975), who were unable to find any relationship between antioxidant activity and color development.

In addition to color intensity, the development of fluorescence during the browning reaction was reported over 30 years ago by Overby and Frost (1950). The fluorescence developed more readily when sugars were incubated with amino compounds (Burton *et al.*, 1962). It was later suggested that nitrogenous compounds containing the chromophoric group -N=CH-CH=CH-NH was probably responsible for this fluorescence (Chio and Tappel, 1961). Adhikari and Tappel (1972) reported that chromophore-containing compounds produced from glucose–glycine browning mixtures were responsible for fluorescent properties. Park and Kim (1983) examined the relationship between color, fluorescence of glycine–glucose systems, and the antioxidant activity of their ethanol extracts. The increase in fluorescence paralleled that of absorbance during the early stages of browning, although the change in fluorescence was much greater (Figure 5.14). Thus fluorescence provided a sensitive method for monitoring the browning reaction. Antioxidant activity as measured by decrease in peroxide value, was detected in ethanol extracts obtained during the early stages of browning but changed very little compared to the increase in fluorescence (Figure 5.15). The formation of active antioxidants occurred prior to the development of any color

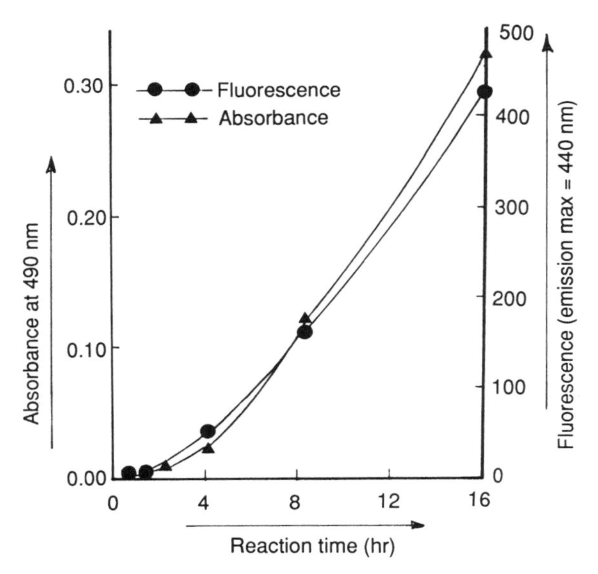

FIG. 5.14. Variations of the absorbance and fluorescence of the 0.2 *M* glucose + 0.1 *M* glycine browning mixture with reaction time. The relative fluorescence of quinine sulfate at 1 μg/ml 0.1*N* H_2SO_4 = 100 (Park and Kim, 1983).

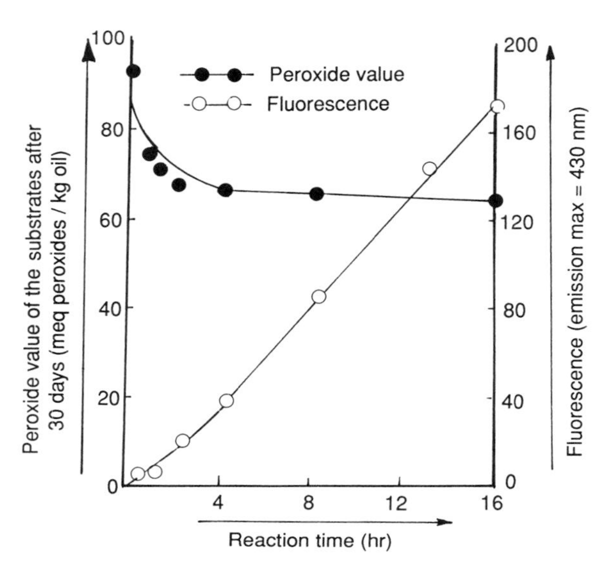

FIG. 5.15. Relationship between fluorescence and absorbance of the ethanolic extracts obtained at successive stages of the 0.1 *M* glucose + 0.1 *M* glycine browning mixture. The relative fluorescence of quinine sulfate at 1 μg/ml 0.1 *N* H_2SO_4 = (Park and Kim, 1983).

or irrespective of the presence of high or low fluorescence activity. Thus, antioxidant activity was not due to chromophore systems involved in fluorescence but involved low-molecular-weight compounds and Amadori compounds formed by the Maillard reaction (Farag *et al.*, 1982).

Studies by Yano and co-workers (1976) showed that dehydroascorbic acid reacted with amino acids or amines to produce a stable free radical species during the early stages of the reaction. They suggested that the antioxidant effect could be due to the radical scavenging action of this species. Namiki *et al.* (1982) later reported antioxidant activity from dehydroascorbic acid and tryptophan comparable to that of the antioxidant BHA (butylated hydroxyanisole).

X. Inhibition of Nonenzymatic Browning

A major concern of food technologists is to control or minimize nonenzymatic browning reactions in food processing. The particular method used must be adapted to the each food product. A variety of methods have been proposed for controlling these reactions.

A. Temperature

An increase in temperature or time of heat treatment accelerates the rate of these reactions (Labuza and Shapiro, 1978). Thus lowering the temperature during processing and storage can lengthen the lag phase, that is, the period needed for the formation of brown-colored products.

B. Moisture Content

The dependency of browning reactions on moisture content provides a convenient method for control. A reduction of moisture content in solid food products by dehydration reduces the mobility of reactive components (Eichner and Karel, 1972; Fox *et al.*, 1983; Labuza and Saltmarch, 1981; Loncin *et al.*, 1968). In the case of solutions, an increase in water activity will diminish the reaction velocity. This not only dilutes the effect of the reactants but water also represents the first reaction product of the condensation step in the Maillard browning reaction. This is illustrated in Figure 5.16, where the browning rates of an avicel/glucose/glycine system were low at both high or low a_W (McWeeny, 1973).

C. pH

The Maillard reaction is generally favored under more alkaline conditions so that lowering the pH provides a useful method of control (Fox *et al.*, 1983). This

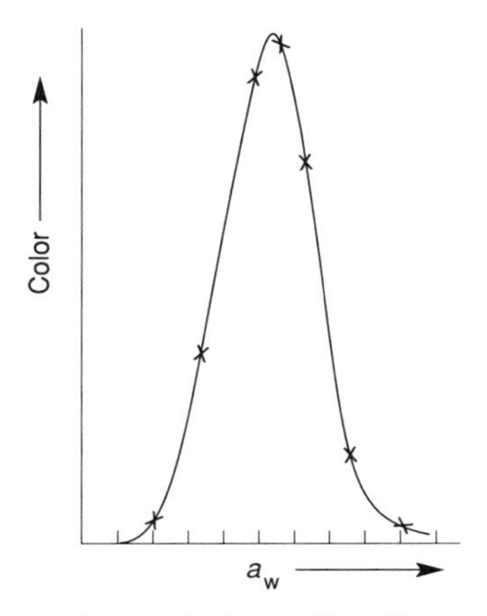

FIG. 5.16. Browning of avicel–glucose–glycine as affected by water activity after 8 days at 38°C (McWeeney, 1973).

method has been utilized in the production of dried egg powder in which acid is added prior to the dehydration process to lower the pH. The pH is restored by addition of sodium bicarbonate to the reconstituted egg.

D. Gas Packing

Gas packing excludes oxygen by packing under an inert gas. This reduces the formation of lipid oxidation products capable of interacting with amino acids. The exclusion of oxygen is thought to affect those reactions involved in the browning process and not the initial carbonylamino reaction step.

E. Biochemical Agents

Removal or conversion of one of the reactants in the sugar–amino acid interactions forms the basis of the biochemical method. For instance, in the commercial production of egg white, glucose is removed by yeast fermentation prior to drying. The direct application of enzymes such as glucose oxidase and catalase mediates the conversion of glucose to gluconic acid, which is no longer capable of combining with amino acids. This enzyme has been used for many years to remove glucose from egg prior to spray-drying (Lightbody and Fevold, 1948). Glucose oxidase has the additional advantage of removing any residual oxygen and is used to reduce headspace oxygen during the production of bottled products.

F. Chemical Inhibitors

A variety of chemical inhibitors have been used to limit browning during the production and storage of a number of different food products. The most widely used are sulfur dioxide and sulfites, although thiols, calcium salts, and aspartic and glutamic acids have also been studied. The use of thiols as inhibitors is limited because of their unpleasant properties.

1. *Sulfur Dioxide/Sulfites*

Sulfur dioxide is unique in its ability to inhibit the Maillard reaction and can be applied as a gas or in solution as sulfite/bisulfite. Sulfur dioxide is not only capable of partially bleaching chromophores already formed but also inhibits color formation at the beginning of the reaction (McWeeny, 1984). The mechanism involves the binding of sulfur dioxide/sulfite with glucose to form hydroxysulfonate and other compounds from which sulfur dioxide/sulfite can be reversi-

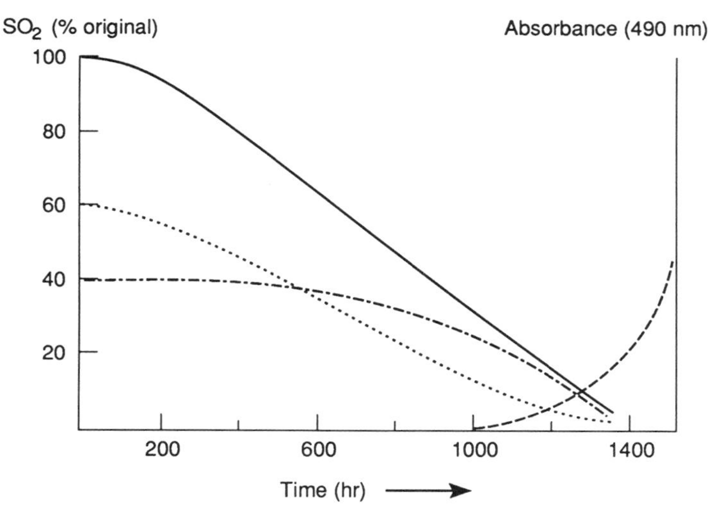

FIG. 5.17. Color production and loss of SO_2 from glucose–glycine–SO_2 during incubation at 55°C. (···) "free" SO_2; (—·—·) "bound" SO_2; (——) total SO_2; (----) absorbance at 490 nm (McWeeny, 1969). Copyright © by Institute of Food Technologists.

bly released. This results in the blocking of the carbonyl group of the sugar, rendering it unavailable for interaction in the typical Maillard reaction. As the reaction proceeds, sulfur dioxide/sulfite becomes irreversibly bound. This permits monitoring of the progress of browning by measuring the amount of sulfur dioxide or sulfite in the "bound" or "free" forms. For example color development for a glucose–glycine model system was not observed by McWeeny (1969) until all the "free" sulfite was depleted, while the ratio of "bound" to "free" sulfite increased as the reaction proceeded (Figure 5.17).

Inhibition of nonenzymatic browning by sulfite appears to involve the formation of stable sulfonates. In the case of ascorbic acid browning, 3-deoxy-4-sulfopentulose is formed with the corresponding 6-carbon compound, 6-deoxy-4-sulfohexulose, for the sulfited inhibition of the Maillard reaction (Wedzicha and McWeeny, 1974a,b; Knowles, 1971). Wedzicha and McWeeny (1975) monitored the formation of several organic sulfonates from sulfited foods. These compounds possess a dicarbonyl group, which makes them particularly reactive and at elevated temperatures can lead to the formation of sulfur compounds with other food components. McWeeny (1984) pointed out that future research on the role of sulfur dioxide/sulfite in inhibiting the Maillard reaction should focus on the nature of those precursors with which it reacts. In addition, further information is needed on the effect of time, temperature, pressure, pH, and additives on the quantitive and qualitative formation of compounds formed in foods to which

TABLE 5.7

Hunter L Values of Control and Treated Potato Chips[a]

Replicate	Control	Dipped in aspartic acid	Dipped in glutamic acid
1	39.0[b]	47.6	44.2
2	38.3	44.6	44.7
3	38.2	45.3	43.0
4	39.0	47.0	43.0
5	37.9	46.8	43.7
6	36.7	46.6	43.9
7	39.5	47.2	43.6
Mean	38.4 ± 0.57[c]	46.4 ± 1.00	43.7 ± 0.57

[a] The treatment consisted in freeze-drying potato slices and dipping them in 0.04 M aspartic or glutamic acid before frying. From Nafisi and Markakis (1983).

[b] Each L value is the average of three readings obtained by rotating the sample 120° angles.

[c] The differences between control and treated samples are significant at the 99% probability level. The difference between treated samples is not significant.

sulfur dioxide/sulfite has been added. This work is particularly important in light of the current trend to limit intake of sulfur dioxide/sulfites in foods.

2. *Aspartic and Glutamic Acids*

A study by Nafisi and Markakis (1983) indicated the potential of aspartic and glutamic acids for inhibiting the Maillard browning reaction. Using model systems containing lysine–glucose/lysine–fructose (pH 8.0, 60°C for 58 hr), these researchers found that L-aspartic acid or L-glutamic acid decreased the rate of browning as seen in Table 5.7. Dipping specially prepared potato chips into either aspartic or glutamic acid solutions prior to frying was accompanied by less darkening as measured by Hunter L values. This corresponded to a mean value of 38.4 for the untreated potato chips compared to 46.4 and 43.7 for the potato chips dipped in aspartic and glutamic acids, respectively.

While the various chemical inhibitors discussed can, with varying degrees of success, limit the progress of browning, the nutritional value of the food could still be reduced. For instance, the initial stage of the Maillard reaction, the carbonylamino step, could still render the amino acids unavailable without any visible browning during this stage. To be certain this step is prevented is extremely difficult to ensure.

Bibliography

Abrams, A., Lowy, P. H., and Borsook, H. (1955). Preparation of 1-amino-1-deoxy-2-keto hexoses from aldohexoses and amino acids. *J. Am. Chem. Soc.* **77,** 4794.

Adhikari, H. R., and Tappel, A. L. (1972). Fluorescent products in a glucose–glycine browning reaction. *J. Food Sci.* **38,** 486.

Aeschbacher, H. V., Chappus, C., Manganel, M., and Aesbach, R. (1981). Investigation of Maillard products in bacterial mutogenicity test systems. *Prog. Food Nutr. Sci.* **5,** 279.

Anderson, R. H., Moran, D. H., Huntley, T. E., and Holahan, J. L. (1963). Responses of cereals to antioxidants. *Food Technol.* **17,** 115.

Anet, E. F. L. J. (1960). Degradation of carbohydrates. I. Isolation of 3–deoxyhexosones. *Aust. J. Chem.* **13,** 396.

Anet, E. F. L. J. (1962). Degradation of carbohydrates. III. Unsaturated hexosones. *Aust. J. Chem.* **15,** 503.

Anet, E. F. L. J. (1964). 3-Deoxyglycosuloses (3-deoxyglycosones) and the degradation of carbohydrates. *Adv. Carbohyd. Chem.* **19,** 181.

Anet, E. F. L. J., and Reynolds, T. M. (1957). Chemistry of nonenzymic browning. 1. Reactions between amino acids, organic acids, and sugar in freeze-dried apricots and peaches. *Aust. J. Chem.* **10,** 182.

Baltes, W. (1973). Nichtenzymatische Veraenderungen bei der Beund Verarbeitung von Lebens. *Z. Ernaehrungswiss. Suppl.* **16,** 34–49.

Barbetti, P., and Chiappini, I. (1976a). Fractionation and spectroscopic characterization of melanoidic pigments from a glucose–glycine non-enzymic browning system. *Ann. Chim. (Rome)* **66,** 293.

Barbetti, P., and Chiappini, I. (1976b). Identification of the furan group in the melanoidic pigments from a glucose–glycine Maillard reaction. *Ann. Chim. (Rome)* **66,** 485.

Bauernfriend, J. C. (1953). The use of ascorbic acid in processing food. *Adv. Food Res.* **4,** 359.

Bauernfriend, J. C., and Pinkert, D. M. (1970). Food processing with ascorbic acid. *Adv. Food Res.* **18,** 219.

Beckel, R. W., and Waller, G. R. (1983). Antioxidative arginine–xylose Maillard reaction products: Conditions for synthesis. *J. Food Sci.* **48,** 996.

Benzing-Purdie, L. M., Ripmeester, J. A., and Ratcliffe, C. I. (1985). Effects of temperature on Maillard reaction products. *J. Agric. Food Chem.* **33,** 31.

Bobbio, P. A., Bobbio, F. O., and Trevisan, L. M. V. (1973). Estudos sobre a meacao de Maillard. 1. Efeitos da temperaturia e do pH. *An. Acad. Bras. Cienc.* **45,** 419.

Bobbio, P. A., Imasato, H., and Leite, S. R. de A. (1981). Maillard reaction. Preparation and characterization of melanoidins from glucose and fructose with glycine. *An. Acad. Bras. Cienc.* **53,** 83.

Bohart, G. S., and Carson, J. F. (1955). Effects of trace metals, oxygen, and light on the glucose–glycine browning reaction. *Nature (London)* **175,** 470.

Braverman, J. B. S. (1963). "Introduction to the Biochemistry of Foods." Elsevier, Amsterdam.

Bryce, D. J., and Greenwood, C. T. (1963). Thermal degradation of starch. *Staerke.* **15,** 166.

Burton, H. S. and McWeeney, D. J. (1963). Non-enzymic browning reactions. Consideration of sugar stability. *Nature (London)* **197,** 266.

Burton, H. S., and McWeeney, D. J. (1964). Non-enzymatic browning: Routes to the production of melanoidins from aldoses and amino compounds. *Chem. Ind. (London)* p. 462.

Burton, H. S., McWeeney, D. J., and Biltcliffe, D. O. (1962). Sulfates and aldose–amino reactions. *Chem. Ind. (London),* p. 219.

Buttery, R. G., and Ling, L. C. (1974). Alkyl thiazoles in potato products. *J. Agric. Food Chem.* **22,** 912.

Buttery, R. G., Ling, L. C., and Juliano, B. O. (1982). 2-Acetyl-1-pyrroline: An important aroma component of cooked rice. *Chem. Ind. (London),* p. 958.

Buttery, R. G., Ling, L. C., Juliano, B. O., and Turnbaugh, J. G. (1983). Cooked rice aroma and 2-acetyl-1-pyrroline. *J. Agric. Food Chem.* **31,** 826.

Buttkus, H. (1967). The reaction of myosin with malonaldehyde. *J. Food Sci.* **32,** 432.

Chang, S. S., Hirai, C., Reddy, B. R., Herz, K. O., Kato, A., and Sipma, G. (1968). Isolation and identification of 2, 4, 5-trimethyl-3-oxazoline and 3, 5-dimethyl 1, 2, 4-trithiolane in the volatile flavour compounds of boiled beef. *Chem. Ind. (London)* p. 1639.

Chio, K. S., and Tappel, A. L. (1961). Synthesis and characterization of the fluorescent products derived from malonaldehyde and amino acids. *Biochemistry* **8,** 2821.

Clegg, K. M. (1964). Non-enzymic browning of lemon juice. *J. Sci. Food Agric.* **15,** 878.

Clegg, K. M. (1969). Citric acid and the browning of solutions containing ascorbic acid. *J. Sci. Food Agric.* **17,** 566.

Coleman, E. C., Ho, C.-T., and Chang, S. S. (1981). Isolation and identification of volatile compounds from baked potatoes. *J. Agric. Food Chem.* **29,** 42.

Curl, L. A. (1949). Ascorbic acid losses and darkening on storage at 69°C (120°F) of synthetic mixtures analogous to orange juice. *Food Res.* **14,** 9.

Danehy, J. P. (1986). Maillard reaction: Nonenzymic browning in food systems with special reference to the development of flavor. *Adv. Food Res.* **30,** 77.

Dawes, I. W., and Edwards, R. A. (1960). Methyl substituted pyrazines as volatile reaction products of heated aqueous aldose, amino acid mixtures. *Chem. Ind. (London),* p. 2203.

Dworshak, E. (1980). Nonenzymic browning and its effect on protein nutrition. *CRC Crit. Rev. Food Sci. Nutr.* **13,** 1.

Dworschak, E., and Hegedüs, M. (1974). Effect of heat treatment on the nutritive value of proteins in milk powder. *Acta Aliment. Acad. Sci. Hung.* **3,** 337.

Dworschak, E., and Orsi, F. (1981). A derivatographical study in the reaction between some essential amino acids (methionine, lysine, tryptophan) and glucose. *Prog. Food Nutr. Sci.* **5,** 415.

Eichner, K., and Karel, M. (1972). The influence of water content and water activity on the sugar–amino browning reaction in model systems under various conditions. *Food Chem.* **20,** 218.

Ellis, G. P. (1959). The Maillard reaction. *Adv. Carbohydr. Chem.* **14,** 63.

Eriksson, C., ed. (1981). "Maillard Reactions in Food," Prog. Food Nutr. Sci., Vol. 5, Pergamon, Oxford.

Farag, R. S., Ghali, Y. and Rashed, M. M. (1982). Linoleic acid oxidation catalyzed by Amadori compounds in aqueous media. *Can. Inst. Food Sci. Technol. J.* **15,** 174.

Feather, M. S., and Harris, J. F. (1973). Dehydration reactions of carbohydrates. *Adv. Carbohydr. Chem.* **28,** 161.

Feather, M. S., and Harris, J. F. (1975). Studies on the mechanism of the interconversion of D-glucose, D-mannose, and D-fructose in acid solution. *J. Am. Chem. Soc.* **97,** 178.

Feather, M. S., and Huang, R. D. (1986). Some studies on a Maillard polymer derived from L-alanine and D-glucose. *Dev. Food Sci.* **13,** 183.

Feather, M. S., and Nelson, D. (1984). Maillard polymers derived from D-glucose, D-fructose, 5-(hydroxymethyl)-2-furaldehyde, and glycine and methionine. *J. Agric. Food Chem.* **32,** 1428.

Feather, M. S., Harris, D. W., and Nichols, S. B. (1972). Routes of conversion of D-xylose, hexuronic acids and L-ascorbic acid to 2–furaldehyde. *J. Org. Chem.* **37,** 1606.

Feeney, R. E. (1980). Overview on the chemical deteriorative changes of proteins and their consequences. *In* "Chemical Deterioration of Proteins" (J. R. Whitaker, and M. Fujimaki, eds.), pp. 1–49. Am. Chem. Soc., Washington, D. C.

Feeney, R. E., Blankenhorn, G., and Dixon, H. B. F. (1975). Carbonylamine reactions in protein chemistry. *Adv. Protein Chem.* **29,** 135.

Ferretti, A., and Flanagan, V. P. (1971). The lactose–casein (Maillard) browning system: Volatile components. *J. Agric. Food Chem.* **19,** 245.

Ferretti, A., and Flanagan, C. P. (1973). Characterization of volatile constituents of an *N*-formyl-L-lysine–D-lactose browning system. *J. Agric. Food Chem.* **21,** 35.

Finot, P. A., and Mauron, J. (1972). Le blocage de la lysine par la reaction de Maillard. II. Propriétés chimiques des dérives *N*-(desoxy-1-D-fructosyl-1) et *N*-(desoxy-1-lactulosyl-1) de la lysine. *Helv. Chim. Acta* **55,** 1153.

Finot, P. A., Bricout, J., Viani, R., and Maurion, J. (1968). Identification of a new lysine derivative obtained upon acid hydrolysis of heated milk. *Experientia* **24,** 1097.

Flink, J. M. (1983). Nonenzymic browning of freeze-dried sucrose. *J. Food Sci.* **48,** 539.

Forsskahl, I., Popoff, T., and Theander, O. (1976). Reactions of D-xylose and D-glucose in alkaline aqueous solutions. *Carbohydr. Res.* **48,** 13.

Fox, M., Loncin, M., and Weiss, M. (1983). Investigations into the influence of water activity, pH and heat treatment on the velocity of the Maillard reactions in foods. *J. Food Qual.* **6,** 103.

Friedman, M. (1982). Chemically reactive and unreactive lysine as an index of browning. *Diabetes* **31,** 5.

Gazzani, G., Vagnarelli, P., Cazzani, M. T., and Bazzay, P. G. (1987). Mutagenic activity of the Maillard reaction products of ribose with different amino acids. *J. Food Sci.* **52,** 757.

Gianturco, M., Giammarino, A. S., and Pitcher, R. G. 1963 The structures of five cyclic diketones isolated from coffee. *Tetrahedron* **19,** 2051.

Gornhardt, L. (1955). Die nichtenzymatische Braunung von Lebensmitteln. 1. *Fette, Seifen, Anstrichm.* **57,** 270.

Grant, N. H., Clark, D. E., and Alburn, H. E. (1966). Accelerated polymerization of *N*-carboxyamino acid anhydrides in frozen dioxane. *J. Amer. Chem. Soc.* **88,** 4071.

Griffith, T., and Johnson, J. A. (1957). Relation of the browning reaction to storage stability of sugar cookies. *Cereal Chem.* **34,** 159.

Grivas, S., Nyhammar, T., Olson, K., and Jogerstad, M. (1985). Formation of a new mutagenic D. Mel Qx compound in a model system by heating creatine, alanine and fructose. *Mutat. Res.* **151,** 177.

Handwerk, R. L., and Coleman, R. L. (1988). Approaches to the citrus browning problem: A review. *J. Agric. Food Chem.* **36,** 231.

Harris, D. W., and Feather, M. S. (1973). Evidence for a C-2 → C-1 intramolecular hydrogen transfer during the acid-catalyzed isomerization of D-glucose to D-fructose. *Carbohydr. Res.* **30,** 359.

Harris, D. W., and Feather, M. S. (1975). Studies on the mechanism of the interconversion of D-glucose, D-mannose and D-fructose in acid solution. *J. Am. Chem. Soc.* **97,** 178.

Harris, J. F. (1972). Alkaline decomposition of D-xylose-1-C, D-glucose-1-C and D-glucose-6-C. *Carbohydr. Res.* **23,** 207.

Harris, R. L., and Mattil, H. A. (1940). The effect of hot alcohol on purified animal proteins. *J. Biol. Chem.* **132,** 477.

Hashiba, H. (1978). Isolation and identification of Amadori compounds from soy sauce. *Agric. Biol. Chem.* **42,** 763.

Hashiba, H. (1979). Oxidative browning of soy sauce. I. The roles of cationic fraction from soy sauce. *Nippon Shoyu Kenkyusho Zasshi* **5**(4), 169.

Hashiba, H. (1986). Oxidative browning of Amadori compounds color formation by iron with Maillard reaction products. *Dev. Food Sci.* **13,** 155.

Hashiba, H., Okura, A., and Iguchi, N. (1981). Oxygen-dependent browning of soy sauce and some brewed product. *Prog. Food Nutr. Sci.* **5,** 93.

Hayashi, T., and Namiki, M. (1986). Role of sugar fragmentation: An early stage browning of amino-carbonyl reaction of sugar and amino acid. *Agric. Biol. Chem.* **50,** 1965.
Hayashi, T., Ohata, Y., and Namiki, M. (1977). Electron spin resonance spectral study of the structure of the novel free radical products formed by the reactions of sugars with amino acids or amines. *J. Agric. Food Chem.* **25,** 1282.
Hendel, C. E., Silveira, V. G., and Harrington, W. O. (1955). Rates of non-enzymatic browning of white potato during dehydration. *Food Technol.* **9,** 433.
Herrmann, J., and Andrae, W. (1963). Oxydative Abbauprodukte der L-Ascorbinsaure. I. Paperchromatographischer Nachweis. *Nahrung* **7,** 243.
Heyns, K., and Breuer, H. (1958). Darstellung und Verhalten weiterer N-substituierter 2-Amino-2-desoxy-aldosen aus D-fructose und Aminosauren. *Chem. Ber.* **91,** 2750.
Heyns, K. and Klier, M. (1968). Braunungsreaktionen und Fragmentierungen von kohlenhydaten. *Carbohydr. Res.* **6,** 436.
Heyns, K., and Noack, H. (1962). Die Umsetzung von D-Fructose mit L-Lysin und L-Histidine mit Hexosen. *Chem. Ber.* **95,** 720.
Heyns, K., and Noack, H. (1964). Die Umsetzung von L-Tryptophan und L-Histidine mit Hexosen. *Chem. Ber.* **97,** 415.
Ho, C. T., and Hartman, G. J. (1982). Formation of oxazolines and oxazoles in Stecker degradation of DL-alanine and L-cysteine with 2, 3-butanedione. *J. Agric. Food Chem.* **30,** 793.
Hodge, J. E. (1953). Chemistry of browning reactions in model systems. *J. Agric. Food Chem.* **1,** 928.
Hodge, J. E., and Osman, E. M. (1976). Carbohydrates. *In* "Principles of Food Science. Part 1. Food Chemistry" (O. R. Fennema, ed.). Dekker, New York.
Hodge, J. E., Fisher, B. E., and Nelson, E. C. (1963). Dicarbonyls, reductones and heterocyclics produced by the reactions of reducing sugars with secondary amine salts. *Proc. Am. Soc. Brew. Chem.* 84.
Hodge, J. E., Mill, F. D., and Fisher, B. E. (1972). Compounds of browned flavor derived from sugar–amine reactions. *Cereal Sci. Today.* **17,** 34.
Holtermand, A. (1966). The browning reaction. *Staerke* **16,** 319.
Homma, S., and Fujimaki, M. (1981). Growth response in rats fed a diet containing non-dialyzable melanoidin. *In* "Maillard Reactions in Food: Chemical, Physiological and Technological Aspects" (C. Eriksson, ed.). Pergamon, Oxford.
Horn, J. M., Lichtenstein, H., and Womack, M. (1968). Availability of amino acids. A methionine–fructose compound and its availability to microorganisms and rats. *J. Agric. Food Chem.* **16,** 741.
Huelin, P. E. (1953). Studies on the anaerobic decomposition of ascorbic acid. *Food Res.* **15,** 78.
Huelin, P. E., Coggiola, I. M., Sidhu, G. S., and Kennett, B. H. (1971). The anaerobic decomposition of ascorbic acid in the pH range of foods and in more acid solutions *J. Sci. Food Agric.* **22,** 540.
Huffman, C. F. (1974). Treatment of fruit juices with ion-exchange resins. U.S. Patent 3,801,717.
Hurrell, R. F., and Carpenter, K. J. (1974). Mechanism of heat damage in proteins. IV. The reactive lysine content of heat-damaged material as measured in different ways. *Br. J. Nutr.* **32,** 589.
Hurrell, R. F., and Carpenter, K. J. (1981). The estimation of available lysine in foodstuffs after Maillard reaction. *Prog. Food Nutr. Sci.* **5,** 159.
Hwang, C.-J., and Kim, D. H. (1973). The antioxidant activity of some extracts from various stages of a Maillard-type browning reaction mixture. *Hanguk Sikp'um Kwahakoe Chi.* **5**(2), 84 (cited in *Chem. Abstr.* **79,** 114189e.
Imasato, H., Leite, S. R. de A., and Bobbio, P. A. (1981). Maillard reaction. VI. Structural determinations in melanoidin from fructose and glycine. *An. Acad. Bras. Cienc.* **53,** 87.
Ingles, D. L., and Gallimore, D. (1985). A new method for the isolation of melanoidin from the Maillard reaction of glucose and glycine. *Chem. Ind. (London),* p. 194.

Isbell, H. S. (1944). Interpretation of some reactions in the carbohydrate field in terms of consecutive electron displacement. *J. Res. Natl. Bur. Stand. (U.S.)* **32,** 45.

Ito, H. (1977). The formation of maltol and isomaltol through degradation of sucrose. *Agric. Biol. Chem.* **41,** 1307.

Jirousova, J., and Davidek, J. (1975). Reactions of *n*-hexanal with glycine in model systems. *Z. Lebensm.-Unters. -Forsch.* **157,** 269.

Johnson, R. R., Alford, E. D., and Kinzer, G. W. (1969). Formation of sucrose pyrolysis products. *J. Agric. Food Chem.* **17,** 22.

Joslyn, M. A. (1957). Role of amino acids in the browning of orange juice. *Food Res.* **22,** 1.

Karel, M., and Labuza, T. P. (1968). Nonenzymatic browning in model systems containing sucrose. *J. Agric. Food Chem.* **16,** 717.

Katchalsky, A. (1941). Interaction of aldoses with α-amino acids or peptides. *Biochem. J.* **35,** 1024.

Kato, H. (1962). Chemical studies on amino-carbonyl reaction. I. Isolation of 3-deoxypentosone and 3-deoxyhexosones formed by browning degradation of *N*-glycosides. *Agric. Biol. Chem.* **26,** 187.

Kato, H. (1963). Chemical studies on amino-carbonyl reaction. II. Identification of D-glucosone formed by oxidative browning degradation of *N*-D-glucoside. *Agric. Biol. Chem.* **27,** 461.

Kato, H. (1973). Antioxidative activity of amino-carbonyl reaction products. *Shokuhin Eiseigaku Zasshi* **14,** 343.

Kato, H., and Fujimaki, M. (1968). Formation of N-substituted pyrrole-2-aldehydes in the browning reaction between D-xylose and amino compounds. *J. Food Sci.* **33,** 445.

Kato, H., Kim, S. S., and Hayase, F. (1986). Amino–carbonyl reactions in food and biological systems. *Dev. Food Sci.* **13,** 215.

Kato, K., Watanabe, K., and Sato, Y. (1978). Effect of the Maillard reaction on the attributes of egg white proteins. *Agric. Biol. Chem.* **42,** 2233.

Kato, K., Watanabe, K. and Sato, Y. (1981). Effect of metals on the Maillard reaction of ovalbumin. *J. Agric. Food Chem.* **29,** 540.

Kawashima, K., Itoh, H., and Chibata, I. (1977). Antioxidant activity of browning products prepared from low molecular carbonyl compounds and amino acids. *J. Agric. Food Chem.* **25,** 202.

Kefford, J. F. (1959). The chemical constituents of fruits. *Adv. Food Res.* **9,** 285.

Kirigaya, N., Kato, H., and Fujimaki, M. (1968). Studies on antioxidant activity of nonenzymic browning reaction products. Part I. Relations of color intensity and reductones with antioxidant activity of browning reaction products. *Agric. Biol. Chem.* **32,** 287.

Kirigaya, N., Kato, H., and Fujimaki, M. (1969). Studies on antioxidant activity of nonenzymic browning reaction products. Part II. Antioxidant activity of nondialyzable browning reaction products. *Nippon Nogei Kagaku Kaishi* **43,** 484.

Kirigaya, N., Kato, H., and Fujimaki, M. (1971). Studies on antioxidant activity of nonenzymic browning reaction products. Part III. Fractionation of browning reaction solution between ammonia and D-glucose and antioxidant activity of the resulting fractions. *Nippon Nogei Kagaku Kaishi* **45,** 292.

Knowles, M. E. (1971). Inhibition of non-enzymic browning by sulphites: Identification of sulphonated products. *Chem. Ind. (London),* p. 110.

Koehler, P. E., and Odell, G. V. (1970). Factors affecting the formation of pyrazine compounds in sugar–amine reactions. *J. Agric. Food Chem.* **18,** 895.

Koehler, P. E., Mason, M. E., and Newell, J. A. (1969). Formation of pyrazine compounds in sugar–amine model systems. *J. Agric. Food Chem.* **17,** 393.

Kort, M. J. (1970). Reactions of free sugars with aqueous ammonia. *Adv. Carbohydr. Chem. Biochem.* **25,** 311.

Kuster, B. F. M., and Temmink, H. M. G. (1977). The influence of pH and weak-acid anions on the dehydration of D-fructose. *Carbohydr. Res.* **54,** 185.

Kuster, B. F. M., and Van der Bean, H. S. (1977). The influence of the initial and catalyst concentrations on the dehydration of D-fructose. *Carbohydr. Res.* **54,** 165.

Kwon, T. W., Menzel, D. B., and Olcott, H. S. (1965). Reactivity of malonaldehyde with food constituents. *J. Food Sci.* **30,** 808.

Labuza, T. P., and Saltmarch, M. (1981). The nonenzymic browning reaction as affected by water in foods. *In* "Water Activity Influences Food Quality" (L. B. Rockland and B. S. Stewart, eds.). Academic Press, New York, p. 605.

Lea, C. H. (1958). Deteriorative reactions involving phospholipids and lipoproteins. *J. Sci. Food Agric.* **8,** 1.

Lea, C. H., and Hannan, R. S. (1949). Studies of the reaction between proteins and reducing sugars in the "dry" state. I. The effect of activity of water, pH and of temperature on the primary reaction between casein and glucose. *Biochim. Biophys. Acta* **3,** 313.

Lea, C. H., and Hannan, R. S. (1950). Biochemical and nutritional significance of the reaction between proteins and reducing sugars. *Nature (London)* **165,** 438.

Ledl, F. (1982a). Formation of coloured products in browning reactions: Reaction of hydroxyacetone with furfural. *Z. Lenbensm.-Unters. -Forsch.* **175,** 203.

Ledl, F. (1982b). Formation of coloured products in browning reactions: Reaction of dihydroxyacetone with furfural. *Z. Lebensm.-Unters. -Forsch.* **175,** 349.

Ledl, F., and Fritsch, G. (1984). Formation of pyrroline reductones by heating hexoses with amino acids. *Z. Lebensm.-Unters. -Forsch.* **178,** 41.

Ledl, F., and Severin, T. (1982). Investigations of the Maillard reaction. XVI (1). Formation of coloured compounds from hexoses. *Z. Lebensm.-Unters. -Forsch.* **175,** 262.

Lee, C. M., Lee, T.-C., and Chichester, C. O. (1979). Kinetics of the production of biologically active Maillard browned products in apricot and glucose–L-tryptophan. *J. Agric. Food Chem.* **27,** 478.

Lee, C. M., Sherr, B., and Koh, Y.-N. (1984). Evaluation of kinetic parameters for a glucose–lysine Maillard reaction. *J. Agric. Food Chem.* **32,** 379.

Lee, M.-H., Ho, C., and Chang, S. S. (1981). Thiazoles, oxazoles and oxazolines identified in the volatile flavor of roasted peanuts. *J. Agric. Food Chem.* **26,** 1049.

Lee, S. S., Rhee, C., and Kim, D. H. (1975). Comparison of the antioxidant activity of absolute ethanol extracts and 90% ethanol extracts obtained at successive stages of a Maillard-type browning reaction mixture. *Korean J. Food Sci. Technol.* **7,** 37.

Lee, T.-C., Pintauro, S. J. and Chichester, C. O. (1982). Nutritional and toxicological effects of nonenzymic browning. *Diabetes* **31,** 37.

Lightbody, H. D., and Fevold, H. L. (1948). Biochemical factors influencing the shelf life of dried whole eggs and means of their control. *Adv. Food Res.* **1,** 149.

Lingnert, N. (1980). Antioxidative Maillard reaction products. III. Application in cookies. *J. Food Process. Preserv.* **4,** 219.

Lingnert, N., and Eriksson, E. (1980). Antioxidative Maillard reaction products from sugars and free amino acids. *J. Food Process. Preserv.* **4,** 161.

Lingnert, N., and Hall, G. (1986). Formation of antioxidative Maillard reaction products during food processing. *Dev. Food Sci.* **13,** 273.

Lingnert, N., and Waller, G. R. (1983). Stability of antioxidants formed from histidine and glucose by the Maillard reaction. *J. Agric. Food Chem.* **31,** 27.

Loncin, M., Bimbenet, J. J., and Lenges, J. (1968). Influence of the activity of water on the spoilage of foodstuffs. *J. Food Technol.* **3,** 131.

Maga, J. A. (1978). Oxazoles and oxazolines in food. *J. Agric. Food Chem.* **26,** 1049.

Maga, J. A. (1981). Pyrroles in foods. *J. Agric. Food Chem.* **29,** 691.

Maga, J. A. (1982). Pyrazines in foods: An update. *CRC Crit. Rev. Food Sci. Nutr.* **16,** 1.

Maga, J. A., and Sizer, C. E. (1973). Pyrazines in foods. *CRC Crit. Rev. Food Technol.* **4,** 39.

Maillard, L. C. (1912). Action des acides amines sur les sucres; formation des mélanoidines par voie méthodique. *C. R. Hebol. Seances Acad. Sci.* **154,** 66.

Markuze, Z. (1963). Effects of traces of metals on the browning of glucose–lysine solutions. *Rocz. Panstw. Zakl. Hig.* **14,** 65, *Chem. Abstr.* **59,** 4980).

Mauron, J. (1970). Nutritional evaluation of proteins by enzymatic methods. *Wenner-Gren Cent. Int. Symp. Ser.* **14.**

Mawhinney, T. P., Madson, M. A., and Feather, M. S. (1980). The isomerization of D-glucose in acidic solution. *Carbohydr. Res.* **86,** 147.

McWeeny, D. J. (1969). The Maillard reaction and its inhibition by sulfite. *J. Food Sci.* **34,** 641.

McWeeny, D. J. (1973). The role of carbohydrates in non-enzymatic browning. *In* "Molecular Structure and Functions of Food Carbohydrates" (G. G. Birch and F. Green, eds.). Applied Science Publishers, London.

McWeeny, D. J. (1984). Sulfur dioxide and the Maillard reaction in food. *Proc. Food Nutr. Sci.* **5,** 395.

Milic, B. L. (1987). CP-mass carbon-13 NMR spectral studies of the kinetics of melanoidin formation. *Analyst* **112,** 783.

Milic, B. L., and Piletic, M. V. (1984). The mechanism of pyrrole, pyrazine and pyridine formation in non-enzymic browning. *Food Chem.* **13,** 165.

Milic, B. L., Piletic, M. V., Grujic-Injac, B., and Premovic, P. I. (1978). A comparison of the chemical composition of boiled and roasted aromas of heated beef. *J. Agric. Food Chem.* **25,** 113.

Milic, B. L., Piletic, M. V., Cembic, S. M., and Odavic-Josic, J. (1979). Free radical formation kinetics in D(+)-glucose and amino acid model systems. *Proc. IUPAC Congr, 27th, 1979.* p. 296.

Milic, B. L., Piletic, M. V., Cembic, S. M., and Odavic-Josic, J. (1980). Kinetic behaviour of free radical formation on the nonenzymatic browning reaction. *J. Food Process. Preserv.* **4,** 13.

Moll, N., Gross, B., That, V., and Moll, M. (1982). A fully automated high-performance liquid chromatographic procedure for isolation and purification of Amadori compounds. *J. Agric. Food Chem.* **30,** 782.

Montgomery, M. W., and Day, E. A. (1965). Aldehyde–amine condensation reaction, possible fate of carbonyls in foods. *J. Food Sci.* **30,** 828.

Nafisi, K., and Markakis, P. (1983). Inhibition of sugar–amine browning by aspartic and glutamic acids. *J. Agric. Food Chem.* **31,** 1115.

Nagy, S. (1980). Vitamin C contents of citrus fruit and their products: A review. *J. Agric. Food Chem.* **28,** 8.

Nagy, S., and Smoot, J. M. (1977). Temperature and storage effects on percentage retention and percentage U.S. recommended dietary allowance of vitamin C in canned single-strength orange juice. *J. Agric. Food Chem.* **25,** 135.

Namiki, M., and Hayashi, T. (1974). Free radical products formed by the reaction of dehydroascorbic acid with amino acids. *Chem. Lett.* **2,** 125.

Namiki, M., and Hayashi, T. (1975). Development of novel free radicals during the amino-carbonyl reaction of sugars with amino acids. *J. Agric. Food Chem.* **23,** 487.

Namiki, M., and Hayashi, T. (1981). Formation of free radical products in an early stage of the Maillard reaction. *In* "Maillard Reactions in Food" (C. Eriksson, ed.), Progr. Food Nutr. Sci., Vol. 5. Pergamon, Oxford.

Namiki, M., and Hayashi, T. (1983). A new mechanism of the Maillard reaction involving sugar fragmentation and free radical formation. Chapt. 2. *In* "The Maillard Reaction in Foods and Nutrition" (G. R. Walter and M. S. Feather, eds.). *ACS Symp. Ser.* **215,** 21.

Namiki, M., Hayashi, T., and Kawakishi, S. (1973). Free radicals developed in the amino-carbonyl reaction of sugars with amino acids. *Agric. Biol. Chem.* **37,** 2935.

Namiki, M., Shigeta, A., and Hayashi, T. (1982). Antioxidant effect of dehydroascorbic acid with tryptophan. *Agric. Biol. Chem.* **46,** 1199.

Newell, J. A., Mason, M. E., and Matlock, R. S. (1967). Precursors of typical and atypical roasted peanut flavor. *J. Agric. Food Chem.* **15,** 767.

Nursten, H. (1986). Maillard browning reactions in dried foods. *In* "Concentration and Drying of Foods" (D. McCarthy, ed.). Elsevier.

Ohno, Y. and Ward, K., Jr. (1961). Acid epimerization of D-glucose. *J. Org. Chem.* **26,** 3928.

Olsson, K., Peremalm, P. A., and Theander, O. (1982). *In* "Maillard Reactions in Food" (C. Eriksson, ed.), Prog. Food Nutr. Sci., Vol. 5. Pergamon, New York.

Otsuka, M., Kurata, T., and Arakawa, N. (1986). Isolation and characterization of an intermediate product in the degradation of 2, 3-diketo-L-gulonic acid. *Agric. Biol. Chem.* **50,** 531.

Overby, L. R., and Frost, D. V. (1950). The effects of heat on the nutritive value of protein hydrolysates with dextrose. *J. Nutr.* **46,** 539.

Park, C. K., and Kim, D. H. (1983). Relationship between fluorescence and antioxidant activity of ethanol extracts of a Maillard browning mixture. *J. Am. Oil Chem. Soc.* **60,** 22.

Partridge, S. M., and Brimley, R. C. (1952). Displacement chromatography on synthetic ion-exchange resins. A systematic method for the separation of amino acids. *Biochem. J.* **51,** 628.

Patton, A. R., Salander, R. C., and Piano, M. (1954). Lysine destruction in casein-glucose interactions measured by quantitative paper chromatography. *Food Res.* **19,** 444.

Peterson, R. J., Izzo, H. J., Jungerman, E., and Chang, S. S. (1975). Changes in volatile flavor compounds during the retorting of canned beef stew. *J. Food Sci.* **40,** 948.

Pigman, W., and Anet, E. F. L. J. (1972). Mutarotation and actions of acids and bases. *In* "The Carbohydrates" (W. Pigman, and D. Horton, eds.), 2nd ed., Vol. 1A, pp. 165–194. Academic Press, New York.

Plakas, S. M., Lee, C.-T., and Walke, K. E. (1988). Bioavailability of lysine in Maillard browned protein as determined by plasma response in rainbow trout *(Salmo gairdneri). J. Nutr.* **118,** 19.

Pokorny, J., Svoboda, H., and Janick, G. (1977). Reaction of lower alkanals with protein. Correlation of flavor changes and nonenzymic browning during storage of model systems. *Sb. Vys. Sk. Chem.-Technol. Praze, Potraviny E49,* 5–21.

Popoff, T., and Theander, O. (1976). Formation of aromatic compounds from carbohydrates. Part III. Reaction of D-glucose and D-fructose in slightly acidic, aqueous solution. *Acta Chem. Scand.* **30,** 397.

Ranganna, S., and Setty, L. (1968). Non-enzymatic discoloration in dried cabbage. Ascorbic acid–amino acid interaction. *J. Agric. Food Chem.* **16,** 529 (commun.).

Ranganna, S., and Setty, L. (1974). Nonenzymatic discoloration in dried cabbage. II. Red condensation product of dehydroascorbic acid and glycine ethyl ester. *J. Agric. Food Chem.* **22,** 719.

Rao, N. M., and Rao, M. N. (1972). Effect of non-enzymatic browning on the nutritive value of casein–sugar system. *J. Food Sci. Technol.* **9,** 66.

Reyes, F. G. R., Poocharoen, B., and Wrolstad, R. E. (1982). Maillard browning reaction of sugar–glycine model systems: Changes in sugar concentration, color and appearance. *J. Food Sci.* **47,** 1376.

Reynolds, T. M. (1963). Chemistry of nonenzymic browning. I. *Adv. Food Res.* **12,** 1.

Reynolds, T. M. (1965). Chemistry of non-enzymatic browning. II. *Adv. Food Res.* **14,** 168.

Reynolds, T. M. (1969). Nonenzymic browning. Sugar–amine interactions. *In* "Foods, Carbohydrates and Their Roles" (H. W. Schultz, R. F. Cain, and R. W. Wrolstadt, eds.). Avi Publ. Co., Westport, Connecticut.

Reynolds, T. M. (1970). Flavours from nonenzymic browning reactions. *Food Technol. Austr.* **22**(11), 610.

Riester, D. W., Braun, O. G., and Pearce, W. E. (1945). Why canned citrus juices deteriorate in storage. *Food Ind.* **17,** 742.

Rizzi, G. P. (1969). Formation of tetraethyl pyrazaine and 2-isopropyl-4,5-dimethyl-3-oxazoline in the Strecker degradation of DL-valine with 2,3–butanedione. *J. Org. Chem.* **34,** 2002.

Rizzi, G. P. (1974). Formation of *N*-alkyl-2-acylpyrroles and aliphatic aldimines in model non-enzymic browning reactions. *J. Agric. Food Chem.* **22,** 279.
Salter, L. J., Mottram, D. S., and Whitfield, F. B. (1988). Volatile compounds produced in Maillard reactions involving glycine, ribose and phospholipid. *J. Sci. Food Agric.* **46,** 227.
Schoebel, T., Tannenbaum, S. R., and Labuza, T. P. (1969). Reaction at limited water concentration. 1. Sucrose hydrolysis. *J. Food Sci.* **34,** 324.
Schonberg, A., and Moubacher, R. (1952). The Strecker degradation of α-amino acids. *Chem. Rev.* **50,** 260.
Schonberg, A., Moubacher, R., and Mostafa, A. (1948). Degradation of α-amino acids to aldehydes and ketones by interaction with carbonyl compounds. *J. Chem. Soc.* p. 176.
Shallenberger, R. S., and Birch, C. G. (1975). "Sugar Chemistry," p. 189. Avi Publ Co., Westport, Connecticut.
Shaw, P. E., and Berry, R. E. (1977). Hexose–amino acid degradation studies involving formation of pyrroles, furans and other low molecular weight products. *J. Agric. Food Chem.* **25,** 641.
Shaw, P. E., Tatum, J. H., and Berry, R. E. (1968). Base-catalyzed fructose degradation and its relation to nonenzymic browning. *J. Agric. Food Chem.* **16,** 979.
Shibamoto, T., and Bernard, R. A. (1977). Investigation of pyrazine formation pathways in sugar–ammonia model systems. *J. Agric. Food Chem.* **25,** 609.
Shigematsu, H., Kurata, T., Kato, H., and Fujimaki, M. (1972). Volatile compounds formed on roasting DL-α-alanine with D-glucose. *Agric. Biol. Chem.* **36,** 1631.
Shinohana, K., Wu, R.-T., Juhan, N., Tanaka, M., Morinaga, N., Murakami, H., and Omura, H. (1980). Mutagenicity of the browning mixtures by amino-carbonyl reactions on *Salmonella typhymurium* TA 100. *Agric. Biol. Chem.* **44,** 671.
Simon, H., and Heubach, G. (1965). Formation of alicyclic and open-chain nitrogenous reductones by reaction of secondary amine salts on monosaccharides. *Chem. Ber.* **98,** 3703.
Smolnik, von H. D. (1987). Herstellung und Andwebdung von Zucker-Kolor aus Starkeprodukten. *Staerke* **39,** 28.
Spark, A. A. (1969). Role of amino acids in non-enzymic browning. *J. Sci. Food Agric.* **20,** 308.
Spingarn, N. E., and Garvie, C. T. (1979). Formation of mutagens in sugar–ammonia model systems. *J. Agric. Food Chem.* **27,** 1319.
Spingarn, N. E., Jahan, N., Tonaka, M., Yamamoto, K., Wu, R.-T., Murakani, H., and Omura, H. (1983). Formation of mutagens in sugar–amino acid model systems. *J. Agric. Food Chem.* **31,** 301.
Stoeffelsman, J., and Pypker, J. (1968). Some new constituents of roasted coffee. *Rec, Trav. Chim. Pays-Bas* **87,** 241.
Stoll, M., Dietrich, P., Sundt, E., and Winter, M. (1967a). Sur l'aroma de café. I. *Helv. Chim. Acta* **50,** 628.
Stoll, M., Winter, M., Gautschi, F., Filament, I., and Willhalm, B. (1967b). Sur l'aroma du cacoa. II. *Helv. Chim. Acta* **50,** 2065.
Svadlenka, I., Davidkova, E., and Rosmus, J. (1975). Interaction of malonaldehyde with collagen. III. Binding site characteristic of malonaldehyde with respect to collagen. *Z. Lebensm.-Unters. -Forsch.* **157,** 263.
Tannenbaum, S. R., Barth, H., and Le Roux, J. P. (1969). Loss of methionine in casein during storage with autoxidizing methyl linoleate. *J. Agric. Food Chem.* **17,** 1353.
Tatum, J. H., Shaw, P. E., and Berry, R. E. (1967). Some compounds formed during nonenzymic browning of orange powder. *J. Agric. Food Chem.* **15,** 773.
Theander, O. (1981). Novel developments in caramelization. *Prog. Food. Nutr. Sci.* **5,** 471.
Tonsbeek, C. H. T., Copier, H., and Plancken, A. J. (1971). Components contributing to beef flavor. Isolation of 2-acetyl-2-thiazoline from beef broth. *J. Agric. Food Chem.* **19,** 1014.

Tsuchuda, H., Tachibana, S., and Kamoto, M. (1976). Identification of heterocyclic compounds produced by pyrolysis of the nondialyzable melanoidins. *Agric. Biol. Chem.* **40,** 2051.

Tu, A., and Eskin, N. A. M. (1973). The inhibitory effect of reducing sugars on the hydrolysis of casein by trypsin. *Can. Inst. Food Sci. Technol. J.* **6,** 50.

Underwood, J. C., Lento, H. G., and Willits, C. O. (1959). Browning of sugar solutions. 3. Effect of pH on the colour produced in dilute glucose solutions containing amino acids with the amino group in different positions in the molecule. *Food Res.* **24,** 181.

Van Praag, M., Stein, H. S., and Tibbetts, M. S. (1968). Steam volatile aroma constituents of roasted cocoa beans. *J. Agric. Food Chem.* **16,** 1005.

Velisek, J., and Davidek, K. (1976a). Reactions of glyoxal with amino acids. I. Formation of brown pigments. *Sb. Vys. Sk. Chem.-Technol. Praze, Potraivin* **E46,** 35.

Velisek, J. and Davidek, K. (1976b). Reactions of glyoxal with amino acids. II. Analysis of the brown pigments. *Sb. Vys. Sk. Chem.-Technol. Praze, Potraivin.* **E46,** 51.

Walradt, J. P., Pittet, A. O., Kinlin, T. E., Muralidhara, R. and Sanderson, A. (1971). Volatile components of roasted peanuts. *J. Agric. Food Chem.* **19,** 972.

Watanabe, K., Kato, Y. and Sato, Y. (1980). Chemical and conformational changes of ovalbumin due to the Maillard reaction. *J. Food Process Preserv.* **3,** 263.

Wedzicha, B. L. and McWeeny, D. J. (1974a). Non-enzymic browning reactions of ascorbic acid and their inhibition. The production of 3-deoxy-4-sulphopentulose in mixtures of ascorbic acid, glycine and bisulphite ion. *J. Sci. Food Agric.* **25,** 577.

Wedzicha, B. L., and McWeeny, D. J. (1974b). Non-enzymic browning reactions of ascorbic acid, glycine and bisulphite ion. The identification of 3-deoxy-4-sulphopentulose in dehydrated, sulphited cabbage after storage. *J. Sci. Food Agric.* **25,** 584.

Wedzicha, B. L. and McWeeny, D. J. (1975). Concentrations of some sulphonates derived from sulphite in certain foods. *J. Sci. Food Agric.* **26,** 327.

Weygand, F. (1940). Uber *N*-Glykoside. II. AmadoriUmlagerungen. *Ber. Dtsch. Chem. Ges.* **73,** 1259.

Wolfrom, M. L., and Rooney, C. C. (1953). Chemical interactions of amino compounds and sugars. VIII. Influence of water. *J. Am. Chem. Soc.* **75,** 5435.

Wolfrom, M. L., Schuetz, R. D., and Calvalieri, L. F. (1948). Discoloration of sugar solutions and 5-(hydroxymethyl)furfural. *J. Am. Chem. Soc.* **70,** 514.

Wong, J. M., and Bernhard, R. A. (1988). Effect of nitrogen source on pyrazine formation. *J. Agric. Food Chem.* **36,** 123.

Yamaguchi, N., and Fujimaki, M. (1974). Studies on browning reaction products from reducing sugars and amino acids. XIV. Antioxidative activities of purified melanoidins and their comparisons with those of legal antioxidants. *Nippon Shokuhin Kogyo Gakkaishi* **21,** 6.

Yamaguchi, N., Naito, S., Yokoo, Y., and Fujimaki, M. (1980). Antioxidative activity of soybean protein hydrolysates in dried model food containing lard. *Nippon Shokuhin Kogyo Gakkaishi* **27,** 51.

Yano, M., Hayashi, T., and Namiki, M. (1974). Structures of the free radical products formed by the reaction of dehydroascorbic acid with amino acids. *Chem. Lett.* **10,** 1193.

Yano, M., Hayashi, T., and Namiki, M. (1976). Formation of free-radical products by the reaction of dehydroascorbic acid with amino acids. *J. Agric. Food Chem.* **24,** 815.

Yu, S. H., and Karel, M. (1978). Reaction of histidine with methyl linoleate: Characterization of the histidine degradation products. *J. Am. Oil Chem. Soc.* **55,** 352.

Yuichiro, T. (1972). Antioxidant activity of amine-carbonyl reaction products. 3. Antioxidant activity of the reaction products of various sugars or aldehydes with tryptophan. *Kagoshima Daigaku Nogakubu Gakujutsu Hokoku* **22,** 99.

6

Biochemistry of Food Processing: Brewing

I. Introduction

The production of beer, like that of wine, dates back over 5000 years. Today brewing still remains an important industry although the amount of beer produced worldwide has stabilized since 1981 at approximately 970 million hectoliters (Schildbach, 1986). A number of cereals can be malted to produce beer but it is the flavor imparted by malted barley that makes beer so acceptable. Of the worldwide production of barley only 10.4% is needed to meet the requirements of the brewing industry for raw brewing barley. The major barley and malt producers are Canada, the United States, Great Britain, France, Federal Republic of Germany, Denmark, Australia, and more recently Turkey. Brewing has undergone major technological changes over the past half century which have transformed it from an art into a well-regulated series of operations. Beer of consistent high quality can now be produced irrespective of the barley variety or adjunct used (Atkinson, 1987). Three major steps are involved in brewing: malting, mashing, and fermentation.

II. Malting

The primary step in brewing is malting, during which barley is allowed to germinate under carefully controlled conditions. This process involves steeping

TABLE 6.1

INFLUENCE OF THE BARLEY HUSK CONTENT ON MALT EXTRACT[a]

Husk content of barley (%)	Malt extract (%)
-10	80.7
10.1–11	80.2
11.1–13	78.6
12.1–13	78.1

[a] Adapted from Schildbach (1986).

the cleaned barley in a shallow bed of water at a temperature of 10–15°C. The time needed to accomplish this phase has been reduced over the past 25 years from 14 days to 4–5 days. A good-quality brewing barley must have high extraction yields as well as good proteolytic and cytolytic activities. This is provided by barley varieties which have been specially bred to meet the demands of the brewing industry. Table 6.1 illustrates the relationship between hull content of barley and malt extract. It is evident that the hull content should not exceed 11% in order to achieve an extract yield greater than 80%. Further improvements in malting quality are also associated with plumper grain size and a high germination capacity (Schildbach, 1986). These qualities are generally provided by 2-row barleys, although in more northerly European countries, Canada, and the United States multirowed varieties are primarily grown. These varieties are very rich in enzymes, making them particularly useful when using unmalted cereals.

A. Germination

The barley grain consists of a small embryo and a large storage tissue, the endosperm, surrounded by the husk (Figure 6. 1). Approximately 90% of the endosperm is starch localized in large dead cells packed with starch granules and storage proteins and surrounded by a living aleurone layer composed of small, thick-walled cells (Enari and Sopanen, 1986). The aleurone layer provides the enzymes, α-amylase, glucanases, and proteases, which are responsible for hydrolyzing the endosperm (Atkinson, 1987). These enzymes are released in response to a message sent out from the embryo in the form of a hormone, gibberellic acid. This hormone is normally sprayed on the steeped barley (0.1–0.2 ppm) to accelerate the germination process (Hudson, 1986).

The cell walls consist of a number of different nonstarch structural polysac-

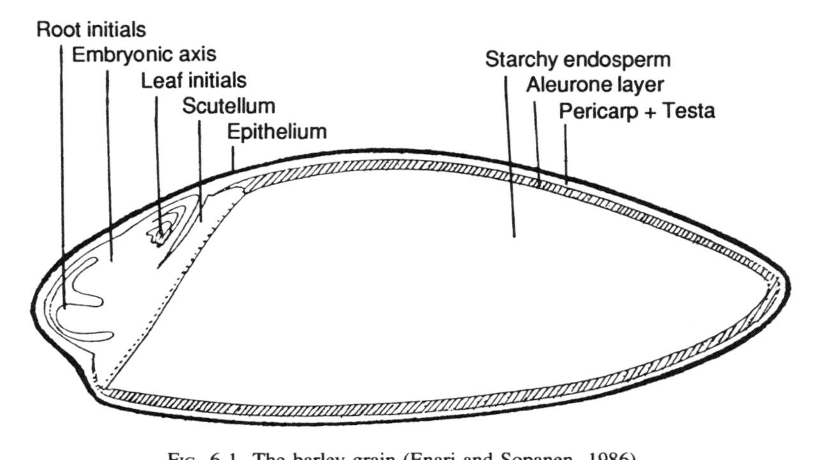

FIG. 6.1. The barley grain (Enari and Sopanen, 1986).

charides. The most prominent of these, β-glucan, a linear polymer of glucose joined by β (1 → 3) and β (1 → 4) linkages, accounts for 70% of the cell wall components together with arabinoxylan (15%). The latter consists of β (1 → 3) linked xylose chains to which are attached arabinose units. Several minor components identified in the cell wall include glucomannan (3%) as well as protein (5%) (Forrest and Wainwright, 1977; Thompson and LaBerge, 1981). The overall result is an insoluble matrix in which β-glucan complexes with protein.

For germination to occur the cell wall surrounding the endosperm must be degraded to allow access for the hydrolytic enzymes to attack the starch endosperm (Briggs and MacDonald, 1983; Fretzdorff *et al.*, 1982). The degradation of the cell wall during malting, referred to as modification, is crucial for the development of a good malt (Mundy *et al.*, 1983). A low modification of the endosperm tissue is undesirable as it produces a lower yield of extract as well as problems in wort separation, beer filtration, and clarity of the final beer product (Brunswick *et al.*, 1987). Thus malting is designed to restrict the development of the barley seedling while at the same time ensuring adequate modification of the endoosperm. Recent studies by Selvig *et al.* (1986) confirmed earlier work by Gram (1982b) that cell wall modification was initiated at the endosperm–embryo junction. Fluorescence microscopy using Calcofluor, which reacts specifically with β-glucan, indicated complete breakdown of the cell wall. A closer examination of the modified endosperm, however, using scanning and transmission electron microscopy, showed remnants of the cell wall remaining after attack by the cell wall-degrading enzymes. These remnants appeared to consist of the middle lamella, which is thought to be composed of pectinlike substances.

The degradation of β-glucans is particularly important as insoluble β-glucans are converted to soluble glucans by the glucanolytic enzymes. Soluble glucans are highly viscous and can lead to brewing problems such as slow filtration of beer unless adequately degraded (Enari and Sopanen, 1986).

B. β-Glucanases

The development of β-glucanases during barley germination is responsible for solubilizing the insoluble β-glucans. Several endo-β-glucanases have been reported in germinating barley grain, including endo-1, 3-β-D-glucanase (EC 3.2.1.39) and endo-1, 3; 1, 4-β-D-glucanase (EC 3.2.1.73) (Manners and Marshall, 1969; Wilson, 1972). Endo-1, 3-β-D-glucanase apparently hydrolyses any β-1, 3-glucan present at the aleurone/endosperm interface as well as some of the consecutive β-1, 3 linkages in the β-1, 3; 1, 4-D-glucans located in the cell walls (Brunswick *et al.*, 1987). Endo-1, 3; 1, 4-β-glucanase specifically hydrolyzes the cell wall β-1, 3;to 1, 4-D-glucans and was co-chromatographed with arabinoxylanase activity following isolation by molecular sieving, anion exchange and isoelectric focusing (Hall, 1978).

The mechanism of β-glucan degradation is complex, initially involving the hydrolysis of the ester linkage between β-glucan and protein by an acidic enzyme, β-glucan solubilase. This enzyme is thought to be a carboxypeptidase which releases the soluble form of β-glucan (Bamforth *et al.*, 1979). The soluble β-glucan is further degraded by at least three enzymes: endo-1, 3-glucanase, endo-1, 4-glucanase, and endo-1, 3; 1, 4-β-glucanase (Briggs, 1964; Manners and Wilson, 1974; Thompson and LaBerge, 1981). β-Glucan solubilase is synthesized during the initial germination phase followed by the other endo-β-glucanases (Bamforth and Martin, 1981; Bourne and Pierce, 1970; MacLeod *et al.*, 1964; Preece and Hoggan, 1956). The order of synthesis was consistent with the sequence of hydrolysis in which the final products were soluble in either cold or hot water and not precipitated in the beer.

C. Proteases

The major storage proteins in barley, the water-insoluble hordeins and glutelins, are found primarily in the starchy endosperm. These are degraded to amino acids during germination and used for new protein synthesis during the growth of the barley seedling. In addition, the protein matrix surrounding the starch granules must be degraded before the α-amylases can hydrolyze starch. Germinating barley contains two major groups of proteinases or endopeptidases: those containing sulfhydryl groups (SH) and those containing metallo-enzymes (Enari and Mikola, 1967; Enari *et al.*, 1968). In addition, there are a variety of

TABLE 6.2

PEPTIDASES IN BARLEY AND MALT[a]

Enzyme	pH optimum	Marker substrate
Carboxypeptidases		
I	5.2	Z-Phe-Ala[b]
II	5.6	Z-Ala-Arg
III	5	Z-Phe-Pe
IV	4.8	Z-Pro-Ala
V	4.8	Z-Gly-Pro-Ala
Neutral aminopeptidases		
I	7.2	Phe-β-NA
II	7.2	Arg-β-NA
III	7.2	Leu-β-NA
IV	7	Leu-β-NA
Alkaline peptidases		
Leucine aminopeptidase	8–10	Leu-Tyr
Dipeptidase	8.8	Ala-Gly

[a] Adapted from Enari and Sopanen (1986).
[b] Z = *N*-carbobenzoxy-; -β-NA = -naphthylamide.

peptidases or exopeptidases, including five carboxypeptidases, four neutral peptidases, and two peptidases with alkaline pH optima. These enzymes exhibit a broad range of pH optima and differ in the particular sequence of amino acids they will attack at the end of the protein chain (Table 6. 2). Only acid carboxypeptidases and proteinases with pH optima of 5 are found in the starchy endosperm, compared to the scutellum, which contains both carboxypeptidases and neutral or alkaline peptidases (Enari and Sopanen, 1986).

D. Protein Mobilization in Germinating Barley Seed

The mobilization of protein during the germination of barley can be divided into three phases: The initial phase involves the degradation of the protein bodies in the scutellum and aleurone layers involving both acid proteinases and carboxypeptidases (Gram, 1982a; Enari and Sopanen, 1986). The free amino acids and small peptides produced apparently enter the cytosol, where the peptides are further degraded by neutral and/or alkaline peptidases with the resultant amino acids used to synthesize the hydrolytic enzymes involved in starch hydrolysis.

During the second phase the proteins present in the starch endosperm are degraded. This was attributed initially to acid proteinases secreted by the scutellum followed by synthesis of proteinases in the aleurone layer in response to gibberellic acid from the embryo (Okamoto *et al.*, 1980; Jacobsen and Varner, 1967).

The final phase of protein mobilization involves the uptake of amino acids and small peptides by the scutellum, with the peptides hydrolyzed by neutral and/or alkaline peptidases (Mikola and Kolehmainen, 1972).

E. Lipids

The lipid content of barley varies from 0.8 to 4.8% on a dry weight basis. These differences were attributed in part to varietal and agronomic factors as well as to methodologies used to determine total lipid content (Ben-Tal, 1975; Bhatty and Rossnagel, 1978; Fedak and de la Roche, 1977; Hernandez *et al.*, 1967; Morrison, 1978; Welch, 1975). Anness (1984) calculated the lipid content of nine barley varieties, when measured as total fatty acids, to be 3.4 to 4.4% (34–44 mg fatty acids/g dry weight) on a dry weight basis. The higher levels obtained in this study were attributed to the milder conditions of hydrolysis (60°C for 1 hr with 6 *M* HCl) used prior to extraction and methylation of the fatty acids for analysis by gas chromatography. Since Price and Parsons (1975) showed that fatty acids accounted for at least 75% of the lipid in barley, Anness (1984) suggested his values might underestimate the total lipid content. The fatty acid composition was similar for all barley varieties and their corresponding malts with the average level of linoleic acid (C18 : 2), palmitic acid (C16 : 0), oleic acid (C18 : 1), linolenic acid (C18 : 3), and stearic acid (C18 : 0) being 58, 20, 12. 9, and 0.8%, respectively. During germination of barley over 30% of lipids (primarily triacyl glycerols) were lost after 5 days for Weah, a fast-germinating Australian variety, compared to 14% for Sonja, a slower-germinating seed. Neutral lipids and free fatty acids accounted for over 80% of the total fatty acids in barley and malt of both these varieties with the remaining constituents being phospholipids (8–10%) and glycolipids (5–8%). Two separate lipases were identified in barley by Baxter (1984), with the more abundant one located primarily in the embryo, while the second lipase was found in the endosperm. These increased rapidly during the course of germination as discussed in Chapter 3 (Fig. 3. 11).

F. Starch Degradation

Approximately two-thirds of the dry weight of the barley grain is starch, which undergoes hydrolysis to glucose during germination. The major polymer in barley grain starch is amylopectin, which consists of glucose units joined by α-1, 4 bonds interspersed by β-1, 6 branch points. The minor polymer, amylose, is a straight-chained glucose polymer as shown in Figure 6. 2. There appears to be a preference for hydrolyzing amylopectin compared to amylose. The enzymes

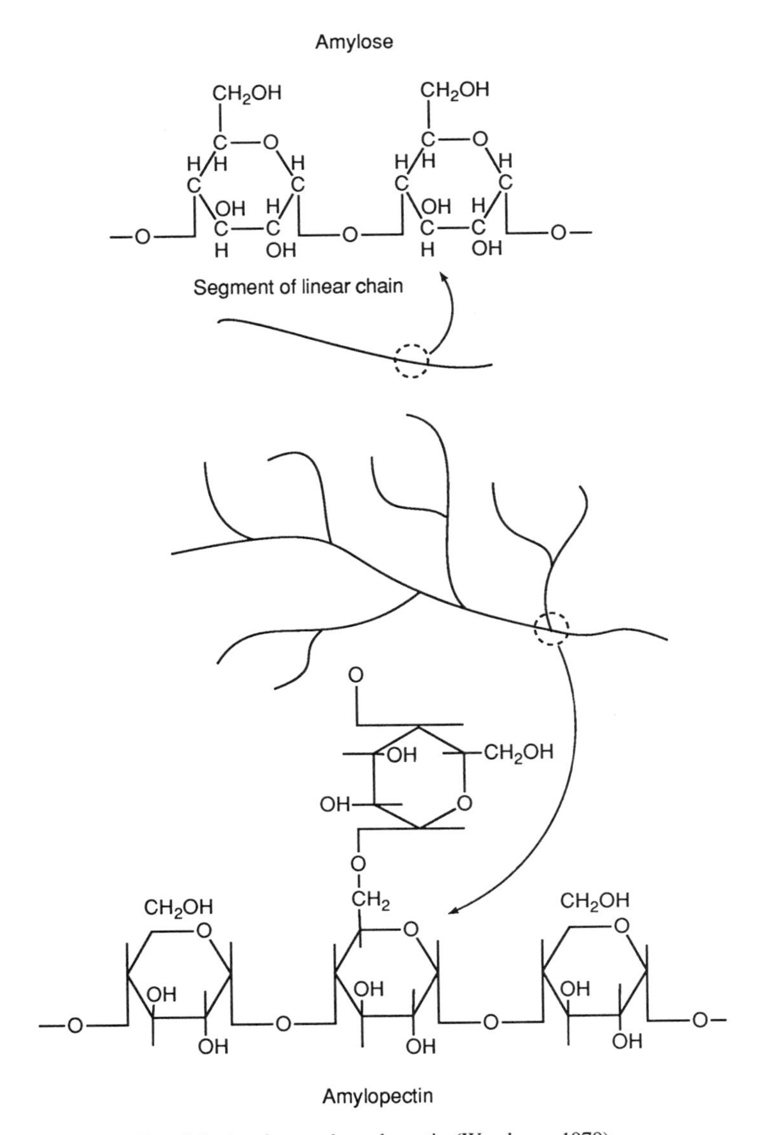

FIG. 6.2. Amylose and amylopectin (Wurzburg, 1970).

involved in degrading nongelatinized barley starch include phosphorylase, β-glucosidase, α- and β-amylases, and debranching enzymes. The most important of these enzymes for malting and brewing are α- and β-amylases.

G. α-Amylase

α-Amylase, a metallo-enzyme, attacks the starch molecule by randomly hydrolyzing any of the α-1,4 linkages. The terminal sugar remaining is in the α-configuration, which is the reason for naming this enzyme α-amylase. The only exceptions are any α-1, 4 linkages close to a branching point or the end of the starch molecule. This endo-enzyme is present in a number of isoenzyme forms in germinating barley (Grabar and Daussant, 1964). Several isoenzymes are recognized, including α-amylases I and II, each shown to be products from two separate gene families (Daussant *et al.*, 1974; Jacobsen and Higgins, 1982; MacGregor and Balance, 1980; Rogers, 1985). α-Amylase only appears during germination following its *de novo* synthesis triggered by gibberellic acid (Filner and Varner, 1967).

The hydrolysis of starch by α-amylase can be divided into two separate phases, an initial rapid dextrinization followed by a slower saccharification (Bathgate and Palmer, 1973; Greenwood and MacGregor, 1965). During dextrinization the linear amylose is randomly degraded to smaller dextrins while saccharification results in mainly maltose and glucose together with some short oligosaccharide polymers. Bertoft and Henriksnas (1982) monitored the initial stages of dextrinization in intact large and small barley granules compared with their respective gelatinized starches at 70°C. Based on molecular weight distribution using gel chromatography, the hydrolyzed products for gelatinized starches from either the large or small starch granules were almost identical. This suggested a similarity in the molecular structure of the individual amylose and amylopectin components for the two starches. Substantive differences in the molecular weight distribution of the products released during the α-amylolysis of the intact large and small starch granules were evident as seen in Figure 6.3. These differences were attributed to the way in which the components were arranged or packed in the different starch granules. Based on molecular weight distribution patterns, dextrinization of gelatinized starch produced a well-defined sequence of fractions which included *a* ($MW > 1 \times 10^6$), *b* ($1 \times 10^6 > MW > 1 \times 10^5$), *c* ($1 \times 10^5 > MW > 2 \times 10^4$), and *d* ($MW < 2 \times 10^4$). These were released in the sequence $a \rightarrow b \rightarrow c \rightarrow d$, in which *a* was decreasing steadily, *d* increasing steadily, while the intermediate fractions *b* and *c* increased initially to a maximum level and then declined. The formation of well-defined intermediate products suggested to these researchers that the dextrinization of starch was not a random process. This conflicted with work by Greenwood and MacGregor (1965), who showed quite clearly that the degradation of amylose by α-amylase from barley and barley malt was a random process. Bertoft and Henriksnas (1982), however, attributed the nonrandom hydrolysis of barley starch observed by Greenwood and MacGregor (1965) to the amylopectin fraction. This starch component was reported previously by Manners and Bathgate (1969) to be the

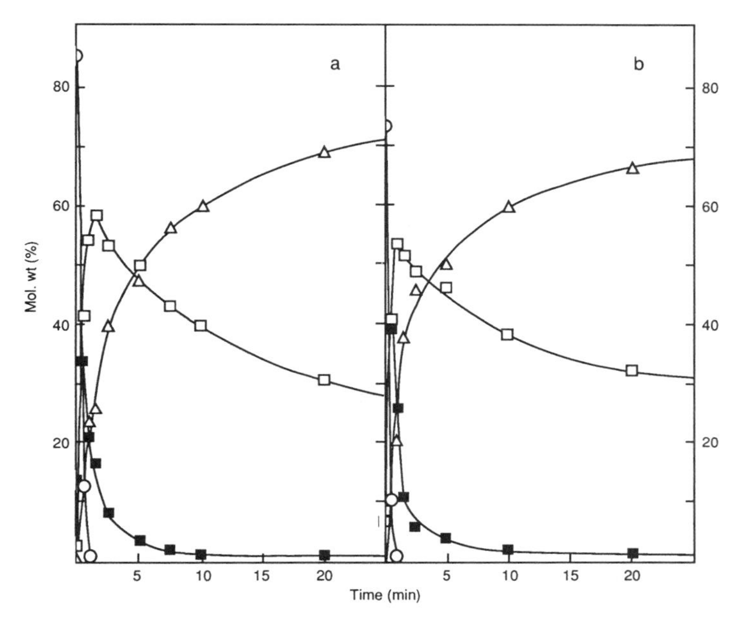

FIG. 6.3. Percentage distribution of molecular weight as a function of time for α-amylolysis at 70°C of large starch granules (a) and small starch granules (b) (Bertoft and Henrksnas, 1982). Fraction a = MW > 1 × 10^6 (○); b = 1 × 10^6 > MW > 1 × 10^5 (■); c = 1 × 10^5 > MW > 2 × 10^4 (□); d = MW < 2 × 10^4 (△).

major one affected during the degradation of the intact starch granule. On the basis of their results, Bertoft and Henriksnas (1982) suggested a highly ordered structure for amylopectin with well-defined glycosidic bonds preferentially attacked during the initial period of starch dextrinization. It was thought to be composed of individual cluster units of molecular weight 3 × 10^4, consistent with the cluster hypothesis of amylopectin proposed by Manners and Matheson (1981) and Robin *et al.* (1974). Several cluster models were proposed to explain the differences in mode of packing between the large and small starch granules. One such model, illustrated in Figure 6.4, shows how the amylopectin chains form in clusters in which the branch points are close together (Manners, 1985).

H. β-AMYLASE

β-Amylase, a thiol-containing exo-enzyme, attacks the starch molecule from the nonreducing end and liberates disaccharide β-maltose units. Maltose, a reducing disaccharide, is the main fermentable sugar of wort. At least four

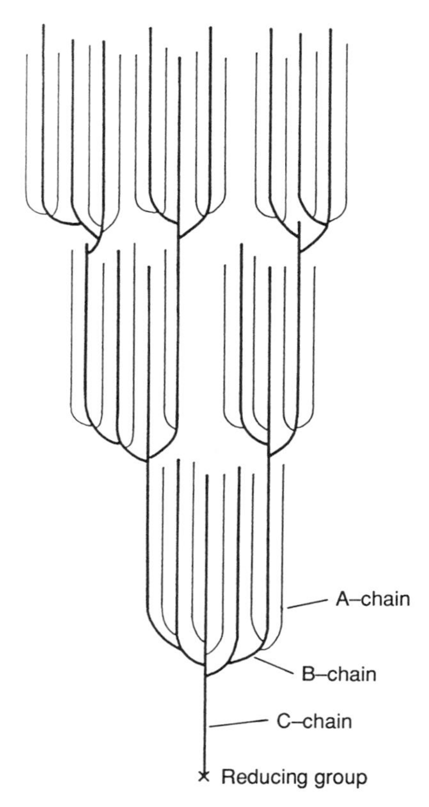

FIG. 6.4. The cluster model of an amylopectin molecule. The differences between A-, B-, and C-chains are illustrated (Eliasson *et al.*, 1987).

β-amylase components with similar antigenic properties have been identified in barley using gel chromatography (Daussant *et al.*, 1965). Subsequent researchers attributed the heterogeneity of this enzyme in part to the reactive monomer's ability to form polymers *via* their thiol groups (Nummi *et al.*, 1965; Visuri and Nummi, 1972). In contrast to α-amylase, β-amylase is synthesized in the starch endosperm during the development of the barley grain (Lauriere *et al.*, 1985). As the barley grain ripens, β-amylase becomes ineffective because it is bound to protein. During the course of germination, however, the bound enzyme is apparently released by endospermal proteinases resulting in an increase in the free and active forms of β-amylase (Figure 6.5) (Daussant and MacGregor, 1979).

The presence of debranching enzymes hydrolyzing β-1,6 linkages provides further substrate for amylolytic attack. Several debranching enzymes have been

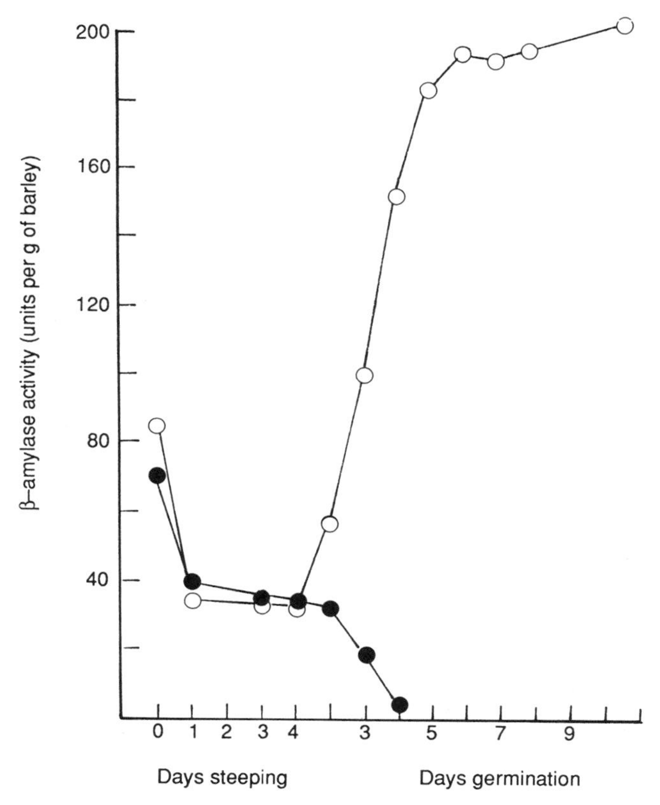

FIG. 6.5. Variation of bound (●) and free (○) β-amylase during malting (Daussant and McGregor, 1979).

reported in barley grain, one appears to be synthesized in a bound form during grain development and released during germination while the other is synthesized in the aleurone layer in response to gibberellic acid (Lauriere *et al.*, 1985; Hardie, 1976). The combined effect of α- and β-amylases is the production of maltose, which is further degraded to glucose by α-glucosidase, also secreted by the aleurone layer (Hardie, 1976).

I. KILNING

The final stage of malting involves arresting the germination process. This is achieved by simply drying out the malt corns by a process referred to as kilning, which reduces the moisture content of the malt from around 45% to 5%. To preserve the enzyme activity in the malt essential for brewing this is generally carried out in several stages. During the first stage of kilning an airflow of 50 to 60°C is used to reduce the moisture content of a bed of barley malt from approximately 60% to 23%. Under these conditions the enzymes are more heat stable so that the temperature can be increased stepwise to around 71°C and the airflow decreased to slowly reduce the moisture content to around 12%. In the final drying or curing step the airflow is reduced even further while the temperature is raised from 71 to 92°C until the required malt color and moisture content are obtained. This step takes at least 2–4 hr as the moisture remaining is in the bound form and therefore more difficult to remove. It is during this step that the typical malt aroma is produced, which is essential in the production of good beer. By changing the humidity, time, and temperature conditions during kilning the final color and aroma of the malt can be manipulated (Runkel, 1975). Lager malts, for example, are normally dried to a moisture content of 4–5% compared to 2–3% for ale malts (Hough, 1985).

III. The Brewing Process

Once the malted barley is dried by kilning and cleaned it is ready to proceed through the brewing process. This involves a number of steps, including crushing/milling the malted barley; preparation of the mash; separation of the aqueous extract or wort, boiling wort with hops; preparation of the wort for the fermentation process; maturation and clarification of the beer; with the final step being the bottling or packaging of the beer.

A. Milling of Malt

The main function of milling is to reduce the particle size in the malt to form a grist (ground or milled grain). The reduction in particle size facilitates the extraction of soluble components, mainly sugars and nitrogenous compounds, from the endosperm. Sufficient husk fragments must be retained, however, to form a permeable filter bed which will facilitate adequate wort separation from the mash. This can be achieved by using dry or wet milling processes common in most breweries. Crescenzi (1987) showed that such factors as modification, moisture, kilning, and abrasion could have a significant influence on the milling performance of the malt if they fell outside commercial malt specifications. Possible synergistic effects between these factors could also lead to significant differences in milling commercial malt. Crescenzi (1987) demonstrated the beneficial effect that steam conditioning of the malt had on milling, wort separation,

TABLE 6.3

EFFECT OF CONDITIONING ON WORT SEPARATION AND EXTRACT RECOVERY IN SINGLE-PASS MILLING[a]

Malt[b]	Moisture increase (%)	Run-off time (min)	Extract (liters/kg)	Extract recovery (%)
1	0	47.5	289.5	95.5
2	1.5	15	246.4	81.3
3	1.1	23.5	265.4	87.6
4	0.8	26.5	273.1	90.1
5	0.5	30	282.7	93.3
6	0.35	35.5	287.2	94.8

[a] From Crescenzi (1987).
[b] Sample 1: unconditioned malt; samples 2–6: steam-conditioned malts.

and extract recovery. The results, summarized in Table 6.3, clearly show that such conditioning produced a grist that varied with respect to both wort separation and extraction. The batch of malt used had a moisture content of 3–4%, typical of an ale type, portions of which were subjected to steam conditioning just prior to milling on a single-pass mill. A moisture content of 0.35% in malt no. 6 reduced the wort separation time compared to the unconditioned malt (1) without affecting the extract recovery. Increasing the moisture content decreased the wort separation time as well as extract recovery. Careful control of the moisture content as well as the mill gap settings is required to maximize the beneficial effects that steam conditioning could have on mashing/wort separation systems.

B. Preparation of Mash

Mashing, the first step in wort production, involves extracting soluble materials from the milled malt. This is accomplished by feeding the grist through a Steel's masher, a hydrator consisting of a large-bore tube bent at right angles (Figure 6.6). During its passage through the vertical portion of the tube the grist is sprayed with hot water (65°C) and then mixed with a revolving screw in the horizontal part of the tube (Briggs *et al.*, 1981). The aerated porridgelike material produced then falls into the tun, which is already heated and partially filled with hot water just above the filter plates. Under these conditions the floating endosperm particles hydrate and undergo further amylolytic attack by α- and β-amylases. These two enzymes exhibit differences in stability, with α-amylase being more thermostable and β-amylase exhibiting optimum activity at a higher pH (6.7). Processors will adjust the pH and temperature conditions to permit both

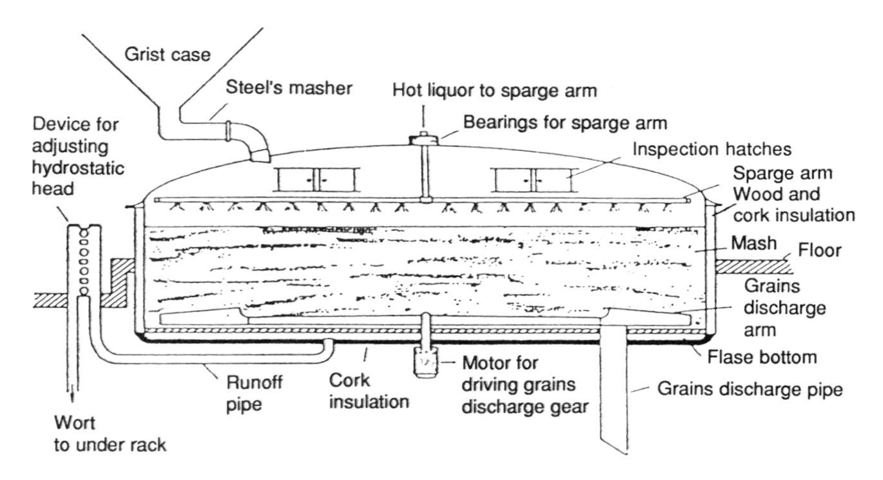

FIG. 6.6. Vertical section through an infusion mash tun equipped with Steel's masher (Hough, 1985).

enzymes to work effectively. This system, involving a single vessel in which the mash temperature is held constant, is referred to as infusion mashing.

The well-modified nature of American malts, which are rich in enzymes and nitrogenous compounds, however, necessitates the addition of adjuncts. These include maize or rice grits, which are first heated in cookers at 65°C to render their starch accessible to attack by the malt enzymes prior to boiling. The main mash is heated at 45°C to permit some protein and starch degradation with the temperature rising to 67°C following the addition of the contents of the cereal cooker. This process, known as the double mash system, is used by most breweries around the world.

Bertoft and Henriksnas (1983) examined the progress of starch hydrolysis during the mashing of three barley malt varieties. The changes in the molecular weight distribution of starch products over time during mashing are summarized in Figure 6.7. During the initial phase at 45°C, fraction *a*-II is unaffected with the exception of a minor change in Pomo, a malt with high enzyme activity. A definite degradation of fraction *a*-I was observed, however, in all the malt fractions, suggesting this phase was a continuation of the later stages of malting. In addition, fraction *b* was markedly degraded. As the temperature increased during the mashing process, fraction *a*-II underwent rapid hydrolysis, resulting in the immediate formation of large fractions *c* and *d* in the case of the enzyme-rich Pomo malt. A much more gradual production of these fractions was evident, however, for the other malts. As the gelatinization temperature of 70°C was reached very little high-molecular-weight material remained in any of the mashes. Similar yields of dextrins were found in fractions *c* and *d* for all three worts produced irrespective of enzyme activities of the starting barley malts. The

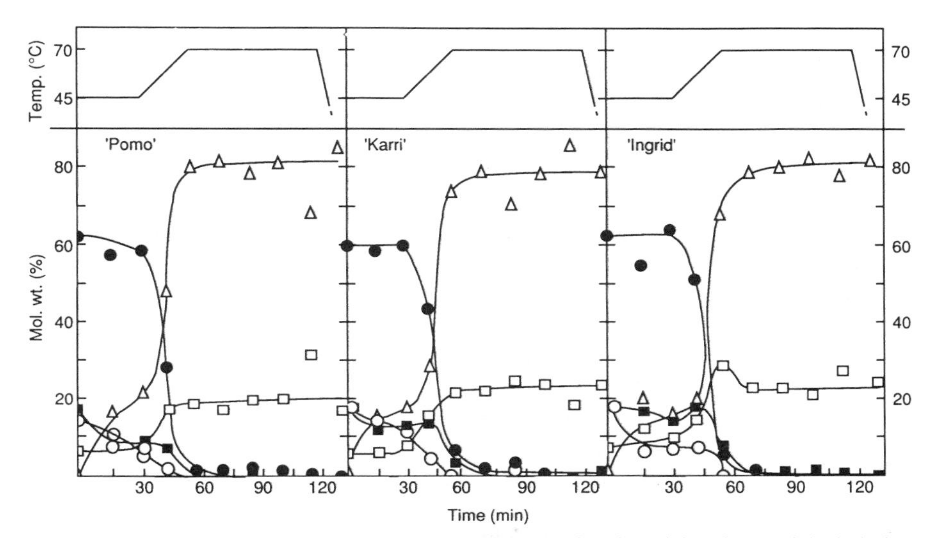

FIG. 6.7. Percentage distribution of molecular weight as a function of time for starch hydrolysis during mashing. Fraction *a*-II = MW $> 2 \times 10^6$ (●); fraction *a*-I = $2 \times 10^6 >$ MW $> 1 \times 10^6$ (○); fraction *b* = $1 \times 10^6 >$ MW $> 1 \times 10^5$ (■); fraction *c* = $1 \times 10^5 >$ MW $> 2 \times 10^4$ (□); fraction *d* = MW $< 2 \times 10^4$ (△). The temperature of the mashing program is indicated in the upper parts of the graphs (Bertoft and Henriksnas, 1983).

suggested sequence of reactions during mashing (shown below) appeared identical with that reported previously for the hydrolysis of gelatinized barley starch by purified α-amylase (Bertoft and Henriksas, 1982):

$$a \rightarrow b \rightarrow c \rightarrow d$$

C. Brewing Water

The major component of beer is water, which accounts for 95% of the fermented beverage. Thus the quality of the water used is a major factor affecting the quality of the beer. The availability of good water supplies has become increasingly rare because of the extensive pollution of both surface and underground waters by pesticides, herbicides, and industrial wastes. This has led to the development of strict water control standards instituted by breweries in which water is filtered through activated carbon as well as ion-exchange resins to remove these impurities. Two ions of particular importance in water are calcium and carbonate/bicarbonate, which control the pH during brewing. Calcium also protects α-amylase from heat destruction, thereby permitting liquefaction of starch during mashing. There are many other roles that calcium plays in the brewing process in the production of a stable and acceptable beer product. Thus breweries will adjust the level of calcium in their water as well as a variety of other ions by the addition of salt mixtures referred to as "Burtonizing the brewing water" (Stewart and Russell, 1985).

D. Adjuncts

The addition of cereal-based adjuncts in brewing provides benefits in both extract costs and beer qualities (Lloyd, 1986). In addition to being cheaper sources of extract than malt they can dilute a malt high in nitrogenous compounds and reduce the tendency for the beer to form a haze. Adjuncts can be classified based on whether the major ingredient is starch-rich or sugar-rich (Table 6.4). The total amount of adjunct presently used in the United States accounts for 35–40% with a ratio of 4 : 1 for starch-rich : sugar-rich. The most commonly used adjuncts are maize and rice grits, which have to be pregelatinized or cooked within a brewery. These are moistened to loosen the husks, which are then removed by steam treatment, and the endosperm is milled to degerm the grits. These grits are then used in brewery cookers particularly in North America.

E. Separation of the Aqueous Extract or Wort

Two systems are available for separating the wort produced during mashing, the lauter tun, and the mash filter. The most widely used system employed in

TABLE 6.4

Classification of Some Brewing Adjuncts[a]

	Starch-rich	Sugar-rich
Product form	Grits	Syrups
	Flakes	
	Raw cereals	Solid sugars
	Torrefied cereals	
	Flours	
	Roasted barley	
	Caramelized malts	
Agricultural source	Cereals	Cane sugar
	Root starches	Beet sugar
		Refined starches
		Malt
		Raw cereals

[a] From Lloyd (1986).

North American breweries is the lauter tun, although the mash filter is still used by many brewers as it is smaller and capable of handling the finer ground malt (Hough, 1985). The lauter tun consists of a vertical cylinder with a large diameter to depth ratio (Stewart and Russell, 1985). This vessel is normally rinsed thoroughly with a sparging or hot water delivery system before receiving the mash, which sinks in the lauter tun and comes to rest on a very flat floor of slotted stainless-steel or brass plates. At the center of the tun is a lautering machine, which is a shaft containing arms to which are attached rakes. These are rotated to facilitate draining of the wort in a collecting vessel called a grant, leaving behind the spent grain. The wort is recirculated through the lauter tun until it has attained a certain degree of clarity, whereupon it is diverted to the kettle.

F. Boiling of the Wort

This is a fairly straightforward procedure during which wort is boiled for up to 2 hr at atmospheric pressure following addition of hops (Miedaner, 1986). The quality of the beer produced, however, can be seriously affected by any changes in boiling technology such as the shape of the copper, boiling time, and temperature. Current energy costs have led brewers to improve the energy efficiency of wort boiling by incorporating modern exchange systems to achieve greater heat recovery. The recent introduction of pressurized boiling in high-temperature wort boiling plants has shortened the wort boiling time to 4 min at 140°C. The most common plant design involves heating the wort from 70°C to 130–140°C. The final temperature is attained through the use of live steam. The reduction in boiling time together with heat exchangers makes this technology energy efficient. The major objectives of wort boiling are:

1. Sterilization of wort and enzyme inactivation.
2. Extraction of bitter and other substances from hops and formation of flavor compounds.
3. Evaporation of excess water and concentration of wort.
4. Evaporation of undesirable flavor volatiles.

1. *Sterilization of Wort and Enzyme Inactivation*

The enzymes in the malt are inactivated to stabilize the wort composition. Thus any residual enzyme activity remaining after separation of the mash is destroyed together with any microorganisms present. High-molecular-weight proteins readily coagulate during wort boiling to an insoluble precipitate and are eliminated as "hot break" or "trub" (Hough *et al.*, 1982; Narziss, 1985). Protein coagulation is thought to involve chemical denaturation followed by colloid-physical coagulation of the denatured proteins (Miedaner, 1986). Their removal is important in terms of the taste and colloidal stability of the finished beer product. The level of coagulable nitrogen is reduced from 35–70 ppm in the unboiled wort to 15–25 ppm during boiling.

2. *Extraction of Bitter and Other Substances from Hops and Formation of Flavor Compounds*

The addition of hops during the wort boiling procedure plays a critical role in the development of bitterness in beer. The bitter taste and flavor imparted by hops became popular only in the nineteenth century but has since become an essential ingredient in the brewing process. Hops (*Humulus lupulus*) are flowering plants grown in many countries around the world although restricted to more temperate climates. It is the female cones of the flowers that are harvested because of the microscopic lupulin glands scattered at the base of the flowers or bracteoles. The harvested cones, once separated from leaves and other debris, are generally air-dried to around 7% moisture (Hough, 1985). These are graded according to appearance, aroma, and bitter resins on account of their contribution to the bitterness and flavor in beer. Since the α-acids in the hops largely contribute to the bitterness in beer it has been used in the following equation to calculate the utilization of potential bitter substances (Hudson, 1983):

$$\%\ \text{Utilization} = \frac{\text{Bitterness units in beer} \times 100}{\text{Concentration of } \alpha\text{-acids added to wort}}$$

A brief discussion of the chemistry of hops is necessary to more fully understand its importance in the brewing process.

The chemical composition of air-dried hops is summarized in Table 6.5 in order of decreasing importance to the production of beer. The major resins are referred to as soft resins because of their solubility in light petroleum ether. The major components of soft resins are α- and β-acids, which together are responsible for the formation of bitter compounds in beer. It is now generally recognized that the α-acids are primarily responsible for bitterness in beer. The α-acids or humulones are a mixture of homologues and analogues composed of six-membered rings differing only in the side chain at carbon-2 (Figure 6.8). The different members of these α-acids, based on the different side chains, and their relative proportions are shown in Table 6.6. These bitter compounds are chemically transformed into the corresponding iso-α-acids or isohumulones during wort boiling. Isohumulones are even more bitter than the original α-acids and form the bitter substances in beer. The mechanism involves isomerization to 5–membered ring compounds with the same side chains, the structures of which are shown in Figure 6.9. This reaction is important in understanding hop chemistry and its contribution to beer quality. A recent study by Verzele and Van de Velde

TABLE 6.5

CHEMICAL COMPOSITION OF AIR-DRIED HOPS[a]

Component	Amount (%)	Relative importance
α-Acids	2–12	XXX
β-Acids	1–10	XX
Essential oils	0.5–1.5	XX
Polyphenols	2–5	XX
Oil and fatty acids	Traces–25	X
Wax and steroids	—	X
Protein	15	
Cellulose	40–50	
Water	8–12	
Chlorophyll	—	
Pectins	2	
Salts (ash)	10	

[a] From Verzele (1986).

TABLE 6.6

CONSTITUENT α-ACIDS IN HOPS[a]

α-Acid	Side chain (R)		Amount (%)
Humulone	$-CH_2CH(CH_3)_2$	(isovaleryl)	35–70
Cohumulone	$-CH(CH_3)_2$	(isobutyryl)	20–55
Adhumulone	$-CH(CH_3)CH_2CH_3$	(2-methylbutyryl)	10–15
Prehumulone	$-CH_2CH_2CH(CH_3)_2$	(4-methylpentanyl)	1–10
Posthumulone	$-CH_2CH_3$	(ethyl)	1–5

[a] Adapted from Verzele (1986).

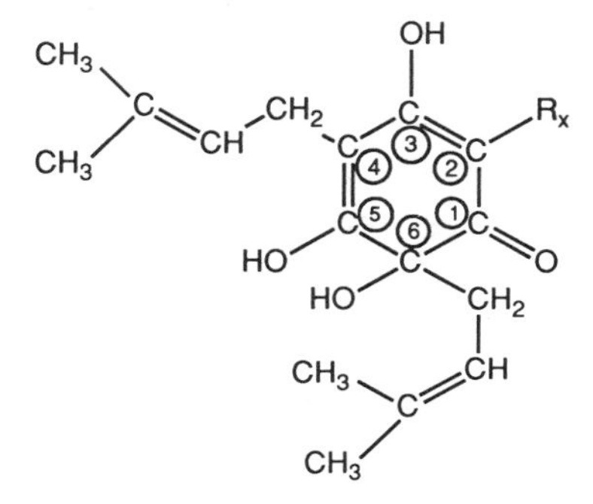

FIG. 6.8. Structure of α-acids or humulones.

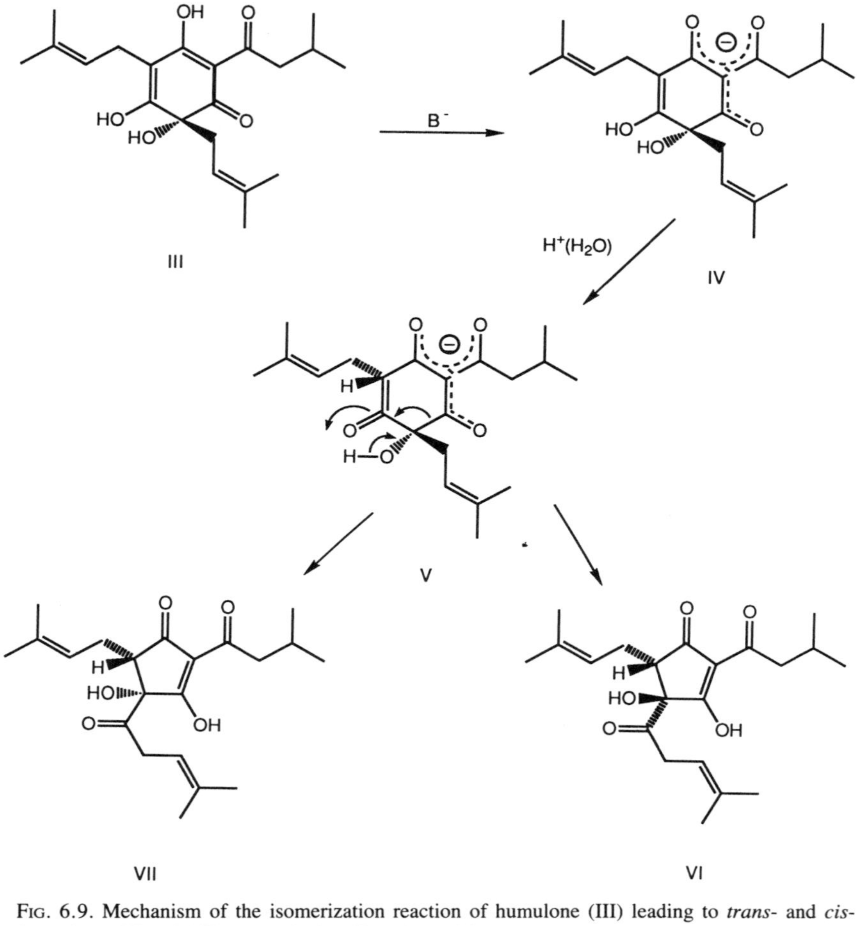

III

IV

V

VII

VI

FIG. 6.9. Mechanism of the isomerization reaction of humulone (III) leading to *trans*- and *cis*-isohumulone (VI and VII, respectively) (Verzele, 1986).

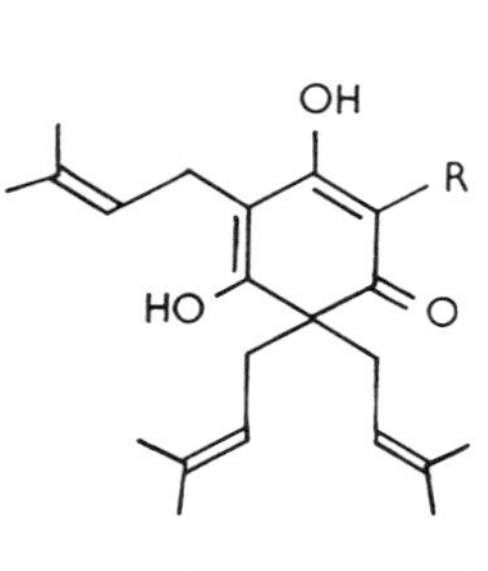

FIG. 6.10. Structure of β-acids.

TABLE 6.7

CONSTITUENT β-ACIDS IN HOPS[a]

β-Acid	Side chain (%)	Amount (%)
Lupulone	$-CH_2CH(CH_3)_2$	30–55
Colupulone	$-CH(CH_3)_2$	20–55
Adlupulone	$-CH(CH_3)CH_2CH_3$	5–10
Prelupulone	$-CH_2CH_2CH(CH_3)_2$	1–3
Postlupulone	$-CH_2CH_3$	?

[a] Adapted from Verzele (1986).

(1987) identified several bicyclic isomerization products (I,II) of humulone and cohumulone in an iso-octane extract of beer using HPLC. These hop-derived bitter acids accounted for several micrograms per milliliter in the beer examined:

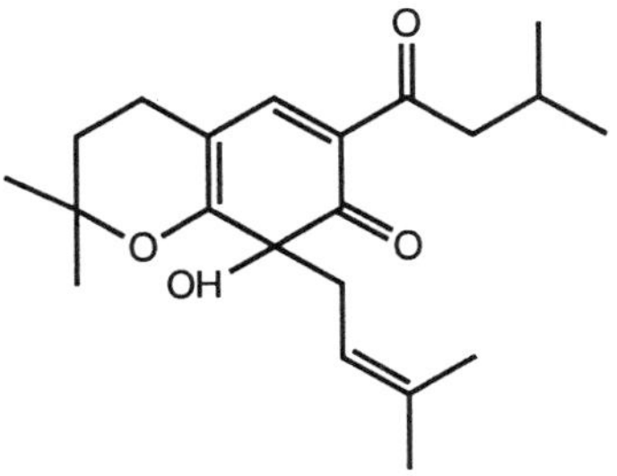

I 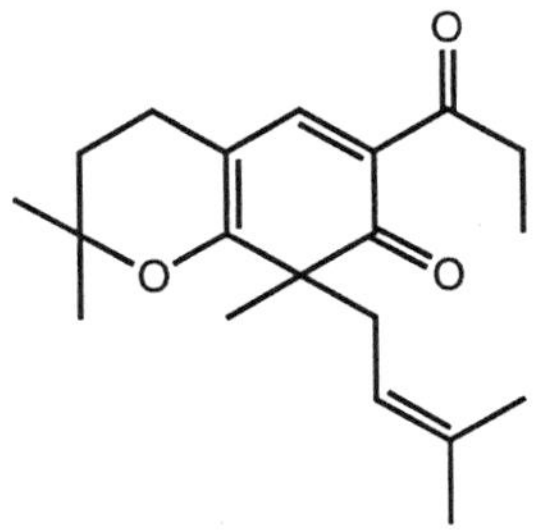 II

The β-acids or lupulones include a similar family of 6-membered ring compounds having identical side chains as the corresponding α-acids (Figure 6.10), however, their contribution to beer bitterness is less important. The members of the family are summarized in Table 6.7. Unlike α-acids, β-acids tend to oxidize during wort boiling, producing a variety of bitter and nonbitter derivatives (Hough, 1985).

a. Essential Oils. The major terpene hydrocarbons in the essential oils of hops are myrcene, humulene, and caryophyllene, which together account for 80–90% of total oil:

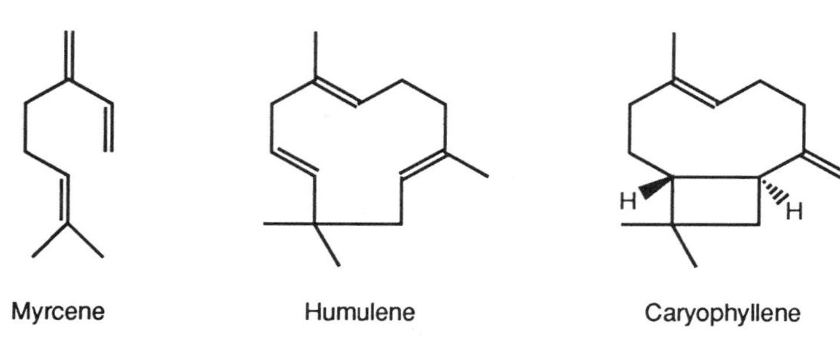

Myrcene Humulene Caryophyllene

Although these components are quite labile, how they influence the aroma of beer is not clear. Peacock and Deinzer (1981) isolated a number of humulene oxidation products in hop oil thought to be responsible for hop aroma using gas chromatography. Those identified included humuladienone, 2-humulene epoxides, and decenoic and decadienoic methyl esters, none of which in the pure state had any of the hoppy aromas. The addition of cold hops to top fermentation beers does impart a strong hoppy flavor if prepared under extremely mild conditions as in the case of vacuum-steam distillation (Picket *et al.*, 1981). Thus the individual contribution by these compounds remains unclear unless there is a combined effect of each compound present at the subthreshold level (Sandra and Verzele, 1975).

b. Polyphenols. Polyphenols play a role in the flavor and color of beer. Hops contain a complex mixture of polyphenols including flavanols and flavanediols (Tressl *et al.*, 1975). Although sparingly soluble, some of these polyphenols pass into the wort and are still present in the final beer product. The role which these compounds play in haze formation of beer will be discussed later in this chapter. During the course of wort boiling these polyphenols form red- to brown-colored anthocyanidins referred to as phlobaphens.

c. Oil and Waxes. Unseeded hops are essentially oil-free but do contain around 0.1% free fatty acids. These are mainly in the form of linoleic and linolenic acids, some of which may be transferred to the beer (Verzele, 1986). The level of free fatty acids found in beer, however, is around 1 ppm, most of which is released during the metabolism of malt and yeast. Studies reported by Narziss (1986), however, showed that hopped worts were considerably lower in fatty acids compared to worts without hops added. It is the oxidation products of these fatty acids, however, such as nonenal and nonadienal, that produce off-flavors. These components have low thresholds and are responsible for the poor flavor stability of beer. For example, *trans*-2-nonenal, which is responsible for the "cardboard flavor," has a flavor threshold of 0.5–0.1 ppb (Jamieson *et al.*, 1969). The precursor of *trans*-2-nonenal is 9, 10, 13-trihydroxy-11-octadecenoic acid (Drost *et al.*, 1971). It appears to be present in three isomeric forms all of which are initial oxidation products of linoleic acid (Esterbauer and Schauerstein, 1977). A rapid capillary gas chromatographic method developed by Verzele *et al.* (1987) for measuring these trihydroxyoctadecenoic acids found that the levels of these acids (measured as the total of the three isomeric forms) ranged from 12.37 to 16.04 ppm. This was considerably higher than the levels of corresponding free fatty acids (0.2–1.2 ppm). The presence of wax in hops, however, does not appear to be involved in the brewing process.

d. Extraction of Hops. Dried hop cones are traditionally added to the wort in their natural state but suffer from oxidative deterioration during storage. Consequently many hops can be either pelleted or solvent extracted and packaged in a form which is relatively stable during storage. In the case of hop extracts, organic solvents such as methylene chloride or ethanol are used to extract resins and essential oils. It is almost impossible to remove the extractant totally so there is always a solvent residue remaining in the extract. The use of liquid carbon dioxide to extract the brewing constituents of hops was first investigated about 10 years ago. This process, operated at a pressure of 800 psig and a temperature range of -5 to 20°C, involved a column of ground hops through which a stream of liquid carbon dioxide was allowed to percolate upward and enter an evaporator (Figure 6.11). By venting the carbon dioxide a residue is obtained consisting primarily of α-acids, β-acids, and essential hop oil (Tidbury, 1986). Carbon dioxide can also be used to extract hops at supercritical temperatures which involve very high pressures. The product obtained with supercritical carbon dioxide was much poorer compared to that from extraction with liquid carbon dioxide, as under the more extreme conditions contaminants such as tannins, polyphenols, and hard resin were present (Atkinson, 1987). Daoud and Kusinski (1986) examined the time course for the extraction of α-acids using liquid carbon dioxide. Essentially two zones were identified where the extraction rates differed substantially. In the first zone, the rate of extraction was dependent on the flow rate of the liquid carbon dioxide, while the extraction time appeared to be the major factor affecting extraction rate. These researchers noted that the main criterion for satisfactory extraction performance was to ensure that the flow rate of liquid carbon dioxide was above that required for the specified hop processing rate. Once this was established it was possible to calculate the hop processing rate attainable, based on the α-acid content of the hops, using the liquid carbon dioxide flow rate available. A number of commercial plants using liquid carbon dioxide are presently operating in Australia, Britain, and Germany for preparing hop extracts (Tidbury, 1986). This process avoids the problems of organic solvent residue as carbon dioxide is itself a product of fermentation.

e. Evaporation of Excess Water and Volatiles. The boiling of wort not only concentrates the wort but also eliminates any unwanted volatiles produced at this

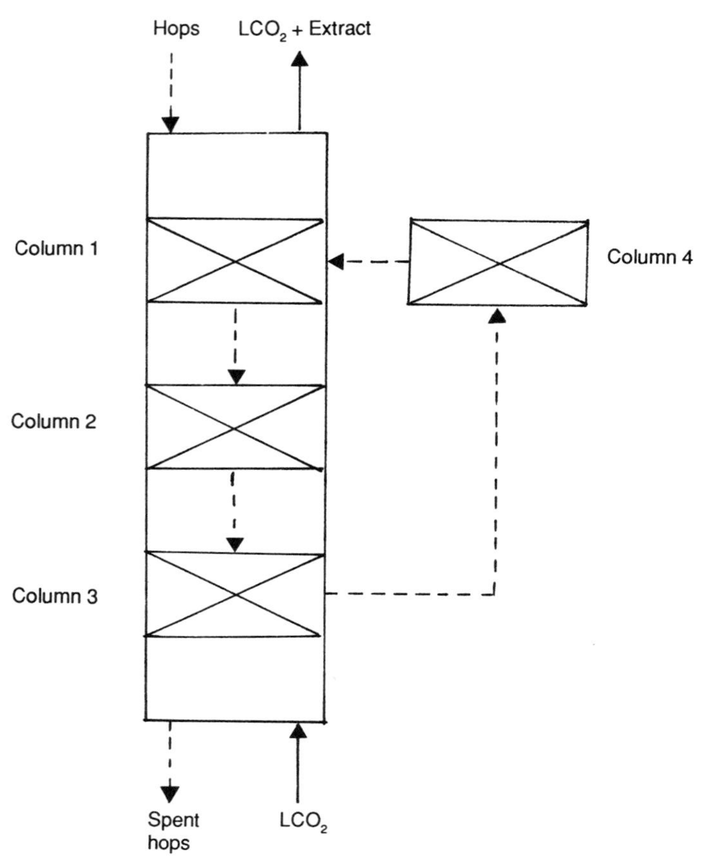

FIG. 6.11. Diagrammatic representation of liquid carbon dioxide hop extraction plant (Daoud and Kusinski, 1986).

stage of brewing. In addition to the more traditional method of wort boiling carried out at 90–120 min under normal atmospheric conditions, the use of boiling under partial vacuum is now employed by many breweries (Miedaner, 1986). The major energy costs incurred during wort boiling are the evaporation of the wort volume. It is possible, however, to reduce total evaporation from 15 to 2% of the wort volume without affecting the flavor. Nevertheless, most breweries still prefer to maintain around 7% evaporation during wort boiling to minimize energy costs while still maintaining beer quality.

The volatiles produced during wort boiling are derived from the malt, hop oils, or from the interaction of the different constituents present in the wort (Miedaner, 1986). The removal of most of the steam-volatile hop oils during wort boiling ensures that the flavor contributed by these oils is not too strong in the final beer product (Stewart and Russell, 1985). Once wort boiling is completed, the wort is clarified using a whirlpool tank to remove the spent hops and trub. The wort is fed tangentially into a rotating vertical cylindrical vessel which deposits the suspended solids into a compact cone located at the center of its base. This permits the wort to be transferred to the cooler and then into the fermentation vessel (Tidbury, 1986).

G. Fermentation

Fermentation is concerned primarily with the production of ethanol by the action of the fermenting yeasts. The availability of yeast genetics for the production of brewing strains with superior amylolytic activity or anticontaminating activity for use in the brewing industry is discussed in Chapter 11. The biochemical process involved in fermentation is glycolysis mediated through the Embden–Meyerhoff–Parnas pathway outlined in Figure 6.12.

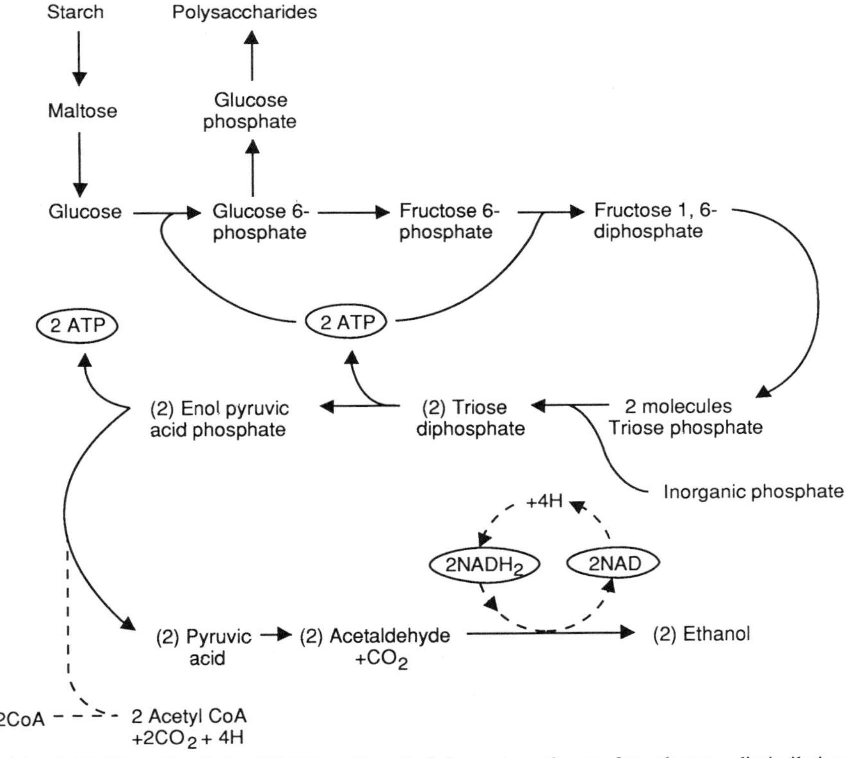

Fig. 6.12. The glycolytic (Embden–Meyerhof–Parnas) pathway for glucose dissimilation (Hough, 1985).

At one time different yeasts were employed in the production of lagers or ales. For instance, lagers were produced using bottom-fermenting yeasts, *Saccharomyces uvarum (Carlsbergenesis)*, at a fermentation temperature ranging between 7 and 15°C. These flocculate and remain at the bottom of the fermenting tank at the conclusion of the fermentation period, where they can be collected or cropped. This method was quite distinct from the production of ales, which used top-fermenting yeasts, *Saccharomyces cerevisiae*, at a higher fermentation temperature range of 18 to 22°C. In this case the yeasts accumulate on the surface of the fermenting wort where they can be skimmed (Stewart and Russell, 1985; Atkinson, 1987). Lagers tend to be darker in color as well as less bitter than ales, which are much paler in color. The development of ales with *Saccharomyces cerevisiae* was found to utilize wort more effectively than top-fermenting yeasts in addition to reducing residual sweetness (Tidbury, 1986). Lagers and ales are no longer separated on the basis of top- or bottom-fermenting yeasts as the modern brewery fermenters consist of vertical cylinders with a conical base and centrifuges. These permit the use of either yeast in the production of ales or lagers as the nonflocculant (top-fermenting) yeasts can be separated from the fermenting medium by centrifugation prior to sedimenting. The bottom-fermenting yeasts, which settle at the bottom of the fermenter, can be removed for further use in fermentation systems. The improved efficiency provided by these cylindrico-conical vessels has reduced the period of fermentation for ales and lager from 7 to 2–3 days and from 10 to 7 days, respectively (Atkinson, 1987).

1. *Flavor Compounds*

a. Alcohols and Esters. During fermentation a large number of alcohols, in addition to ethanol, are formed that affect beer flavor. These are produced from wort carbohydrates via oxo-acids or from transamination and deamination of amino acids in wort. They include both the aromatic alcohol 2-phenylethanol as well as such aliphatic alcohols as butanol, propanol, and hexanol.

The most important group of beer volatiles providing strong fruity flavors are esters. Their formation is associated with lipid metabolism by yeast in which acetyl-CoA reacts enzymatically with different alcohols, as shown here for ethyl acetate:

$$\underset{\text{Acetyl-CoA}}{CH_3COCoA} + \underset{\text{Ethanol}}{CH_3CH_2OH} \rightarrow \underset{\text{Ethyl acetate}}{CH_3COOCH_3CH_2OH} + \underset{\text{Coenzyme A}}{CoA}$$

The amount of esters formed during fermentation depends on the yeast strain, fermentation temperature, and aeration of wort (Nordstrom, 1964).

b. Carbonyl Compounds. The most important carbonyl compounds formed during fermentation are diketones such as diacetyl and pentane-2,3-dione. The flavor attributed to both of these compounds has been described as "buttery," "honey-or toffee-like," or "butterscotch" (Stewart and Russell, 1985). These volatiles are formed from oxidative decarboxylation of oxaloacetate and acetohydroxybutyrate, intermediates in the biosynthesis of leucine and valine, both of which are excreted by yeast into the wort. Once formed, diacetyl and pentane-2, 3-dione can be reduced to the corresponding alcohols, acetoin and pentane-2, 3-diol, by yeast reductases. The presence of diacetyl in lager above a threshold level of 0.10–0.14 ppm can produce a flavor defect. This can arise in the absence of or inability of the yeast to reduce diacetyl to the corresponding alcohol, which is relatively flavorless. Diacetyl can be reduced to an acceptable level in beer using acetolactate decarboxylase from *Enterobacter aerogenes* which converts it to acetoin (Stewart and Russell, 1985).

c. Sulfur Compounds. Volatile sulfur compounds originate from sulfur-containing amino acids or proteins present in the wort. These compounds elicit strong flavors and aromas in the beer and include hydrogen sulfide and dimethyl sulfide, formed during active yeast growth. Dimethyl sulfide (DMS) is a thioether which contributes to the flavor and aroma of such food products as tea, cocoa, milk, wines, and cooked vegetables (Keevil *et al.*, 1979; Lopez and Quesnel, 1976; Ronald and Thomson, 1964; Loubser and du Plessis, 1976).

$$H_3C{-}S{-}CH_3$$

Dimethyl sulfide (DMS)

The flavor of lager beers has been attributed in large part to the presence of dimethyl sulfide (White, 1977; Dickenson, 1983). At concentrations above its flavor threshold (0.03 ppm) it imparts a desirable aroma and taste to the beer, whereas above 0.1 ppm it elicits an undesirable "cooked sweet-corn" or "black currant-like" flavor (Anderson *et al.*, 1975). The precursors of dimethyl sulfide are *S*-methyl-methionine and (SMM) and dimethyl sulfoxide (DMSO), both of which are present in the malt.

The formation of dimethyl sulfide during brewing involves its conversion from dimethyl sulfoxide by yeast or through the heat degradation of *S*-methyl-methionine (Anness and Bamforth, 1982). The enzyme involved in the reduction of dimethyl sulfoxide to dimethyl sulfide in yeast appears to be methionine sulfoxide reductase (Bamforth, 1980; Bamforth and Anness, 1981). Purification of this enzyme from yeast showed it to consist of three proteins. In addition to the methionine sulfoxide-reducing protein are thioredoxin and thioredoxin reductase, which facilitate the transfer of electrons from NADPH to the methionine reductase protein. The sequence of reactions involved is shown in Scheme 6.1. The relative contribution of this pathway to the formation of dimethyl sulfide in beer is unclear as a result of conflicting reports from different breweries (Dickenson and Anderson, 1981; Booer and Wilson, 1979). This discrepancy appears to

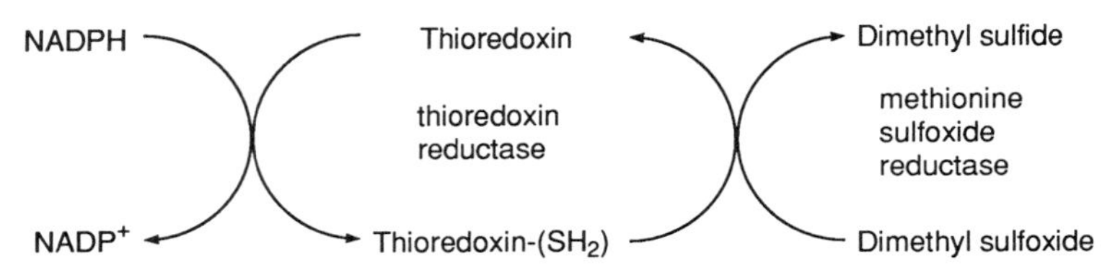

SCHEME 6.1. Enzymatic reduction of dimethyl sulfoxide by yeast.

reflect the amounts of dimethyl sulfoxide in the wort. Yeast cells (*S. cerevisae* NCYC 40), for example, reduced 2–5% of this compound in worts with concentrations of dimethyl sulfoxide ranging from 0.22 to 0.60 ppm.

The thermal degradation of *S*-methyl-methionine to dimethyl sulfide occurs during kilning, although most of it is lost with the exhaust gases. This reaction can be controlled by ensuring that the temperature used for well-modified malts (which generally contain high levels of *S*-methyl-methionine) never exceeds 65°C (Anness and Bamforth, 1982). Thermal decomposition of *S*-methyl-methionine during wort boiling can result in significant levels of dimethyl sulfide in beer. This process was shown to be a first-order reaction in which the half-life at pH 5.4 was 35 min (Wilson and Booer, 1979; Dickenson, 1979).

H. Continuous versus Batch Fermentation Systems

The ultimate goal of fermentation technology is the development of a continuous system. This has not been achieved so far because of many technical and commercial problems encountered compared to the batch system. The only continuous system at present is in the brewing industry in New Zealand (Atkinson, 1987).

IV. Maturation and Clarification of Beer

Maturation includes all those changes occurring between the end of primary fermentation to the final filtration of beer (Masschelein, 1986). Ale is matured at relatively warm temperatures of 12–20°C while lagers are held under much cooler conditions. The warmer temperatures under which ale is matured allow for the rapid metabolism of any residual and priming sugars as well as loss of green flavors within 1 to 2 weeks depending on beer type, yeast strain, wort composition, and primary fermentation conditions. In the case of lager beer, and Canadian ales, the beer used to be held at refrigerated temperatures for up to several months after fermentation (Stewart and Russell, 1985). This process, referred to as "aging," "storing," or "lagering," now only involves several weeks. The flavor of lager beer is affected by the length of the low-temperature

(2–6°C) maturing process as well as the amount of yeast and fermentable sugar present in the beer. Lagers are normally stored for 2 to 4 weeks during which time the flavor matures. Other factors such as the stability of the beer are also affected during this period, particularly haze formation.

A. Clarification of Beer

Beer produced during fermentation is quite turbid and must be clarified before it can be marketed. This turbidity is due to the presence of yeasts and proteinaceous materials associated with carbohydrates and polyphenols (Morris, 1986). The formation of these protein precipitates results from the cold temperature, low pH (4.2), and poor solubility in alcoholic solutions (Stewart and Russell, 1985). As discussed earlier, most of the yeasts remaining at the end of the fermentation period can be readily removed by centrifugation. Nevertheless this technique could result in the concentration of very fine haze particles which are difficult to filter. These can be removed with fining agents both in a prefiltration step as well as in sedimenting the beer in cask (Morris, 1986). The presence of a wide spectrum of charged species in wort and beer has led to the use of a variety of finings. For instance, copper and auxiliary finings are negatively charged polysaccharides which interact with the positively charged protein present. Negatively charged polyphenolic compounds and yeast, on the other hand, can only be removed by interacting with isinglass, which is composed of a positively charged high-molecular-weight protein collagen. Copper finings prevent the continued suspension of much of this material, present in wort as cold break, from remaining in the beer. Morris (1986) reported that while copper finings substantially reduced the amount of material suspended, a more rapid clarification as well as smaller sedimentation volume was obtained using a smaller amount of a mixture of auxiliary and isinglass finings.

1. *Polyphenols and Haze Formation*

Haze formation is well documented in the brewing literature and involves the reaction between malt proteins and malt and hop polyphenols. The polyphenols are present in the malt and hops as catechins and proanthocyanidins (Gardner and McGuiness, 1977).

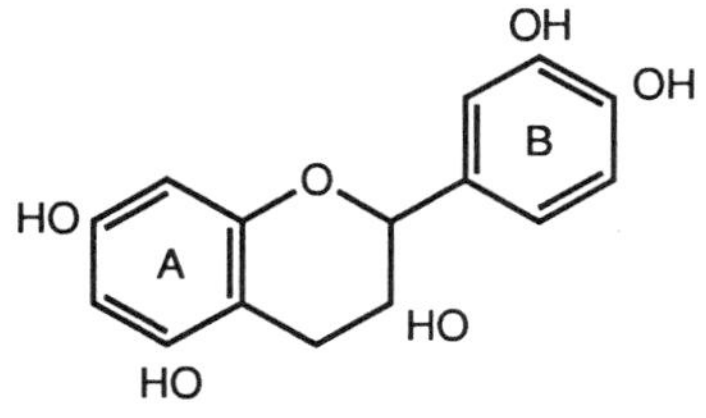

Catechin

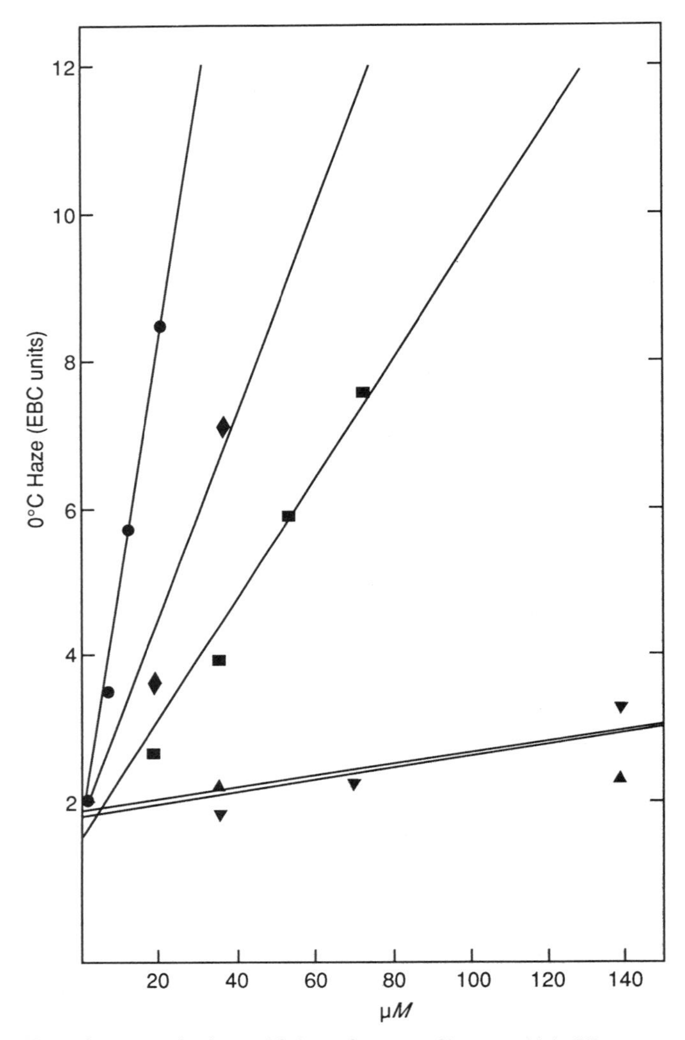

FIG. 6.13. Haze after pasteurization and 2 days of storage of beer to which different concentrations of flavanoids were added. (◆) (−)-Epicatechin; (▼) (+)-catechin; (■) procyanidin B3; (▲) procyanidin B6; (●) trimeric and tetrameric procyanidins (Delcour *et al.*, 1984).

The formation of this colloidal haze is affected by temperature as well as the presence of metal ions and oxygen. While the addition of simple anthocyanogens can hasten the onset of haze, the effect by polymeric polyphenols is immediate. Intensive research by the Carlsberg Research Centre resulted in the isolation of over 500 barley mutants which were proanthocyanidin-free (Erdal, 1986). The development of proanthocyanidin-free barley enabled the production of a proanthocyanidin-free beer when fermented with tannin-free hop extracts. Delcour *et al.* (1984), for example, brewed an all-malt Pilsner beer using proanthocyanidin-free malt together with tannin-free hop extracts. Different levels of catechins and procyanidins were added to the bottled beer and haze formation was monitored with a Radiometer Haze meter. As seen in Figure 6.13, a linear increase in haze formation occurred with increasing levels of phenolics added. Increasing the molecular weight of polyphenolics produced a corresponding increase in haze formation with the smallest change for (−)-epicatechin and the largest for the trimeric and tetrameric procyanidins. On the basis of the levels of the different catechins reported in a number of beers by earlier researchers, Delcour and coworkers (1984) implicated trimeric catechins as the major contributor to haze formation in beer. Baxter *et al.* (1987) examined the malting performance of Galant barley, a low-anthocyanogen variety, in the production of beer. While the malt met commercial specifications, it developed excessive coloring during kilning. This necessitated lowering the kilning temperature to obtain the same color as the control malt made from the malting variety Triumph. The green beer was treated with a range of fining and filtration tests normally carried out in the production of cask-conditioned and chilled and filtered beer. The results in Table 6.8 indicate that fewer particles (<4 μm) were present in beer from Galant compared to Triumph prior to fining. The overall clarity of the fined beer, based on haze value, was much greater in Galant beer fined at 13°C. A marked reduction of chill haze was also observed in Galant beer when fined at 4°C compared to the control. This study demonstrated the improved chill haze stability of beer brewed from Galant beer due to the substantial reduction of nonbiological haze in the finished product.

Proteases and Chill Haze Reduction. The application of proteases for reducing chill haze was first described some 70 years ago by Wallerstein (1911). Proteases used today are sulfhydryl proteases such as papain, ficin, and bromelain, all obtained from tropical plants (see Chapter 11). The major one used, papain, while relatively cheap, has a number of disadvantages including hydrolysis of other proteins in beer. It decreases head retention or foam capacity of beer as it is thermostable and survives pasteurization and hydrolyzes the barley proteins responsible for this phenomenon (Gray *et al.*, 1963; Jones *et al.*, 1967). In addition there have been increased cases of allergic reactions to papain in response to its use in meat tenderizing. Nelson and Young (1986) examined extracellular proteases from 119 strains of yeasts for their potential chill-proofing activity. The most prolific protease producers were those secreted from the genus *Candida*. The extracellular protease secreted by *Candida olea* 148 appeared to have a number of advantages over papain, including total inactivation during pasteurization as well as little effect on beer head retention.

TABLE 6.8

RESULTS OF FININGS TEST CONDUCTED AT 13°C[a]

			Galant	Triumph
Unfined beer[b]				
Particles present ($\times 10^{-6}$ ml/ml)		1–4 μm	9	36
		4–12 μm	180	145
Haze value (EBC units)			>12	>12
Fined in 5-gallon cask[c]				
Haze value (EBC units)		4 hr	12	>12
		7 hr	2.25	6.4
	After	48 hr	0.80	1.50
Fined in 2-liter aspirators				
Haze value[c] (EBC units)	After	4 hr	5.4	5.5
		24 hr	1.0	1.55
Haze value (EBC units)[d]	After	4 hr	2.05	3.8
		24 hr	0.80	1.25

[a] Adapted from Baxter *et al.* (1987).
[b] Volume of particles as measured by a Coulter counter.
[c] Fined with 0.5 pints/barrel (0.175 liter/hecaliter) alginate auxiliary finings and 2.5 pints/barrel (0.875 liter/hecaliter) isinglass finings.
[d] Fined and 1 pint/barrel (0.35 liter/hecaliter) alginate auxiliary finings and 4 pints/barrel (1.4 liter/hecaliter) isinglass finings.

2. *Pasteurization of Beer*

Pasteurization is carried out to ensure the shelf life of the beer over a period of months. This was accomplished by the development of tunnel pasteurization at the end of the nineteenth century in which the beer bottle was subjected to 60°C for 20 min (Fricker, 1984). The metal can, with its superior heat transfer properties, was exploited by Del Vecchio *et al.* (1951) for packaging beer. Beer is currently subjected to flash pasteurization at 75°C for several seconds. It is important, however, to avoid overpasteurization as this can adversely affect beer flavor (Fricker, 1984). The plate heat exhanger remains the most critical item in the system. The chilled, filtered beer containing 2.2 volumes of carbon dioxide per volume of beer is pumped through the heat exchanger, where it undergoes preheating by the hot, pasteurized beer returning from the holding tube. On entering the holding tube it is then heated to pasteurization temperatures, after which it is cooled and passed to a presterilized buffer tank prior to packaging.

V. Bottling and Canning of Beer

Bottle-blowing machines were introduced around 1870 and produced very inexpensive bottles. The bottles used today in the production of beer are dark in color to ensure flavor stability. Small packaged beer, mostly bottle returnables, increased from 64–90% of the total packaged beer in the United States since World War II (Lowe and Elkin, 1986). The introduction of the beer can was reviewed by Johnstone (1986) in which he highlights the many remarkable developments in its design since being introduced successfully by the brewing industry in 1935. The major problem with the use of cans was to prevent the transfer of metal ions from the tin or steel. This resulted in the development of metallic flavors and haze or "metal turbidity." To avoid this, many linings were tried but it was not until the introduction of epoxy linings in the 1960s that these problems were overcome. The present aluminum beer can is lightweight, attractive, and easy to open but necessitates strict quality control procedures to maintain the highest standards. It is important to transfer beer aseptically following flash pasteurization to prevent microbial growth. Thus sterile filling methods have been developed, including hot filling, prefilling pasteurization, and sterile filtration.

VI. Flavor Stability of Beer

Flavor stability is critical for the development of a quality beer and has been a subject of great concern in the brewing industry. The development of good beer flavor is a complex process involving many different reactions, many occurring simultaneously. These include the Maillard reaction, involving the Strecker degradation reactions of amino acids during kilning, mashing, and wort boiling; oxidative degradation of isohumolones; oxidation of alcohols to aldehydes; autoxidation of fatty acids; enzymatic degradation of lipids; and secondary autoxidation of lipids (Narziss, 1986). The development of stale flavor during the storage of beer, which is attributed to the formation of aldehydes, remains a major problem for the brewing industry (Hashimoto and Kuroiwa, 1975; Hashimoto and Eshima, 1977). A number of measures can be imposed by the brewer to limit the development of those compounds responsible for the "aged" or "stale" flavor. The most important of these is to limit the amount of oxygen present as this could increase lipid oxidation reactions in the beer as well as lower the reducing substances present during wort production. Narziss (1986) stored bottled beer containing a low (0.27 mg O_2/liter) and high (5.4 mg 0_2/liter) oxygen level for a week at 70°C and compared them to a fresh bottle of beer with low oxygen content. The results using a taste panel showed that the original beer with low oxygen content was of excellent quality but deteriorated in quality during storage. Compared to the bottled beer with high oxygen content, however, the stored beer with low oxygen still retained some of the hoppy aroma of the corresponding fresh beer. In the case of the beer bottled with a high oxygen content, its flavor and aroma were totally changed as reflected by its inferior quality (Narziss, 1986). It was evident from subsequent work that limiting aeration during mashing, wort production, and lautering of beer as well as replacing copper or mild steel with stainless-steel equipment minimized flavor deterioration in beer. The use of brown bottles provides additional protection against the development of light-struck flavor in beer, which is attributed to the formation of 3-methyl-2-buten-1-thiol from hydrogen sulfide.

Lynch and Seo (1987) developed a method for monitoring the development of staling in beer based on headspace analysis of ethylene. Unlike aldehydes, which undergo secondary reactions and may disappear during storage, ethylene was stable and simple to measure. These researchers found that the increase in ethylene during storage of three brands of lager beer at 40°C coincided with increases in both TBA (oxidation products) and browning. The origin of ethylene in the staling of beer was attributed to a combination of three pathways, including breakdown of aldehydes, oxidative degradation of polyunsaturated fatty acids, and methionine breakdown.

Bibliography

Anderson, R. J., Clapperton, J. F., Crabb, D., and Hudson, J. R. (1975). Dimethyl sulphide as a feature of lager flavour. *J. Inst. Brew.* **81,** 208.

Anness, B. J. (1984). Lipids of barley, malt and adjuncts. *J. Inst. Brew.* **90,** 315.

Anness, B. J., and Bamforth, C. W. (1982). Dimethyl sulphide: A review. *J. Inst. Brew.* **88,** 244.

Atkinson, B. (1987). The recent advances in brewing technology. *In* "Food Technology: International Europe," pp. 142–145. Lavenham Press Ltd., U. K.

Bamforth, C. W. (1980). Dimethyl sulfoxide reductase of *Saccharomyces* spp. *FEMS Microbiol. Lett.* **7,** 55.

Bamforth, C. W., and Anness, B. J. (1981). The role of dimethyl sulphoxide reductase in the formation of dimethyl sulphide during fermentation. *J. Inst. Brew.* **87,** 30.

Bamforth, C. W., and Martin, H. L. (1981). The development of β glucan solubilase during barley germination. *J. Inst. Brew.* **87,** 81.

Bamforth, C. W., Martin, H. L., and Wainwright, T. (1979). A role for carboxypeptidase in the solubilization of barley β-glucan. *J. Inst. Brew.* **85,** 334.

Bathgate, G. N., and Palmer, G. H. (1973). *In vivo* and *in vitro* degradation of barley and malt starch grannules. *J. Inst. Brew.* **79,** 402.

Baxter, E. D. (1984). Recognition of two lipases from barley and green malt. *J. Inst. Brew.* **90,** 277.

Baxter, E. D., Morris, T. M., and Picksley, M. A. (1987). Low proanthocyanidin malts for production of chilled and filtered or fired beer. *J. Inst. Brew,* **93,** 387.

Ben-Tal, Y. (1975). Activation of phosphorylcholine glyceride transferase by gibberellic acid in barley aleurone cells. *Diss. Abstr.* **35,** 43.

Bertoft, E. and Henriksnas, H. (1982). Initial stages in α amylolysis of barley starch. *J. Inst. Brew.* **88,** 261.

Bertoft, E., and Henriksnas, H. (1983). Starch hydrolysis in malting and mashing. *J. Inst. Brew.* **89,** 279.

Bhatty, R. S., and Rossnagel, B. G. (1980). Lipids and fatty acid composition of Riso 1508 and normal barley. *Cereal Chem.* **57,** 383.

Booer, C. D., and Wilson, R. J. H. (1979). Synthesis of dimethyl sulphide during fermentation by a route not involving the heat-labile DMS precursor of malt. *J. Inst. Brew.* **85,** 35.

Bourne, D. T., and Pierce, J. S. (1970). β Glucan and β glucanase in brewing. *J. Inst. Brew.* **76,** 328.

Briggs, D. E. (1964). Origin and distribution of α amylase in malt. *J. Inst. Brew.* **70,** 14.

Briggs, D. E., and MacDonald, J. (1983). Patterns of modification in malting barley. *J. Inst. Brew.* **89,** 260.

Briggs, D. E., Hough, J. S., Stevens, R., and Young, T. (1981). "Malting and Brewing Science," 2nd. ed., Vol. 1. Chapman & Hall, London.

Brunswick, P., Manners, D. J., and Stark, J. R. (1987). The development of β glucanases during the germination of barley and the effect of kilning on individual isoenzymes. *J. Inst. Brew.* **93,** 181.

Crescenzi, A. M. (1987). Factors governing the milling of malt. *J. Inst. Brew.* **93,** 193.

Daoud, I. S., and Kusinski, S. (1986). Process aspects of the extraction of hops with liquid carbon dioxide. *J. Inst. Brew.* **92,** 559.

Daussant, J., Grabar, P., and Nummi, M. (1985). *Eur. Brew. Conv., Proc. Congr.*, p. 165.

Daussant, J., and MacGregor, A. N. (1979). Evolution of barley β amylase during the first days of germination. Qualitative and quantitative aspects. *Eur. Brew. Conv., Proc. Congr.* **17,** 663.

Daussant, J., Skakoun, A., and NikuPaavola, M. C. (1974). Immunochemical study on barley α amylases. *J. Inst. Brew.* **80,** 55.

Delcour, J. A., Schoeters, M. M., Meysman, E. W., and Dondeyne, P. (1984). The intrinsic influence of catechins and procyanidins on beer haze formation. *J. Inst. Brew.* **90,** 381.

Del Vecchio, H. W., Dayharsh, C. A., and Baselt, F. C. (1951). *Proc. Am. Soc. Brew. Chem.* 50.

Dickenson, C. J. (1979). The relationship of dimethyl sulphide levels in malt, wort and beer. *J. Inst. Brew.* **85,** 235.

Dickenson, C. J. (1983). Dimethyl sulfide, its origin and control in brewing. *J. Inst. Brew.* **89,** 41.

Dickenson, C. J., and Anderson, R. G. (1981). The relative importance of *S*-methylmethionine and dimethyl sulfoxide as precursors of dimethyl sulfoxide. *Proc. Congr.—Eur. Brew. Conv., 18th,* p. 481.

Drost, B. W., van Eerde, P., Hoekstra, S. F., and Strating, J. (1971). Fatty acids and staling of beer. *Eur. Brew. Conv., Proc. Congr.* **14,** 451.

Eliasson, A.-C., Larsson, K., Anderson, S., Hyde, T., Nesper, R., and von Schnering, H.-G. (1987). On the structure of native starch—An analogue to the quartz structure. *Starke.* **39**(5), 147.

Enari, T.-M., and Mikola, J. (1967). Characterization of the soluble proteolytic enzymes of green malt. *Eur. Brew. Conv., Proc. Congr.*, p. 9.

Enari, T.-M., and Sopanen, T. (1986). Centenary review. Mobilization of endospermal reserves during the germination of barley. *J. Inst. Brew.* **92,** 25.

Enari, T.-M., Puputti, E., and Mikola, J. (1968). Fractionation of proteolytic enzymes of barley and malt. *Eur. Brew. Conv., Proc. Congr.*, p. 39.

Erdal, K. (1986). Proanthocyanidinfree barley-malting and brewing. *J. Inst. Brew.* **92,** 220.

Esterbauer, H., and Schauerstein, C. (1977). Isomeric trihydroxyoctadecenoic acids in beer, evidence for their presence and quantitative determination. *Z. Lebensm-Unters-Forsch.* **164,** 1155.

Fedak, G., and de la Roche, I. (1977). Lipid and fatty acid composition of barley kernels. *Can. J. Plant Sci.* **57,** 257.

Filner, P., and Varner, J. E. (1967). A test for *de novo* 18 synthesis of enzymes, density labelling of H_2O of barley α amylase induced by gibberellic acid. *Proc. Nat Acad. Sci. U.S.A.* **58,** 1520.
Forrest, I. S., and Wainwright, T. (1977). The mode of binding of β glucans and pentosans in barley endosperm cell walls. *J. Inst. Brew.* **83,** 279.
Fretzdorff, B., Pomeranz, Y. and Bechtel, D. B. (1982). Malt modification assessed by histochemistry, light microscopy and transmission scanning electron microscopy. *J. Food Sci.* **47,** 786.
Fricker, R. (1984). The flash pasteurization of beer. *J. Inst. Brew.* **90,** 146.
Gardner, R. J., and McGuiness, J. D. (1977). Complex phenols in brewing: A critical survey. *Tech. Q.—Master Brew. Assoc. Am.* **14,** 250.
Grabar, P., and Daussant, J. (1964). Study of barley and malt amylases by immunochemical methods. *Cereal Chem.* **41,** 523.
Gram, N. H. (1982a). The ultrastructure of germinating barley seeds. I. Changes in the scutellum and the aleurone layer in Nordal barley. *Carlsberg Res. Commun.* **47,** 143.
Gram, N. H. (1982b). The ultrastructure of germinating barley seeds. II. Breakdown of starch granules and cell walls of endosperm in three barley varieties. *Carlsberg Res. Commun.* **47,** 173.
Gray, P. P., Salentan, L. T., and Gantz, C. S. (1963). Survival of enzyme activity in stabilized beer. *Wallerstein Lab. Commun.* **26,** 161.
Greenwood, C. T., and MacGregor, A. W. (1965). Isolation of α amylase from barley and malted barley, and a study of the properties and action patterns of the enzymes. *J. Inst. Brew.* **71,** 405.
Hall, R. S. (1978). Ph.D. Thesis, Herriot Watt University, Edinburgh, U. K.
Hardie, D. G. (1976). Control of carbohydrase formation by gibberellic acid in barley endosperm. *Phytochemistry* **14,** 1719.
Hashimoto, N., and Eshima, T. (1977). Composition and pathway of formation of stale aldehydes in bottled beer. *Proc. Am. Soc. Brew. Chem.* **33,** 104.
Hashimoto, N., and Kuroiwa, Y. (1975). Proposed pathways for the formation of volatile aldehydes during storage of bottled beer. *Proc. Am. Soc. Brew. Chem.* **33,** 106.
Hernandez, H. H., Banasik, O. J., and Gilles, K. A. (1967). Changes in lipase activity and lipid content resulting from malting. *Proc. Am. Soc. Brew. Chem.* **24**.
Hough, J. S. (1985). "The Biotechnology of Malting and Brewing," Cambridge Stud. in Biotechnol. 1. Cambridge Univ. Press, London and New York.
Hough, J. S., Briggs, D. E., Stevens, R., and Young, T. W. (1982). "Malting and Brewing Science," 2nd ed., Vol. 2. Chapman & Hall, London.
Hudson, O. P. (1983). Horace Brown Memorial Lecture. Expanding brewing technology. *J. Inst. Brew.* **89,** 189.
Hudson, O. P. (1986). Centenary review. Malting technology. *J. Inst. Brew.* **92,** 115.
Jacobsen. J. V., and Varner, J. E. (1967). Gibberellic acid induced synthesis of protease by isolated aleurone layers of barley. *Plant Physiol.* 42, 1596.
Jacobsen, M., and Higgins, T. J. V. (1982). Characterization of the α amylases synthesized by aleurone layers of *Himalaya* barley in response to gibberellic acid. *Plant Physiol.* **70,** 1647.
Jamieson, A. M., Chen, E. and Van Cheluve, J. E. A. (1969). Study of the cardboard flavor in beer by gas chromatography. *J. Am. Soc. Brew. Chem.*, p. 123.
Johnstone, D. I. H. (1986). Centenary review. Beer packaging in cans. *J. Inst. Brew.* **92,** 529.
Jones, M., Woof, B., and Pierce, J. S. (1967). Peptidase activity in wort and beer. *Proc. Am. Soc. Brew. Chem.* **14**.
Keevil, W. J., Hough, J. S., and Cole, J. A. (1979). The influence of a coliform bacterium on fermentation by yeast. *J. Inst. Brew.* **85,** 99.
Lauriere, C., Mayer, C., Renard, H., MacGregor, A. W., and Daussant, J. (1985). Ripening of barley grain, variations of isoforms of α and β amylases, debranching enzyme and α amylase inhibitor. *Proc. Congr.—Eur. Brew., Conv., 20th,* p. 52.

Lloyd, W. J. W. (1986). Centenary review. Adjuncts. *J. Inst. Brew.* **92,** 336.
Lopez, A. S., and Quesnel, V. C. (1976). Methyl-*S*-methionine sulphonium salt: A precursor of dimethyl sulphide in cacao. *J. Sci. Food Agric.* **27,** 85.
Loubser, G. J., and du Plessis, C. S. (1976). The quantitative determination and some values of dimethyl sulfoxide in white table wine. *Vitis* **15,** 248.
Lowe, C. M., and Elkin, W. I. (1986). Centenary review. Beer packaging in glass and recent developments. *J. Inst. Brew.* **92,** 517.
Lynch, P. A., and Seo, C. W. (1987). Ethylene production in staling beer. *J. Food Sci.* **52,** 1270.
MacGregor, A. W., and Ballance, D. (1980). Quantitative determination of α amylase enzymes in germinated barley after separation by isoelectricfocusing. *J. Inst. Brew.* **86,** 131.
MacLeod, A., Duffus, J. H., and Johnston, C. S. (1964). Development of hydrolytic enzymes in germinating grain. *J. Inst. Brew.* **70,** 521.
Manners, D. J. (1985). Some aspects of the structure of starch. *Cereal Foods World* **30,** 461.
Manners, D. J., and Bathgate, G. N. (1969). α-1, 4-Glucans. XX. Molecular structure of the starches from oats and malted oats. *J. Inst. Brew.* **75,** 169.
Manners, D. J., and Marshall, J. J. (1969). Carbohydrate metabolizing enzymes. XXII. The β glucanase system of malted barley. *J. Inst. Brew.* **75,** 550.
Manners, D. J., and Matheson, N. K. (1981). α(1 → 4)D-GLUCANS. PART XXV. THE FINE STRUCTURE OF AMYLOPECTIN. *Carbohydr. Res.* **90,** 99.
Manners, D. J., and Wilson, G. (1974). Purification and properties of an endo (1 → 3)-β-D-glucanase from malted barley. *Carbohydr. Res.* **37,** 9.
Masschelein, C. A. (1986). Centenary review. The biochemistry of maturation. *J. Inst. Brew.* **92,** 213.
Miedaner, H. (1986). Centenary review. Wort boiling today—old and new aspects. *J. Inst. Brew.* **92,** 330.
Mikola, J., and Kolehmainen, L. (1972). Localization and activity of various peptidases in germinating barley. *Planta* **104,** 167.
Morris, T. M. (1986). The effect of cold break on the fining of beer. *J. Inst. Brew.* **92,** 93.
Morrison, W. R. (1978). Cereal lipids. *Adv. Cereal Sci. Technol.* **2,** 221.
Mundy, J., Svedsen, I., and Heijgaard, J. (1983). Barley α amylase/subtilisin inhibitor. I. Isolation and characterization. *Carlsberg Res. Commun.* **48,** 81.
Narziss, L. (1985). "Technologie der Würzebeitung." Enke, Stuttgart.
Narziss, L. (1986). Centenary review. Technological factors of flavor stability. *J. Inst. Brew.* **92,** 346.
Nelson, G., and Young, T. W. (1986). Yeast extracellular proteolytic enzymes for chillproofing beer. *J. Inst. Brew.* **92,** 599.
Nordstrom, K. (1964). Formation of esters from alcohols by brewers yeast. *J. Inst. Brew.* **70,** 328.
Nummi, M., Vilhunen, R., and Enari, T.-M. (1965). Exclusion chromatography of barley β amylase on Sephadex G 75. *Acta Chem. Scand.* **19,** 1793.
Okamoto, K., Kitano, H., and Akazawa, T. (1980). Biosynthesis and excretion of hydrolases in germinating cereal seeds. *Plant Cell Physiol.* **21,** 201.
Peacock, V., and Deinzer, M. (1981). Nonbitter contributors to beer flavor. *ACS Symp. Ser.* 119.
Preece, I. A., and Hoggan, J. (1956). Enzymic degradation of cereal hemicelluloses. I. Observation in the β glucosanase system and its development during malting. *J. Inst. Brew.* **62,** 486.
Price, P. B., and Parsons, J. G. (1975). Lipids of seven cereal grains. *J. Am. Oil Chem. Soc.* **52,** 490.
Robin, J. P., Mercier, C., Charbonniere, R., and Guilbot, A. (1974). Linterized starches. Gel filtration and enzymatic studies of insoluble residues. *Cereal Chem.* **51,** 389.
Rogers, J. C. (1985). Two barley α amylase gene families are regulated differently in aleurone cells. *J. Biol. Chem.* **260,** 3731.
Ronald, A. P., and Thomson, W. A. B. (1964). Volatile sulfur compounds of oysters. *J. Fish. Res. Board Can.* **21,** 1481.

Runkel, U. D. (1975). Malt kilning and its influence on malt and beer quality. *Eur. Brew. Conv., Monogr.* **11,** 222.

Sandra, P., and Verzele, M. (1975). Contribution of hop-derived compounds to beer aroma. *Eur. Brew. Conv., Proc. Congr.* **15,** 107.

Schildbach, R. (1986). Centenary review barley worldwide. *J. Inst. Brew.* **62,** 486.

Selvig, A., Aarnes, H., and Lie, S. (1986). Cell wall degradation in endosperm of barley during germination. *J. Inst. Brew.* **92,** 185.

Stewart, G. G., and Russell, I. (1985). Modern brewing technology. *Comp. Biotechnol.* **3,** 375–381.

Thompson, R. G., and LaBerge, D. E. (1981). Barley endosperm cell walls morphological characteristics and chemical composition. *Tech. Q.—Master Brew. Assoc. Am.* **18,** 116.

Tidbury, C. H. (1986). Sixth Richard Seligman Memorial Lecture "Technical Progress in Brewing." *J. Inst. Brew.* **92,** 147.

Tressl, R., Kossa, T., Renner, R., and Koppler, H. (1975). Gas chromatographischmassenspelkmetrische und deren Genese I. Organische sauren in Wurse und Bier. *Monatschr. Brau..* **28,** 109.

Verzele, M. (1986). Centenary review. 100 years of hop chemistry and its relevance to brewing. *J. Inst. Brew.* **92**, 32.

Verzele, M., and Van de Velde, N. (1987). Bicyclic isomerization products of humulone and cohumulone in beer. *J. Inst. Brew.* **93,** 190.

Verzele, M., Scuddinck, G., Proot, M., and Sandra, P. (1987). Quantitative capillary gas chromatography of trihydroxyoctadecenoic acids in beer. *J. Inst. Brew.* **93,** 26.

Visuri, K. and Nummi, M. (1972). Purification and characterization of crystalline β amylase from barley. *Eur. J. Biochem.* **28,** 555.

Wallerstein, L. (1911). Brewing. U. S. Patent 995, 820.

Welch, R. W. (1975). Fatty acid composition of grain from winter and spring sown oats, barley and wheat. *J. Sci. Food Agric.* **26,** 429.

White, F. H. (1977). The origin and control of dimethylsulfide in beer. *Brewer's Dig.* **52,** 38.

Wilson, G. (1972). Ph.D. Thesis, Herriot Watt University, Edinburgh, U. K.

Wilson, R. J. H., and Booer, C. D. (1979). Control of the dimethyl sulphide content of beer by regulation of the copper boil. *J. Inst. Brew.* **85,** 144.

7

Biochemistry of Food Processing: Baking

I. Introduction

The major wheat-based food product worldwide is bread. It is one of the least expensive yet most important staples in the world today. The baking of bread has undergone dramatic changes during the past two decades. The growth of large supermarkets and shopping malls has led to the establishment of "in-store" bakeries and hot bread shops. Nowadays frozen doughs and part-baked flour products are prepared in centralized locations and transported to "in-store" bakeries for final baking (Hoover, 1984; Chamberlain, 1987). Variety breads have also increased in popularity as reflected by the growth and expansion of specialty and ethnic breadstuffs. This chapter will discuss those biochemical changes essential for the development of bread.

II. Baking Technology

When wheat flour is mixed with water it forms a viscoelastic dough which is critical for the production of high-quality bread. Four major components can be separated from wheat flour: gluten, water-soluble fractions, starch, and lipids. This section will focus on the contribution by each of these components to the functional properties of dough. A good-quality loaf is obtained only when there is an optimized combination of constituents, ingredients, and processing (Bushuk, 1985). The three steps involved in the production of bread are mixing and dough development, dough aeration, and oven baking.

A. Mixing and Dough Development

The formation of a cohesive and elastic dough is unique to wheat flour (Hoseney, 1985). This occurs by mixing or stirring the wheat flour–water mixture until the dough is fully developed. Under these conditions the dough becomes quite resistant to extension, which is characteristic of its cohesive and elastic properties. This is illustrated by mixogram curves of wheat flour mixes which record the resistance to extension by the height of the curve (Figure 7.1). The longer a dough is mixed, the more resistant to extension it becomes, as shown by the increasing height of the curve. During the development or mixing of the dough the flour particles become hydrated. Thus when a dough is optimally developed all the protein and starch become fully hydrated. Hoseney (1985) examined an optimally mixed freeze-dried dough by electron microscopy

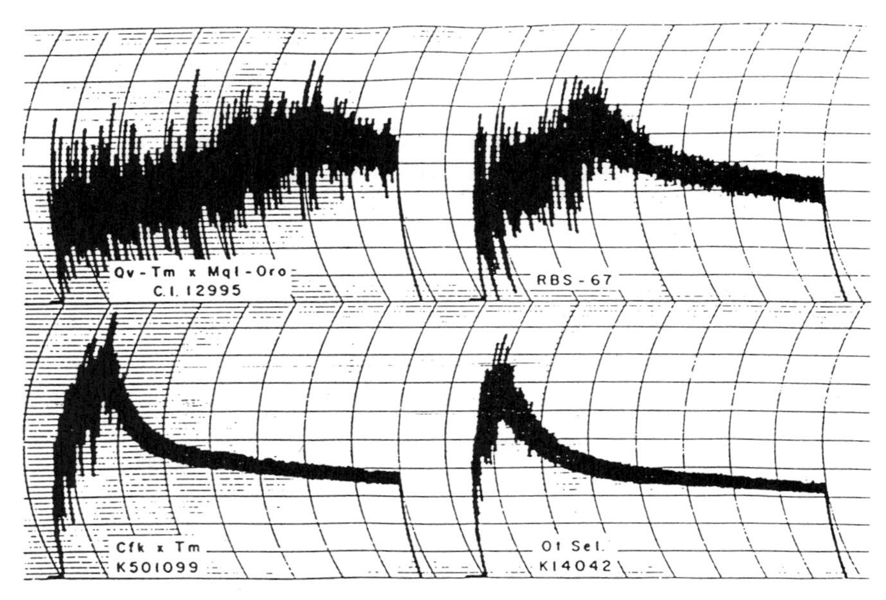

Fig. 7.1. Mixogram curves of hard wheat flours, showing long (C.I. 12995), medium (RBS-67), short (K501099), and very short (K14042) mixing times (Hoseney, 1985).

and found it to consist of a random mixture of protein fibrils with adhering starch.

In addition to hydration, development of the cohesive nature of the dough involves incorporation of air into the dough, which decreases its density. Junge *et al.* (1981) reported that about one-half of the total air possible was incorporated into the dough at optimal mix. Nitrogen was particularly important, as several research groups showed that it produced cells in which carbon dioxide diffuses (Chamberlain and Collins, 1979; Mahdi *et al.*, 1981).

1. *Proteins and Dough Development*

Wheat flour must be high in both protein content and quality to develop the functional properties essential to bread baking. The protein content of the flour affects the mixing time as low-protein flours (less than 12%) have been shown to require longer mixing times (Finney and Shogren, 1972). When mixed with water, flour forms a viscoelastic substance called gluten, which is composed of two protein fractions, glutenin and gliadin. The control of mixing time has also been associated with the glutenin fraction of the wheat flour (Hoseney *et al.*, 1969b). Differences between wheat varieties in bread-making properties appear to be a function of the nature of gluten. A study conducted on 104 flour samples by Orth and Bushuk (1972) showed that loaf volume was inversely related to the amount of acid-soluble (in 0.05 *N* acetic acid) glutenin and directly related to the proportion of insoluble glutenin (Figure 7.2). In a study of 26 wheat varieties, Orth and Bushuk (1973) concluded that glutenin was responsible for loaf volume, which was recently confirmed by MacRitchie (1980, 1985). This differed from studies by Hoseney *et al.* (1969 a,b) and Finney *et al.* (1978), who implicated the gliadin fraction as the loaf volume controlling factor in wheat flour. The controversy regarding the roles of glutenin and gliadin in controlling loaf volume still remains.

A closer examination of the two protein fractions showed them to be quite different. Glutenin was composed of high-molecular-weight proteins with values reported from 150,000 to 3 million, while gliadins were low-molecular-weight proteins ranging in molecular weight from 25,000 to 100,000 daltons (Jones *et al.*, 1961; Nielsen *et al.*, 1962; Taylor and Cluskey, 1962). Considerable efforts have been made to elucidate the structure of glutenin, as variations in molecular architecture might explain varietal differences in the bread-making quality of wheats.

2. *Structure of Glutenin*

The reduction in viscosity of glutenins, following the addition of reducing agents, led Pence and Olcott (1952) to suggest that glutenins were composed of a system of polypeptide subunits held together by disulfide bonds. A marked

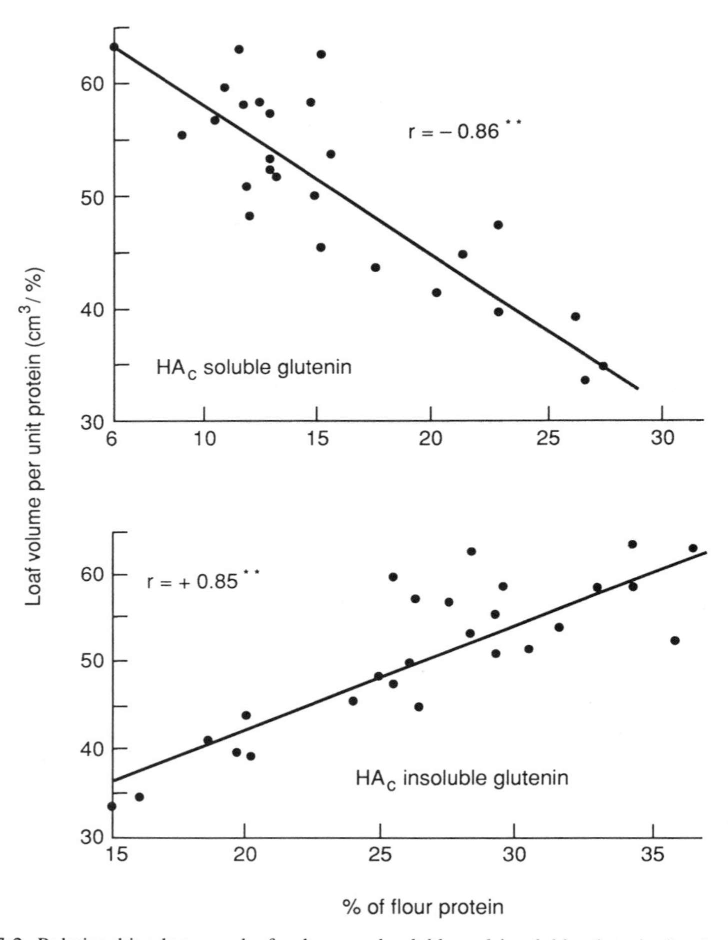

FIG. 7.2. Relationships between loaf volume and soluble and insoluble glutenin fractions (Orth and Bushuk, 1973).

decrease in viscosity immediately following the addition of reducing agents to glutenin was attributed by Beckwith and Wall (1966) to the breakdown of the interpolypeptide disulfide bonds. On standing, however, a slight increase in viscosity was observed, which was attributed to the unfolding of the polypeptide chains of glutenin as a consequence of a slower reduction of intrapolypeptide disulfide bonds or by noncovalent aggregation of the products. Oxidation of reduced glutenin at 5% (w/v) resulted in a product similar in viscosity, molecular weight, and elasticity characteristics to native glutenin. Beckwith and Wall (1966) concluded that intra- and interpolypeptide disulfide bonds must be present in glutenin in a certain ratio to provide the necessary viscoelastic properties for bread baking. Attempts to differentiate between inter- and intrasubunit disulfide bonds in glutenin using specific chemical reagents proved unreliable as they were based on the premise that only the intersubunit disulfide bonds were reactive (Kasarda *et al.*, 1979).

Studies on glutenin have been difficult to compare, as each researcher has tended to develop his or her own method of extraction. The original Osborne extraction involved dilute acid or alkali, which only partially extracted the glutenin (Kasarda *et al.*, 1976). Thus each researcher has tended to develop his or her own particular extractant, including NaCl; acetic acid; a combination of acetic acid, urea, and cetyltrimethylammonium bromide then separation of extract on Sephadex G-200 or precipitation of dispersed glutenin with ethanol at pH 6.4; and ammonium sulfate (Ewart, 1972a,b; Jones *et al.*, 1959; Meredith and Wren, 1966; Wasik and Bushuk, 1974; Woychik *et al.*, 1964). A novel procedure was developed by Graveland *et al.* (1979, 1982) in which sodium dodecyl sulfate (SDS) was used to separate flour proteins into SDS-insoluble and SDS-soluble fractions. These researchers were able to identify four glutenin fractions, I, II, III, and IV, using both SDS and chromatography. In an attempt to clarify the confusion, Laszity (1984) recommended that "glutenin" should only refer to the original protein prepared by the Osborne method (Osborne, 1907). It was suggested that other glutenins prepared by different extraction procedures should use the method of preparation as a suffix, such as "glutenin prepared according to the method of—".

3. *Glutenin and Bread Making Quality*

Pomeranz (1965) noted that good bread-making wheat cultivars were higher in urea-unextractable proteins especially insoluble glutenins. This was subsequently confirmed in baking studies with reconstituted flours by Shogren and coworkers (1969). On the basis of the salting out response of glutenins from 11 wheat cultivars varying in baking performance, Huebner (1970) observed a greater proportion of high-molecular-weight glutenins in the good bread quality cultivars. Lee and MacRitchie (1971) reported that flour proteins extracted with 4 *M* urea and sodium hydroxide, primarily glutenins, enhanced mixing stability and strength of the dough. An examination of the relationship between the protein fractions obtained from 26 diverse wheat cultivars using the Osborne procedure and bread-making quality was carried out by Orth and Bushuk (1972) and Orth *et al.* (1972). These researchers found loaf volume to be inversely related to the amount of soluble glutenin and directly related to the level of insoluble glutenin. Glutenin content was identified as the major factor responsible for baking differences between bread wheat varieties. Two glutenin fractions, I and II, were separated from wheat cultivars by Huebner and Wall (1976) using agarose gel filtration. A better bread-making cultivar tended to have a higher amount of fraction I, a high-molecular-weight (HMW) glutenin fraction, relative to fraction II, a low-molecular-weight glutenin fraction.

Attempts by Orth and Bushuk (1973) to correlate SDS–PAGE patterns of glutenin with baking quality of hard red spring wheat varieties proved unsuccessful. A definite relationship has since been reported, however, between baking quality and high-molecular-weight glutenin subunits (>80,000) for European wheats (Payne *et al.*, 1979; Burnouf and Bouriquet, 1980).

4. *High Molecular Weight Glutenins and Bread Baking Quality*

Studies conducted at the Plant Breeding Institute of Cambridge, England, have established a strong relationship between HMW glutenin subunits and bread-making quality. While these HMW subunits account for less than 1% of the dry weight of the endosperm, they make a major contribution to the bread-making quality of wheat. For example, "Maris Widgeon," a good bread-making wheat variety, contained two HMW glutenin subunits not found in "Maris Ranger," a poor bread-making wheat variety (Payne *et al.*, 1979). By crossing these two varieties a good bread-making wheat variety was selected after several generations and referred to as "Maris Freeman." This variety contained only one of the two HMW glutenin subunits originally present in "Maris Widgeon" which had a molecular weight of 145,000. This glutenin subunit was identified in 31% of the 67 wheat varieties screened and correlated with their bread-making quality using the SDS–sedimentation test (Axford *et al.*, 1978). A similar correlation was reported by Burnouf and Bouriquet (1980) between bread-making quality and two HMW glutenin subunits for a stock of 47 genetically related wheat varieties examined.

Payne and Corfield (1979) fractionated 12 glutenin fractions ranging in molecular weight from 31,000 to 136,000, which they attributed to the elasticity of glutenin. Further studies by Payne and co-workers (1980a,b) identified 12 HMW glutenin subunits from 7 wheat varieties ranging in molecular weight from 95,000 to 145,000 and numbered them from 1 to 12. Several additional HMW glutenin subunits were subsequently reported by these researchers (Payne *et al.*, 1980a).

The relationship between bread-making qualities and HMW glutenin subunits for 70 wheat cultivars was studied by Branlard and Dardevet (1985). A number of flour quality characteristics were examined, including Chopin Alveograph (W), tenacity (P), swelling (G), and extensibility (L). A stepwise regression analysis to predict each of these flour quality characteristics for the HMW glutenin subunits showed that 40% of the variation in W, 41% of the variation in Z, and 46% of the variation in P could be attributed to eight, seven, and six subunits, respectively. Ng (1987) identified 13 different HMW glutenin subunits from SDS–PAGE patterns of 26 wheat varieties ranging from 90,000 to 147,000.

Varieties with superior bread-making qualities were high in specific subunits, with 128,100 and 91,600 HMW subunits highly correlated with bread-making quality. Seven prediction equations were developed between HMW glutenin subunits and such bread-making criteria as dough development time (DT), mixing tolerance index (MTI), maximum resistance (R), and extensibility (E). The prediction equations for all these parameters, with the exception of E, had a high predictive power, which suggested that other factors were responsible for dough extensibility.

5. *Genetics and Functional Properties of Dough*

Payne *et al.* (1984) showed the HMW glutenin subunits were coded by genes located at three genetically unlinked loci, Glu-A1, Glu-B1, and Glu-D1, on the long arms of chromosomes 1A, 1B, and 1D, respectively (Payne, 1987). Any allelic variation at each locus was distinguished by separation of the allelic protein subunits using SDS–PAGE. Branlard and Dardevet (1985) pointed out the difficulty in assessing the contribution of allelic variation in a group of proteins, relative to variation in other flour components, on bread-making quality. This resulted in the creation of a Glu-1 quality score based on the relationship between individual HMW subunits of glutenin and quality, as measured by the SDS–sedimentation test (Payne *et al.,* 1987). Using this procedure, the Glu-1 score for seed proteins of 84 home-grown wheat varieties were recorded and ranged from a maximum of 10 to a minimum of 3. A positive relationship was observed between Glu-1 quality score and bread-making quality, which supported its usefulness as an indicator of potential elastic development or dough strength. These researchers identified somewhat low Glu-1 values for British-grown wheats, with a mean score of 6.8, compared to the highest-quality West Germany wheat varieties, Monopol, Severin, and Rektor, with scores of 9, 9, and 7, respectively (Anonymous, 1983). To develop wheat varieties with improved bread-making qualities plant breeders must select progeny with high Glu-1 quality scores (Payne *et al.,* 1987). A recent study by Payne and coworkers (1988) fractionated the high-molecular-weight glutenin subunits of 33 Spanish-grown wheat varieties by SDS–PAGE. A strong correlation was observed between Glu-1 score, Alveograph *w* value, and Zeleny volume, all measurements of dough strength, and Falling Number values for 32 of the varieties were also examined. The authors attributed these results to the fact that the breeder has consistently selected progeny with high Falling Number values and strong mixing doughs, both of which are critical for good bread-making wheat varieties.

6. *Protein Functionality and Structure*

Protein functionality is dependent on surface characteristics such as hydrophobicity. Using reversed-phase high-performance liquid chromatography (RP-

HPLC), gluten proteins were bonded to hydrophobic packings in the presence of polar solvents and then eluted by increasing the solvent hydrophobicity (Bietz, 1983, 1984). Further improvements in this technique were reported by Bietz and Cobb (1985) in which they proposed a hydrophobicity index to characterize wheat proteins based on alkylphenone standards. The latter vary widely in hydrophobicity and could be used to identify wheat proteins rather than elution time. This technique has considerable potential for identifying varietal differences in wheat glutenin as well as for characterizing these proteins.

During mixing of dough, the gluten forms a flexible film structure capable of retaining the carbon dioxide gas produced by fermentation during the proofing and initial baking period. This structure is then set and stabilized by denaturation of the gluten proteins during baking, providing the necessary framework for the loaf of bread (Bushuk, 1985). This development of the gluten network led Atkins (1971) to describe bread as essentially foamed gluten. Several models were proposed to explain the unique viscoelastic properties of gluten, including the "linear glutenin hypothesis." This model, proposed by Ewart (1968), suggested that glutenin was a linear molecule consisting of polypeptide chains attached to one another by interpolypeptide disulfide bonds. Ewart (1972a), however, was unable to explain the rheological properties of the dough using this model. He later suggested a more complex, concatenated structure for glutenin in which extensively folded polypeptide chains were joined by two interchain disulfide bonds to each of the neighboring polypeptides (Ewart, 1972b). A modification of this structure by Greenwood and Ewart (1975) suggested that polypeptides of glutenins were linked by disulfide bonds into linear chains, or concatenations. The dramatic reduction in viscosity of glutenins observed with disulfide-breaking agents could be explained by a decrease in the degree of polymerization of this linear model. The linear model could also explain the increase in dough resistance observed during the initial period of mixing. For instance, the dough's tensile strength increases because the concatenations become oriented in the direction of stress, thus permitting secondary forces such as hydrogen bonding and hydrophobic forces to come into play (Pyler, 1983a). These concatenations may become entangled during dough mixing and provide the cross-links in the gluten network which are essential for dough elasticity.

Early studies on protein structure led to the discovery of the α-helix, in which the polypeptide chain was arranged in a helical coil, and the sheet structure consisting of an extended polypeptide chain was held in a parallel arrangement by hydrogen bonds (Schultz and Schirmer, 1979). Venkatachalam (1968), using theoretical calculations, identified three arrangements where the polypeptide chain could undergo a 180° change in direction. This led to identification of the β-turn, which differed from the α-helix as hydrogen bonding occurred between the carbonyl group of the first residue and the amide nitrogen on the fourth residue rather than the fifth residue. Studies on wheat gluten proteins showed the dominant conformation to be the β-turn (Tatham *et al.*, 1984; Tatham and Shewry, 1985). These researchers proposed a model in which the repeated β-turns in the disulfide-bonded high-molecular-weight subunits were responsible for the elasticity of gluten. Using circular dichroism spectroscopy and computer predictions of amino acid sequences, Tatham *et al.* (1985) suggested that the high-molecular-weight glutenin subunits were the major elastic components of gluten. This was based on the regular distribution of β-turns forming a spiral arrangement in a similar manner similar to that observed for elastin. The absence of denaturation when gluten was heated in a differential scanning calorimeter (DSC) led Hoseney and co-workers (1986) to conclude that wheat gluten has no long-range order. Wheat gluten appeared to be an amorphous random polymer, inconsistent with the β-turn conformation proposed earlier by Tatham *et al.* (1985).

III. The Water-Soluble Fraction of Wheat Flour

The water-soluble fraction of wheat flour consists of proteins, pentosans, and dialyzable components. These can be separated into a number of fractions as shown in Scheme 7.1. The dialyzable fraction provides growth factors for yeast fermentation and thus contributes to the production of gas. The residual superna-

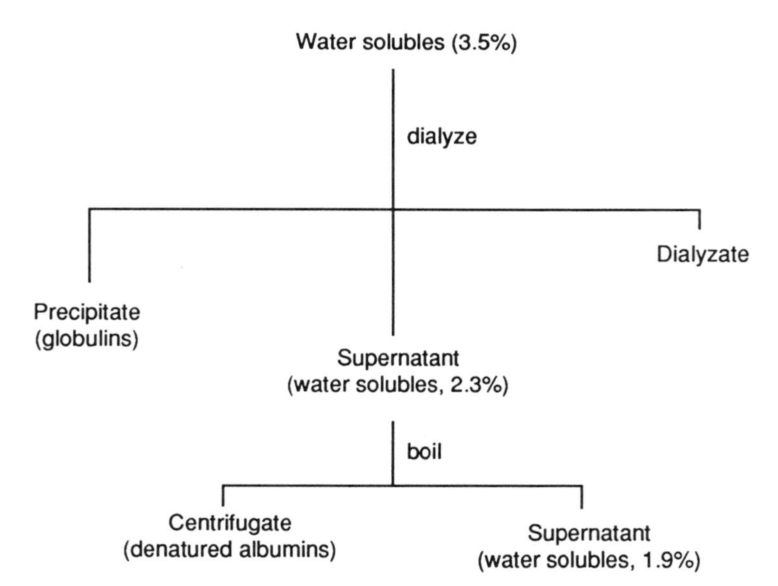

SCHEME 7.1. Fractionation scheme of water-soluble fractions of flour (percentages based on total flour weights) (Hoseney *et al.*, 1969a).

tant solution, containing the soluble pentosans and glycoproteins, is important for the development of gluten and its ability to retain gas in fermented doughs (Pence *et al.*, 1951).

A. Pentosans

The predominant polymer in the water-soluble pentosans (Figure 7. 3) consists of a straight chain of anhydro-D-xylopyranosyl residues linked β-1, 4 with an anhydro-L-arabinofuranosyl residue at the 2, 3 position (Perlin, 1951). Pentosans are generally classified on the basis of solubility in cold water, which is determined by the extent of arabinose branching of the xylose chain. The presence of more branching on the xylan chain renders the xylan chain less soluble. Kundig *et al.* (1961a) obtained five fractions by separating water-soluble pentosans on DEAE-cellulose (borate form). One fraction appeared to be pure arabinoxylan while the remainder were glycoproteins. Using this technique other researchers demonstrated that this method produced overlapping fractions (Lin and Pomeranz, 1968; Medcalf *et al.*, 1968; Wrench, 1965). Fincher and Stone (1974) clarified the situation by obtaining two distinct fractions when treating water-soluble pentosans with saturated ammonium sulfate. The soluble fraction was found to be arabinogalactan covalently linked to a peptide. The precipitate, arabinoxylan contaminated with free protein, was later shown by Yeh *et al.*, (1980) to contain ferulic acid (Figure 7.4).

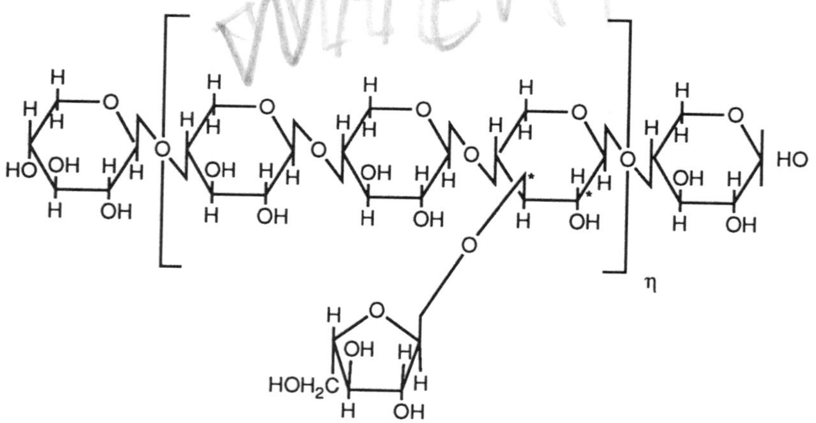

Fig. 7.3. Structure of wheat endosperm pentosans. Possible arabinose branching points at C_2 and C_3 are designated by asterisks (Shelton and D'Appolonia, 1985).

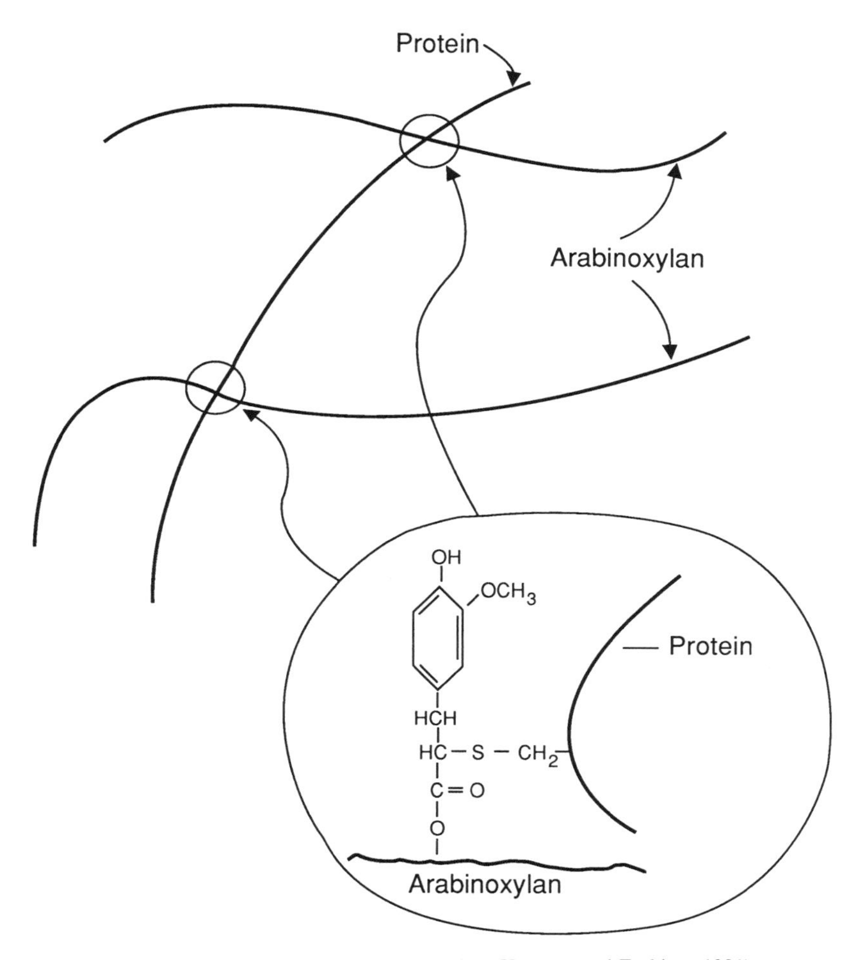

FIG. 7.4. Mechanism of oxidative gelation (Hoseney and Faubion, 1981).

B. FUNCTIONAL PROPERTIES OF PENTOSANS

Water-soluble pentosans form very viscous solutions because of their much higher intrinsic viscosity compared to soluble proteins (Udy, 1956). A unique property of water-soluble pentosans is their ability to form gels in the presence of oxidizing agents. This process of oxidative gelation was monitored by Kundig *et al.* (1961b), who noted the disappearance of a peak at 320 nm in the ultraviolet spectrum of pentosans following the addition of oxidizing agents. The loss of this peak was attributed to involvement of ferulic acid in the gelation process. A

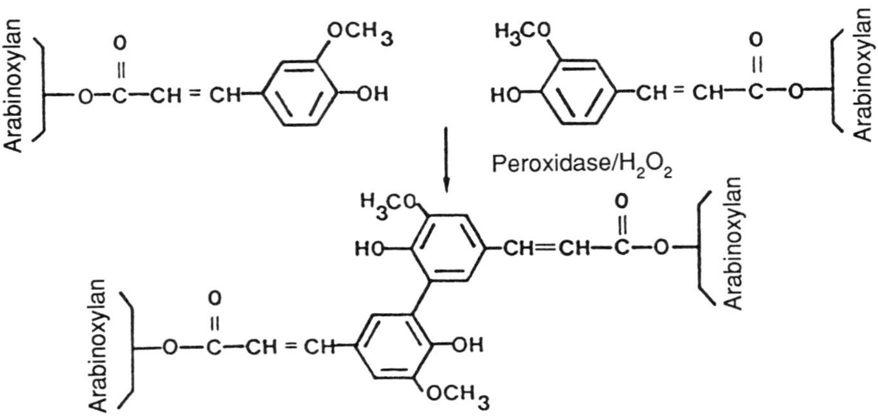

FIG. 7.5. Suggested mechanism for the oxidative coupling of two ferulic acid residues (Neukom and Markwalder, 1978).

mechanism was subsequently proposed by Neukom and Markwalder (1978) in which a dimer of ferulic acid formed a cross-link between pentosans (Figure 7.5). A variety of oxidants, including potassium bromate, potassium iodate, and ascorbic acid, were examined by Hoseney and Faubion (1981) with respect to their effect on the viscosity of water-soluble pentosans. None of these caused any increase in viscosity with the exception of hydrogen peroxide in the presence of peroxidase, ammonium persulfate, and formamimidine disulfide. Two active centers in ferulic acid were thought to account for the increase in viscosity via cross-linking. One was the aromatic nucleus while the other was the activated double bond (Neukom and Markwalder, 1978; Sidhu *et al.*, 1980a,b). Hoseney and Faubion (1981) found that fumaric acid, with no aromatic nucleus, stopped gelation in the presence of hydrogen peroxide compared to vanillin, with no activated double bond, which had no effect. Based on these results it was evident that only the activated double bond in ferulic acid was involved in the gelation process. Ferulic acid and cysteine added to the water-soluble fraction prior to addition of hydrogen peroxide also stopped gelation. Based on these results the following pathway (Fig. 7.4) was proposed by Hoseney and Faubion (1981) to explain gelation. This involved addition of a protein thiyl radical to the activated double bond of ferulic acid, which was itself esterified to the arabinoxylan fraction. The formation of such cross-links through covalent binding of protein and polysaccharide chains could affect the rheology of the dough.

The contribution of pentosans to loaf volume has been somewhat confused by the conflicting results in the literature. Shelton and D'Appolonia (1985) attributed this to differences in isolation procedures, purity of the pentosans, and baking procedures reported by various researchers. An early study by Pence and co-workers (1950) associated increased loaf volume with the residual protein in the pentosan extract. Hoseney (1984), however, reported that reduced loaf volume occurred only after the water-soluble pentosans were removed and was not affected by the water-soluble proteins, albumin and globulin. These results were in direct conflict with an earlier study by D'Appolonia and co-workers (1970), who found little change in loaf volume following the addition of water-soluble pentosans to gluten–starch loaves. Kim and D'Appolonia (1977) suggested that pentosans decreased the rate of bread staling by decreasing the amount of starch components available for crystallization.

IV. Lipids in Wheat Flour

Lipids, although a minor component of wheat flour, play an important role in bread making. Of particular interest are the polar lipids, which contribute to mixing requirements and loaf volume potential (Pomeranz, 1985). Hoseney *et al.* (1969a,b) demonstrated that the addition of polar lipids rich in glycolipids to defatted flour restored loaf volume.

A. Polar Lipids and Bread Baking

The amount and type of lipid present in wheat flour influences the functionality of the dough. Nonpolar lipids, including stearyl esters, glycerides, and free fatty acids, all have a detrimental effect on bread making (Chung and Pomeranz, 1981). Polar lipids, on the other hand, have been shown to be effective improvers in bread making by several researchers. Hoseney and co-workers (1970) reported that the addition of small amounts of free polar lipids (rich in glycolipids) to petroleum ether-defatted wheat flour completely restored loaf volume compared to bread baked with original flour and 3% shortening. This differed from the addition of bound polar lipids, containing the equivalent levels of phospholipids and glycolipids, which proved totally ineffective. These researchers also found that the addition of small amounts of polar lipids to almost completely defatted flour proved detrimental to loaf volume. Increasing the levels of either free or bound polar lipids, however, had an improving effect. This improving effect of higher levels of polar lipids to chloroform-defatted flour without added shortening was confirmed by MacRitchie and Gras (1973). Fractionation of polar lipids into phospholipids and glycolipids was performed by Dafraty *et al.* (1968). Supplementation of petroleum ether-defatted wheat flour containing no shortening with 0.2% glycolipids proved beneficial to bread mak-

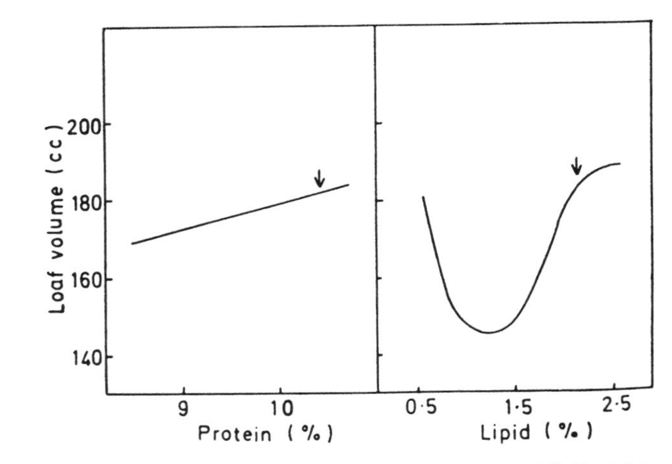

FIG. 7.6. Variation of loaf volume with changing protein (left) and lipid (right) contents of the same flour, redrawn from results of MacRitichie (1978). Arrows indicate natural values for the flour. Lipids are expressed as hydrolysate lipid. Starch contains approximately 0.5% hydrolysate lipid; the curve shows effects of the addition of chloroform-extracted lipid (MacRitchie, 1981).

ing while phospholipids were slightly detrimental. The detrimental effect of phospholipids added to petroleum ether-defatted wheat flour was also shown by Lin *et al.* (1974). A slight decrease in loaf volume was reported by MacRitchie (1977) for chloroform-defatted flour in the presence of up to 0.6% phospholipids. Addition of much higher levels of phospholipids, however, resulted in a marked increase in loaf volume. Wheat lipids were shown by MacRitchie (1981) to produce the greatest changes in loaf volume and texture. Figure 7.6 illustrates decrease in loaf volume with decreasing levels of proteins or lipids. While this decrease was linear with respect to protein, a minimum curve was evident for lipids, after which the loaf volume increased on further removal of lipids. Unlike defatted flour, the complete removal of proteins destroys dough-forming properties and bread-baking capacity. This was not the case for defatted flour.

B. Lipid-Protein Interaction and Shortening Effect

Phospholipids, particularly glycolipids, are effective improvers when added to petroleum ether-defatted flours. Most baking formulations use added shortenings or surfactants so that the role of the indigenous flour lipids tend to be overlooked. On the other hand, the presence of lipids in flour makes it difficult to delineate the effects of added lipids to untreated flour (Pomeranz, 1985). It is now clear that lipid–protein interactions play an important role in shortening. Once the flour is wetted with water and mixed into a dough, the free lipids become "bound" by the gluten proteins. Of the polar lipids, glycolipids appeared to be bound to the gliadin proteins hydrophilically and to the gluten proteins hydrophobically (Hoseney *et al.*, 1970). The simultaneous binding of glycolipids to gliadin and gluten proteins was thought to contribute structurally to the gas-retaining ability of the gluten proteins. Chung *et al.* (1978) reported that as much as one-half to two-thirds of the free lipids, all the polar lipids as well as some nonpolar lipids, are bound during gluten formation. Chung *et al.* (1980a) clearly demonstrated the role that wheat flour lipids play in shortening by differentially defatting a composite hard red winter wheat flour with good loaf volume potential and medium mixing and oxidation requirements. These solvents (Skellysolve B, benzene, acetone, and 2-propanol) totally removed the nonpolar lipids but left different amounts of polar lipids in the flour. In the absence of shortening the loaf volume of bread baked from the untreated flour was 69.3 cc compared to 84 cc for defatted flour reconstituted with polar lipids. This difference could be due, in part, to the detrimental effect exerted on loaf volume by the nonpolar lipids. The addition of shortening appeared to compensate for the nonpolar lipids as there was no significant difference between loaf volume obtained for the untreated flour (85.5 cc) compared to flour reconstituted with polar lipids (87 cc). Further studies by Chung and co-workers (1980b) attempted to explain lipid–protein interactions in terms of good- and poor-quality bread wheat flours. These researchers defatted 11 wheat flours of different bread-making qualities with Skelysolve B or 2-propanol. The defatted flours contained negligible amounts of nonpolar lipids, although flours defatted with Skellysolve B had more residual bound polar lipids compared to flours defatted with 2-propanol. Based on bread baking of these flours, loaf volume was found to increase linearly with free polar lipids but not with total polar lipids in the flour. Differences in loaf volume potential between flours were attributed to interaction of free polar lipids in the flour with added shortening. In the absence of free polar lipids, shortening may provide a mechanical barrier which interferes with the formation of protein–protein complexes. Such interference could lead to a reduction in loaf volume for bread baked with defatted flours in the presence of 3% shortening. In the absence of shortening, however, protein–protein interactions may well be enhanced by lipid removal.

C. Lipoxygenase and Bread Making

The addition of the enzyme lipoxygenase (EC 1. 13. 11. 12) to bread making is related to a number of important functions it plays in the bread-making process (Eskin *et al.*, 1977; Faubion and Hoseney, 1981; Nicolas and Drapron, 1983). These include increasing the amount of free lipids in the dough, destruction of essential fatty acids and bleaching of carotenoids, and increasing mixing tolerance and dough stability.

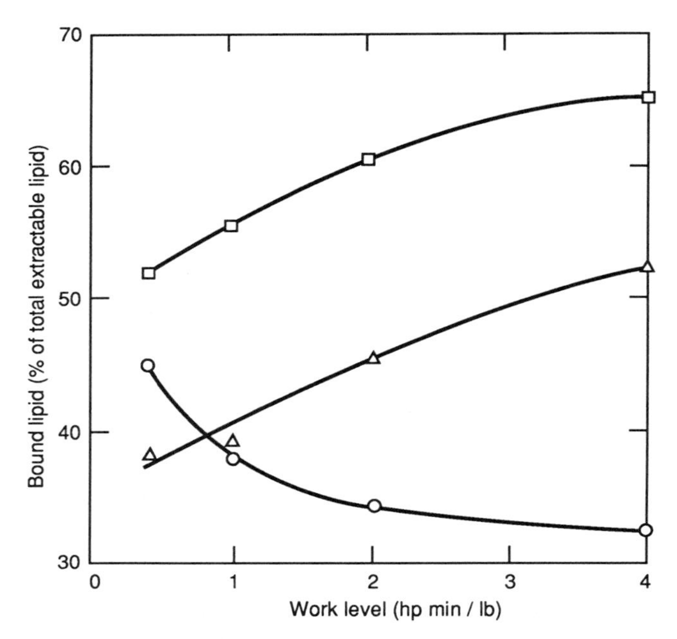

FIG. 7.7. Effect of peroxidized lipid on lipid binding in nitrogen-mixed doughs (Daniels *et al.*, 1970). (□) Control doughs mixed in nitrogen; (○) control doughs mixed in air; (△) nitrogen-mixed doughs with added peroxidized lipid in place of shortening fat.

1. *Increasing the Amount of Free Lipids in the Dough*

Once the dough is formed there is a decrease in the amount of extractable or "free" lipid, which further decreases during mixing. In the presence of air, however, the formation of bound lipid is halted resulting in an increase in "free" lipid (Davies *et al.*, 1969). This was shown by Daniels and co-workers (1970) to occur in the presence of only 1% oxygen in the mixing chamber. This phenomenon, illustrated in Figure 7.7, was attributed to lipid peroxidation, in which decreased lipid binding only occurred in doughs mixed in air. The absence of any effect following the addition of peroxidized lipids suggested that the process of peroxidation in the aerobically mixed doughs rather than lipid peroxides exerted an influence on lipid binding (Eskin *et al.*, 1977; Frazier, 1979; Faubion and Hoseney, 1981). One possible mechanism proposed to explain the release of bound lipids during the mixing of dough is the following. The oxidized intermediates generated by the action of lipoxygenase on polyunsaturated fatty acids are thought to enter the nonaqueous region of gluten, where they oxidize sulfhydryl groups. This results in a reversal of the electrical charge on the protein surface which causes an inversion of the lipoprotein micelle hydrophilic binding sites. Following this, water now enters the protein structure with the release of bound lipid (see Ch. 11, Scheme 11.4). Lipoxygenase is particularly important as prevention of lipid binding ensures the effectiveness of added shortening fat in enhancing the loaf volume and soft crumb of bread (Coppock, 1974).

2. *Destruction of Essential Fatty Acids and Bleaching of Carotenoids*

Lipoxygenase improves the color of flour. Its use dates back over half a century when Haas and Bonn (1934) patented the use of ground soybeans to replace chemical agents to bleach the flour pigments. A detailed discussion of the ability of lipoxygenase to bleach the flour pigments can be found in Chapter 11.

3. *Increasing Mixing Tolerance and Dough Stability*

Mixing tolerance is very important in commercial baking as it implies that the dough is resistant to overmixing and breakdown after reaching peak development (Faubion and Hoseney, 1981). The tolerance of the dough can be monitored using a mixograph, in which breakdown is measured by the height and width of the tail after optimum development is attained. Koch (1956) reported that lipoxygenase caused an increase in mixing tolerance of dough, which was confirmed in later work by Frazier and co-workers (1973). These researchers used an enzyme-active soy flour on dough mixed in air at a rate of 20 kJ/kg /min. This dough had a relaxation time of 45 sec at 260 kJ/kg as compared to 31 sec at 160 kJ/kg for untreated flours (Figure 7.8). Consequently addition of enzyme-active soy flour improved the dough's mixing tolerance. Weak and co-workers (1977) showed that fast-acting oxidants such as potassium iodate (KIO) decreased mixing tolerance of dough which was reversed by the addition of enzyme-active soybean flour. This reversal in mixing tolerance by lipoxygenase was not dependent on oxygen as it still took place in an atmosphere of nitrogen (Hoseney *et al.*, 1980). The final bread product has a much whiter crumb, enhanced loaf volume and modified aroma (Drapron *et al.*, 1974; Frazier *et al.*, 1977). The source of lipoxygenase in North America is soybean flour, while in France the horse bean or *Vicia faba* L. is permitted. A further discussion of lipoxygenase is found in Chapter 11.

V. Role of Starch in Breadmaking

The contribution of starch to bread making is related to its three important properties: water absorption, gelatinization, and retrogradation.

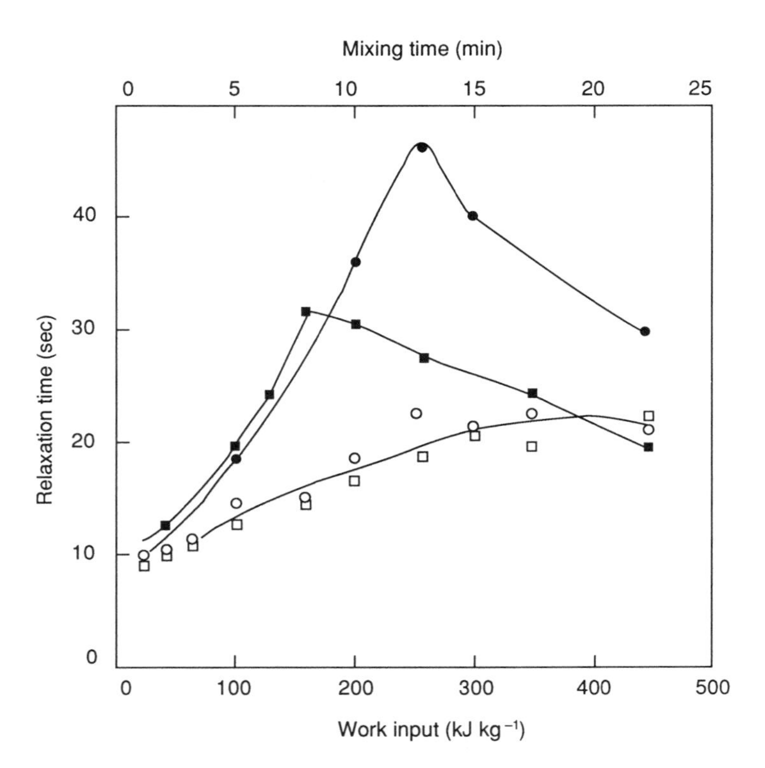

FIG. 7.8. Effect of enzyme-active soya flour on the mechanical development of doughs at 20 kg/min (Frazier *et al.*, 1973). (□, ■) Flour–salt–water dough; (○, ●) flour–salt–water dough plus enzyme-active soya flour.

A. WATER ABSORPTION

Starch granules are normally insoluble in cold water but when added to warm water they readily absorb water and swell. This process appears to be reversible at room temperature, but at gelatinization temperatures (>50°C) water absorption increases, resulting in an expansion of the starch granule volume (Hoseney *et al.*, 1983). Intact starch granules absorb only half their weight of cold water compared to damaged starch granules, which absorb up to twice their own weight of water (Tipples, 1982). The importance of damaged starch to bread making is related to its ability to increase the amount of water absorbed by the flour, which in turn increases the bread yield. Increasing the amount of damaged starch requires more water to be added to the flour to produce a dough of constant consistency (Kulp, 1973). The degree of starch damage can be controlled during the milling process as the optimum level required depends on the wheat variety used, protein content of the flour, amylase activity, and the particular baking process used.

B. Gelatinization

Gelatinization of starch occurs when starch granules are heated in aqueous medium. Starches normally exhibit birefringence in polarized light, which implies a high degree of molecular organization within the granules (Lineback, 1984). In the case of wheat starch this is characterized by the presence of concentric rings. When heated between 58 and 64°C, the gelatinization temperature range for wheat starch, birefringence properties are lost because of the loss of molecular organization (Lineback, 1984; Dengate, 1984). During heating, hydrogen bonds are broken or weakened in the starch, resulting in increased water absorption. This first takes place within the amorphous region of the starch granule but as the temperature rises the more crystalline areas start to melt. The importance of starch gelatinization to bread making is its effect on protein–starch interactions, which influence the dough mixing properties (Kulp and Lorenz, 1981). Until starch is gelatinized, water added to the flour associates primarily with the protein. During the baking process, partial gelatinization occurs as a result of limited availability of water. The starch granules swell, particularly the lenticular-shaped ones, resulting in a change in configuration which allows them to elongate during expansion of the gas cell. Thus starch granules appear to be involved in formation of the film surrounding the gas cells (Dennet and Sterling, 1979). During swelling, wheat starch granules transform into a saddle shape, which provides a larger surface area without increase in thickness (French, 1984). The formation of this unique saddle shape pattern of the wheat starch granule contributes to the formation of the gas cell film and the crumb structure of bread.

C. Retrogradation of Starch

Retrogradation refers to the physical change in starch from a gel-like state to a more crystalline structure (Krog *et al.*, 1989). In fresh bread, starch is mostly amorphous but undergoes recrystallization during storage. This process is responsible for bread staling and was shown by Katz (1928), using x-ray diffraction, that starch returned to a semicrystalline state in stale bread. Retrogradation takes place during aging and results in decreased starch solubility and increased rigidity of the starch system (D'Appolonia *et al.*, 1971). The mechanism of recrystallinity has tended to implicate the linear amylose fraction, as only pure solutions of amylose and not branched amylopectin underwent any retrogradation (Knightly, 1971). Subsequent research, however, indicated that amylopectin also played a role in the retrogradation process (Kulp and Ponte, 1981; D'Appolonia and Morad, 1981). The development of differential scanning colorimetry (DSC) enabled the thermal properties of starch to be explored and crystallinity to be quantified (Longton and LeGrys, 1981; Fearn and Russell, 1982). Using DSC enthalpy values (the energy needed to melt the starch crystals) as an index of crystallinity in aged gels, Longton and LeGrys (1981) found that crystallinity reached a maximum in 50% gels and disappeared in either too dilute (10%) or too concentrated (80%) gels. These changes were consistent with x-ray diffraction studies by Hellman and co-workers (1954), who reported the most intense x-ray patterns with 50% gels which decreased with higher or lower concentration gels. Thus water appeared to play an integral role in the retrogradation process, as shown in more recent studies by Zeleznak and Hoseney (1986). These researchers found that moisture content of the starch gel influenced the degree of retrogradation as illustrated in Figure 7.9. DSC enthalpy values were strongly dependent on the moisture content during the aging of starch gels. Solubilized amylopectin was shown to crystallize after aging for 14 days at 50 and 70% water at temperatures typical of granular starch. This indicated that amylopectin retrograded to the nongranular form with the major endotherm in bread crumb and starch gels consistent with the melting rate of amylopectin. The effect of moisture on the recrystallization rate of starch in bread samples with and without added antistaling agents was found to be identical. This suggested that the mechanism of antistaling agents to change the rate of starch recrystallization by altering the moisture relationships during aging was not the correct one to explain their function in bread making.

Zeleznak and Hoseney (1987) studied the effect of storage temperature on retrogradation of starch. Bread was stored at 4, 25, and 40°C for up to 5 days to see if the starch crystals annealed, that is, whether a more perfect crystal was formed at temperatures closer to the crystal melting temperature. Several conflicting results were reported previously, one on starch gels by Colwell *et al.* (1969), who found a negative correlation between storage temperature and recrystallization. This was confirmed by Longton and LeGrys (1981), who also found that the onset temperature (T_0) of melting increased for retrograded starch stored at elevated temperatures. Dragsdorf and Verriano-Marston (1980), however, reported the formation of different crystal structures in retrograded starch in breads treated with amylases ("V"-hydrate and "B") compared to the corresponding control breads ("V"-hydrate and "A"). The growth of "A" and "B" crystals was previously shown by Wright (Knight and Wade, 1971) to correlate with storage temperature as well as crumb moisture and firmness. Zeleznak and

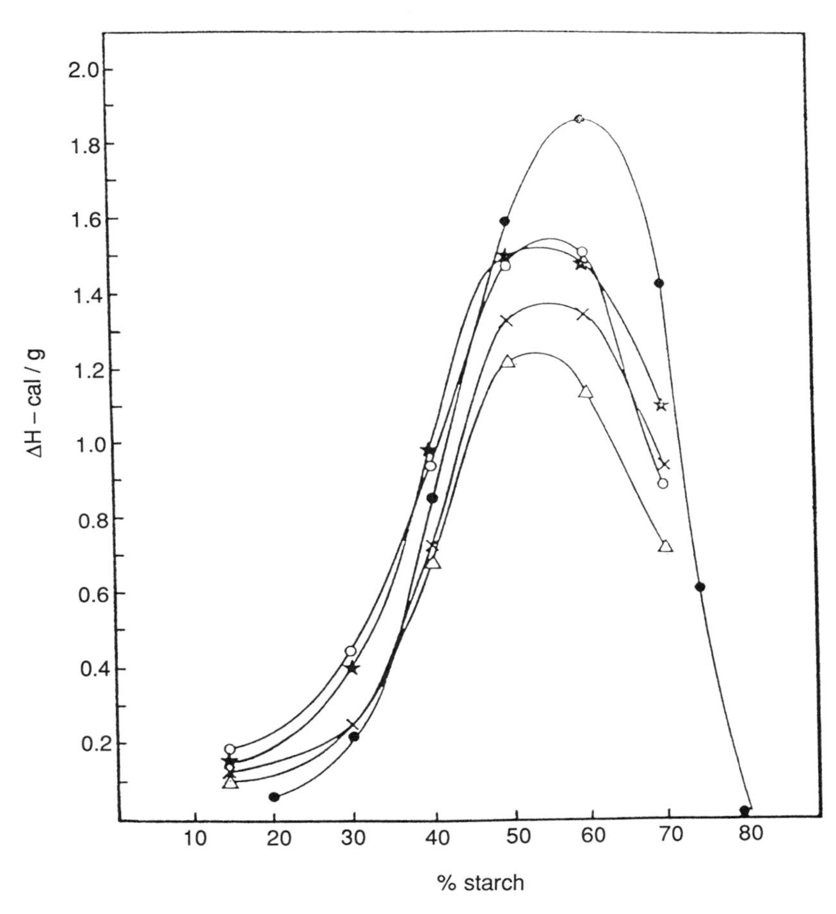

FIG. 7.9. Effect of the moisture present during aging on the enthalpy (ΔH in calories per gram of starch) of retrograded starch in starch gels and in bread baked with antistaling agents. Starch gels (●), control bread (★), bread with Crisco shortening (○), bread with Durkee D-10 shortening (△), bread with Durkee, Dur-em monoglyceride (×) (Zeleznak and Hoseney, 1986).

Hoseney (1987) reported an increase in both onset temperature (T_0) and maximum temperature (T_p) for the DSC thermograms of bread crumbs stored at increasing temperatures. X-ray diffraction patterns of starch isolated from these breads showed identical patterns, suggesting similar crystalline structures. These results pointed to the annealing behavior of the starch crystals as reflected by the increase in T_p during storage of bread at elevated temperatures.

VI. Fermentation

The essential ingredients of fermentation are yeast, flour, and water. During fermentation, yeast undergoes anaerobic metabolism, producing carbon dioxide gas which aerates the dough. In addition to this leavening effect, yeast also imparts flavor to the baked product.

A. Leavening Effect

The anaerobic fermentation of carbohydrates by yeast is responsible for the production of carbon dioxide and ethanol. The reaction involved is summarized in the following equation in which carbon dioxide was attributed to leavening of the dough:

$$C_6H_{12}O_6 \xrightarrow{\text{Yeast}} 2C_2H_5OH + 2CO_2$$

Moore and Hoseney (1985) examined the loaf volume during the bread making of a pup loaf (from 100 g of flour). Their results are summarized in Scheme 7.2, which shows an expansion of dough volume from 145 cc to 515 cc by the end of the proofing. During baking the dough was heated from 28 to 80°C, further loaf

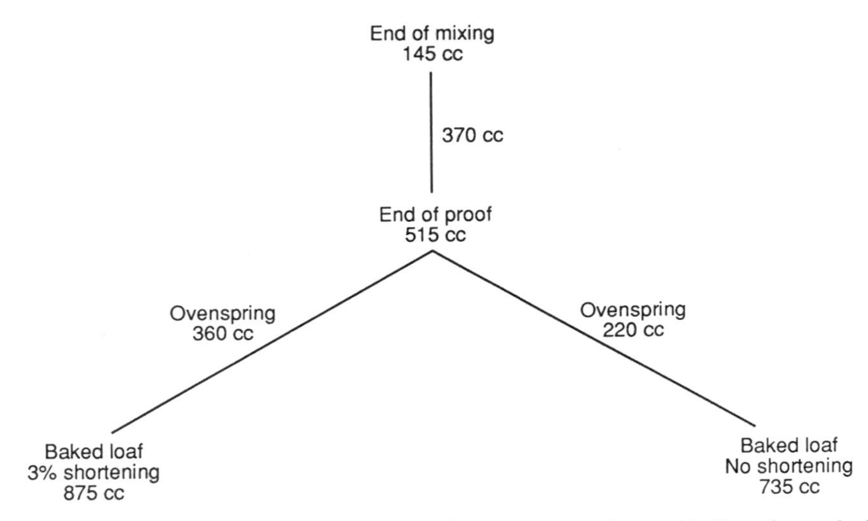

Scheme 7.2. Changes in dough volume during fermentation, proofing, and baking of a pup loaf (Moore and Hoseney, 1985).

expansion (ovenspring) of 360 cc occurred in the presence of shortening compared to 220 cc without added shortening. Moore and Hoseney (1985) found that only 40% of the total carbon dioxide gas produced by yeast fermentation was retained in fully proofed dough. The remaining 60% appeared to be lost during fermentation, punching, molding, and proofing of the dough. The increase in dough volume during baking could not be attributed totally to either expansion of carbon dioxide gas trapped in the air cells or its contribution from the aqueous phase. The additional expansion of dough during heating was attributed primarily to the vaporization of ethanol, with a small amount contributed by water vaporization.

B. Commercial Processes

White bread still accounts for the majority of bread eaten in North America. Different methods are used for developing the dough, including yeast fermentation, mechanical development, and chemical development.

1. *Yeast Fermentation*

Yeast fermentation involves mixing flour, water, yeast, fat, and salt to produce the dough, which is then fermented at 27°C prior to baking. Several variations of this method include the straight dough system and the sponge and dough system. In North America the sponge and dough system is used to produce 60% of all bread (Ponte, 1985). This involves mixing only a part of the flour with yeast and water, which is then fermented to produce a spongelike dough. After mixing, the remainder of the flour, water, salt, fat, etc., is added to the dough for a short fermentation period before proofing and baking. This process differs from the straight dough system, used in England, in which all the ingredients are mixed prior to fermentation. Bread produced by the sponge and dough system has good loaf volume, fine grain and texture, and fuller flavor. The major disadvantages of this method are the high costs of production and equipment involved in a process which takes 7 to 8 hr to complete. Efforts have been made to develop short-time doughs which require no more than 2 hr from mixing the dough to baking. This has been successfully developed for the production of hearth-baked bread and rolls which must be consumed within a couple of days. The production of excellent-quality white pan bread was reported by Ponte (1985) using short-time dough technology in which added dough improvers compensated for the absence of yeast fermentation.

2. *Mechanical Development*

Mechanical methods are designed to bypass the fermentation period by subjecting the dough to intense mechanical mixing. One particular method developed in England is the Chorleywood Bread Process (Chamberlain, 1984). This is a batch or continuous system for developing the dough by subjecting it to intense mechanical work (11 Wh/kg) for a period of up to 5 min. The mechanically developed dough required a relatively high level of slow oxidizing agents such as ascorbic acid or potassium bromate (75 ppm) as well as hard fat or solid monoglycerides (0.7% on flour weight) to promote gas retention during baking (French and Fish, 1983). Bread produced by this process was considered indistinguishable from bread produced by the bulk fermentation and did not stale as fast (Axford *et al.*, 1968).

The main advantages of the Chorleywood Bread Process include reduction of the total bread-making process from 4.5 to less than 2 hr, increased bread yield from a combination of increased absorption and elimination of fermentation losses, and better controlled conditions for dough development. These advantages are offset by the much higher energy costs to run this process and the twofold increased requirement in yeast. The Chorleywood Bread Process is particularly popular in the United Kingdom as it enables wheat flour of lower protein content to be used effectively in bread making (French and Fish, 1983).

3. *Chemical Development*

Chemical methods involve breaking disulfide bonds in the dough protein with a reducing agent such as L-cysteine or an oxidizing agent such as potassium bromate together with ascorbic acid. One such process is the Activated Dough Development Method, which accomplishes dough development without the need for fermentation or mechanical development.

VII. Baking

Baking is the climax of the bread-making process when the dough is finally transformed into bread. The effectiveness of the baking process is determined by heat, enzyme activity, water, starch, and protein content (Marston and Wannan, 1976).

The primary factor governing baking is the condition within the oven. In a conventional oven there is considerable variation in the temperature between the surface and interior of the dough. This is due to the cellular structure of the dough which makes it a poor conductor of heat (Marston and Wannan, 1976). For example, the temperature–time profiles for a dough piece baked at 235°C is shown in Figure 7.10. Toward the interior of the dough, the temperature rises slowly because of poor conductivity of the dough. The water content of the dough is important since its translocation from free liquid and protein-held liquid

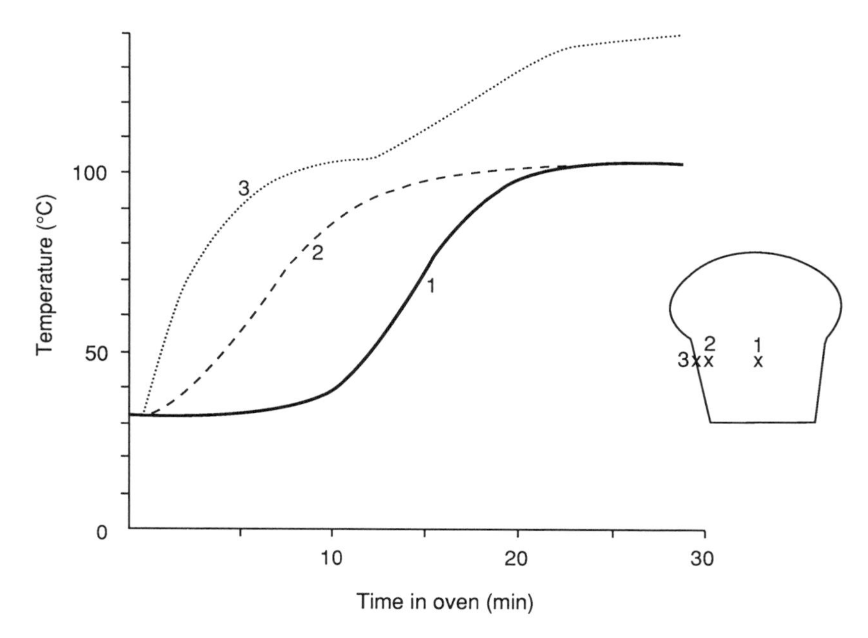

Fig. 7.10. Temperature–time profiles from a dough piece baked at 235°C (Marston and Wannan, 1983).

to starch at 60°C permits gelatinization. As gelatinization progresses at 70°C, various enzymes begin to be inactivated. Differences in thermal stabilities of cereal, fungal, and bacterial amylases are discussed in Chapter 11. In addition, irreversible denaturation and rupturing of gluten films occurs rapidly once the internal loaf temperature rises above 75°C. The overall effect is swelling of starch and buildup of the internal pressure from the gas and vapor, which results in the development of the dough structure. During the oven baking of bread, water is rapidly lost from the surface, which undergoes browning via the Maillard reaction (see Chapter 5) under the higher oven temperature conditions. The final result is a freshly baked loaf with an attractive resilient crumb and delicious flavor.

Bibliography

Anonymous. (1983). Descriptive list of varieties of cereals, maize, and oilseeds. Government Varieties Office, Alfred Strothe Verlag, Hannover.

Atkins, J. H. C. (1971). Mixing requirements of baked products. *Food Manuf.* **46**(2), 47.

Axford, D. W. E., Colwell, K. H., Cornford, S. J., and Elton, G. A. H. (1968). Effect of loaf specific volume on the rate and extent of staling in bread. *J. Sci. Food Agric.* **19,** 95.
Axford, D. W. E., McDermott, E. E., and Redman, D. G. (1978). Small-scale tests of breadmaking quality. *Milling Feed Fertil.* **161,** 18.
Beckwith, A. C., and Wall, J. S. (1966). Reduction and reoxidation of wheat glutenin. *Biochim. Biophys. Acta* **130,** 155.
Bietz, J. A. (1983). Separation of cereal proteins by reversed-phase high performance liquid chromatography. *J. Chromatogr.* **255,** 219.
Bietz, J. A. (1984). Analysis of wheat gluten proteins by high pressure liquid chromatography. Part II. *Baker's Dig.* **58**(2), 22.
Bietz, J. A., and Cobb, L. A. (1985). Improved procedures for rapid wheat varietal identification by reversed-phase high-performance liquid chromatography of gliadin. *Cereal Chem.* **62,** 332.
Branlard, G., and Dardevet, M. (1985). Diversity of grain proteins and bread wheat quality. I. Correlation between gliadin bands and flour quality characteristics. *J. Cereal Sci.* **3,** 329.
Burnouf, T., and Bouriquet, R. (1980). Glutenin subunits of genetically related European hexaploid wheat cultivars: Their relationship to breadmaking quality. *Theor. Appl. Genet.* **58,** 107.
Bushuk, W. (1985). Flour proteins: Structure and functionality in dough and bread. *Cereal Foods World* **30,** 447.
Chamberlain, N. (1984). The Chorleywood bread process: International prospects. *Cereal Foods World* **29,** 656.
Chamberlain, N. (1987). Recent developments in baking technology. *In* "Food Technology International Europe," pp. 117–119. Lavenham Press Ltd., U. K.
Chamberlain, N., and Collins, T. H. (1979). The Chorleywood bread process: the roles of oxygen and nitrogen. *Bakers Dig.* **53**(1), 18.
Chung, O. K., and Pomeranz, Y. (1981). Recent research on wheat lipids. *Baker's Dig.* **55**(5), 38.
Chung, O. K., Pomeranz, Y., and Finney, K. F. (1978). Wheat flour lipids in breadmaking. *Cereal Chem.* **55,** 598.
Chung, O. K., Pomeranz, Y., Jacobs, R. M., and Howard, B. G. (1980a). Lipid extraction conditions to differentiate among hard red winter wheats that vary in breadmaking. *J. Food Sci.* **45,** 1168.
Chung, O. K., Pomeranz, Y., and Finney, K. F. (1980b). Wheat flour lipids in breadmaking. *Cereal Chem.* **55,** 598.
Colwell, K. H., Axford, D. W. E., Chamberlain, N., and Elton, G. A. H. (1969). Effect of storage temperature on the aging of concentrated wheat starch gels. *J. Sci. Food Agric.* **20,** 550.
Coppock, J. B. M. (1974). Selling food technology. *Chem. Ind. (London),* p. 358.
Dafraty, R. D., Pomeranz, Y., Shogren, M. D., and Finney, K. F. (1968). Functional breadmaking properties of lipids. II. The role of flour lipid fractions in breadmaking. *Food Technol.* **22,** 327.
Daniels, N. W. R., Richmond, J. J., Russell Eggitt, P. W., and Coppock, J. B. M. (1970). Studies on lipids of flour. V. Effect of air on lipid binding. *J. Sci. Food Agric.* **20,** 129.
D'Appolonia, B. L., and Morad, M. M. (1981). Bread staling. *Cereal Chem.* **58,** 186.
D'Appolonia, B. L., Gilles, K. A., and Metcalf, D. G. (1970). Effect of water-soluble pentosans on gluten-starch loaves. *Cereal Chem.* **47,** 194.
D'Appolonia, B. L., Gillies, K. A., Osman, E. M., and Pomeranz, Y. (1971). Carbohydrates. *Monogr. Ser.—Am. Assoc. Cereal Chem.* **3**(rev.), 301–392.
Davies, R. J., Daniels, N. W. R., and Greenshields, R. N. (1969). An improved method of adjusting flour moisture in studies on lipid binding. *J. Food Technol.* **4,** 117.
Dengate, H. N. (1984). Swelling, pasting and gelling of wheat starch. *Adv. Grain Sci. Technol.* **6,** 49–82.
Dennet, K., and Sterling, C. (1979). Role of starch in bread formation. *Staerke* **31,** 305.

Dragsford, R. D., and Verriano-Marston, E. (1980). Bread staling, X-ray diffraction studies on bread supplemented with α-amylases from different sources. *Cereal Chem.* **57,** 310.

Drapron, R., Beaux, Y., Cormier, M., Geffrey, J., and Adrian, J. (1974). Lipoxygenase action during bread making. Destruction of essential fatty acids, carotenoids, and tocopherols. Deterioration of flavor of bread. *Ann. Technol. Agric.* **23,** 353.

Eskin, N. A. M., Grossman, S., and Pinsky, C. (1977). Biochemistry of lipoxygenase in relation to food quality. *CRC Crit. Rev. Food Sci. Nutr.* **9,** 1.

Ewart, J. A. D. (1968). A hypothesis for the structure and rheology of glutenin. *J. Sci. Food Agric.* **23,** 687.

Ewart, J. A. D. (1972a). A modified hypothesis for the structure and rheology of glutenins. *J. Sci. Food Agric.* **23,** 687.

Ewart, J. A. D. (1972b). Isolation of an albumin from Cappelle-Des prez and Manitou wheats. *J. Sci. Food Agric.* **23,** 701.

Faubion, J. M., and Hoseney, R. C. (1981). Lipoxygenase, its biochemistry and role in breadmaking. *Cereal Chem.* **58,** 175.

Fearn, T., and Russell, P. L. (1982). A kinetic study of bread staling by differential scanning calorimetry. The effect of loaf specific volume. *J. Sci. Food Agric.* **33,** 537.

Fincher, G. B., and Stone, B. A. (1974). A water-soluble arabinogalactan-peptide from wheat endosperm. *J. Biol. Sci.* **27,** 117.

Finney, K. F., and Shogren, M. D. (1972). A ten gram mixograph for determining and predicting functional properties of wheat flours. *Baker's Dig.* **46**(2), 32.

Finney, K. F., Heyne, E. G., Shogren, M. D., Bolte, L. C., and Pomeranz, Y. (1978). Functional properties of high yielding European wheats grown at Manhattan, Kansas. *Cereal Foods World.* **23,** 479.

Frazier, P. J. (1979). Lipoxygenase action and lipid binding in breadmaking. *Baker's Dig.* **53**(6), 8.

Frazier, P. J., Leigh-Dugmore, F. A., Daniels, N. W. R., Russell Eggitt, P. W., and Coppock, J. B. M. (1973). The effect of lipoxygenase action on the mechanical development of wheat flour doughs. *J. Sci. Food Agric.* **24,** 421.

Frazier, P. J., Brimblecombe, F. A., Daniels, N. W. R., and Russell Eggitt, P. W. (1977). The effect of lipoxygenase action on the mechanical development of doughs from fat-extracted and reconstituted wheat flours. *J. Sci. Food Agric.* **28,** 247.

French, D. (1984). Organization of starch granules. *Adv. Cereal Sci. Technol.* **7,** 321–334.

French, F. D., and Fish, A. R. (1983). High speed mechanical dough development. *Baker's Digest.***57**(4), 94.

Graveland, A., Dazer, P., and Bosveld, P. (1979). Extraction and fractionation of wheat flour protein. *J. Sci. Food Agric.* **30,** 71.

Graveland, A., Boosveld, P., Lichtendank, N. J., Moonen, H. G., and Scheepstra, A. (1982). Extraction and fractionation of wheat flour protein. *J. Sci. Food Agric.* **33,** 1117.

Greenwood, C. T., and Ewart, J. A. D. (1975). Hypothesis for the structure of gluten in relation to rheological properties of gluten and dough. *Cereal Chem.* **52,** 146.

Haas, L. W., and Bonn, R. M. (1934). Bleaching bread dough. U. S. Patent 1,957,333.

Hellman, N. N., Fairchild, B., and Sentl, F. R. (1954). The bread staling problem. Molecular organization of starch upon aging of concentrated starch gels at various moisture levels. *Cereal Chem.* **31,** 495.

Hoover, W. (1984). Baking—The future looks bright. *Cereal Foods World* **29,** 644.

Hoseney, R. C. (1984). Gas retention in bread doughs. *Cereal Foods World* **29,** 305.

Hoseney, R. C. (1985). The mixing phenomenon. *Cereal Foods World* **30,** 453.

Hoseney, R. C., and Faubion, H. (1981). A mechanism for the oxidative gelation of wheat flour water-soluble pentosans. *Cereal Chem.* **58,** 421.

Hoseney, R. C., Finney, K. F., Shogren, M. D., and Pomeranz, Y. (1969a). Functional (breadmaking) and biochemical properties of wheat flour components. II. Role of water solubles. *Cereal Chem.* **46,** 117.

Hoseney, R. C., Finney, K. F., Pomeranz, Y., and Shogren, M. D. (1969b). Functional (breadmaking) and biochemical properties of wheat flour components. IV. Gluten protein fractionation by solubilizing in 70% ethyl alcohol and in dilute lactic acid. *Cereal Chem.* **46,** 495.

Hoseney, R. C., Finney, K. F., Pomeranz, Y., and Shogren, M. D. (1969c). Functional (breadmaking) and biochemical properties of wheat flour components. V. Roles of total extractable lipids. *Cereal Chem.* **46,** 606.

Hoseney, R. C., Finney, K. F., and Pomeranz, Y. (1970). Functional (breadmaking) and biochemical properties of wheat flour components. VI. Gliadin–lipid–glutenin interactions in wheat gluten. *Cereal Chem.* **47,** 135.

Hoseney, R. C., Rao, H., Faubion, J., and Sidhu, J. S. (1980). Mixograph studies. IV. The mechanism by which lipoxygenase increases mixing tolerance. *Cereal Chem.* **7,** 163.

Hoseney, R. C., Lineback, P. R., and Seib, P. (1983). Role of starch in baked goods. *Baker's Digest.* **57**(4), 65.

Hoseney, R. C., Zeleznak, K., and Lai, C. S. (1986). Wheat gluten: A glassy polymer. *Cereal Chem.* **63,** 285.

Huebner, F. R. (1970). Comparative studies on glutenins from different classes of wheat. *J. Agric. Food Chem.* **18,** 256.

Huebner, F. R., and Wall, J. S. (1976). Fractionation and quantitative differences of glutenin from wheat varieties varying in baking quality. *Cereal Chem.* **53,** 228.

Jones, R. W., Taylor, N. W., and Senti, F. R. (1959). Electrophoresis and fractionation of wheat gluten. *Arch. Biochem. Biophys.* **84,** 363.

Jones, R. W., Babcock, G. E., Taylor, N. W., and Senti, F. R. (1961). Molecular weight of wheat gluten fractions. *Arch. Biochem. Biophys.* **94,** 483.

Junge, R. C., Hoseney, R. C., and Varriano-Marston, E. (1981). Effect of surfactants on air incorporation in dough and crumb grain of bread. *Cereal Chem.* **58,** 338.

Kasarda, D. D., Bernardin, J. C., and Nimmo, C. C. (1976). Wheat protein. *Adv. Cereal Sci. Technol.* **1,** 158–236.

Kasarda, D. D., Nimmo, C. C., and Kohler, G. O. (1979). Proteins and the amino acid composition of wheat fractions. In "Wheat, Chemistry and Technology" (Y. Pomeranz, ed.), 2nd ed., pp. 227–299. Am. Assoc. Cereal Chem., St. Paul, Minnesota.

Katz, J. R. (1928). The X-ray spectrography of starch and gelatinization and retrogradation of starch in the staling process. *In* "A Comprehensive Survey of Starch Chemistry" (R. P. Walton, ed.), pp. 68–76. Chemical Catalog Co., New York.

Kim, S. K., and D'Appolonia, B. L. (1977). Bread staling studies. III. Effect of pentosans on dough, bread, and bread staling rate. *Cereal Chem.* **54,** 225.

Knight, R. A., and Wade, P. (1971). Starch: Granule structure and technology. *Chem. Ind. (London),* p. 568.

Knightly, W. H. (1977). The staling of bread. A review. *Baker's Dig.* **51**(5), 52.

Koch, R. B. (1956). Mechanism of fat oxidation. *Baker's Dig.* **30,** 48.

Krog, N., Olesen, S. K., Toernaes, H., and Joensson, T. (1989). Retrogradation of the starch fraction in wheat bread. *Cereal Foods World* **34,** 281.

Kulkarni, R. G., Ponte, Jr., J. G., and Kulp, K. (1987). Significance of gluten content as an index of flour quality. *Cereal Chem.* **4,** 1.

Kulp, K. (1973). Physiochemical properties of wheat starch as related to bread. *Baker's Dig.* **47,** 34.

Kulp, K., and Lorenz, K. (1981). Starch functionality in white pan breads: New developments. *Baker's Dig.* **55,** 24.

Kulp, K., and Ponte, J. G. (1981). Staling of white pan bread: Fundamental causes. *CRC Crit. Rev. Food Sci. Nutr.* **15,** 1.

Kundig, W., Neukom, H., and Deuel, H. (1961a). Untersuchungen uber Getriedesleimstoffe. I. Chromatographische Fraktionierung von wasserloslichen Weizenmehlpentosanen and Diethylaminoethyl cellulose. *Helv. Chim. Acta* **44,** 823.

Kundig, W., Neukom, H., and Deuel, H. (1961b). Untersuchungen uber Getreideschleimstoffe. II. Uber die Gelierung Wasseriger von Weizenmehlpentosanen durch. Oxydationsmittel. *Helv. Chim. Acta* **44,** 969.

Laszity, R. (1984). "The Chemistry of Cereal Proteins." CRC Press, Boca Raton, Florida.

Lee, J. W., and MacRitchie, F. (1971). The effect of gluten protein fractions on dough properties. *Cereal Chem.* **48,** 620.

Lin, F. M., and Pomeranz, Y. (1968). Characterization of water-soluble flour pentosans. *J. Food Sci.* **33,** 599.

Lin, F. M., D'Appolonia, B. L., and Youngs, V. L. (1974). Hard red spring and durum wheat polar lipids. II. Effect on quality of bread and pasta products. *Cereal Chem.* **51,** 34.

Lineback, D. R. (1984). The starch granule: Organization and properties. *Baker's Dig.* **58**(3), 16.

Longton, J., and LeGrys, G. A. (1981). Differential scanning calorimetry studies on the crystallinity of aging wheat starch gels. *Staerke* **33,** 410.

MacRitchie, F. (1977). Flour lipids and their effects in baking. *J. Sci. Food Agric.* **28,** 79.

MacRitchie, F. (1980). Physicochemical aspects of some problems in wheat research. *Adv. Cereal Sci. Technol.* **3,** 271–321.

MacRitchie, F. (1981). Flour lipids: Theoretical aspects and functional properties. *Cereal Chem.* **58,** 156.

MacRitchie, F. (1985). Studies of the methodology for fractionation and reconstitution of wheat flours. *J. Cereal Sci.* **3,** 221.

MacRitchie, F., and Gras, P. W. (1973). The role of flour lipids in baking. *Cereal Chem.* **50,** 292.

Mahdi, J. G., Varriano-Marston, E., and Hoseney, R. C. (1981). The effect of mixing atmosphere and fat crystal size on dough structure and bread quality. *Baker's Dig.* **55**(2), 28.

Marston, P. E., and Wannan, T. L. (1983). Bread baking, the transformation from dough to bread. *Baker's Dig.* **50**(4), 24.

Medcalf, D. G., D'Appolonia, B. L., and Gilles, K. A. (1968). Comparison of chemical composition and properties between hard red spring and durum wheat endosperm pentosans. *Cereal Chem.* **45,** 539.

Meredith, O. B., and Wren, J. J. (1966). Determination of molecular weight distribution in wheat flour proteins by extraction and gel filtration in a dissociating medium. *Cereal Chem.* **43,** 169.

Moore, W. R., and Hoseney, R. C. (1985). The leavening of bread dough. *Cereal Foods World* **30,** 791.

Neukom, H., and Markwalder, H. U. (1978). Oxidative gelation of wheat flour pentosans: A new way of cross-linking polymers. *Cereal Foods World* **23,** 374.

Ng, P. (1987). Relationship between high molecular weight subunits of glutenin and breadmaking quality of Canadian grown wheats. Ph. D. Thesis, University of Manitoba, Winnipeg, Canada.

Nicolas, J., and Drapron, R. (1983). Influence of some biochemical parameters on lipoxygenase activity during breadmaking. *In* "Progress in Cereal Chemistry and Technology. Proceedings of the VIIth World Cereal and Bread Congress, Prague" (J. Holas and J. Kratochvil, eds.). Elsevier.

Nielsen, H. C., Babcock, G. E., and Senti, F. R. (1962). Molecular weight studies on glutenin before and after disulfide-bond splitting. *Arch. Biochem. Biophys.* **96,** 252.

Orth, R. A., and Bushuk, W. (1972). A comparative study of the proteins of wheats of diverse baking qualities. *Cereal Chem.* **49,** 268.

Orth, R. A., and Bushuk, W. (1973). Studies of glutenin. II. Relation of variety, location of growth, and baking quality to molecular weight distribution of subunits. *Cereal Chem.* **50,** 191.
Orth, R. A., Baker, R. J., and Bushuk, W. (1972). Statistical evaluation of techniques for predicting baking quality of wheat cultivars. *Can. J. Plant Sci.* **52,** 139.
Osborne, T. B. (1907). The proteins of the wheat kernel. *Carnegie Inst. Washington Publ.* **84.**
Payne, P. I. (1987). Genetics of wheat storage proteins and the effect of allelic variation on bread-making quality. *Annu. Rev. Plant Physiol.* **38,** 141.
Payne, P. I. and Corfield, K. G. (1979). Subunit composition of wheat glutenin proteins isolated by gel filtration in a dissociating medium. *Planta* **145,** 83.
Payne, P. I., Corfield, K. G., and Blackman, J. A. (1979). Identification of a high molecular weight subunit of glutenin whose presence correlates with breadmaking quality in wheats of related pedigree. *Theor. Appl. Genet.* **55,** 153.
Payne, P. I., Harris, P. A., Law, C. N., Holt, L. M., and Blackman, J. A. (1980a). The high-molecular weight subunits of glutenin: Structure, genetics and relationships to breadmaking quality. *Ann. Technol. Agric.* **29,** 309.
Payne, P. I., Law, C. N., and Mudd, F. E. (1980b). Control by homeologous group 1 chromosomes of the high-molecular weight subunits of glutenin, a major protein of wheat endosperm. *Theor. Appl. Genet.* **58,** 113.
Payne, P. I., Holt, L. M., Johnson, E. A., and Law, C. N. (1984). Wheat storage proteins, their genetic and their potential for manipulation by plant breeding. *Trans. R. Soc. London. Ser. B* **304,** 359.
Payne, P. I., Nightingale, M. A., Krattiger, A. F., and Holt, L. M. (1987). The relationship between HMW glutenin subunit composition and the bread-making quality of British-grown wheat varieties. *J. Sci. Food Agric.* **40,** 51.
Payne, P. I., Holt, L. M., Krattinger, A. F., and Carrilo, J. M. (1988). Relationships between seed quality characteristics and HMW glutenin subunit composition determined using wheats grown in Spain. *J. Cereal Sci.* **7,** 229.
Pence, J. W., and Olcott, H. S. (1952). Effect of reducing agents on gluten protein. *Cereal Chem.* **29,** 292.
Pence, J. W., Elder, H. A., and Mecham, D. K. (1950). Preparation of wheat flour pentosans for use in reconstituted doughs. *Cereal Chem.* **27,** 60.
Pence, J. W., Elder, H. A., and Mecham, D. K. (1951). Some effects of soluble flour components on baking behaviour. *Cereal Chem.* **28,** 94.
Perlin, A. S. (1951). Structure of soluble pentosans of wheat flours. *Cereal Chem.* **28,** 382.
Pomeranz, Y. (1965). Dispersibility of wheat proteins and aqueous urea solutions: A new parameter to evaluate breadmaking potentialities of wheat flour. *J. Sci. Food Agric.* **16,** 586.
Pomeranz, Y. (1985). Wheat flour lipids—What they can and cannot do in bread. *Cereal Foods World* **30,** 443.
Ponte, J. G. (1985). Short time doughs simplify pan bread processing. *Baker's Dig.* **59**(1), 24.
Pyler, E. J. (1983a). Flour proteins: Role in baking performance. I. *Baker's Dig.* **57**(May), 24.
Pyler, E. J. (1983b). Flour proteins: Role in baking performance. II. *Baker's Dig.* **57**(September), 44.
Schultz, G. E., and Schirmer, R. H. (1979). "Principles of Protein Structure," Chapter 5. Springer-Verlag, New York.
Shelton, D. R., and D'Appolonia, B. L. (1985). Carbohydrate functionality in the baking process. *Cereal Foods World* **30,** 437.
Shogren, M. D., Finney, K. F., and Hoseney, R. C. (1969). Functional (breadmaking) and biochemical properties of wheat flour components. I. Solubilizing gluten and flour proteins. *Cereal Chem.* **46,** 93.
Sidhu, J. S., Nordin, P., and Hoseney, R. C. (1980a). Mixograph studies. III. Reaction of fumaric acid with gluten proteins during mixing. *Cereal Chem.* **57,** 159.

Sidhu, J. S., Hoseney, R. C., Faubion, J. M., and Nordin, P. (1980b). Reaction of C^{14}–cysteine with wheat flour water solubles under ultraviolet light. *Cereal Chem.* **57,** 380.

Tatham, A. S., and Shewry, P. R. (1985). The conformation of wheat gluten proteins. The secondary structures and thermal stabilities of α, β, γ and ω-gliadins. *J. Cereal Sci.* **3,** 103.

Tatham, A. S., Shewry, P. R., and Miflin, B. J. (1984). Wheat gluten elasticity: A similar molecular basis to elastin. *FEBS Lett.* **177,** 205.

Tatham, A. S., Miflin, B. J., and Shewry, P. R. (1985). The beta configuration in wheat gluten proteins: Relationship to gluten elasticity. *Cereal Chem.* **62,** 405.

Taylor, N. W., and Cluskey, J. E. (1962). Wheat gluten and its glutenin component: Viscosity, diffusion and sedimentation analysis. *Arch. Biochem. Biophys.* **97,** 399.

Tipples, K. H. (1982). Breadmaking technology. *In* "Grains and Oilseeds: Handling, Marketing, Processing," 3rd ed., Ch. D6, pp. 601–635. Canadian International Grains Institute, Winnipeg, Canada.

Udy, D. C. (1956). The intrinsic viscosities of the water-soluble components of wheat flour. *Cereal Chem.* **33,** 67.

Venkatachalam, C. M. (1968). Stereochemical criteria for polypeptides and proteins. V. Conformation of a system of three linked peptide units. *Biopolymers* **6,** 1425.

Wasik, R. J., and Bushuk, W. (1974). Studies of glutenin. V. Note on additional preparative methods. *Cereal Chem.* **51,** 112.

Weak, E. D., Hoseney, R. C., Seib, P. A., and Barg, M. (1977). Mixograph studies. I. Effect of certain compounds on mixing properties. *Cereal Chem.* **54,** 794.

Woychik, J. H., Hu, R., and Dimler, R. J. (1964). Reduction and electrophoresis of wheat gliadin and glutenin. *Biochim. Biophys. Acta* **105,** 155.

Wrench, P. M. (1965). The role of wheat pentosans in baking. III. Enzymatic degradation of pentosans. *J. Sci. Food Agric.* **16,** 51.

Yeh, Y. F., Hoseney, R. C., and Lineback, D. R. (1980). Changes in wheat flour pentosans as a result of dough mixing and oxidation. *Cereal Chem.* **57,** 144.

Zeleznak, K. J., and Hoseney, R. C. (1986). The role of water in the retrogradation of wheat starch gels and bread crumb. *Cereal Chem.* **63,** 407.

Zeleznak, K. J., and Hoseney, R. C. (1987). Characterization of starch from bread aged at different temperatures. *Staerke* **39,** 231.

8

Biochemistry of Food Processing: Cheese and Yoghurt

I. Introduction

The production of cheese originated thousands of years ago in the Middle East (Scott, 1981). Cheese making was subsequently introduced to Europe during the period of the Roman Empire, where it was produced either in monasteries or on farms. It was not until the middle of the nineteenth century, however, that cheese was first produced in factories. Today over 800 different types of cheeses are produced worldwide, many differing only in shape, size, degree of ripening, type of milk, condiments, packaging, and geographical area of production (Irvine and Hill, 1985). In addition to cheese, a large number of fermented milk products are also consumed around the world. These differ in the process of manufacture and starter organism used (International Dairy Federation, 1983). The main cultured product consumed in Australia, Canada, England, and the United States is yoghurt, while various cultured buttermilk and cream products are very popular in Scandinavian and Eastern European countries. This chapter will discuss those biochemical changes taking place during the production of cheese and yoghurt.

TABLE 8.1

CLASSIFICATION OF CHEESE

Classification	Type of cheese
Hard	Cheddar, Swiss, Romano
Soft or semisoft	Gouda, Camembert, Brick
Fresh	Cottage, Cream, Quarg

II. Cheese

Cheese is made by the concentration of all or part of the milk components by coagulation of the milk protein with enzymes, acid-producing bacteria, or acid. The classification of cheese is based on the method of production used (Table 8.1). For instance, the production of hard cheeses requires both high acid development and high temperatures, while soft or semisoft cheeses are associated with slow acid development, washing to control lactose, and minimal cooking temperatures. Fresh cheese, on the other hand, involves development of high acidity by bacterial action, while processed cheese is characterized by the use of high temperatures to arrest ripening (Irvine and Hill, 1985). This section will focus on the biochemistry of cheese production.

A. Milk Quality

The raw material for cheese is milk, the quality of which has a great influence on the quality of the final cheese product. Any flavor defects developing in milk, as discussed in Chapter 10, will have a deleterious effect on the finished cheese product. In addition, milk high in bacteria or antibiotics is not permitted for use in the production of cheese. It is important to use milk of the highest quality as well as sanitary equipment for the manufacture of cheese products. The major milk quality tests include:

1. organoleptic evaluation of odor and taste;
2. bacterial plate counts (ICMSF, 1978);
3. inhibitory substances test (ICMSF, 1978);
4. fermentation and curd tests (Irvine, 1982).

The type of cheese produced determines whether cream and milk are both used. Cheddar cheese, for example, uses whole milk, while only skim milk is used for the production of cottage cheese. In the case of creamed cottage cheese, cream is added as a dressing. Swiss and Edam cheeses, however, are made from a mixture of whole and skim milk.

B. Cheese Production

The production of cheese commences with the formation of the curd and concludes with the ripening of the final cheese product. Essentially five steps are involved in the cheese-making process, including acidification, coagulation, dehydration, molding/shaping, and salting.

1. *Acidification*

In the process of cheese making a "starter" culture of lactic acid bacterium, *Streptococcus lactis*, is added to the milk. The "starter" culture ferments the carbohydrates via the hexose diphosphate pathway to pyruvate, which is then reduced to lactic acid (Adda *et al.*, 1982). The fermentation of lactose to lactic is fundamental to cheese making and is carried out by both *Streptococcus* and *Lactobacillus* bacteria. While *Streptococcus* bacteria are used in the production of Cheddar, Gouda, and cottage cheese, *Lactobacillus* organisms such as *L. bulgaricus* and *L. acidophilus* are used in the production of both Swiss and Grana cheeses (Sharpe, 1978, 1979). This step is crucial for cheese making and has been the subject of considerable research activity to develop improved starter organisms by genetic manipulation to increase phage resistance as well as acid production and proteolytic activity (Fox, 1987). In the Netherlands, mixed-strain starters or P-starters are used in the production of cheese (Stadhousers, 1986). These are concentrates of phage-resistant starters selected on the basis of desirable acidity and final cheese flavor, which are stored at neutral pH under liquid nitrogen at the Netherlands Institute of Dairy Research (NIZO). The increased production of lactic acid lowers the pH, which in turn facilitates rennet action, curd formation, volatile formation, and enzymatic action and helps to maintain the shelf life and maturation of the cheese products (McKay *et al.*, 1971).

2. *Milk Coagulation*

a. Rennin or Chymosin. Coagulation of milk by chymosin or rennin involves several steps, both enzymatic and nonenzymatic. The calcium-sensitive casein fractions (α_{s1}, α_{s2}, and β) in the core of the micelles are protected from precipitation by κ-casein located primarily on the surface. Hydrolysis of κ-casein by rennet during the initial phase, however, releases a highly negatively charged macropeptide. The overall result is a 50% reduction of the negative charge on the casein micelles, which causes destabilization (Green and Grutchfield, 1971; Pearce, 1976). This is followed by the second phase in which the casein micelles aggregate (Lindqvist, 1963a,b) by hydrophobic forces or possibly electrostatic interactions (Payens, 1966, 1977).

b. Phase One: Enzymatic Hydrolysis. Proteolysis of κ-casein by rennet follows the typical Michaelis–Menten kinetics (Castle and Wheelock, 1972; Chaplin and Green, 1980; Garnier, 1963). The availability of different types of rennets for cheese making is discussed in detail in Chapter 11. Rennet hydrolyzes an essential peptide bond (Phe–Met) of κ-casein (Delfour *et al.*, 1965; Waugh and von Hippel, 1956), liberating a soluble casein macropeptide and para-κ-casein. The optimum pH for rennet action on κ-casein is 5.1 to 5.3 (Humme, 1972). In fact it is possible to separate enzymatic activity from the secondary phase of coagulation since clotting does not occur at low temperatures or high pH (Berridge, 1942). The activation energy required for this reaction was reported to be around 10 kcal/mole (Nitschmann and Bohren, 1955).

While rennet specifically attacks the κ-casein fraction, it is also capable of rapidly hydrolyzing the peptide bond between Phe 23 and Phe 24 from isolated α_{s1}-casein and to a much slower degree that of β-casein (Mullvihill and Fox, 1977). These proteins do not become accessible to enzymatic attack until the micelle structure is disrupted following proteolysis of κ-casein. The eventual hydrolysis of α- and β-casein fractions is particularly important during the ripening of cheese and affects its body, texture, and flavor.

The rate equation for the enzymatic hydrolysis of κ-casein by chymosin/rennet based on a random Bi–Bi mechanism is summarized in Scheme 8.1. This is based on random binding of chymosin to κ-casein or water with the conversion of the enzyme–substrate complex to enzyme–product complex and the subse-

$$E + H_2O \underset{}{\overset{K_1}{\rightleftharpoons}} E - H_2O \quad \xrightleftharpoons[-\kappa]{+\kappa,\ K_2} \quad E - \kappa H_2O \underset{k_{cr}}{\overset{k_{cf}}{\rightleftharpoons}} E - P - M \quad \xrightleftharpoons[+M]{-M,\ K_5} \quad E - P \overset{K_7}{\rightleftharpoons} E + P$$

$$E + \kappa \underset{K_2}{\rightleftharpoons} E - \kappa \quad \xrightleftharpoons[-H_2O]{+H_2O,\ K_4} \quad E - \kappa H_2O \qquad E - P - M \quad \xrightleftharpoons[-P]{+P,\ K_6} \quad E - M \underset{K_3}{\rightleftharpoons} E + M$$

SCHEME 8.1. The complete reaction scheme proposed for κ-casein hydrolysis by renneting enzymes. The reaction proposed is a random Bi–Bi mechanism (Cleland, 1963). E, free enzyme; κ, κ-casein. The *K* values are various equilibrium constants which are related to the Michaelis–Menten constants (Mahler and Cordes, 1971; Carlson *et al.*, 1987a).

quent release of the enzyme and product. A series of experiments were recently carried out by Carlson and co-workers (1987a–d) on the kinetics of milk coagulation. They found that the rate of hydrolysis of κ-casein by rennet was affected by temperature, requiring an activation energy of 9.9 kcal/mole, while pH had a relatively minor effect. By carrying out enzymatic hydrolysis at temperatures above 50°C or at pH levels greater than 7.0, Carlson *et al.* (1987a) could denature the milk-clotting enzymes. While this is undesirable from a cheese production point of view it permitted an opportunity to study micelle flocculation following the completion or partial completion of the enzymatic phase of the milk-clotting process.

c. Phase Two: Clotting. Following rennet action the casein micelles aggregate to form a coagulum from liquid milk. Payens (1977) predicted that these two processes would occur concurrently. Several other researchers demonstrated that aggregation was not initiated in fat-free milk at 25 and 30°C until 85–90% of rennet action was complete (Dagleish, 1979; Green *et al.*, 1978). Chaplin and Green (1980) reported that at least 90% of the total κ-casein in skim milk was hydrolyzed at pH 6.6 and 30°C before clotting was observed. These results were consistent with the theory proposed by Dagleish (1979), using soluble and immobilized chymosin, that micelle aggregation only occurred when the majority of κ-casein (97%) had been hydrolyzed (Figure 8.1). The formation of para-κ-

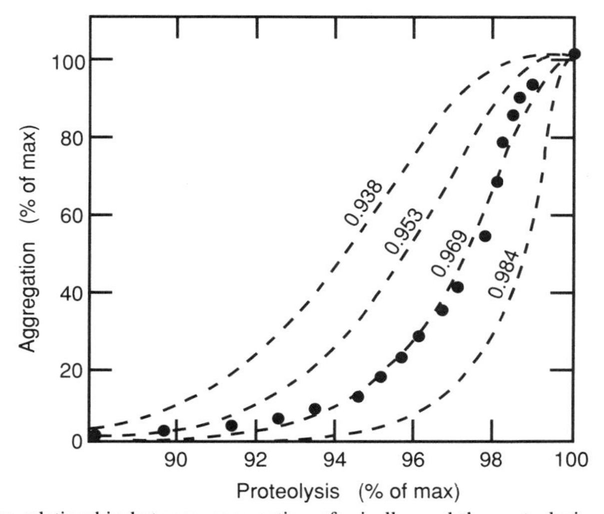

FIG. 8.1. The relationship between aggregation of micelles and the proteolysis of κ-casein, for insolubilized enzyme, based on the probability of micelles requiring greater than a given extent of κ-casein destruction to allow aggregation. Experimentally determined points (●) are compared with calculated values for four different values of the critical extent of reaction (Dagleish, 1979).

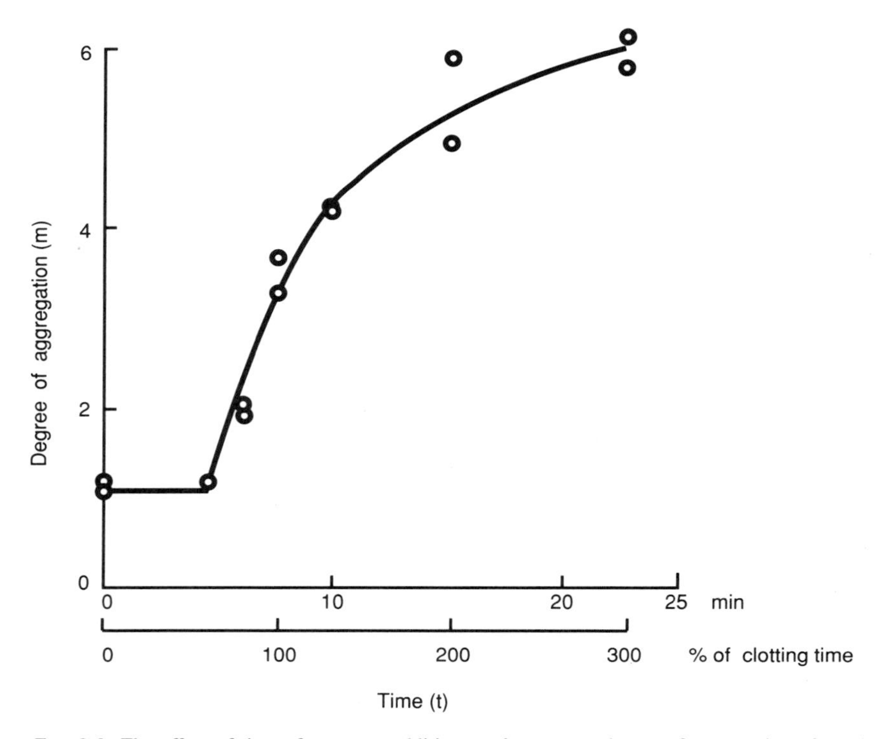

FIG. 8.2. The effect of time after rennet addition on the average degree of aggregation of casein micelles in skim milk. Each point is deduced from the count of one stereopair of micrographs (Green and Morant, 1981).

casein by rennet action was necessary for micellar aggregation while it was inhibited by κ-casein (Walstra, 1979). Green and Morant (1981) studied the mechanism of casein micelle aggregation. Several models were examined to establish the mechanism of aggregation including the von Smoluchowski mechanism. This model defined the rate of disappearance of primary particles by their diffusion rates and collision efficiencies. The frequency distribution of the casein micelle aggregate in rennet treated skim milk was examined in electron micrographs. Their results, shown in Figure 8.2, indicate no change in the degree of aggregation between 0 and 60% of the rennet clotting time, but this increased linearly to 100% of the coagulation time and then leveled off. This suggested random diffusion of particles during aggregation consistent with the von Smoluchowski model. Carlson and co-workers (1987b) suggested that it was not possible in this study to determine whether or not the primary or secondary phase of agglomeration limited the observed rate. These researchers examined the kinetics of flocculation by following turbidity under conditions designed to deac-

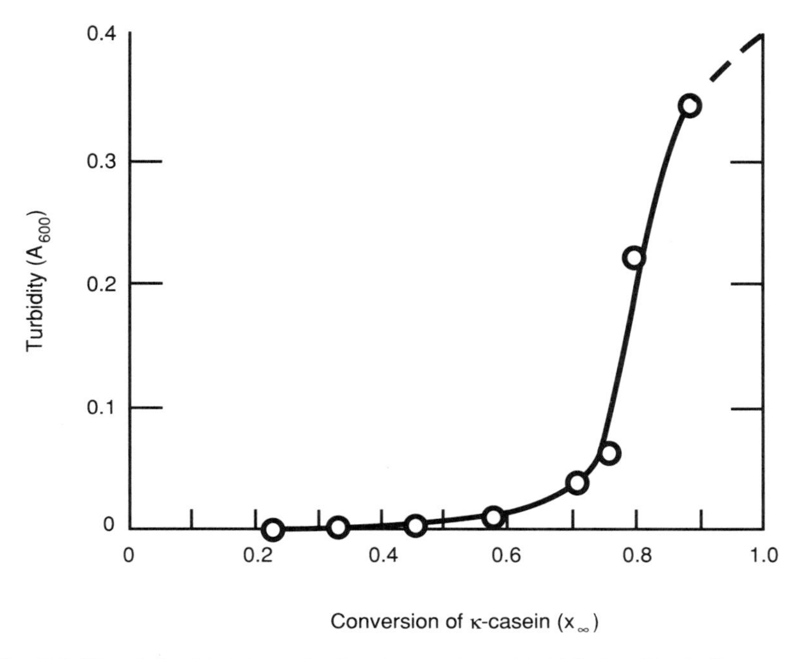

FIG. 8.3. The relationship between fractional conversion and turbidity in milk micelle suspensions (Carlson *et al.*, 1987b).

tivate rennet. Their results, shown in Figure 8.3, indicate that the rate of agglomeration was affected by the degree of κ-casein hydrolysis. This study provided the first experimental measurement of micelle agglomeration in the presence of partially hydrolyzed κ-casein in which the secondary phase was the rate-limiting step. It was evident that a marked increase in turbidity occurred once 80% of the κ-casein had been hydrolyzed. A kinetic model was proposed by Carlson *et al.* (1987b) to explain the agglomeration of milk micelles following κ-casein hydrolysis in which their stability was attributed to the surface potential on the casein micelles. Once κ-casein is hydrolyzed the change in surface potential leads to an unstable state. Thus micelles agglomerate through "patches" on their surface so that two particles form a single entity affected by second-order kinetics:

$$ds/dt = 2k'_2 s,$$

where s = patch concentration in solution at any time t and K' is a second-order rate constant.

Carlson and co-workers (1987c) presented a step functional mathematical model for milk micelle agglomeration to explain the observed kinetics of milk clotting by rennet. Both primary and secondary phases of milk coagulation were responsible for coagulation time, as determined by viscosity increase, visible flocculation, or gelation. In the presence of low enzyme concentrations, however, the observed coagulation was controlled by the primary phase. Micelle flocculation did not occur until at least 75% of the κ-casein was hydrolyzed with the rate-limiting step at this stage being κ-casein hydrolysis. Extrapolation of the t (coagulation time) vs. E_0 (enzyme concentration) curve to infinite enzyme concentration provided the average rate for the secondary phase of agglomeration. Further research by Carlon and co-workers (1987d) examined the kinetics of gel development during milk coagulation. The primary enzymatic phase resulted in the formation of gel cross-link sites through hydrolysis of κ-casein while the nonenzymatic phase was associated with the formation of cross-links and depletion of active sites. This model appeared to predict gel firmness over a temperature range of 31–45°C in the presence of different enzyme concentrations. For a detailed discussion of this work the articles published by Carlson and co-workers (1987a–d) are strongly recommended.

In addition to the factors discussed, the coagulation of renneted bovine casein micelles is also affected by cationic material (Green and Marshall, 1977; Marshall and Green, 1980). Dagleish (1983) showed that at low and moderate temperatures the rate of coagulation increased with rise in calcium ion concentration. Besides the formation of hydrophobic bonds, Payens (1977) proposed the formation of calcium bridges between micelles. Grandison *et al.* (1984) noted an increase in coagulum strength in renneted Friesen milk following change from winter rations to spring grazing, which could be related to the levels of casein, citrate, and minerals. The ability of casein to form a network characteristic of the milk coagulum depends on the pH (Kowalchyk and Olson, 1977), association with calcium (Green and Marshall, 1977; Kowalchyk and Olson, 1979), and higher casein concentration (Storry and Ford, 1982).

3. *Dehydration*

Curd formation is a critical step in the cheese-making process, the strength of which affects both yield and rate of dehydration or syneresis. Many factors affect the strength of the rennet gel, including amount of milk components, calcium, and protein, pH, rennet source, and heat treatment of milk (Fox, 1987). The cheese maker still relies on organoleptic methods to assess when the curd is firm enough for cutting. When this point is reached the curd is cut vertically and horizontally into cubes of uniform size using wired curd knives to ensure uniform release of whey when heated (Irvine and Hill, 1985). It is during dehydration that the individual characteristics of cheese varieties start to be acquired. Once the curd is ready for cutting it is heated in a steam jacket to different temperatures depending on the type of cheese produced. For example, Cheddar cheese is normally heated to 38°C and Swiss cheese is cooked to 52°C, while soft or semisoft cheeses are not heated at all. The dehydration of the curd occurs through loss of whey during cutting with further syneresis taking place when cooked. This expulsion of whey out of the curd results in shrinkage and may be accompanied by some further degradation of κ-casein. The latter may lead to the formation of new cross-links in the rennet gel or curd and development of a firmer or more rubbery curd (Dagleish, 1982).

4. *Molding or Shaping*

Once the curds are cooked they are separated from the whey, placed in molds, and pressed. This is a highly mechanized process which in some plants may be partially automated. In the case of Cheddar cheese, however, the curd is "cheddared" to allow for greater acidification as well as removal of whey before pressing. Many cheese-making plants will use cheddaring belts or towers to ensure that the desired acidity (pH 5.35–5.45) and fibrous texture are attained for Cheddar cheese products. The curd is then comminuted, mixed with salt, and shaped in a mold using a large block or hydraulic press with further loss of whey.

5. *Salting*

Salting is carried out for all cheeses although the amount added varies with the particular cheese produced (Table 8.2). Salting is accomplished by either immersing the cheese in a brine tank or by surface application with dry salt. This process depends on the rate of diffusion of salt, which depends on the moisture content. In addition to causing further dehydration of the cheese through whey removal, salt also slows down the rate of acid development as well as inhibits undesirable bacteria (Irvine and Hill, 1985).

TABLE 8.2

AMOUNT OF SALT IN DIFFERENT CHEESES[a]

Variety	Salt (%)
Emmental	0.7
Cheddar	1.7
Gouda	2.5
Blue cheeses	4
Feta	5–6

[a] Data taken from Fox (1987).

C. Biochemistry of Cheese Ripening

Once manufactured, many cheeses are ripened to develop the required texture and flavor. There are a great variety of cheeses but the ripened cheese can be grouped on the basis of texture as follows:

1. hard grating cheese (Parmesan type);
2. hard cheese (Swiss, Cheddar, Gouda);
3. semihard cheese (Roquefort, Blue Limburger);
4. soft cheese (Camembert).

During the ripening period the major constituents, proteins, carbohydrates, and fats, undergo a variety of chemical and physical changes, including:

1. fermentation of lactose to lactic acid and small quantities of acetic acid, propionic acid, and carbon dioxide;
2. proteolysis, affecting the flavor, texture, and body of cheese;
3. lipolysis, affecting the flavor of cheese.

These changes are brought about by enzymes released by microorganisms added as starters or present in the milk, having survived pasteurization or other heat treatments, by rennet, and by enzymes present in the milk itself (Webb *et al.*, 1974).

1. *Fermentation of Lactose*

The formation of lactic acid from lactose during cheese production and ripening is brought about by the microorganisms added as starters. Lactic acid is essential for proper manufacture, flavor development, normal ripening, and longer shelf life of the cheese. It is particularly important in repressing any undesirable microorganisms, including those belonging to the *coli-aerogenes* group, and those which could produce acetic acid, butyric acid, carbon dioxide, and water (Webb *et al.*, 1974). For example, in the production of Camembert cheese the lactic acid lowers the pH to around 4.6 (Adda *et al.*, 1982).

2. *Proteolysis*

Proteolysis of milk proteins occurs during the ripening of cheese and converts para-casein as well as other minor protein components into proteoses, peptones, amino acids, and ammonia. Para-casein accounts for the majority (95%) of the protein in cheeses with only 1% contributed by whey proteins (O'Keeffe *et al.*, 1978).

a. Proteolysis and Texture. Since the network of cheese consists of protein in which fat is entangled, any modification in the nature and amount of protein in cheese would have major implications on texture. For instance, the protein matrix of soft and semihard cheeses is converted to a smoother, more homogeneous structure during ripening (deJong, 1978; Knoop and Peters, 1971). The firmness of both soft and hard cheeses is related to the moisture content (deJong, 1978; Ruegg *et al.*, 1980), which influences the rate of hydrolysis. DeJong (1977) reported that hydrolysis of α_{s1}-casein by residual coagulant was responsible for the softening of soft Meshanger cheese. Proteolysis of both α_{s1}- and β-casein fractions by rennet occurs gradually and has been termed the tertiary phase in the coagulation of milk (Alais *et al.*, 1953). Coagulant enzymes appear to be primarily responsible for the breakdown of α_{s1}-casein in semihard and hard cheeses, which occurs at a much faster rate than breakdown of β-casein. The bond of α_{s1}-casein which is susceptible to hydrolysis by chymosin was reported by Hill *et al.* (1974) and Resmini *et al.* (1976, 1977) to be Phe 23–Phe 24. Creamer and Richardson (1974) suggested that this bond was Phe 24–Val 25. It was apparent from studies by Pelissier and co-workers (1974) that chymosin was capable of hydrolyzing both of these peptide bonds. This occurred during the early stages of ripening resulting in the formation of an α_{s1}-I-casein fraction, which has been identified as the principal degradation product of a number of different cheeses (Creamer and Richardson, 1974; Resmini *et al.*, 1977). β-casein is degraded much more slowly, which may be related to the salt content in cheese (O'Keeffe *et al.*, 1976; Visser and de Groot-Mostert, 1978). This appears to inhibit the degradation of β-casein to a much greater extent compared to that of α_{s1}-casein (Fox and Walley, 1971).

Approximately 10% of rennet is retained by the curd during cheese manufacture and is responsible for the release of large-molecular-weight peptides during the ripening period (O'Keeffe *et al.*, 1978). The production of free amino acids and smaller peptides is brought about by the starter bacteria (El Soda *et al.*, 1978; Thomas and Mills, 1981). The starter bacteria provide exo- and endopeptidases, although primarily exopeptidases, which result in selective amino acid production (Adda *et al.*, 1982). The major pathway for proteolysis during cheese ripening involves the primary action of rennet, which results in a limited degradation of casein followed by the proteolytic action of the starter bacteria, producing free amino acids and small peptide fragments (O'Keeffe *et al.*, 1978; Visser, 1977). Studies by Visser and de Groot-Mostert (1977) on Gouda cheese concluded that rennet was responsible for the degradation of α_{s1}-casein and β-casein during the first month of ripening with the starter bacteria contributing to further degradation of proteins, particularly β-casein, over the longer ripening period. While milk protease could degrade β-casein, its effect was insignificant compared to rennet and starter bacteria proteinase action. The major indigenous protease in milk, alkaline protease or plasmin, is extremely thermostable, hydrolyzing β-casein and α-caseins. Its action was influenced by both pH and salt concentration, although in the case of plasmin pH appeared to be more important (Noomen, 1978). In the case of Gouda cheese, Visser and de Groot-Mostert (1977) found that starter (*S. cremoris*) protease was responsible for 10% of the degradation of β-casein compared to 40% for rennet. Several aspartyl-proteinases from *P. caseicolum* and *P. roqueforti* were also capable of splitting three specific bonds in β-casein (Trieu-Cuot *et al.*, 1982a,b). The complex changes taking place during cheese ripening are due to the combined action of proteinases from milk, rennet or rennet substitutes, and to a lesser degree the starter cultures and other specialized microorganisms (Grappin *et al.*, 1985).

b. Proteolysis and Flavor. The development of cheese flavor is due to the production of a wide range of compounds formed during the ripening period. Of these, amino acids were among the first compounds examined for their contribution to cheese flavor. Early research by Kosikowski (1951) reported that an increase in cheese flavor development was accompanied by an increase in total free amino acids. Confirmation was provided by Kristoffersen and Gould (1960), who suggested a possible relationship between the overall flavor intensity and total amount of amino acids, but were unable to correlate this with any of the individual amino acids produced. Some evidence for the contribution of glutamic acid to the "brothy" flavor of cheese was subsequently presented by Mulder (1952) and Harper (1959). The increase in free amino acids in cheese can be estimated by the amount of phosphotungstic acid (PTA) soluble nitrogen (N) levels in cheese (Jarrett *et al.*, 1982). Studies on the acceleration of Cheddar cheese ripening by Aston (1981) showed strong correlations between PTA-soluble (N) levels and flavor, the latter based on taste panel assessment. Nevertheless, there has been a substantial body of research indicating that little or no contribution is made by free amino acids to Cheddar cheese flavor other than background (Dacre, 1953; Mabbit, 1955; Law *et al.*, 1976). This has been questioned since the publication of a paper by McGugan *et al.* (1979) showing that the water-soluble extract, containing amino acids, salts, and peptides, isolated from mild and aged cheese contributed to the intensity of cheese flavor as distinct from the volatiles, which contributed to the quality of the cheese flavor.

c. Bitter Peptides. The development of bitter flavors in cheese is caused by the release of low-molecular-weight peptides. These are formed during the curing of cheese by enzymatic hydrolysis of casein (Harwalkar and Elliott, 1965). This proteolytic activity is due to rennet and starter bacteria proteases *(Streptococci)*. A worldwide shortage of rennet has resulted in the use of rennet substitutes which are higher in proteolytic activity. This has been demonstrated for microbial (Arima, 1972; Babbar *et al.*, 1965; Sardinas, 1972) and vegetable rennet substitutes (Gupta and Eskin, 1977).

Stadhousers and co-workers (1983) found that only certain starter bacteria produced bitter peptides in cheese. The enzymes responsible, however, were present in both bitter and nonbitter starter cultures. The inability of the nonbitter

TABLE 8.3

THE EFFECT OF SALT ON THE FORMATION OF BITTER FLAVOR IN GOUDA CHEESE[a]

Starter	Salt in dry matter (%)	Intensity of bitter flavor[b] after			
		6	8	12	16
		(weeks)			
S. cremoris HP	3.76	1.8			
	2.07	2.0			
	0.10	3.2			
S. cremoris TR	2.53			1.3	
	0.10			2.3	
S. cremoris Eg	2.62			1.2	
	1.5				0.10
S. cremoris HP	3.43		1.9		2.1
	0.12		3.0		3.5

[a] Taken from Stadhousers *et al.* (1983).
[b] Bitter flavor intensity was estimated in pastes made from various cheeses in which the salt content was raised to the highest value determined. On a scale of 0 to 4, a higher value indicates greater bitterness.

starter culture to produce bitter peptides suggested a different location for these enzymes in the culture cells. It appeared that those proteolytic enzymes responsible for bitter peptide production were located primarily on the membranes. These researchers also showed that salt inhibited the production of bitter flavor in Gouda cheese. Their results, summarized in Table 8.3, indicate that a lower salt content in Gouda cheese produced a greater incidence of bitter flavor.

Umetsu and associates (1983) isolated a bitter peptide fraction by peptic digestion of casein. When treated with wheat carboxypeptidase, the bitterness of this fraction was reduced, which correlated with an increase in free amino acids (Figure 8.4). The reduction in bitterness was attributed to the release of hydrophobic amino acids from the carboxyl end of the bitter peptide. These results were in agreement with earlier work reported by Clegg and co-workers (1974) who observed a close relationship between the bitterness of peptides and the content and sequence of hydrophobic amino acids. Bitter peptides isolated from casein were shown by Ney (1971) to have large amounts of such hydrophobic amino acids, leucine and phenylalanine. The high content of hydrophobic side chains in bitter peptides was also reported by Guigoz and Solms (1976) and Matoba *et al.* (1970, 1977). Using high-pressure liquid chromatography (HPLC), Champion and Stanley (1982) separated the bitter peptides from Cheddar cheese coagulated with chicken pepsin. Two bitter fractions isolated by HPLC had a slightly higher

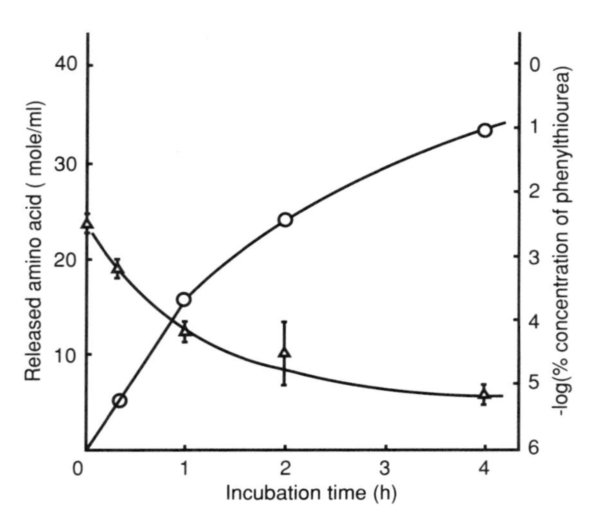

FIG. 8.4. Correlation between released amino acids (○) and the bitterness (△) during the hydrolysis of the bitter peptide fraction by wheat carboxypeptidase. The substrate was hydrolyzed with the enzyme under the following condition: substrate concentration, 2%; enzyme/substrate ratio, 1 : 50 (w/w); pH 5.1; temperature, 30°C; incubation time, 0.5, 1, 2, and 4 hr. After the scheduled periods of time, the reaction was stopped to measure free amino acids and bitterness (Umetsu *et al.*, 1983).

average hydrophobicity than the corresponding nonbitter fractions, although a bitter fraction isolated by gel filtration on Sephadex G-50 did not show a similar relationship. This discrepancy was attributed to the fact these were mixtures of compounds, so that those present in larger quantities would affect the hydrophobicity but not necessarily contribute as much to bitterness.

d. Amino Acid Catabolism and Cheese Ripening. As discussed in previous sections, the formation of free amino acids and peptides is the result of the proteolytic activity of rennet, starter organisms, and contaminating microorganisms. This section will discuss the overall contribution of microorganisms to amino acid catabolism during cheese ripening. They are responsible for the variety of amino acids and degradation products which play a crucial role in flavor development during the ripening process. These reactions are summarized in Scheme 8.2 and involve decarboxylation, transamination, deamination, and hydrolytic cleavage of amino acid side chains (Hemme *et al.*, 1982).

Decarboxylation of amino acids is carried out by decarboxylases, which require pyridoxal phosphate as a coenzyme. These enzymes exhibit optimal activity at around pH 5.5, corresponding to the pH of cheese (Schormuller, 1968). A wide range of decarboxylases have been reported in starter organisms as well as in microorganisms in general (Nakazawa *et al.*, 1977). Transaminases and

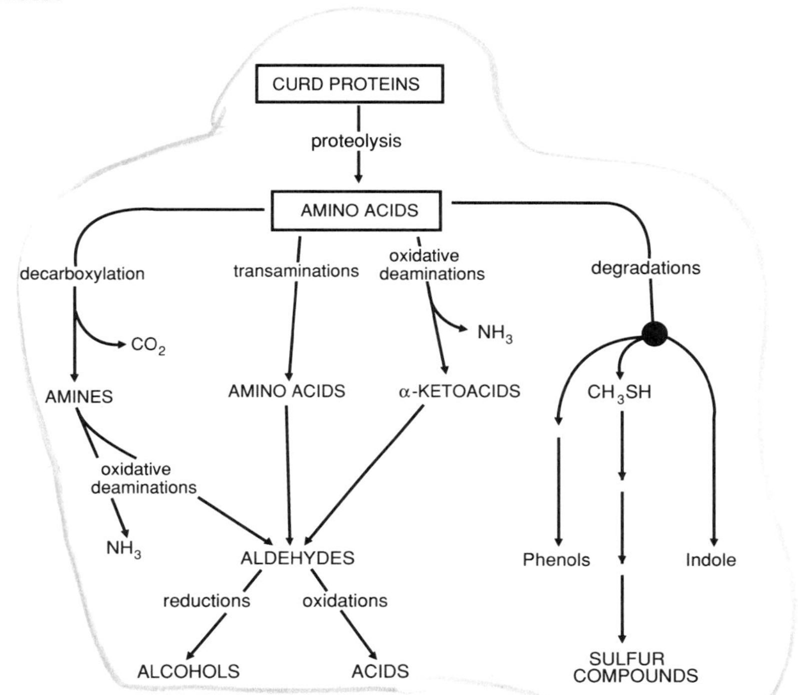

SCHEME 8.2. General scheme of microbial catabolism of amino acids during cheese ripening (Hemme *et al.*, 1982).

deaminases are widely distributed in microorganisms catalizing oxidative deamination of amino acids to form aldehydes, which can be oxidized or reduced to the corresponding alcohols or acids. The degradation reactions outlined in Scheme 8.2 include the hydrolysis of amide groups and cleavage of indole or phenol groups from tryptophan or tyrosine by lyases.

Microorganisms can also degrade methionine by cleaving the C–S bond. This process, referred to as a gamma elimination, usually involves deamination by a deaminating methionine-demethiolase (EC 4. 4. 1. 11), producing methanethiol (CH_3SH) (Kreiss and Hession, 1973; Laakso, 1979: Law and Sharpe, 1978). Sulfur compounds, such as hydrogen sulfide, methanethiol, methyl sulfide, and dimethyl sulfide, all play an important role in the flavor of foods including cheese (Aston and Dulley, 1982; Boelens *et al.*, 1974; Maga, 1976; Panouse *et al.*, 1972). Methanethiol has been shown to be one of the characteristic compounds associated with Cheddar cheese flavor (Manning, 1979). It also appears to be the precursor of a wide range of sulfur compounds responsible for the garlic flavor associated with such surface-ripened cheeses as Liverot and Munster and other French cheeses (Dumont and Adda, 1978). Cuer and co-workers (1979) synthesized a number of sulfur compounds, including 2, 4-dithiapentane, 2, 3, 4-trithiapentane, 2, 3, 5-trithiapentane, and a range of *S*-methylthioesters, thought to contribute to cheese aroma. All of these compounds appeared to contribute to cheese aroma, although only two of them exhibited strong cheese odors. These included 2, 3, 4-trithiapentane, which had an overripened cheese aroma with an experimental threshold value of 0.1 ppm, and *S*-methylthiopropanoate, with a cheesy aroma and a much higher experimental threshold value of 200 ppm.

$$CH_3\text{–}S\text{–}S\text{–}S\text{–}CH_3 \qquad CH_3\text{–}CH_2\text{–}\underset{\underset{O}{\|}}{C}\text{–}S\text{–}CH_3$$

2,3,4-Trithiapentane *S*-Methylthiopropanoate

The formation of some of these thioesters appeared likely as several strains of microorganisms isolated were capable of esterifying methanethiol. Further discussion on the contribution of proteinases to cheese ripening can be found in Chapter 11.

3. *Lipolysis*

The production of free fatty acids is extremely important for the development of flavors in both soft and hard cheeses. They are derived by either lipolysis of fats or from the metabolism of carbohydrates and amino acids. Lipolysis, however, appears to be mainly responsible for free fatty acid production in which the enzyme involved is lipase. A detailed discussion on the role of lipase in cheese ripening can be found in Chapter 11.

D. Accelerating Cheese Ripening

Cheese ripening is a very costly process in which cheese is held under controlled storage conditions for an extended period of time. This has resulted in attempts to accelerate the ripening period and reduce labor storage costs. A number of methods have received attention, including storage at elevated temperatures, addition of exogenous enzymes, and new developments in cheese ripening technology.

1. *Storage of Cheese at Elevated Temperatures*

Several trials were successfully carried out in which high temperatures were found to accelerate the ripening rate of Cheddar cheese without adversely affecting the quality of the cheese (Aston *et al.*, 1983a,b; Fedrick *et al.*, 1983). Attempts to further enhance the ripening of Cheddar cheese were conducted by Aston and co-workers (1985). These researchers subjected Cheddar cheese to a variety of temperature/time storage conditions and followed the progress of proteolysis as well as flavor development using a trained panel. It was evident from their research that Cheddar cheese could be stored up to a maximum of 15°C for 32 weeks without any deterioration in quality. Under these conditions the cheese was found to contain similar levels of proteolytic products as controlled cheese stored at 8°C for 32 weeks (Figure 8.5). Estimation of cheese age for the 20-week-old cheese stored at 15°C was equivalent to that of the control cheese stored at 8°C for 32 weeks (Figure 8.6). A substantial reduction in the cheese ripening period for Cheddar was clearly demonstrated using controlled high-temperature storage conditions.

2. *Addition of Exogenous Enzymes*

The addition of exogenous enzymes to enhance the ripening of cheese is discussed in Chapter 11, Section V.

3. *New Developments in Cheese Ripening Technology*

New developments in cheese ripening have been introduced during the past few years using techniques developed in medicine (El Soda, 1986). For example,

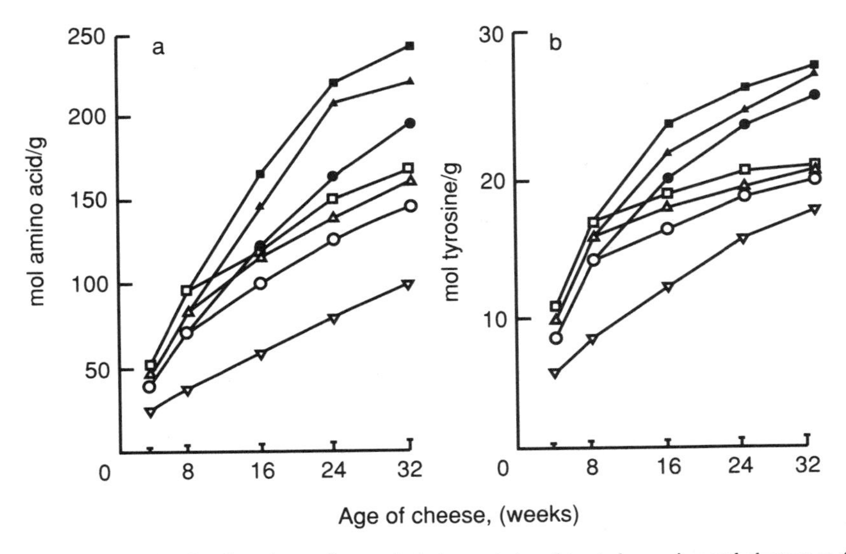

Fig. 8.5. Mean levels of products of proteolysis in control and treated experimental cheeses at 4, 8, 16, 24, and 32 weeks of age. (a) PTA-soluble amino N, (b) TCA-soluble tyrosine. Standard deviations are indicated by heights of bars at each age (Aston *et al.*, 1985). (▽) Control cheese ripened at 8°C for 32 weeks; (○) cheese ripened at 15°C for 8 weeks, 8°C for 24 weeks; (△) cheese ripened at 17.5°C for 8 weeks, 8°C for 24 weeks; (□) cheese ripened at 20°C for 8 weeks, 8°C for 24 weeks; (●) cheese ripened at 15°C for 32 weeks; (▲) cheese ripened at 17.5°C for 32 weeks; (■) cheese ripened at 20°C for 32 weeks.

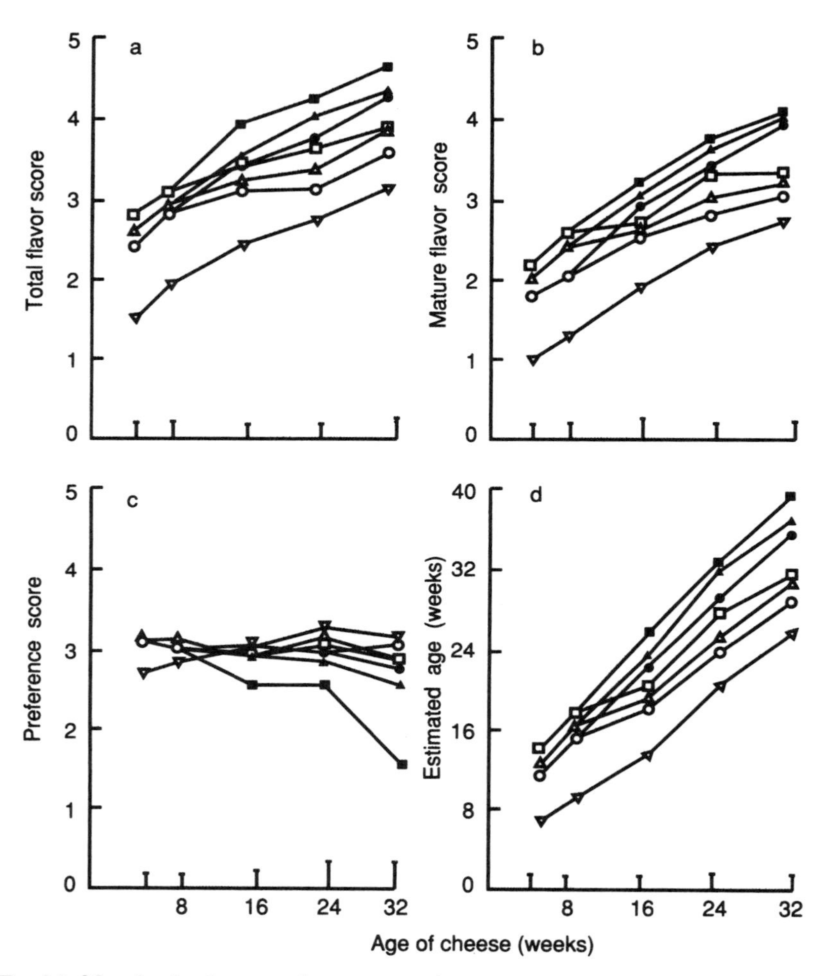

FIG. 8.6. Mean levels of taste panel assessments of control and treated experimental cheeses at 4, 8, 16, 24, and 32 weeks of age. (a) Total flavor score, (b) mature flavor scores, (c) preference scores, and (d) estimated ages. Standard deviations are indicated by heights of bars at each age (Aston *et al.*, (1985a,b). (▽) Control cheese ripened at 8°C for 32 weeks; (○) cheese ripened at 15°C for 8 weeks, 8°C for 24 weeks; (△) cheese ripened at 17.5°C for 8 weeks, 8°C for 24 weeks; (□) cheese ripened at 20°C for 8 weeks, 8°C for 24 weeks; (●) cheese ripened at 15°C for 32 weeks; (▲) cheese ripened at 17.5°C for 32 weeks; (■) cheese ripened at 20°C for 32 weeks.

genetic modification of starter cultures by Grieve and Dulley (1984) and Grieve *et al.* (1983) successfully increased the population of *Streptococcus lactis,* which enhanced flavor development in Cheddar cheese without increasing the rate of acid development. Other developments include microencapsulation of the cell-free extracts of selected microorganisms by Braun and Olson (1983, 1984), which enhanced the catalytic products formed compared to those of the uncapsulated extracts. The use of liposomes to carry enzymes for enhancing ripening is a recent development which has shown considerable promise in enhancing the ripening of Cheddar cheese (Law and King, 1985; El Soda *et al.*, 1983, 1984).

III. Fermented Milk-Cultured Products

A large number of fermented milk products are consumed worldwide which differ in the process of manufacture and nature of the starter organism used (International Dairy Federation, 1983). The main cultured product consumed in Australia, Canada, England, and the United States is yoghurt, while various cultured buttermilk and cream products are predominant in Scandinavian and East European countries. This section will discuss the biochemistry of yoghurt, the main fermented dairy product in North America.

A. Yoghurt

Yoghurt is a product manufactured from milk by fermentation with a mixed starter culture composed of *Streptococcus thermophilus* and *Lactobacillus bulgaricus* (Tamime and Robinson, 1985). The initial step in yoghurt production is homogenization of the milk, which splits the fat into small globules. These become coated with new membranes of casein submicelles, which reduces agglutination of the fat globules. The milk is then preheated at 85°C for 30 min or 90–95°C for 5–10 min to destroy pathogenic and spoilage organisms. This heating also affects the physicochemical structure of the protein and the stability of the yoghurt gel. The majority of the whey proteins are denatured by the heat treatment, with α-lactalbumin completely denatured (Parry, 1974). This protein fraction interacts with casein to increase its hydrophilic properties, thus causing formation of a stable coagulum (Grigorov, 1977a,b,c). The milk is then cooled to around 42–43°C and the yoghurt starter bacteria added. The mixture is held at this temperature until the desired acidity is reached. During this period the starter organisms are involved in the development and production of flavor. A number of studies showed that *L. bulgaricus* provided essential amino acids for the growth of *S. thermophilus,* while the latter organism provided formic acid and related compounds required to stimulate *L. bulgaricus* growth (Botazzi *et al.*, 1971; Higashio *et al.*, 1977a,b; Veringa *et al.*, 1968). The formation of lactic acid during the metabolism of the starter organism is crucial to yoghurt production. It destabilizes the casein micelle, causing coagulation of the milk protein and formation of a yoghurt gel (Dyatchenko, 1971). The symbiotic relationship between the two starter cultures is extremely important to the yoghurt process and was first observed over half a century ago by Orla-Jensen (1931).

1. *Starter Culture Metabolism*

The metabolism of the starter cultures requires energy which is provided by the only available carbohydrate in milk, lactose. To be utilized by the intracellular enzymes, lactose must be transported through the cell membrane. The transport mechanism involved in *S. thermophilus* or *L. bulgaricus* is probably similar to that reported for other organisms in which the phosphoenolpyruvate (PE) dependent phosphotransferase system is involved (Saier, 1977). Once inside the cell, lactose is hydrolyzed to glucose and galactose by the enzyme β-D-galactosidase (Rao and Dutta, 1977, 1978). The ability to hydrolyze lactose is considerably greater in *L. bulgaricus* compared to *S. thermophilus* (Kilara and Shahani, 1976). A second enzyme, β-D-phosphogalactosidase, was also reported in both starter cultures and it hydrolyzes lactose phosphate to glucose and galactose 6-phosphate (Somkuti and Steinberg, 1978a,b). The glucose is then metabolized via the Embden–Meyerhoff cycle to pyruvate, which is then converted to lactic acid by lactate dehydrogenase. The configuration of the final product is D(−)- or L(+)-lactate, depending on the enzymes present in the bacterial culture. Several studies demonstrated that *S. thermophilus* preferentially produces the L(+) form of lactic acid while *L. bulgaricus* produces the D(-) form (Benner, 1976; Garvie, 1978). Blumenthal and Helbing (1972) reported that the rate of L(+)-lactate production was much faster than the corresponding rate of D(−)-lactate, which corresponded to the respective growth rate patterns previously reported for these two organisms (Figure 8.7).

Storage of yoghurt was found to increase the D(−) form of lactate as a result of the greater tolerance of *L. bulgaricus* (Abrahamsen, 1978). In fact good-quality yoghurt was reported to have a ratio of L(+)-lactate : D(−)-lactate of 2 or less as an excess of D(−)- lactate produced a sharper flavor (Blumenthal and Helbing, 1974). In addition to improving the flavor of yoghurt, lactate also acts as a preservative.

2. *Yoghurt Flavor*

A large number of flavor components have been identified in yoghurt with the major ones being carbonyls such as acetaldehyde, acetone, acetoin, and diacetyl. Several researchers have implicated acetaldehydes as prime contributors to yoghurt flavor development (Divivedi, 1973; Sandine and Elliker, 1970; Sandine *et al.*, 1972). Several pathways have been described for acetaldehyde production, including decarboxylation of pyruvate, reduction of acetyl-CoA, and cleavage of

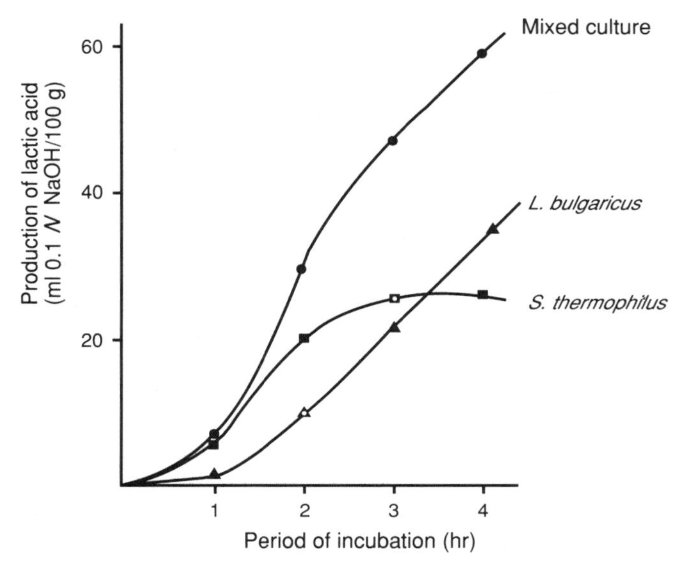

FIG. 8.7. Production of lactic acid in milk by yoghurt starter cultures (Pette and Lolkema, 1950).

threonine to glycine and acetaldehyde (Sandine and Elliker, 1970). *Streptococcus thermophilus* has been reported to produce greater amounts of acetaldehyde compared to *L. bulgaricus* in the presence of necessary stimulatory growth factors (Shankar, 1977; Shankar and Davies, 1978). In addition to carbonyl compounds, volatile fatty acids and amino acids have also been shown to enhance yoghurt flavor (Dumont and Adda, 1973; Groux, 1976). The major flavor components in yoghurt, however, are formed from the microbial fermentation of lactose (Tamime and Deeth, 1980).

3. *Proteolysis*

Proteolytic activity by lactic acid bacteria is responsible for the hydrolysis of milk protein in cultured dairy products such as yoghurt. The increase in free amino acids in cultured products was attributed by Miller *et al.* (1964) to proteolysis by *L. bulgaricus*. This organism appeared to have greater proteolytic activity compared to *S. thermophilus* with respect to casein hydrolysis (Maret *et al.*, 1974; Yu and Nakanishi, 1975a,b). *Streptococcus thermophilus*, however, had more peptidase activity, which degraded the intermediate products of casein proteolysis by *L. bulgaricus* (Carini and Loda, 1974; Torneur, 1974). This further exemplifies the synergistic relationship between these two organisms which determines the degree of proteolysis of casein. The maximum level of free amino acids released was reported by Luca (1974) in the presence of a 1 : 1 ratio of *L. bulgaricus* and *S. thermophilus*. The production of bitter peptides in yoghurt has been attributed to the proteolytic activity of *L. bulgaricus* during storage (Renz and Puhan, 1975).

4. *Lipolysis*

Only a small degree of lipolysis occurs in yoghurt, although it makes an important contribution to the development of flavor. Since natural lipase of milk is inactivated by pasteurization, any lipolytic activity must be provided by the starter cultures. These are intracellular enzymes located in the cytoplasm of the cultures with little activity present in the cell membranes (Formisano *et al.*, 1974). An increase in low-molecular-weight volatile acids has been reported in yoghurt with *L. bulgaricus* making a greater contribution compared to *S. thermophilus* (Yu and Nakanishi, 1975c; Yu *et al.*, 1974). The amount produced by these organisms varies depending on whether the starter bacteria are cultured in whole or skim milk.

B. Nutritional Value of Yoghurt

The nutritional value of yoghurt is similar to that of milk. Some confusion has surrounded the vitamin content of yoghurt versus milk in which yoghurt was considered to be a superior source of B vitamins. This idea of yogurt being a "B vitamin factory" was discounted by Acott and Labuza (1972), who showed, with the exception of nicotinic acid, that the levels of vitamins in yoghurt were considerably lower compared to those in whole milk (Table 8.4). A more recent

TABLE 8.4

Vitamin Content of Whole Milk and Plain Yoghurt[a,b]

	Whole fluid milk	Plain yoghurt	Difference from milk (%)
Vitamin A (IU/100 g)	156	69	−56
Thiamine (mg/liter)	0.44	0.37	−16
Riboflavin (mg/liter)	1.75	1.40	−20
Nicotinic acid (mg/liter)	0.94	1.3	+12
Biotin (mg/liter)	0.031	0.012	−61
Vitamin B (mg/liter)	0.0043	0.0012	−72
Choline (mg/liter)	121	6	−95
Vitamin C (mg/liter)	21.1	6.2	−71

[a] Taken from Acott and Labuza (1972).
[b] From data by Hartman and Dryden (1965).

paper by Gurr (1984) found that only folates were higher in yoghurt as compared to milk. The vitamin content could be increased by the addition of fresh fruit or stabilizers, and in the case of the latter this could be folic acid from carageenan (Reddy *et al.*, 1976).

C. Lactose Intolerance and Yoghurt

Sufferers of lactose intolerance who are unable to drink milk can tolerate yoghurt. This is due to the presence of active lactose-hydrolyzing enzymes in the yoghurt microflora in which *S. thermophilus* exhibits three times the activity of *L. bulgaricus* (Kilara and Shahani, 1976). Broussalain and Westhoff (1983) demonstrated the deleterious effects that high levels of lactose in milk have on the yoghurt microflora. Improved weight gains by lactose-intolerant individuals was attributed to the reduction of lactose during fermentation by microbial lactase activity.

D. Antitumor Activity in Yoghurt

Yoghurt has been reported to inhibit both the rate and extent of tumor development (Ayebo *et al.*, 1981; Farmer *et al.*, 1975; Reddy *et al.*, 1973). Several compounds were isolated from the cell wall of *L. bulgaricus* which inhibited the growth of certain tumors in laboratory animals (Bogdanov *et al.*, 1963, 1975). A later study by Ayebo and co-workers (1981) isolated antitumor components from yoghurt by dialysis which were low-molecular-weight (>14,000) compounds. The precise nature of these compounds remains to be characterized. More recent studies by Friend and Shahani (1984) showed that lactic acid bacteria exhibited antibacterial, antiatherosclerotic, and anticancerous properties. The antimutagenic activity of the cell walls of the lactic acid bacteria may account, in part, for the low risk of colon cancer among populations in which cultured dairy products are consumed (IARC, 1977).

Bibliography

Abrahamsen, R. K. (1978). The content of lactic acid and acetaldehyde in yoghurt stored at different temperatures. *Int. Dairy Congr., 20th, 1978* Vol. E. 829.

Acott, K. M., and Labuza, T. P. (1972). Yogurt: Is it truly Adelle's B vitamin factory? *Food Prod. Dev.* **6**(7), 50.

Adda, J., Gripon, J. C., and Vassal, L. (1982). The chemistry of flavour and texture generation in cheese. *Food Chem.* **9**, 115.

Alais, C., Mocquot, G., Nitschmann, H., and Zahler, P. (1953). Rennet and its action on the casein

of milk. VII. Hydrolysis of nonprotein nitrogen and its relation to the primary reaction of rennet-induced curdling of milk. *Helv. Chim. Acta* **36,** 1955.
Arima, K. (1972). Mucor rennin. *Proc. Int. Symp. Convers. Manuf. Foodst. Microorg., 1971,* 99–100.
Aston, J. W. (1981). Master of Applied Science Thesis, Queensland Institute of Technology, Australia.
Aston, J. W., and Dulley, J. R. (1982). Cheddar cheese flavour. *Aust. J. Dairy Technol.* **37,** 59.
Aston, J. W., Fedrick, I. A., Durward, I. G., and Dulley, J. R. (1983a). The effect of elevated ripening temperatures on proteolysis and flavor development in Cheddar cheese. I. Higher initial storage temperatures. *N. Z. J. Dairy Sci. Technol.* **18,** 143.
Aston, J. W., Grieve, P. A., Durward, I. G. and Dulley, J. R. (1983b). Proteolysis and flavor development in Cheddar cheeses subjected to accelerated ripening treatments. *Aust. J. Dairy Technol.* **38,** 59.
Aston, J. W., Giles, J. E., Durward, I. G., and Dulley, J. R. (1985). Effect of elevated ripening temperatures on proteolysis and flavor development in Cheddar cheese. *J. Dairy Res.* **52,** 565.
Ayebo, A. D., Shahani, K. M., and Dam, R. (1981). Antitumor component(s) of yogurt: Fractionation. *J. Dairy Sci.* **64,** 2318.
Babbar, L. J., Srinivason, R. A., Chakravotty, S. C., and Dudani, A. T. (1965). Microbial rennet substitutes. A review. *Indian J. Dairy Sci.* **18,** 89.
Benner, J. (1976). Production of Dor L-lactate in yoghurt cultured milk and kefir. *Dairy Sci. Abstr.* **38,** 544.
Berridge, N. J. (1942). The second phase of rennet coagulation. *Nature (London)* **149,** 194.
Blumenthal, A., and Helbing, J. (1972). Contents of L(+) and D-lactic acid in various cultured milks. *Diary Sci. Abstr.* **34,** 342.
Blumenthal, A., and Helbing, J. (1974). The L(+) and D(−) lactic acid concentration in yoghurts of different fat content. *Dairy Sci.Abstr.* **36,** 331.
Boelens, M., Van der Linde, L. M., De Valois, P. J., Van Dort, H. M., and Takken, H. J. (1974). Organic sulfur, compounds from fatty aldehydes, hydrogen sulfide, thiols, and ammonia as flavor constituents. *J. Agric. Food Chem.* **23,** 1047.
Bogdanov, I. G., Popkhristov, P., and Marinov, L. (1963). In "Anticancer effect of *Antibioticum bulgaricum* in Sarcoma 180 and on the solid form of Ehrlich carcinoma. *Abstr., Int. Cancer Congr., 8th, 1962,* p. 366.
Bogdanov, I. G., Daler, P. G., Gurevich, A. T., Kolosov, M. N., Mal'kova, V. P., Plemyannikova, L. A., and Sorokina, I. B. (1975). Antitumor glycopeptides from *Lactobacillus bulgaricus* cell wall. *FEBS Lett.* **57,** 259.
Botazzi, V., Battistotti, B., and Vesovo, M. (1971). Continuous production of yoghurt cultures and stimulation of *Lactobacillus bulgaricus* by formic acid. *Milchwissenshaft* **26,** 214.
Braun, S. D. and Olson, N. F. (1983). Microencapsulation of multiple bacterial cell free extracts to demonstrate feasibility of heterogenous enzyme systems of cofactor recycling. *J. Dairy Sci.* **66,** Suppl. 1, 77.
Braun, S. D., and Olson, N. F. (1984). Stability of fat microencapsules for use in cheese ripening. *J. Dairy Sci.* **67,** Suppl. 1, 60.
Broussalain, J. and Westhoff, D. (1983). Influence of lactose concentration of milk and yogurt on growth rate in rats. *J. Dairy Sci.* **66,** 438.
Carini, S. and Loda, R. (1974). The lactic acid bacteria, their proteolytic, reducing and acidifying activities in the ripening of cheese. *International Dairy Congr., 19th, 1974* Vol. 1E, p. 433.
Carlson, A., Hill, C. G., Jr., and Olson, N. F. (1987a). Kinetics of milk coagulation. I. The kinetics of kappa casein hydrolysis in the presence of enzyme deactivation. *Biotechnol. Bioeng.* **29,** 582.
Carlson, A., Hill, C. G., Jr., and Olson, N. F. (1987b). Kinetics of milk coagulation. II. Kinetics of the secondary phase: Micelle flocculation. *Biotechnol. Bioeng.* **29,** 590.

Carlson, A., Hill, C. G., Jr., and Olson, N. F. (1987c). Kinetics of milk coagulation. III. Mathematical modeling of the kinetics of curd formation following enzymatic hydrolysis of κ-casein—parameter estimation. *Biotechnol. Bioeng.* **29,** 601.

Carlson, A., Hill, C. G., Jr., and Olson, N. F. (1987d). The kinetics of milk coagulation. IV. The kinetics of the gelfirming process. *Biotechnol. Bioeng.* **29,** 612.

Castle, A. V., and Wheelock, J. V. (1972). Effect of varying enzyme concentration on the action of rennin in whole milk. *J. Dairy Res.* **39,** 15.

Champion, H. M., and Stanley, D. W. (1982). HPLC separation of bitter peptides from cheddar cheese. *Can. Inst. Food Sci. Technol. J.* **15,** 283.

Chaplin, B., and Green, M. L. (1980). Determination of the proportion of κ-casein hydrolyzed by rennet on coagulation of skim milk. *J. Dairy Res.* **47,** 351.

Clegg, K. M., Lim, G. L., and Manson, W. (1974). The structure of a bitter peptide derived from casein by digestion with papain. *J. Dairy Sci.* **41,** 283.

Cleland, W. W. (1963). The kinetics of enzyme-catalyzed reactions iwth two or more substitutes or products. *Biochim. Biophys. Acta* **67,** 104.

Creamer, L. K., and Richardson, B. C. (1974). Identification of the primary degradation product of α-casein in cheddar cheese. *N. Z. J. Dairy Sci.* **9,** 9.

Cuer, A., Dauphin, G., Kergomard, A., Roger, S., Dumont, J. P., and Adda, J. (1979). Flavour properties of some sulphur compounds isolated from cheeses. *Lebensm.-Wiss. Technol.* **12,** 258.

Dacre, J. C. (1953). Amino acids in New Zealand cheddar cheese: Their possible contribution to flavor. *J. Sci. Food Agric.* **4,** 604.

Dagleish, D. G. (1979). Proteolysis and aggregation of casein micelles treated with immobilized or soluble chymosin. *J. Dairy Res.* **46,** 653.

Dagleish, D. G. (1982). The enzymatic coagulation of milk. *In* "Developments in Dairy Chemistry" (P. F. Fox, ed.), pp. 157–188. Applied Science Publishers, London, U. K.

Dagleish, D. G. (1983). Coagulation of renneted bovine casein micelles: Dependence on temperature, calcium concentration and ionic strength. *J. Dairy Res.* **50,** 331.

deJong, L. (1977). Protein breakdown in soft cheese and its relation to consistency. 2. The influence of the rennet concentration. *Neth. Milk Dairy J.* **31,** 314.

deJong, L. (1978). The influence of the moisture content on the consistency and protein breakdown of cheese. *Neth. Milk Dairy J.* **32,** 1.

Delfour, A., Jollès, J., Alais, C., and Jollès, P. (1965). Casein-glycopeptides. Characterization of a methionine residue of the N-terminal sequences. *Biochem. Biophys. Res. Commun.* **19,** 452.

Divivedi, B. K. (1973). The role of enzymes in food flavors. Part 1. Dairy products. *CRC Crit. Rev. Food Technol.* **3,** 457.

Dumont, J. P. and Adda, J. (1973). Rapid method for highly volatile flavour compounds in dairy products. Application to yoghurt. *Lait* **53,** 12.

Dumont, J. P. and Adda, J. (1978). Flavor formation in dairy products. *In* "Progress in Flavor Research" (D. G. Land, and H. E. Nursten, eds.), pp. 245–262. Applied Science Publishers, London.

Dyatchenko, P. (1971). "Chemistry of Milk," p. 13. Ministry of Meat and Milk Industry, Tallinn, USSR.

El Soda, M. (1986). Acceleration of cheese ripening: Recent advances. *J. Food Prot.* **49,** 395.

El Soda, M., Bergere, J.-L., and Desmazeaud, J. (1978). Detection and localization of peptide hydrolases in *Lactobacilli* casein. *J. Dairy Res.* **45,** 519.

El Soda, M., Fathallah, S., and Ezzat, N. (1983). Acceleration of Cheddar cheese ripening with liposome trapped extracts from *Lactobacillus* casein. *J. Dairy Sci.* **66** Suppl. 1, 78. (Abstr.).

El Soda, M., Korayem, M., Ezzat, N., and Ismail, A. (1984). Acceleration of Cheddar cheese ripening with liposome trapped extracts from *Lactobacillus helveticus. Proc. Egypt. Conf. Dairy Sci. Technol. 2nd,* p. 28.

Farmer, R. E., Shahani, K. M., and Reddy, G. V. (1975). Inhibitory effect of yogurt components. *J. Dairy Sci.* **58,** 787.

Fedrick, I. A., Aston, J. W., Durward, I. G., and Dulley, J. R. (1983). The effect of elevated ripening temperatures on proteolysis and flavor development in Cheddar cheese. II. High temperature storage midway through ripening. *N. Z. J. Dairy Sci. Technol.* **18,** 253.

Formisano, M., Coppola, S., Percuoco, G., Percuoco, S., Zoina, A., Germano, S., and Capriglione, I. (1974). *Ann. Microbiol Enzymol.* **24,** 281.

Fox, P. F. (1987). New developments in cheese production. *In* "Food Technology: International Europe, pp. 112–115. Lavenham Press Ltd., U. K.

Fox, P. F., and Walley, B. F. (1971). Influence of sodium chloride on the proteolysis of casein by rennet and by trypsin. *J. Dairy Res.* **38,** 260.

Friend, B. A., and Shahani, K. M. (1984). Nutritional and therapeutic aspects of lactobacilli. *J. Appl. Nutr.* **36,** 125.

Garnier, J. (1963). Etude cinétique d'une protéolyse limitée action de la presure sur la caséine κ. *Biochim. Biophys. Acta* **66,** 366.

Garvie, E. I. (1978). Lactate dehydrogenases of *Streptococcus thermophilus. J. Dairy Res.* **45,** 515.

Grandison, A. S., Ford, G. D., Owen, A. J., and Millard, D. (1984). Chemical composition and coagulating properties of renneted Friesian milk during the transition from winter rations to spring grazing. *J. Dairy Res.* **51,** 69.

Grappin, R., Rank, T. C., and Olson, N. F. (1985). Primary proteolysis of cheese proteins during ripening. A review. *J. Dairy Sci.* **68,** 531.

Green, M. L., and Grutchfield, G. (1971). Density gradient electrophoresis of native and of rennet-treated casein micelles. *J. Dairy Res.* **38,** 151.

Green, M. L., and Marshall, R. J. (1977). The acceleration by cationic materials of the coagulation of casein micelles. *J. Dairy Res.* **44,** 521.

Green, M. L., and Morant, S. V. (1981). Mechanism of aggregation of casein micelles in rennet-treated milk. *J. Dairy Res.* **48,** 57.

Green, M. L., Hobbs, D. G., and Morant, S. V. (1978). Intermicellar relationships in rennet-treated separated milk. 2. Process of gel assembly. *J. Dairy Res.* **45,** 413.

Grieve, P., and Dulley, J. R. (1984). Use of *Streptococcus lactis lac.* mutants for accelerating Cheddar cheese ripening. 2. Their effect on the rate of proteolysis and flavor development. *Aust. J. Dairy Technol.* **38,** 49.

Grieve, P., Lockie, B., and Dulley, J. R. (1983). Use of *Streptococcus lactis. lac* mutants for accelerating Cheddar cheese ripening. 1. Isolation, growth and properties of a C2 lac. variant. *Aust. J. Dairy Technol.* **38,** 10.

Grigorov, H. (1967a). Effect of heat treatment of cow's milk on the hydrophilic properties of the protein in Bulgarian yoghurt. *Int. Dairy Congr. [Proc.], 17th, 1966* Vol. E/F, p. 643.

Grigorov, H. (1967b). Effect of various types of heat processing of cow's milk on the duration of the coagulation process and on the pH and acidometric titration values of Bulgarian sour milk (yoghurt). *Int. Dairy Congr. [Proc.], 17th, 1966* Vol. E/F, p. 649.

Grigorov, H. (1967c). Interrelation between heat treatment and homogenization of cow's milk and the quality and hydrophilic properties in Bulgarian yoghurt. *Int. Dairy Congr. [Proc.], 17th, 1966* Vol. E/F, p. 655.

Groux, M. (1976). Study of the components of yoghurt flavour. *In* "Nestle Research News" (C. Boella, ed.), p. 50. Nestles Products Technical Assistance Co. Ltd.

Guigoz, Y., and Solms, J. (1976). Bitter peptides, occurrence and structure. *Chem. Senses Flavour* **2,** 71.

Gupta, C. B., and Eskin, N. A. M. (1977). Potential use of vegetable rennet in the production of cheese. *Food Technol.* **31,** 62.

Gurr, M. I. (1984). The nutritional role of cultured dairy products. *Can. Inst. Food Sci. Technol. J.* **17,** 57.

Harper, W. J. (1959). Our industry today. Chemistry of cheese flavors. *J. Dairy Sci.* **42,** 207.

Hartman, A. M., and Dryden, L. P. (1965). Milk and milk products: A review. American Dairy Sci. Assoc., U.S.D.A. Beltsville, Maryland, pp. 15, 21, 35, 46, 61, and 70.

Harwalkar, V. R., and Elliot, J. A. (1965). Isolation and partial purification of bitter components from Cheddar cheese. *J. Dairy Sci.* **48,** 784.

Hemme, D., Bouillanne, C., Metro, F., and Desmazeaud, M.-J. (1982). Microbial catabolism of amino acids during cheese ripening. *Sci. Aliment.* **2,** 113.

Higashio, K., Yoshioka, Y., and Kikuchi, T. (1977a). Symbiosis in yoghurt culture. 1.Isolation and identification of a growth factor for *Streptococcus thermophilus* produced by *Lactobacillus bulgaricus. J. Agric. Chem. Soc. Jpn.* **51,** 203.

Higashio, K., Yoshioka, Y. and Kikuchi, T. (1977b). Symbiosis in yoghurt culture. II. Isolation and identification of a growth factor for *Lactobacillus bulgaricus* produced by *Streptococcus thermophilus. J. Agric. Chem. Soc. Jpn.* **51,** 209.

Hill, R. D., Lahav, E. and Givol, D. (1974). A rennin sensitive bond in α, β-casein. *J. Dairy Sci.* **41,** 147.

Humme, H. E. (1972). The optimum pH for the limited specific hydrolysis of kappa-casein by rennin (primary phase of milk clotting). *Neth. Milk Dairy J.* **26,** 180.

IARC International Microecology Group (1977). Dietary fiber, transit time, fecal bacteria, steriods and colon cancer in two Scandinavian populations. *Lancet* **2,** 207.

ICMSF (1978). "Microorganisms in Foods. 1. Their Significance and Methods of Enumeration." Univ. of Toronto Press, Toronto, Canada.

International Dairy Federation (1983). "Cultured Dairy Foods in Human Nutrition," Doc. No. 159. IDF.

Irvine, D. M. (1982). "Cheddar Cheese," Bull. Ont. Dep. Agric. Food, Toronto, Ontario, Canada.

Irvine, D. M., and Hill, A. R. (1985). Cheese technology. *In* "Comprehensive Biotechnology" (H. W. Blanch, S. Drew, and D. I. C. Wang, eds.), Vol. 3, pp. 523–566. Pergamon, Oxford.

Jarrett, W. D., Aston, J. W., and Dulley, J. R. (1982). A simple method for estimating free amino acids in Cheddar cheese. *Aust. J. Dairy Technol.* **37,** 55.

Kilara, A., and Shahani, K. M. (1976). Lactase activity of cultured and acidified dairy products. *J. Dairy Sci.* **59,** 2031.

Knoop, A. M., and Peters, K. H. (1971). Sunmikroskopische Strukturveranderungen in Camembert Kase wahrend der Reifung. *Milchwissenschaft* **26,** 193.

Kosikowski, F. V. (1951). The liberation of free amino acids in raw and pasteurized milk cheddar cheese during ripening. *J. Dairy Sci.* **34,** 235.

Kowalchyk, A. W., and Olson, N. F. (1977). Effects of pH and temperature on the secondary phase of milk clotting by rennet. *J. Dairy Sci.* 60, 1256.

Kowalchyk, A. W., and Olson, N. F. (1979). Milk clotting and curd firmness as affected by type of milk-clotting enzyme, calcium chloride, concentration and season of year. *J. Dairy Sci.* **62,** 1233.

Kreiss, W., and Hession, C. (1973). Isolation and purification of L-methionine-deamino-mercapto-methane lyase (L-methioninase) from *Clostridium sporogenes. Cancer Res.* **33,** 1862.

Kristoffersen, T., and Gould, I. A. (1960). Cheddar cheese flavor. II. Changes in flavor quality and ripening products of commercial cheddar cheese during controlled curing. *J. Dairy Sci.* **41,** 1202.

Laakso, S. (1979). Evidence of multiple demethiolation of methionine by a methionine utilizing mutant of *Pseudomonas fluorescens UK1. FEMS Microbiol. Let.* **5,** 407.

Law, B. A., and King, J. (1985). The use of liposomes for proteinase addition to Cheddar cheese. *J. Dairy Res.* **52,** 183.
Law, B. A., and Sharpe, M. E. (1978). Formation of methanethiol by bacteria isolated from raw milk and Cheddar cheese. *J. Dairy Res.* **46,** 267.
Law, B. A., Castanon, M., and Sharpe, M. E. (1976). The contribution of *Streptococci* to flavour development in Cheddar cheese. *J. Dairy Res.* **43,** 301.
Lindqvist, B. (1963a). Casein and the action of rennin. Part 1. *Dairy Sci. Abstr.* **25,** 257.
Lindqvist, B. (1963b). Casein and the action of rennin. Part 2. *Dairy Sci. Abstr.* **25,** 265.
Luca, C. (1974). Improvement of yoghurt quality: Lactic acid bacteria and proteolysis of nitrogenous compounds. *Dairy Sci. Abstr.* **36,** 633.
Mabbit, L. A. (1955). Quantitative estimation of the amino acids in Cheddar cheese and their importance in flavor. *J. Dairy Res.* **22,** 224.
Maga, J. A. (1976). The role of sulfur compounds in food flavor. Part III. Thiols. *CRC Crit. Rev. Food Sci. Nutr.* 147.
Mahler, H. R., and Cordes, E. H. (1971). "Biological Chemistry," 2nd ed. Harper & Row, New York.
Manning, D. J. (1979). Cheddar cheese flavor studies. II. Relative flavor contributions of individual volatile components. *J. Dairy Res.* **46,** 523.
Manning, D. J. (1979). Chemical production of essential Cheddar flavour compounds. *J. Dairy Sci.* **46,** 531.
Maret, R., Sozzi, T., and Bohren, H. (1974). Proteolytic activity of different strains of lactic acid bacteria. *Int. Dairy Congr., 19th, 1974* Vol. 1E, p. 371.
Marshall, R. J., and Green, M. L. (1980). The effect of chemical structure of additives on the coagulation of casein micelle suspensions by rennet. *J. Dairy Res.* **47,** 359.
Matoba, T., Hayashi, R., and Hata, T. (1970). Isolation of bitter peptides from tryptic hydrolysate of casein and their chemical structure. *Agric. Biol. Chem.* **34,** 1234.
McGugan, W. A., Emmons, D. B., and Larmond, E. (1979). Influence of volatile and nonvolatile fractions on the intensity of Cheddar cheese flavors. *J. Dairy Sci.* **62,** 398.
McKay, L. L., Sandine, W. E., and Elliker, P. R. (1971). Lactose utilization by lactic acid and bacteria. A review. *Dairy Sci. Abstr.* **33,** 493.
Miller, I., Martin, H., and Kindler, O. (1964). Spectrum of amino acids in yoghurt. *Milchwissenschaft* **19,** 18.
Mulder, H. (1952). Taste and flavour forming substances in cheese. *Neth. Milk Dairy J.* **6,** 157.
Mullvihill, D. M., and Fox, P. F. (1977). Proteolysis of β-casein by chymosin: Influence of pH and urea. *J. Dairy Res.* **44,** 533.
Nakazawa, H., Sano, K., Kumagai, H., and Yamada, H. (1977). Distribution and formation of aromatic L-amino acid decarboxylase in bacteria. *Agric. Biol. Chem.* **41,** 2241.
Ney, K. H. (1971). Voraussage der bitterkeit von peptiden aus deren aminosaurezusammenstezung. *Z. Lebensm.-Unters -Forsch.* **147,** 64.
Nitschmann, H. and Bohren, H. U. (1955). Rennin and its action on the casein of milk. X. A method for the direct determination of the velocity of the primary reaction of rennin coagulation. *Helv. Chim. Acta* **38,** 1953.
Noomen, A. (1978). Activity of proteolytic enzymes in simulated soft cheeses (Meshanger type). 1. Activity of milk protease. *Neth. Milk Dairy J.* **32,** 26.
O'Keeffe, R. B., Fox, P. F., and Daly, C. (1976). Contribution of rennet and starter proteases to proteolysis in Cheddar cheese. *J. Dairy Res.* **43,** 97.
O'Keeffe, R. B., Fox, P. F., and Daly, C. (1978). Proteolysis in cheddar cheese: Role of coagulant and starter bacteria. *J. Dairy Res.* 45, 465.
Orla-Jensen, S. (1931). "Dairy Bacteriology" (Engl. Transl.), 2nd ed., pp. 1–53 and 109–129. Amp & Churchill, London.

Panouse, J. J., Masson, J. D., and Thanh Tong, T. (1972). Cheese flavors. *Ind. Aliment. Agric.* **89,** 133.

Parry, R. M. (1974). Milk coagulation and protein denaturation. In "Fundamentals of Dairy Chemistry" (B. H. Webb, A. H. Johnson, and J. A. Alford, eds.), 2nd ed., pp. 621–629. Avi Publ. Co., Westport, Connecticut.

Payens, T. A. J. (1966). Association of caseins and their possible relation to structure of the casein micelle. *J. Dairy Sci.* **49,** 1317.

Payens, T. A. J. (1977). On enzymatic clotting processes. 2. The colloidal instability of chymosin-treated casein micelles. *Biophys. Chem.* **6,** 263.

Pearce, L. E. (1976). Moving boundary electrophoresis of native and rennet-treated casein in micelles. *J. Dairy Res.* **43,** 27.

Pelissier, J. P., Mercier, J. C., and Ribadeau Dumas, B. (1974). Proteolysis of bovine α and β-caseins by s1 rennin. Proteolytic specificity of the enzyme and the bitter peptides released. *Ann. Biol. Anim. Biochim. Biophys.* **14,** 343.

Pette, J. W., and Lolkema, H. (1950). Acid production and aroma formation in yoghurt. *Neth. Milk Dairy J.* **4,** 261.

Rao, M. V. R., and Dutta, S. M. (1977). Production of betagalactosidase from *Streptococcus thermophilus* grown in whey. *Appl. Environ. Microbiol.* **34,** 185.

Rao, M. V. R., and Dutta, S. M. (1978). Factors influencing production of lactase by *Streptococcus thermophilus. Int. Dairy Congr., 20th, 1978* Vol. E, p. 495.

Reddy, G. V., Shahani, K. M., and Banergee, M. R. (1973). Inhibitory effect of yogurt on Erhlich ascites tumor cell proliferation. *J. Natl. Cancer Inst. (U.S.)* **50,** 815.

Reddy, G. V., Shahani, K. M. and Kulkarni, S. M. (1976). B-complex vitamins in cultured and acidified yogurt. *J. Dairy Sci.* **59,** 191.

Renz, U., and Puhan, Z. (1975). Factors promoting bitterness in yoghurt. (Bietrag zur Kennfis von Faktoren, die Bitterkeitlm Im Joghurt begunstigen). *Milchwissenschaft* **30,** 265.

Resmini, P., Saracchi, C., Pazzaglia, C., and Bernardi, G. (1976). Amino acid sequence of the first peptide with high electrophoretic mobility (α_{s1},-I) formed during cheese ripening. *Sci. Tech. Latt.-Casearia* **27,** 7.

Resmini, P., Saracchi, C. and Pazzaglia, C. (1977). Some physicochemical characteristics of, -I and B-I peptides split from α and β-casein by rennin action. *Sci. Tec. Latt.-Casearia.* **28,** 405.

Ruegg, M., Eberhard, P., Moor, U., Fluckiger, E., and Blanc, B. (1980). Relationships between texture and composition of cheese. *Schweiz. Milchwirtsch. Forsch.* **9,** 3.

Saier, M. H., Jr. (1977). Bacterial phosphoenolpyruvate : sugar phosphotransferase systems: Structural, functional and evolutionary interrelationships. *Bacteriol. Rev.* 41, 856.

Sandine, W. E. and Elliker, P. R. (1970). Microbially induced flavors and fermented foods. *J. Agric. Food Chem.* **18,** 557.

Sandine, W. E., Daly, C., Elliker, P. R., and Vedamuthu, E. R. (1972). Causes and control of culture-related flavor defects in cultured dairy products. *J. Dairy Sci.* **55,** 1030.

Schormuller, J. (1968). The chemistry and biochemistry of cheese ripening. *Adv. Food Res.* **16,** 231.

Scott, R. (1981). "Cheesemaking Practice." Applied Science Publishers, London.

Shankar, P. A. (1977). Inter-relationship of *S. thermophilus* and *L. bulgaricus* in Yoghurt Cultures. Ph.D. Thesis, University of Reading, Berkshire, U. K.

Shankar, P. A., and Davies, F. L. (1978). Proteinase and peptidase activities of yoghurt starter bacteria. *Int. Dairy Congr. 20th, 1978* Vol. 3, p. 467.

Sharpe, M. E. (1978). Lactic acid bacteria (including *Leuconostocs*). *Int. Dairy Congress., 20th, 1978. Moscow.*

Sharpe, M. E. (1979). Lactic acid bacteria in the dairy industry. *J. Soc. Dairy Technol.* **32,** 9.

Snoeren, T. H. M., and Both, P. (1981). Proteolysis during the storage of UHT-sterilized whole milk. II. Experiments with milk heated by the indirect system for 4s at 1422°. *Neth. Milk Dairy J.* **35,** 113.

Somkuti, G., and Steinberg, D. H. (1978a). Adaptability of *Streptococcus thermophilus* to carbohydrates. *J. Dairy Sci.* **61,** Suppl.1, 118.

Somkuti, G. and Steinberg, D. H. (1978b). Lactose catabolism in *Thermophilus streptococci. J. Dairy Sci.* **61,** Suppl. 1, 118.

Stadhousers, J. (1986). The control of cheese starter activity. *Neth. Milk Dairy J.* **40,** 155.

Stadhousers, J., Hupo, G., Exterkate, F. A., and Visser, S. (1983). Bitter flavour in cheese. 1. Mechanism of the formation of the bitter flavour defect in cheese. *Neth. Milk Dairy J.* **37,** 157.

Storry, J. E., and Ford, G. D. (1982). Development of coagulum firmness in renneted milk—A two-phase process. *J. Dairy Res.* **49,** 343.

Tamime, A. Y., and Deeth, H. C. (1980). Yogurt: Technology and biochemistry. *J. Food Prot.* **43,** 939.

Tamime, A. Y., and Robinson, R. K. (1985). "Yoghurt Science and Technology." Pergamon, Oxford.

Thomas, E. L., and Mills, D. E. (1981). Proteolytic enzymes of starter bacteria. *Neth. Milk Dairy J.* **35,** 255.

Torneur, C. (1974). The proteolytic activity of *lactobacilli. Int. Dairy Congr., 19th, 1974* Vol. 1E, p. 366.

Trieu-Cuot, P., Archieri-Haze, M. J., and Gripon, J. Ç. (1982a). Effect of aspartyl proteinases of *Penicillium roqueforti* on caseins. *J. Dairy Res.* **49,** 487.

Trieu-Cuot, P., Archieri-Haze, M. J., and Gripon, J. C. (1982b). A study of proteolysis during Camembert cheese ripening using isoelectric focussing and two-dimensional electrophoresis. *J. Dairy Res.* **49,** 501.

Umetsu, H., Matsuoka, H., and Ichishima, E. (1983). Debittering mechanisms of bitter peptides from milk casein by wheat carboxypeptidase. *J. Agric. Food Chem.* **31,** 50.

Veringa, H. A., Galesloot, Th. E. and Davelaar, H. (1968). Symbiosis in yoghurt. II. Isolation and identification of a growth factor for *Lactobacillus bulgaricus* produced by *Streptococcus thermophilus. Neth. Milk. Dairy J.* **22,** 114.

Visser, F. M. W. (1977). Contribution of enzymes from rennet, starter bacteria and milk to proteolysis and flavour development in Gouda cheese. 2. Development of bitterness and cheese flavour. *Neth. Milk Dairy J.* 31, 188.

Visser, F. M. W., and de Groot-Mostert, A. E. A. (1977). Contribution of enzymes from rennet, starter bacteria and milk to proteolysis and flavour development in Gouda cheese. 4. Protein breakdown: A gel electrophoretical study. *Neth. Milk Dairy J.* **31,** 247.

Walstra, P. (1979). The voluminosity of bovine casein micelles and some of its implications. *J. Dairy Res.* **46,** 317.

Waugh, D. F., and von Hippel, P. H. (1956). κ-Casein and the stabilization of casein micelles. *J. Am. Chem. Soc.* **78,** 4576.

Webb, D. H., Johnson, A. H., and Alford, J. A. (1974). "Fundamentals of Dairy Chemistry," 2nd ed. Avi Publ. Co., Westport, Conn.

Yu, J. H., and Nakanishi, T. (1975a). Studies on the production of flavor constituents by various lactic acid bacteria. II. Effect of milkfat on formation of volatile carbonyl compounds by various lactic acid bacteria. *Jpn. J. Dairy Sci.* **74,** A27.

Yu, J. H., and Nakanishi, T. (1975b). Studies on the production of flavor constituents by various lactic acid bacteria. IV. Effect of milkfat on production of free amino acids by various lactic acid bacteria. *Jpn. J. Dairy Sci.* **74,** A63.

Yu, J. H., and Nakanishi, T. (1975c). Studies on the production of flavor constituents by various lactic acid bacteria. IV. Effect of milkfat on production of free volatile fatty acids by mixed lactic acid bacteria. *Jpn. J. Dairy Sci.* **74,** A79.

Yu, J. H., Kakanishi, T., and Suyama, K. (1974). Production of flavor substances by various lactic acid bacteria. 1. Effects of milk fat on the formation of lactic acid, nonprotein nitrogen, amino nitrogen and free volatile fatty acids by various lactic acid bacteria. *Jpn. J. Dairy Sci.* **23,** A195.

Part III

Biochemistry of Food Spoilage

9

Biochemistry of Food Spoilage: Enzymatic Browning

I. Introduction

Enzymatic browning is a phenomenon which occurs in many fruits and vegetables, such as potatoes, mushrooms, apples, and bananas. When the tissue is bruised, cut, peeled, diseased, or exposed to any abnormal conditions it rapidly darkens on exposure to air as a result of the conversion of phenolic compounds to brown melanins.

The international nomenclature for the enzymes involved in the browning reaction has changed. The first enzyme, monophenol monoxygenase or tyrosinase (EC 1.14.18.1), initiates the browning reaction, which later involves diphenol oxidase or catechol oxidase (EC 1.10.3.2) and laccase (EC 1.10.3.1). Catecholase oxidase will be referred to as "polyphenol oxidase" in this chapter. This enzyme requires the presence of both a copper prosthetic group and oxygen. The copper is believed to be monovalent in the case of the mushroom polyphenol oxidase and divalent in the case of the potato enzyme (Bendall and Gregory, 1963). Polyphenol oxidase is classified as an oxidoreductase, with oxygen functioning as the hydrogen acceptor. The enzyme is widely distributed in higher plants and fungal and animal tissue and was reviewed by Swain (1962), Mathew and Parpia (1971), Mayer and Harel (1979), Vamos-Vigyazo (1981), and recently by Mayer (1987).

II. Mechanism of Reaction

Polyphenol oxidase has been studied extensively in mushrooms (Bouchilloux *et al.*, 1963; Jolley and Mason, 1965; Smith and Krueger, 1962) and potatoes (Patil and Zucker, 1965). Subsequent work examined the isoenzymes of polyphenol oxidase (Constantinides and Bedford, 1967; Cheung and Henderson, 1972) and the intracellular location of the enzyme (Craft, 1966). Recent studies by Jayaraman and Ramanuja (1987) identified 9–13 isoenzymes by polyacrylamide gel electrophoresis of partially purified banana polyphenol oxidase obtained from five commercial varieties. Each one exhibited different specificities toward phenolic substrates, catechol, dopamine, D- and L-dopa, protocatechuic acid, chlorogenic acid, catechin, caffeic acid, and pyrogallol. A recent study by Variyar and co-workers (1988) identified *o*-diphenolase oxidase activity to be the cause of blackening in green pepper berries (*Piper nigrum*). This enzyme was shown to catalyze the oxidation of 3, 4-dihydroxyphenylethanol glycoside, which is naturally present in the fruit. This reaction occurred immediately following damage to the skin tissues with the subsequent development of black oxidized products.

Polyphenol oxidase catalyzes two types of reaction, the cresolase activity in which monophenols are hydroxylated to *o*-diphenols and catecholase activity. The catecholase or diphenolase type of reaction is best illustrated by the oxidation of catechol, an *o*-diphenol which is a commonly used laboratory substrate:

$$\text{Catechol} + E\text{–}2Cu^{2+} + {}^{1}/{}_{2}O_2 \xrightarrow{-2e^-} o\text{-Benzoquinone} + E\text{–}2Cu^{+} + H_2O$$

Catechol o-Benzoquinone

The cresolase or monophenolase activity involves hydroxylation of monophenols to *o*-diphenols as shown by the oxidation of L-tyrosine to 3, 4-dihydroxyphenylalanine, which occurs in potatoes (Mapson *et al.*, 1963; Schwimmer and Burr, 1967):

$$\text{L-Tyrosine} + E\text{–}2Cu^{+} + O_2 + 2H^{+} \xrightarrow{+2e^-} \text{3,4-Dihydroxyphenylalanine} + E\text{–}Cu^{2+} + H_2O$$

$CH_2CH(NH_2)COOH$ (OH) $CH_2CH(NH_2)COOH$ (HO, OH)

L-Tyrosine 3,4-Dihydroxyphenylalanine

The overall mechanism of plant polyphenol oxidases is still incompletely understood although the mechanism for fungal polyphenol oxidase (*Neurospora crassa*) was presented by Lerch (1981). A model was proposed in which the site of interaction with the phenolic substrate, whether mono- or diphenol, was based on the binuclear center of copper. The basic functional molecular unit for the fungal enzyme appeared to be a single-chain protein with two copper atoms per molecule, liganded in part by histidine (Lerch, 1981; Solomon, 1981). The active site of copper appeared to be binuclear and occurred in different functional states: met, oxy, and deoxy. One of the Cu^{2+} atoms is bound to monophenols while diphenols bind to both of them. As shown in the reaction mechanisms, monophenolase activity is intimately coupled to diphenolase activity and produces two electrons. These are required to incorporate one oxygen atom into the monophenol substrate.

The product of this reaction is subsequently oxidized by the catecholase reaction to form *o*-quinone phenylalanine. This is followed by the formation of a red compound, dopachrome (5, 6-quinone indole-2-carboxylic acid), which contains a heterocyclic ring derived from the closure of the aminocarboxylic acid side chain. Dopachrome then undergoes polymerization to form brown melanins (Kang *et al.*, 1983).

Quinone formation is both enzyme and oxygen dependent. Once this has taken place, the subsequent reactions occur spontaneously and no longer depend on the presence of polyphenol oxidase or oxygen. Joslyn and Ponting (1951) summarized those chemical reactions which may account for the formation of brown melanins.

The first reaction is thought to be a secondary hydroxylation of the *o*-quinone or of excess *o*-diphenol:

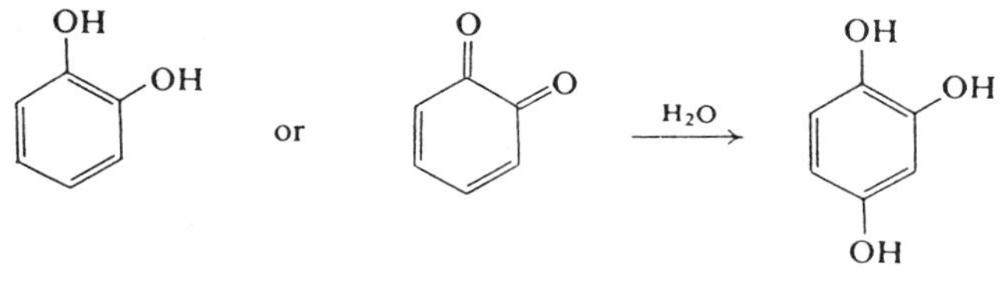

The resultant compound (triphenolic trihydroxybenzene) interacts with *o*-quinone to form hydroxyquinones:

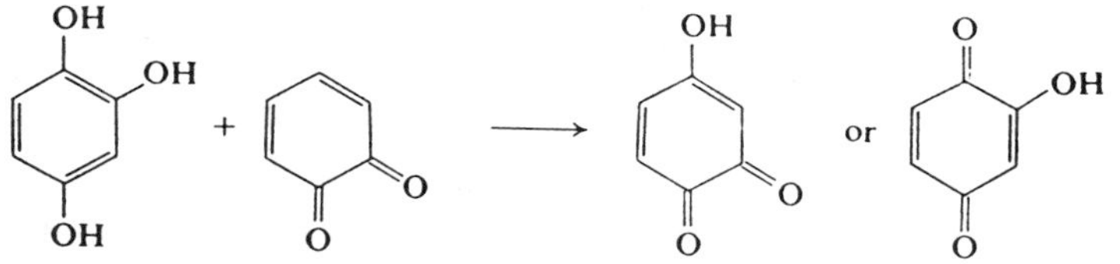

Hydroxyquinones undergo polymerization and are progressively converted to red and red-brown polymers, and finally to the brown melanins which appear at the site of plant tissue injury (Whitaker, 1972; Matheis and Whitaker, 1984).

The ratio of diphenolase : monophenolase activity depends on the plant source and can vary from 1 : 10 to 1 : 40 (Vamos-Vigyazo, 1981). Sanchez-Ferrer *et al.* (1988) partially purified polyphenol oxidase activity in Monastrel grapes (*Vitis vinifera* L. *cv Monastrell*) and identified both cresolase and catecholase activity. A number of fruits, however, have been found to be devoid of monophenolase activity, including mango (Joshi and Shiralkar, 1977), cherry (Lanzarini *et al.*, 1972), banana (Palmer, 1963), and pear (Rivas and Whitaker, 1973).

A. Historical Aspects of Polyphenol Oxidase

The earliest work is attributed to Lindet, who in 1895 recognized the enzymatic nature of browning while working on cider. At the same time Bourquelot and Bertrand commenced studies on mushroom tyrosine oxidase. Subsequently Mrs. Onslow in 1920 showed that enzymatic browning of plant tissue in air was attributed to the presence of *o*-diphenolic compounds, such as catechol, protocatechuic acid, and caffeic acid, plus the appropriate enzymes (oxygenases). The product of this reaction was thought to be a peroxide which reacted with a "chromogen" to form a brown pigment. Many fruits and vegetables, including apples, pears, apricots, and potatoes, were all found to be rich in phenolic compounds and oxygenases. Other fruits, such as citrus, pineapples, and red currants, lacked these substances and were thus termed "peroxidase plants."

This distinction was eliminated when it was shown that peroxidase and catalase were present in both groups of plants and in fact were ubiquitous in plant tissues. The term "oxygenase" was subsequently replaced by "phenolase" or "polyphenol oxidase." Kubowitz in 1937 demonstrated that polyphenol oxidase was a copper-containing enzyme.

B. Biological Significance of Polyphenol Oxidase in Plants

The role of polyphenol oxidase in the living intact cell has until recently remained somewhat obscure. Early studies suggested its involvement as a terminal oxidase in respiration (James, 1953) or in the biosynthesis of lignin (Mason *et al.*, 1955). This was later discounted in studies by Nakamura (1967), who examined the role of three enzymes isolated from the latex of the Japanese lacquer tree (*Rhis vermicifera*), phenolase, peroxidase, and laccase. Of these enzymes only peroxidase was involved in lignification. The exclusive involvement of peroxidase in polymerization of soluble phenolics to insoluble lignin was confirmed by later researchers (Egley *et al.*, 1985; Harkin and Obst, 1973; Stafford, 1974). More recent studies on polyphenol oxidase have shown it to be restricted to the plastids (Bar-Nun and Mayer, 1983; Martyn *et al.*, 1979; Vaughn and Duke, 1981a,b, 1982, 1984). These are cytoplasmic organelles in the chloroplasts where there is intense photochemical activity. The enzyme appeared to be in a latent form and bound to the thylakoid membrane, where the photochemical reactions of photosynthesis occur. In this form it had little activity toward phenolics, which are located in the vacuole somewhat isolated from the plastid.

A possible role for polyphenol oxidase appeared to be in pseudocyclic photophosphorylation (Tolbert, 1973). This is a noncyclic electron transport system in which oxygen instead of $NADP^+$ is the terminal acceptor, also referred to as the "Mehler reaction." The inhibition of polyphenol oxidase by KCN was shown by Siegenthaler and Vaucher-Bonjour (1971) to result in an increase in the photosynthetic evolution of oxygen. Later studies by Meyer and Biehl (1982) noted an increase in oxygen evolution as latent polyphenol oxidase activity was activated during the aging of isolated chloroplasts from spinach. It was clear from these studies that an increase in polyphenol oxidase activity can be correlated with loss of Mehler reaction function (Vaughn and Duke, 1984). Since the enzyme functions normally only when the cells are damaged or undergo senescence, it may also have a protected role as proposed originally by Craft and Audia (1962). The activation of polyphenol oxidase was previously reported during aging or senescence (Goldbeck and Cammarata, 1981), or in the presence of fatty acids (Siegenthaler and Vaucher-Boniour, 1971), detergents (Sato and Hasegawa, 1976), or trypsin (Tolbert, 1973). Polyphenol oxidase apparently becomes involved in phenolic metabolism when the plastid and vacuole contents are mixed. This occurs during senescence, when the integrity of the cell is disrupted, and is responsible for the pigmentation of black olives (Ben-Shalom *et al.*, 1977). Enzymatic browning also results following injury to the fruit or vegetable, causing a disruption of the plastid, activation of latent PPO, and catalysis of phenolics released from the vacuole. Activation of polyphenol oxidase into an active protein was confirmed by Vaughn and Duke (1984), who found that tentoxin prevented this from occurring, resulting in the accumulation of nonactive polyphenol oxidase in the plastid envelope membranes of isolated *Vicia faba* L. chloroplasts.

Al-Barazi and Schwabe (1984) reported that polyphenol oxidase was present in both the secondary xylem and pith, close to the sites of root initiation in *Pistacia vera* cuttings. They suggested the enzyme may be involved in root formation and development.

The enzyme appears to play an important role in the resistance of plants to infection by viruses, bacteria, fungi, or mechanical damage. Under these conditions the activity of the enzyme increases with the production of insoluble polymers which serve as a barrier to the spread of infection in the plant (Rubin and Artsikhovskaya, 1960). Alternatively, some of the intermediates in the oxidative polymerization of polyphenols prevent or reduce infection by inactivating or binding some labile plant enzymes or viruses (Pierpoint, 1966). The bacteriostatic properties of melanin and quinones have been shown by their ability to partially inactivate a potato virus (Pierpoint *et al.*, 1977).

C. Phenolic Compounds in Food Material

The phenolic compounds which occur in food material and which participate in browning may be classified into three groups, namely, simple phenols, cinnamic acid derivatives, and flavonoids.

1. *The Simple Phenols*

The simple phenols include monophenols such as L-tyrosine and *o*-diphenols such as catechol. Gallic acid is present in an esterified form in tea flavonoids.

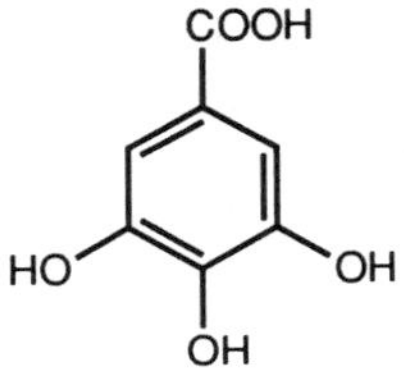

Gallic acid

2. *The Cinnamic Acid Derivatives*

The most important member of this group of compounds in food material is chlorogenic acid, which is the key substrate for enzymatic browning, particularly in apples and pears (Weurman and Swain, 1953; Hulme, 1958). It is thought to play a major role in the after-blackening phenomenon sometimes observed in potatoes (Bate-Smith *et al.*, 1958) in which, on standing, parts of the cooked tuber flesh turn a gray-black color. This discoloration is attributed to oxidation of complexes formed between iron and caffeic and chlorogenic acids.

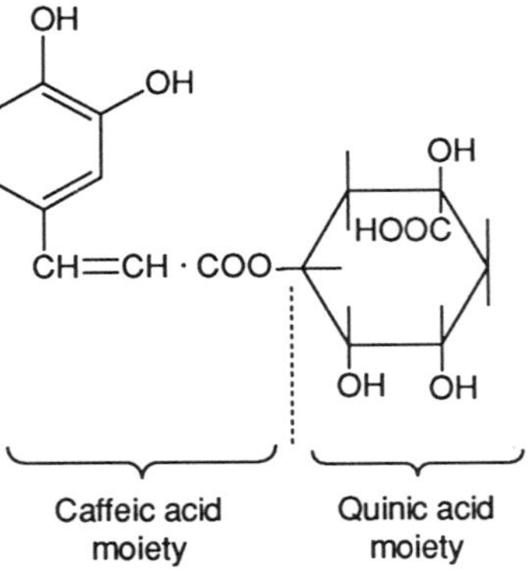

Caffeic acid moiety — Quinic acid moiety

Other members of this group of compounds include *p*-coumaric, caffeic, ferulic, and sinapic acids.

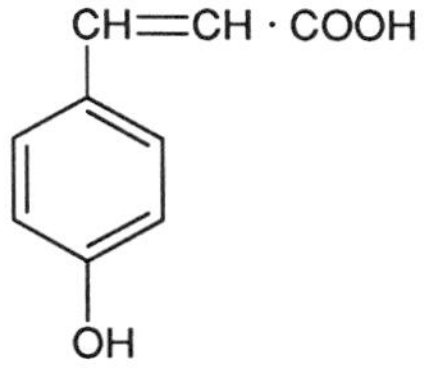

p-Coumaric acid

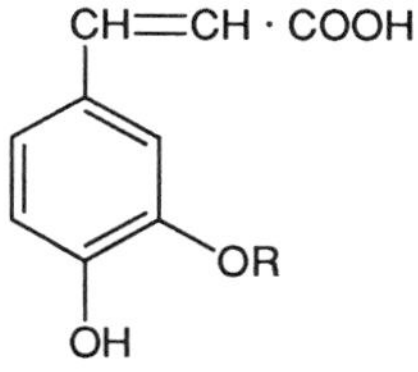

R = H, caffeic acid
R = CH_3, ferulic acid

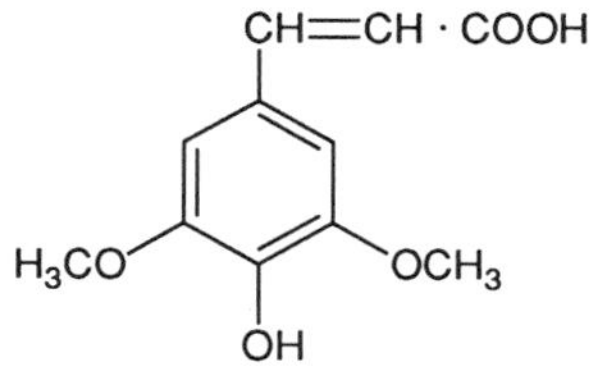

Sinapic acid

3. *The Flavonoids*

All members of this group of compounds are structurally related to flavone:

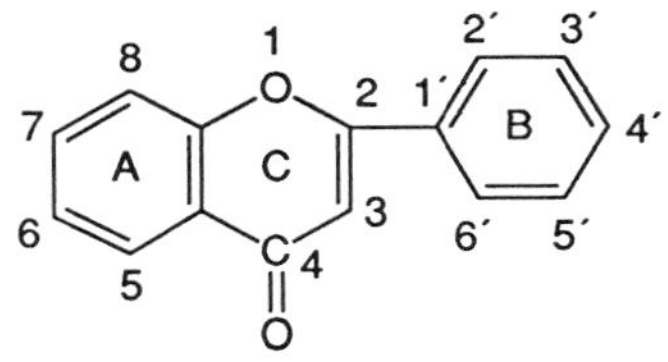

In food material, the important flavonoids are the catechins and leucoanthocyanidins (sometimes called the food tannins), anthocyanins (described in Chapter 2), and the flavonols. Catechin has the following structure:

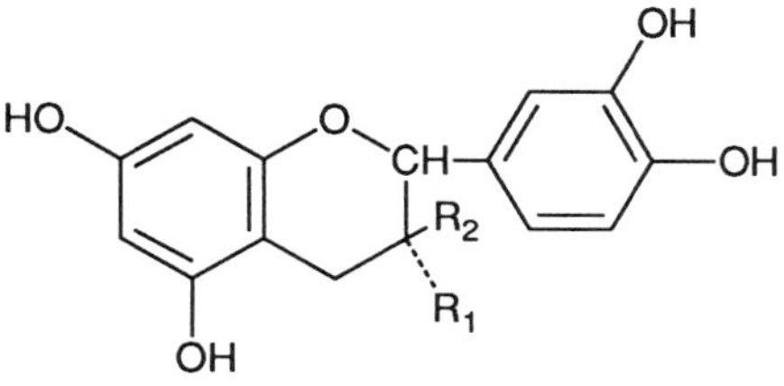

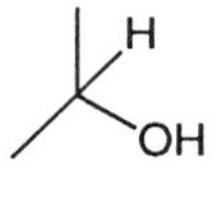

Catechin 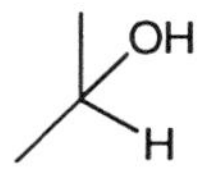 Epicatechin

(+)–Catechin (R_1 = H; R_2 = OH)
(–)–Epicatechin (R_1 =OH; R_2 = H)

An extra hydroxyl group attached to the 5′ position on the B ring of catechin and epicatechin give rise to gallocatechin and epigallocatechin, respectively. The catechin gallates are esters of catechins and gallic acid, the ester linkage being formed from the carboxyl group of gallic acid and the hydroxyl group attached to position 3 of the catechin C ring. An example is (−)-epigallocatechin gallate, which is the major polyphenol in dried tea leaves.

Leucoanthocyanidins are widely distributed in food material and the accepted chemical structure is that of a flavan-3, 4-diol:

Leucoanthocyanidin (R_1 = OH; R_2 = H)
Leucodelphinidin (R_1 = R_2 = OH)
Leucopelargonidin (R_1 = R_2 = H)

Such compounds may polymerize, as evident for dimeric leucoanthocyanidins isolated from cacas. A general definition of leucoanthocyanidins is any compound which yields anthocyanidin on heating in the presence of mineral acid.

The flavonols also participate in browning reactions and are widely distributed in plant tissues. The most commonly occurring flavonols are kaempferol, quercetin, and myricetin:

Kaempferol (R_1 = R_2 = H)
Quercetin (R_1 = OH; R_2 = H)
Myricetin (R_1 = R_2 = OH)

Flavonols posses a light yellow color and are particularly important in fruits and vegetables in terms of the astringency they impart to the particular food. They occur naturally as glycosides, examples of which are rutin and the glycosides of quercetin, the latter occurring in tea leaves and in apple skins (Hulme, 1958).

All the compounds discussed so far are substrates for polyphenol oxidase. Such oxidation reactions are important in tea fermentation, in the maturation of dates (Maier and Metzler, 1965), in the browning of cling peaches (Luh *et al.*, 1967), and in the drying stage of the curing of fresh cacao seeds (Roelofsen, 1958). It appears to be an important step in the development of the final color, flavor, and aroma of cacao and chocolate. Polyphenol oxidase plays a beneficial role in the case of tea and cacao fermentation, in contrast to its perhaps more familiar role in the browning of fruits and vegetables.

D. Inhibitors of Polyphenol Oxidase

A number of compounds possess chemical structures closely related to *o*-diphenols but do not function as substrates of polyphenol oxidase. These include methyl-substituted derivatives such as guaiacol and ferulic acid (Finkle and Nelson, 1963), and *m*-diphenols such as resorcinol and phloroglucinol, which have been shown to have an inhibitory effect, possibly competitive (Bendall and Gregory, 1963).

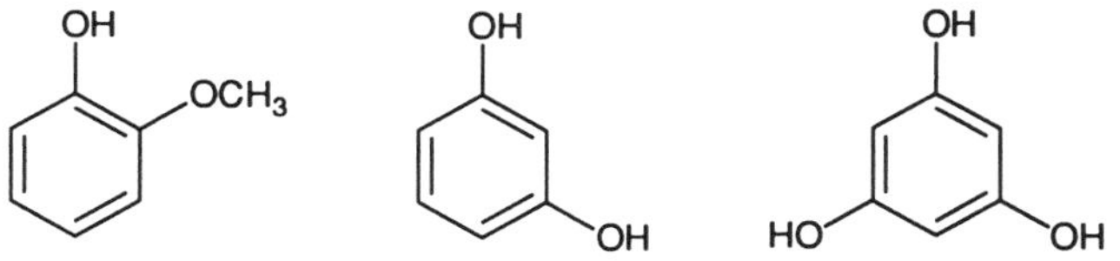

Guaiacol Resorcinol Phloroglucinol

Since polyphenol oxidase is a metalloprotein in which copper is the prosthetic group, it is inhibited by a variety of chelating agents, including sodium diethyldithiocarbamate (DIECA), sodium azide, potassium ethylxanthate, and ethylenediaminetetraacetate (EDTA). While sodium azide and EDTA inhibit polyphenol oxidase, they appear to be less specific chelators compared to DIECA or potassium ethylxanthate (Abukhara and Woolhouse, 1966; Wong *et al.*, 1971). The latter compounds were reported by Anderson (1968) to combine with the quinones produced by polyphenol oxidase in addition to chelating copper.

In addition to sulfites, discussed in Section IV,E, L-cysteine has also been shown to be an effective inhibitor of polyphenol oxidase by acting as a quinone-coupler as well as reducing agent. Kahn (1985) reported L-cysteine to be the most effective inhibitor of avocado, banana, and mushroom polyphenol oxidase. Zawistowski and co-workers (1987) studied the effect of metabisulfite and L-cysteine on the activity of Jerusalem artichoke polyphenol oxidase stored over a 2-week period. Their results (Figure 9.1) showed complete inhibition of enzyme activity in the presence of 5 or 10 m*M* sodium metabisulfite during the 15-day storage period compared to inhibition for only 3 days in the presence of 5 ppm L-cysteine. The resumption of polyphenol oxidase activity after 3 days was attributed to the consumption of L-cysteine through quinone-coupling (Pierpoint, 1966).

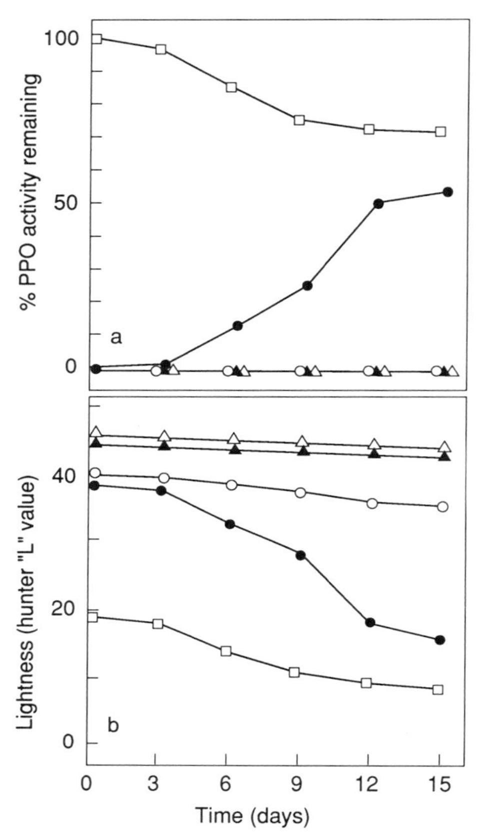

FIG. 9.1. Effect of inhibitors on polyphenol oxidase activity (a) and color development-lightness (b) in artichoke extracts stored at 4°C. Extracts contained L-cysteine at concentrations of 10 m*M* (●) or 25 m*M* (○) or sodium metabisulfite at concentrations of 5 m*M* (▲) and 10 m*M* (△) or no inhibitor (□) (Zawistowski *et al.*, 1987).

E. Laccase

Early studies on enzymatic browning of phenolic compounds identified two types of activity which were initially termed tyrosinase and laccase. The respective systematic names formerly used were *o*-diphenol: oxygen oxidoreductase (EC 1. 10. 3. 1) and p-diphenol : oxygen oxidoreductase (EC 1. 10. 3. 2), which are now combined as monophenol monooxygenase (EC 1. 14.18.1). The diagnostic feature of laccase is its ability to oxidize *p*-diphenols, a property not possessed by tyrosinase or polyphenol oxidase (Mayer and Harel, 1979). Research on the reaction mechanism of laccase has been restricted to one higher plant enzyme from *Rhus* and two fungal enzymes from *Polyporus* and *Podospora*. Much research has focused on three different types of copper sites involved in substrate binding (Morris *et al.*, 1983) or responsible for the reduction of oxygen (Goldberg *et al.*, 1980). The recent proposal by Allendorf and coworkers (1985) for the existence of a trinuclear Cu site capable of binding and bridging small molecules further illustrates the complexity of the mechanism involved in the laccase reaction. The basic laccase-catalyzed reaction responsible for the oxidation of *p*-diphenol is as follows:

HO–C$_6$H$_4$–OH + [O] ⟶ O=C$_6$H$_4$=O + H_2O

p-Diphenol (quinol) ⟶ *p*-Quinone

or

o-Diphenol + [O] ⟶ *o*-Quinone + H_2O

F. Specificity of Polyphenol Oxidase

Polyphenol oxidase, as discussed previously, catalyzes two different reactions, either the hydroxylation of monophenols to *o*-dihydroxyphenols or the oxidation of *o*-dihydoxyphenols to *o*-quinones (Robb, 1984). The most appropriate chemical structure with respect to polyphenol oxidase activity, when the reaction rate is at a maximum, appears to correspond to the *o*-dihydroxy structure as evident in such compounds as catechol, caffeic acid, and the catechins (Halim and Montgomery, 1978). Oxidation of *o*-diphenols to the corresponding *o*-quinones is a general reaction of all known polyphenol oxidases, irrespective of whether the source material is mushroom, potato, sweet potato (Hyodo and Uritani, 1965), tobacco leaf (Clayton, 1959), apple (Harel *et al.*, 1966), tomato (Hobson, 1967), banana (Palmer and Roberts, 1967), broad bean (Robb *et al.*, 1964, 1965), tea leaf (Gregory and Bendall, 1966), avocado (Kahn, 1975, 1976, 1977), or green olives (Ben-Shalom *et al.*, 1977). Monophenols are more slowly acting

TABLE 9.1

SUBSTRATE SPECIFICITY OF CRUDE ARTICHOKE POLYPHENOL OXIDASE[a]

Substrate	Specific activity[b]
Monohydroxyphenols	
DL-Tyrosine	0.08
p-Cresol	0.08
Dihydroxyphenols	
Chlorogenic acid	3.91
Catechol	1.70
DL-Dopa[c]	1.57
Caffeic acid	0.74

[a] Adapted from Zawistowski *et al.* (1986).
[b] Specific activity was defined as the amount of oxygen (μmole) consumed per minute at 30°C per mg of protein.
[c] 3, 4-dihydroxyphenylalanine.

substrates as they have to be hydroxylated prior to their oxidation to the corresponding *o*-quinones. The oxidation of monophenols is less widespread than that of diphenols, being catalyzed, for example, by potato and mushroom enzyme preparations but not by those from tea, tobacco leaves, or sweet potatoes. It seems that the relationship between cresolase and catecholase activities is not yet fully understood. It appears that many polyphenol oxidases are specific to a high degree in that they only attack *o*-diphenols. Zawistowski and co-workers (1986), in studies on artichoke polyphenol oxidase, found that it exhibited the highest specificity toward dihydroxyphenolic substrates, as shown in Table 9. 1. Cholorogenic acid was the most reactive, followed by catechol, DOPA, and caffeic acid.The specific activity of the enzyme ranged from 50 times greater for chlorogenic acid to 10 times greater for caffeic acid compared the the corresponding monophenols. The optimum substrate for artichoke polyphenoloxidase was chlorogenic acid. This was consistent with studies reported by Lanzarini and co-workers (1972), who showed that the presence of an electron-donating group in position four, as in chlorogenic acid, increased the reactivity of the substrate.

III. Polyphenol Oxidase in Foods and Food Processing

A. Role in Tea Fermentation

The production of black tea is dependent on the oxidative changes that tea leaf polyphenols undergo during processing. Such changes are particularly important for the development of color as well as reduction of the bitter taste associated with unoxidized tannin (polyphenol compound). This subject is reviewed by Swain (1962), Bokuchava and Skobeleva (1969), Mathew and Parpia (1971), and recently Jain and Takeo (1984).

The main tea leaf polyphenols, determined by partition chromatography, include (+)-catechin, (−)-epicatechin, (+)-gallocatechin, (−)-epigallocatechin, (−)-epicatechin gallate, and (−)-epigallocatechin gallate. Of these compounds (−)-epigallocatechin gallate is the major component in the tea shoot. During fermentation (−)-epigallocatechin and its gallate appear to be the only substrates oxidized by tea polyphenol oxidase.

The production of black tea, the most popular form of the beverage, is carried out in four stages. The first stage is called withering, when the shoots from the tea plant are allowed to dry out. This is followed by rolling with the roller, which disrupts the tea leaf tissue and causes cell damage, providing the necessary conditions for the development of the oxidative processes. The next step is the fermentation of the fragmented tea leaves which are held at room temperature in a humid atmosphere with a continuous supply of oxygen. These conditions are optimal for polyphenol oxidase action on the tea leaf tannins, which in addition to reducing astringency also converts the the green color of the rolled tea leaves to give coppery-red and brown pigments. Fermentation is terminated by firing, where the tea is dried at 90–95°C and the moisture reduced to 3–4%.

The critical biochemical reaction during tea fermentation is oxidation of catechins by polyphenol oxidase to the corresponding *o*-quinones. These quinones are intermediate compounds which are subjected to secondary oxidation leading to the production of theaflavin and theaflavin gallate, the yellow-orange pigments in black tea, and to a group of compounds referred to as thearubigins. These thearubigins are dark brown and the main contributors to the familiar color of black tea, and they are the oxidative products of theaflavins. A simplified scheme for the oxidative reactions occurring during tea fermentation is outlined in Scheme 9.1. The theaflavin content of tea was shown by Hilton and Ellis (1972) to correlate with the tea taster's evaluation. This was consistent with earlier studies by Roberts (1952) and Sanderson (1964), who noted a positive correlation between tea quality and polyphenol oxidase activity. Polyphenol oxidase was later purified by Hilton (1972). The oxidative degradation of phlo-

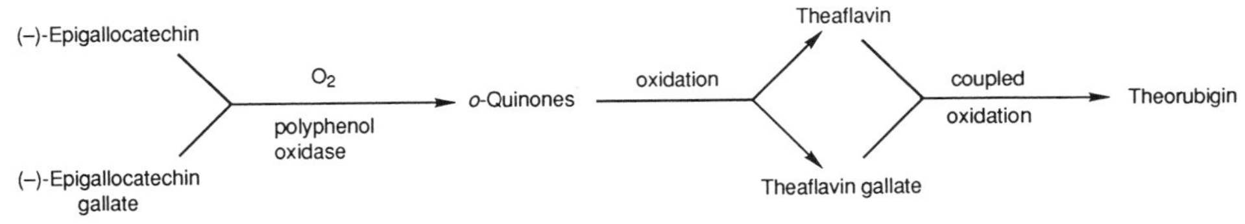

SCHEME 9.1. Oxidative transformations of (−)-epigallocetechin and its gallate during tea fermentation.

TABLE 9.2

SPECIFIC ACTIVITIES OF MT12 AND SEEDLING TEA LEAVES[a,b]

Clone	Peroxidase[c]	Polyphenol oxidase[d]
M12	1.821	0.055
Seedling	0.735	0.020

[a] From Van Lelyveld and de Rooster (1986).
[b] Specific activity in Δ OD/min/mg protein.
[c] Significant at $P < 0.05$.
[d] Significant at $P < 0.01$.

roglucinol rings of the theaflavins by peroxidase caused a loss of theaflavins and a decline in tea quality (Cloughley, 1980a,b). Consequently the presence of both these enzymes affects the quality of tea. Van Lelyveld and de Rooster (1986) examined the browning potential of black tea clones and seedlings. They found a much higher level of polyphenol oxidase in a high-quality hybrid clone (MT12) compared to a low-quality seedling tea (Table 9.2). The reverse was true for peroxidase, in which lower-quality tea had more than double the activity of peroxidase. This suggested that the combination of higher theaflavin levels and polyphenol oxidase activity was responsible for the better quality associated with the MT12 clone.

Green tea is particularly popular in Oriental countries, such as Japan. It is an unfermented tea with a light color and a characteristic degree of astringency. This is achieved by the application of heat during the early stages of tea manufacture which inhibits or prevents oxidation. Red and yellow teas are intermediate between black and green teas and are semifermented products (partially fermented prior to firing). An example of the latter is the Chinese variety Oolong.

B. SHRIMP AND CRUSTACEANS

Enzymatic browning, while studied extensively in fruits and vegetables, has also been implicated in the discoloration of shrimp and other crustaceans (Savagaon and Sreenivasan, 1978). In the latter case it expresses itself as melanosis or black spot, which renders these products unattractive to the consumer as well as lowers their market value. The involvement of phenol oxidase in the formation of melanosis remained somewhat obscure because of lack of any attempt to purify it. Recent studies by Simpson and co-workers (1987, 1988) purified this enzyme from the heads of fresh white shrimps (*Penaeus setiferus*) using affinity chromatography. A single enzyme band of molecular weight 30,000 was identified which was capable of oxidizing dihydroxyphenylalanine. This enzyme, as observed with other phenolases, was also found to be activated by copper. Detailed studies on other phenolases remain to be carried out in addition to developing methods for retarding melanosis in crustaceans.

IV. Methods for Controlling or Inhibiting Enzymatic Browning

A number of methods have been proposed for inhibiting phenolase activity although relatively few can be used in food material (Vamos-Vigyazo, 1981; Walker, 1977). This is illustrated by such toxic inhibitors as cyanide and hydrogen sulfide. Inhibitors such as diethyldithiocarbamate (DIAE) suffer from objectionable flavors at the levels required to inhibit phenolase (Muneta and Walradt, 1968). Others, including naturally occurring materials, ATP, or cysteine, are unsuitable as they are too costly to be economically feasible. Inhibitors of polyphenol oxidase activity can be categorized into four groups based on their mode of action:

1. exclusion of reactants such as oxygen;
2. denaturation of enzyme protein;
3. interaction with the copper prosthetic group;
4. interaction with phenolic substrates or quinones.

A. Exclusion of Oxygen

The simplest method of controlling enzymatic browning is by immersing the peeled product such as potato in water prior to cooking. This can be done very easily in the home to limit access of oxygen to the cut potato tissue. This procedure is used on a large scale for the production of potato chips and French fries (Talburt and Smith, 1967). The method is limited as the fruit or vegetable will brown on reexposure to air or via the oxygen occurring naturally in the plant tissues. The removal of oxygen from fruit or vegetable tissue could lead to anaerobiosis if they are stored for extended periods, which in turn could lead to abnormal metabolites and tissue breakdown. In the case of frozen sliced peaches, the surfaces are treated with excess ascorbic acid to use up the surface oxygen.

B. Application of Heat

The inactivation of polyphenol oxidase as well as other spoilage enzymes can be achieved by subjecting the food article to high temperatures for an adequate

TABLE 9.3

OPTIMAL TEMPERATURES FOR POLYPHENOL OXIDASE FROM DIFFERENT PLANT SOURCES

Source	Optimal temperature (°C)
Apricots[a]	25
Bananas[b]	37
Apples[a]	25,30
Grapes[c]	10–15, 20, 25–30
Potatoes[d]	22

[a] Mihalyi *et al.* (1978).
[b] From Palmer (1963).
[c] From Montedoro (1969), Cash *et al.* (1976), and Lee *et al.* (1983), respectively.
[d] From Schaller (1972).

length of time to denature the protein. The pretreatment of vegetables for canning, freezing preservation, or dehydration or in the manufacture of fruit juices and purees by blanching and high-temperature short-time (HTST) pasteurization essentially achieves this goal. Caution must be exercised when using heat treatment to avoid cooking the fruit or vegetable as this could result in unfavorable texture changes and off-flavor development (Mapson and Tomalin, 1961; Ponting, 1960). This could occur particularly with the preprocessing of potatoes, apples, pears, and peaches.

The heat inactivation of enzymes in foods is not only dependent on time but is also affected by pH. The optimum temperature for polyphenol oxidase varies considerably for different plant sources as well as among cultivars, as shown in Table 9.3. The temperature optimum is also influenced by the substrate used in the assay. For example, Schaller (1972) reported that polyphenol oxidase from potato exhibited maximum activity at 22°C with catechol, whereas an almost linear increase in activity was observed by Mihalyi and coworkers (1978) between 15 and 35°C for potato polyphenol oxidase with pyrogallol. This compared to studies with DeChaunac grapes by Lee *et al.* (1983), who found that the enzyme exhibited maximum activity at 25°C with caffeic acid and little activity when catechol was the substrate.

The effect of temperature on the activity of polyphenol oxidase in pear purée at a constant heating time of 8 sec is shown in Fig. 9.2. A temperature of 80°C was necessary to reduce the enzyme activity by 50%, with almost total inactivation at 90°C. The use of high temperature to inactivate this enzyme requires careful control to insure the desired flavor and texture.

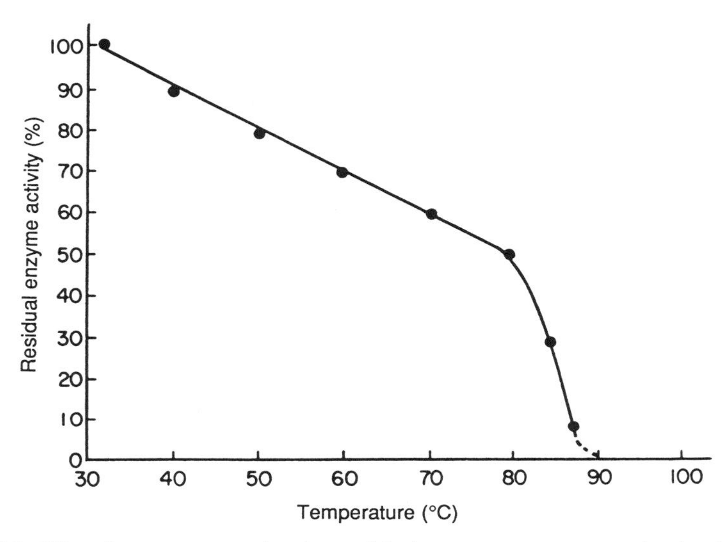

FIG. 9.2. Effect of temperature on phenolase activity in pear puree at a constant heating time of 8 sec. Adapted from Dimick *et al.* (1951).Copyright © by Institute of Food Technologists.

Hot water blanching remains the major unit operation in processing whole potatoes for French fries, potato chips, hash browns, and flakes. This process retards enzymatic browning but also results in the leaching of soluble materials, particularly reducing sugars. This reduces the sugar content, which in turn produces a lighter and more attractive product. In this method the heat penetrates rather slowly, which is a problem since it is important to heat the core sufficiently to inactivate the enzymes. Consequently this operation could result in a cooked potato with associated off-flavors. An alternative method using microwave heating to inactivate polyphenol oxidase in whole potatoes was proposed by Collins and McCarty (1969). They demonstrated a rapid inactivation of the enzyme compared to use of boiling water with no significant difference in texture between the products. Further studies have also confirmed these observations, although this technology has not yet been adopted by the potato industry.

C. pH TREATMENT

The application of acids to control enzymatic browning is used extensively. The acids employed are those found naturally in plant tissues, including citric, malic, phosphoric, and ascorbic acids. This method is based on the fact that lowering the tissue pH will reduce or retard the development of enzymatic browning. The optimum pH of most polyphenol oxidases lies between pH 4.0

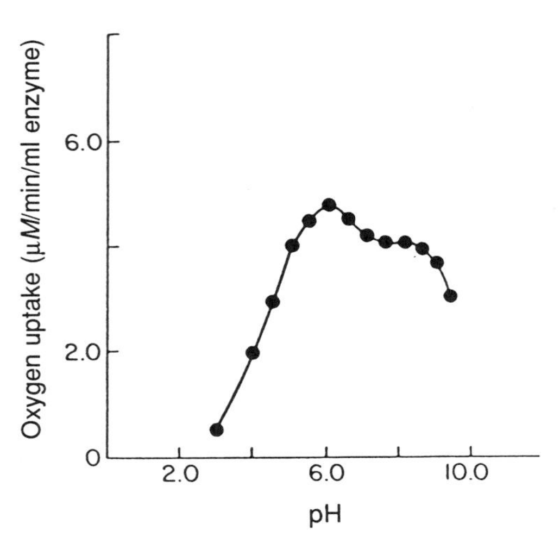

FIG. 9.3. pH optimum of crude polyphemol oxidase using catechol as substrate (Zawistowski *et al.*, 1986).

and 7.0 with little activity below pH 3.0, as illustrated in Fig. 9.3 for Jerusalem artichoke (Zawistowski *et al.*, 1986). Muneta (1977) examined the effect of pH on the development of melanins during enzymatic browning. Although, as discussed earlier, the initial reaction involving the formation of quinone is enzyme catalyzed, polymerization of these quinones to the brown or brown-black melanins is essentially nonenzymatic. While both are pH dependent, Muneta (1977) studied the effect of pH on the nonenzymatic reactions leading to the formation of melanins in potatoes. The formation of dopachrome from dopaquinone was monitored in phosphate buffers at pH 5.0, 6.0, and 7.0. A very rapid melanin development occurred at pH 7.0 compared to a rather slow process at pH 5.0. This observation could be important to the processor, particularly if the lye-peeled potatoes are inadequately washed, resulting in a high surface pH that facilitates enzymatic browning and melanin formation.

Citric acid has been used in conjunction with ascorbic acid or sodium sulfite as a chemical inhibitor of enzymatic browning (Joslyn and Ponting, 1951; Ponting, 1960; Schwimmer and Burr, 1967). Cut fruit, such as peaches, is often immersed in dilute solutions of these acids just prior to processing. This is particularly important in the case of lye-peeled cling peaches, where the acid dip counteracts the effect that any residual lye might have on enzymatic browning. Citric acid also inhibits polyphenol oxidase by chelating the copper moiety of the enzyme. Compared to citric acid, however, malic acid, the principal acid in apple juice, is a much more effective inhibitor of enzymatic browning.

A particularly effective inhibitor of polyphenol oxidase is ascorbic acid (AA)

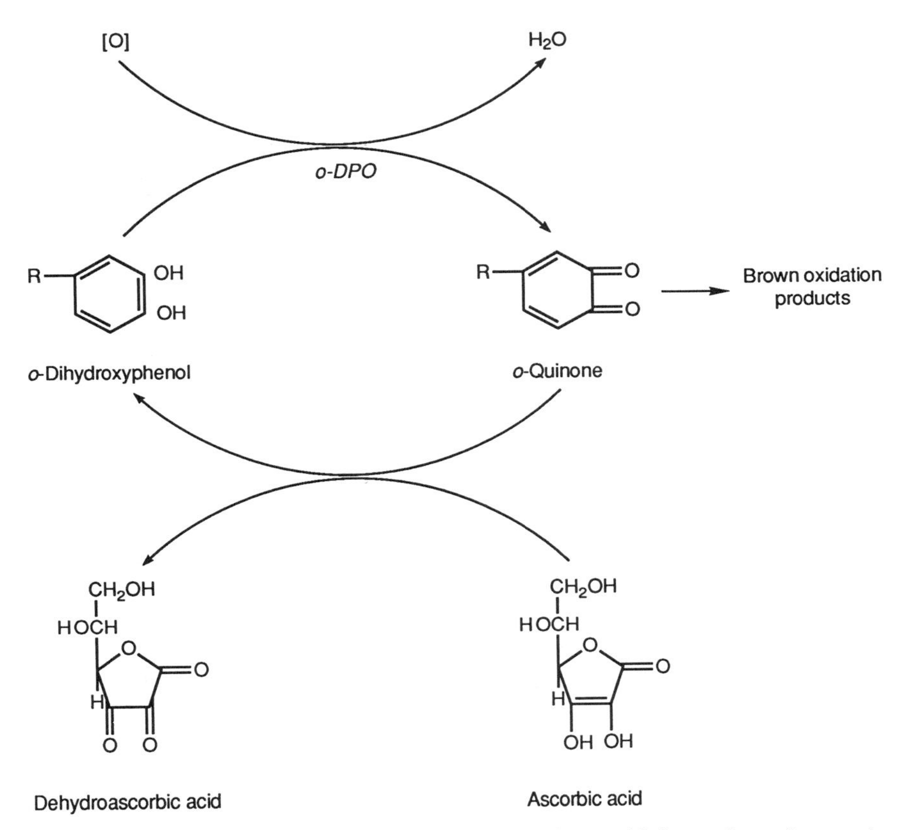

SCHEME 9.2. Reduction by ascorbic acid of the primary quinone oxidation products of enzymatic browning (Walker, 1976).

(Taeufel and Voigt, 1964). It does not have detectable flavor at the level used to inhibit this enzyme nor does it have a corrosive action on metals. In addition to these advantages, it is an important vitamin. The mode of action of ascorbic acid is outlined in Scheme 9.2 (Walker, 1976). The quinone is converted back to its precursor diphenol by oxidation of the ascorbic acid which prevents polymerization from taking place. The mode of action of ascorbic acid on this enzyme system is still unclear but it has been postulated that it could inactivate apple polyphenol oxidase by itself. Golan-Goldhirsh and Whitaker (1985) reported a K-type interaction between polyphenol oxidase and ascorbic acid in which the product of the reaction formed an inactive covalent enzyme derivative. A recent study by Hsu and co-workers (1988) compared the effectiveness of several ascorbic acid derivatives, including dehydroascorbic acid, isoascorbic acid, ascorbic

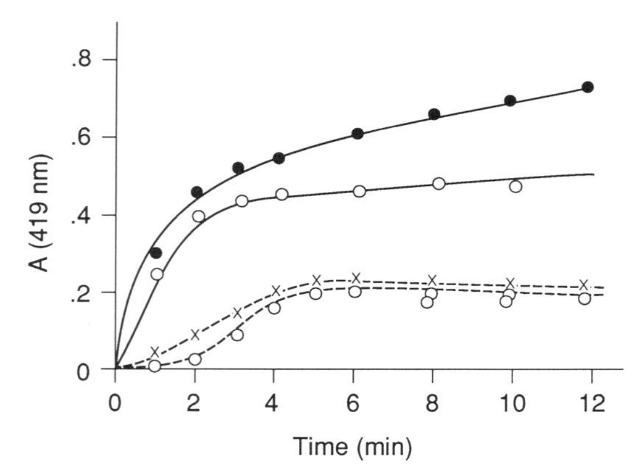

FIG. 9.4. Comparison of ascorbic acid derivatives on inhibition of mushroom polyphenol oxidase (PPO). AA (○---○), isoAA (×——×), dehydroAA (○——○), AA-2-PO_4 (●——●), and AA-2-SO_4 (●——●) at 0.25 m*M* concentration were incubated with 20 μg of mushroom PPO and 0.25 m*M* dihydroxyphenylalanine in 50 m*M* sodium phosphate buffer (pH 6.8) at 25°C. The control experiment (●——●) was carried out without the above inhibitors under the same conditions. The activity of PPO was measured spectrophotometrically at 419 nm (Hsu *et al.*, 1988). Copyright © by Institute of Food Technologists.

acid-2-phosphate, and ascorbic acid-2-sulfate, with that of ascorbic acid. On the basis of kinetic studies they found ascorbic acid and isoascorbic acid to be the most effective inhibitors of mushroom polyphenol oxidase followed by dehydroascorbic acid (Fig. 9.4). No inhibitory activity was evident in the presence of either of the other ascorbate derivatives. Using electron spin resonance a new inhibitory mechanism was proposed in which the Cu^{2+} of polyphenol oxidase was reduced to Cu^{+} in the presence of ascorbate.

An adequate amount of ascorbic acid must be added to food material to delay enzymatic browning. In fruit juices treated with ascorbic acid autooxidation of ascorbic acid, or natural ascorbic acid oxidase activity, will use up any dissolved oxygen in the fruit juice. Thus oxygen would become the limiting factor determining the rate of enzymatic browning. The addition of ascorbic acid at a concentration of 300 mg per pound of fruit was shown by Hope (1961) to control browning as well as reduce headspace oxygen in canned apple halves. This was a particularly effective method for controlling enzymatic browning in spite of the thickness of the apple tissue and its comparatively high oxygen content.

Because of the concern regarding the use of sulfite to control enzymatic browning, discussed later in this chapter, Langdon (1987) reported the development of an alternative chemical treatment for packaging potatoes. Immersion of the peeled and sliced potatoes in a solution of ascorbic and citric acids followed by vacuum packaging the drained product in a polyolefin multilayer bag extended the shelf life for over 14 days. This technique has considerable promise for extending the shelf life of other fresh fruits and vegetables.

Muneta (1977) suggested sodium acid pyrophosphate (SAPP) as an alternative inhibitor of enzymatic browning. It has several advantages over the organic acids, being much less sour than citric acid as well as minimizing after-cooking blackening of potatoes by complexing iron. An additional benefit of chelating iron is to limit its role in the catalysis of rancidity in fried or dehydrated potatoes.

D. Cinnamic Acids

Studies on potatoes, apples, pears, and sweet cherries demonstrated the potential of *o*-diphenol oxidase inhibitors in controlling enzymatic browning (Macrae and Duggleby, 1968; Walker, 1969; Rivas and Whitaker, 1973; Pifferi *et al.*, 1974). Subsequent research by Walker (1975) reported that cinnamic and *p*-coumaric acids were potent inhibitors of enzymatic browning. Walker (1976) found that these substituted cinnamic acids were also very effective in preventing enzymatic browning. It appeared that the site of substitution of mono- and dihydroxyphenols was a crucial factor in determining polyphenol oxidase activity. For example, monophenols were hydroxylated as long as the para-substituent was greater than CH_2, or para-substituted 3,4-dihydroxyls were ox-

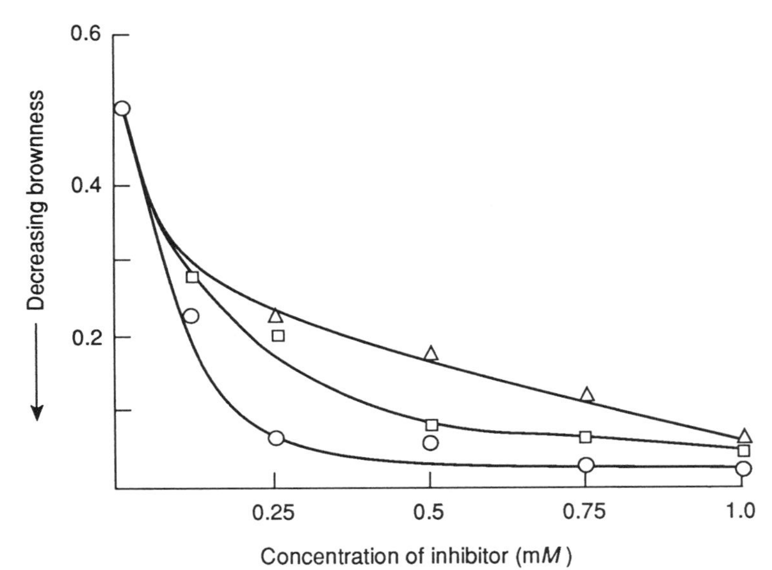

Fig. 9.5. Effect of concentration of inhibitor on the browning of juice from Granny Smith apples. Cinnamic acid (○), *p*-coumaric acid (□), and ferulic acid (△) (Walker, 1976).

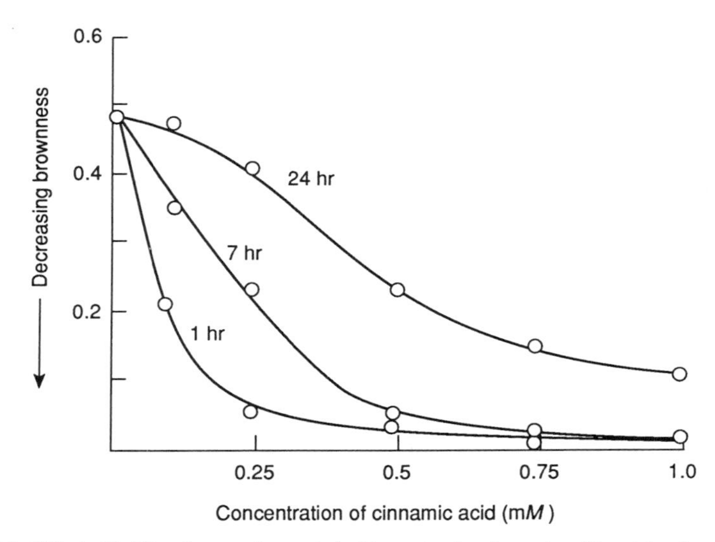

FIG. 9.6. Effect of holding time on the control of browning by cinnamic acid on juice from Granny Smith apples (Walker, 1976).

idized at a faster rate compared to the corresponding 2,3-dihydroxyphenols. The results in Fig. 9.5 illustrate the effect of adding cinnamic, *p*-coumaric, and ferulic acids to freshly prepared apple juice. Of these acids, cinnamic acid was the most potent, requiring less than 0.01% to prevent browning. The effect of cinnamic acid was long term as seen in Fig. 9.6, indicating considerable potential for these compounds in preventing this type of browning in fruit juices (Walker, 1977).

E. APPLICATION OF SULFUR DIOXIDE AND SULFITES

Sulfur dioxide and sulfites are powerful inhibitors of polyphenol oxidase (Ponting, 1960; Mapson and Wager, 1961; Mapson, 1965; Golan-Goldhirsh and Whitaker, 1984; Zawistowski *et al.*, 1987). Sodium bisulfite by itself or in combination with citric acid is commonly used commercially to inhibit enzymatic blackening of prepeeled potatoes or in the processing of apples and peaches (Feinberg *et al.*, 1967; Ponting, 1960). It can be employed as either gaseous sulfur dioxide or a dilute solution of the sulfite. The gas will penetrate at a faster rate into the fruit or vegetable but sulfite solutions in the form of a dip in the processing plant or as a spray on newly harvested potatoes are much easier to handle.

The use of sulfur dioxide and sulfites is particularly advantageous where heating would result in unfavorable texture and off-flavor changes in a product. In addition to having antiseptic properties it also preserves vitamin C. Several disadvantages are evident, however, when they are used in foods, including the development of objectionable flavor and odor, the bleaching of the natural food pigments, and the hastening of can corrosion. Sulfites are toxic at high levels, with concentrations above 0.01 *M* being readily detected organoleptically in cooked potato slices (Mapson and Swain, 1961). A major problem associated with the use of sulfur dioxide or sulfites in foods is their destructive effect on vitamin B_1 or thiamine (Mapson and Swain, 1961; Mapson, 1965). In spite of these drawbacks, this group of phenolase inhibitors is widely used in food processing, largely because of their effectiveness and low cost. Nevertheless, in recent years there has been considerable concern regarding their safety.

Sodium bisulfite appears to be most effective in inhibiting tyrosine oxidation to 3,4-dihydroxyphenylalanine (DOPA) but far less effective in preventing the oxidation of DOPA (Muneta, 1966). Inhibition of enzymatic browning has also been attributed to the formation of colorless addition products between bisulfite and quinones (Bouchilloux, 1959; Embs and Markakis, 1965). The effectiveness

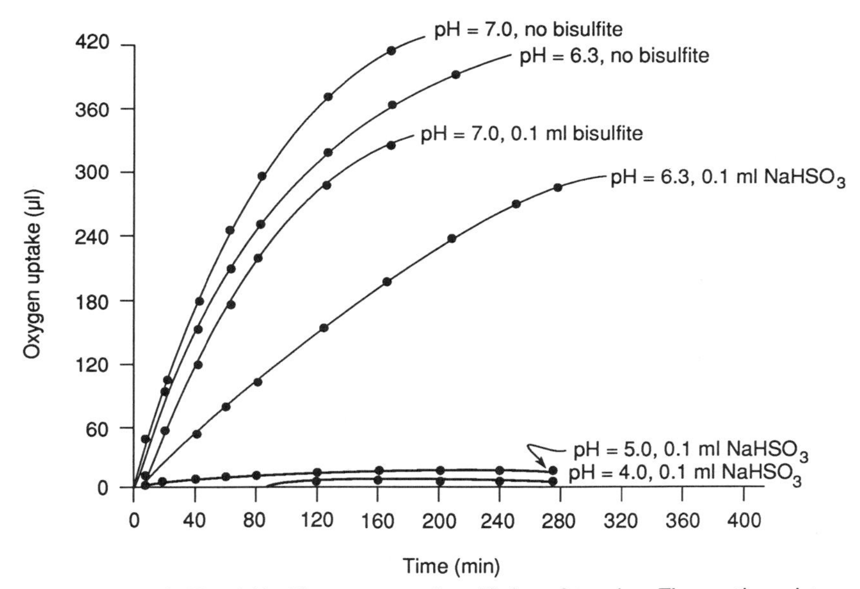

FIG. 9.7. Effect of pH and bisulfite on enzymatic oxidation of tyrosine. The reaction mixture contained 0.8 ml enzyme, 0.8 ml tyrosine (1×10^{-2} *M*), 0.1 ml $NaHSO_3$ (5×10^{-3} *M*), H_2O to a total reaction volume of 2.55 ml, and 0.2 ml 20% KOH for CO_2 absorption (30°C) (Muneta and Wang, 1977).

of bisulfite as an inhibitor of enzymatic browning was pH dependent. Mapson (1965) reported that a pH of 6.0 was quite satisfactory for sulfite inhibition of enzymatic browning of potato slices. Muneta and Wang (1977) found bisulfite to be a very effective inhibitor of enzymatic browning at a pH below 5.0 and was less effective at high pH conditions. Their results, shown in Fig. 9.7, indicate this clearly for the enzymatic oxidation of tyrosine. Their studies demonstrated that at pH 4.0 the pH alone was effective in inhibiting enzymatic browning, with little effect at pH 5.0 or above. At pH 5.0, however, bisulfite was primarily responsible for the inhibition of enzymatic browning. These researchers recommended the use of a bisulfite dip at pH 4.0 for prepeeled potatoes to rapidly inactivate polyphenol oxidase followed by rinsing with water to remove acid and bisulfite solution. Any residual enzyme activity could be further reduced by dipping in a more dilute solution of bisulfite. This treatment could be used to prevent enzymatic blackening while at the same time minimize exudation, softening, and flavor problems normally associated with the combined use of high concentrations of bisulfite and low pH (<5.0) (Amla and Francis, 1961; Muneta and Wang, 1977).

Muneta (1981) compared bisulfite inhibition of tyrosine oxidation at pH 6.3 with that by cysteine, dithioreitol, sodium diethyldithiocarbamate, and ascorbic and dihydroxyfumaric acids. Of these chemical inhibitors, bisulfite was most preferred by food processors. Its ability to react with disulfide bonds in proteins to cause enzyme inactivation was of considerable practical importance in inhibiting the enzymatic blackening of potatoes. Kang *et al.* (1983) inhibited the production of dopachrome, the initial pigment formed during enzymatic browning, by addition of sodium sulfite to potato disks. The formation of dopachrome and its subsequent conversion to black melanin was delayed at low pH. A high concentration of sulfite (80 μg/ml) was required, as evident from Fig. 9.8, to bleach any dopachrome which had already been formed.

Sulfur dioxide in food was considered by Wedzicha and co-workers (1987) to be a mixture of oxospecies of sulfur in oxidation state IV (SO_2, HSO_3^-, and SO_3^{2-}) in which the proportions present varied with the pH. Wedzicha (1984) suggested that 4-sulfocatechol was the initial reaction product formed when these sulfur oxospecies reacted with a polyphenol oxidase–catechol system. Wedzicha *et al.* (1987) confirmed the formation of monosulfonates when sulfite was added to a mushroom tyrosinase–catechol system in which the main product was shown to be 4-sulfocatechol. This reaction was irreversible with the products being unreactive to mushroom tyrosinase and not exerting any inhibitory effect on the enzyme. This work was consistent with work reported by Embs and Markakis (1965) in which the ability of sulfite to react with the phenolic substrates was primarily responsible for the inhibition of enzymatic browning.

Concern regarding the allergenic reactions to sulfite has led to banning its use in some parts of the United States. Ingestion of sulfites has recently been associ-

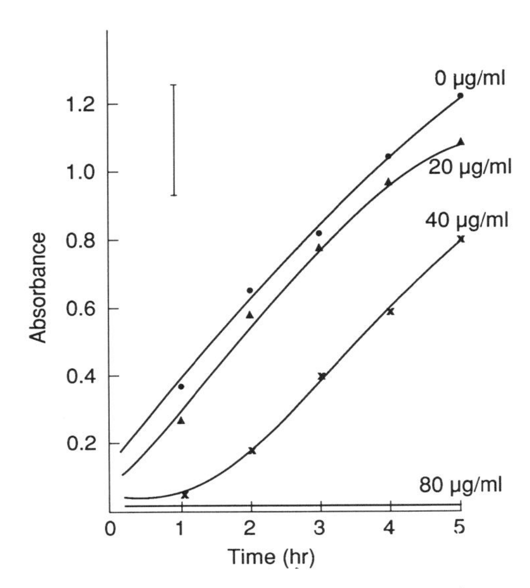

FIG. 9.8. The pattern of dopachrome formation as influenced by various concentrations of sulfur dioxide in the presence of sliced disks (Kang *et al.*, 1983).

ated with initiation of asthmatic attacks among a small number of asthmatics (Baker *et al.*, 1981; Bush *et al.*, 1985; Delohery *et al.*, 1984). As a consequence the Food and Drug Administration (1986) banned the use of sulfites for preserving fresh fruits and vegetables. Its present use in potatoes is allowed but remains under close scrutiny by regulatory bodies (Taylor and Bush, 1986). This necessitated the development of new and alternative methods for controlling enzymatic browning, as described earlier. To accomplish this an understanding of the precise mechanism of action is required. Several mechanisms were proposed, including the formation of quinone–sulfite complexes as well as polyphenol oxidase inactivation (Embs and Markakis, 1965; Haisman, 1974). Sayavedra-Soto and Montgomery (1986) found that the main mode of action of polphenol oxidase activity by sulfite involved modification of its protein structure, resulting in inactivation.

F. Effect of Sprout Inhibitors on Polyphenol Oxidase

To minimize discoloration in potatoes caused by polyphenol oxidase, several treatments were examined, including sprout suppression and ventilation during high-temperature storage. Randhawa and co-workers (1986) studied the effect of a chemical sprout suppressant, isopropyl-*N*-(3-chlorphenyl)carbamate (CIPC),

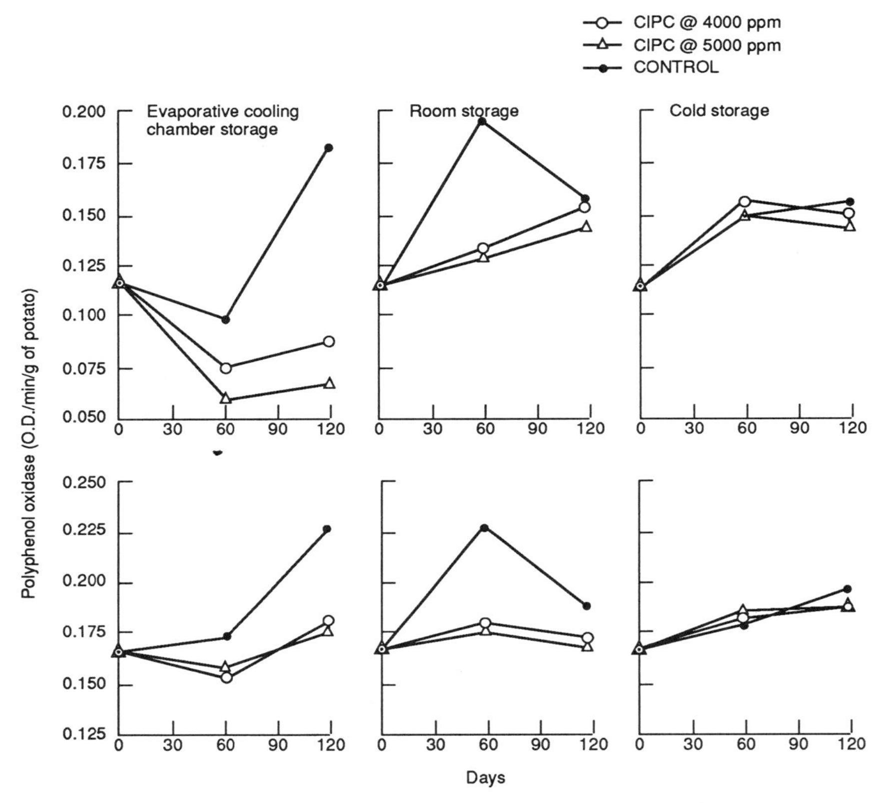

FIG. 9.9. Polyphenol oxidase activity in CIPC-treated potatoes of Kufri Chandramukhi (top) and Kufri Sindhuri (bottom) varieties stored under different storage conditions (Randhawa *et al.*, 1986).

on two Indian potato varieties stored at room temperature, in cold storage, and in a cool evaporative chamber for up to 4 months. Their results in Fig. 9.9 show a substantial decrease in polyphenol oxidase activity for both potato varieties during evaporative cooling chamber storage in the presence of increasing levels of suppressant compared to the control. Little effect on enzyme activity was found during cold storage, with room storage being intermediate. It was clear from this study that polyphenol oxidase could be substantially reduced by postharvest application of CIPC in evaporative cooling chambers and room storage. Potato slices produced from Kufri Chandramukhi, the variety with a lower activity of polyphenol oxidase, proved to be superior to slices prepared from Kufri Sindhuri.

Bibliography

Abukhara, R. H., and Woolhouse, H. W. (1966). The preparation and properties of *o*-diphenol, oxygen oxidoreductase from potato tubers. *New Phytol.* **65,** 477.

Al-Barazi, Z., and Schwabe, W. W. (1984). The possible involvement of polyphenol-oxidase and the auxin-oxidase system in root formation and development in cuttings of *Pistacia vera*. *J. Hortic. Sci.* **59,** 453.

Allendorf, M. D., Spira, D. J., and Solomon, E. I. (1985). Low-temperature magnetic circular dichroism studies of native laccase, spectroscopic evidence for exogenous ligand bonding at a trinuclear copper active site. *Proc. Natl. Acad. Sci. U.S.A.* **82,** 3063.

Amla, B. L., and Francis, F. J. (1961). Effect of pH of dipping solution on the quality of prepeeled potatoes. *Am. Potato J.* **38,** 121.

Anderson, J. W. (1968). Extraction of enzyme and subcellular organelles from plant tissues. *Phytochemistry* **7,** 1973.

Anderson, R. H., Moran, D. H., Huntly, T. E., and Holahan, J. L. (1963). Responses of cereals to antioxidants. *Food Technol.* **17,** 1587.

Aylward, F., and Haisman, D. R. (1969). Oxidation systems in fruits and vegetables:Their relation to the quality of preserved products. *Adv. Food Res.* **17,** 1.

Baker, G. J., Collett, P., and Allen, D. H. (1981). Bronchospasm induced by metabisulfite-containing foods and drugs. *Med. J. Aust.* **2,** 614.

Bar-Nun, N., and Mayer, A. M. (1983). Suppression of catechol oxidase by norleucine in plant suspension cultures. *Phytochemistry* **22,** 1329.

Bate-Smith, E. C., Hughes, J. C., and Swain, T. (1958). After-cooking discolouration in potatoes. *Chem. Ind.* (*London*), p. 627.

Bendall, D. S., and Gregory, R. P. F. (1963). Purification of phenol oxidases. *Enzyme Chem. Phenolic Compd., Proc. Plant Phenolics Group Symp., 1962,* pp. 7–24.

Ben-Shalom, N., Kahn, V., Harel, E., and Mayer, A. M. (1977). Olive catechol oxidase—Changes during fruit development. *J. Sci. Food Agric.* **28,** 545.

Bokuchava, M. A., and Skobeleva, G. A. (1969). The chemistry and biochemistry of tea and tea manufacture. *Adv. Food Res.* **17,** 215.

Bouchilloux, S. (1959). Sur la décoloration par le bisulfite des milieux d'oxydation chimique on enzymatique de divers phenols. Formation de dérives bisulfitiques. *C. R. Seances Soc. Biol. Ses Fil.* **153,** 642.

Bouchilloux, S., McMahill, P., and Mason, H. S. (1963). The multiple forms of mushroom tyrosinase: Purification and molecular properties of the enzymes. *J. Biol. Chem.* **238,** 1699.

Bush, R. K., Taylor, S. L., and Busse, W. W. (1985). A critical evaluation of clinical trials in reactions to sulfites. *J. Allergy Clin. Immunol.* **78,** 191.

Cash, J. N., Sistrunk, W. A., and Stutte, C. A. (1976). Characteristics of Concord grape polyphenoloxidase involved in juice color loss. *J. Food Sci.* **41,** 1398.

Cheung, K. W. K., and Henderson, H. M. (1972). Effect of physiological stress on potato polyphenol oxidase. *Phytochemistry* **11,** 1255.

Clayton, R. A. (1959). Properties of tobacco polyphenol oxidase. *Arch. Biochem. Biophys.* **81,** 404.

Cloughley, J. B. (1980a). The effect of fermentation on the quality parameters and price evaluation of Central African black teas. *J. Sci. Food Agric.* **31,** 911.

Cloughley, J. B. (1980b). The effect of temperature on enzyme activity during the fermentation phase of black tea manufacture. *J. Sci. Food Agric.* **31,** 920.

Collins, J. L., and McCarty, I. E. (1969). Comparison of microwave energy with boiling water for blanching whole potatoes. *Food Technol.* **23,** 337.

Constantinides, S. M., and Bedford, C. L. (1967). Multiple forms of phenol oxidase. *J. Food Sci.* **32,** 446.

Craft, C. C. (1966). Localization and activity of phenolase in the potato tuber. *Am. Potato J.* **43,** 112.

Craft, C. C., and Audia, W. V. (1962). Phenolic substances associated with wound-barrier formation in vegetables. *Bot. Gaz. (Chicago)* **123,** 211.

Delohery, J., Simmul, R., Castle, W. D., and Allen, D. H. (1984). The relationship of inhaled sulfur dioxide reactivity to ingested metabisulfite sensitivity with asthma. *Am. Rev. Respir. Dis.* **130,** 1027.

Dimick, K. P., Ponting, J. D., and Makower, B. (1951). Heat inactivation of polyphenolase in fruit purees. *Food Technol.* **5,** 237.

Egley, G. H., Paul, R. N., Jr., Duke, S. O., and Vaughn, K. C. (1985). Peroxidase involvement in lignification in water impermeable seed coats by weedy leguminous and malvaceous species. *Plant Cell Environ.* **8,** 253.

Embs, R. J., and Markakis, P.(1965). The mechanism of sulfite inhibition of browning caused by polyphenol oxidase. *J. Food Sci.* **30,** 753.

Feinberg, B., Olson, R. L., and Mullins, W. R. (1967). Prepeeled potatoes. *In* "Potato Processing" (W. F. Talburt, and O. Smith, eds.), p. 491. Avi Publ. Co., Westport, Connecticut.

Finkle, B. J., and Nelson, R. F. (1963). Enzyme reactions with phenolic compounds: Effects of *o*-methyltransferase on a natural substrate of fruit polyphenol oxidase. *Nature (London)* **197,** 902.

Food and Drug Administration (1986). "Chemical Preservatives," Title 21, Part 182, Part 101. FDA, Code of Federal Regulations, Washington, D.C.

Golan-Goldhirsh, A., and Whitaker, J. R. (1984). Effect of ascorbic acid, sodium bisulfite and thiol compounds on mushroom polyphenol oxidase. *J. Agric. Food Chem.* **32,** 1003.

Golan-Goldhirsh, J. R., and Whitaker, J. R. (1985). k_{CAT} inactivation of mushroom polyphenol oxidase. *J. Mol. Catal.* **32,** 141.

Goldbeck, J. H., and Cammarata, K. V. (1981). Spinach thylakoid polyphenoloxidase. Isolation, activation, and properties of the native chloroplast enzyme. *Plant Physiol.* **67,** 977.

Goldberg, M., Farver, O., and Pecht, J. (1980). Interaction of *Rhus* laccase with dioxygen and its reduction. *J. Biol. Chem.* **255,** 2057.

Gornhardt, L. (1955). Die nichtenzymatische Braunung von Lebensmitteln. 1. *Fette, Seifen, Anstrichm.* **57,** 270.

Grant, N. H., Clark, D. E., and Alburn, H. E. (1966). Accelerated polymerization of *N*-carboxyamino acid anhydrides in frozen dioxane. *J. Am. Chem. Soc.* **88,** 4071.

Gregory, R. P. F., and Bendall, D. S. (1966). The purification and some properties of the polyphenol oxidase from tea *(Camellia sinensis* L.*). Biochem. J.* **101,** 569.

Gross, G. G. (1977). The structure, biosynthesis and degradation of wood. *Recent Adv. Phytochem.* **11,** 141.

Haisman, D. R. (1974). The effect of sulphur dioxide on oxidising systems in plant tissues. *J. Sci. Food Agric.* **25,** 803.

Halim, D. H., and Montgomery, M. W. (1978). Polyphenol oxidase of d'Anjou pears *(Pyrus commumis* L.*). J. Food Sci.* **43,** 603.

Harel, E., Mayer, A. M., and Shain, Y. (1966). Catechol oxidases, endogenous substrates and browning in developing apples. *J. Sci. Food Agric.* **17,** 389.

Harkin, P. M., and Obst, J. R. (1973). Lignification in trees: Indication of exclusive peroxidase participation. *Science* **180,** 296.

Hilton, P. J. (1972). *In vitro* oxidation of flavonols from tea leaf. *Phytochemistry* **11,** 1243.

Hilton, P. J., and Ellis, R. T. (1972). Estimation of the market value of Central African tea by theaflavin analysis. *J. Sci. Food Agric.* **23,** 227.

Hobson, G. E. (1967). Phenolase activity in tomato fruit in relation to growth and to various ripening disorders. *J. Sci. Food Agric.* **18,** 523.

Hope, G. W. (1961). The use of antioxidants in canning apple halves. *Food Technol.* **15,** 548.

Hsu, A. F., Shieh, J. J., Bills, D. D., and White, K. (1988). Inhibition of mushroom polyphenoloxidase by ascorbic acid derivatives. *J. Food Sci.* **53,** 765.

Hulme, A. C. (1958). Some aspects of the biochemistry of apple and pear fruits. *Adv. Food Res.* **8,** 297.

Hyodo, H., and Uritani, I. (1965). Purification and properties of *o*-diphenol oxidases in sweet potato. *J. Biochem. (Tokyo)* **58,** 388

Jain, J. C., and Takeo, T. (1984). A review: The enzymes of tea and their role in tea making. *J. Food Biochem.* **8,** 243.

James, W. O. (1953). The terminal oxidases of plant respiration. *Biol. Rev. Cambridge Philos. Soc.* **28,** 245.

Jayaraman, K. S., and Ramanuja, M. N. (1987). Studies on multiple forms of polyphenoloxidase from some banana cultivars differing in browning rate. *Lebensm.-Wiss. Technol.* **20,** 16.

Jolley, R. L., Jr., and Mason, H. S. (1965). The multiple forms of mushroom tyrosinase, interconversion. *J. Biol. Chem.* **240,** PC^{14}89.

Joshi, P. R., and Shiralkar, N. D. (1977). Polyphenoloxidases of a local variety of mango. *J. Food Sci. Technol.* **14,** 77.

Joslyn, M. A., and Ponting, J. D. (1951). Enzyme-catalyzed oxidative browning of fruit products. *Adv. Food Res.* **3,** 1.

Kahn, V. (1975). Polyphenol oxidase activity and browning of three avocado varieties. *J. Sci. Food Agric.* **26,** 1319.

Kahn, V. (1976). Polyphenoloxidase isoenzymes in avocado. *Phytochemistry* **15,** 267.

Kahn, V.1977. Some biochemical properties of polyphenoloxidase from two avocado varieties differing in their browning rates. *J. Food Sci.* **42,** 38.

Kahn, V. (1985). Effect of protein hydrolyzates and amino acids on *o*-dihydroxyphenolase activity of polyphenol oxidase of mushroom, avocado, and banana. *J. Food Sci.* **50,** 111.

Kang, K.-J., Cho, Y.-O., and Kim, M-S, L. (1983). Enzymatic discoloration of raw potatoe tubers with special reference to the formation and inhibition of dopachrome. *Am. Potato J.* **60,** 451.

Langdon, T. T. (1987). Preventing browning in fresh prepared potatoes without the use of sulfiting agents. *Food Technol.* **41(5),** 64.

Lanzarini, G., Pifferi, P. G., and Zamorani, A. (1972). Specficity of an *o*-diphenoloxidase from *Prunus avium* fruits. *Phytochemistry* **11,** 89.

Lee, C. Y., Smith, N. L., and Pennesi, A. P. (1983). Polyphenoloxidase from DeChaunac grapes. *J. Sci. Food Agric.* **34,** 987.

Lerch, K. (1981). Copper monooxygenases, tyrosinases and dopamine-monooxygenase. *In* "Metal Ions in Biological Systems. Copper Proteins." (H. Siegel, ed.), p. 143. Dekker, New York.

Luh, B. S., Hsu, E. T., and Stachowicz, K. (1967). Polyphenolic compounds in canned cling peaches. *J. Food Sci.* **32,** 251.

Macrae, A. R., and Duggleby, R. G. (1968). Substrates and inhibitors of potatoe tuber phenolase. *Phytochemistry* **7,** 855.

Maier, V. P., and Metzler, D. M. (1965). Quantitative changes in date polyphenols and their relation to browning. *J. Food Sci.* **30,** 80.

Mapson, L. W. (1965). Enzymic browning of pre-peeled potato tissue. *Nutrition (London)* **19,** 123.

Mapson, L. W., and Swain, T. (1961). Oxidation of ascorbic acid and phenolic constituents. *SCI Monogr.* **11,** 121.

Mapson, L. W., and Tomalin, W. (1961). Preservation of peeled potato. III. The inactivation of phenolase by heat. *J. Sci. Food Agric.* **12,** 54e.

Mapson, L. W., and Wager, H. G. (1961). Preservation of peeled potato. I. Use of sulfite and its effect on the thiamine content. *J. Sci. Food Agric.* **12,** 43.

Martyn, R. D., Samuelson, D. A., and Freeman, T. E. (1979). Ultrastructural localization of polyphenoloxidase activity in leaves of healthy and diseased water hyacinth. *Phytopathology* **69,** 1278.

Mason, H. S., Fowlks, W. L., and Peterson, E. (1955). Oxygen transfer and electron transport by the phenolase complex. *J. Am. Chem. Soc.* **77,** 2914.

Matheis, G., and Whitaker, J. R. (1984). Peroxidase-catalyzed crosslinking of proteins. *J. Protein Chem.* **3,** 35.

Mathew, A. G., and Parpia, H. A. B. (1971). Food browning as a polyphenol reaction. *Adv. Food Res.* **19,** 75.

Mayer, A. M. (1987). Polphenol oxidases in plants—Recent progress. *Phytochemistry* **26,** 11.

Mayer, A. M., and Harel, E. (1979). Polyphenol oxidase in plants (review). *Phytochemistry* **18,** 193.

Meyer, H. V., and Biehl, B. (1982). Relation between photosynthetic and phenolase activities in spinach chloroplasts. *Phytochemistry* **21,** 9.

Mihalyi, K, Vamos-Vigyazo, L., Kiss-Kutz, N., and Babos-Szebenyi, E. (1978). The activities of polyphenol oxidase in fruits and vegetables as related to pH and temperature. *Acta Aliment. Acad. Sci. Hung.* **7,** 57.

Montedoro, G. F. (1969). Grape polyphenoloxidases. 1. Distribution and evolution of the enzymic activity of grapes of diverse varieties. *Ind. Agrar.* **7,** 197.

Morris, M. C., Hauenstein, B. L., Jr., and McMillin, D. R. (1983). Metal-replacement studies of copper proteins. The removal and replacement of copper (1) from Chinese laccase. *Biochim. Biophys. Acta* **743,** 389.

Muneta, P. (1966). Bisulfite inhibition of enzymatic blackening caused by tyrosine oxidation. *Am. Potato J.* **43,** 397.

Muneta, P. (1977). Enzymatic blackening in potatoes: Influence of pH on dopachrome oxidation. *Am. Potato J.* **54,** 387.

Muneta, P. (1981). Comparisons of inhibitors of tyrosine oxidation in the enzymatic blackening of potatoes. *Am. Potato J.* **58,** 85.

Muneta, P., and Walradt, J. (1968). Cysteine inhibition of enzymatic blackening with polyphenol oxidase from potatoes. *J. Food Sci.* **33,** 606.

Muneta, P., and Wang, H. (1977). Influence of pH and bisulfite on the enzymatic blackening reaction in potatoes. *Am. Potato J.* **54,** 73.

Nakamura, W. (1967). Studies on the biosynthesis of lignin. I. Disproof against the catalytic activity of laccase in the oxidation of coniferyl alcohol. *J. Biochem. (Tokyo)* **62,** 54.

Palmer, J. K. (1963). Banana polyphenoloxidase. Preparation and properties. *Plant Physiol.* **38,** 508.

Palmer, J. K., and Roberts, J. B. (1967). Inhibition of banana polyphenoloxidase by 2-mercaptobenzothiazole. *Science* **157,** 200.

Patil, S. S., and Zucker, M. (1965). Potato phenolases: Purification and properties. *J. Biol. Chem.* **240,** 3938.

Pierpoint, W. S. (1966). The enzymatic oxidation of chlorogenic acid and some reactions of the quinone produced. *Biochem. J.* **98,** 567.

Pierpoint, W. S., Ireland, R. J., and Carpenter, J. M. (1977). Modification of proteins during the oxidation of leaf phenols, reaction of potato virus X with chlorogenoquinone. *Phytochemistry* **16,** 29.

Pifferi, R., Baldassan, C., and Cultera, R. (1974). Inhibition by carboxylic acids of an *o*-diphenol oxidase of *Prunus avium* fruits. *J. Sci. Food Agric.* **25,** 263.

Ponting, J. D. (1960). The control of enzymatic browning in fruits. *In* "Food Enzymes" (H. W. Schultz, ed.), p. 105. Avi Pub. Co., Wesport, Connecticut.

Randhawa, K. S., Khurana, D. S., and Bajaj, K. L. (1986). Effect of CIPS [isopropyl-*N*-(3-

chlorophenyl)carbamate] on total nitrogen and polyphenol oxidase activity in relation to processing of potatoes *(Solanum tuberosum* L.*)* stored under different conditions. *Qual. Plant.—Plant Foods Hum. Nutr.* **36,** 207.
Rivas, N. deJ., and Whitaker, J. R. (1973). Purification and some properties of two polyphenol oxidases from Bartlett pears. *Plant Physiol.* **52,** 501.
Robb, D. A. (1984). Tyrosinase. Chapter 7. *In* "Copper Proteins and Copper Enzymes" (R. Lontie, ed.), p. 207. CRC Press, Baco Raton, Florida.
Robb, D. A., Mapson, L. W., and Swain, T. (1964). Activation of the latent tyrosinase of broad bean *(Vicia faba* L.*). Phytochemistry* **43,** 731.
Robb, D. A., Mapson, L. W., and Swain, T. (1965). On the heterogeneity of the tyrosinase of broad bean. *Nature (London)* **201,** 503.
Roberts, E. A. H. (1952). Chemistry of tea fermentation. *J. Sci. Food Agric.* **23,** 227.
Roelofsen, P. A. (1958). Fermentation, drying and storage of cocoa beans. *Adv. Food Res.* **8,** 225.
Rubin, B. A., and Artsikhovskaya, E. V. (1960). "Biokhimiya i fiziologiya Immuniteta rastenni (Biochemistry and Physiology of Plant Immunity)." Izd. Akad. Nauk SSR, Moscow (*Chem. Abstr.* **55,** 8557d).
Sanchez-Ferrer, A., Bou, R., Cabanes, J., and Garcia-Carmona, F. (1988). Characterization of catecholase and cresolase activities of Manartrell grape polyphenol oxidase. *Phytochemistry* **27,** 319.
Sanderson, G. W. (1964). The chemical composition of fresh tea flush as affected by clone and climate. *Tea Q.* **35,** 101.
Sato, M., and Hasegawa, M. (1976). The latency of spinach chloroplast phenolase. *Phytochemistry* **15,** 61.
Savagaon, K. A., and Sreenivasan, A. (1978). Activation of pre-phenoloxidase in lobster and shrimp. *Fish. Technol.* **15,** 49.
Sayavedra-Soto, L. A., and Montgomery, M. W. (1986). Inhibition of polyphenol oxidase by sulfite. *J. Food Sci.* **51,** 1531.
Schaller, K. (1972). Zur Bestimmung der Polyphenoloxidaseaktivitat in Kartoffelknollen. *Z. Lebensm.-Unters. -Forsch.* **150,** 211.
Schwimmer, S., and Burr, H. K. (1967). Structure and chemical composition of the potatoe tuber. *In* "Potato Processing" (W. F. Talburt, and O. Smith, eds.), p. 12. Avi Pub Co., Westport, Connecticut.
Siegenthaler, P.-A., and Vaucher-Bonjour, P. (1971). Vieillessement de l'appareil photosynthétique. III. Variations et caractéristiques de l'activité *o*-diphenyloxydase (polyphenoloxydase) au cours du viellissement *in vitro* de chloroplastes isolés dépinard. *Planta* **100,** 106.
Simpson, B. K., Marshall, M. R., and Otwell, W. S. (1987). Phenoloxidase from shrimp *(Penaeus setiferus):* Purification and some properties. *J. Agric. Food Chem.* **35,** 918.
Simpson, B. K., Marshall, M. R., and Otwell, W. S. (1988). Phenoloxidases from pink and white shrimp: Kinetic and other properties. *J. Food Biochem.* **12,** 205.
Smith, J. L., and Krueger, R. C. (1962). Separation and purification of the phenolases of the common mushroom. *J. Biol. Chem.* **237,** 1121.
Solomon, E. I. (1981). *In* "Copper Proteins: Metal Ions in Biology" (T. G. Spiro, ed.), Vol. 3, p. 183. CRC Press, Boca Raton, Florida.
Stafford, H. A. (1974). The metabolism of aromatic compounds. *Annu. Rev. Plant Physiol.* **25,** 459.
Swain, T. (1962). Economic importance of flavonoid compounds. Foodstuffs. *In* "The Chemistry of Flavonoid Compounds" (T. A. Geissman, ed.), pp. 513–552. Pergamon, Oxford.
Taeufel, K., and Voigt, J. (1964). On the inhibiting action of ascorbic acid on the polyphenol oxidase of apples. *Z. Lebensm.-Unters. -Forsch.* **126,** 19.
Talburt, W. F., and Smith, O., eds. (1967). "Potatoe Processing." Avi Publ. Co., Westport, Connecticut.

Taylor, S. L., and Bush, R. K. (1986). Sulfites as food ingredients. A scientific status summary by the Institute of Food Technologists Expert Panel on Food Safety and Nutrition. *Food Technol.* **40 (6),** 47.

Tolbert, N. E. (1973). Activation of polyphenol oxidase of chloroplasts. *Plant Physiol.* **51,** 234.

Vamos-Vigyazo, L. (1981). Polyphenol oxidase and peroxidase in fruits and vegetables. *CRC Crit. Rev. Food Sci. Nutr.* **15,** 49.

Van Lelyveld, L. J., and de Rooster, K. (1986). Browning potential of tea clones and seedlings. *J. Hortic. Sci.* **61,** 545.

Variyar P. S., Pendharkar, M. B., Banerjee, A., and Bandyopadhyay, C. (1988). Blackening in green pepper berries. *Phytochemistry* **27,** 715.

Vaughn, K. C., and Duke, S. O. (1981a). Tissue localization of polyphenol oxidase in sorghum. *Protoplasma* **108,** 319.

Vaughn, K. C., and Duke, S. O. (1981b). Tentoxin-induced loss of plastidic polyphenol oxidase. *Plant Physiol.* **53,** 421.

Vaughn, K. C., and Duke, S. O. (1982). Tentoxin effects on sorghum: The role of polyphenol oxidase. *Protoplasma* **110,** 48.

Vaughn, K. C., and Duke, S. O. (1984). Functioning of polyphenol oxidase in higher plants. *Physiol. Plant.* **60,** 106.

Walker, J. R. L. (1969). Inhibition of the apple phenolase system through infection by *Penicillium expansum. Phytochemistry* **8,** 561.

Walker, J. R. L. (1975). Studies on the enzymic browning of apples. Inhibition of apple *o*-diphenol oxidase by phenolic acids. *J. Sci. Food Agric.* **26,** 1825.

Walker, J. R. L. (1976). The control of enzymic browning in fruit juices by cinnamic acids. *J. Food Technol.* **11,** 341.

Walker, J. R. L. (1977). Enzymic browning in foods. Its chemistry and control. *Food Technol. N. Z.* **12,** 19.

Wedzicha, B. C. (1984). "Chemistry of Sulphur Dioxide in Foods," pp. 219–228. Elsevier Applied Science, Amsterdam.

Wedzicha, B. C., Goddard, B. L., and Garner, D. N. (1987). Enzymic browning of sulphocatechol. *Int. J. Food Sci. Technol.* **22,** 653.

Weurman, C., and Swain, T. (1953). Chlorogenic acid and the enzymic browning of apples and pears. *Nature (London)* **172,** 678.

Whitaker, J. R. (1972). "Principles of Enzymology for the Food Sciences." pp. 571–582. Dekker, New York.

Wong, T. C., Luh, B. S., and Whitaker, J. R. (1971). Isolation and characterization of polyphenol oxidase isoenzymes of Clingstone peach. *Plant Physiol.* **48,** 19.

Zawistowski, J., Blank, G., and Murray, E. D. (1986). Polyphenol oxidase activity in Jerusalem artichokes *(Helianthus tuberosum* L.). *Can. Inst. Food Sci. Technol. J.* **19,** 210.

Zawistowski, J., Blank, G., and Murray, E. D. (1987). Inhibition of enzymatic browning in extracts of Jerusalem artichoke *(Helianthus tuberosus* L.). *Can. Inst. Food Sci. Technol. J.* **20,** 162.

10

Biochemistry of Food Spoilage: Off-Flavors in Milk

I. Introduction

Fresh milk has a pleasing, slightly sweet flavor, little aroma, and a pleasant mouthfeel and aftertaste. Since it is a rather bland product any off-flavor can be readily detected by the consumer. Being a naturally produced beverage it is influenced by a variety of genetic and environmental factors from production to consumption. Improvements in milking, transportation, storage, and marketing have all helped to maintain and preserve the flavor quality of milk. Once the milk has been removed from the cow, however, a variety of biochemical reactions are initiated which could lead to the deterioration of flavor and eventual spoilage. The Committee on Flavor Nomenclature and Reference Standards of the American Dairy Science Association reviewed off-flavors in milk and categorized them as shown in Table 10.1 (Shipe *et al.*, 1978). Those off-flavors that are of economic importance to the dairy industry will be emphasized as well as possible methods for minimizing or eliminating them altogether.

TABLE 10.1

CATEGORIES OF OFF-FLAVORS IN MILK[a]

Cause	Descriptive or associated terms
Heated	Cooked, caramelized, scorched
Light-induced	Light, sunlight, activated
Lipolyzed	Rancid, butyric, bitter, goaty[b]
Microbial	Acid, bitter, fruity, malty, putrid, unclean
Oxidized	Papery, cardboard, metallic, oily, fishy
Transmitted	Feed, weed, cowy, barny
Miscellaneous	Flat, chemical, foreign, lacks freshness, salty

[a] From Shipe *et al.* (1978).

[b] Bitter flavor may arise from a number of different causes. If a specific cause is unknown it should be classified under miscellaneous.

II. Off-Flavors in Milk

A. LIPOLYZED FLAVORS

Lipolyzed flavors are produced by the enzymatic hydrolysis of milk fat triacylglycerols by lipase, resulting in the accumulation of free fatty acids (FFA) as major degradation products. At one time this type of flavor defect was described as rancid, which caused considerable confusion because of its association with lipid oxidation. This was eventually resolved by referring to it as hydrolytic rancidity as distinct from oxidative rancidity, which more closely described "oxidized flavors." Increased mechanization in harvesting milk on the dairy farm has resulted in a greater incidence of lipolyzed flavor (Pillay *et al.*, 1980; Thomas, 1981). This was attributed to the installation of bulk tanks and pipeline milking systems which mechanically transfer milk to the farm bulk tanks, causing considerable agitation and foaming of the milk. While improvements in design and operations have reduced the incidence of this problem it still remains a major concern. Several surveys conducted by Barnard (1974) and Bandler *et al.* (1975) in the United States showed hydrolytic rancidity to be a serious problem in the commercial milk supply. A more recent study by Shipe *et al.* (1980) reported that 25% of randomly collected commercial milks were criticized by consumers for lipolyzed flavor at the sell-by date. Raw milk delivered to the processor which is borderline in free fatty acids would require very little lipolysis to produce a fatty acid level which exceeds the flavor threshold for lipolyzed flavor (American Public Health Association, 1978; Thomas, 1956).

The impairment of milk flavor by lipolysis appears to be a general problem throughout the world and has been the subject of discussion by the International Dairy Federation in a number of recent documents (International Dairy Federation, 1974, 1975, 1980). Downey (1974a,b) noted that for milk of good (5×10^3 counts/ml) or reasonable (10^5 counts/ml) microbiological quality it was the intrinsic milk lipolytic enzymes which hydrolyzed milk fat during cold storage. A microbiological count in excess of 10^6 or 10^7/ml was required before microbial lipases could hydrolyze lipids sufficiently to impart a detectable off-flavor in the milk. Mabbitt (1981), however, noted the development of severe sensory problems in raw milk with total counts of less than 10^6/ml when psychrotrophic bacteria predominated. The intrinsic milk enzymes are present in sufficient amounts to cause extensive fat hydrolysis with concomitant flavor impairment (Herrington, 1954). The level of enzymes, however, was not the critical factor in determining the susceptibility of milk to lipolysis. The critical factor is the milk fat globule membrane (MFGM), which protects the micelle triacylglycerols from lipolytic attack, which is further impaired by their association and/or occlusion with the casein micelles (Downey and Murphy, 1975) as well as possible inhibitors of lipolysis in the milk itself (Deeth and FitzGerald, 1975). Raw milk or cream requires excessive agitation or turbulence to disrupt the milk fat globule membrane thereby permitting the lipolytic enzymes to attack the triacylglycerols, which is referred to as induced lipolysis (Kirst, 1986). This occurs during the transfer of the milk to the farm bulk tanks, where considerable agitation and turbulence are generated (Pillay *et al.*, 1980; Kirst, 1986). Once lipolysis occurs it is accompanied by the rapid accumulation of free fatty acids together with mono- and diglycerides, resulting in a rancid or lipolyzed flavor in the milk. Lipolysis continues for a relatively short period of time and then levels off in spite of the presence of excess amounts of triacylglycerol substrates. Tolle and Heeschen (1975) bubbled gas through warm milk to induce turbulence and foaming. This caused lipolysis which leveled off after 10–15 min when the free fatty acid level approached 3.5 mequiv./liter (Fig. 10.1). These results confirmed earlier work by Nilsson and Willart (1961a) which showed that incubation of a mixture of raw and pasteurized milk at 16°C (3 : 2 v/v) caused lipolysis to proceed for about 2 hr, stopping when the free fatty acid (FFA) level had reached approximately 3 mequiv./liter. In addition to the holding temperature and time of incubation, these researchers found that the rate of lipolysis was affected by the degree of agitation. Increasing the homogenization pressure from 250 to 2500 psi or use of short (30-sec) successive homogenization treatments increased the level of free fatty acids (Fig. 10.2). The reinitiation of lipolysis showed that neither irreversible loss of enzyme activity nor the complete hydrolysis of triacylglycerols was responsible for the leveling off of lipolysis over time. This phenomenon also occurred in the presence of purified lipolytic enzymes from milk and artificial triglyceride emulsions (Patel *et al.*, 1968).

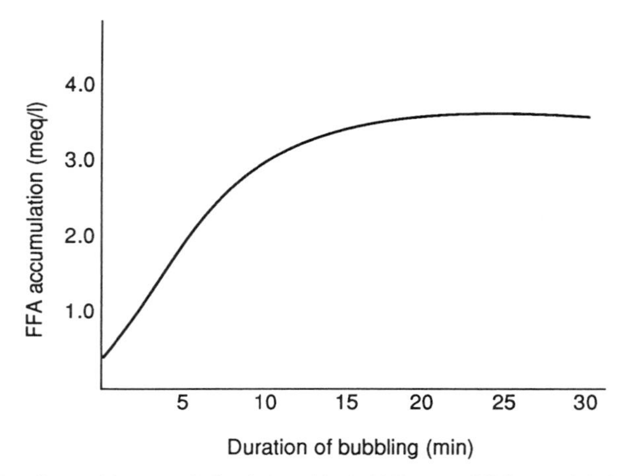

FIG. 10.1. Free fatty acid accumulation induced by bubbling gas (30 liters/min) through raw milks (Tolle and Heeschen, 1975).

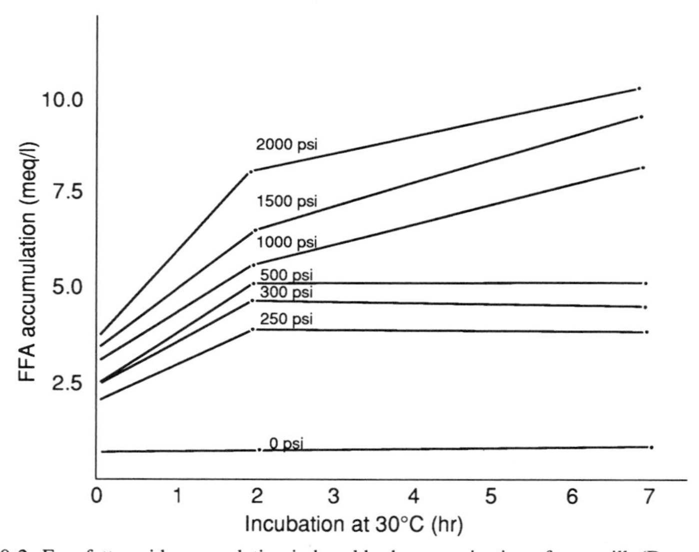

FIG. 10.2. Free fatty acid accumulation induced by homogenization of raw milk (Downey, 1980).

The peculiar feature of lipolytic enzymes is that the locus of attack on the emulsified substrate is the ester bond at the oil–water interface to which the enzymes must be reversibly absorbed. The importance of bile salts in lipolysis is well established (Desnuelle, 1961), with a number of theories proposed to explain its stimulatory effect. The current theory is that rather than affecting the enzyme per se, the bile salts stimulate lipolysis by sweeping the oil–water interface of liberated free fatty acids by complex formation and thereby favoring reversibility of enzyme desorption at the interface (Benzonana and Desnuelle, 1965). In the absence of bile salts, free fatty acids accumulate at the oil–water interface, causing noncovalent interaction with hydrophobic moieties of the substrate which could result in the unfolding of the enzyme and the subsequent blockage of the oil–water interface. Based on this theory the leveling off with time of induced lipolysis in milk could be explained by the accumulation of free fatty acids at the fat globule membrane and inability of the enzyme to de-adsorb from the interface. The result is a slowdown and eventual inhibition of lipolysis. The sweeping of free fatty acids from the oil–water interface by homogenization thus permits the resumption of new enzyme–substrate interactions until the interface becomes blocked again (Downey, 1980). A reduction of free fatty acids during the holding of lipolyzed milk has also been attributed to the presence of very labile enzymatic activity capable of synthesizing triacylglycerols in freshly secreted milk (Christie, 1974; McCarthy and Patton, 1964). Part of this activity may have originated from fragments derived from mammary gland secretory cells (Christie and Wooding, 1975). Compared to the interfacial blocking, however, the synthesis of triacylglycerols represents only a minor effect on the cessation of lipolysis over time.

Bhavadasan and co-workers (1982) noted a much greater degree of lipolysis after agitation of buffalo and cow's milk at 15°C compared to at 10 or 25°C. Disruption of the milk fat globule membrane was monitored by following the

TABLE 10.2

Activity of Membrane-Bound Xanthine Oxidase from Skim Milk of Cow and Buffalo Milk Agitated at Different Temperatures[a]

Agitation temperature (°C)	Increase in activity (%)	
	Cow	Buffalo
10	2.1	5.5
15	90.5	116.6
25	38.1	27.7

[a] From Bhavadasan *et al.* (1982).

release of bound xanthine oxidase. This enzyme constituted 8% of the intrinsic proteins of the milk fat globule membrane and was only released when the membrane was ruptured (Briley and Eisenthal, 1975). Table 10.2 shows an increased release of xanthine oxidase when milk was agitated at 10, 15, and 25°C, with the maximum corresponding to the highest degree of lipolysis at 15°C. This study confirmed that enhanced lipolysis occurred, with greater disruption of the globule membrane increasing the susceptibility of milk fat to lipolytic attack.

In addition to induced lipolysis, which requires agitation, some naturally active milks when cooled develop lipolysis spontaneously. This phenomenon, referred to by Jensen (1964) as spontaneous lipolysis, was influenced by the rapidity and extent of cooling (Johnson and von Gunter, 1962). The incidence of spontaneous lipolysis in milk can vary from 3 to 35%, although the current term is used to describe milks that develop lipolysis without any apparent agitation. The earlier stipulation that it should be initiated by cooling is no longer adhered to.

1. *Lipases in Milk*

Bovine milk contains considerable lipolytic activity of which the β-type esterases are the predominant ones. These include glycerol tricarboxyl esterases, aliphatic esterases, diesterases, and lipases (EC 3.1.1.3) with a pH optimum of 8–9. The majority of the lipase activity is associated with casein, of which 70% is bound to micellar casein (Downey and Andrews, 1966). The remainder of lipase is present as a soluble casein–enzyme complex in the milk serum (Anderson, 1982; Hoynes and Downey, 1973). One particular lipase in milk is lipoprotein lipase (EC 3.1.1.34), which was isolated by Egelrud and Olivecrona (1972b). This enzyme plays an important role in removing lipids from the blood to the mammary gland and its presence may be due to leakage from the tissue (Mendelssohn *et al.*, 1977; Shirley *et al.*, 1973). It is a particularly efficient enzyme with a turnover number of more than 3000 per sec (1, 3-diolein, pH 8.5, 25°C) and it hydrolyzes long- and short-chain triacylglycerols, partial glycerides and phospholipids (Egelrud and Olivecrona, 1973; Scow and Egelrud, 1976).

The physicochemical characteristics of lipoprotein lipase are very similar to those of lipase and it has been concluded that both enzymes reside in the same enzyme in bovine milk, although this has since been questioned. Only the primary ester groups are hydrolyzed in the triacylglycerols (Nilsson-Ehle *et al.*, 1973) and it preferentially attacks *sn*-1-ester (Morley *et al.*, 1974). The short-chain fatty acids are in the *sn*-3 position in milk triglycerides and these are readily attacked by the enzyme. Triacylglycerols, formerly referred to as triglycerides, are esters of glycerol in which all the hydroxyl groups are esterified with fatty acids. Glycerol, while appearing to be a symmetric molecule, is in fact asymmetric, possessing neither rotational symmetry nor the ability to superimpose on itself by rotation. As a consequence, the two primary hydroxyl groups in glycerol are not identical, requiring a nomenclature system capable of distinguishing between them. This was provided by a stereochemical numbering system recommended by the IUPAC/IUCB Commission on Biochemical Nomenclature (1967). In this system the carbon atom on the top in the Fischer projection which shows a vertical chain with the secondary hydroxyl group to the left is numbered C-1, while the one below is C-3. This numbering system uses the prefix *sn* (stereochemical number) before the name of the compound as shown in the following:

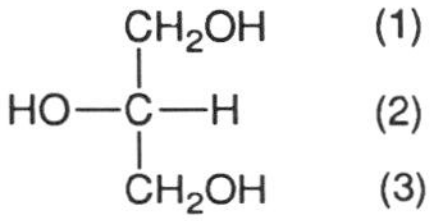

CH_2OH (1)
HO—C—H (2)
CH_2OH (3)

Glycerol

$CH_2OC(=O)R'$ (1)
$R'C(=O)OCH$ (2)
$CH_2OC(=O)R'$ (3)

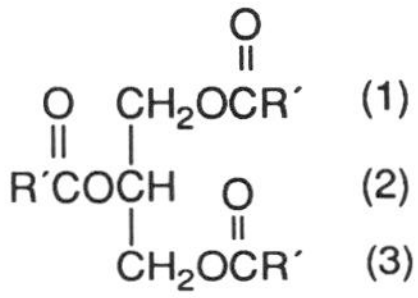

1, 2, 3-triacyl-*sn*-glycerol

(sn numbering system in parentheses)

The disproportionate amount of short-chain fatty acids released by lipolysis of milk fat pointed to some fatty acid specificity for lipase. This may be due to the physical property of the substrate, as ester groups with short fatty acids are less hydrophobic and more likely to be exposed at the oil–water interface compared to ester groups with longer fatty acids. Lipoprotein lipases are activated by a cofactor protein which appears to orient the enzyme and/or lipid substrate to facilitate lipolysis. Partial isolation of lipoprotein lipase and a bovine cofactor were reported by Lim and Scanu (1976) and Liesman *et al.* (1977). Enhanced lipolysis following addition of 1–5% bovine blood serum to milk was reported by Castberg and Solberg (1974) and Jellema (1975) to be due to leakage of the blood components into the milk. Further research is required before placing too much importance on blood constituents, as spontaneous lipolysis appears to be affected by late lactation, poor nutrition, low milk yield, hormonal controls, and cell count (Deeth and FitzGerald, 1975; Salih, 1978).

Lipoprotein lipase is considered responsible for spontaneous lipolysis (Olivecrona, 1980). The milk fat globules in normal milk appear to be resistant to lipoprotein lipase as the majority of the enzyme is bound to the casein micelles (Anderson, 1982; Hohe *et al.*, 1985; Sundheim and Bengtsson-Olivecrona, 1986). Cooling milk may increase the amount of lipoprotein lipase bound to the fat globule membrane, which is correlated with spontaneous lipolysis (Sundheim and Bengtsson-Olivecrona, 1985, 1986). Using isolated fat globules from fresh warm milk, Sundheim and Bengtsson-Olivecrona (1987) found them to be quite resistant to attack by purified bovine lipoprotein lipase. Cooling the milk fat globules, however, rendered them susceptible to lipolysis by lipoprotein lipase,

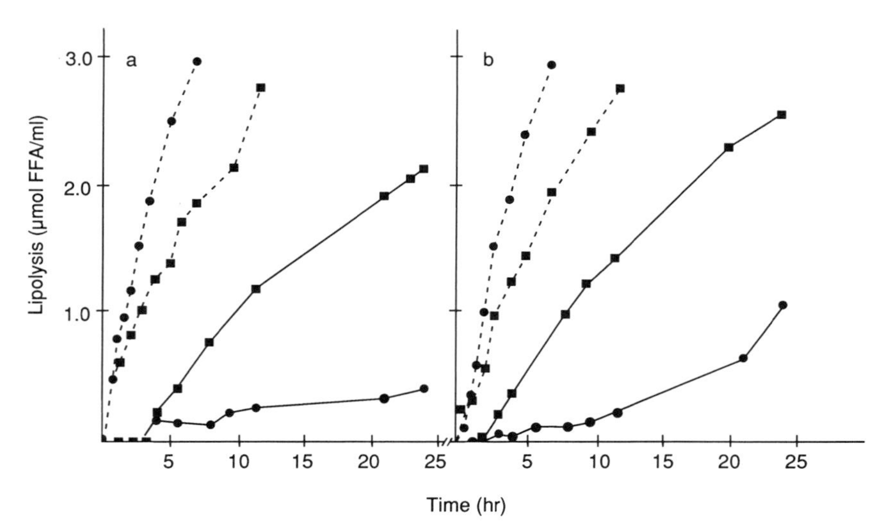

FIG. 10.3. Milk fat globules as substrate for lipoprotein lipase (LPL). Effect of temperature on milk fat globules (MFG) isolated from normal milk (a) or milk with spontaneous lipolysis (b). Purified LPL was added to warm MFG (——) or to MFG precooled for 16 hr at 4°C (-----) before incubation at 25°C (●) or 4°C (■) under the conditions described. The degree of lipolysis in whole milk incubated at 4°C for 24 hr was 0.90 μmole free fatty acids (FFA) /ml for the milk used in (a) and 1.98 μmole FFA/ml for the milk used in (b) (Sundheim and Bengtsson-Olivecrona, 1987).

as observed in the case of milk fat globules isolated from milk with spontaneous lipolysis (Fig. 10.3). The milk fat globules from the most stable milk exhibited the longest lag phase in the presence of lipoprotein lipase with half the amount of free fatty acids released after 24 hr. Recent studies by Sundheim (1988) found that lipoprotein lipase was consistently higher in evening milk samples compared to in morning samples. This was also the case with respect to cold storage lipolysis, although the amount of free fatty acids produced per unit of lipase remained the same. Several researchers reported the inability of lipoprotein lipase to bind to the milk fat globule in normal milk (Olivecrona, 1980; Olivecrona and Bengtsson, 1984). The factors in milk which prevent the binding of lipoprotein lipase to the milk fat globule, however, remain to be identified, although addition of serum lipoproteins or of apolipoprotein causes rapid lipolysis (Bengtsson and Olivecrona, 1982; Castberg and Solberg, 1974; Jellema, 1975; Sundheim *et al.*, 1983). The spontaneous lipolysis during cold storage of milk remains a complex problem in which the properties of the milk fat globules, skim milk, and lipoprotein lipase are all contributory factors (Sundheim, 1988).

2. *Inactivation of Lipase*

It is generally assumed that milk lipase is sufficiently inactivated during pasteurization. Shipe and Senyk (1981), however, noted that research published over 20 years ago, now apparently overlooked, showed that minimum pasteurization did not completely destroy milk lipase. In this study, milk heated using a laboratory tubular pasteurizer at 71.11 and 87.78°C for 17.6 sec showed residual lipase activities of 15 and 2%, respectively (Harper and Gould, 1959). A second study by Nilsson and Willart (1961b), however, could not detect lipolysis in milk heat-treated at 73°C for 30 sec when incubated for 3 hr at 22°C. Incubation of the same heat-treated milk for 24 hr, however, resulted in the detection of 10% of the original lipolytic activity. Shipe and Senyk (1981) recognized the inadequacy of minimal pasteurization to prevent lipolysis in milk. They also noted that the variation in the lipolytic activity of milk, as determined by acid degree value, from milk obtained from different herds pointed to the need for higher pasteurization temperatures. Experiments conducted using combinations of different pasteurization temperatures and times supported their recommendation of pasteurization at 76.7°C for 16 sec to prevent lipolysis in most milk samples for up to 7 days postpasteurization (Table 10.3). To extend the shelf life of susceptible raw milk samples they recommended a higher pasteurization temperature be used.

TABLE 10.3

EFFECT OF PASTEURIZATION TIME AND TEMPERATURE ON LIPOLYSIS[a,b]

Temperature[c] (°C)	Holding time[d]		
	16 sec	20 sec	24 sec
72.2	2.1	1.7	1.3
74.4	1.0	1.0	0.9
76.7	0.9	0.9	0.9
78.9	0.9	0.8	0.8
81.1	0.8	0.8	0.8

[a] From Shipe and Senyk (1981).

[b] Average acid degree values (ADVs) for six pasteurized–homogenized milk samples after storage for 7 days at 5°C. Average raw milk ADV was 0.7.

[c] ADV means for 72.2°C were significantly different from all others ($P < 0.01$); values for 74.4°C differed from those for 81.1°C ($P < 0.05$); no significant differences were observed between others.

[d] ADVs for different holding times were different ($P < 0.05$) at temperatures below 74.4°C.

B. Microbial Flavors

Milk is an ideal medium for microbial growth so it is particularly important to utilize the most thorough sanitizing procedures as well as proper cooling and holding temperatures to ensure that the raw milk on the farm is of good quality. Most microbial spoilage and associated off-flavors are due to postpasteurization contamination generally involving psychrotrophic bacteria. In commercial pasteurized milk, contamination is due to gram-negative psychrotrophic recontaminants (Thomas and Druce, 1969). The temperature recommended by the Institute of Food Technology in 1981 for low-temperature storage of pasteurized milk is 0–5°C.

Psychrotrophs were originally referred to as a group of microorganisms which grow best at low temperatures, that is, they are cold loving. However, objections were raised by a number of researchers to the use of this term, with its implied preference for growth at low temperatures, since in fact they grow better at 20°C or higher. These microorganisms were defined at a meeting of the International Dairy Federation (1976) as "microorganisms" which can grow at 7°C or less irrespective of their optimal temperature of growth. Of particular concern to the dairy technologist are those psychrotrophs which are sufficiently resistant to cause problems in pasteurized milk. The use of low-temperature storage provides a convenient control method for prolonging the shelf life of pasteurized milk. Milk with a high psychrotrophic count has a much shorter shelf life than milk with a low psychrotrophic count.

The majority of psychrotrophic organisms belong to the genus *Bacillus* and are probably variants of mesophilic bacilli that have adapted to lower growth temperature ranges (Collins, 1981). In addition to this genus, others belong to *Clostridium* as well as *Arthrobacter, Microbacterium, Streptococcus,* and *Corynebacterium* (Bhadsavle *et al.*, 1972). These microorganisms biochemically alter the constituents of milk and cause spoilage. The initial changes during the early growth phase cause a lack of freshness or stale taste. Subsequent changes lead to the development of definite off-flavors and odors caused primarily by their lipolytic and/or proteolytic activity.

Monika and co-workers (1982) examined the keeping quality of HTST-pasteurized milk from five dairies in the United Kingdom. Milk spoilage was evident at total colony counts of 10 cfu/ml (colony-forming units) and had a bitter or fruity component that was not detected in the laboratory-pasteurized milk. The degree of postpasteurization contamination resulted in corresponding changes in keeping quality at 5°C. This could be minimized by proper sanitation of milk-contacting surfaces and proper filling practices. The use of high-efficiency particulate air filters to reduce contamination by airborne microbial populations was reported by Wainess (1981) to extend the shelf life of pasteurized milk.

TABLE 10.4

SHELF LIFE (IN DAYS) OF MILK STORED RAW AT 4°C AND PASTEURIZED AT 8 AND 16°C[a]

Days of raw milk storage	Shelf life of pasteurized milk: 8°C storage		16°C storage	
	LP	Control	LP	Control
4	12	9	6	5
6	11	7	5	1
8	1	0	0	0

[a] From Martinez *et al.* (1988).

Activation of the lactoperoxidase system (LP) was proposed by several researchers to control psychrotrophic bacteria present in milk (Reiter and Harnulv, 1982; Zajak *et al.*, 1983). Lactoperoxidase in milk produces an antibacterial compound, hypothiocyanate ($OSCN^-$) from thiocyanate (SCN^-) in the milk, which has limited stability. Martinez and co-workers (1988) examined the potential of reactivating the LP system during the storage of raw milk and on the quality of the pasteurized milk. Activation of the LP system was achieved by adjusting the level of SCN^- in raw milk to 0.25 m*M*/ml with a solution of NaSCN and to 0.25 m*M* peroxide with a freshly prepared solution of hydrogen peroxide. The raw milk was reactivated every 48 hr during 8 days of storage at 4°C, then pasteurized, reactivated, and stored at 8 or 16°C. The improved shelf life obtained by this treatment is shown in Table 10.4 compared to that of the untreated milk. These researchers recommended that reactivation of LP be used when raw milk is stored longer than 2 days to extend the shelf life of pasteurized milk.

C. Lipolytic Enzymes

Lipolytic activity by psychrotrophs varies with the species as well as optimum temperature, pH, and specificity. The development of fruity flavor in milk held at low temperatures was described in 1902 by Eicholz as well Gruber (Morgan, 1976). The name of the organism involved was proposed by Hussong (1932) to be *Pseudomonas fragi* which he isolated from defective Iowa dairy products. This organism hydrolyzed milk fat at 7°C, producing short-chain fatty acids apparently responsible for the development of fruit flavors in cottage cheese (Reddy *et al.*, 1971). This enzyme was shown by Alford *et al.* (1964) and Mencher and Alford (1967) to have a 1,3 position specificity for triacylglycerols.

Reddy and co-workers (1968) isolated six esterases from milk cultures of *P. fragi* responsible for fruity aroma.

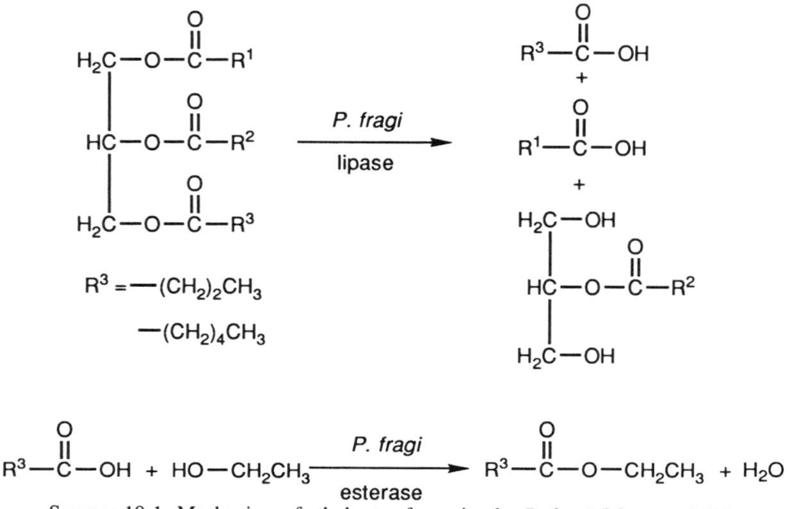

$$\begin{array}{l} H_2C-O-\overset{O}{\overset{\|}{C}}-R^1 \\ HC-O-\overset{O}{\overset{\|}{C}}-R^2 \\ H_2C-O-\overset{O}{\overset{\|}{C}}-R^3 \end{array} \xrightarrow[\text{lipase}]{P.\ fragi} R^3-\overset{O}{\overset{\|}{C}}-OH + R^1-\overset{O}{\overset{\|}{C}}-OH + \begin{array}{l} H_2C-OH \\ HC-O-\overset{O}{\overset{\|}{C}}-R^2 \\ H_2C-OH \end{array}$$

$$R^3 = -(CH_2)_2CH_3$$
$$-(CH_2)_4CH_3$$

$$R^3-\overset{O}{\overset{\|}{C}}-OH + HO-CH_2CH_3 \xrightarrow[\text{esterase}]{P.\ fragi} R^3-\overset{O}{\overset{\|}{C}}-O-CH_2CH_3 + H_2O$$

SCHEME 10.1. Mechanism of ethyl ester formation by *P. fragi* (Morgan, 1976).

The major fruity volatiles isolated were ethyl butyrate and ethyl hexanoate (Scheme 10.1).

D. PROTEOLYTIC ENZYMES

Proteolytic enzymes degrade proteins and release a range of nitrogenous compounds. Those proteases, which attack either casein or whey proteins, result in the production of bitter flavors and coagulation of milk. The increased shelf life of milk is approximately 2 weeks in the store plus 1 week at home if refrigerated properly. While bacteria may be killed by pasteurization, certain heat-resistant enzymes may cause proteolytic and lipolytic action.

Pseudomonas and *Acinetobacter* spp. were reported by Law *et al.* (1977) to modify the polyacrylamide and starch gel electrophoretic patterns of casein. Skean and Overcast (1960) reported proteolytic activity in three *Pseudomonas* spp. (*P. fluorescens, P. fragi,* and *P. putrefaceins*) which were capable of degrading casein and whey proteins in pasteurized milk held at low temperatures. Nine raw milk isolates were shown by Adams and co-workers (1976) to degrade κ-casein. A greater degree of degradation was observed by these psychrotroph strains on β-casein compared to α-casein. A particular problem associated with microbiologically spoiled milk is malty flavor, which is caused by the organism *Streptococcus lactis*. This problem was particularly prevalent in Connecticut in the 1940s and was attributed by Jackson and Morgan (1954) to the formation of

3-methylbutanal by the organism *S. lactis* var. *maltigenes*. The mechanism of aldehyde formation was revealed in studies by Hosono and co-workers (1974) and Hosono and Elliot (1974) to originate from amino acids:

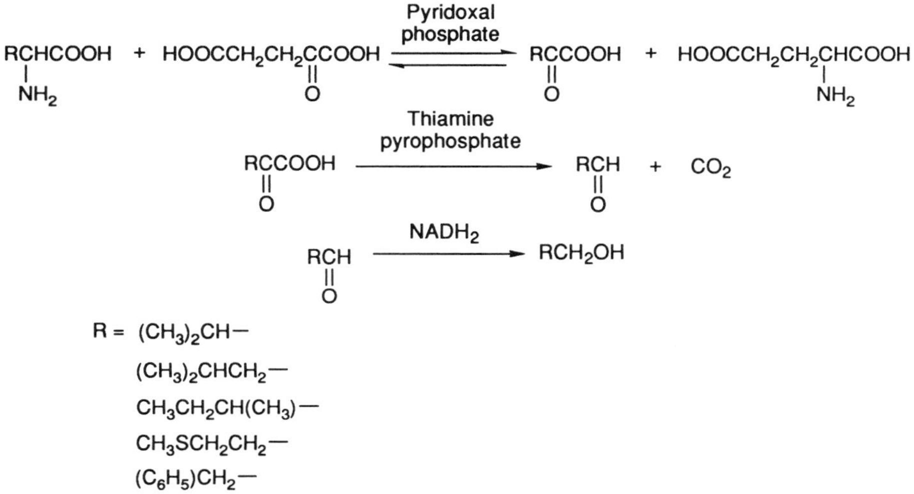

$$\underset{NH_2}{RCHCOOH} + HOOCCH_2CH_2\underset{\|\\O}{C}COOH \xrightleftharpoons{\text{Pyridoxal phosphate}} R\underset{\|\\O}{C}COOH + HOOCCH_2CH_2\underset{NH_2}{CH}COOH$$

$$R\underset{\|\\O}{C}COOH \xrightarrow{\text{Thiamine pyrophosphate}} R\underset{\|\\O}{CH} + CO_2$$

$$R\underset{\|\\O}{CH} \xrightarrow{NADH_2} RCH_2OH$$

$R = (CH_3)_2CH-$
$(CH_3)_2CHCH_2-$
$CH_3CH_2CH(CH_3)-$
$CH_3SCH_2CH_2-$
$(C_6H_5)CH_2-$

SCHEME 10.2. Mechanism of aldehyde and alcohol formation from amino acids by *S. lactis* var. *maltigenes* (Morgan, 1976).

MacLeod and Morgan (1955, 1956) carried out a series of studies on the production of aldehydes from amino acids in which *S. lactis* and its malty variant required leucine, isoleucine, and valine for growth in a synthetic medium. The resting cells of both organisms were able to convert α-ketoisocaproic and α-ketoisovaleric acids to leucine and valine in the presence of glutamic acid. Only *S. lactis* var. *maltigenes* cells concomitantly decarboxylated the keto acids to the corresponding aldehydes. Dialyzed acetone powders showed transaminase activity for both malty and nonmalty cells although only the malty cells possessed TPP-mediated α-ketodecarboxylase. The resting cells of *S. lactis* var. *maltigenes* were also found to convert phenylalanine and methionine to their respective aldehydes by the same reaction mechanism.

Tucker and Morgan (1967) observed that C_5 and C_6 keto acids branched at the penultimate carbon atom were rapidly converted to their respective aldehydes by *S. lactis* var. *maltigenes* as shown in Fig. 10.4. A new species, *Lactobacillus maltaromicus,* identified by Miller *et al.* (1974), also produced 3-methylpropanal and 3-methylbutanal and their corresponding alcohols by a mechanism similar to that found with *S. lactis* var. *maltigenes*.

In 1932 Levine and Anderson reported an organism which caused mustiness in eggs. An identical organism was also reported by Olson and Hammer (1932) which produced a musty potato aroma in milk. The organism involved was

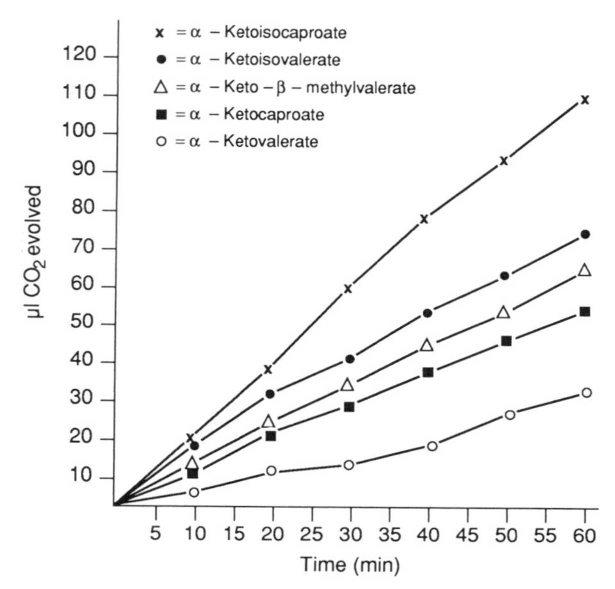

FIG. 10.4. Rates of decarboxylation of α-keto acids by *S. lactis* var. *maltigenes* resting cells. Each flask contained 0.0025 *N* neutralized keto acid, 5.0 ml of resting cells (12.5 mg dry wt. per flask), 1.0 ml ellvaine's buffer (pH 5.2), and 1.0 ml water. Total reaction volume was 3.0 ml and the temperature was 30°C (Tucker and Morgan, 1967).

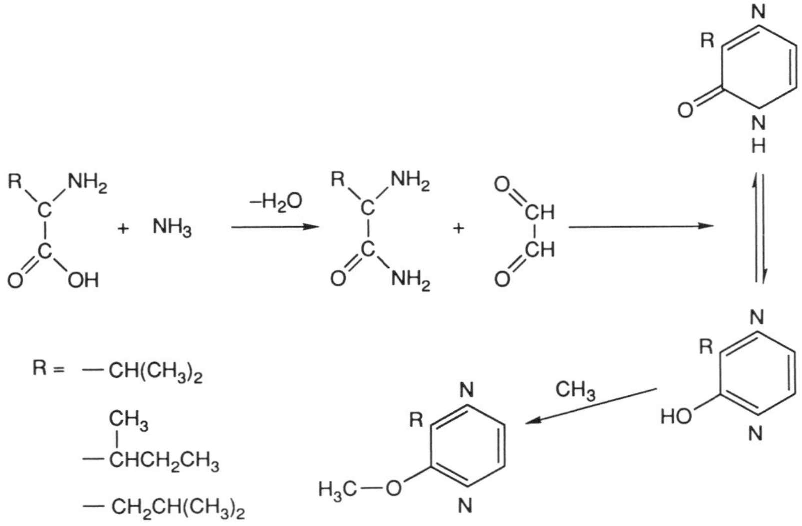

R–C(NH2)–C(=O)OH + NH_3 $\xrightarrow{-H_2O}$ R–C(NH2)–C(=O)NH2 + OHC–CHO

R = —$CH(CH_3)_2$

—$CH(CH_3)CH_2CH_3$

—$CH_2CH(CH_3)_2$

CH_3

H_3C—O

SCHEME 10.3. Biosynthesis of 2-methoxyalkylpyrazines (Morgan, 1976).

Pseudomonas taetrolens and the compounds produced were identified by Morgan *et al.* (1972) to be 2, 5-dimethylpyrazine and 2-methoxy-3-isopropylpyrazine by GLC using headspace vapor analysis. The latter compound was demonstrated earlier by Seifert *et al.* (1970) to have a potent potatolike aroma. The mechanism involved biological amidation of amino acids and condensation with α- and β-dicarbonyl, followed by methylation of the hydroxyl groups to produce substituted 2-methoxypyrazine derivatives of isobutyl and secondary butyl alcohols (Scheme 10.3). These have been reported in a number of vegetables, possibly involving valine, isoleucine, and leucine and two carbon compounds (Murray *et al.*, 1970; Murray and Whifield, 1975).

E. Oxidized Flavors

Oxidized flavors result from oxidation of unsaturated fatty acids by molecular oxygen. The flavors produced by oxidation of dairy products have been described as oxidized, cardboard, papery, metallic, tallowy, oily, and fishy. Shipe *et al.* (1978) recommended "oxidized flavor" as the generic term to describe all these flavors. The initial products of lipid oxidation, lipid hydroperoxides, are quite bland but degrade rapidly to hydrocarbons, acids, alcohols, aldehydes, and ketones which elicit undesirable flavors. The precise mechanism of hydroperoxide formation remained unclear because from an energetic consideration it was unlikely to occur between an unsaturated fatty acid (RH) and oxygen. The oxygen must be in an excited singlet state to react readily with the singlet state lipid to produce a singlet state peroxide. Aurand *et al.* (1977), using a singlet oxygen quencher (1,4-diazobicyclo[2.2.2]octane) or scavenger (1,13-diphenyl isobenzofuran), suggested that O′ (singlet oxygen) was the intermediate source of hydroperoxides catalyzed by copper, light, and enzymes. The reagents used by Aurand and co-workers (1977), however, were not really specific for singlet oxygen as other oxygen radicals gave similar results. The polyunsaturated fatty acids in the phosphatides at the interface of the milk fat globule were thought to be the precursors of flavor (Shipe, 1964). Further research by Foote and co-workers (1980) on the spontaneous dismutation of superoxide, however, showed that singlet oxygen was not, in fact, the product. Consequently Korycka-Dahl and Richardson (1980) and Richardson and Korycka-Dahl (1983) proposed a number of alternative ways for generating singlet oxygen, including:

1. Chemical reaction between any residual hypochlorite and hydrogen peroxide in milk.
2. Chemical or enzymatic catalyzed reactions involving metalloproteins, for example, peroxidase.
3. Photochemical oxidation in the presence of a sensitizer, for example, riboflavin (see Section I).

4. Interaction between secondary peroxy radicals based on studies by Boveris *et al.* (1981).
5. Oxidation of superoxide by selected oxidizing agents (Nanni *et al.*, 1981).

Hill *et al.* (1977) suggested that two systems were responsible for lipid oxidation in milk. One of these was dependent on Cu^{2+} and ascorbic acid for the generation of $^{\cdot}OH$ radicals while the other involved the generation of superoxide radicals and singlet oxygen by oxidases (xanthine oxidase and lactoperoxidase). The catalytic effect of Cu^{2+} on lipid oxidation in milk was first recognized by King and Dunkley (1959). Using protected supplements to feed dairy cows permits the polyunsaturated fatty acid content to increase from the normal 3% to over 20% (Scott *et al.*, 1970). This increased level of polyunsaturated fatty acids (High-Lin) in milk made it particularly susceptible to oxidation (Sidhu *et al.*, 1975). This was evident as the addition of antioxidant BHA to milk prevented this type of oxidation. Sidhu *et al.* (1976) noted, however, that the destruction of ascorbic acid in milk by treating with hydrogen peroxide prior to pasteurization resulted in greater resistance to oxidation. This suggested that the catalysis of lipid oxidation in milk involved Cu^{2+} and ascorbic acid. Aurand *et al.* (1977) found Cu(II) to be the primary catalyst in the generation of superoxide anion. This was confirmed by the partial inhibitory effect of superoxide dismutase (SOD), which catalyzes the dismutation of superoxide anion to triplet oxygen and hydrogen peroxide, thus having a protective role in milk.

$$2(O_2^{\cdot -}) + SOD \rightarrow O_2 + H_2O_2$$

Hill *et al.* (1977) found that addition of SOD and catalase to 80°C pasteurized milk inhibited oxidation (as measured by TBA). Oxidation of milk induced by Cu^{2+} could not be inhibited by SOD and catalase in the presence of formate, since the latter converts $^{\cdot}OH$ radicals to superoxide ($O_2^{\cdot -}$) radicals. These results point to one pathway which depends on the presence of $^{\cdot}OH$ radicals and added copper while the other pathway depends on the generation of superoxide radicals by enzymes.

F. Superoxide Dismutase and Xanthine Oxidase.

Superoxide dismutase was first reported in milk by Hill (1975) and Hicks *et al.* (1975). It appeared to be similar to an enzyme found in bovine blood containing Cu/Zn (Korycka-Dahl *et al.*, 1979; Hicks *et al.*, 1979). Later studies by Allen and Wrieden (1982a) used a milk-related model to examine the effect of milk proteins on lipid oxidation. They confirmed the ability of SOD to reduce the development of oxidized flavor in heat-treated (80°C) high-linolenic acid (High-Lin) milk. Addition of 0.1 ppm Cu^{2+} to the heated milk, however, essentially removed the protective effect of SOD. This system differed from that proposed

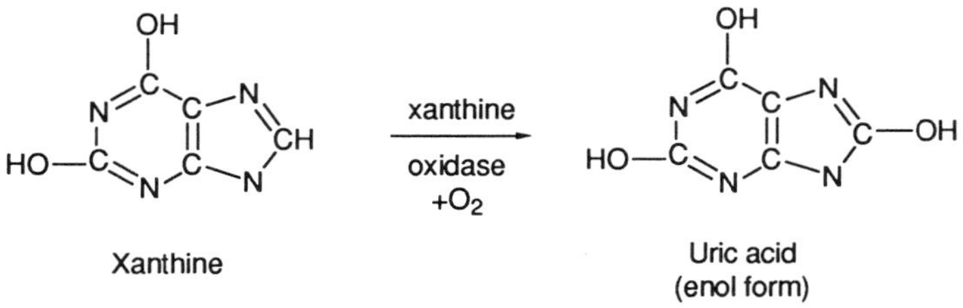

Xanthine —xanthine oxidase, $+O_2$→ Uric acid (enol form)

SCHEME 10.4. Oxidation of xanthine to uric acid by xanthine oxidase.

by Hill (1975) since no oxidases were present to generate superoxide radicals to be scavenged by SOD. It appeared, therefore, that SOD was exerting a non-enzymatic protective effect against lipid oxidation. The reduction of SOD by Cu^{2+} suggested that it was involved in converting superoxide radicals to a lipid reactive species $\cdot OH$. Hasegawa and Peterson (1978) reported that superoxide radicals did not react directly with polyunsaturated fatty acids or the corresponding hydroperoxides, thus supporting the conversion to $\cdot OH$ as the operative process (Bors *et al.*, 1979). Allen and Wrieden (1982b) also found that the addition of Cu^{2+} removed the antioxidant effect of SOD but not the enzyme activity. Cu^{2+} appeared to compete with the enzyme for superoxide radicals converting them to a lipid-reactive species.

In addition to the metal-catalyzed oxidation of milk flavor, oxidases also play an important role. Of these enzymes, xanthine oxidase, an enzyme which catalyzes the oxidation of the purines hypoxanthine and xanthine to uric acid (Scheme 10.4), plays a definite role. Bovine milk is a particularly rich source of this enzyme, which accounts for approximately 8% of the intrinsic proteins of the milk fat globule membrane (Briley and Eisenthal, 1975). This enzyme follows the fat into the cream when milk is separated. The molecular weight of xanthine oxidase is 275,000 with two moles of FAD, 8 moles of nonheme iron, 2 moles of molybdenum, and 8 labile sulfide groups. It was generally thought that xanthine oxidase promoted lipid oxidation in milk in the presence of xanthine or acetaldehyde as substrate by generating superoxide anions which then initiated fat oxidation (Aurand *et al.*, 1977; Kellog and Fridovich, 1975; Lynch and Fridovich, 1979). Studies by Fridovich and Porter (1981) distinguished between the rapid co-oxidation of micellar lipid by xanthine oxidase/acetaldehyde, which is inhibited by catalase or SOD and accelerated by Fe^{2+} but independent of the presence of hydroperoxides. This differed from the slow autoxidation of linolenic acid (C18:3) by xanthine oxidase as described by Thomas *et al.* (1978), which depended on the presence of hydroperoxides. These were thought to react with intermediate superoxide anion radicals ($O_2^{\cdot -}$) to form alkoxyl radicals which initiated oxidation according to the following equation:

$$O_2^{\cdot -} + LOOH \rightarrow O_2 + LO^{\cdot} + HO^{-}$$

TABLE 10.5

OXIDATIVE CAPABILITY OF XANTHINE OXIDASE-CONTAINING PROTEIN FROM MILK-FAT GLOBULE MEMBRANE[a]

	Rate of conjugated diene formation (m*M*/day)
Control (no protein added)	0.121
Non-xanthine oxidase protein	0.143
Active xanthine oxidase	0.173
Denatured xanthine oxidase	0.279

[a] From Allen and Humphries (1977).

[b] The xanthine oxidase-containing fractions from the Sephadex chromatography were collected and subjected to isoelectric focusing. The enzymatically active fractions were combined, and a portion was heat-denatured. The effects of these, and an equivalent amount of non-xanthine oxidase protein from the isoelectric focusing experiment, on the oxidation of 6 m*M* linoleic acid were determined in 0.1 *M* HEPES at pH 7.5 containing 0.1% (w/v) Empigen BB at 25°C. Protein concentration = 20 ± 4 μg/ml in each assay.

Allen and Wrieden (1982b) observed an increase in lipid oxidation in the presence of Cu^{2+} and xanthine oxidase. This suggested an alternative mechanism as xanthine oxidase was inactivated by Cu^{2+}, which reflected the ability of xanthine oxidase to bind nonnative metals causing denaturation and inactivation (Bergel and Bray, 1959). This could explain why Allen and Humphries (1977) found a xanthine oxidase fraction in the milk fat globule membrane which was more effective as a lipid pro-oxidant when heat-denatured, and increased 11-fold in the presence of Cu^{2+} as shown in Table 10.5. Their results suggested that under normal conditions xanthine oxidase may not be a significant cause of lipid oxidation unless contaminated with Cu^{2+}. Allen and Wrieden (1982a,b) also noted that lactoperoxidase exerted a pro-oxidative effect which was reduced by heating to 80°C. This was consistent with the loss of lactoperoxidase reported in heat-treated whole milk by Hill *et al.* (1977). It was apparent to Allen and Wrieden (1982b) that the two lipid-oxidizing systems proposed by Hill (1979) needed to be modified as the generation of the superoxide anion radical could not be attributed solely to oxidases, particularly as these radicals have no direct activity on lipid oxidation. In addition to these effects Allan and Wrieden (1982a) also found that casein at a concentration equivalent to that in bovine milk had antioxidant properties while whole whey protein was less effective in retarding lipid oxidation.

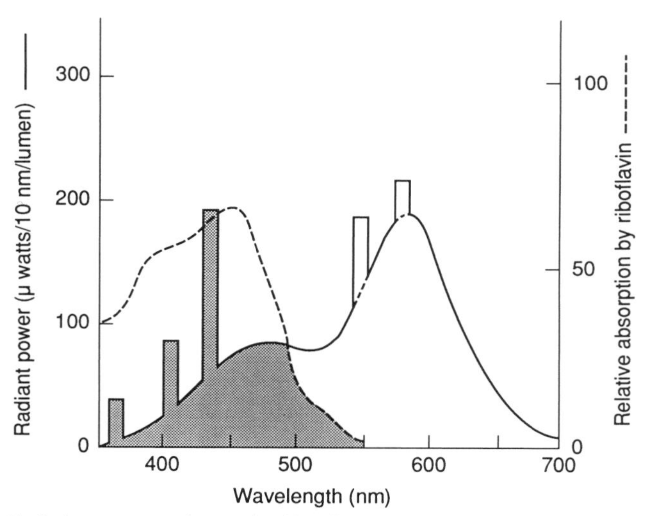

FIG. 10.5. Emission spectra of a cool white fluorescent lamp compared with absorption of riboflavin (Dunckley *et al.*, 1962). Copyright © by Institute of Food Technologists.

G. LIGHT-INDUCED FLAVORS

A major cause of milk deterioration is due to light (International Dairy Federation, 1970, 1976). The decline in home delivery of milk in North America means that most milk is sold through supermarkets, corner groceries, or convenience stores (Milk Industry Foundation, 1977). This has resulted in a longer period of time between processing and consumption of fluid milks, which accounts for as much as 9 days after processing (Dimick, 1982). Of particular concern is the use of fluorescent lights to illuminate dairy display cases, which is responsible for flavor deterioration and loss of nutrient quality in the milk (Bradley, 1980; Dimick, 1982; Hoskin and Dimick, 1979; Sattar and deMan, 1975). Exposure of milk to fluorescent light 24 hr/day for 1–4 days is not uncommon. The use of "white" fluorescent light in supermarkets is widespread, and this light has a spectral output of 350–750 nm and peaks at 470 and 600 nm as shown in Fig. 10.5. The radiant energy emitted by the fluorescent light is absorbed by and interacts with the milk components, resulting in photochemical reactions that lead to off-flavor development.

The susceptibility of milk to light-induced flavor changes depends on the nature and composition of the milk container. A variety of milk containers are available on the market and these are listed in Table 10.6 together with their light transmission values. Evaluation of these containers on exposure to fluorescent light has been carried out by many researchers. Dimick (1973) reported that

TABLE 10.6

DESCRIPTION AND CHARACTERISTICS OF ONE-GALLON MILK CONTAINERS[a]

Container	Type	Trade name[b]	Thickness (nm)	Transmission[c] (%)
Clear flint glass	Returnable	—	3.4	92
Clear polycarbonate	Returnable	Merlon	1.5	90
Tinted polycarbonate	Returnable	Merlon	1.5	75
High-density polyethylene	Returnable	Polytrip	1.7	58
Unprinted fiberboard	Nonreturnable	Pure-Pak	0.7	4

[a] From Hoskin and Dimick (1979).

[b] The use of trade names is for description only and does not imply endorsement by The Pennsylvania State University.

[c] Light transmission with 40-W cool white fluorescent lamp source measured inside and outside of container using Model 756 Weston meter.

homogenized milk packaged in imprinted fiber board container protected it from the development of light-induced flavor for up to 48 hr exposure to 100 foot-candles, while in plastic containers off-flavors were evident after only 12 hr exposure (Fig. 10.6). Sattar and deMan (1973) found that exposure of milk in clear polyethylene pouches to 100 and 200 foot-candles of fluorescent light for 3 and 6 hr resulted in rapid flavor deterioration. This was in sharp contrast to milk packaged in fiberboard containers, where exposure for 12 hr was needed before off-flavors were detected. The rapid deterioration of the flavor of milk held in transparent or translucent containers was confirmed in subsequent studies by

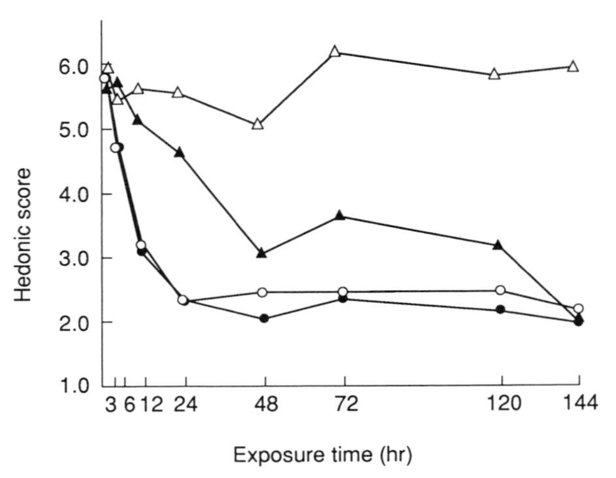

FIG. 10.6. Mean hedonic scores from expert panel for milk exposed to fluorescent light in various containers for 144 hr at 7± 1°C (Dimick, 1973). △, Control; ▲, fiberboard; ○, plastic; ●, glass.

Bradley (1980), Christy *et al.* (1981), Hansen *et al.* (1975), Hoskin and Dimick (1979), and deMan (1978).

One method to reduce light transmission through the milk container is by incorporation of colored materials to block out or absorb that part of the visible spectrum (440–488 nm) detrimental to milk flavor. Dunckley *et al.* (1962) reported a reduction in the transmission of fluorescent light at 480 nm through yellow, turquoise, red, and blue colored fiberboard cartons corresponding to 20, 50, 10, and 25%, respectively, compared to the uncolored carton. Using the uncolored fiberboard carton as the 100% reference, Bradfield and Duthie (1965) observed transmission values of 54, 27, 27, and 18% for red, blue, black, and green colored containers exposed to 400 foot-candles of fluorescent light. Sattar and deMan (1973) noted transmission of 0% at 400 nm and 13% at 700 nm for imprinted paperboard compared to over 60% transmission at 400 nm and over 80% at 700 nm for plastic 3-quart jugs. The latter produced light-induced flavor in milk after 3 hr exposure to 200 foot-candles compared to 12 hr for the corresponding fiberboard container. When exposed to 100 foot-candles for 12 hr both the plastic- and fiberboard-contained milk developed off-flavors compared to the unexposed milk. Christy *et al.* (1981) demonstrated that co-extruded black/white polyethylene totally blocked diffuse transmission at 400–500 nm and with aluminum ink only 1–3% transmission occurred, thus protecting the milk from off-flavor development. Fanelli and co-workers (1985) demonstrated that opacification of high-density polyethylene milk packages with visible and UV light screens provided an alternative way of shielding the milk from harmful radiation. Of these, FD&C yellow #5 proved quite effective in protecting

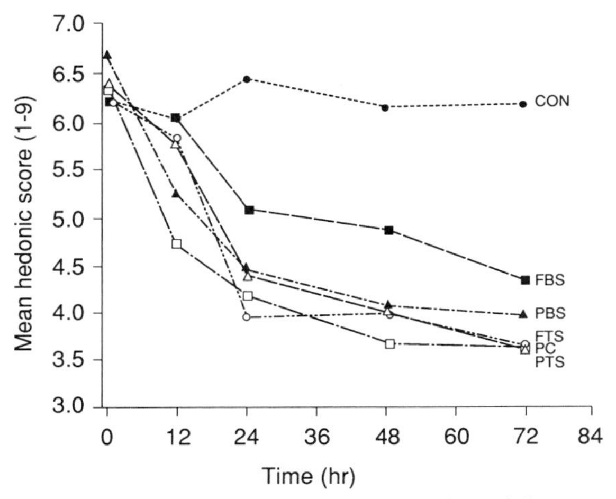

FIG. 10.7. Mean hedonic score of milk by treatment over time of fluorescent light exposure, average of trials 1 and 2 (Hoskin, 1988). CON = unexposed control; FBS = foil, bottom shield; PBS = polypropylene, bottom shield; FTS = foil, top shield; PC = unmodified; PTS = polypropylene, top shield.

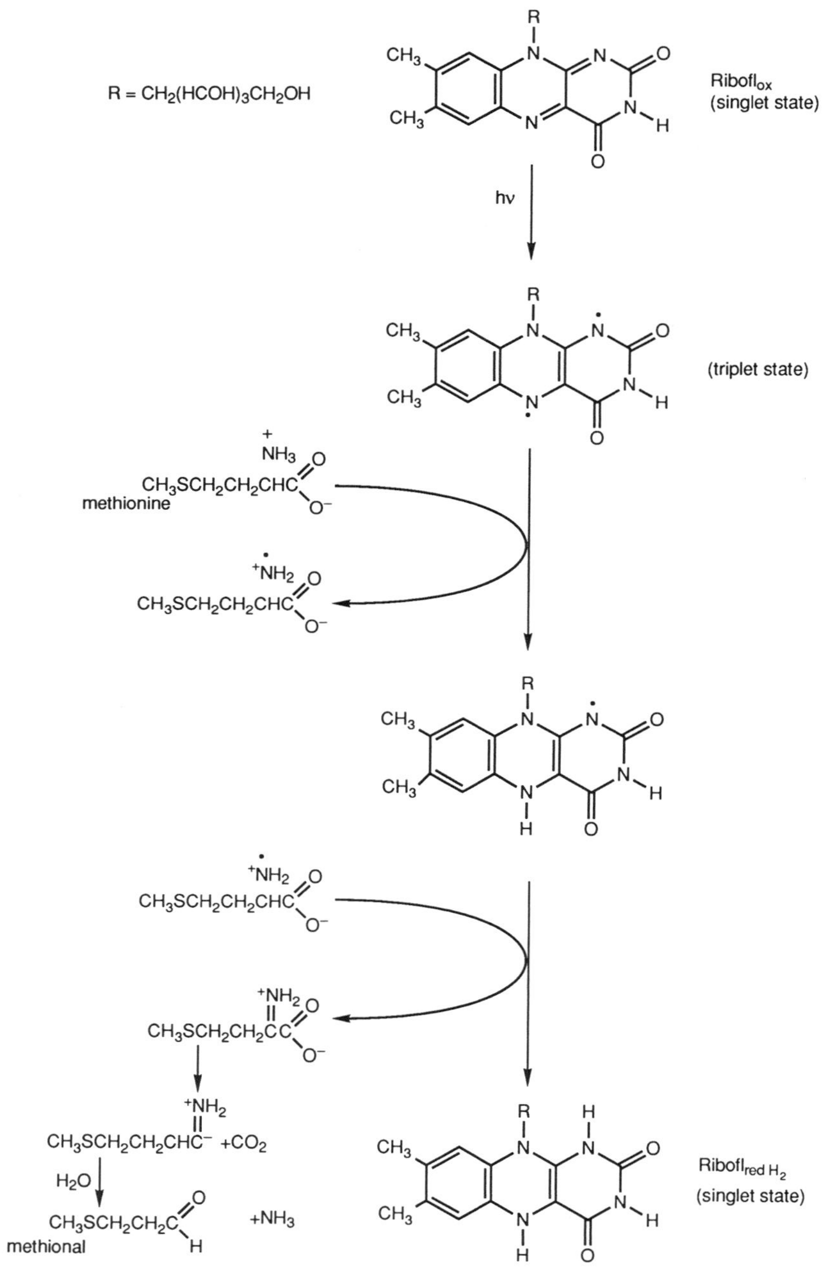

$R = CH_2(HCOH)_3CH_2OH$

Ribofl$_{ox}$ (singlet state)

hν

(triplet state)

methionine

methional

$+NH_3$

$+CO_2$

H_2O

Ribofl$_{red}$ H_2 (singlet state)

SCHEME 10.5. Proposed Strecker degradation-like reaction of methionine catalyzed by riboflavin (Hoskin, 1979).

vitamin A and riboflavin from photodegradation. A recent study by Hoskin (1988) found that milk packaged in half-gallon containers was protected from fluorescent light when aluminum foil and oriented polypropylene were placed on the outer surface of the container. The protection afforded by this technique was far more effective in retaining vitamins and flavor compared to just shielding the top part of the container with these materials. Aluminum foil proved more effective in protecting milk from fluorescent light compared to polypropylene as a result of its greater opacity (Fig. 10.7).

The origin of light-induced flavor has been examined for over 20 years. Aurand and co-workers (1964) detected a loosely bound complex of milk proteins and riboflavin which required the presence of tryptophan and was oxygen dependent. Further research by Aurand *et al.* (1966) implicated riboflavin as the primary factor responsible for the development of light-induced oxidized flavor in milk with ascorbic acid playing a secondary role. The destruction of riboflavin was mediated by light of the same wavelength which induces off-flavor. Velander and Patton (1955) showed that removal of riboflavin from milk essentially eliminated off-flavor development. Patton (1954) reported the formation of 3-methylthiopropanal (methional) when riboflavin and methionine were exposed to light. This compound has an odor similar to that of light-exposed milk and was probably formed via the Strecker degradation reaction (Scheme 10.5). Methional elicited the light-induced flavor at levels as low as 50 parts per billion. Absorption of light (max. 450 nm) causes riboflavin to be excited to a singlet and then to a triplet state, thus attracting methionine which is oxidized to methional while riboflavin is reduced. Several competing pathways can operate depending on the reactants and environmental conditions in the food. For example, under conditions of high oxygen concentration, singlet oxygen is formed which reacts with the unsaturated lipids to produce rancid compounds, while the sensitizer reverts to the ground state to be reactivated again. Alternatively the sensitizer abstracts a hydrogen atom from the fatty acid to produce a free radical which then reacts with ground-state oxygen (O_2^3) to form a peroxy radical. In the case of riboflavin it is univalently reduced to form a stable flavin semiquinone (Richardson and Korycka-Dahl, 1983). The reduced riboflavin can be reoxidized to a flavin radical in the presence of oxygen to produce superoxide anions ($O_2^{\bar{}}$) as shown in Scheme 10.2 (Ballou *et al.*, 1969). Korycka-Dahl and Richardson (1978) found that cysteine, histidine, tyrosine, and tryptophan also supported the photogeneration of superoxide anions when exposed to fluorescent light and that photoreaction occurred even if the terminal alpha amino and carboxyl groups of methionine were blocked. These results suggested that the photogeneration of superoxide anions could occur with polypeptides or proteins containing methionine. Allen and Parks (1975) previously showed the presence of methional in skim milk exposed to direct sunlight for 10 to 15 min periods.

In addition to the degradation of serum protein, lipid oxidation is another important factor in the development of light-induced flavor. The effect of radiation on the oxidation of fatty acids is well established, leading to peroxides and then carbonyl compounds by a free radical reaction mechanism. Compared to the oxidation of proteins, the rate of lipid degradation is relatively slower (Aurand *et al.*, 1966). This was confirmed by Dimick (1973), who monitored TBA values, as a measure of lipid oxidation, and compared it to the development of light-induced flavor. Wishner and Keeney (1963) reported that C_6 to C_{10} alk-2-enals were formed in milk exposed to sunlight but were absent in the unexposed sample. Bassette (1976) found increases in acetaldehyde, *n*-pentanal, and *n*-hexanal in milk exposed to sunlight and fluorescent light. Bray *et al.* (1977) noted that this type of induced flavor was detected early by consumers with more than 73% surveyed preferring the nonexposed milk.

A recent study by Olsen and Ashoor (1987) did not find any damaging effects on the flavor, odor, appearance, or riboflavin content of milk whether packaged in plastic or fiberboard containers displayed in dairy cases of four grocery stores. These researchers attributed their results to the fact that most of the previous studies were conducted under controlled laboratory conditions and not under actual display conditions, where lower light intensities may be used (Dillman *et al.*, 1971; Hankin and Dillman, 1972; Reif *et al.*, 1983). Three of the four grocery stores studied had much lower light intensities in their display cases compared to the national average. A nationwide survey of 105 supermarkets conducted in the United States in 1973 found that the average illumination was

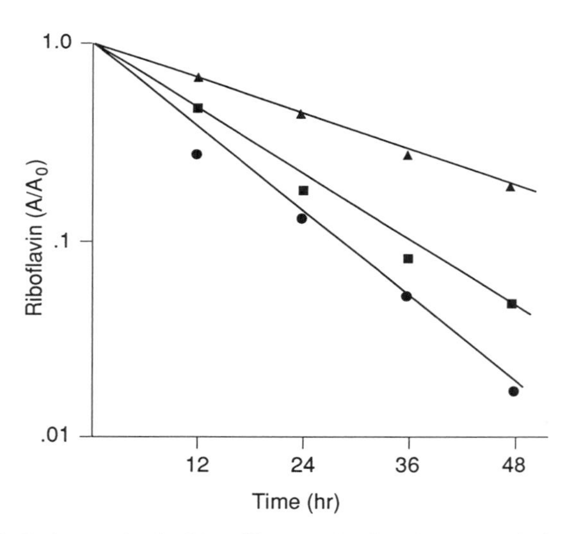

FIG. 10.8. Riboflavin retention in skim milk exposed to light intensities of (▲) 150, (■) 300, and (●) 450 foot-candles at 4°C. A = Concentration at a given time of light; A_0 = initial concentration (Gaylord *et al.*, 1986).

186 foot-candles (Dimick, 1973). Sattar and deMan (1973) reported that the average intensity of fluorescent light most common in display cases in Canada ranged from 100 to 200 foot-candles. Deger and Ashoor (1987) surveyed seven supermarkets and three convenience stores and reported that the light intensity in the dairy cases ranged from 10 to 180 foot-candles. Previous studies by Reif and co-workers (1983) did not detect any differences in the riboflavin content of milk purchased in the retail market in California irrespective of whether packaged in paper or plastic containers. These researchers did find, however, that those milk samples packaged in plastic containers were criticized for light-induced flavor over ten times more frequently compared to milk packaged in paper containers. No information was presented, however, on either the intensity of the fluorescent light or the previous history of the milk. The rate of riboflavin destruction has been shown previously to be affected by the intensity of the incident light. Singh and co-workers (1975) reported a threefold increase in the first-order rate constant for riboflavin destruction in milk when exposed to 450 foot-candles at 4°C in glass containers compared to in milk exposed to 150 foot-candles. Studies by Gaylord *et al.* (1986) found riboflavin destruction in skim milk to be far greater at higher light intensities as shown in Figure 10.8. Their results together with those by Olsen and Ashoor (1987) suggest that the light-induced flavor problem could be minimized or eliminated by regulating the intensity of fluorescent light in dairy display cases. The use of yellow fluorescent lamps with yellow shields was shown by several researchers to restrict maximum emission between 400 and 500 nm (Bradley, 1980; Hansen *et al.*, 1975). Under these conditions the lighting system emitted very little energy at wavelengths below 540 nm and reduced the incidence of light-induced off-flavors in the milk.

Bibliography

Adams, D. M., Barach, J. T., and Speck, M. L. (1976). Effect of psychotrophic bacteria from raw milk on milk proteins and stability of milk proteins to UHT-treatment. *J. Dairy Sci.* **59,** 823.

Alford, A. J., Pierce, D. A., and Suggs, F. G. (1964). Activity of microbial lipases on natural fats and synthetic triglycerides. *J. Lipid Res.* **5,** 390.

Allen, J. C., and Humphries, C. (1977). The oxidation of lipids by components of bovine milk-fat globule membrane. *J. Dairy Res.* **44,** 495.

Allen, J. C., and Parks, O. W. (1975). Evidence for methional in skim milk exposed to sunlight. *J. Dairy Sci.* **58,** 1609.

Allen, J. C., and Wrieden, W. L. (1982a). Influence of milk proteins on lipid oxidation in aqueous emulsion. I. Casein, whey protein and α-lactalbumin. *J. Dairy Res.* **49,** 239.

Allen, J. C., and Wrieden, W. L. (1982b). Influence of milk proteins on lipid oxidation in aqueous emulsion. II. Lactoperoxidase, lactoferrin, superoxide dismutase and xanthine oxidase. *J. Dairy Sci.* **49,** 249.

American Public Health Association (1978). "Standard Methods for the Examination of Dairy Products," 14th ed., APHA, New York.
Anderson, M. (1982). Factors affecting the distribution of lipoprotein lipase activity between serum and casein micelles in bovine milk. *J. Dairy Res.* **49,** 51.
Aurand, L. W., Singleton, J. A., and Matrone, G. J. (1964). Sunlight flavor in milk. II. Complex formation between milk proteins and riboflavin. *J. Dairy Sci.* **47,** 827.
Aurand, L. W., Singleton, J. A., and Noble, B. W. (1966). Photooxidation reactions in milk. *J. Dairy Sci.* **49,** 138.
Aurand, L. W., Boone, N. H., and Giddings, G. G. (1977). Superoxide and singlet oxygen in milk lipid peroxidation. *J. Dairy Sci.* **60,** 363.
Ballou, D., Palmer, G., and Massey, V. (1969). Direct demonstration of superoxide anion production during the oxidation of reduced flavin of its catalytic decomposition by erythro-caprein. *Biochem. Biophys. Res. Commun.* **36,** 898.
Bandler, D. K., Brown, R. O., and Wolff, E. T. (1975). Milk quality in the New York public school system. *J. Milk Food Technol.* **38,** 223
Barnard, S. E. (1974). Flavor and shelf life of fluid milk. *J. Milk Food Technol.* **37,** 346.
Bassette, R. (1976). Effects of light on concentrations of some volatile materials in milk. *J. Milk Food Technol.* **39,** 10.
Bassette, R., Fung, D. Y. C., and Mantha, V. R. (1986). Off-flavors in milk. *CRC CRit. Rev. Food Sci. Nutr.* **24,** 1.
Bengtsson, G., and Olivecrona, T. (1982). Activation of lipoprotein lipase by apoprotein C. II. Demonstration of an effect of the activator on the binding of the enzyme to milk fat globules. *FEBS Lett.* **147,** 183.
Benzonana, G., and Desnuelle, P. (1965). Etude cinétique de la lipase pancréatique sur des triglycerides en emulsion. Essai d'une enzymologie en milieu hétérogène. *Biochim. Biophys. Acta* **105,** 121.
Bergel, F., and Bray, R. C. (1959). The chemistry of xanthine oxidase. 4. The problems of enzyme inactivation and stabilization. *Biochem. J.* **73,** 182.
Bhadsavle, C. H., Shehata, T. E., and Collins, E. B. (1972). Isolation and identification of psychrophilic species of *Clostridium* from milk. *Appl. Microbiol.* **24,** 699.
Bhavadasan, M. K., Abraham, M. J., and Ganguli, N. C. (1982). Influence of agitation on milk lipolysis and release of membrane-bound xanthine oxidase. *J. Dairy Sci.* **65,** 1692.
Bors, W., Michel, C., and Saran, M. (1979). Superoxide anions do not react with hydroperoxides. *FEBS Lett.* **107,** 403.
Boveris, A., Cadenas, E., and Chance, B. (1981). Ultraweak chemiluminescence: A sensitive assay for oxidative radical reactions. *Fed. Proc., Fed. Am. Soc. Exp. Biol.* **40,** 195.
Bradfield, A., and Duthie, A. H. (1965). Protecting milk from fluorescent light. *Am. Dairy Rev.* **27,** 110.
Bradley, R. L. (1980). Effect of light on alteration of nutritional value and flavor of milk: A review. *J. Food Prot.* **43,** 314.
Bray, S. L., Duthie, A. H., and Rogers, R. P. (1977). Consumers can detect light-induced flavor in milk. *J. Food Prot.* **40,** 586.
Briley, S., and Eisenthal, R. (1975). Association of xanthine oxidase with the bovine milkfat-globule membrane. Nature of the enzyme-membrane association. *Biochem. J.* **147,** 417.
Castberg, H. B., and Solberg, P. (1974). The lipoprotein lipase and the lipolysis in bovine milk. *Meieriposten* **51–52,** 961.
Christie, W. (1974). Biosynthesis of triglycerides in freshly secreted milk from goats. *Lipids* **9,** 876.
Christie, W., and Wooding, F. B. P. (1975). Site of triglyceride biosynthesis in milk. *Experientia* **31,** 1445.

Christy, G. E., Amantea, G. F., and Irwin, R. E. T. (1981). Evaluation of effectiveness of polyethylene overwraps in preventing light-induced oxidation of milk in pouches. *Can. Inst. Food Sci. Technol. J.* **14,** 135.

Collins, E. B. (1981). Heat resistant psychotrophic microorganisms. *J. Dairy Sci.* **64,** 157.

Deeth, H. C., and FitzGerald, C. H. (1975). Factors governing the susceptibility of milk to spontaneous lipolysis. *Proc. Lipolysis Symp., Int. Dairy Fed., Doc.* **86,** 24–34.

Deger, D., and Ashoor, S. H. (1987). Light-induced changes in taste, appearance, odor, and riboflavin content of cheese. *J. Dairy Sci.* **70,** 1371.

deMan, J. M. (1978). Possibilities of prevention of light induced quality loss of milk. *Can. Inst. Food Sci. Technol. J.* **11,** 152.

Desnuelle, P. (1961). Pancreatic lipase. *Adv. Enzymol.* **23,** 129.

Dillman, W. F., Anderson, E. O., and Hankin, L. (1971). An assessment of the flavour quality of whole milk available at commercial outlets. *J. Milk Food Technol.* **34,** 244.

Dimick, P. S. (1973). Effect of fluorescent light on the flavor and select nutrients of homogenized milk held in conventional containers. *J. Milk Food Technol.* **36,** 383.

Dimick, P. S. (1982). Photochemical effects on flavor and nutrients of fluid milk. *Can. Inst. Food Sci. Technol. J.* **15,** 247.

Downey, W. K. (1974a). Enzymic aspects of hydrolytic rancidity. *Int. Dairy Cong., 19th, 1974* Vol. 1E, p. 351.

Downey, W. K. (1974b). Enzyme systems influencing processing and storage of milk and milk products. *Int. Dairy Congr., 19th, 1974* Vol. II, p. 323.

Downey, W. K. (1980). Risks from pre- and post-manufacture lipolysis. *In* "Flavour Impairment of Milk and Milk Products due to Lipolysis," Doc. No. 118, pp. 4–19. Int. Dairy Fed., Brussels, Belgium.

Downey, W. K., and Andrews, P. (1966). Studies on the properties of cow's-milk tributyrinases and their interaction with milk proteins. *Biochem. J.* **101,** 651.

Downey, W. K., and Murphy, R. F. (1975). Partitioning of the lipolytic enzymes in bovine milk. *Proc. Lipolysis Symp., Int. Dairy Fed., Doc.* **86,** 19.

Dunckley, W. L., Franklin, J. D., and Pangborn, R, M. (1962). Effects of fluorescent light on flavor, ascorbic acid and riboflavin in milk. *Food Technol.* **16,** 112.

Egelrud, T., and Olivecrona, T. (1972a). A comparative study of lipase in bovine colostrum and in bovine milk. *Neth. Milk Dairy J.* **30,** 186.

Egelrud, T., and Olivecrona, T. (1972b). The purification of a lipoprotein lipase from bovine skim milk. *J. Biol. Chem.* **247,** 6212.

Egelrud, T., and Olivecrona, T. (1973). Purified bovine milk lipoprotein lipase: Activity against lipid substrates in the absence of exogenous serum factors. *Biochim. Biophys. Acta* **306,** 115.

Fanelli, A. J., Burlew, J. V., and Gabriel, M. K. (1985). Protection of milk packaged in high density polyethylene against photodegradation by fluorescent light. *J. Food Prot.* **48,** 112.

Foote, C. S., Shook, F. C., and Abakerli, R. A. (1980). Chemistry of superoxide ion. 4. Singlet oxygen is not a major product of dismutation. *J. Am. Oil Chem. Soc.* **102,** 2503.

Fridovich, S. E., and Porter, N. A. (1981). Oxidation of arachidonic acid in micelles by superoxide and hydrogen peroxide. *J. Biol. Chem.* **256,** 260.

Gaylord, A. M., Warthesen, J. J., and Smith, D. E. (1986). Influence of milk fat, milk solids and light intensity on vitamin A and riboflavin in low fat milk. *J. Dairy Sci.* **69,** 2779.

Hankin, L., and Dillman, W. F. (1972). Further studies on the flavor quality of retail milk in Connecticut. *J. Milk Food Technol.* **35,** 710.

Hansen, A. P., Turner, L. G., and Aurand, L. W. (1975). Fluorescent light-activated flavor in milk. *J. Milk Food Technol.* **38,** 388.

Harper, W. J., and Gould, I. A. (1959). Some factors affecting the heat-inactivation of the milk lipase enzyme system. *Int. Dairy Congr., Proc., 15th, 1959* Vol. 1, p. 455.

Hasegawa, K., and Peterson, L. K. (1978). Pulse radiolysis studies in lipid model systems: Formation and behaviour of peroxy radicals in fatty acids. *Photochem. Photobiol.* **28,** 817.
Herrington, B. L. (1954). Lipase: A review. *J. Dairy Sci.* **37,** 775.
Hicks, C. L., Korycka-Dahl, M., and Richardson, T. (1975). Superoxide dismutase in bovine milk. *J. Dairy Sci.* **58,** 796.
Hicks, C. L., Bucy, L., and Stofer, W. (1979). Heat inactivation of superoxide dismutase in bovine milk. *J. Dairy Sci.* **62,** 529.
Hill, R. D. (1975). Superoxide dismutase activity in bovine milk. *Aust. J. Dairy Technol.* **30,** 26.
Hill, R. D. (1979). Oxidative enzymes and oxidative processes in milk. *CSIRO Food Res. Q.* **39,** 33.
Hill, R. D., Van Leeuwen, V. V., and Wilkinson, R. A. (1977). Some factors influencing the autoxidation of milks rich in linoleic acid. *N. Z. Dairy Sci. Technol.* **12,** 69.
Hohe, K. A., Dimick, P. S., and Kilara, A. (1985). Milk lipoprotein lipase distribution in the major fractions of bovine milk. *J. Dairy Sci.* **68,** 1067.
Hoskin, J. C. (1979). Sensory evaluation and riboflavin analysis of milk held in light-exposed one-gallon containers. MS Thesis. The Pennsylvania State University, University Park, Pennsylvania.
Hoskin, J. C. (1988). Effect of fluorescent light on flavor and riboflavin content of milk held in modified half-gallon containers. *J. Food Prot.* **51,** 19.
Hoskin, J. C., and Dimick, P. S. (1979). Evaluation of fluorescent light on flavor and riboflavin content of milk held in gallon returnable containers. *J. Food Prot.* **42,** 105.
Hosono, A., and Elliot, J. A. (1974). Properties of crude ethylester-forming enzyme preparations from some lactic acid and psychrotrophic bacteria. *J. Dairy Sci.* **57,** 1432.
Hosono, A., Elliot, J. A., and McGugan, W. A. (1974). Production of ethyl esters by some lactic acid and psychotrophic bacteria. *J. Dairy Sci.* **57,** 535.
Hoynes, M. C. T., and Downey, W. K. (1973). Relationship of the lipase and lipoprotein lipase activities of bovine milk. *Biochem. Soc. Trans.* **1,** 256.
Hussong, R. V. (1932). Ph.D. Thesis, Iowa State College, Ames.
International Dairy Federation (1970). "Annual Bulletin. Part III. Influence of Packaging Materials on Stability and Organoleptic Quality of Milk." IDF, Brussels, Belgium.
International Dairy Federation (1974). "Lipolysis in Cooled Bulk Milk," Doc. No. 82. IDF, Brussels, Belgium.
International Dairy Federation (1975). "Proceeding of Lipolysis Symposium, Cork, Ireland," Doc. No. 86. IDF, Brussels, Belgium.
International Dairy Federation (1976). "Technical Guide for the Packaging of Milk and Milk Products," Doc. No. 92. IDF, Brussels, Belgium.
International Dairy Federation (1980). "Flavour Impairment of Milk and Milk Products due to Lipolysis," Doc. No. 118. IDF, Brussels, Belgium.
IUPAC/IUCB Commission on Biochemical Nomenclature. (1967). The nomenclature of lipids. *Biochem. J.* **105,** 897.
Jackson, H. W., and Morgan, M. E. (1954). Identity and origin of the malty aroma from milk culture of *Streptococcus lactis* var. *maltigenes*. *J. Dairy Sci.* **37,** 1316.
Jellema, A. (1975). Note on susceptibility of bovine milk to lipolysis. *Neth. Milk Dairy J.* **29,** 145.
Jensen, R. G. (1964). Lipolysis. *J. Dairy Sci.* **47,** 210.
Jensen, R. G., Smith, A. C., McLeod, R., and Dowd, L. (1957). The acid degree of milk obtained with pipeline and bucket milkers. *J. Milk Food Technol.* **20,** 352.
Johnson, P. E., and von Gunter, R. L. (1962). A study of factors involved in the development of rancid flavor in milk. *Bull.—Okla. Agric. Exp. Stn.* **B-593.**
Kellog, E. W., and Fridovich, I. (1975). Superoxide, hydrogen peroxide, and singlet oxygen in lipid peroxidation by xanthine oxidase system. *J. Biol. Chem.* **250,** 8812.

King, R. L., and Dunkley, W. L. (1959). Relation of natural copper in milk to incidences of spontaneous oxidized flavor. *J. Dairy Sci.* **42,** 420.

Kirst, E. (1986). Lipolytic changes in the milk fat of raw milk and their effects on the quality of milk products. *Food Microstruc.* **5,** 265.

Korycka-Dahl, M., and Richardson, T. (1978). Photogeneration of superoxide anion in serum of bovine milk and in model systems containing riboflavin and amino acids. *J. Dairy Sci.* **61,** 400.

Korycka-Dahl, M., and Richardson, T. (1980). Initiation of oxidative changes in foods. *J. Dairy Sci.* **63,** 1181.

Korycka-Dahl, M., Richardson, T., and Hicks, C. L. (1979). Superoxide dismutase activity in bovine milk serum and its relevance for oxidative stability of milk. *J. Dairy Sci.* **62,** 183.

Law, B. A., Andrews, A. T., and Sharpe, M. E. (1977). Gelation of UHT-sterilized milk by protease from a strain of *Pseudomonas fluorescens* isolated from raw milk. *J. Dairy Res.* **44,** 145.

Levine, M., and Anderson, D. Q. (1932). Two new species of bacteria causing mustiness in eggs. *J. Bacteriol.* **23,** 343.

Liesman, J., Emery, R. S., and Daitch, J. (1977). Different activation of lipoprotein lipase by different sera. *Fed. Proc., Fed. Am. Soc. Exp. Biol.* **36,** 1114.

Lim, C. T., and Scanu, A. M. (1976). Apoproteins of bovine serum high density lipoproteins: Isolation and characterization of the small-molecular-weight components. *Artery (Fulton, Mich.)* **2,** 483.

Lynch, B. A., and Fridovich, I. (1979). Autoinactivation of xanthine oxidase. The role of superoxide radical and hydrogen peroxide. *Biochim. Biophys. Acta* **571,** 195.

Mabbitt, L. A. (1981). Metabolic activity of bacteria in raw milk. *Kiel. Milchwirtsch. Forschungsber.* **33,** 273.

MacLeod, H. P., and Morgan, M. E. (1955). Leucine metabolism of *Streptococcus lactis* var. *maltigenes*. 1. Conversion of alpha-ketoisocaproic acid to leucine and 3-methyl butanal. *J. Dairy Sci.* **38,** 1208.

MacLeod, H. P., and Morgan, M. E. (1956). Leucine metabolism of *Streptococcus lactis* var. *maltigenes* conversion of alpha keto isocaproic acid to leucine and 3–methyl butanal. *J. Dairy Sci.* **37,** 1316.

Martinez, C. E., Mendoza, P. G., Alarcon, F. J., and Garcia, H. S. (1988). Reactivation of the lactoperoxidase system during raw milk storage and its effect on the characteristics of pasteurized milk. *J. Food Prot.* **51,** 558.

McCarthy, R. D., and Patton, S. (1964). Biosynthesis of glycerides in freshly secreted milk. *Nature (London)* **202,** 347.

Mencher, J. R., and Alford, J. A. (1967). Purification and characterization of the lipase of *Pseudomonas fragi*. *J. Gen. Microbiol.* **48,** 317.

Mendelssohn, C. R., Zinder, O., Blanchette-Mackie, E., Chernick, S., and Scow, R. O. (1977). Lipoprotein lipase and lipid metabolism in mammary gland. *J. Dairy Sci.* **60,** 666.

Milk Industry Foundation (1977). "Milk Facts." Milk Ind. Found., Washington, D. C.

Miller, A., III, Morgan, M. E., and Libbey, L. M. (1974). *Lactobacillus maltaromicus,* a new species producing a malty aroma. *Int. J. Syst. Bacteriol.* **24,** 346.

Monika, J., Schroder, A., Cousins, C. M., and McKinnon, C. H. (1982). Effect of psychotrophic post-pasteurization contamination on the keeping quality at 11 and 5°C of HTST pasteurized milk in the UK. *J. Dairy Res.* **49,** 619.

Morgan, M. E. (1976). The chemistry of some microbially induced flavor defects in milk and dairy foods. *Biotechnol. Bioeng.* **18,** 953.

Morgan, M. E., Libbey, L. M., and Scanlan, R. A. (1972). Identity of the musty potatoe aroma compound in milk cultures of *Pseudomonas faetrolleus*. *J. Dairy Sci.* **55,** 666.

Morley, N. H., Kuksis, A., and Buchnea, D. (1974). Hydrolysis of synthetic triacylglycerols by pancreatic and lipoprotein lipase. *Lipids* **9,** 481.

Murray, K. E., and Whitfield, F. B. (1975). The occurrence of 3–alkyl-2-methoxy pyrazines in raw vegetables. *J. Sci. Food Agric.* **26,** 973.

Murray, K. E., Shipton, J., and Whitfield, F. B. (1970). 2-Methoxypyrazines and the flavour of green peas (*Pisum sativuum*). *Chem. Ind. (London)* **27,** 897.

Nanni, E. J., Jr., Birge, R. R., Hubbard, L. M., Morrison, M. M., and Sawyer, D. T. (1981). Oxidation and dismutation of superoxide ion solutions to molecular oxygen. Singlet *vs* triplet state. *Inorg. Chem.* **20,** 737.

Nilsson, R., and Willart, S. (1960). Lipolytic activity in milk. 1. The influence of homogenization on the fat splitting of milk. *Milk Dairy Res., Rep.* **60.**

Nilsson, R., and Willart, S. (1961). Lipolytic activity in milk. II. The heat inactivation of the fat-splitting in milk. *Milk Dairy Res., Rep.* **64.**

Nilsson-Ehle, P., Egelrud, T., Belfrage, P., Olivecrona, T., and Borgstrom, B. (1973). Positional specificity of purified milk lipoprotein lipase. *J. Biol. Chem.* **248,** 6734.

Olivecrona, T. (1980). Biochemical aspects of lipolysis in bovine milk. *In* "Flavour Impairment of Milk and Milk Products due to Lipolysis," Doc. No. 118, p. 19. Int. Dairy Fed., Brussels, Belgium.

Olivecrona, T., and Bengtsson, G. (1984). Lipases in milk. *In* "Lipases" (B. Borgstrom, and H. L. Brockman, eds.). Elsevier, Amsterdam.

Olsen, J. R., and Ashoor, S. H. (1987). An assessment of light induced off-flavors in retail milk. *J. Dairy Sci.* **70,** 1362.

Olson, H. C., and Hammer, B. W. (1932). Organisms producing a potato odor in milk. *Iowa State Coll. J. Sci.* **9,** 125.

Patel, C. V., Fox, P. F., and Tarassuk, N. P. (1968). Bovine milk lipase. 11. Characterization. *J. Dairy Sci.* **51,** 1879.

Patton, T. A. J. (1954). The mechanism of sunlight flavor formation with special reference to methionine and riboflavin. *J. Dairy Sci.* **37,** 446.

Pillay, V. T., Myhr, A. N., and Gray, J. I. (1980). Lipolysis in milk. 1. Determination of free fatty acid and threshold value for lipolyzed flavor detection. *J. Dairy Sci.* **63,** 1213.

Reddy, M. C., Lindsay, R. C., and Montgomery, M. W. (1971). Ester production by *Pseudomonas fragi.* IV. *Int. Dairy Congr., Congr. Pap., 18th, 1970* Vol. E, p. 495.

Reif, G. D., Franke, A. A., and Bruhn, J. C. (1983). Retail dairy foods quality—An assessment of the incidence of off-flavors in California milk. *Dairy Food Sanit.* **3,** 44.

Reiter, B., and Harnulv, G. (1982). The preservation of refrigerated and uncooled milk by its natural lactoperoxidase system. *Dairy Ind. Int.* **47,** 13.

Richardson, T., and Korycka-Dahl, M. (1983). Lipid oxidation. Chapter 7. *Dev. Dairy Chem.* **2,** 241.

Salih, A. M. A. (1978). Factors affecting lipolytic activity in cow's milk. Ph.D. thesis, University of Reading, Berkshire, U.K.

Sattar, A., and deMan, J. M. (1973). Effect of packaging material on light-induced quality deterioration of milk. *Can. Inst. Food Sci. Technol. J.* **6,** 170.

Sattar, A., and deMan, J. M. (1975). Photooxidation of milk and milk products: A review. *CRC Crit. Rev. Food Sci. Nutr.* **7,** 13.

Scott, T. W., Cook, L. J., Ferguson, K. A., McDonald, I. W., Buchanan, R. A., and Lofthus Hills, G. (1970). Production of polyunsaturated milk fat in domestic ruminants. *Aust. J. Sci.* **32,** 291.

Scow, R. D., and Egelrud, T. (1976). Hydrolysis of chylomicron phosphatidylcholine *in vitro* by lipoprotein lipase, phospholipase A and phospholipase C. *Biochim. Biophys. Acta* **431,** 538.

Seifert, R. M., Buttery, D. G., Guadagni, D. G., Block, D. R., and Harris, J. G. (1970). Synthesis of some 2-methoxy-3-alkyl pyrazines with strong bell pepper-like odors. *J. Agric. Food Chem.* **18,** 246.

Shipe, W. F. (1964). Oxidation in the dark. *J. Dairy Sci.* **47,** 221.

Shipe, W. F., and Senyk, G. F. (1981). Effects of processing conditions on lipolysis in milk. *J. Dairy Sci.* **64,** 2146.

Shipe, W. F., Bassette, R., Deane, D. D., Dunkley, W. L., Hammond, E. G., Harper, W. J., Kleyn, D. H., Morgan, M. E., Nelson, J. H., and Scanlan, R. A. (1978). Off flavors of milk: Nomenclature, standards, and bibliography. *J. Dairy Sci.* **61,** 855.

Shipe, W. F., Senyk, G. E., Ledford, R. A., Bandler, D. K., and Wolff, E. T. (1980). Flavor and chemical evaluations of fresh and aged market milk. *J. Dairy Sci.* **63,** Suppl. 1, 43.

Shirley, J. E., Emery, R. S., Convey, E. M., and Oxender, W. D. (1973). Enzymic changes in bovine adipose and mammary tissue. Serum and mammary tissue hormonal changes with initiation of location. *J. Dairy Sci.* **56,** 569.

Sidhu, G. S., Brown, M. A., and Johnson, A. R. (1975). Autoxidation in milk rich in linoleic acid. 1. An objective method for measuring autoxidation and evaluating antioxidants. *J. Dairy Res.* **42,** 185.

Sidhu, G. S., Brown, M. A., and Johnson, A. R. (1976). Autoxidation in milk rich in linoleic acid. II. Modification of the initiation system and control of oxidation. *J. Dairy Res.* **43,** 239.

Singh, R. P., Heldman, D. R., and Kirk, J. R. (1975). Kinetic analysis of light induced riboflavin loss in whole milk. *J. Food Sci.* **40,** 164.

Skean, J. D., and Overcast, W. W. (1960). Changes in the paper electrophoretic protein patterns of refrigerated skim milk accompanying growth of three *Pseudomonas* species. *Appl. Microbiol.* **8,** 335.

Sundheim, G. (1988). Spontaneous lipolysis in bovine milk: Combined effects of cream, skim milk and lipoprotein lipase activity. *J. Dairy Sci.* **71,** 620.

Sundheim, G., Zimmer, T. L., and Astrup, H. N. (1983). Induction of milk lipolysis by lipoprotein components of bovine blood serum. *J. Dairy Sci.* **66,** 400.

Sundheim, G., and Bengtsson-Olivecrona, G. (1985). Lipolysis in milk induced by cooling or by heparin: Comparisons of amounts of lipoprotein lipase in the cream fraction and degree of lipolysis. *J. Dairy Sci.* **68,** 589.

Sundheim, G., and Bengtsson-Olivecrona, G. (1986). Iodine-125–labelled lipoprotein lipase as a tool to detect and study spontaneous lipolysis in bovine milk. *J. Dairy Sci.* **69,** 1776.

Sundheim, G., and Bengtsson-Olivecrona, G. (1987). Isolated milk fat globules as substrates for lipoprotein lipase: Study of factors relevant to spontaneous lipolysis in milk. *J. Dairy Sci.* **70,** 499.

Thomas, E. L. (1956). The problem of off-flavored milk in the bulk tank era. *Proc. 49th Annu. Conf. Milk Ind. Found., Milk Supplies Sect.,* p. 11.

Thomas, E. L. (1981). Trends in milk flavors. *J. Dairy Sci.* **64,** 1023.

Thomas, M. J., Mehl, K. S., and Pryor, W. A. (1978). The role of the superoxide anion in the xanthine oxidase-induced autoxidation of linoleic acid. *Biochem. Biophys. Res. Commun.* **83,** 927.

Thomas, S. B., and Druce, R. G. (1969). Psychotrophic bacteria in refrigerated pasteurized milk. A review. *Dairy Ind.* **34,** 430.

Thompson, J. N., and Erdody, P. (1974). Destruction by light of vitamin A in milk. *Can. Inst. Food Sci. Technol. J.* **7,** 157.

Tolle, A., and Heeschen, W. (1975). Free fatty acids in milk in relation to flow conditions in milking plants. *Proc. Lipolysis Symp. Int. Dairy Fed., Doc.* **86,** 136.

Tucker, J. S., and Morgan, M. E. (1967). Decarboxylation of alpha-keto acids by *Streptococcus lactis* var. *maltigenes. Appl. Microbiol.* **15,** 694.

Velander, H. J., and Patton, S. (1955). Prevention of sunlight flavor in milk removal of riboflavin. *J. Dairy Sci.* **38,** 376.

Wainess, H. (1981). Factors affecting the keeping quality of heat treated milk. Shelf life and environmental aspects. *Bull.—Int. Dairy Fed.* **130,** 71.

Wishner, L. A., and Keeney, M. (1963). Carbonyl pattern of sunlight exposed milk. *J. Dairy Sci.* **46,** 785.

Zajak, M., Gladys, M., Skarzinska, M., Harnülv, G., and Bjorck, L. (1983). Changes in bacteriological quality of raw milk stabilized by activation of its lactoperoxidase system and stored at different temperatures. *J. Food Prot.* **46,** 1065.

Part IV

Biotechnology

11

Biotechnology: Enzymes in the Food Industry

I. Introduction

Living organisms carry out their life processes within the physical limits of temperature, pH, and pressure. In the case of plants these changes are primarily endothermic, in which kinetic energy from the sun is converted into potential energy, while those in animals are exothermic, in which potential energy is liberated as kinetic energy in the form of movement and work. Thus green plants carry out photosynthesis, whereby energy is stored in complex products which can then be utilized by animals. These chemical changes are taking place continuously in nature under somewhat mild conditions as living cells cannot tolerate high temperatures or extreme pH conditions. Such changes involve a myriad of biochemical reactions which are brought about by biological catalysts or enzymes, defined as "proteins with catalytic properties due to their powers of specific activation."

Over the past decade research on enzymes has focused on isoenzymes (Latner and Skillen, 1968; Brewer, 1970), immobilization (Pitcher, 1980; Zaborsky, 1973), allosteric properties (Monod *et al.*, 1963, 1965; Koshland *et al.*, 1966; Stadtman, 1966, Sanwal, 1970), mechanisms of enzyme action (Storm and Koshland, 1970; Port and Richards, 1971), and the chemical synthesis of enzymes (Gutte and Merrifield, 1969; Denkswater *et al.*, 1969). Exciting developments are taking place in the microbial production of enzymes, with the new emphasis on biotechnology (Kirsop, 1981). The application of genetic engineering, particularly recombinant DNA technology, is having a major impact on the development of new sources of industrial enzymes for the food industry (Lawrence, 1987). This chapter will focus on those enzymes presently used in the production of foods or food-related products.

II. Historical Highlights

In 1915 Zemplen postulated the colloidal nature of enzymes produced from living cells. The first pure crystalline enzyme, urease, was isolated by Sumner in 1926 from jack beans (*Canavalia ureiformis*). This enzyme was found to be a protein which was catalytically active. Subsequent work by Northrop in 1930 provided a pure crystalline form of pepsin which has since been followed by the isolation and purification of many enzymes from animal, plant, and microbial sources. It soon became necessary to develop a system for the classification of these enzymes. This lead to the Commission on Enzymes by the International Union of Biochemistry, which convened over 1956–1961 to classify enzymes.

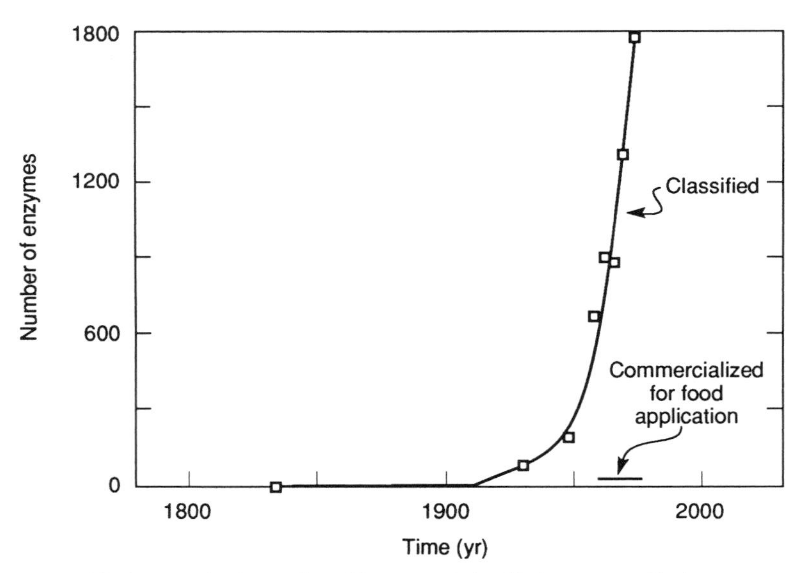

FIG. 11.1. The increase in the number of known and classified enzymes with time, compared to their commercialization in food systems (Beck and Scott, 1974).

Thus all known enzymes were renamed and classified systematically according to the reactions they catalyzed (Anonymous, 1961). There have been a number of revisions of the enzyme classification table, which now includes over two thousand enzymes (Enzyme Nomenclature, 1976). Of these enzymes a very small number are used on a commercial scale in the food industry, only three dozen enzymes in total (Fig. 11.1) (Beck and Scott, 1974). This number has not increased substantially over the past sixteen years although significant progress has been made in their production from microbial sources.

III. Industrial Enzymes and Their Applications

Enzymes are particularly suited to the food industry for several reasons. They occur naturally in biological material, are nontoxic, and have specific action. Control of enzymatic reactions is achieved quite easily by adjusting the temperature, pH, or the amount of the enzyme itself. Individual enzymes in their varied roles in the food industry are listed in Table 11.1. The majority of these enzymes are hydrolases, which indicates the importance that hydrolytic changes have in the production of foods.

IV. Carbohydrases

Carbohydrases hydrolytically split glycosidic linkages and are therefore concerned with the hydrolysis of polysaccharides and oligosaccharides. The amylases, in particular, are used widely in the food industry, although the other members within this group, pectic enzymes, cellulase, lactase, invertase, and hemicellulase, all have important functions in the commercial production of foods.

A. Amylases

The substrate for amylases is starch, a polysaccharide consisting of two fractions, amylose and amylopectin. Amylose is a straight-chain polysaccharide in which glucose units are joined by α-1, 4–glycosidic bonds, and the chain length can be up to 350 units long. Amylopectin, on the other hand, has branching points joining other linear chains through α-1, 6-glycosidic linkages, resulting in much shorter linear regions of approximately 30 glucose units.

TABLE 11.1

CLASSIFICATION OF ENZYMES SIGNIFICANT IN FOOD AND IN THE FOOD INDUSTRY

Trivial name(s)	Systematic name	Enzyme Commission number	Reaction (as significant in food material)
Oxidoreductases			
Glucose oxidase	β-D-Glucose : O_2 oxidoreductase	1.1.3.4	β-D-Glucose + $O_2 \rightarrow$ D-glucono-δ-lactone + H_2O_2
Phenolase (polyphenol oxidase)	*o*-Diphenol : O_2 oxidoreductase	1.10.3.1	2 *o*-Diphenol + $O_2 \rightarrow$ 2 *o*-quinone + 2 H_2O
Ascorbic acid oxidase	L-Ascorbate : O_2 oxidoreductase	1.10.3.3	2 L-ascorbate + $O_2 \rightarrow$ 2 dehydroascorbate + 2 H_2O
Catalase	H_2O_2 : H_2O_2 oxidoreductase	1.11.1.6	$H_2O_2 + H_2O_2 \rightarrow O_2 + 2\ H_2O$
Peroxidase	Donor : H_2O_2 oxidoreductase	1.11.1.7	Donor + $H_2O_2 \rightarrow$ oxidized donor + 2 H_2O
Lipoxygenase	Linoleate : oxygen oxidoreductase	1.13.11.12	Unsaturated fat + $O_2 \rightarrow$ a peroxide of the unsaturated fat
Hydrolases			
Lipase	Glycerol ester hydrolase	3.1.1.3	Triglyceride + $H_2O \rightarrow$ glycerol + fatty acids
Pectin methylesterase	Pectin pectyl-hydrolase	3.1.1.11	Pectin + *n* $H_2O \rightarrow$ pectic acid + *n* MeOH
Chlorophyllase	Chlorophyll chlorophyllido-hydrolase	3.1.1.14	Chlorophyll + $H_2O \rightarrow$ phytol + chlorophyllide
Phosphatase (acid or alkaline)	Orthophosphoric monoester phos-phohydrolase	3.1.3.(1,2)	An orthophosphoric monoester + $H_2O \rightarrow$ an alcohol + HPO_4
α-Amylase	α-1,4-Glucan 4-glucanohydrolase	3.2.1.1	Hydrolysis of α-1,4-glucan links: Internal random hydrolysis
β-Amylase	α-1,4-Glucan maltohydrolase	3.2.1.2	Hydrolysis of α-1,4-glucan links: Successive maltose units removed
Glucoamylase	α-1,4-Glucan 4-glucohydrolase	3.2.1.3	Hydrolysis of α-1,4-glucan links: Successive glucose units removed
Cellulase	β-1,4-Glucan 4-glucanohydrolase	3.2.1.4	Hydrolyzes β-1,4-glucan links in cellulose
Polygalacturonase	Polygalacturonide glycanohydrolase	3.2.1.15	Pectic acid + $(x - 1)$ $H_2O \rightarrow$ *x* α-D-galacturonic acid
Maltase (α-glucosidase)	α-D-Glucoside glucohydrolase	3.2.1.20	Maltose + $H_2O \rightarrow$ 2 α-D-glucose
Lactase	β-D-Galactoside galactohydrolase	3.2.1.23	Lactose + $H_2O \rightarrow$ 2 α-D-glucose + β-D-galactose
Invertase (sucrase)	β-D-Fructofuranoside fructohydrolase	3.2.1.26	Sucrose + $H_2O \rightarrow$ α-D-glucose + β-D-fructose

Pullulanase, R-enzyme, debranching enzyme, amylopectin 6-glucanohydrolase	Pullulan 6-glucanohydrolase	3.2.1.41	Hydrolyzes α-1,6-glucosidic linkages in pullulan, amylopectin and glycogen, and in the α- and β-amylase limit dextrin
Isoamylase	Glycogen 6-glucanohydrolase	3.2.1.68	Hydrolysis of D-glucosidic branch linkages in glycogen, amylopectin and β-limit dextrin
Chymotrypsin	—	3.4.21.1	Hydrolysis of peptide linkages
Trypsin	—	3.4.21.4	
Elastase	—	3.4.21.11	
Bacterial proteinase	—	3.4.21.14	
Papain	—	3.4.22.2	
Ficin	—	3.4.22.3	
Bromelain	—	3.4.22.4	
Chymopapain	—	3.4.22.6	
Pepsin	—	3.4.23.1	
Fungal proteinase	—	3.4.23.6	
Collagenase	—	3.4.24.3	
Rennin	—	3.4.23.4	
L-Amino acid acylase	*N*-Acylamino-acid amido hydrolase	3.5.1.14	DL-Acyl amino acid + $H_2O \rightleftharpoons$ L-amino acid + D-acyl amino acid
Lyases			
Pectin lyase	Poly(methoxygalacturonide)lyase	4.2.2.10	Hydrolysis of α-1-4-glycosidic bond between residues of methylgalacturonate by transelimination
Isomerases			
Glucose isomerase		5.3.1.5 or 5.3.1.9	D-Glucose $\rightleftharpoons$ D-fructose

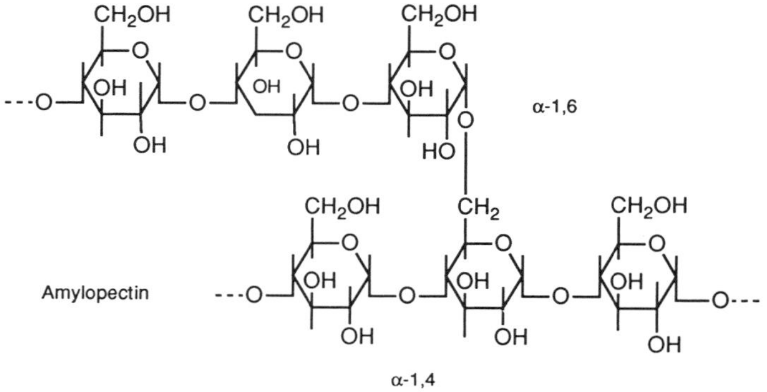

α-1,6

Amylopectin

α-1,4

α-Amylase (endo-amylase), also referred to as the liquefying or dextrinogenic amylase, hydrolyzes the splitting of α-1, 4 bonds of amylose and amylopectin in a random manner. It is unable to cleave the α-1,6 linkages in amylopectin, resulting in the production of low-molecular-weight dextrins.

β-Amylase (exo-amylase), the saccharifying or saccharogenic amylase, hydrolyzes the α-1, 4 bonds in the starch fractions by removing maltose units from the nonreducing end in an orderly fashion. It can completely hydrolyze the amylose fraction to maltose, while its activity is stopped short at the branching points in amylopectin because of the presence of α-1,6 linkages. The resistant residues that remain, high-molecular-weight dextrins, can be made accessible to further hydrolysis by β-amylase by the random action of α-amylase. The joint action of both these enzymes can bring about almost complete hydrolysis of the starch molecule.

1. *Amylases and Baking*

The need for enzyme supplementation in the baking process arose from the low level or total absence of α-amylase activity in flour obtained from mechanically harvested grain. The mechanical harvesters were particularly efficient in producing a grain cut and stored at such speeds that conditions were no longer suitable for germination to occur. Germination of wheat is essential for the development of α-amylase activity which has been reported to increase 1000-fold (Kneen and Hoch, 1945). The lack of α-amylase in such flours results in the production of baked products with inferior color, volume, and texture.

The major commercial sources of α-amylase are fungal (*Aspergillus oryzae*), bacterial (*Bacillus subtilis*), and cereal (malted wheat and barley). They all exhibit good activity over a pH range of 5.0 to 5.5 but differ markedly in their thermal stabilities. This is evident from Figure 11.2, which shows that cereal and bacterial α-amylases are thermally more stable compared to the fungal source.

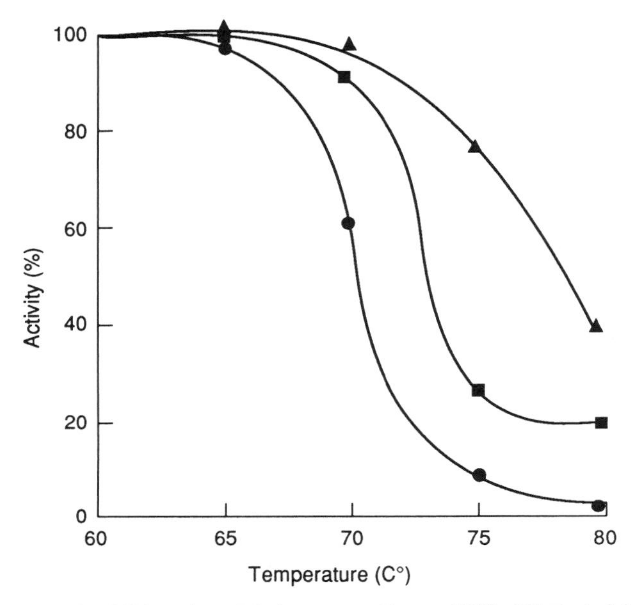

FIG. 11.2. Thermal stabilities of amylolytic enzymes (Amos, 1955). (▲) Bacterial amylase, (■) cereal amylase, (●) fungal amylase.

The first stage in the baking process immediately following selective ball milling of the flour is the fermentation period, during which α-amylase catalyzes the dextrinization of the damaged starch granules. These dextrins are then hydrolyzed by β-amylase and converted to maltose, which can then be fermented by the yeast cells. The only starch granules that are accessible to enzyme attack are those damaged during the milling process. Thus by controlling the milling process the desired percentage of damaged starch granules can be produced, which in turn affects the degree of dextrinization as shown in Table 11.2.

TABLE 11.2

EFFECT OF STARCH DAMAGE ON MALTOSE PRODUCTION[a]

Ball milling (hr)	Starch damage (%)	Maltose value (mg/10 g)
0	7.0	289
5	8.5	324
10	9.9	362
20	14.9	416

[a] From Ponte *et al.* (1961).

The action of amylases during fermentation is limited by the content of the damaged starch in the dough. During the baking period the temperature of the dough rises, however, gelatinizing the remaining undamaged starch granules and rendering them open to hydrolytic attack by amylases. For several minutes in the oven, until heat inactivation of amylases occurs, there is rapid dextrinization and saccharification of starch, which is important for final loaf quality in terms of color, volume, and texture.

The extent of amylase action and subsequent starch breakdown is directly related to the thermostability of the α-amylase supplement used. If during the baking process the activity of amylase continues, it results in liquefaction of starch and produces a wet and sticky bread crumb structure. This is unlikely to occur with the heat-labile fungal α-amylase preparation, which would be destroyed as the oven temperature rises, but could be a problem with cereal α-amylase. It thus becomes important to estimate the level of α-amylase in flour in order to assess the level of enzyme supplement needed. A number of viscometric methods have been proposed, including the use of the amylograph (Ranum *et al.*, 1978) and the falling number test (Sprossler, 1982). Perten (1984) developed a modified falling number procedure capable of monitoring both cereal and fungal α-amylase activity in flour with 50% gelatinized starch. The procedure was performed at fermentation (30°C) and oven temperatures (100°C) and the modified falling number values were highly correlated with enzyme activity for both sources of the enzyme. These enzyme supplements can be added at the mill or by the individual baker with the level dependent on the amount of fermentable sugar present. Flours containing sugar levels of 3–5% need little enzyme supplementation while flours with less than 1% fermentable sugars require such enzymes to produce the fermentable sugars essential for fermentation. Fungal α-amylase is not used at the mill in the United States but is permitted in Canada, Sweden, the United Kingdom, and some South American countries. The major source of the enzyme is *Aspergillus oryzae* (Vedernikova *et al.*, 1962).

In addition to producing fermentable sugar from damaged starch granules, α-amylase also releases water absorbed in the hydrated starch, which assists in softening the dough. In addition, the detrimental effect that damaged starch granules have on the extensibility of the dough is eliminated by the action of α-amylase. The relatively heat-stable bacterial α-amylase is used as the supplement in the dough of fruit cakes, where a moist, somewhat sticky crust is desirable.

2. *Amylases in Brewing*

The main raw ingredient in brewing beer is barley grain. This is converted into mash by a combination of steeping, germination, and kilning as shown in Scheme 11.1. During germination, hydrolytic enzymes are synthesized, particu-

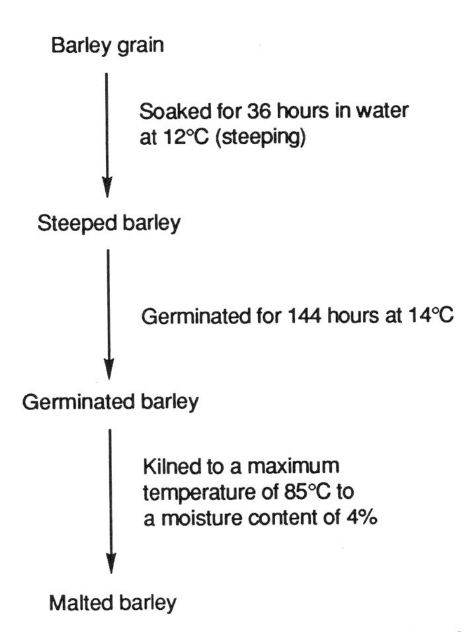

SCHEME 11.1. Outline of process involved in the production of malt barley.

larly carbohydrases and proteolytic enzymes. Unmalted barley contains considerable β-amylase activity although little activity appears until germination commences (Satyanarayana Rao and Narasimham, 1976). Thus malt provides fermentable sugar through the action of its amylolytic enzymes, α- and β-amylases, and glucoamylase for yeast fermentation in beer production.

The high cost of malting barley combined with considerable malting losses led to the replacement of the indigenous barley enzymes by microbial enzymes. Amylases and proteases from bacterial (*Bacillus subtilis*) and fungal (*Aspergillus oryzae*) sources are now used by the brewing industry. The malt, bacterial, or fungal amylases are essential for the liquefaction of starch adjuncts added during mashing. Of these sources the bacterial α-amylases have the highest optimum temperature (75°C at pH 7.0) compared to the fungal (49–54°C at pH 5.1) or malt (50–55°C at pH 4.7–5.4; 70–75°C at pH 5.6–5.8) α-amylases (Saletan, 1968). The higher inactivation temperature for bacterial α-amylase makes it very useful during the mashing process, where starch is hydrolyzed to maltose, the principal substrate for yeast fermentation. The more heat-labile fungal α-amylase is used for treating cold storage beer to reduce any residual starch.

An enzyme complex called "Brewnzyme," used in the Netherlands for beer production, contains bacterial proteases and α-amylases together with barley β-amylase for use in mashing unmalted starch materials. Obi and Odibo (1984) suggested the need to develop an enzyme complex which is derived totally from microbial sources. This required identification of a microbial β-amylase which was more heat stable than the barley enzyme. These researchers reported a β-amylase from *Thermactinomyces sp.* isolated from the soil which exhibited maximum stability over 60–70°C with an optimum temperature of 60°C. It was apparent from this study that this enzyme had considerable potential as an alternative to barley β-amylase in the brewing industry.

3. *Debranching Enzymes and Brewing*

In addition to β-amylase, the presence of the debranching enzymes amyloglucosidase (EC 3.2.1.3) and glucoamylase (EC 3.2.1.20) have been reported in *Aspergillus oryzae* (Saletan, 1968). These enzymes hydrolyze both α-l, 4 and α-l, 6 linkages in the starch molecule although the rate at which the α-l,6 linkages are cleaved is somewhat slower. The ability to hydrolyze the α-l,6 linkages makes these debranching enzymes particularly attractive to the brewing industry. Of particular interest were several other microbial debranching enzymes, pullulanase (EC 3.2.1.41) and isoamylase (EC 3.2.1.68). These enzymes were quite distinct from amyloglucosidase or glucoamylase in hydrolyzing only α-l,6 branch points. Pullulanase hydrolyzes α-D-(1 → 6) linkages in amylopectin, requiring a minimum of two D-glucosyl units in the group attached to the molecule through an α (1 → 6) bond. This differed from the action of isoamylase, in which a minimum of three D-glucosyl units were required (Abdullah *et al.,* 1966; Gunja-Smith *et al.,* 1971). Pullulanase, also referred to as R-enzyme, obtained its name from its ability to hydrolyze pullulan, a linear α-glucan composed predominantly of maltotriose units linked head to tail by α-l,6 linkages produced by *Pullularia pullulans.* The best-known source of pullulanase is *Aerobacter aerogenes,* while isoamylase is obtained from *Pseudomonas* species. Isoamylase is also a debranching enzyme but is incapable of hydrolyzing pullu-

Generalized structure of pullulan (Norman, 1982)
○ = D-glucose
↓ = 1, 6-α-linkage
— = 1, 4-α-linkage

lan. The main problem with pullulanase from *Aerobacter aerogenes* is that it is inactivated above 63°C, necessitating reduction of the temperature to 45–50°C following gelatinization of the starch granules prior to its addition (Enevoldsen, 1975). Thus identification of a microbial pullulanase with improved heat stability characteristics is needed to provide more efficient and effective debranching enzymes for use in brewing and related industries.

The critical phase in the production of beer and other alcoholic beverages is fermentation by yeast cells. To carry out this process a constant supply of fermentable sugars must be provided by treatment of starch with malt, bacterial, or fungal amylases. A more efficient and cost-saving process would be the use of yeast cells which also produce their own β-amylase and glucoamylase capable of generating fermentable sugars. Wilson and Ingledew (1982) isolated extracellular amylolytic enzymes from the yeast *Schwaniomyces alluvius* which contained both β-amylase and glucoamylase. The combined effect of these enzymes on soluble starch resulted in the production of mainly glucose and isomaltose together with traces of maltose and maltotriose. This organism is one of the few yeast species identified so far which both hydrolyzes starch as well as possesses some fermentative ability to produce alcohol.

B. Starch Processing with Enzymes

The production of glucose syrups from starch can be accomplished by hydrolysis with mineral acids. The industry prefers the more economical enzyme-catalyzed route, however, which produces specific end products (Pannell, 1970). In some processes starch may be degraded by a dual acid–enzyme conversion catalyzed system. The degree of conversion is defined by the "dextrose equivalent" (DE), which is the percentage of the theoretical maximum hydrolysis which has actually occurred (Bucke, 1982). The principal enzymes involved are α-amylase and glucoamylase. The enzyme which initiates starch hydrolysis is α-amylase, obtained from *Bacillus subtilis*. This enzyme is extremely stable at high temperatures (82.2°C), which results in the corn starch becoming hydrated and accessible to hydrolysis by amylase (MacAllister, 1979). A heat-stable bacterial α-amylase was reported by Madsen *et al.* (1973) which tolerated much higher temperatures of 10–15°C compared to other bacterial α-amylases. The enzyme, whose trade name is Thermamyl, could function at 100°C, thus eliminating the need for cooling following enzymatic liquefaction. This enzyme is now obtained from *Bacillus licheniformis* and was reported to produce higher levels of glucose (8–10%) compared to the equivalent enzyme from *Bacillus subtilis* (Barfoed, 1978).

The other enzyme used is glucoamylase, which cleaves α-1, 4-linked D-glucose units from the nonreducing end of amylopectin up to the α-1, 6 branch

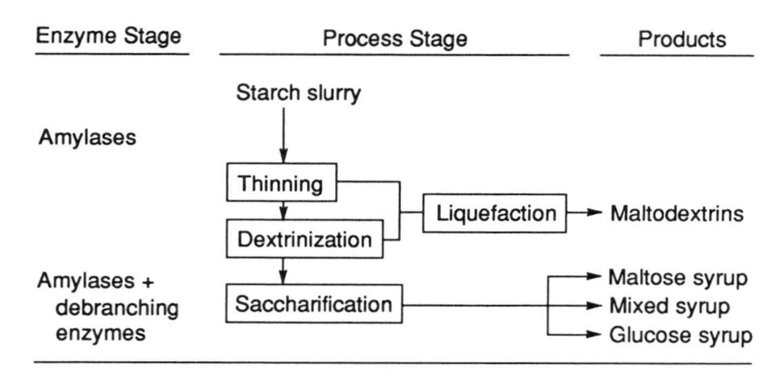

SCHEME 11.2. Enzymatic starch processing (Godfrey, 1978).

points, which it can hydrolyze but at a slower rate. Abdullah *et al.* (1966) separated glucoamylase isoenzymes from *Aspergillus niger*, one of which could effectively hydrolyze the α-1, 6 linkages of amylopectin. The process involved in the industrial degradation of starch is outlined in Scheme 11.2.

Most of the starches used in the manufacture of glucose syrups are composed of 75–85% amylopectin so that they contain 4–5% α (1 → 6) linkages. These branch points must be hydrolyzed to maximize the efficiency of the saccharifying amylases. This could be achieved through the combined action of pullulanase and isoamylase enzymes although their use in the syrup industry is limited. This is because isoamylase and pullulanase enzymes obtained from microbial sources are thermolabile and of little use at 60°C. Following an intensive screening program, however, Norman (1982) isolated a *Bacillus* species which possessed a thermostable pullulanase. This enzyme exhibited remarkable heat stability properties and retained 80% activity when incubated for up to 3 days at pH 4.24–5.2 and at 60°C. This enzyme could be used during the saccharifying stages, where temperatures of 60°C are maintained to minimize microbial contamination. A combination of pullulanase and *Aspergillus niger* glucoamylase during the saccharifying stage was recommended, which should increase the amount of D-glucose formed. The effect of different pullulanase : glucoamylase ratios required for the formation of 96% D-glucose syrup from enzyme-liquefied corn starch is shown in Figure 11.3. A glucoamylase level of 0.2 units/DS was needed for the production of 96% D-glucose, which could be halved when supplemented with 0. 15 pullulanase units/DS. Several advantages were noted from this study, including the reduction in the level of glucoamylase. This was particularly important as it reduced the polymerization of D-glucose to isomaltose by this enzyme, which would lower glucose yields. Further advantages from the use of these branching enzymes were their ability to use higher substrate concentrations and substantially reduce the saccharification time from around 80 to 30 hr.

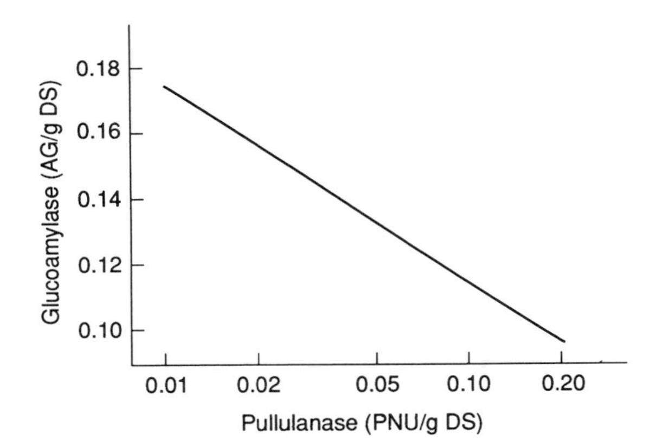

FIG. 11.3. Glucoamylase/pullulanase dosage required to obtain 96% D-glucose in 96 hr under the given conditions (substrate: DE 7 enzyme-liquefied corn starch, 30% DS, pH 4.8–4.3, 60°C, reaction time 96 hr) (Norman, 1982).

C. MALTOSE SYRUP PRODUCTION BY ENZYMES

The increased demand for maltose syrups, particularly in Japan, has led to improved methods for its production (Maeda and Tsao, 1979). The particular attraction of maltose syrups is its mild sweet flavor as well as its ability to replace acid glucose syrups in many formulations. Of particular pharmaceutical interest is the absence of blood sugar elevation by maltose, which makes it attractive for intravenous feeding. The traditional method for maltose production involves cereal β-amylase, which removes maltose units from the outer chains of starch until the α-1,6 branch point linkages are reached, leaving behind limit dextrins. The maximum yield possible for maltose from starch using this enzyme is 60%. Using debranching enzymes, however, higher maltose syrups should be attainable although the technology needs to be developed. Maeda *et al.* (1978) commented on the potential of such debranching enzymes as well as maltotriase to increase the amount of maltose. A combination of α-amylase, β-amylase, debranching enzymes, and maltotriase in various proportions could result in the production of high-maltose syrups in the foreseeable future.

D. INVERTASE

Invertase from yeast has been used for the production of cast cream centers by the confectionery industry for over 60 years (Janssen, 1960). The enzyme was first reported by Dumas and Boullay in 1828 and it catalyzes the hydrolysis of sucrose.

There are two classes of invertase, each characterized by its ability to cleave sucrose at a different point in the molecule (Myrbäck, 1960; Fukui *et al.*, 1974). One is α-D-glucoside glucohydrolase (EC 3.2.1.20) with the same group specificity as maltose, which attacks the sucrose molecule from the glucose side of the oxygen glucosidic ring (site A), while the other is β-D-fructofuranoside fructohydrolase (EC 3.2.1.26), which attacks the molecule from the fructose end (site B):

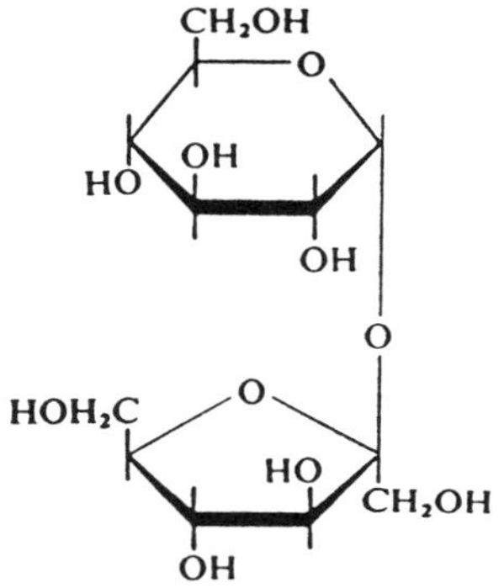

Sucrose

α-D-Glucopyranosyl-β-D-fructofuranoside

$$\underset{[\alpha]_D = +66.5^\circ}{\text{Sucrose} + H_2O} \longrightarrow \underbrace{\underset{[\alpha]_D = +52.5^\circ}{\text{D-glucose}} \quad + \quad \underset{[\alpha]_D = -92^\circ}{\text{D-fructose}}}_{[\alpha]_D = -20^\circ}$$

Yeast invertase, a fructohydrolase, also hydrolyzes raffinose, releasing fructose and mellobiose, while fungal invertase (a glucohydrolase) exerts no action on raffinose (a trisaccharide consisting of sucrose attached to galactose). The hydrolysis of sucrose to an equimolar mixture of glucose and fructose is referred to as the inversion of sucrose, and is accompanied by a change in optical rotation from positive *(dextro)* to the negative *(levo)* as a result of the highly levoratory property of fructose. The resulting glucose–fructose mixture, referred to as invert sugar, has a sweeter taste than the original sucrose. The improved sweetness of invert sugar together with its higher solubility and decreased tendency to crystallize makes it invaluable to the confectionery industry.

Invertase is available commercially from several sources, including baker's and brewer's yeast as well as molds. It is used in the production of cream and liquid centers by liquefying creams or fondant-containing materials. Creams are biphasic systems composed of a syrup (liquid) fraction and a sugar (crystalline) fraction. The concentration of syrup determines the fluidity of the cream. The main criterion in this process is to obtain a sugar level in the cream or liquid center which will not support microbial fermentation. A solid content of 79% or greater was shown by Janssen (1960) to be unable to support fermentation. The action of invertase was found to meet this requirement as it hydrolyzed sucrose in the solid phase to fructose and glucose, the latter compounds dissolving in the liquid or syrup fraction. This process continues until a solid content of approximately 82% is reached in the syrup fraction at which point the activity of invertase is terminated. Under normal casting temperature of 65°C the activity of baker's yeast (*Saccharomyces cerevisiae*) invertase is reduced by 20% after 60 min, although Janssen (1964) reported complete loss of activity at higher casting temperatures of 80°C. Woodward and Wiseman (1975) identified an invertase from *Candida utilis* which was far more stable than baker's yeast invertase at 80°C in the presence of 60% sucrose. These researchers suggested that *Candida utilis* invertase could be particularly useful at the higher casting temperature.

In the production of candies the crystalline sucrose is coated with chocolate followed by addition of invertase to the crystalline portion. The enzyme liquefies the sucrose portion, resulting in the formation of a soft center.

Invertase is also used commercially for the production of invert sugar syrup and artificial honey. Much of the recent research has focused on immobilization of invertase for the continuous hydrolysis of sucrose (Woodward and Wiseman, 1978; Yamazaki *et al.*, 1984).

E. Lactase

Lactase (β-D-galactosidase, EC 3.2.1.23) is used to reduce the lactose content of milk and dairy products. The presence of lactose is generally undesirable because of its low sweetness, limited solubility, as well as its intolerance in some individuals (Katz and Speckmann, 1978). A number of methods are available for the production of lactose-free milk, including physical removal by ultrafiltration, although the biochemical method using lactase is preferred. The main sources of lactase are neutral lactase from *Saccharomyces lactis* (Olling, 1972; Guy and Bingham, 1978; Kosikowski and Wierzbicki, 1973) and acid lactase from *Aspergillus niger* (Wierzbicki and Kosikowski, 1973a,b). Of these the *Saccharomyces* (or *Kluyveromyces*, based on the latest nomenclature) *lactis* enzyme is most widely used with a pH optimum of 6.6–6.8 compared to 3.5–4.5 for *Aspergillus niger* lactase. The enzyme hydrolyzes the β-galactosidic bond between galactose and glucose in the lactose molecule, releasing the constituent monosaccharides. The decrease in lactose results in a substantial improvement in sweetness of milk or whey as well as eliminates the grittiness in ice cream or frozen milk concentrates associated with lactose crystallization.

Lactase-deficient individuals can use milk in which over 90% of the lactose has been hydrolyzed (Payne-Bose *et al.*, 1977; Cheng *et al.*, 1979). Lactase is available in a powdered form in the United States, where it is added to the milk a day prior to its consumption (McCormick, 1976). A liquid form of the enzyme was recently introduced which is more stable and convenient to use (Kligerman, 1981). The use of lactase to modify milk in the home, however, is both expensive and time-consuming compared to the commercially produced modified milk. Dahlqvist and co-workers (1977) reported a method for reducing the lactose content of UHT sterilized milk using lactase from *Saccharomyces lactis*. Only minimal amounts of the sterilized enzyme were added to UHT milk in sterilized packages which then hydrolyzed lactose at room temperature for 10 days before the product was released in the retail market. Kligerman (1981) reported a continuous method whereby lactase is added to the milk carton by an automatic electric syringe followed by addition of the milk. The sealed milk carton was then refrigerated for a minimum of 24 hr to facilitate lactose hydrolysis. Lactase has also been added to milk for the production of cottage cheese (Gyuricsek and Thompson, 1976), Cheddar cheese (Thompson and Brower, 1974), and low-lactose yoghurt (O'Leary and Woychik, 1976). The use of lactase-treated whey has considerable potential in the food industry for the production of confectionery, bakery products, and syrups (Holsinger, 1981; Rand, 1981).

F. Pectic Enzymes

Pectic enzymes are involved in the cloud stability of fruit juices and concentrates (Krop, 1974; Termote *et al.*, 1977) as well as in the clarification of wine, grape juice, and cider (Braddock and Kesterson, 1979; Kilara, 1982). In all cases, the primary objective is the production of juices with good color, flavor, and appearance.

The primary substrates for pectic enzymes are collectively referred to as pectic substances of which the basic polymer is (1 → 4)-D-polygalacturonic acid. Pectin itself is a colloid composed of this polygalacturonic acid chain in which approximately two-thirds of the carboxyl groups are esterified with methanol (Kertesz, 1951). The known pectic enzymes acting on pectic substances are either depolymerizing or de-esterifying. The depolymerizing enzymes include polygalacturonase or pectin transeliminase. The classification of pectic enzymes used in the food industry is shown in Table 11.3.

Pectin lyases (poly(methoxygalacturonide)lyase, EC4.2.2.10) are endo-enzymes in nature which cause the rapid depolymerization of esterified pectin in a random manner, resulting in a dramatic decrease in viscosity. Low-methoxyl pectin or pectic acid is depolymerized by pectate endolyases (poly[(1 → 4)α-D-galacturonide]lyase, EC 4.2.2.2), which randomly break the chain, as distinct from the exolyase, which releases unsaturated dimers from the reduced end of

TABLE 11.3

INDUSTRIAL PECTIC ENZYMES[a]

A. Depolymerizing Pectic Enzymes
(1) Acting mainly on pectin
(a) Polymethylgalacturonases
(i) endo-
(ii) exo-
(b) Pectin lyases (PL)
(i) endo-
(ii) exo-
(2) Acting mainly on pectic acid
(a) Polygalacturonases
(i) endo-
(ii) exo-
(b) Pectate lyases (PAL)
(i) endo-
(ii) exo-
B. Pectinesterases
(1) Pectin methylesterase (PME)

[a] Taken from Kulp (1975).

the chain. Neither exo-PMG nor exo-PAL has been reported in nature. Endo-polygalacturonase (EC 3.2.1.15) depolymerizes pectic acid while exo-polygalacturonase (EC 3. 2. 1. 67) removes galacturonobiose units in a stepwise manner from the end of the chain.

The pectin esterases (pectin methyl esterase, EC 3.1.1.11) hydrolyze methyl esters from the pectin chain, resulting in the formation of low-ester pectin and pectic acid.

1. *Fruit Juice Clarification with Pectolytic Enzymes*

Commercial pectolytic enzymes are mixtures of pectin methyl esterase and polygalacturonase obtained mainly from molds of the genus *Aspergillus* (Sidler and Zuber, 1978). They are used to supplement the normal pectolytic activity of apples and grapes for fruit juice and wine clarification. These preparations also contain traces of other enzymes such as cellulase, amylase, protease, arabinase, and xylanase. The pectolytic activity of these preparations causes the hydrolysis of the soluble pectin and removal of its colloidal properties. This is accompanied by a rapid flocculation of the cloud-forming particles, which can then be removed by sedimentation, filtration, or centrifugation (Kilara, 1982). The amount of enzyme preparation needed varies with the particular fruit variety as well as the ratio of soluble to insoluble pectic substances present. An increase of 12–17% in insoluble solids was reported by Bieleg and co-workers (1971) following

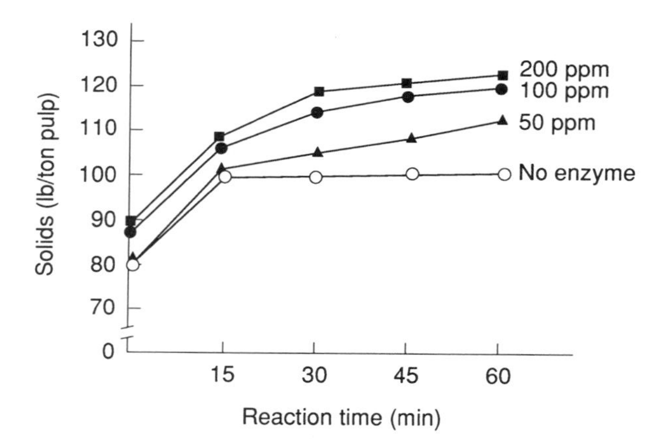

FIG. 11.4. Yield of soluble solids from orange juice pulp versus reaction time at various enzyme concentrations (Braddock and Kesterson, 1979). Copyright © by Institute of Food Technologists.

the addition of pectolytic enzymes to apple juice. Addition of pectic enzymes to orange juice also improved the recovery of soluble solids as evident in Figure 11.4 (Braddock and Kesterson, 1979).

2. *Fruit Juice Concentrates*

In contrast to fruit juice clarification, the production of products such as fruit juice concentrates and tomato purée or ketchup requires the presence of cloud stability. The loss of cloud in citrus juices was attributed to the de-esterification of the pectin by pectinesterase and the subsequent precipitation of the calcium pectinate or pectate (Joslyn and Pilnik, 1961). Consequently, considerable emphasis has been focused on pectinesterase inhibition in order to retain cloud stability. Baker and Bruemmer (1969) attributed the destabilizing effect by pectinesterase to the formation of pectin low in methyl esters rather than to the direct role of pectin. These researchers (Baker and Bruemmer, 1972) showed that depolymerization of orange juice pectin by a commercial pectinase preparation produced soluble pectates which did not precipitate in the presence of calcium ions. Krop and Pilnik (1974) found that addition of pure yeast polygalacturonase to a pectinesterase-active orange juice rapidly depolymerized the pectate, preventing its coagulation as calcium pectate. Different pectinesterase inhibitors have been suggested to maintain cloud stability although the most viable appeared to be the use of pectate hydrolysates (Termote *et al.*, 1977). These hydrolysates were produced enzymatically with the degree of polymerization ranging from 8 to 30. Of these the pectate hydrolysate with degree of polymeri-

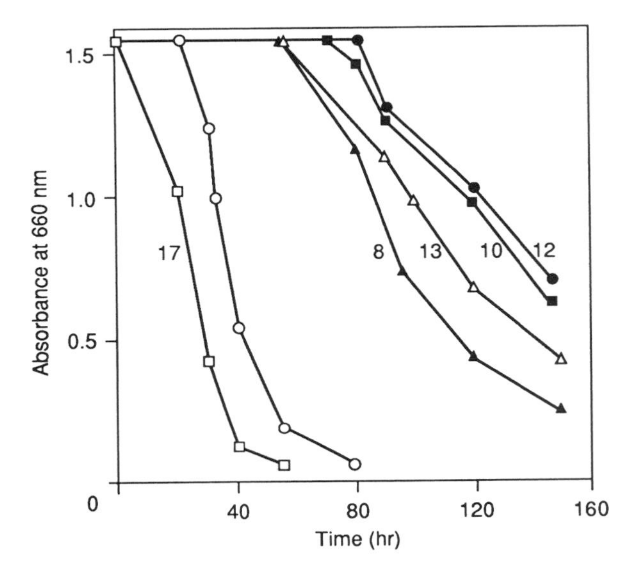

FIG. 11.5. Cloud stability of reconstituted orange juice. The juice contained 1 U/ml of pectinesterase, no inhibitor (○) or 1.0 mg/ml of enzymatically prepared pectic acid hydrolysates with varying degrees of polymerization (indicated with number in the figure). The pH of the juice was adjusted to 4.0 and the juice was incubated at 30°C. Cloud loss was measured as absorbance at 660 nm of the supernatant after centrifugation of the juice for 10 min at 320*g* (Termote *et al.*, 1977).

zation 8–15 prevented any clarification of the fruit juice as shown in Figure 11.5. These results were substantiated by Chandler and Robertson (1983), although they found that the residual activity of the inhibited enzymes prior to orange juice pasteurization as well as pectic hydrolysis by the acidic conditions of the juice also affected cloud stability.

Pectic enzyme preparations are also used for the production of low-methoxy pectins for diabetic foods as well as in the initial steps of coffee manufacture. In the latter case the enzymes are used to digest the mucilage which surrounds the coffee bean (Johnston and Foote, 1951).

G. CELLULASE

Cellulase, as discussed previously in Chapter 2, is an enzyme complex consisting of C_1,C_x, and β-glucosidase, which catalyzes the following reaction sequence:

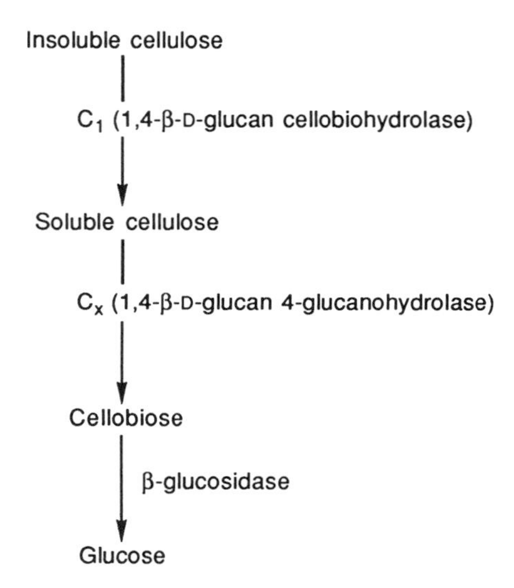

This complex is present in bacteria and fungi and its identification in higher plants was reported in tomato fruit by Sobotka and Stelzig (1974). Fungi excrete large amounts of cellulase enzymes, making them the preferred source, while bacterial cellulases are mainly cell bound. The three main enzymes produced by fungi are endo-1,4-β-glucanase (endo-1,4-β-D-glucanohydrolase, EC 3.2.1.4), cellobiohydrolase, (1,4-β-D-glucan cellobiohydrolase, EC 3.2.1.91), and β-glucosidase (EC 3.7.1.21) (Wood and McCrae, 1979). The main enzyme in this complex is cellobiohydrolase, an exo-glucanase capable of hydrolyzing crystalline cellulose. It releases cellobiose units from the nonreducing end of the cellulose chain (Wood and McCrae, 1973; Berghem and Petersson, 1973). The endo-glucanases or C_x degrade CM-cellulose randomly, causing a sharp decrease in chain length and slowly increasing the number of reducing groups. The third main enzyme in the fungal complex, β-glucosidase, hydrolyzes cellobiose as well as cello-oligosaccharide intermediates produced by the action of C_x. In addition to these enzymes several others are present but only in trace amounts.

Cellulase has seen limited use by the food industry, for example, in the modification of fibrous vegetables, clarification of citrus juice as a result of cellulose cloud formation, and in the production of fermentable sugars during the mashing procedure in the brewing of beer. One the main reasons for its limited use has been the unavailability of adequately active cell-free enzyme preparations. This situation was changed at the Natick Laboratories in the United States, where a highly active cell-free cellulase was obtained from a mutant of the fungus *Trichoderma viride* (Spano *et al.*, 1976; Mandels and Sternberg, 1976). The enzyme present in the culture filtrate was capable of saccharifying ball-milled waste cellulose to produce a glucose syrup. Since then cell-free culture filtrates prepared from *Trichoderma koningii, Fusarium solani,* and *Penicillium funiculosum* were all shown to hydrolyze cellulose (Wood and McCrae, 1978). The availability of commercial high-active cellulase is expected to have considerable impact on the food industry. Pilnik and co-workers (1975) liquefied fruit pulps using pectinolytic and cellulolytic enzymes. A similar process was accomplished on tropical fruits using a technical cellulase preparation obtained from *Trichoderma viride* by Kittsteiner-Eberle *et al.* (1985) and Schreier and Idstein (1984).

Considerable interest has been shown in the cellulase complex from the mesophilic fungus *Trichoderma resie,* which effectively converts crystalline cellulose to glucose. Improved mutant strains have been identified which produce more active cellulase preparations, although there is considerable room for improvement (Merivuori *et al.,* 1985). The development of improved strains of *Trichoderma resei* will hopefully result in a biochemical conversion system capable of competing with the starch to glucose process (Mandels *et al.,* 1981).

Cellulase has considerable potential for utilizing cellulose waste material accumulated by the pulp and paper industry. This could be hydrolyzed to glucose as demonstrated by Andren and co-workers (1976). The production of glucose from such waste materials could be used in the food industry or for microbial fermentation.

V. Proteases

Proteases degrade protein by hydrolyzing peptide linkages. There are many proteolytic enzymes with different degrees of specificity, and no single protease is capable of hydrolyzing all the peptide bonds in the protein molecule. The presence or utilization of these enzymes is of fundamental importance in many food processes.

A. Meat Tenderization by Proteases

Papain is the most widely used protease in the food industry with primary applications in the chill proofing of beer as well as artificial tenderization of meat. Chymopapain is not used as a food enzyme per se,but accompanies papain in crude preparations from papaya latex. A process developed by Jones and Mercier (1974) produced a more active and stable refined papain extract from papaya latex.

The major problem with artificial meat tenderization is to ensure a uniform distribution of the enzyme within the meat, without having to grind the meat into

TABLE 11.4

ENZYMES IN MEAT TENDERIZATION

Protease	Source	Muscle protein substrate
Papain	Papaya latex	
Bromelain	Pineapple	Collagen, elastin (connective tissue)
Ficin	Fig	
Microbial proteases	*Aspergillus oryzae*	
	Actin, myosin	
	Bacillus subtilis	

a paste. To tenderize those meat cuts of inferior quality, the enzyme hydrolyses one or more muscle tissue components, in particular the sarcolemma, without excessive degradation of the muscle fibers. The latter can result in a rather mushy tissue with an undesirable hydrolysate flavor. Those main enzymes permitted as meat tenderizers are listed in Table 11.4.

All the proteases listed in Table 11.4 hydrolyze the sarcolemma, which holds the muscle fibers together. The use of enzymes in this process was first reported over 400 years ago by Hernando Cortez, a Spanish conquistador, who noted that Mexican Indians tenderized their tough meat by wrapping it in papaya leaves overnight. It was not until 1949 that the proteolytic enzyme native to this plant, papain, was promoted commercially in the United States as a meat tenderizer. The application of proteases requires the uniform distribution of low enzyme concentrations to ensure limited proteolysis to attain a particular degree of tenderness. The meat may either be sprinkled with powdered enzyme preparation or immersed into a liquid preparation. The use of a fork to pierce the steak before or after the enzyme is applied was reported by Mier and co-workers (1962). By dipping thin cuts of meat in a meat tenderizer solution Bernholdt (1969) reported a significant improvement in meat tenderness. The development of a spray tenderizer procedure by Wattenbarger (1961) used a combination of gaseous and liquid phases to atomize the enzyme for uniform distribution into those meat cuts to be tenderized. By tailoring the enzyme tenderizing mixture to the individual cuts and grades of meat he was able to achieve the desired degree of tenderness. Papain was used for tenderizing steaks while a mixture of papain and bromelain was found to be more suitable for tenderizing roasts. The use of bromelain was important as it was inactivated at a lower temperature (68.5°C). This prevented excessive tenderizing when meat cuts were held on a steam table at 60–70°C for several hours, as is the practice in hotels, restaurants, and institutions. The potential of actinidin, a proteolytic enzyme found in kiwi fruit (*Actinidia chimemsis*), was recently investigated by Lewis and Luh (1988). Although this enzyme is not as active as papain, treatment with it resulted in more tender steaks.

The postmortem injection of enzymes for tenderizing meat has also been reported, including direct injection into a specific area of the carcass (Silberstein, 1966). Improvement in tenderness was noted which probably depended on whether the muscle was in the pre- or postrigor state. An alternative approach to meat tenderization was by injecting the meat tenderizer into the live animal prior to slaughter (Goeser, 1961; Robinson and Goeser, 1962). The enzyme was distributed by the vascular system and remained dormant until the meat was cooked. This was confirmed in papain-treated mutton, which increased in tenderness when cooked at temperatures greater than 65°C. This process was approved by the United States Department of Agriculture for the production of a beef product with the trademark ProTen Beef in 1959.

The use of proteolyic enzymes is permitted by the USDA for tenderizing certain turkey parts, baking hens, and roasters. A process was patented by Murphy and Murphy (1964) involving the antemortem injection of hyalluronidase together with seasoning under the skin of the fowl to enhance both tenderness and flavor. Hogan and Bernholdt (1964) injected proteolytic enzymes into the vascular system of fowl just prior to slaughter to improve tenderness. A later study by Cunningham and Tiede (1981) injected poultry muscles with a tenderizing marinade containing 0.05% papain. Those poultry products treated with the enzyme were all significantly more tender compared to the untreated samples irrespective of whether they were cooked conventionally or in a microwave oven. Their results in Figure 11.6 show that much lower shear press values were obtained for the enzyme-treated baking hen thighs consistent with the sensory data. Similar results were reported for both turkey drumsticks and baking hen thighs and breasts. The benefit to the consumer by improving the tenderness of poultry parts normally considered to be tough is obvious.

B. Chill Proofing of Beer

The development of beer haze is due to either the proliferation of infecting microorganisms (biological haze) or chemical reactions in the beer (nonbiological haze). While biological haze has been eliminated by proper manufacturing practices, chemical haze still remains a serious problem.

The formation of nonbiological haze (chill haze) results from the combination of polypeptide and tannin molecules occurring when beer is cooled below 10°C. Chill haze formation also involves the participation of carbohydrates, including glucose, arabinose, and xylose, as well as metal ions such as copper and iron. Its formation involves the following steps:

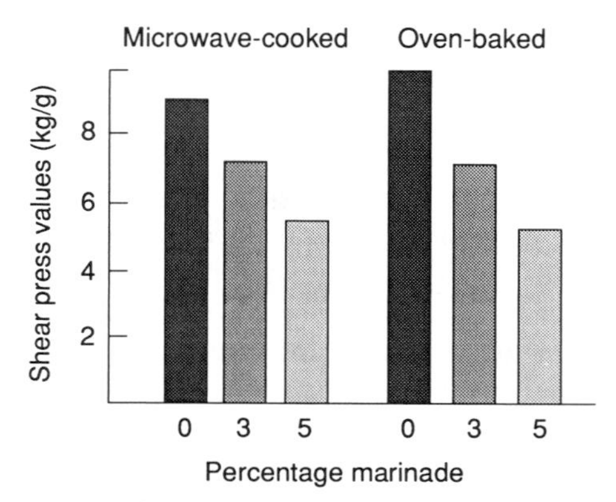

FIG. 11.6. Shear press values for cooked baking hen thighs after injection with a marinade containing 0.05% papain (Cunningham and Tiede, 1981).

1. polymerization of the hop-derived tannin molecules, including caffeic acid, gallic acid, and leucoanthocyanins, to form active polymers, and
2. the reaction between the active polymers and the polypeptide molecules.

The reaction scheme is illustrated in the following:

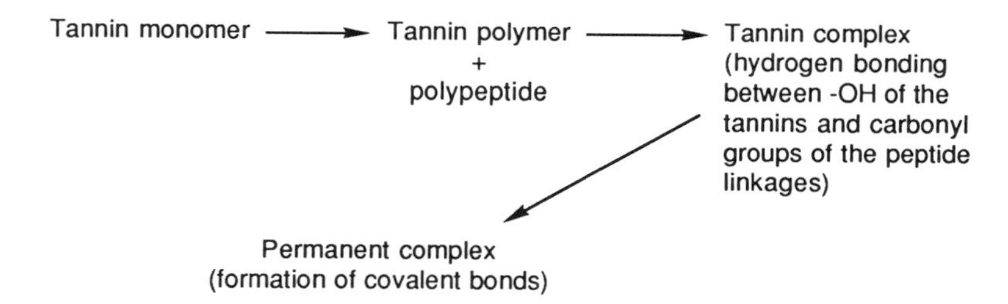

Proteolytic enzymes are used to prevent beer haze formation (chill proofing), since polypeptides are the major components of chill haze. These enzymes include papain, pepsin, ficin, bromelain, and bacterial protease and must be active at the normal acid pH of beer. Chill haze may not be a problem in draft beer but can appear in bottled or canned beer stored at cool temperatures. Papain was first used commercially for chill proofing beer by Wallerstein (1911), who patented the process. Chill proofing is used in beers produced in North, South, and Central America as well as in Europe, although in Germany it is only used for export beers (Jones and Mercier, 1974).

The chill proofing activity of papain has been attributed to its hydrolytic cleavage of proteins in beer during pasteurization. An alternative theory was proposed by Horie (1964), who suggested that papain clotted the beer proteins. Support for this theory was provided by Jones and co-workers (1967), who observed an increase in high-molecular-weight material following papain treatment. Later studies by Segura *et al.* (1980) also noted an increase in higher-molecular-weight protein characteristic of plastein formation or protein precipitation. They subsequently attributed the chill proofing of beer to the proteolytic activity of papain (Segura *et al.*, 1981). Previous research by Stone and Saletan (1968) noted that the loss of proteolytic activity of papain preceded its chill proofing property. This distinction between proteolytic and chill proofing activity was further clarified by Kennedy and Pike (1981), who reported that *S*-carboxymethylpapain exhibited only chill proofing activity. The separation of these two properties in papain was examined by Fukal and Kas (1984), who followed changes in molecular sizes of proteins and polypeptides, release of free amino acids, and changes in the colloidal stability of beer treated with both active and inactive forms of this enzyme. Figure 11.7 shows the chromatographic profiles of proteins in beer obtained after separation on Sephadex G-25 with and without papain treatment using the Folin reagent. Marked changes in two peaks were evident and were also accompanied by increases in the amino acids released. Increasing the level of papain 10-fold from 5 to 50 μg/ml resulted in a much greater hydrolysis of the beer proteins, although above a certain level colloidal stability decreased. This was attributed to changes in the protein–polyphenol

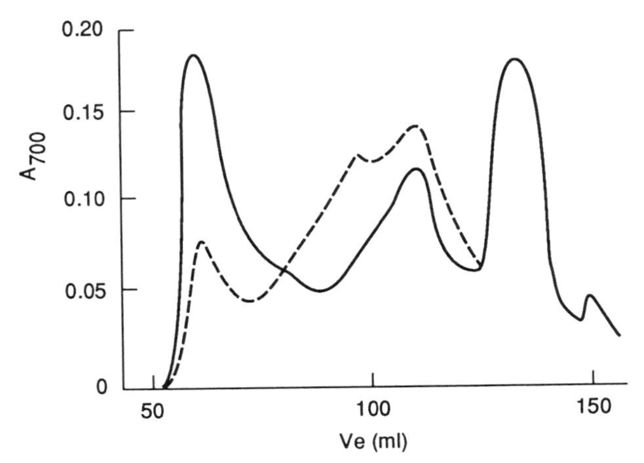

FIG. 11.7. Comparison of chromatographic profiles of control beer and beer treated with papain (5 μg papain/ml, 24 hr at 40°C). Solid line, chromatographic profile at 700 nm without papain; dashed line, chromatographic profile at 700 nm treated with papain. V_e = elution volume (Fukal and Kas, 1984).

equilibrium. The addition of inactivated papain, by aeration or copper and ascorbic acid, did not effect the protein pattern in the beer but did significantly improve the colloidal stability of beer. On the basis of their results Fukal and Kas (1984) suggested that two different mechanisms were involved in the chill proofing of beer by papain:

1. As long as sufficient proteolysis of the high-molecular-weight protein fraction (involved in the chill haze) occurred, papain was no longer needed for colloidal stability of beer.
2. In the absence of proteolysis a steady-state equilibrium occurs between proteins and polyphenols. Under such conditions both active and inactive forms of papain are necessary for colloidal stability.

C. Rennin and Rennet Substitutes in Cheese Making

The clotting of milk by rennin is the primary step in the manufacture of cheese. Rennin is the proteolytic agent of milk clotting found in rennet, the general name for commercial powders or extracts produced from the stomaches of calves, lambs, or young goats. The enzyme was first crystallized and purified by Berridge and Woodward (1953), who found that the preparation consisted of a number of fractions each exerting a specific activity (Foltmann, 1960).

Rennin or chymosin (EC 3.4.23.4) hydrolyses peptide linkages in casein which brings about the clotting of milk. As discussed in Chapter 4, casein is composed of several components with rennin specifically attacking the κ-casein fraction by limited proteolysis. The specific cleavage of the phenylalanine (105)–methionine (106) bond of κ-casein by rennin releases a highly negatively charged macropeptide, the micelle-stabilizing segment of κ-casein. This exposes the other casein fractions, which undergo nonenzymatic coagulation in the presence of calcium ions at temperatures below 20°C.

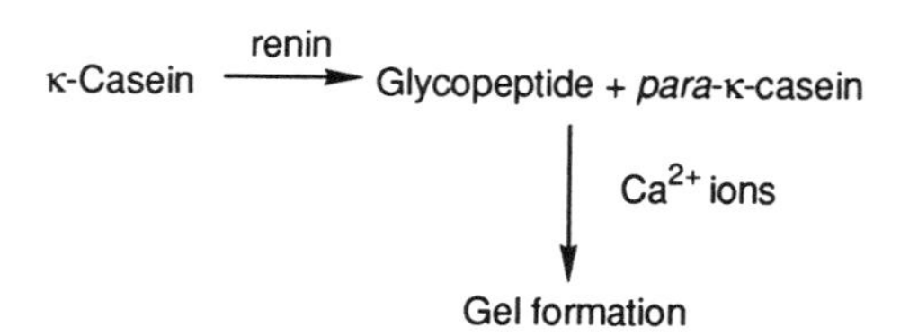

The nonenzymatic phase, during which the unstable micelles coagulate, starts to take place during the enzymatic reaction (Payens, 1984; Johnston, 1984; Schmidt *et al.*, 1973). The coagulation of milk consists of the primary enzymatic phase, for which the temperature coefficient Q_{10} is approximately 2. 0, and the secondary nonenzymatic coagulation phase, for which the Q_{10} value is 13.1

(Weestall, 1975). The Q_{10} value indicates that the enzymatic phase may occur at relatively low temperatures while coagulation takes place upon subsequent warming.

An important criterion for the suitability of a protease is that the enzyme should clot milk, but have little or no action upon the curd. Most proteases will bring about the initial clot formation but will continue to act upon peptide linkages in all casein fractions. This is termed general proteolysis, which results in the formation of bitter peptides, weakening and eventual dissolution of the curd, and reduced yields (Sternberg, 1972). For the production of cheese, a protease should have as high a milk-clotting to general proteolysis ratio as possible. This property is best exhibited by rennin, although other rennet substitutes have been found to be acceptable for cheese production, including pepsin, and fungal rennets from *Mucor pusillus, M. michei,* and *Endothia parasitica.*

Calf rennet, the best source of rennin for cheese manufacture, is becoming extremely scarce as a result of the increased world production of cheese (Food and Agriculture Organization, 1968). In 1973 calf rennet was only able to meet 34% of the cheese-making demands in the United States (Nelson, 1975). Consequently there has been a continuing search for alternative sources of milk-clotting enzymes. This triggered the reintroduction of swine pepsin in 1960 as well as the development of some microbial rennet preparations. This is reflected by the change in the composition of rennet preparations from high rennin or chymosin activity to equivalent amounts of rennin and bovine pepsin activities in modern preparations (Rothe *et al.,* 1977). Chicken pepsin was studied by Gordin and Rosenthal (1978) for the commercial production of Emmental (Swiss) and Kashkaval-type soft cheeses. This enzyme was found to be quite suitable for the production of soft cheeses. This contrasted with the poor results of Green (1972), who found that chicken pepsin produced a Cheddar cheese of inferior quality which exhibited soft body, weak flavor, and intense off-flavor due to excessive proteolysis. The production of good-quality soft cheeses with chicken pepsin was attributed by Gordin and Rosenthal (1978) to the different processing conditions used. In the case of Emmental and Kashkaval-type cheeses, a high temperature of 52 and 85°C, respectively, is required compared to 38°C during cheddaring. The higher temperatures involved in soft cheese manufacture resulted in partial inactivation of the chicken pepsin, thus limiting proteolysis compared to in the Cheddar cheese process.

Although the various rennet substitutes discussed are suitable for the production of cheeses they are not as satisfactory as calf rennet (Green, 1977). The development of recombinant DNA technology for the production of a number of important compounds has led to the application of this technique for cloning the calf rennet gene in *Escherichia coli* (Emtage *et al.,* 1983). A discussion of the recombinant enzyme will be found in Section XII,B,1.

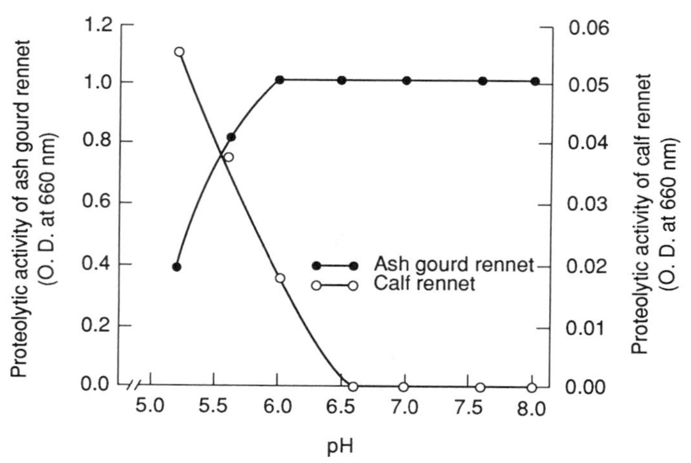

FIG. 11.8. Proteolytic activity of ash gourd rennet and commercial calf rennet over pH range 5.0–8.0 (Gupta and Eskin, 1977). Copyright © by Institute of Food Technologists.

1. *Vegetable Rennets*

Ficin was first crystallized from fig latex by Walti in 1938 and has been used for the production of cheese. Religious considerations in India and Israel have led to the search for other vegetable rennets. Plant proteases, however, have been found to suffer from excessive proteolytic activity (Green, 1977). Nevertheless, the use of plant coagulants has been traditionally used in some countries as coagulants for cheese production. The flowers of Cardo *(Cyanara cardunculus)* in Portugal, for example, provide a rennet substitute for soft-body cheeses from sheep milk, for example, Serra. Its high proteolytic activity, however, presented a problem with respect to the production of Edam, although it made a satisfactory Roquefort cheese (Vieira de Sá and Barbosa, 1972). A milk-clotting enzyme was partially characterized by Eskin and Landman (1975) from ash gourd *(Benincasa cerifera).* This enzyme produced an acceptable Cheddar cheese although its high proteolytic activity compared to calf rennet (Fig. 11.8) required some modifications in the cheese-making process. A more mature flavor, however, was apparent in the unripened cheese (Gupta and Eskin, 1977). These researchers pointed to the potential of this enzyme for accelerating the cheese-ripening process.

A traditional cheese-making process in parts of West Africa such as Nigeria and the Republic of Benin utilizes the juice from the leaves of the sodom apple *(Calotropis procera).* One of the cheeses produced in Nigeria is Wara, a white cheese product (Ogundiwin and Oke, 1983). Aworh and Nakai (1986) partially purified the enzyme responsible for clotting from sodom apple leaves. A more detailed study was conducted by Aworh and Muller (1987), who compared the cheese produced with the vegetable and calf rennets. The cheese-making procedure using the sodom apple leave extract was based on the traditional West African method (Aworh and Egounlety, 1985) and differed from the calf rennet process with respect to cooking temperature and pH. These researchers found that cheese produced with the vegetable rennet was lower in soluble nitrogen in spite of its greater proteolytic activity in casein solution. The cheese produced with vegetable enzyme was harder, much less cohesive, and far gummier than cheese produced with calf rennet. These differences in textural characteristics were reflected by the corresponding variation in cheese composition, although how it affected acceptability remained unclear. The overall yield and recovery of milk solids in cheese using the sodom leaf extract indicated considerable potential for this vegetable rennet in cheese production (Aworh and Muller, 1987).

D. Proteinases and Cheese Ripening

The evolution of cheese production from the farmhouse to large commercial production is reflected by the need to reduce overall production costs. The need to reduce the cheese storage period required for ripening led to examination of exogenous enzymes or modification of the starter bacteria (Law, 1980). Addition of food-grade proteinases to American Cheddar cheese was shown by Sood and Kosikowski (1979) to cause the development of strong flavors over a much shorter period without any flavor defects. Law (1981) examined the ripening of English Cheddar cheese using a commercial fungal proteinase (R: *Aspergillus oryzae*) and bacterial proteinase (N: *Bacillus subtilis*). The fungal proteinase resulted in a very bitter flavor compared to the neutral bacterial proteinase because of the more extensive degradation of α- and β-caseins. Addition of low levels of the neutral proteinase resulted in an acceptable and more mature cheese flavor although it still caused textural problems in the cheese, particularly a crumbly body. Further studies by Law and Wigmore (1982) confirmed the advantage of bacterial neutral proteinase over the fungal enzyme for accelerating the ripening of cheddar cheese. However, in spite of the improvement in cheese flavor the texture of the cheese was much weaker when treated with the bacterial enzyme. Thus more effective proteinases that do not cause textural problems need to be identified if the period of cheese ripening is to be reduced commercially.

The role of *Penicillium roqueforti* in the production of blue cheeses is associated with proteolysis during ripening. Compared to the proteolytic activity of the other flora, including lactic acid bacteria, micrococci, yeast, and corynebacteria, *Penicillium roqueforti* provides the major proteolytic activity in the blue cheese (Desmazeaud *et al.*, 1976). *Penicillium roqueforti* produces two endo-peptidases and three exo-peptidases (Zevaco *et al.*, 1973; Modler *et al.*, 1974; Gripon and

Hermier, 1974; Gripon and Debest, 1976; Gripon, 1977a,b). One of the endopeptidases is the acid proteinase aspartyl proteinase, while the other is a metalloproteinase, with optimum pHs of 4.0 and 6.0, respectively. Le Bars and Gripon (1981) attempted to characterize the extracellular proteolytic activity of this mold in relation to blue cheese ripening. Aspartyl proteinase appeared to hydrolyze α_{s1}- and β-caseins quite differently. The main products obtained from the hydrolysis of β-casein by aspartyl proteinase and metalloproteinase were identified by electrophoresis as the peptides β-Prap1 and β-Prmp1, respectively. These could be used as an index of their proteinase activity. In spite of the extent of the proteolysis of blue cheese, based on the large amount of solubilized nitrogen, the texture and cohesion were still retained. Creamer (1976) pointed to the role of the C-terminal part of the β-casein fraction in polymer formation with β- or α-casein fractions. Le Bars and Gripon (1981) suggested that the peptides β-Prap1 and β-Prap2 formed from β-casein by aspartyl proteinase represented the C-terminal half of the β-casein molecule.These fractions exert this polymerization effect and maintain the texture of blue cheese.

E. Proteases in Baking

The rheological properties of wheat flour doughs or batters are due to the state of the flour protein gluten. Hydrated gluten, a major component of dough, is responsible for the viscoelastic properties of the dough and the strength of flours. For instance, strong flours produce doughs which can tolerate extensive mixing. Oxidizing agents such as bromates, peroxides, and iodates are improving agents used to strengthen flours, while reducing agents, cysteine or glutathione, have the opposite effect by weakening the flour. The effect of reducing agents is attributed to their influence on sulfhydryl or disulfide groups in gluten, although the system is actually far more complex (Blocksma and Hlynka, 1964).

Wheat normally contains a very low level of protease activity, so the milled flour will reflect this deficiency. This necessitates protease supplementation in the baking process. The fungal protease used in baking is derived from *Aspergillus oryzae* although a protease from *Bacillus subtilis* is also utilized (ter Haseborg, 1981). These proteases disrupt the gluten network by hydrolyzing the peptide bonds thereby controlling the condition of the dough, resulting in the modification of the final quality of the baked loaf. A similarity exists between protease action and that of reducing agents insofar as both reduce dough viscosity and mixing time, rendering the dough more pliable and extensible. Unlike the action of reducing agents, the softening effect by protease cannot be reversed by oxidizing agents (Sandsted and Mattern, 1958). The controlled addition of protease improves the loaf characteristics by allowing greater gas retention, resulting in increased volume, better symmetry, and improved texture, flavor, and storage life. The release of free amino groups by protease action facilitates the

Maillard reaction, which is responsible for the improved browning and flavor of the enzyme-treated loaf.

The action of protease is limited to the dough mixing and fermentation stages as its thermolability renders it inactive during baking. An optimum amount of the enzyme supplement is required as determined by the quality of the flour and baking conditions, with enzyme supplements containing different α-amylase–protease ratios prepared to the baker's own specifications. Excess proteolysis, as a result of too high a concentration of the enzyme, must be avoided, as this could lead to an unmanageable sticky or slack dough, resulting in poor loaf characteristics.

Unlike the production of breads, which is based on gluten-strong flours, biscuits and waffles require gluten-weak flours. The increased production of gluten-strong wheats necessitates their modification for production of biscuits and waffles. This can be accomplished by treating the batter with protease to reduce the viscosity or by the addition of the enzyme at the mill (ter Haseborg, 1981).

F. Proteases and Natural Meat Tenderization

Tenderness in meat is developed by the action of the endogenous proteases during the conversion of muscle to meat and subsequent storage. The phenomenon of postrigor tenderness in meat is described in detail in Chapter 1.

Cathepsins, a group of intracellular proteolytic enzymes present in animal tissues, for example, liver and muscle, appear to have an indirect effect on the development of postrigor tenderness. The main function of the cathepsins is degradation of sarcoplasmic proteins with the production of peptides and amino acids which act as flavor precursors in meat.

Current theory regarding the mechanism of postrigor tenderness in meat centers around changes occurring within the myofibrillar structure. The disintegration of the Z-line material is initiated by a calcium-activated neutral protease,and at low pH lysosomal cathepsins B and D both contribute to myofibrillar breakdown. The intramuscular connective tissues may be weakened by the release of cathepsins B and N (collagenases) into the extracellular spaces of the muscle tissues. There is a direct relationship between enzymatic activity and morphological changes in postrigor muscle tissues, and the significance of the activities of endogenous proteases in meat is now recognized.

G. Proteases and Oil Extraction

The additional release of oil from Nigerian melon seeds treated with proteolytic enzymes indicated its potential for improving oil extraction (Fullbrook, 1983).

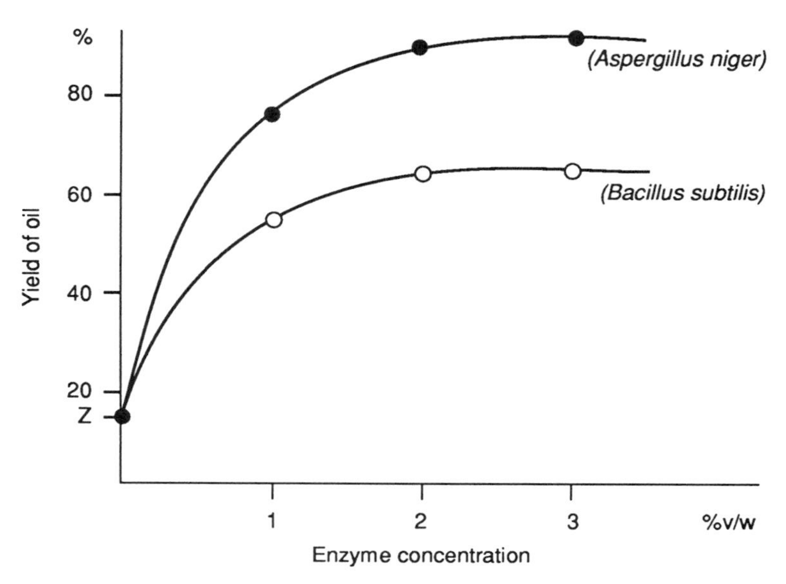

FIG. 11.9. Use of the two enzyme systems in extraction of oil from soybean (Fullbrook, 1983).

Melon seeds contained around 30% w/w fatty material and 50% crude protein, although not all the oil was extractable with solvents. Treatment of the seeds with proteolytic enzymes, however, somehow enhanced extraction of the oil. The proteinase from *Bacillus licheniformis* solubilized around 35% of the crude protein, and hydrolyzed 2.8% of solubilized protein, while at the same time enhancing total oil extraction by 16%. A comparison of two enzyme systems, one from *Bacillus sp.* and the other from *Aspergillus niger,* was carried out by Fullbrook (1983) to assess their potential for extracting oil from soybean. As seen in Figure 11.9, the preparation from *Aspergillus niger* gave higher yields of oil. An increase in oil extraction was obtained on increasing the enzyme supplements, which leveled off at the higher enzyme levels. Sosulski and co-workers (1988) recently noted both a decrease in extraction time and an increase in oil yield when three canola cultivars were incubated with different crude carbohydrase preparations.

VI. Esterases

This group of enzymes is responsible for the splitting of ester linkages with the introduction of a molecule of water:

$$\underset{\text{Ester}}{RCO\text{–}OR'} + H_2O \rightarrow \underset{\text{Acid}}{RCOOH} + \underset{\text{Alcohol}}{R\text{–}OH}$$

These enzymes are considered hydrolases and are therefore classified as EC 3.1.1. in the enzyme classification. The alcohol moiety of the ester may be monohydric or polyhydric and aliphatic or aromatic, and the acid moiety may be organic or inorganic. An esterase is specific for the ester linkage, and in addition may be specific for the acid or alcohol moiety. Of the many esterases found in biological material, relatively few are important in food production. This activity has also been identified with trypsin, chymotrypsin, and papain, all of which are capable of hydrolyzing simple fatty acid esters.

A. Lipases

Glycerol ester hydrolases (EC 3.1.1.3) or lipases are enzymes widely distributed in animals, plants, and microorganisms responsible for the hydrolysis of insoluble fats and oils (triacylglycerols, diacylglycerols, and in some cases monoacylglycerols). The reaction is complex as the fats and oils are present as a separate nonaqueous phase of an emulsion with the enzyme action taking place at the fat–water interface. Lipases are subdivided into two types based on the positional specificity for the primary ester. For example, those obtained from *Penicillium roqueforti* are 1,3-lipases while 2-lipases are obtained from *Aspergillus flavus*. The 1,3-lipase preferentially hydrolyzes the ester linkages at the 1 and 3 positions in the triacylglycerol while 2-lipase preferentially attacks the ester linkage at the 2 position:

$$\begin{array}{ll} (1) & CH_2OOCR \\ (2) & R'COOCH \\ (3) & CH_2OOCR'' \end{array} \xrightarrow[+\,H_2O]{\text{1,3-lipase}} \begin{array}{l} CH_2OH + RCOOH \\ R'COOCH \\ CH_2OH + R''COOH \end{array}$$

or

$$\begin{array}{ll} (1) & CH_2OOCR \\ (2) & R'COOCH \\ (3) & CH_2OOCR'' \end{array} \xrightarrow[+\,H_2O]{\text{2-lipase}} \begin{array}{l} CH_2OOCR \\ HOCH \quad + R'COOH \\ CH_2OOCR'' \end{array}$$

B. Lipases and Cheese Ripening

Cheese ripening is a dynamic process which involves a large number of biochemical changes such as lipid and protein hydrolysis. These reactions are mediated by the enzymes produced by microorganisms growing in the cheese. Different types of bacteria or molds predominate at different stages of cheese ripening.

There are two main types of cheese, hard and soft. In hard cheeses, for example, Cheddar and Swiss, the ripening is brought about by bacterial action in the cheese itself. In Cheddar cheese the organism responsible for this is *Lactobacillus citrovorum*. In the case of the soft cheeses Camembert and Limburger, however, ripening is carried out by yeasts, slime molds, or bacteria on the cheese surface. This limits the final product to a small size to ensure penetration by the surface microorganisms, which results in a uniform flavor throughout the cheese.

During cheese ripening there is extensive proteolysis and lipolysis of the butterfat. Controlled lipolysis during cheese production is necessary for the development of characteristic flavors, which is particularly important in Italian-type cheeses. Supplementation of the lipolytic activity of microorganisms with enzymes has been successful using oral or pregastric lipases. These lipases have a preference for liberating free fatty acids below C_{10} from the cream, which is especially important in Italian-type cheeses (Harper, 1957). Pregastric esterase supplements produce the desired picante flavor in Romano and Provolone cheeses while a combination of lamb gastric enzymes and pregastric calf esterase produces a Provolone-type flavor (Richardson and Nelson, 1967). Pregastric lipases are used, with ordinary rennet for coagulation, for the production of Italian-type cheeses. The pregastric lipase cannot be replaced with lipases from other sources owing to the production of different ratios of free fatty acids, which result in atypical, soapy, or rancid flavors. The production of lower chain fatty acids by the action of seven different lipases on butter oil is shown in Table 11.5. The concentration of butyric acid was shown by Long and Harper (1956) to be critical for the production of a desirable flavor in Provolone and Romano cheese. This was confirmed in a later study by Shahani *et al.* (1976) based on flavor

TABLE 11.5

EFFECT OF DIFFERENT LIPASES ON THE RELEASE OF LOW-CHAIN FATTY ACIDS FROM BUTTER OIL[a]

Lipase	Percentage of lower-chain fatty acid
Kid	42
Kid–lamb	40
Calf	31
Milk	19
Bovine pancreatic	17
Penicillium roqueforti	38
Achromobacter	22

[a] Adapted from Shahani *et al.* (1976).

evaluation and free fatty acids released in commercially produced Romano cheese using kid–lamb lipase mixture or kid lipase alone. A more desirable flavor was obtained in Romano cheese produced with the kid–lamb lipase mixture, which also resulted in higher levels of the short-chain fatty acids acetic, butyric, and propionic acids. Of these butyric acid accounted for over 60% of the total free fatty acids in the most desirable cheese samples.

Penicillium roqueforti, the organism involved in the production of Roquefort, Gorgonzola, and Stilton cheeses, possesses a very high lipase activity resulting in the accumulation of free short-chain saturated fatty acids, especially caprylic acid, which gives rise to a sharp, peppery flavor (Currie, 1914). Caprylic acid can then be converted by the action of lipolytic microorganisms to methylketone, which contributes to the final flavor of mold-ripened cheeses.

$$RCH_2CH_2CH_2COOH \qquad\qquad RCH_2\overset{\overset{\displaystyle O}{\|}}{C}-CH_3$$

Caprylic acid Methylketone

The formation of methylketones from milkfat by *P. roqueforti* has been studied extensively in relation to its role in the fermentation of blue cheese flavor (Nelson, 1969). Large fluctuations in ketones during the ripening of blue cheese were attributed to an interconversion mechanism, including metabolism into secondary alcohols by the mold. These fluctuations were found by Fan *et al.* (1976) to vary with the physiological stage of the mold and the amount of ketones present.

C. Lipases and Flavor Deterioration of Dairy Products

Hydrolytic rancidity is ascribed to the development of off-flavors because of the liberation of free fatty acids from fats by lipase action. These fatty acids undergo either oxidative rancidity or β-oxidation to produce volatile short-chain saturated fatty acids with distinctive flavors. Lipases in milk are responsible for hydrolytic rancidity in improperly pasteurized or unpasteurized milk, as well as in cream and butter (see Chapter 10). This type of rancidity may also occur in stored grains and flour, including oats (Hutchinson and Martin, 1952; Urquardt *et al.,* 1983) and faba beans (Dundas *et al.,*1978), through indigenous lipase activity.

D. Lipases and Oil Processing

The application of lipases in oleochemical processing has received increasing attention in recent years (Posorske, 1984). The three main areas of interest include:

1. the enzymatic hydrolysis of fats for the production of fatty acids;
2. the synthesis of lipids by reversal of hydrolysis;
3. enzymatic modification of lipids by interesterification.

Intensive work is being carried out in these areas for possible commercial application. The production of free fatty acids with lipases has obvious economic advantage over the current high-temperature and pressure conditions needed to liberate the fatty acids. Lipase from *Candida cylindracea* was found to rapidly hydrolyze olive oil with almost complete hydrolysis within 4 hr (Nielsen, 1985). These results were identical to those obtained using the standard chemical hydrolysis but required much milder conditions. *Candida cylindracea* lipase is now used by a large Japanese company to release free fatty acids which have a better color and odor compared to those produced by the chemical process (Kilara, 1985).

The use of lipase for synthesis of fats is achieved by decreasing water activity, thereby shifting the reaction in favor of esterification (Strobel *et al.*, 1983). Using lipase from *Rhizopus arrhizus,* which is specific for long-chain fatty acids, a greater than 80% yield of ester was obtained. By careful selection of the enzyme it is possible to produce very specific products.

Interesterification of fats is used to modify composition and change the physical properties of triacylglycerol mixtures (Vaisey-Genser and Eskin, 1987). In this process the fatty acid in one triacylglycerol molecule is exchanged for another fatty acid present in the reaction medium. The final result is new species of triacylglycerols with completely changed fatty acid arrangements. Unlike chemical interesterification, which is a random process, the use of specific lipase permits careful control over this process and produces the triacylglycerol molecule shown in Scheme 11.3. A patent was obtained by Coleman and MacRae (1981) to produce cocoa butter equivalents by interesterification with lipase. Under normal physiological conditions triacylglycerols are hydrolyzed by lipase, releasing glycerols and fatty acids. This reaction can be reversed under certain conditions so that the predominant reaction is ester synthesis (MacRae, 1983). Thus by reversing this reaction, interesterification of fats can be achieved (Eigtved *et al.*, 1985). This was demonstrated in a recent study by Thomas and co-workers (1988) in which canola oil was successfully interesterified with lauric acid/trilaurin or fully hydrogenated high erucic rapeseed oil in the presence of porcine pancreatic lipase.

E. Chlorophyllase

Chlorophyllase (EC 3.1.1.14) catalyzes the first step in the degradation of chlorophyll during senescence or storage of fruits and vegetables:

$$\text{Chlorophyll} + H_2O \xrightarrow{\text{Chlorophyllase}} \text{Phytol} + \text{methyl chlorophyllide}$$

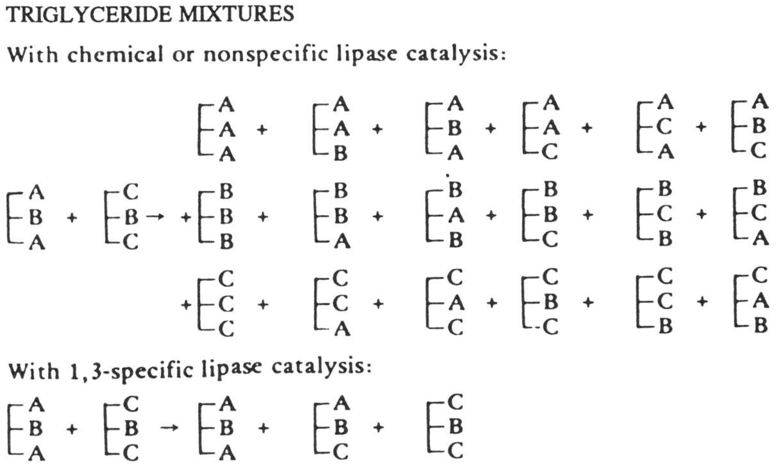

TRIGLYCERIDE MIXTURES

With chemical or nonspecific lipase catalysis:

AAA + AAB + ABA + AAC + ACA + ABC

ABA + CBC → + BBB + BBA + BAB + BBC + BCB + BCA

+ CCC + CCA + CAC + CBC + CCB + CAB

With 1,3-specific lipase catalysis:

ABA + CBC → ABA + ABC + CBC

SCHEME 11.3. Products formed by interesterification of mixtures of fats (MacRae, 1983).

This enzyme is widely distributed in plant tissues, although its precise physiological role in plants is not known. It is important to inactivate this enzyme to maintain the desirable green color of fresh vegetables in the processed product (Eskin,1979).

The importance of chlorophyllase in the degreening of citrus fruit is discussed in Chapter 2. The activity of this enzyme is enhanced by treatment with ethylene to accelerate the breakdown of chlorophyll and expose the carotenoids.

F. PHOSPHATASES

Phosphatases are present in bacteria although the physiological role of alkaline phosphatase is poorly understood. The enzyme is nonspecific in action and is generally cell bound with the exception of an extracellular phosphatase by *Bacillus sp.* (Fogarty *et al.*, 1974). The main application of phosphatases is found in the commercial production of 5′-mononucleotide, guanosine-5′-monophosphate (GMP), and inosine-5′-monophosphate (IMP). These are particularly useful as flavor enhancers in foods which are released from ribonucleic acid (RNA) by the action of 5′-phosphodiesterase. This enzyme can be produced from the organism *Bacillus subtilis* and hydrolyzes yeast RNA to yield 5′-nucleotides.

Phosphatases are important in milk as their inactivation is used to assess the adequacy of high-temperature short-time (HTST) pasteurization. This process destroys any pathogenic bacteria as well as inactivates many of the milk enzymes. Consequently a negative phosphatase test confirms the effectiveness of the pasteurization treatment. The reaction catalyzed by phosphatase is shown as follows:

$ROPO_3^{2-}$	$+ H_2O \rightarrow$	R–OH +	$HOPO_3^{2-}$
Orthophosphoric monoester		Alcohol	Orthophosphoric acid

Under certain conditions a negative phosphatase test after pasteurization may result in the regeneration of phosphatase activity during storage. This phenomenon is referred to as phosphatase reactivation, with the reactivated enzyme having identical biochemical properties to that of the original or raw milk enzyme (Wright and Tramer, 1953). Reactivation is enhanced by a number of factors, including high fat content, pasteurization temperatures in excess of 77.8°C, and elevated storage temperatures (Eddleman and Babel, 1958; Fram, 1957; McFarren *et al.*, 1960). Tests established to differentiate between the residual and reactivated phosphatases have proved to be inconsistent. Kwee (1983) examined these tests on pasteurized cream and found that prewarming the cream samples or prolonged storage at elevated temperatures caused false-positive results for phosphatase activity. By ensuring that the cream samples are not exposed to either of these conditions the anomolous reactivation of phophatase could be prevented.

VII. Oxidoreductases

Oxidoreductases are enzymes which catalyze oxidation–reduction reactions. The application of these enzymes to foods is extremely limited compared to the hydrolases (Schmid, 1979). Of these, glucose oxidase (EC 1.1.3.4) is the main one, while catalase (EC 1.11.1.6) is used only to a small extent. Another enzyme, lipoxygenase (EC 1.13.11.12), is added in the form of a soybean flour supplement in the baking of bread (Eskin *et al.*, 1977). These enzymes will be discussed in this section, while a detailed description of polyphenol oxidase can be found in Chapter 9.

A. Glucose Oxidase

Glucose oxidase catalyzes the oxidation of β-D-glucose with molecular oxygen to form D-gluconic acid:

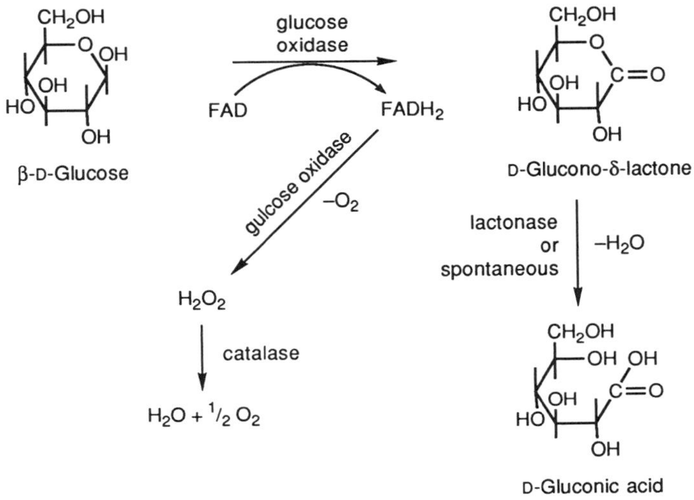
CH2OH
OH
OH
HO
OH
β-D-Glucose
glucose
oxidase
FAD
FADH2
gulcose oxidase
–O2
H2O2
catalase
H2O + 1/2 O2
CH2OH
O
C=O
OH
HO
OH
D-Glucono-δ-lactone
lactonase
or
spontaneous
–H2O
CH2OH
OH OH
C=O
OH
HO
OH
D-Gluconic acid

B. Removal of Glucose by Glucose Oxidase

Glucose oxidase shows a high specificity for β-D-glucose and is used as an analytical reagent for the specific determination of glucose in biological material. The commercial enzyme is obtained from *Aspergillus niger,* which also contains catalase. This enzyme is used for the removal of glucose from food material to minimize the Maillard reaction. It is added in small quantities, for example, to egg albumin or dried egg powder, to oxidize the aldehyde group of glucose to gluconic acid. This suppresses the Maillard reaction, which requires the presence of an aldehyde group to react with the amino group of an amino acid or protein (see Chapter 5). The enzyme is generally added before drying to prevent the deterioration of powdered egg products during storage by the Maillard reaction.

C. Removal of Oxygen by Glucose Oxidase

In the process of oxidizing glucose, glucose oxidase removes oxygen. This property is used to remove traces of oxygen from such products as beer, wine, fruit juices, or mayonnaise which prevents deterioration occurring as a result of enzymatic browning or oxidative rancidity (Underkofler, 1968).

D. Catalase

Catalase contains heme as the prosthetic group and catalyzes the decomposition of hydrogen peroxide to water and oxygen. It is thought to be involved in the oxidative deterioration of vegetables during storage. The function of catalase *in vivo* is not fully understood but it is unlikely that its main function in the living intact organism is the decomposition of hydrogen peroxide. A catalase preparation which was resistant to pH and temperature changes as well as to inactivation by hydrogen peroxide was obtained from *Aspergillus niger*. The reaction catalyzed by catalase is shown as follows:

$$H_2O_2 \xrightarrow{\text{Catalase}} H_2O + \tfrac{1}{2}O_2$$

E. Removal of Sterilant Hydrogen Peroxide by Catalase

Catalase is used for removing hydrogen peroxide when added to preserve milk on the farm. The H_2O_2–catalase treatment of milk is preferred to regular heat pasteurization as it destroys only the pathogenic organisms and leaves intact the lactic acid-forming microorganisms as well as many of the indigenous milk enzymes. These play an important role in the development of a number of cultured products. However, it is crucial that all traces of hydrogen peroxide be removed as it could interfere with the growth of starter microorganisms in dairy products, and with the final quality of the cheese. An immobilized catalase has been reported for the removal of hydrogen peroxide in dairy products (Chu *et al.*, 1975).

F. Catalase and Glucose Oxidase as Oxygen Scavengers

Catalase is used in conjunction with glucose oxidase to break down hydrogen peroxide as described previously. A co-immobilized glucose oxidase–catalase system has been used for the production of glucose-free dietetic drinks, fructose-free invert sugar, and glucose-free maltose. The effectiveness of this enzyme combination as oxygen scavengers has led to many patents for the removal of oxygen from foods (Schmid, 1979)

G. Peroxidase

Peroxidase (EC 1.11.1.7) also contains a heme prosthetic group and catalyzes the following reaction:

$$AH_2 + R\text{–}OOH \rightarrow A + H_2O + R\text{–}OH$$

where A= a hydrogen donor such as benzidine, guaiacol, pyrogallol, flavonoids, or tyrosine, and R–OOH represents hydrogen peroxide or an organic peroxide such as methyl or ethyl hydrogen peroxide.

TABLE 11.6

RECOMMENDED RESIDUAL PEROXIDASE LEVELS IN SOME FINISHED VEGETABLE PRODUCTS[a]

Vegetable	Peroxidase activity (% of original)
Peas	2.0–6.3
Green beans	0.7–3.2
Cauliflower	2.9–8.2
Brussels sprouts	7.5–11.5

[a] From Bottcher (1975).

H. PEROXIDASE AS AN INDEX OF BLANCHING

Peroxidase is believed to play a role in the oxidative deterioration of vegetables during storage. It is a highly thermostable enzyme and is frequently used as an index of the effectiveness of blanching treatments. The heat stability of peroxidase is apparent, for if held at 85°C for 32 min half the original activity still remains. The corresponding time required to retain 50% enzyme activity at 145°C is 0.4 min (Reed, 1975). The loss of peroxidase activity in a blanched food product indicates a corresponding loss of activity for other deteriorative enzymes. Peroxidase is self-generating because of reversible denaturation so this must be taken into account when determining the efficiency of the blanching procedure. Nevertheless, Bottcher (1975) concluded that complete inactivation may not be necessary to indicate overblanching. He suggested that residual peroxidase levels could be tolerated in different vegetables without adversely affecting the quality of the product while at the same time minimizing the effects of blanching. The level of residual peroxidase recommended for peas, green beans, cauliflower, and brussel sprouts is shown in Table 11.6. The different ranges reflect differences among varieties with respect to the stability of peroxidase. This has been attributed to the variability in the stability of the different isoenzymes of peroxidase to heat (Winter, 1969; Delincee and Schaefer, 1975).

I. ASCORBIC ACID OXIDASE

The oxidation of ascorbic acid may involve either direct attack by molecular oxygen (autoxidation) or the action of the copper-dependent enzyme ascorbic acid oxidase (EC 1.10.3.3). This enzyme is particularly significant in such fruit and vegetable products as lemon and grapefruit juices and concentrates, where it is responsible for the initiation of browning and loss of vitamin C activity during storage (see Chapter 5). The following reaction illustrates the chemical events which initiate ascorbic acid browning:

Ascorbic acid $\rightleftharpoons$ Dehydroascorbic acid $\longrightarrow$ 2,3-Diketogulonic acid

Ascorbic acid: $O{=}C-C(OH){=}C(OH)-HC(-O-)-HOCH-CH_2OH$ (lactone ring through O)

Dehydroascorbic acid: $O{=}C-C{=}O-C{=}O-HC(-O-)-HOCH-CH_2OH$ (lactone ring through O)

2,3-Diketogulonic acid: $COOH-C{=}O-C{=}O-HCOH-HOCH-CH_2OH$

The extent of ascorbic acid browning can be minimized by steam blanching or exclusion of oxygen. Food processed in plain tin cans and processing equipment should be copper-free. The rate of ascorbic acid oxidation increases markedly in the presence of metallic ions, especially copper and iron. While the loss of ascorbic acid cannot be prevented completely, it can be reduced to a minimal level during processing.

J. Lipoxygenase

Lipoxygenase (linoleate : oxygen oxidoreductase, EC 1.13.11.12) is widely distributed in plants (Axelrod, 1974; Eskin *et al.*, 1977; Pinsky *et al.*, 1971). This enzyme catalyzes the oxidation of lipids containing *cis, cis*-1,4-pentadiene groups to conjugated *cis, trans*-hydroperoxides in the presence of molecular oxygen. Substrates of lipoxygenase include linoleic, linolenic, and arachidonic acids in both the free and esterified forms, but not oleic acid. The reactions catalyzed by lipoxygenase are shown as follows:

$$R_1-\underset{cis}{CH{=}CH}-CH_2-\underset{cis}{CH{=}CH}-R_2 + O_2$$

$$\downarrow$$

$$(R_1-CH{=}CH-\dot{C}H-CH{=}CH-R_2 + \dot{O}OH)$$

$$\downarrow$$

$$(R_1-\underset{cis}{CH{=}CH}-\underset{trans}{CH{=}CH}-\dot{C}H-R_2 + \dot{O}OH)$$

$$\downarrow$$

$$R_1-\underset{cis}{CH{=}CH}-\underset{trans}{CH{=}CH}-\underset{\displaystyle OOH}{\underset{|}{CH}}-R_2$$

Lipoxygenase functions as a pro-oxidant and initiates the oxidative rancidity of plant lipids containing a high proportion of polyunsaturated fatty acids. This is a problem associated with raw legumes such as peas, lentils, and soybeans (Haydar and Hadziyev, 1973; Haydar *et al.*, 1973; Kon *et al.*,1961; Arai *et al.*, 1970). A possible relationship between lipoxygenase and off-flavor development in ground-stored faba beans was proposed by Hinchcliffe *et al.* (1977).

K. Lipoxygenase and Baking Technology

Lipoxygenase plays an important role in baking by improving the baking quality of the flour (Pringle, 1974). It is normally added in the form of an enzyme-active soy flour supplement in Canada, the United States, and the United Kingdom. In France, however, faba bean flour is used to supplement this enzyme because of the unavailability of soybean flour. The presence of lipoxygenase in faba beans was established by Eskin and Henderson (1974a,b). The application of ground soybeans to flour was first patented by Haas and Bohn (1934) to bleach the flour pigments and produce a whiter crumb. It has since been established that lipoxygenase oxidizes the flour protein gluten, resulting in improved crumb structure (Wood, 1967).

L. Lipoxygenase and Bleaching of Pigments

The reaction involved in the bleaching of the natural flour pigments is a coupled oxidation of polyunsaturated fatty acids and pigments:

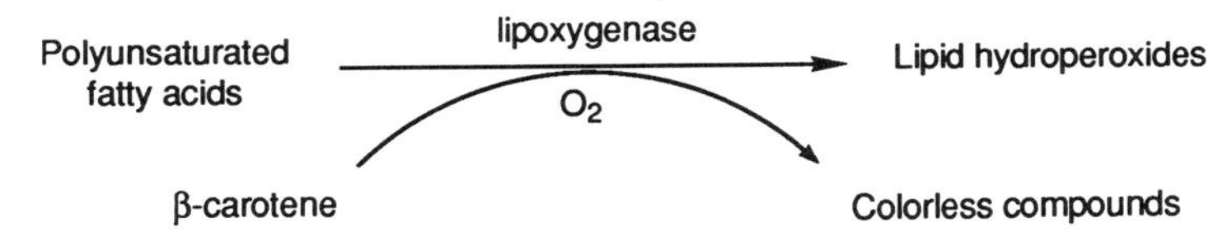

This controlled reaction is employed for the production of a white crumb.

M. Lipoxygenase and Dough Development

The importance of lipoxygenase in the development of the dough is related to the release of the bound lipid (Daniels *et al.*, 1970). This is accomplished by a coupled oxidation reaction in which the bound lipid is released possibly by oxidation of thiol groups at hydrophobic binding sites in the dough protein, as shown in Scheme 11.4.

The oxidized gluten produces the desirable rheological properties in the dough. In baking, the lipoxygenase preparation used of defatted soybean flour is approximately 0.5–1.0% in order to achieve the desired effects. Lipoxygenase

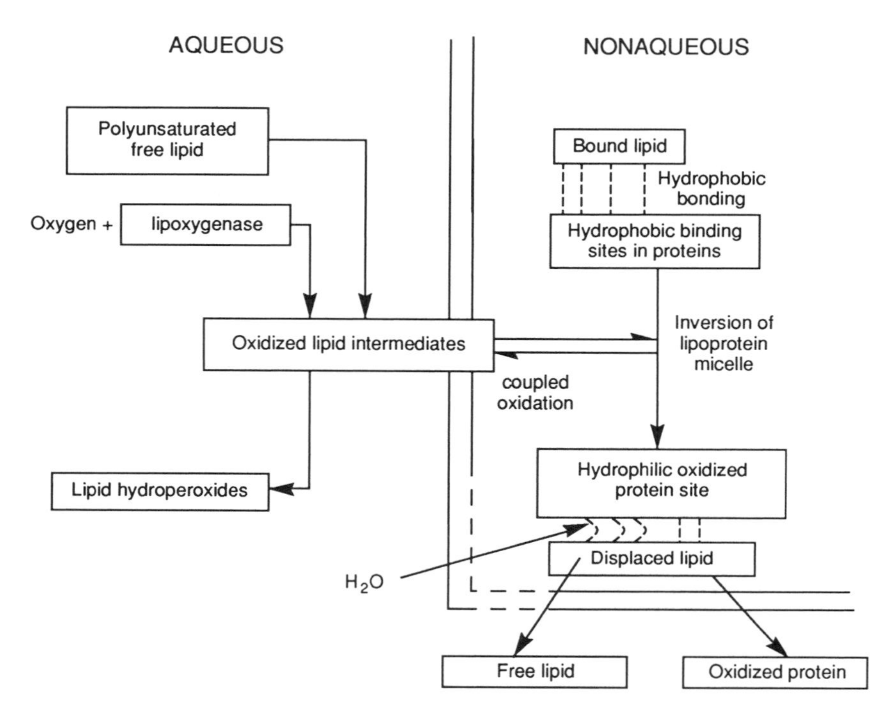

SCHEME 11.4. Proposed mechanism for the release of bound lipid during mixing dough in the presence of air (Daniels *et al.*, 1970).

affects crumb structure and produces hexanal, which alters the bread flavor (Frazier, 1979; Eskin *et al.*, 1977; Wood, 1980).

VIII. Miscellaneous Enzymes

In this context the term "miscellaneous enzymes" refers to those enzymes which cannot be classified as either hydrolases or oxidoreductases. The most important one is glucose isomerase, which is used extensively in the food industry for the production of sweet syrups.

A. Glucose Isomerase

Glucose isomerase (EC 5.3.1.5) is used for the large-scale production of sweet syrups as it catalyzes the reversible isomerization of glucose to fructose. This enzyme has been reported in bacteria and actinomycetes but not in molds or yeasts (Chen, 1980). In most cases the glucose isomerase preparations can also isomerize xylose.

Until recently, the high cost of glucose isomerase limited its use for converting glucose to fructose as it was not competitive with the production of invert sugar from sucrose. This changed, however, when Japanese researchers found a strain of *Streptomyces sp.* in which glucose isomerase could be induced by the presence of xylan or xylose (Takasaki, 1972, 1974; Takasaki *et al.*, 1969; Takasaki and Tanabe, 1971). *Streptomyces albus* YT-4 and YT-5 were both found to be extremely efficient producers (Cotter *et al.*, 1971). A large number of *Streptomyces species* have since been identified as sources of glucose isomerase and used commercially, including *S. wedmorensis, S. venezuella, S. olivochromogenes, S. olivaceus,* and *S. glaucescenes* (Chen, 1980). The conversion of glucose to fructose is accompanied by a marked increase in sweetness as well as more desirable physical and chemical properties. The importance of this is reflected by the use of glucose isomerase to convert glucose in corn syrup to fructose. The enzyme is used as a continuous system, in an immobilized form, for the production of high-fructose corn syrup (HFCS). This is composed of 42% fructose with the remainder being glucose together with traces of maltose and oligosaccharides. HCFS possesses sweetness comparable to that of sucrose or medium invert sugar, and can replace them in many foods at a much lower cost (Mermelstein, 1975). The major plants for HFCS production are located in the United States with facilities expanding in Europe and the Far East. The anticipated market for HFCS in 1985 is 6.8 million pounds, which is approximately three times that produced worldwide in 1978 (Chen, 1980).

IX. Immobilized Enzymes

The versatility of enzymes in food processing is evident by the many examples cited in this chapter. In all the processes discussed the enzyme(s) are added to the reaction mixture, either in a liquid or powder form,with the products removed at the termination of the reaction. A number of problems are inherent in this procedure, particularly the lack of a purification process for removing the enzyme from the accrued end products. Preparing the enzyme in an insoluble form would not only facilitate recovery of the enzyme or enzymes but permit them to be used repeatedly, which has obvious economic advantages.

A. Preparation of Immobilized Enzyme Systems

Immobilized or bound enzymes have been prepared by physically or chemically binding the enzyme to an insoluble support. Such enzymes represent one of the most rapidly growing areas of applied enzymology, where the field is open to applications in food processing, food analysis, pharmaceutics, medicine, and the detection and elimination of environmental pollution (Kilara and Shahani, 1979; Munnecke, 1978; Enfors and Molin, 1978; Skogberg and Richardson,1980; Weetall, 1975). The application of immobilized enzymes has resulted in the development of the new technology of enzyme engineering (Wingard, 1972).

One of the earliest attempts to bind enzymes was carried out in 1916 by Nelson and Griffin, who adsorbed invertase on charcoal and alumina and noted that the bound enzyme still exhibited some activity. Grubhofer and Schlieth (1954) were the first to covalently link amylase, pepsin, ribonuclease, and carboxypeptidase to an insoluble matrix and then observe residual enzyme activity. Bar-Eli and Katchalski (1960) chemically bound trypsin to a copolymer of *p*-aminophenyl-alanine and reported that the bound enzyme was more stable in storage than the corresponding free enzyme preparation. In 1962, Manecke was successful in insolubilizing a nonhydrolytic enzyme, alcohol dehydrogenase.

Attempts have been made to bind enzymes to cellulose-derivative supports. Tosa *et al.* (1966a) investigated a number of possible adsorbents for aminoacyl-ase and found that DEAE-Sephadex and DEAE-cellulose were the most satisfactory. Wiseman and Gould (1968) devised a method for linking an enzyme chemically to a carboxymethylcellulose support by diazotization. Kay and co-workers (1968) insolubilized chymotrypsin, ribonuclease, and lactase using cyanuric chloride to couple the enzymes to a cellulose support. Mason and Weetall (1972) successfully immobilized several enzymes to glass. Glass has a number of advantages as an enzyme support, including rigidity, ready access, economic cost, and it is relatively inert chemically. Other enzyme supports which have been examined in recent years are polyacrylamide and collagen.

The use of immobilized enzymes in the food industry will ultimately be governed by economic considerations. Enzymes are ideal reagents for carrying out chemical reactions because of their specific action, however, their use in batch food processes is limited by the costs incurred. Soluble enzymes can only be used once as it is too costly or difficult to recover or remove soluble enzymes from reaction mixtures after the enzymatic reaction is completed. Immobilized enzymes, on the other hand, can be used in a batch process and then recovered by filtration or centrifugation for further use. Application of this technology facilitates repeated use of the same enzyme preparation in a batch process.

Immobilized enzymes can be used in continuous processes in specially designed reactor vessels. The most widely used reactor designs include the continuous-feed stirred tank, the packed-bed reactor, and the fluidized bed reactor, shown diagrammatically in Figure 11.10. These systems are characterized by the continuous input of the substrate into the reactor and the continuous output of the products.

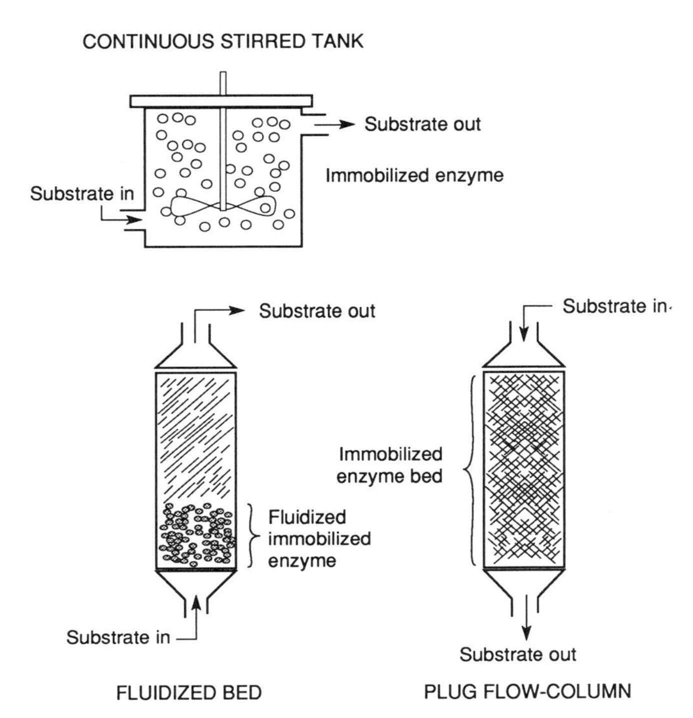

FIG. 11.10. Reactor designs for immobilized enzymes (Weetall, 1973).

B. Properties of Immobilized Enzymes

Immobilized enzymes possess several properties that confer a number of advantages over their water-soluble counterparts. For example, bound enzymes are generally more stable to heat than soluble enzymes, as illustrated for cellobiase and cellobiase bound to dextrose (CDC) or amylose (CAC) in Figure 11.11. A 50% residual enzyme activity was still evident at 75.5°C for the immobilized enzyme compared to 65°C for the corresponding soluble enzyme (Lenders *et al.*, 1985). The immobilized enzyme can also be stored and used for several months without any significant loss of activity even at room temperature. This compensates for the loss of activity which results from immobilization. Of the various methods used to attach enzymes to solid matrices, covalent attachment remains the preferred method for enzyme applications in food processing and food analysis. It is relatively stable and not reversed by changes in pH and ionic strength, or by the addition of substrate, although it significantly affects the activity of the enzyme. This may be due to steric hindrance with respect to both the matrix and the active site of the enzyme or through the reaction of amino acid residues

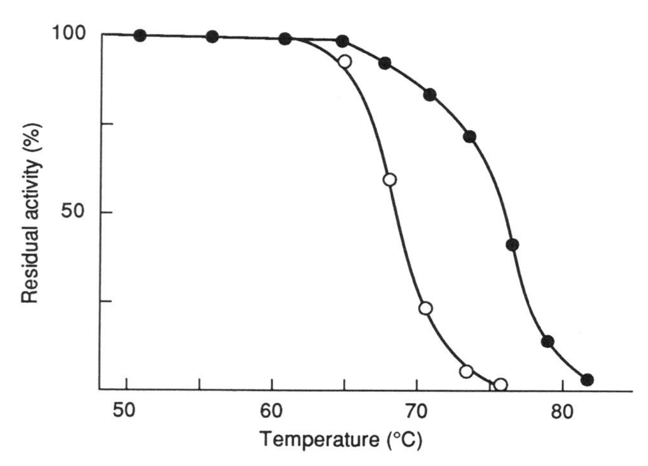

FIG. 11.11. Thermal stabilities of cellobiase (○) and conjugates CDC and CAC (●). Enzymes were maintained at the indicated temperature for 10 min prior to determining residual activity (Lenders *et al.*, 1985). CDC, Cellobiase-dextran conjugate; CAC, cellobiase-amylase conjugate.

important for enzyme activity and the chemical groups used in the immobilization process.

C. APPLICATION OF IMMOBILIZED ENZYMES IN THE FOOD INDUSTRY

The old concept of one enzyme preparation for one process, which is both uneconomical and inefficient, can now be replaced by immobilized enzymes in a continuous catalytic process. Immobilized enzymes are finding increasing use in the food and beverage industry as indicated by the following examples (Weetall, 1975; Kilara and Shahani, 1979; Pitcher, 1980).

1. L-Amino acylase

This enzyme was the first to be used successfully on a commercial scale in an immobilized form in Japan in 1966 as a result of research conducted by Tosa and co-workers (1966a,b, 1967, 1969, 1971a,b). This enzyme catalyzed the resolution of racemic mixtures of amino acids produced by microbial fermentation as follows:

$$\underset{\text{(racemic mixture)}}{\text{DL-Acyl amino acid}} \xrightarrow[\text{Amino acylase}]{H_2O} \text{L-Amino acid} + \text{D-acyl amino acid}$$

L-Amino acylase was bound to the ion-exchange resin DEAE-Sephadex A-25 at pH 7.0 and could be operated continuously at 50°C for over a month with 60%

of the original activity retained. The lost activity could then be supplemented by addition of fresh enzyme to the Sephadex column (Weetall, 1975). This process reduced the cost of production of L-amino acids by almost half compared to the batch process using the soluble enzyme. The covalent attachment of L-amino acylase to cellulose was later reported by Sato *et al.* (1971) to be as efficient as the Sephadex bound enzyme system. In this process the racemic mixture is poured down the column containing the immobilized enzyme at 50°C and eluent, containing the L-amino acids, is crystallized and separated. The remaining D-acyl amino acids then undergo chemical racemization to form DL-acyl amino acids which are then passed through the immobilized enzyme system for resolution.

Since the L-form of the amino acid is the only one that can be utilized by humans, the application of L-amino acylase is essential for the production of L-amino acids such as L-methionine. The latter is used to supplement legume proteins, which are deficient in sulfur amino acids, particularly methionine (Eskin *et al.*, 1985).

2. *Glucose Isomerase*

The isomerization of glucose to fructose by glucose isomerase, as discussed previously, is accompanied by improvements in both sweetness and solubility properties.The importance of this reaction is in the production of high-fructose corn syrups, which are much sweeter as well as less liable to crystallization during transport and storage. The enzyme is produced by the organism *Streptomyces albus* and could be fixed in the *Streptomyces* cells by first heating to 65°C for 15 min and then entrapping the cells in a filter bed (Takasaki *et al.*, 1969; Lloyd and Logan, 1972). In the United States, glucose isomerase immobilized on fibrous DEAE-cellulose has been in commercial use since 1968. In the continuous process the corn glucose syrup is pumped through the filter bed reactor, resulting in the production of corn syrup containing approximately 42% fructose. Sweeter syrups may be obtained through the development of chromatographic techniques for separating fructose from the other components, resulting in the production of 55 and 90% fructose syrups (Bucke, 1980). A new process by Hashimoto *et al.* (1983) using a combination of selective fructose adsorption and an immobilized glucose isomerase produced a syrup containing 45–65% fructose. This product was referred to as a higher-fructose syrup, which is sought after by the food industry. Antrim and co-workers (1986) developed a new immobilized glucose isomerase for continuous isomerization of glucose. The enzyme, obtained from *Streptomyces rubiginosus,* was purified and electrostatically adsorbed onto a granular, inert, food-grade carrier composed of positively charged DEAE-cellulose and titanium oxide agglomerated with polystyrene. The particular advantage of this system was its ability to regenerate the enzyme after use by reimmobilization with fresh glucose isomerase. The high activity of glucose isomerase was reported to produce over 9 metric tons of 42%

fructose syrup solids per kilogram of immobilized enzyme on a commercial scale.

3. *Lactase (β-Galactosidase)*

A number of immobilized lactase systems have been developed over the past ten years. These include adsorption on porous glass beads and stainless steel, covalent linkage to organic polymers, entrapment within gels or fibers, and microencapsulation within nylon or cellulose microcapsules (Greenberg and Mahoney, 1981). In addition to producing low-lactose products for consumption by lactose-intolerant individuals, the hydrolysis of lactose to galactose and glucose results in improved sweetness, solubility, humectant properties, and reduced tendency to crystallize (Shah and Nickerson, 1978a,b,c).

The production of low-lactose milk using immobilized lactase is carried out on a small scale in Italy using sterilized milk (Greenberg and Mahoney, 1981). The immobilized enzyme has potential application for the production of sweet syrups from cheese whey and could compete with corn syrup and high-fructose syrup. Moore (1980) reported the cost of producing syrups from deproteinized whey in the United States using an *Aspergillus niger* lactase immobilized to glass beads to be 14–20 cents per pound. This was cheaper than the 20–25 cents per pound reported for the production of corn syrup in Europe, which is almost twice as high as equivalent production costs in the United States.

Lactase-hydrolyzed milk is much more susceptible to microbial spoilage than unhydrolyzed milk This problem was resolved, however, by Kaul and coworkers (1984), who immobilized lactase to hen egg white powder by cross-linking with glutaraldehyde. The presence of lysozyme in the hen egg white, which is co-immobilized, minimized microbial spoilage as a result of its bacteriolytic activity. The stability of lyophilized immobilized lactase powder and used immobilized enzyme suspension at refrigerated temperatures made it suitable for domestic use as a cheaper and safer alternative to the soluble enzyme (Richmond *et al.*, 1981).

4. *Glucoamylase*

Glucoamylase produces dextrose from starch by hydrolyzing the α-1,4-glucosidic linkages from the nonreducing end of the starch chains as well as slowly hydrolyzing the α-1,6-glucosidic linkages in amylopectin (Pazur and Ando, 1960). It is used in the corn wetmilling industry for the saccharification of corn starch to D-glucose. Cereal starches are initially converted to linear and branched dextrins by the action of α-amylase followed by their rapid and almost complete hydrolysis to D-glucose by glucoamylase. The use of glucoamylase immobilized to alkylamine porous silica was reported by Lee *et al.* (1976). They reported conversions of up to 94% when freshly prepared solutions of α-amylase-hydrolyzed dextrins of 25–30 DE were passed through the immobilized glucoamylase at concentrations of 27%. The poorer conversions of liquefied starch by immobilized glucoamylase compared to the soluble enzyme were attributed to the lower expression of activity in the bound enzyme (Thompson *et al.*, 1978). This could be overcome by supplementation with saccharifying α-amylases. The major impediments with respect to the development of immobilized glucoamylase technology were reported to be a combination of the low costs of glucoamylase combined with the difficulty in using this enzyme at sufficiently high temperatures to minimize microbial contamination. Hausser *et al.* (1983), however, developed a two-enzyme immobilized system composed of fungal α-amylase and glucoamylase for the continuous production of high-conversion maltose-containing corn syrup. The enzymes were each chemically attached to separate reactors made from Microporous Plastic Sheets. The reactors were operated separately at 50°C with a pH of 4.3 for immobilized glucoamylase and 5.5 for fungal α-amylase. By altering the temperature and flow rate of the corn syrup feedstock it was possible to obtain syrups with varying glucose/maltose levels in a fraction of the time it normally takes. These researchers clearly demonstrated the viability of the dual immobilized enzyme reactor system for use by industry.

5. *Proteases in Cheese Manufacture*

The use of immobilized proteases for the continuous coagulation of milk is feasible based on the different Q_{10} values for each phase of the coagulation process. The coagulation of milk can be selectively retarded in an enzyme reactor by lowering the temperature to inhibit the nonenzymatic phase while permitting completion of the enzymatic phase. Following this the temperature is increased to allow the second phase, the clotting of the milk, to occur. One possible advantage of this process is that immobilized protease could be used for cheese production, while avoiding general proteolysis responsible for the release of bitter peptides and curd weakening. However, immobilized rennin cannot completely replace soluble rennin owing to the involvement of the latter in cheese ripening. Nevertheless, a number of immobilized rennin and pepsin preparations have been reported for the continuous coagulation of milk (Green and Crutchfield, 1969; Taylor *et al.*, 1977).

D. Immobilized Proteases and Chill Proofing of Beer

The development of haze in beer, described earlier, involves complexation of protein–tannins and carbohydrates. The core of the haze complex is composed primarily of proteins which can be broken down enzymatically. The use of proteolytic enzymes was first employed 65 years ago by the Wallerstein Company to break down the proteins involved (Finley *et al.*, 1979). Of the enzymes examined, papain has been shown to work most efficiently in eliminating the haze problem without causing any other changes in the beer. The use of immobilized proteases for chill proofing was reported by Wildi and Boyce (1971) and Weinrich *et al.* (1971). The covalent cross-linking of papain using glutaraldehyde was examined by Witt *et al.* (1970) for use in chill proofing. However, papain immobilized to collagen was reported by Venkatasubramanian *et al.* (1975) to be far more effective for chill proofing. Finley and co-workers (1979) reported that papain immobilized to crab chitin was as effective as the corresponding soluble enzyme with no detectable flavor differences. This method appeared to have considerable potential for eliminating this problem in the brewing industry.

In addition to the immobilized enzymes discussed there are others which have been examined for use in the food industry, including polygalacturonase (Pifferi and Preziuso, 1987), pectin esterase (Weibel *et al.*, 1975), invertase (Ooshima *et al.*, 1980a,b), cellobiase (Lenders and Crichton, 1984; Tjerneld *et al.*, 1985), catalase (Altomare *et al.*, 1974; Wang *et al.*, 1974), amyloglucosidase (Park and Lima, 1973), and cellulase (Shimizu and Ishihara, 1987; Takeuchi and Makino, 1987). Kilara *et al.* (1977) examined the kinetics of lactase, papain, and lipase immobilized on a single support. The potential of immobilized lipase for the production of lipolyzed cream and butter was discussed by Kilara (1981). In addition, the application of an immobilized lipase from *Rhizopus arrhius* for interesterification of fats was recently reported by Wisdom and co-workers (1987).

X. Enzyme Electrodes and Food Analysis

The application of immobilized enzymes for the analysis of foods led to the development of enzyme electrodes. These consist of an electrochemical sensor and an immobilized enzyme placed in close proximity to the surface of the sensor as shown in Figure 11.12 (Enfors and Molin, 1978). The probe is placed in a solution containing the substrate, which reacts with the enzyme, releasing a product which can be detected by the electrode. The most common types of electrodes used in enzyme electrodes are potentiometric, polarographic, and potentiometric gas-sensing membrane electrodes (Skogberg and Richardson, 1980). Clark and Lyons (1962) first reported an immobilized glucose oxidase to measure glucose amperometrically by monitoring the uptake of oxygen. The term "enzyme electrode" was first used, however, to describe a potentiometric electrode with glucose oxidase entrapped in a polyacrylamide gel (Updike and Hicks, 1967). This system has been refined and uses either a polarographic platinum electrode to measure the depletion of oxygen in the enzyme layer as a

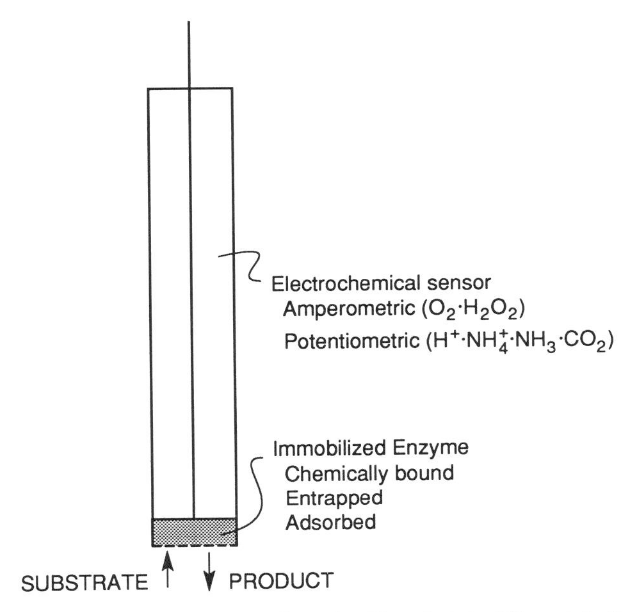

FIG. 11.12. The principle of an enzyme electrode. The substrate diffuses into the immobilized enzyme where it is converted into product. The electrochemical sensor measures the activity of the product or of a cosubstrate of the enzymatic reaction (Enfors and Molin, 1978).

result of glucose oxidation or polarographic measurement of hydrogen peroxide or pH changes as a result of gluconic acid formation:

$$\text{Glucose} + O_2 \xrightarrow{\text{Glucose oxidase}} \text{Gluconic acid} + H_2O_2$$

The preparation of an enzyme electrode using immobilized lysine decarboxylase together with a CO_2 electrode was reported by Skogberg and Richardson (1980) for measuring L-lysine in cereal grains. The results obtained with the enzyme electrode compared favorably with amino acid analysis but provided an inexpensive method for routinely analyzing for L-lysine in grain and feed samples.

It is obvious from the two examples cited that enzyme electrodes have considerable potential for carbohydrate and amino acid analysis. Enfors and Molin (1978) suggested that enzyme electrodes could prove particularly useful for monitoring changes during fermentation processes. These could provide a relatively simple and inexpensive method to implement fermentation control. Mattiasson and Danielsson (1982) described the use of an enzyme thermoresistor for measuring L-ascorbic acid, D-galactose, D-glucose, cellobiose, lactose, and sucrose. This involved a combination of immobilized enzyme (D-galactose oxidase, D-glucose oxidase, invertase, L-ascorbate oxidase, β-glucosidase, and lactase) and a flow calorimeter for measuring the heat produced by the change in enthalpy associated with the enzyme reaction. It could be used to measure discrete samples or in continuous monitoring of systems. A multienzyme electrode for measuring sucrose in food products was recently developed by Nabi Rahni and co-workers (1987) which proved extremely stable and efficient.

XI. Immobilized Cells

Another development in enzyme technology is the use of immobilized microbial cells for carrying out biochemical transformations. This procedure eliminates the necessity of isolating and purifying individual enzymes while at the same time maintaining the natural environment of the enzyme (Bucke and Wiseman, 1981; Cheetham, 1979). The methods involved in binding the microbial cells are essentially the same as those involved in enzyme immobilization although greater care is required to minimize the destruction or inactivation of the particular enzyme activities within the bound organism. The most successful technique has been entrapping the cells within biochemically inert hydrogels of which polyacrylamide is the most common (Bucke and Wiseman, 1981). While this technology is still in its infancy there is considerable potential for the use of immobilized yeast cells in the fermentation of beer or in the conversion of cellulose wastes to ethanol (Kolot, 1980).

Nilsson and co-workers (1983) reported that immobilization of whole cells for complex biochemical conversions and syntheses required immobilization within spherical particles of polymeric matrices. This resulted in optimal activity as spherical particles were homogeneous and easily packed in a column. The formation of these spherical polymer particles containing the entrapped enzymes was achieved by carrying out gel formation in an organic phase of toluene: chloroform (Nilsson *et al.*, 1972). This technique, however, proved detrimental to the viability of the entrapped cells. An improved method for immobilizing cells was developed by Nilsson and co-workers (1983), which used milder conditions and resulted in a beaded uniform catalyst. Using this technique these researchers reported that microbial, algal, plant, and animal cells could be immobilized on a variety of polymer/monomer matrices by suspension in a hydrophobic phase such as soy, paraffin, or silicon oil, tri-*n*-butylphosphate, or dibutylphthalate. Under these conditions the entrapped cells remained fully viable and active.

Fumi *et al.* (1987) examined the potential of immobilizing yeast cells on sodium alginate for the production of sparkling wines in Italy. Two yeast starters

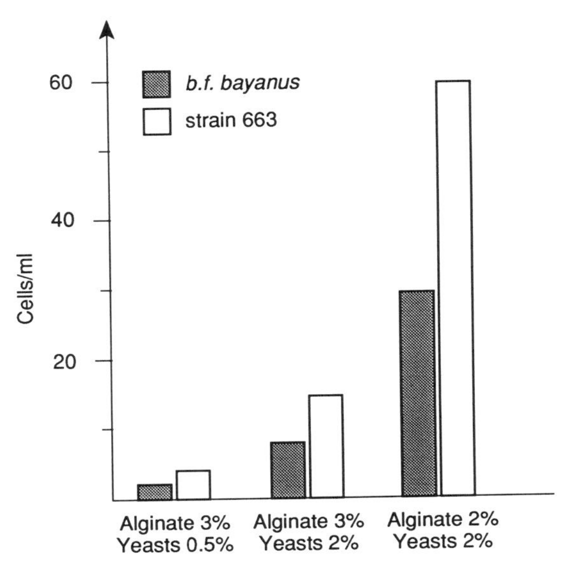

FIG. 11.13. Influence of yeast strain, alginate concentration, and yeast loading on cell release into sparkling wine (Fumi *et al.*, 1987).

were examined in both the free and bound forms. The yeast cells were immobilized by dropping the yeast–sodium–alginate slurry into 0.5 *M* calcium chloride using a peristaltic pump, with the beads formed being washed with sterilized water. A base wine obtained from black Pinot grapes was allowed to undergo secondary fermentation by the different yeast cells. The alginate system was used at 2 or 3% concentrations (weight/weight) and the yeast in concentrations of 0.5 and 2% (dry weight/weight). A calculated amount of beads was added to provide 10 yeast cells/liter to the base wine while the free-cell samples were inoculated into the base wine to give an equivalent concentration of 10 cells/liter. The chemical and physical properties of the sparkling wines produced from the free and bound yeast cells using *Saccharomyces cerevisiae b. f. bayanus* (a commercial product) and an experimental *Saccharomyces cerevisiae* strain 663 were examined. No major differences were observed between the chemical and physical characteristics of wine fermented with either the free or immobilized yeast cells. The structural integrity of the alginate gel was found to change after fermentation and appeared to be related to the concentration of alginate and yeast cells as well as the type of yeast used. An increased amount of yeast cells was released into the sparkling wine which paralleled an increase in alginate and yeast strain levels as shown in Figure 11.13. These differences were markedly higher for *Saccharomyces cerevisiae* strain 663 and was attributed to its high fermentation potential.

The immobilization of *Kluyveromyces marxianus* on alginate beads was successfully reported by Bajpai and Margaritis (1986) for the production of high-fructose syrups from Jerusalem artichoke. This organism contained an active inulase enzyme which hydrolyzed inulin, the polyfructan carbohydrate present in Jerusalem artichoke, to fructose.

Robinson and co-workers (1987) recently reviewed immobilized algae technology and noted that while a large number of reports have been published on their immobilization, considerable basic research remained to be undertaken before any major commercial application was possible.

XII. Genetic Engineering

Current interest in biotechnology is focused on the use of genetic engineering *via* recombinant DNA technology in microbial, plant cell, and tissue culture. These techniques have exciting potential for the development of more effective enzyme systems for the production of improved products. Table 11.7 indicates those food-processing enzymes and food additives which have benefited from genetic engineering technology. Rosen (1987) suggested that the good investment opportunities afforded by biotechnology should lead to its commercialization. This section will examine the impact of this technology on some areas of food production.

TABLE 11.7

FOOD-PROCESSING ENZYMES AND FOOD ADDITIVES USED IN THE U.S. FOOD INDUSTRY THAT BENEFIT FROM GENETIC ENGINEERING[a]

Food industries	Enzymes	
Starch processing	α-Amylase Glucoamylase Pullanase	β-Amylase Glucose isomerase
Dairy industry	Rennin Lactase	Lipase
Brewing	Amylase	Proteases
Wine/fruit/vegetable processing		Pectinases

[a] Adapted from Lin (1986). Copyright © by Institute of Food Technologists.

A. Recombinant DNA Technology

Recombinant DNA technology involves isolating and identifying the gene coding for a particular protein and placing it in a host cell where it can be produced (Moo-Young, 1986). The technique necessitates isolating and separating the messenger RNA (mRNA) responsible for the genetic code of the protein required so that it can be cloned. This requires converting it to a double-stranded DNA molecule with the aid of three enzymes, reverse transcriptase, DNA polymerase, and S1 nuclease. Reverse trancriptase synthesizes a single-stranded DNA molecule which is complementary to mRNA, while DNA polymerase produces the second DNA strand. The third enzyme, S1 nuclease, cuts the DNA molecule at the closed end so that it can be inserted into the expression vector. The primary vehicle or vector for these DNA fragments is bacterial plasmids. These are double-stranded DNA fragments quite distinct from the chromosomal DNA but capable of independent multiplication. To insert the desired genetic material into the plasmid requires splicing or opening it up by restriction enzymes. These enzymes cleave certain sites in the plasmid for insertion of the new genetic material. Once in place the plasmid is sealed with the aid of DNA ligase, thus creating a recombinant DNA. The plasmid incorporating the new DNA can then be introduced into a suitable host organism for expression as protein (Scheme 11.5). Of these, *Escherichia coli* is the most well developed, although other host organisms including *Bacillus spp.* and yeasts are receiving increasing

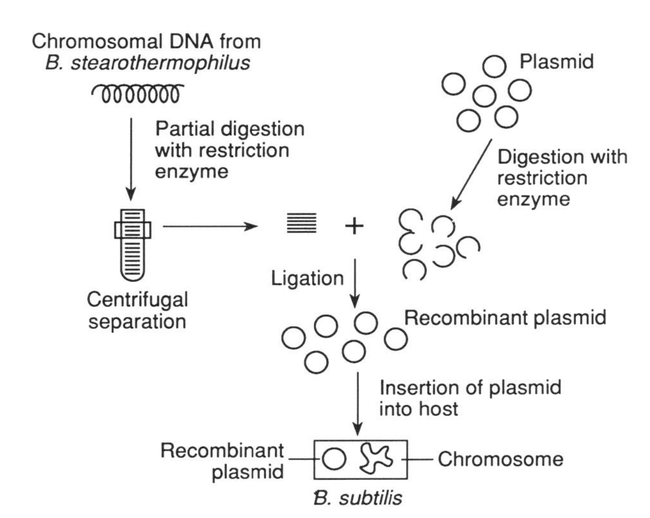

Scheme 11.5. Basic steps in cloning of the maltogenic amylase gene from *B. stearothermophilus* into *B. subtilis* (Andersen *et al.*, 1987).

attention (Botstein and Davies, 1972; Dubnau, 1982; Ganesan *et al.*, 1982). The basic steps involved in cloning α-amylase from *B. stearothermophilus* into *B. subtilis* are shown in Scheme 11.5 (Andersen *et al.*, 1987).

B. Application to Food Production

A wide variety of food products and processes have been the focus of recombinant DNA technology including dairy products, starch processing, brewing, and food additives (Addy and Stuart, 1986; Lawrence, 1987; Lin, 1986; Pitcher, 1986; Saha and Zeikus, 1987).

1. *Cheese Making*

While various microbial rennet substitutes are used for the production of cheeses they still prove inferior to calf rennet (Green, 1977). Recent advances in genetic engineering resulted in the successful cloning of the calf rennet gene into *E. coli* (Emtage *et al.*, 1983; Nishimori, 1986; Nishimori *et al.*, 1982; Harris *et al.*, 1982). This recombinant enzyme was subsequently studied and purified by Marston *et al.* (1984). The cheese-making properties of the recombinant enzyme were compared to that of a standard calf rennet by Green and co-workers (1985). The cheeses were not significantly different although those made with recombinant enzyme were marginally firmer after 5 and 18 weeks (Table 11.8). The cheese produced with recombinant enzyme was judged to be more mature and preferred by the tasters. This was attributed to the presence of trace amounts of proteolytic enzymes in the recombinant enzyme which remained active during cheese ripening. The recombinant rennin from *E. coli* was severely limited in that the enzyme was not secreted but accumulated in the cytoplasm, requiring an expensive and poor yielding extraction process (Pitcher, 1986). This has led to efforts using "supersecretor" yeast strains or filamentous fungi as hosts (Lawrence, 1987).

2. *Starch Processing for Production of Syrups*

The first petition to the FDA in the United States to obtain GRAS (Generally Regarded as Safe) status for a genetically engineered enzyme in food processing was for α-amylase (Lawrence, 1987). This enzyme, as discussed earlier in this chapter, plays a crucial role in the degradation of starch in syrup production. The genetic code for α-amylase was obtained from *Bacillus stearothermophilus*, an organism accepted by FDA as GRAS (Tamuri and Kanno, 1981). The genetic code for α-amylase was inserted into a DNA plasmid vector and introduced into a strain of *Bacillus subtilis* which did not contain this enzyme (see Scheme 11.5). The genetically engineered α-amylase was equivalent in properties to the original

TABLE 11.8

PROTEOLYSIS DURING RIPENING OF CHEESES MADE WITH TWO DIFFERENT COAGULANTS[a]

Cheese no.	Coagulant	Degraded in first 5 weeks of ripening (mg/g dry wt)		Produced in first 5 weeks of ripening (mg/g dry wt)		Produced in first 18 weeks of ripening (mg/g dry wt)
		α_{s1}-Casein	β-Casein	α_{s1}-I-Casein	N soluble in 2.5% TCA[b]	N soluble in 2.5% TCA
1	Rennet	91	13	104	5.3	7.4
2	Recombinant chymosin	86	19	105	5.5	10.5
3	Rennet	96	0	87	4.7	9.8
4	Recombinant chymosin	92	0	88	4.2	10.7
5	Rennet	104	0	98	5.4	10.8
6	Recombinant chymosin	87	0	84	5.8	11.6

[a] From Green *et al.* (1985).
[b] TCA = Trichloroacetic acid.

B. stearothermophilus enzyme. Further developments in enzyme production for syrups include the recent patenting of a genetically engineered pullulanase in *B. subtilis* from *E. coli* strains (Coleman and McAlister, 1986).

3. *Brewing*

A major goal in brewing is the development of a yeast strain which will be able to hydrolyze starch. One particular yeast strain, *Saccharomyces diastaticus* (formally *S. cerevisiae* var. *diastaticus*), was capable of utilizing dextrin or starch by secreting α-1,4-amyloglucosidase during vegetative growth. The limitation of using this organism for brewing is that it produced a phenolic "off-flavor" in the beer (MacQueen, 1987). The genes associated with the synthesis of amylase in *S. diastaticus* were referred to as DEX or STA (Erratt and Stewart, 1978). The thermolability of this enzyme would facilitate its rapid inactivation during post-fermentation pasteurization. This is important in obtaining good flavor stability with low-carbohydrate beer (Tubb, 1987). Consequently efforts were made to isolate the amylase gene from this organism and insert it into brewing yeasts via the plasmid technique. The instability of the gene when transferred in the plasmid vector still requires alternative strategies for its incorporation.

The sensitivity of brewing yeasts to copper ions is a particularly difficult problem because of the general use of copper vessels in the brewing process. Consequently a copper-resistant gene was developed which was introduced by recombinant DNA technology into both ale and lager yeast strains (Henderson *et al.*, 1985). A plasmid was constructed containing both the copper-resistant and the amylase genes and used for fermentation of beer (Meaden and Tubb, 1985). Future research in this area includes introducing genes to encode β-glucanase as well as enzymes capable of hydrolyzing β-1,6 branch points in amylopectin (Tubb, 1987).

C. Application to Agricultural Products

In addition to the more direct use of genetic engineering to produce specific proteins, biotechnology has also been applied to improving the yield, productivity, and quality of plant crops (Jaworski, 1987). The techniques involved include tissue culture, somaclonal variation and gametoclonal variation, somatic cell hybridization, cellular selection procedures, and recombinant DNA (Sharp *et al.*, 1984). All of these techniques are genetic engineering tools for developing new crop varieties for the food industry. Biotechnology appears to have considerable potential for improving the nutritive value of plant foods (Teutonico and Knorr, 1984, 1985).

Current developments in animal biotechnology are focused primarily on improving production (Evans and Hollaender, 1986). The successful introduction of a sheep gene for the milk protein β-lactoglobulin into mice, however, represents a dramatic breakthrough in the development of genetically engineered milk (Simons *et al.*, 1987). These researchers suggested that this could lead to a significant improvement in the productivity of dairy animals.

The recognition of the importance of genetic engineering in the marketplace will lead, ultimately, to the integration of this technology into food production (Lawrence, 1987).

Bibliography

Abdullah, M., Catley, B. J., Lee, E. Y. C., Robyt, J., Wallenfels, K., and Whelan, W. J. (1966). The mechanism of carbohydrase action. XI. Pullanase, an enzyme specific for the hydrolysis of 1 → 6 bonds in amylaceous oligo and polysaccharides. *Cereal Chem.* **43,** 111.

Addy, N. D., and Stuart, D. A. (1986). Impact of biotechnology on vegetable processing. *Food Technol.* **40,** 64.

Altomare, R. E., Kohler, J., Greenfield, P. F., and Kettrell, J. R. (1974). Deactivation of immobilized beef liver catalase by hydrogen peroxide. *Biotechnol. Bioeng.* **16,** 1659.

Amos, J. A. (1955). The use of enzymes on the baking industry. *J. Sci. Food Agric.* **6,** 489.

Andersen, J. R., Diderichsen, B. K., Hjortkjaer, R. K., De Boer, A. S., Bootman, J., West, H., and Ashby, R. (1987). Determining the safety of maltogenic amylase produced by rDNA technology. *J. Food Prot.* **50,** 521.

Andren, R. K., Mandels, M., and Modeiros, J. E. (1976). Production of sugars from waste cellulose by enzymatic hydrolysis: Primary evaluation of substrates. *Process Biochem.* **11,** 1.

Anonymous (1961). "Report of the Commission on Enzymes of the International Union of Biochemistry." Macmillan (Pergamon), New York.

Antrim, R. L., Iowa, C., and Kantwik, A.-L. A. (1986). A new regenerable immobilized glucose isomerase. *Stäerke* **38,** 132.

Arai, S., Noguchi, M., Kaji, M., Kato, H., and Fujimaki, M. (1970). Studies on flavor components in soybean. V. n-Hexanal and some volatile alcohols. Their distribution in raw soybean tissues and formation in crude soy protein concentrate by lipoxygenase. *Agric. Biol. Chem.* **34,** 1420.

Aworh, O. C., and Egounlety, M. (1985). Preservation of West African soft cheese by chemical treatment. *J. Dairy Res.* **52,** 189.

Aworh, O. C., and Muller, H. G. (1987). Cheese-making properties of vegetable rennet from Sodom apple *(Calotropis procera). Food Chem.* **26,** 71.

Aworh, O. C., and Nakai, S. (1986). Extraction of milk clotting enzyme from sodom apple *(Calotropis procera). J. Food Sci.* **51,** 1569.

Axelrod, P. (1974). Lipoxygenases. Chapter 13. *Adv. Chem. Ser.* **136,** 324.

Bajpai, P., and Margaritis, A. (1986). Optimization for production of high fructose syrup from Jerusalem artichoke using calcium alginate immobilized cells of *Kluyveromyces marxianus. Process Biochem.*, **21**(1), 16.

Baker, R. A., and Bruemmer, J. H. (1969). Cloud stability in the absence of various orange juice soluble components. *Proc. Fla. State Hortic. Soc.* **82,** 215.

Baker, R. A., and Bruemmer, J. H. (1972). Pectinase stabilization of orange juice cloud. *J. Agric. Food Chem.* **20,** 1169.

Bar-Eli, A., and Katchalski, E. (1960). A water-insoluble trypsin derivative and its use as a trypsin column. *Nature (London)* **188,** 856.

Barfoed, H. (1978). Enzymes in starch processing. *Cereal Foods* **21,** 588.
Beck, C. I., and Scott, D. (1974). Enzymes in foods—For better or worse. *In* "Food Related Enzymes" (J. R. Whitaker, ed.), Ch. 1. *Adv. Chem. Ser.* **136,** 1. American Chemical Society, Washington, D.C.
Berghem, L. E. R., and Petersson, L. G. (1973). Mechanism of enzymic degradation: Purification of a cellulolytic enzyme from *Trichoderma viride* active on highly ordered cellulose. *Eur. J. Biochem.* **37,** 21.
Bernholdt, H. F. (1969). Enzymic tenderization of meat. *Annu. Rep.—N. Y. State Assoc. Milk Food Sanit.,* **43,** 35.
Berridge, N. J., and Woodward, C. (1953). A simple method for preparing crystalline rennet. *J. Dairy Res.* **20,** 255.
Bieleg, H., Wolf, J., and Balcke, K. (1971). Enzymic clarification of apple pulp for removal by rack and cloth presses or centrifuges. *Fluess. Obst.* **38,** 408.
Blocksma, A. H., and Hlynka, I. I. (1964). Basic consideration in unleavened doughs from normal and defatted wheat flour. *In* "Wheat Chemistry and Technology" (I. Hylinka, ed.), p. 445. Am. Assoc. Cereal Chem., St. Paul, Minnesota.
Botstein, D., and Davies, P. W. (1972). *In* "Molecular Biology of the Yeast *Saccharomyces,* Metabolism and Gene Expression" (S. N. Cohen, and H. W. Boyer, eds.). Cold Spring Harbor Lab., Cold Spring Harbor, New York.
Bottcher, H. (1975). Enzyme activity and quality of frozen vegetables. I. Remaining residues of peroxidase. *Nahrung* **19,** 173.
Braddock, R. J., and Kesterson, J. W. (1979). Use of enzyme in citrus processing. *Food Technol. (Chicago)* **31,** 78.
Brewer, G. J. (1970). "An Introduction to Isoenzyme Techniques." Academic Press, New York.
Bucke, C. (1980). Enzymes in fructose manufacture. *In* "Enzymes in Food Processing" (G. G. Birch, and K. S. Parker, eds.). Applied Science Publishers, London.
Bucke, C. (1982). Practicality of industrial enzymes. *Biochem. Soc. Trans.* **11,** 13.
Bucke, C., and Wiseman, A. (1981). Immobilised enzymes and cells. *Chem. Ind. (London),* p. 234.
Chandler, B. V., and Roberston, G. L. (1983). Effect of pectic enzymes on cloud stability and soluble limonin concentration in stored orange juice. *J. Sci. Food Agric.* **34,** 599.
Cheetham, P. S. J. (1979). Physical studies on the mechanical stability of columns of calcium alginate gel pellets containing entrapped microbial cells. *Enzyme Microb. Technol.* **1,** 183.
Chen, W.-P. (1980). Glucose isomerase (a review). *Process Biochem.* **15(5),** 30.
Cheng, A. H., Brunser, O., Espinoza, J., Jones, H. L., Monckeberg, F., Chichester, C. O., Rand, G., and Hourigan, A. G. 1979. Long-term acceptance of low-lactose milk. *Am. J. Clin. Nutr.* **32,** 1989.
Chu, H. D., Leeder, J. G., and Gilbert, S. G. (1975). Immobilized catalase reactor for use in peroxide sterilization of dairy products. *J. Food Sci.* **40,** 641.
Clark, L. C., and Lyons, C. (1962). Electrode systems for continuous monitoring in cardiovascular surgery. *Ann. N. Y. Acad. Sci.* **102,** 39.
Coleman, R. D., and McAlister, M. P. (1986). Plasmids containing a gene coding for a thermostable pullulanase and pullulanase-producing strains of *Escherichia coli* and *Bacillus subtilis* containing the plasmids. U. S. Patent 4,612,287.
Coleman, M. H., and MacRae, A. R. 1981.Fat process and composition. U. S. Patent 4,275,081.
Cotter, W. P., Lloyd, M. E., and Hinman, C. W. (1971). Method for isomerizing glucose syrups. U. S. Patent 3,623,953.
Creamer, L. K. (1976). A further study of the action of rennin. *N. Z. J. Dairy Sci. Technol.* **11,** 30.
Cunningham, F. E., and Tiede, L. M. (1981). Properties of selected poultry products treated with a tenderizing marinade. *Poult. Sci.* **60,** 2475.
Currie, J. N. (1914). Flavor of Roqueforti cheese. *J. Agric. Res.* **2,** 1.

Dahlqvist, A., Asp., N.-G., Burvall, A., and Rausing, H. (1977). Hydrolysis of lactose in milk and whey with minute amounts of lactase. *J. Dairy Res.* **44,** 541.
Daniels, N. W. R., Wood, P. S., Eggitt, P. W., and Coppock, J. B. M. (1970). Studies on the lipids of flour. V. Effect of air on lipid binding. *J. Sci. Food Agric.* **21,** 377.
Delincee, H., and Schaefer, W. (1975). Influence of heat treatments of spinach at tempereratures up to 100°C on important constituents. Heat inactivation of peroxidase isoenzymes in spinach. *Lebensm.-Wiss. Technol.* **8,** 217.
Denkswater, R. G., Veber, D. F., Holly, F. W., and Hirschmann, R. (1969). Studies on the total synthesis of an enzyme. 1. Objective and strategy. *J. Am. Chem. Soc.* **91,** 502.
Desmazeaud, M. J., Gripon, J. C., Le Bars, D., and Bergere, J. L. (1976). Study of the role of microorganisms and enzymes during cheese ripening. III. Effect of microorganisms *Streptococcus lactis* and *Penicillium caseicolum* and *P. roqueforti. Lait* **56,** 379.
Dubnau, D. A., ed. (1982). "The Molecular Biology of the Bacilli," Vol. 1. Academic Press, New York.
Dundas, D. G. A., Henderson, H. M., and Eskin, N. A. M. (1978). Lipase from *Vicia faba minor Food Chem.* **3,** 171.
Eddleman, T. L., and Babel, F. J. (1958). Phosphatase reactivation in dairy products. *J. Milk Food Technol.* **21,** 126.
Eigtved, P., Hansen, T. T., and Huge-Jensen. (1985). Tech. Rep. No. A-0524A. Novo Industri, Bagsvaerd, Denmark.
Emtage, J. S., Angal, S., Doel, M. T., Harris, T. J. R., Jenkins, B., Lilley, G., and Lowe, P. A. (1983). Synthesis in calf prochymosin (prorennin) in *Escherichia coli. Proc. Natl. Acad. Sci. U.S.A.* **80,** 3671.
Enevoldsen, B. C. (1975). Debranching enzymes in brewing. *Eur. Brew. Conv., Proc. Congr.* **15,** 683.
Enfors, S. O., and Molin, N. (1978). Enzyme electrodes for fermentation control. *Process Biochem.* **13,** 9.
Enzyme Nomenclature (1984). "Recommendations of the Nomenclature Committee of the International Union of Biochemistry on the Nomenclature and Classification of Enzyme-Catalyzed Reaction." Academic Press, Orlando, Florida.
Erratt, J. A., and Stewart, G. G. (1978). Genetic and biochemical studies on yeast strains able to utilize dextrins. *J. Am. Soc. Brew. Chem.* **36,** 151.
Eskin, N. A. M. (1979). "Plant Pigments, Flavors and Textures: The Chemistry and Biochemistry of Selected Compounds." Academic Press, New York.
Eskin, N. A. M., and Henderson, H. M. (1974a). Lipoxygenase in *Vicia faba minor. Phytochemistry* **13,** 2713.
Eskin, N. A. M., and Henderson, H. M. (1974b). A study of lipoxygenase from small faba beans. *Proc. Int. Congr. Food Sci. Technol., 4th, 1974* Vol. 1, p. 263.
Eskin, N. A. M., and Landman, A. D. (1975). Study of milk clotting by an enzyme from ash gourd *(Benincasa cerifera). J. Food Sci.* **40,** 413.
Eskin, N. A. M., Grossman, S., and Pinsky, A. (1977). Biochemistry of lipoxygenase in relation to food quality. *CRC Crit. Rev. Food Sci. Nutr.* **9,** 1.
Eskin, N. A. M., Mokady, S., and Cogan, U. (1985). The potential of legumes in human nutrition. *In* "Advances in Diet and Nutrition" (C. Horowitz, ed.), Chapter 68. John Libbey & Co., U.K.
Evans, J. W., and Hollaender, A., eds. (1986). "Genetic Engineering of Animals." Plenum, New York.
Fan, T. Y., Hwang, D. H., and Kinsella, J. E. (1976). Methyl ketone formation during germination of *Penicillium roqueforti. J. Agric. Food Chem.* **24,** 443.
Finley, J. W., Stanley, W. L., and Watters, G. G. (1979). Chillproofing beer with papain immobilized on chitin. *Process Biochem.* **14,** 12

Fogarty, W. M., Griffin, P. J., and Joyce, A. M. (1974). Enzymes of *Bacillus* species. Part 2. *Process Biochem.* **9,** 27.

Foltmann, B. (1960). Studies on rennin. IV. Chromatographic fractionation of rennin. *Acta Chem. Scand.* **14,** 2059.

Food and Agriculture Organization (1968). "Ad Hoc Consultation in World Shortage of Rennet in Cheesemaking," Rep. No. AN 1968-3. FAO, Rome.

Fram, H. (1957). The reaction of phosphatase in HTST pasteurized dairy products. *J. Dairy Sci.* **40,** 19.

Frazier, P. J. (1979). Lipoxygenase action and lipid binding in breadmaking. *Baker's Dig.* **53**(6), 8.

Fukal, L., and Kas, J. (1984). The role of active and inactivated papain in beer chillproofing. *J. Inst. Brew.* **90,** 247.

Fukui, K., Jukui, Y., and Moriyama, T. P. (1974). Purification and properties of dextran sucrase and invertase from *Streptococcus mutans. J. Bacteriol.* **118,** 796.

Fullbrook, P. D. (1983). The use of enzymes in the processing of oil seeds. *J. Am. Oil Chem. Soc.* **60,** 428A.

Fumi, M. D., Trioli, G., and Colagrande, O. (1987). Preliminary assessment on the use of immobilized yeast cells in sodium alginate for sparkling wine processes. *Biotechnol. Lett.* **9,** 339.

Ganesan, A. T., Chang, S., and Hoch, J. A., eds. (1982). "Molecular Cloning and Gene Regulation in Bacilli." Academic Press, New York.

Godfrey, J. (1978). Enzymes in food processing. *Int. Flavor Fragrance Assoc.,* p. 163.

Goeser, P. A. (1961). Tenderized meat through antemortem vascular injection of proteolytic enzymes. *Am. Meat Inst. Found., Circ.* **64,** 55.

Gordin, S., and Rosenthal, I. (1978). Efficacy of chicken pepsin as a milk clotting enzyme. *J. Food Prot.* **41,** 684.

Green, M. L. (1972). On the mechanism of milk clotting by rennin. *J. Dairy Sci.* **39,** 55.

Green, M. L. (1977). Review of the progress of dairy science: Milk coagulants. *J. Dairy Res.* **44,** 159.

Green, M. L., and Crutchfield, G. (1969). Studies on the preparation of water-insoluble derivatives of rennin and chymotrypsin and their use in the hydrolysis of casein and the clotting of milk. *Biochem. J.* **115,** 182.

Green, M. L., Angal, S., Lowe, P. A., and Marston, F. A. O. (1985). Cheddar cheesemaking with recombinant calf chymosin synthesized in *Escherichia coli. J. Dairy Res.* **52,** 281.

Greenberg, N. A., and Mahoney, R. R. (1981). Immobilization of lactase (β-galactosidase) for use in dairy processing: A review. *Process Biochem.* **16 (2),** 2.

Gripon, J. C. (1977a). Proteolytic system of *Penicillium roqueforti.* IV. Properties of an acid carboxypeptidase. *Ann. Biol. Anim. Biochim., Biophys.* **17,** 283.

Gripon, J. C. (1977b). The proteolytic system of *Penicillium roqueforti.* V. Purification and properties of an alkaline amino peptidase. *Biochimie* **59,** 679.

Gripon, J. C., and Debest, B. (1976). Electrophoretic study of the exocellular proteolytic system of *Penicillium roqueforti. Lait* **56,** 423.

Gripon, J. C., and Hermier, J. (1974). Proteolytic system of *Penicillium roqueforti.* III. Purification, properties and specificity of a protease inhibited by EDTA. *Biochimie* **56,** 1323.

Grubhofer, N., and Schlieth, L. (1954). Coupling of proteins on diazotised polyaminostyrene. *Hoppe-Seyler's Z. Physiol. Chem.* **297,** 108.

Gunja-Smith, Z., Marshall, J. J., and Smith, E. E. (1971). Enzymic determination of the unit chain length of glycogen and related polysaccharides. *FEBS Lett.* **13,** 309.

Gupta, C. B., and Eskin, N. A. M. (1977). Potential use of vegetable rennet in the production of cheese. *Food Technol.* **31,** 63.

Gutte, B., and Merrifield, R. B. (1969). The total synthesis of an enzyme with ribonuclease A activity. *J. Am. Chem. Soc.* **91,** 501.

Guy, E. J., and Bingham, E. W. (1978). Properties of β-galactosidase of *Saccharomyces lactis* in milk and milk products. *J. Dairy Sci.* **61,** 147.

Gyuricsek, D. M., and Thompson, M. P. (1976). Hydrolysed lactose cultured dairy products. II. Manufacture of yoghurt, buttermilk and cottage cheese. *Cult. Dairy Prod.* **11,** 12.

Haas, L. W., and Bohn, R. M. (1934). Bleaching bread dough. U. S. Patents 1,957,333–7.

Harper, W. J. (1957). Lipase systems used in the manufacture of Italian chesse. II. Selective hydrolysis. *J. Dairy Sci.* **40,** 556.

Harris, T. J. R., Lowe, P. A., Lyons, A., Thomas, P. G., Eaton, M. A. W., Millicon, T. A. L., Patel, T. P., Bose, C. C., Carey, N. H., and Doel, M. T. (1982). Molecular cloning and nucleotide sequence of cDNA coding for calf preprochymosin. *Nucleic Acids Res.* **10,** 2177.

Hashimoto, K., Adachi, S., and Noujima, H. (1983). A new process combining adsorption and enzyme reaction for producing higher-fructose syrup. *Biotechnol. Bioeng.* **25,** 2371.

Hausser, A. G., Goldgerg, B. S., and Mertens, J. C. (1983). An immobilized two-enzyme system (fungal–amylase–glucoamylase) and its use in the continuous production of high conversion maltose-containing corn syrups. *Biotechnol. Bioeng.* **25,** 525.

Haydar, M., and Hadziyev, D. (1973). A study of lipoxidase in pea seeds and seedlings. *J. Sci. Food Agric.* **24,** 1039.

Haydar, M., Steele, L., and Hadziyev, D. (1973). Oxidation of pea lipids by pea seed lipoxygenase. *J. Food Sci.* **40,** 808.

Henderson, R. C. A., Cox, B. S., and Tubb, R. S. (1985). The transformation of brewing yeasts with a plasmid containing the gene for copper resistance. *Curr. Genet.* **9,** 133.

Hinchcliffe, J., Vaisey-Genser, M., McDaniel, M., and Eskin, N. A. M. (1977). The flavor of fababeans as affected by heat and storage. *Can. Inst. Food Sci. Technol. J.* **10,** 181.

Hogan, J. M., and Bernholdt, J. F. (1964). Proteolytic enzyme treatment of animals. U.S. Patent 3,163,540.

Holsinger, V. H. (1981). Lactase-modified milk and whey. *Food Technol.* **40,** 35.

Horie, Y. (1964). Protein-clotting activity of papain. *Proc. Am. Soc. Brew. Chem.*, p. 174.

Hutchinson, J. B., and Martin, H. F. (1952). The measurement of lipase activity in oat products. *J. Sci. Food Agric.* **3,** 312.

Janssen, F. M. (1960). Invertase and cast cream centers. *MC, Manuf. Confect.* **40,** 41.

Janssen, F. M. (1964). Effect of high temperatures on the efficiency of invertase. *MC, Manuf. Confect.* **44,** 63.

Jaworski, E. G. (1987). The impact of biotechnology on food production. *Cereal Foods World* **32,** 754.

Johnston, D. E. (1984). Application of polymer cross-linking theory to rennet-induced milk gels. *J. Dairy Res.* **51,** 96.

Johnston, W. R., and Foote, H. G. (1951). Development of a new process for curing coffee. *Food Technol.* **5,** 464.

Jones, J. G., and Mercier, P. L. (1974). Refined papain. *Process Biochem.* **9,** 21.

Jones, M., Woof, J. B., and Pierce, J. S. (1967). Peptidase activity in wort and beer. *Proc. Am. Soc. Brew. Chem.* **14,** 14.

Joslyn, M. A., and Pilnik, W. (1961). Enzymes and enzyme activity. *In* "The Orange: Its Biochemistry and Physiology" (W. B. Sinclair, ed.), pp. 373–475. Univ. of California Press, Berkeley.

Katz, R. S., and Speckmann, F. N. (1978). A perspective on milk intolerance. *J. Food Prot.* **41,** 220.

Kaul, R., D'Souza, S. F., and Nadkani, G. B. (1984). Hydrolysis of milk lactose by immobilized β-galactosidase-hen egg white powder. *Biotechnol. Bioeng.* **26,** 901.

Kay, G., Lilly, M. D., Sharp, A. K., and Wilson, R. J. H. (1968). Preparation and use of porous sheets with enzyme action. *Nature (London)* **217,** 641.

Kennedy, J. F., and Pike, V. W. (1981). Papain, chymotrypsin and related proteins—A comparative study of their chillproofing abilities and characteristics. *Enzyme Microb. Technol.* **3,** 59.

Kertesz, Z. I. (1951). "The Pectic Substances." Wiley (Interscience), New York.
Kilara, A. (1981). Immobilized proteases and lipases. *Process Biochem.* **16**(2), 25.
Kilara, A. (1982). Enzymes and their uses in the processed apple industry. *Process Biochem.* **17,** 35.
Kilara, A. (1985). Enzyme-modified lipid food ingredients. *Process Biochem.* **20**(2), 35.
Kilara, A., and Shahani, K. M. (1979). The use of immobilized enzymes in the food industry: A review. *CRC Crit. Rev. Food Sci. Nutr.* **12,** 161.
Kilara, A., Shahani, K. M., and Wagner, F. W. (1977). The Kinetic properties of alctase, papain and lipase immobilized on a single support. *J. Food Biochem.* **1,** 261.
Kirsop, B. H. (1981). Biotechnology in the food processing industry. *Chem. Ind.* (*London*), p. 218.
Kittsteiner-Eberle, R., Hofelmann, M., and Schreier, P. (1985). Isolation and characterization of 1,4β glucan-4-glucanohydrolases (EC 3.2.1.4) from a technical *Trichoderma viride* cellulase. *Food Chem.* **17,** 131.
Kligerman, A. E. (1981). Development of lactase-reduced milk products. *In* "Lactose Digestion: Clinical and Nutritional Implications" (D. M. Paige, and T. M. Bayless, eds.), pp. 252–256. Johns Hopkins Univ. Press, Baltimore, Maryland.
Kneen, E., and Hads, H. L. (1945). Effects of variety and environment on the amylases of germinated wheat and barley. *Cereal Chem.* **22,** 407.
Kolot, F. B. (1980). New trends in yeast technology—Immobilized cells. *Process Biochem.* **15(7),** 2.
Kon, S., Wagner, J. P., Guadagni, D. C., and Horvat, R. J. (1970). pH adjustment control of oxidative off-flavor during grinding of raw legume seeds. *J. Food Sci.* **35,** 343.
Koshland, D. E., Jr., Nemethy, G., and Filmer, D. (1966). Comparison of experimental binding data and theoretical models in proteins containing subunits. *Biochemistry* **5,** 365.
Kosikowski, F. W., and Wierzbicki, L. E. (1973). Lactose hydrolysis of raw and pasteurized milks by *Saccharomyces lactis* lactase. *J. Dairy Sci.* **56,** 146.
Krop, J. J. P. (1974). The mechanism of cloud loss phenomena in orange juice. Doctoral Thesis, Agricultural University, The Netherlands.
Krop, J. J. P., and Pilnik, W. (1974). Cloud loss studies in citrus juices: Cloud stabilization by a yeast polygalacturonase. *Lebensm.-Wiss. Technol.* **7,** 121.
Kulp, K. (1975). Carbohydrases. *In* "Enzymes in Food Processing" (G. Reed, ed.), Chapt. 6, p. 108. Academic Press, New York.
Kwee, W. S. (1983). Phosphatase reactivation in cream samples. *Aust. J. Dairy Technol.* **38**(4), 160.
Latner, A. L., and Skillen, A. W. (1968). "Isoenzymes in Biology and Medicine." Academic Press, New York.
Law, B. A. (1980). Accelerated ripening of cheese. *Dairy Ind. Int.* **45,** 15.
Law, B. A. (1981). Accelerated ripening of cheddar cheese with microbial proteinases. *Neth. Milk Dairy J.* **35,** 313.
Law, B. A., and Wigmore, A. (1982). Accelerated cheese ripening with food grade proteinases. *J. Dairy Res.* **49,** 137.
Lawrence, R. H., Jr. (1987). New applications of biotechnology in the food industry. *Cereal Foods Today* **32,** 758.
Le Bars, D., and Gripon, J. C. (1981). Role of *Penicillium roqueforti* proteinases during blue cheese ripening. *J. Dairy Res.* **48,** 479.
Lee, D. D., Lee, Y. Y., Reilly, P. J., Collins, E. V., and Tsao, G. T. (1976). Pilot plant production of glucose with glucoamylase immobilized to porous silica. *Biotechnol. Bioeng.* **18,** 253.
Lenders, J. P., and Crichton, R. R. (1984). Thermal stabilization of amylolytic enzymes by covalently coupling to soluble polysaccharides. *Biotechnol. Bioeng.* **26,** 1343.
Lenders, J. P., Germain, P., and Crichton, R. R. (1985). Immobilization of a soluble chemically thermostabilized enzyme. *Biotechnol. Bioeng.* **27,** 572.
Lewis, D. A., and Luh, B. S. (1988). Application of actinidin from kiwi fruit to meat tenderization and characterization of beef muscle protein hydrolysis. *J. Food Biochem.* **12,** 147.

Lin, Y.-L. (1986). Genetic engineering and process development for production of food processing enzymes and additives. *Food Technol.* **40,** 104.

Lloyd, N. E., and Logan, R. M. (1972). Process for isomerizing glucose to fructose. U. S. Patent 3,694,314.

Long, J. E., and Harper, W. J. (1956). Italian cheese ripening. VI. Effect of different types of lipolytic enzyme preparations on the accumulation of various free fatty and amino acids and the development of flavor in Provolone and Romano cheese. *J. Dairy Sci.* **39,** 245.

MacAllister, R. V. (1979). Nutritive sweeteners made from starch. *Adv. Carbohydr. Chem. Biochem.* **36,** 15.

MacQueen, H. (1987). Brewers tap biotechnology. *New Sci.* **116**(1580), 66.

MacRae, A. R. (1983). Lipase-catalyzed interesterification of oils and fats. *J. Am. Oil Chem. Soc.* **60,** 291.

Madsen, G. B., Norman, B. E., and Slott, S. (1973). A new heat stable bacterial amylase and its use in high temperature liquefaction. *Staerke* **25,** 304.

Maeda, H., and Tsao, G. T. (1979). Maltose production. *Process Biochem.* **14(7),** 2.

Maeda,H., Tsao, G. T., and Chen, L. F. (1978). Preparation of immobilized soybean β-amylase on porous cellulose beads and continuous maltose production. *Biotechnol. Bioeng.* **20,** 383.

Mandels, M., and Sternberg, D. (1976). Recent advances in cellulase technology. *Hakko Kogaku Zasshi* **54,** 267.

Mandels, M., Medeires, J., Andreotti, R., and Bisett, R. (1981). Enzymic hydrolysis of cellulose, evaluation of cellulose culture filtrates under use conditions. *Biotechnol. Bioeng.* **23,** 2009.

Manecke, G. (1962). Reactive polymers and their use for the preparation of antibody and enzyme resins. *Pure Appl. Chem.* **4,** 507.

Marston, F. A. O., Lowe, P. A., Doel, M. T., Shoemaker, J. M., White, S., and Angal, S. (1984). Purification of calf prochymosin (prorennin) synthesised in *Escherichia coli. Bio/Technology* **2,** 800.

Mason, P. D., and Weetall, H. H. (1972). Invertase covalently coupled to porous glass. *Biotechnol. Bioeng.* **14,** 637.

Mattiasson, B., and Danielsson, B. (1982). Calorimetric analysis of sugars derivatives with aid of an enzyme thermistor. *Carbohydr. Res.* **102,** 273.

McCormick, R. D. (1976). A nutritious alternative for the lactose-intolerant consumer. *Food Prod Dev.* **10,** 17.

McFarren, E. F., Thomas, R. C., Black, L. A., and Campell, J. E. (1960). Differentiation of reactivated from residual phosphatase in high temperature-short time pasteurized milk and cream. *J. Assoc. Off. Anal. Chem.* **3,** 414.

Meaden, P. G., and Tubb, R. S. (1985). A plasmid vector system for the genetic manipulation of brewing strains. *Proc. Congr.—Eur. Brew. Conv.* **20,** 219.

Merivuori, H., Siegler, K. M., Sands, J. A., and Montenecourt, B. S. (1985). Regulation of cellulose biosynthesis and secretion of fungi. *Biochem. Soc. Trans.* **13,** 411.

Mermelstein, N. H. (1975). Immobilized enzymes produce high fructose corn syrup. *Food Technol.* **29,** 20.

Mier, G., Rhodes, W. J., Mahoney, L. G., Webb, N. S., Rodgers, C., Mangel, M., and Baldwin, P. (1962). Beef tenderization by proteolytic enzymes. The effects of two methods of application. *Food Technol.* **16,** 111.

Modler, H. W., Brunnner, J. R., and Stine, C. M. (1974). Extracellular protease of *Penicillium roqueforti.* II.Characterization of a purified enzyme preparation. *J. Dairy Sci.* **57,** 528.

Monod, J., Changeux, J.-P., and Jacob, F. (1963). Allosteric proteins and cellular control system. *J. Mol. Biol.* **16,** 306.

Monod, J., Wyman, J., and Changeux, J.-P. (1965). On the nature of allosteric transitions: A plausible model. *J. Mol. Biol.* **12,** 815.

Moore, K. (1980). Immobilized enzyme technology commercially hydrolyzes lactose. *Food Prod. Dev.* **14,** 50.

Moo-Young, M., ed. (1986). "Comprehensive Biotechnology: The Principles, Applications and Regulations of Biotechnology in Industry, Agriculture and Medicine," Vol. 3. Pergamon, Oxford.

Munnecke, D. M. (1978). Detoxification of pesticides using soluble or immobilized enzymes. *Process Biochem.* 14,

Murphy, J. F., and Murphy, R. E. (1964). Seasoning composition and hyaluronidase for antemortem injection. U.S. Patent 3,159,489.

Myrbäck, K. (1960). Invertases. *In* "Enzymes" (P. D. Boyer, H. A. Lardy, and K. Myrbäck, eds.), Vol. 4, p. 379. Academic Press, New York.

Nabi Rahni, M. A., Lubrano, G. J., and Guilbault, G. G. (1987). Enzyme electrode for the determination of sucrose in food products. *J. Agric. Food Chem.* **35,** 1001.

Nelson, J. H. (1969). Blue cheese flavor by fermentation. *Food Prod. Dev.* **3,** 54.

Nelson, J. H. (1975). Impact of new milk clotting enzymes on cheese technology. *J. Dairy Sci.* **58,** 1739.

Nelson, J. M., and Griffin, E. G. (1916). Adsorption of invertase. *J. Am. Chem. Soc.* **38,** 1109.

Nielsen, T. (1985). Industrial application possibilities for lipase. *Fette, Seifen, Anstrichm.* **87,** 15.

Nilsson, H., Mosbach, R., and Mosbach, K. (1972). The use of bead polymerization of acrylic monomers for immobilization of enzymes. *Biochim. Biophys. Acta* **268,** 253.

Nilsson, K., Birnbaum, S., Flygare, S., Linse, L., Schroder, V., Jeppsson, V., Larsson, P.-O., Mosbach, K., and Brodelius, P. (1983). A general method for the immobilization of cells with preserved viability. *Eur. J. Appl. Microbiol. Biotechnol.* **17,** 319.

Nishimori, K. (1986). Molecular cloning of calf prochymosin cDNA expression in microorganisms. *Nippon i Kaishi* **60,** 119.

Nishimori, K., Kawaguchi, Y., Hidaka, M., Uozumi, T., and Beppu, T. (1982). Expression of cloned calf prochymosin gene sequence in *Escherichia coli*. *Gene* **19,** 337.

Norman, B. E. (1982). A novel debranching enzyme for application in the glucose syrup industry. *Starke* **10,** 340.

Obi, S. K. C., and Odibo, F. J. C. (1984). Partial purification and characterization of a thermostable *Actinomycete* β-amylase. *Appl. Environ. Microbiol.* **47,** 571.

Ogundiwin, J. O., and Oke, O. L. (1983). Factors affecting the processing of Wara—A Nigerian white cheese. *Food Chem.* **11,** 1.

O'Leary, V. S., and Woychik, J. H. (1976). A comparison of some chemical properties of yogurts made from cereal and lactase-treated milks. *J. Food Sci.* **41,** 791.

Olling, C. C. J. (1972). Lactose treatment in the dairy industry. *Ann. Technol. Agric.* **21,** 343.

Ooshima, H., Sakimoto, M., and Harano, Y. (1980a). Characteristics of immobilized invertase. *Biotechnol. Bioeng.* **22,** 2155.

Ooshima, H., Sakimoto, M., and Harano, Y. (1980b). Kinetic study on thermal stability of immobilized invertase. *Biotechnol. Bioeng.* **22,** 2169.

Pannell, R. J. N. (1970). Glucose syrups and related carbohydrates. *In* (G. G. Birch, and F. Green, eds.). Elsevier, London.

Park, Y. K., and Lima, D. C. (1973). A continuous conversion of starch to glucose by an amyloglucosidase–resin complex. *J. Food Sci.*

Payens, T. A. J. (1984). On the gelpoint in the enzyme triggered clotting of casein micelles: A comment. *Neth. Milk Dairy J.* **38,** 195.

Payne-Bose, D., Welsch, J. D., Gearhart, H. C., and Morrison, R. D. (1977). Milk and lactose-hydrolyzed milk. *Am. J. Clin. Nutr.* **30,** 695.

Pazur, J. H., and Ando, T. (1960). The hydrolysis of glucosyl oligosaccharides with α-D (1 → 4) and α-D (1 → 6) bonds by fungal amyloglucosidase. *J. Biol. Chem.* **235,** 297.

Perten, H. (1984). A modified falling number method suitable for measuring both cereal and fungal alpha-amylase activity. *Cereal Chem.* **61,** 108.

Pifferi, P. G., and Preziuso, M. (1987). Immobilization of endo-polygalacturonase on γ-alumina for the treatment of fruit juices. *Lebensm-Wiss. Technol.* **20,** 137.

Pilnik, W., Voragen, A. G. J., and De Vos, L. (1975). Enzymatische Verflussigung von Obst und Gemise. *Fluess. Obst.* **42,** 448.

Pinsky, A., Grossman, S., and Trap, M. (1971). Lipoxygenase content and antioxidant activity of some fruits and vegetables. *J. Food Sci.* **36,** 571.

Pitcher, W. H. (1980). "Immobilized Enzymes for Food Processing." CRC Press, Boca Raton, Florida.

Pitcher, W. H. (1986). Genetic modification of enzymes used in food processing. *Food Technol.* **40,** 62.

Ponte, J. G., Jr., Titcomb, S. T., Rosen, J., Drakert, W., and Cotten, R. H. (1961). The starch damage of white bread flours. *Cereal Sci. Today* **6,** 108.

Port, G. N. J., and Richards, W. G. (1971). Orbital steering and the catalytic power of enzymes. *Nature (London)* **231,** 312.

Posorke, L. H. (1984). Industrial scale application of enzymes to the fats and oil industry. *J. Am. Oil Chem. Soc.* **61,** 1758.

Pringle, W. (1974). Full-fat soy flour. *J. Am. Oil Chem. Soc.* **51,** 74A.

Rand, A. G., Jr. (1981). Enzyme technology and the development of lactose-hydrolyzed milk. *Lactose Dig. [conf.], 1979,* p. 219.

Ranum, P. M., Kulp, K., and Agasie, F. R. (1978). Modified amylograph test for determining diastatic activity in flour supplemented with fungal α-amylase. *Cereal Chem.* **55,** 321.

Reed, G., ed. (1975). "Enzymes in Food Processing," 2nd ed. Academic Press, London.

Richardson, G. A., and Nelson, J. H. (1967). Assay and characterization of pregastric esterases. *J. Dairy Sci.* **50,** 1061.

Richmond, M. L., Gray, J. J., and Stine, C. M. (1981). Beta-galactosidase: A review of recent research related to technological applications, nutritional concerns, and immobilization. *J. Dairy Sci.* **64,** 1759.

Robinson, H. E., and Goeser, P. A. (1962). Enzymatic tenderization of meat. *J. Home Econ.* **54,** 195.

Robinson, P. K., Mak, A. L., and Trevan, M. D. (1986). Immobilized algae: A review. *Process Biochem.* **21**(4), 122.

Rosen, C.-G. (1987). Biotechnology: It's time to scale up and commercialize. *CHEMTECH.*, p. 812.

Rothe, G. A. L., Harboe, M. K., and Marting, S. C. (1977). Quantitation of milk-clotting enzymes in 40 commercial bovine rennets, comparing rocket immunoelectrophoresis with an activity ratio assay. *J. Dairy Res.* **44,** 73.

Saha, B. C., and Zeikus, J. G. (1987). Biotechnology of maltose syrup production. *Process Biochem.* **22**(3), 78.

Saletan, L. T. (1968). Carbohydrases of interest in brewing with particular reference to amyloglucosidase. *Wallerstein Lab. Commun.* **31,** 33.

Sandsted, P. M., and Mattern, P. J. (1958). The relation of proteolysis to the characteristics of oxidation and reduction in doughs. IV. Evidence obtained through baking procedure. *Baker's Dig.* **32,** 33.

Sanwal, B. D. (1970). Allosteric controls of amphibolic pathways in bacteria. *Bacteriol. Rev.* **34,** 20.

Sato, T., Mori, T., Tosa, T., and Chibata, I. (1971). Studies on immobilized enzymes: Preparation and properties of amino acylase covalently attached to halogen *o*-acetyl celluloses. *Arch. Biochem. Biophys.* **147,** 788.

Satyanaryana Rao, B. A., and Narasimham, V. V. L. (1976). Brewing with enzymes. *J. Food Sci. Technol.* **13,** 119.

Schmid, R. D. (1979). Oxidoreductases—Present and potential applications in technology. *Process Biochem.* **14(5),** 2.
Schmidt, D. G., Walstra, P., and Buchheim, W. (1973). The size distribution of casein micelles in cow's milk. *Neth. Milk Dairy J.* **27,** 128.
Schreier, P., and Idstein, H. (1984). Untersuchungen uber die Aromastoffzusammenstezung enzymatisch verflussigter Guava (*Psidium guajava* L.) und Mango (*Mangifera indica* L. var. *Alphjonso*). *Fruchtpulpen. Dtsch. Lebensm. Rolsch.* **80,** 335.
Segura, R., Perez, J. B., and Martinez, J. L. J. (1980). *EFCE Publ. Ser.* **12,** 637.
Segura, R., Perez, J. B., Martinez, J. L., and Posada, J. (1981). *Proc. Congr.—Eur. Brew. Conv. 18th,* p. 443.
Shah, N. O., and Nickerson, T. A. (1978a). Functional properties of hydrolyzed lactose: Solubility, viscosity and humectant properties. *J. Food Sci.* **43,** 1081.
Shah, N. O., and Nickerson, T. A. (1978b). Functional properties of hydrolyzed lactose crystallization. *J. Food Sci.* **43,** 1085.
Shah, N. O., and Nickerson, T. A. (1978c). Functional properties of hydrolyzed lactose: Relative sweetness. *J. Food Sci.* **43,** 1575.
Shahani, K. M., Arnold, R. G., Kilara, A., and Devivedi, R. K. (1976). Role of microbial enzymes in flavor development in foods. *Biotechnol. Bioeng.* **18,** 891.
Sharp, W. R., Evans, D. A., and Ammirato, P. V. (1984). Plant genetic engineering: Designing crops to meet food industry specifications. *Food Technol.* **38,** 112.
Shimizu, K., and Ishihara, M. (1987). Immobilization of cellulolytic and hemicellulolytic enzymes on inorganic supports. *Biotechnol. Bioeng.* **29,** 236.
Sidler, W., and Zuber, H. (1978). Production of extracellular thermostable neutral proteinase and α-amylase by *Bacillus thermophilus. Eur. J. Appl. Microbiol.* **4,** 255.
Silberstein, O. O. (1966). Tenderizing meat. U.S. Patent 3,276,879.
Simons, J. P., McClenaghan, M., and Clark, A. J. (1987). Alteration of the quality of milk by expression of sheep β lactoglobulin in transgenic mice. *Nature (London)* **328,** 530.
Skogberg, D., and Richardson, T. (1980). Enzyme electrodes for the food industry. *J. Food Prot.* **43,** 808.
Sobotka, F. E., and Stelzig, A. A.1974. An apparent cellulase complex in tomato (*Lycopersicon esculentum* L.) fruit. *Plant Physiol.* **53,** 759.
Sood, V. K., and Kosikowski. F. V. (1979). Accelerated cheddar cheese ripening by added microbial enzymes. J. *Dairy Sci.* **62,** 1865.
Sosulski, K., Sosulski, F. W., and Coxworth, E. (1988). Carbohydrase hydrolysis of canola to enhance oil extraction with hexane. *J. Am. Oil Chem. Soc.* **65,** 357.
Spano, L. A., Madeiros, J., and Mandels, M. (1976). Enzymic hydrolysis of cellulosic wastes to glucose. *Resour. Recovery Conserv.* **1,** 279.
Sprossler, B. (1982). Neu analytische Methoden fur Enzyme in Mehl. *Muehle* **119,** 425.
Stadtman, E. R. (1966). Allosteric regulation of enzymatic activity. *Adv. Enzymol.* **28,** 41.
Sternberg, M. M. (1972). Bond specificity, active site and milk clotting mechanism of the *Mucor miehei* protease. *Biochim. Biophys. Acta* **285,** 383.
Stone, I. M., and Saletan, L. T. (1968). Stability of diluted chillproofing enzyme solutions. *Wallerstein Lab. Commun.* **31,** 45.
Storm, D. R., and Koshland, D. E. Jr. (1970). A source for the special catalytic power of enzymes: Orbital steering. *Proc. Natl. Acad. Sci. U.S.A.* **66,** 445.
Strobel, R. J., Ciavarelli, L. M., Starnes, R. L., and Lanzilotta, R. P. (1983). *Am. Soc. Microbiol., Abstr. Annu. Meet.,* No. 053.
Sumner, J. B. (1926). Note. The recrystallization of urease. *J. Biol. Chem.* **70,** 97.
Takasaki, Y. (1972). Enzymatic method for manufacture of fructose. U. S. Patent 3,689,362.
Takasaki, Y. (1974). Method for separating fructose. U. S. Patent 3,806,363.

Takasaki, Y., and Tanabe, O. (1971). Enzymatic method for converting glucose in glucose syrups to fructose. U. S. Patent 3,616,221.

Takasaki, Y., Kosugi, Y., and Kanbayashi, A. (1969). *Streptomyces* glucose isomerase. *In* "Fermentation Advances" (D. Perlman, ed.), p. 561. Academic Press, New York.

Takeuchi, T., and Makino, K. (1987). Cellulase immobilized on poly-L-glutamic acid. *Biotechnol. Bioeng.* **29,** 160.

Tamuri, M., and Kanno, M. (1981). Heatand acid-stable alpha-amylase enzymes and processes for producing the same. U. S Patent 4,284,722.

Taylor, M. J., Cheryan, M., Richardson, T., and Olson, N. F. (1977). Pepsin immobilised on inorganic supports for the continuous coagulation of serum milk. *Biotechnol. Bioeng.* **19,** 683.

ter Haseborg, E. (1981). Enzymes in flour and baking applications, especially waffle batters. *Process Biochem.* **16,** 16.

Termote, F., Roumbouts, F. M., and Pilnik, W. (1977). Stabilization of cloud in pectinesterase active orange juice by pectic acid hydrolysate. *J. Food Biochem.* **1,** 15.

Teutonica, R. A., and Knorr, D. (1984). Plant cell culture: Food applications and the potential reduction of nutritional stress factors. *Food Technol.* **38,** 120.

Teutonica, R. A., and Knorr, D. (1985). Impact of biotechnology on nutritional quality of food plants. *Food Technol.* **39,** 127.

Thomas, K. C., Magnusen, B., and McCurdy, A. R. (1988). Enzymatic interesterification of canola oil. *Can. Inst. Food Sci. Technol. J.* **21,** 167.

Thompson, K. N., Lloyd, N. E., and Johnson, R. A. (1978). Dextrose using mixed immobilized enzymes. U.S. Patent 4,102,745.

Thompson, M. P., and Brower, D. P. (1974). Manufacture of cheddar cheese from hydrolysed lactose milk. *J. Dairy Sci.* **57,** 598.

Tjerneld, F., Persson, I., Albertsson, P.-A., and Hahn-Hagerdal, B. (1985). Enzymatic hydrolysis of cellulose in aqueous two-phase systems. II. Semicontinuous conversion of a model substrate. Solka Floc. BW 200. *Biotechnol. Bioeng.* **27,** 1044.

Tosa, T., Mori, T., Fuse, N., and Chibata, I. (1966a). Studies on continuous enzyme reactions. I. Screening of carriers for the preparation of water-insoluble amino acylase. *Enzymologia* **31,** 214.

Tosa, T., Mori, T., Fuse, N., and Chibata, I. (1966b). Studies on continuous enzyme reactions. IV. Preparation of a DEAE–cellulose–aminoacylase column and continuous optical resolution of acyl-DL-amino acids. *Biotechnol. Bioeng.* **9,** 603.

Tosa, T., Mori, T., Fuse, N., and Chibata, I. (1967). Studies on continuous enzyme reactions. IV. Preparation of a DEAE–Sephadex aminoacylase column and continuous optical resolution of acyl-DL-amino acids. *Biotechnol. Bioeng.* **9,** 603.

Tosa, T., Mori, T., Fuse, N., and Chibata, I. (1969). Studies on continuous enzyme reactions. V. Kinetics and industrial application of amino acylase column for continuous optical resolution of acyl-DL-amino acids. *Agric. Biol. Chem.* **33,** 1047.

Tosa, T., Mori, T., and Chibata, I. (1971a). Studies on continuous enzyme reactions. 8. Kinetics and pressure drop of amino acylase. *Hakko Kogaku Zasshi* **49,** 522.

Tosa, T., Mori, T., and Chibata, I. (1971b). Studies on continuous enzyme reactions. VII. Activation of water insoluble aminoacylase by protein denaturing agents. *Enzymologia* **40,** 41.

Tubb, R. S. (1987). Gene technology for industrial yeasts. *J. Inst. Brew.* **93,** 91.

Underkofler, L. A. (1968). Enzymes. *In* "Handbook of Food Additives" (T. E. Furia, ed.), Chapter 2. Chem. Rubber Publ. Co., Cleveland, Ohio.

Updike, S. J., and Hicks, G. P. (1967). The enzyme electrode. *Nature* (*London*) **214,** 986.

Urquardt, A. A., Altosaar, I., and Matlashenski, G. J. (1983). Lipase activity in oat grains and milled oat fraction. *Cereal Chem.* **60,** 181.

Vaisey-Genser, M., and Eskin, N. A. M. (1982). "Canola Oil: Properties and Performance," Pub. No. 60. Canola Council of Canada.
Vedernikova, E. I., Lyushinskaya, I. I., Linetskaya, G. N,. and Polak, M. V. (1962). Maltase activity of enzymic preparations of molds for the baking industry. *Mikrobiologiya* **31,** 1087.
Venkatasubramanian, K., Saini, R., and Vieth, W. R. (1975). Immobilization of papain on collagen and the use of collagen–papain membranes in beer chillproofing. *J. Food Sci.* **40,** 109.
Vieira de Sá, F., and Barbosa, M. (1972). Cheese-making with a vegetable rennet from Cardo *(Cyanara cardunculus). J. Dairy Res.* **39,** 335.
Wallerstein, L. (1911). Preventing the clouding of beer on chilling by adding papain. U.S. Patent 995,825.
Walt, A. (1938). Crystalline ficin. *J. Am. Chem. Soc.* **60,** 493.
Wang, S. S., Gallile, G. E., Gilbert, S. G., and Leeder,J. G. (1974). Inactivation and regeneration of immobilized catalase. *J. Food Sci.* **39,** 338.
Wattenbarger, C. J. (1961). U.S. Patent 3,216,826.
Weetall, H. H. (1973). Immobilized enzymes: Some applications in foods. Part II. Applications. *Food Prod. Dev.* **7**(4), 94.
Weetall, H. H. (1975). Application of immobilized enzymes. *In* "Immobilized Enzymes for Industrial Reactors" (R. A. Messing, ed.), p. 201. Academic Press, New York.
Weibel, M. K., Barries, R., Delotto, R., and Humphry, A. E. (1975). Immobilized enzymes: Pectin esterase covalently couples to porous glass particles. *Biotechnol. Bioeng.* **17,** 85.
Weinrich, B. W., Johnson, J. H., Wildi, B. S., and Boyce, D. C. (1971). Chillproofing of beverages using insoluble polymer–enzyme products. U.S. Patent 3,597,220.
Wierzbicki, L. W., and Kosikowski, F. V. (1973a). Food syrups from acid whey treated treated with β-galactosidase of *Aspergillus niger. J. Dairy Sci.* **56,** 1182.
Wierzbicki, L. W., and Kosikowski, F. V. (1973b). Kinetics of lactase hydrolysis in acid whey by β-galactosidase from *Aspergillus niger. J. Dairy Sci.* **56,** 1396.
Wildi, B. S., and Boyce, C. (1971). Chillproofing of beverages using insoluble polymer–enzyme product. U.S. Patent 3,597,219.
Wilson, J. J., and Ingledew, W. (1982). Isolation and characterization of *Schwannomyces alluvius* amylolytic enzymes. *Appl. Environ. Microbiol.* **44,** 301.
Wingard, L. B., Jr. (1972). "Enzyme Engineering," Biotechnol. Bioeng. Symp. No. 3. Wiley (Interscience), New York.
Winter, E. (1969). Behaviour of peroxidase during blanching of vegetables. *Z. Lebensm.-Unters. -Forsch.* **141,** 201.
Wisdom, R. A., Dunnill, P., and Lilly, M. D. (1987). Enzymic interesterification of fats: Laboratory and pilot-scale studies with immobilized lipase from *Rhizopus arrhius. Biotechnol. Bioeng.* **29,** 1081.
Wiseman, A,. and Gould, B. (1968). New enzymes for industry. *New Sci.* **38,** 66.
Witt, R. R., Sair, R. A., Richardson, R. A., and Olson, N. F. (1980). Chill-proofing beer with insoluble papain. *Brewers Dig.* **45**(10), 70.
Wood, J. C. (1967). Soya flour in food products. *Food Manuf.* **42,** 11.
Wood, P. S. (1980). Recent advances in bread improvers. *Process Biochem.* **15**(6), 12.
Wood, T. M., and McCrae, S. I. (1973). The purification and properties of the C component of *Trichoderma Koningii* cellulase. *Biochem. J.* **128,** 1183.
Wood, T. M., and McCrae, S. I. (1978). *In* "Bioconversion of Cellulosic Substances into Energy, Chemicals and Microbial Protein" (T. K. Ghose, ed.), pp. 114–141. IIT, Delhi.
Wood, T. M., and McCrae, S. I. (1979). Enzymes involved in the generation of glucose from cellulose. *Dev. Food Sci.* **2,** 257.
Woodward, J., and Wiseman, A. (1975). Heat-stable invertase. *J. Sci. Food Agric.* **26,** 540.

Woodward, J., and Wiseman, A. (1978). Immobilized and stabilised invertase and its applications. *J. Sci. Food Agric.* **29,** 957.

Wright, R. C., and Tramer, J. (1953). Reactivation of milk phosphatase following heat treatment. 1. *J. Dairy Res.* **20,** 177.

Yamazaki, H., Cheok, R. K. H., and Fraser, A. D. E. (1984). Immobilization of invertase on polyethylenimine-coated cotton cloth. *Biotechnol. Lett.* **6**(3), 165.

Zaborsky, O. R. (1973). "Immobilized Enzymes." CRC Press, Cleveland, Ohio.

Zevaco, C., Hernier, J., and Gripon, J. C. (1973). Proteolytic system of *Penicillium roqueforti*. II. Purification and properties of an acid protease. *Biochimie* **55,** 1353.

Index

D

O

P

Q

R

S

ISBN 0-12-242351-8